DEMCO

Renewable Energy Engineering and Technology

Renewable Energy Engineering and Technology

Principles and Practice

Edited by

V V N Kishore

Senior Fellow, Energy–Environment Technology Division
The Energy and Resources Institute, New Delhi

GOVERNORS STATE UNIVERSITY
UNIVERSITY PARK
IL 60466

London • Sterling, VA

First published by Earthscan in the UK and USA in 2009

Copyright © TERI Press, 2009

ISBN   978-1-84407-699-4

Typeset by TERI Press, New Delhi
Printed and bound by Parangat Offset Pvt Ltd, New Delhi, India
Cover design by Archana Singh

For a full list of publications please contact
Earthscan
Dunstan House
14a St Cross St
London, EC1N 8XA, UK
Tel: +44 (0)20 7841 1930
Fax: +44 (0)20 7242 1474
Email: earthinfo@earthscan.co.uk
Web: **www.earthscan.co.uk**

22883 Quicksilver Drive, Sterling, VA 20166-2012, USA

Earthscan publishes in association with the International Institute for
Environment and Development

A catalogue record for this book is available from the British Library

Library of Congress Cataloging-in-Publication Data has been applied for

Contents

Foreword

As humanity progresses in the 21st century, it would in future encounter major challenges in terms of ensuring adequate and equitable provision of energy. While the 20th century was characterized by growing dependence on fossil fuels, the current century would have to deal with the depletion of reserves of fossil fuels, growing environmental problems as a result of production and use of these fuels as well as the threat of climate change, which results from the emissions of GHGs (greenhouse gases) due to the combustion of fossil fuels. There are, therefore, several reasons for the world to explore with some urgency alternative sources of energy supply. The IPCC (Intergovernmental Panel on Climate Change) has clearly established that warming of the earth is unequivocal and that while adaptation measures in the short run would require to be implemented with a sense of urgency, mitigation measures to reduce the emissions of GHGs are crucial and essential.

This poses a major challenge for those dealing with energy decisions, particularly in the context of bringing about a major transition to renewable sources of energy. In 1998, I was the president of the International Association for Energy Economics, a body then consisting of about 3000 professionals including academics and senior leaders from the energy supply industry as well as other sectors. In my presidential address, on the occasion of the annual international conference of this body, I made the following statement: 'one area where our profession needs to make a much stronger entry than it has thus far achieved is in the field of energy–environment interface issues. Not only are the direct environmental effects of energy, such as air and water pollution and acid rain, serious enough to merit attention, but there is also now a definite basis for concern over the effects of energy use and production on the global climate. We can postpone a deeper interest in the subject only at the risk of a continuing insularity and myopia. Climate changes are already resulting in serious problems in the tropics in the form of frequent droughts and floods.'

Recent statements by several world leaders express the need to bring about deep cuts in the emission of GHGs, particularly carbon dioxide. This would be possible only with a major shift towards renewable energy technologies. The IPCC's *Fourth Assessment Report* has assessed several scenarios. In one of the scenarios, concentration of GHGs can be stabilized to limit temperature increase at an equilibrium of 2.0–2.4 °C. In order to reach this level of stabilization, a reduction of 50%–85% in the current level of emissions would be required by 2050. Needless to say, such a shift towards a low-carbon economy can only be achieved through a major movement towards greater use of renewable energy.

Development of renewable energy systems for various applications, coupled with their implementation, is a challenging task as it involves understanding of a multitude of disciplines. Developing countries including India need diverse renewable energy products and services, which need to be developed locally to make them affordable and sustainable. This book, which covers several aspects of design, sizing, and system integration, attempts to address the basic needs of a renewable energy designer and developer. The emphasis on the basics of engineering design and practical examples will make the task of system designers easy. The book is well timed to fulfil the needs of students as well as practitioners at all levels. It will also help students from diverse backgrounds such as architecture, biochemistry, and physics to understand the required basics of engineering science. I congratulate the editor and authors of *Renewable Energy Engineering and Technology – a knowledge compendium* for publishing such an important and useful piece of work.

R K Pachauri

Preface

This book is the partial outcome of a dream project that aimed at putting 'everything under the sun' in one place. Conceived nearly three years ago, this book was planned to cover various items, including core subjects like solar thermal engineering, photovoltaics, wind energy, biomass energy, along with ancillary subjects like economic/financial analysis, rural energy, techno-social interaction, energy planning, among others. In addition, we thought of covering different manufacturers, standards, policy issues, educational institutes, renewable energy experts, and so on. It was soon realized that it would not be possible to produce such a mammoth work in such a short span of time, and we settled to cover only the core subjects as a first attempt.

Every branch of engineering has its own genesis, growth, and stabilization. For example, as late as the 1960s, chemical engineering was considered as a new and somewhat risky branch with uncertain job markets, compared to conventional branches such as civil, mechanical, and electrical engineering. Soon, however, it established itself as a core branch of engineering. More recently, because of the IT (information technology) revolution, several branches of engineering have emerged, which were hitherto not considered as viable career options.

Renewable energy engineering is yet to establish itself as a viable branch. However, it may happen sooner than later. The Nobel Peace Prize for the IPCC (Intergovernmental Panel on Climate Change) is the ultimate recognition of the fact that climate change is a result of rapid development and that there are no viable solutions other than sustainable energy solutions. There will be a huge interest in the study of philosophy, science, engineering, and technology of sustainable energy in the coming years. In India, for example, the demand for energy professionals is growing rapidly, and the 1000-odd postgraduates coming out of various 'centres of energy' are deemed as inadequate compared to the demands of the growing industry. This book lays the foundation for an entirely new branch of engineering, renewable energy engineering.

There is vast and rich literature available on renewable energy. Several books on renewable energy, published both in India and abroad, are available, but most of these are in the specialization category. Also, many of these books are not targeted at practitioners. It is well known that renewable energy is a multidisciplinary area, requiring knowledge of basic sciences such as physics, chemistry, biology; applied sciences such as material science; and engineering subjects such as mechanical, chemical, and electrical. Many existing books on renewable energy cover only a few basics relevant to the topic. The main aim of this book is to provide a broad background so as to cover as many renewable energy topics as possible.

The other important aspect of the book is that almost all the authors who have contributed to the various chapters are essentially practitioners. Right from its formative years, TERI laid a lot of stress on product development and field implementation, along with basic and applied research. Thus, along with incorporating the scientific and technical content, we endeavoured to capture the rich practical experience gained at TERI during the past two decades.

The first two chapters establish the raison d' être of renewable energy. The third chapter tries to establish a foundation of broad scientific and engineering principles required to understand the design basis of renewable energy systems. This chapter includes topics like properties of matter, material and energy balances, thermodynamics, fluid flow, heat transfer, mass transfer, and is normally covered in the first two years of chemical/mechanical engineering streams. This chapter serves as a bridge material for non-engineering students and as a brief review for engineering students.

Chapters 4–7 cover solar energy basics and applications. The resource characteristics of solar radiation are covered in Chapter 4, followed by photovoltaics, solar thermal engineering, and solar passive architecture in Chapters 5, 6, and 7, respectively.

Wind energy is covered in Chapters 8 and 9, and small hydro is covered in Chapter 10. Four important energy sources, namely, geothermal, tidal, wave, and ocean thermal are covered in Chapter 11. We realize that each of these subjects deserves complete and independent treatment, but we had to limit the coverage owing to two reasons. First, not much work has been done in India and, hence, expertise was found to be lacking. Second, the potential for these resources, either technical or economical, is yet to be established firmly in the Indian context.

Chapters 12–15 cover various topics in bio-energy utilization, which is emerging as a very important subject both in developed and developing economics. The resource characteristics are covered in Chapter 12, followed by thermo-chemical conversion, biochemical methods of conversion, and liquid biofuels in Chapters 13, 14, and 15, respectively.

As mentioned earlier, there is an urgent and huge task of human resource development in renewable energy so as to tackle the challenges of climate change and sustainable development in the current millennium. We hope that this book serves as a pilot for much bigger human resource development initiatives.

V V N Kishore

Acknowledgements

The editor and authors thank Dr R K Pachauri, Director-General, TERI, for his overall support, encouragement, and perseverance during the course of writing this book.

Thanks are also due to Air Commodore M M Joshi (Retd), K P Eashwar, Beena Menon, Richa Sharma, Madhu Singh Sirohi, R K Joshi, Archana Singh, T Radhakrishnan and S Gopalakrishan of TERI Press, all of whom provided excellent support during the production of this book.

The secretarial help provided by C Surendran and administrative support given by R Venkatesan are gratefully acknowledged.

Internationalization support was received by Intelligent Renewable Energy, Freiburg, Germany (<www.intelligent-re.com>).

Contributors

V V N Kishore

Dr V V N Kishore is a Senior Fellow at the Energy–Environment Technology Division, TERI, and Adjunct Faculty at the Centre for Energy and Environment, TERI University.

He is a chemical engineer by training. After graduating from Andhra University, Waltair, he obtained his MTech degree and his doctorate from IIT (Indian Institute of Technology), Kanpur. He started working in the area of renewable energy technologies in 1978, first at IIT Kanpur and later at CSMCRI (Central Salt and Marine Chemicals Research Institute), Bhavnagar. He joined TERI in 1984 and served in various capacities as Area Convener, Dean (Energy–Environment Technology Division), Advisory Board member, and Resource Adviser. Over a long span of about three decades, he has acquired a vast experience in the broad area of renewable energy utilization, covering basic research, product/process development, field testing, technology transfer, execution of turnkey projects, policy research, consultancy, training and capacity building, and teaching. He has trained a large number of professionals in renewable energy in a wide range of topics such as solar thermal energy, biomass gasification, biomass densification, bio-methanation, solar passive building design, techno-economics of renewable energy systems, emission measurements, system integration, and so on. His current activities include distributed power generation through biomass gasification, gasifier thermal applications, high performance biomass cook stoves, lignocellulosic ethanol, and solar thermal–biomass hybrid systems. He has eight books, 160 publications in journals/conference proceedings, numerous reports, 30 products/end-use packages, and eight patents to his credit. He was a visiting fellow at Kennedy School of Government, Harvard University, USA and a consultant for the World Bank and for Winrock International, Washington. He teaches and guides PhD students at TERI University. He was on many expert evaluation/monitoring committees of MNRE (Ministry of New and

Renewable Energy), TIFAC (Technology Information, Forecasting, and Assessment Council), CSIR (Council of Scientific and Industrial Research), among others, and served SESI (Solar Energy Society of India) for many years as secretary and secretary/treasurer. He is a member of ASME (American Society of Mechanical Engineers), AIChE (American Institute of Chemical Engineers), and ACS (American Chemical Society), and a Fellow of World Technology Network. He is the recipient of Dr K S Rao Memorial Award of SESI for 2001 for outstanding contributions to the field of new and renewable sources of energy. He has travelled extensively all over the world on account of professional assignments, conferences, and so on.

Amit Kumar

Amit Kumar is a Senior Fellow and Director, Energy–Environment Technology Division, TERI. He studied mechanical engineering with specialization in thermal engineering from IIT, Roorkee. He has been working on the development and diffusion of renewable energy in India for 25 years. His experience ranges from policy and programme formulation, to project implementation, design and development of renewable energy technologies, and manufacturing of solar energy devices.

At present, he is responsible for research activities in the field of renewable energy, sustainable building design, and resource-efficient process technology applications in TERI. Besides coordinating part of a multi-country GEF-(Global Environment Facility) supported study to assess solar and wind resources in developing countries, he is also actively involved in a project related to the development of Sri Lanka's policy framework for renewable-resource-based electricity generation. At the field level, he managed the construction and commissioning of one of the largest solar ponds at Bhuj in western India from 1987 to 1996, and Asia's first solar-powered cold storage in the early 1980s.

Besides, he is also actively involved in activities related to knowledge management and sharing. He was a member of the Renewable Energy Subcommittee of the International Performance Measurement and Verification Protocol, US Department of Energy and the Climate Change Advisory Group to the Government of India. He was involved in the preparation of the National Renewable Energy Policy Statement and was member, Working Group for Non-conventional Energy Sources for formulation of the Tenth Five-year Plan; besides being an Alternate Member, Working Group on Non-traditional Sources of Energy and on R&D (research and development) in India within the Subgroup on Energy Security set up by the National Security

Council Secretariat, Government of India. He also worked as Renewable Energy Expert for the Asian Development Bank. He has published more than 20 papers in reputed national and international journals.

K V Rajeshwari

K V Rajeshwari is a Fellow at the Energy–Environment Technology Division, TERI. Her areas of interest include research and technology development in the field of waste water treatment and solid waste management. After having completed 10 years of research in the field of application of biochemical process for waste treatment, she has gained expertise in designing of biodigesters and various microbiological and analytical techniques. She has been involved in the development of reactor designs for industrial effluent treatment and solid waste digestion at IIT Delhi and TERI. A patent has been filed for her research resulting in the development of a novel process for degradation of solid wastes.

Jami Hossain

Jami Hossain is the Director, Programmes, and Secretary, Indian Wind Energy Association. A mechanical engineer by training, he is a known international expert on wind energy. Having implemented the very first wind farm projects with grid-connected wind turbines in India, he has more than 20 years of experience with renewable energies, primarily with wind energy. He also conducted the very first studies of assessment of potential for wind farms in India, grid integration of renewables, and assessment of wind energy sites based on geomorphological indicators that provided useful inputs to policy-makers in the early stages of wind power development in the country. He has played a key role in wind turbine technology transfer and was responsible for introducing cutting-edge wind turbine technologies from Europe to India through joint ventures and collaboration in manufacturing. He specializes in wind turbine technology, wind resource assessment, wind farm site selection, project planning and implementation, renewable energy programmes, policy formulation and interpretation, and modern communication approaches and technologies. In India, large wind farms have come up on many sites selected by him.

Jami Hossain has worked closely with various wind turbine manufacturers such as Enercon, Vestas, Micon, and Suzlon, and has been associated with research institutions such as TERI, New Delhi, and DRAL (Duff, Rutherford Appleton Laboratory), UK. He initiated wind energy research at TERI in 1986 and has trained many professionals under him, who are now serving the industry. He has also worked with international NGOs (non-governmental organizations) like Winrock International in their Clean Energy Group and has been a part of World Bank missions to Cambodia and Sri Lanka. He has been an external evaluator for many prestigious wind energy projects. He has published two books and more than 20 papers in international journals and conference proceedings.

In addition to interests in wind energy, Jami Hossain also has interests in journalism and interacts with industry associations. He has worked as Associate Editor in the Business Publications Division of the Indian Express group of newspapers. Currently, he is the editor of a magazine, *In Wind Chronicles*. He has been associated with the Independent Power Producers' Association of India and currently heads the programmes of InWEA (Indian Wind Energy Association). He also works closely with the WWEA (World Wind Energy Association) and is the member and coordinator of the International Programme Committee of the WWEA.

Malini Balakrishnan

Dr Malini Balakrishnan is a Fellow, Energy–Environment Technology Division, TERI. She currently leads the group on Resource Efficient Process Technology Applications. She has nearly 20 years of professional experience and her major interests are development and dissemination of environmentally sound technologies, with emphasis on membrane applications and value-added utilization of waste materials. In addition to research activities, she teaches postgraduate courses and guides doctoral students at TERI University, where she is an Adjunct Faculty. Dr Balakrishnan obtained her doctorate in biochemical engineering from IIT Delhi, masters in chemical engineering from the University of Waterloo, Canada, and a bachelors degree in chemical engineering from BITS (Birla Institute of Technology and Science), Pilani. She has over 70 publications in journals and conferences to her credit.

Sanjay P Mande

Sanjay Mande is a Fellow at the Energy–Environment Technology Division, TERI. After obtaining a bachelors degree in mechanical engineering, he completed his masters from IIT Mumbai, with thermal and fluids engineering as specialization. He has also completed his PhD in environmental sciences. For almost two decades, he has been working on the development and promotion of various renewable and energy-efficient technologies. He has been working on low-temperature solar thermal systems, development of solar double chimney system for thermal conditioning of buildings in composite climate, and development and promotion of biomass gasification technologies (both thermal as well as power applications) for various applications. Recently, he has initiated research activities in the area of liquid biofuels, optimizing bio-diesel production, and utilization technologies. He has one book, over 30 national international project reports, and over 55 publications in several national and international journals/conference proceedings to his credit. He has two patents—'energy-efficient dryer for cash crops' and 'an updraft gasifier use with a large cardamom drying system'. He is also an Adjunct Faculty at TERI University, where he teaches and guides PhD students. He is a life member of SESI, ISREE (Indian Society of Renewable Energy Education), Indian Chapter of the ICTP (International Centre for Theoretical Physics), and member of BUN India, India Section of the International Biomass User's Network. He has served as associate editor for the SESI journal and editor of the SESI newsletter for several years.

Ashish V Kulkarni

Ashish V Kulkarni is a Research Associate, Energy–Environment Technology Division, TERI. An electrical engineer from Nagpur, he has experience of working on different aspects of renewable-resource-based power-generating plants. He is involved in activities ranging from resource assessment, product development, installation, commissioning, and regulatory and policy-related aspects. He has also worked on electrical research and development projects such as development of electronic load controller at IIT Delhi.

Mahesh C Vipradas

Mahesh C Vipradas is a Fellow, Energy–Environment Technology Division, TERI. His core research area is solar thermal devices and he has been involved in design, development, and testing of a solar desalination system. He is presently working on a roadmap for renewable power in the restructured power sector. In addition, he has worked in the areas of climate change and developed baselines for renewable energy project in India. He holds an MSc degree in physics and MTech degree from IIT Delhi in energy studies.

Parimita Mohanty

Parimita Mohanty is an Associate Fellow, Energy–Environment Technology Division, TERI. She holds an MTech degree in energy science and technology from Jadavpur University. She has been involved in activities related to solar PV (photovoltaic) technology such as solar-PV-based product design and customization, testing of various solar PV products, project development and formulation of off-grid electrification project, study on rooftop grid interactive PV system, among others. She is actively involved in developing TERI Solverter ™ (patent applied), a solar-based power supply solution. She has also conducted various feasibility studies for electrification through solar home systems and power plants. She is also involved in sizing and developing appropriate PV system in buildings. She has also carried out studies on assessing various solar PV technologies.

Shirish Garud

Shirish Garud is a Fellow, Energy–Environment Technology Division, TERI. After completing his MTech in Energy Systems Engineering from IIT Mumbai (1986), he has been working in solar thermal and PV technologies for the past 20 years, during which he worked on various projects and technologies. He has wide experience in solar system designing, manufacturing, and project implementation. His areas of specialization include solar energy utilization; solar thermal systems and project development and management; designing of integrated PV project; solar system designing using simulation software; renewable energy project designing software like TRNSYS, RETScreen and HOMER; and hydrogen energy utilization. His outstanding contributions include design and installation of one of the world's largest solar water heating system of 120 000 litres per day capacity, commercialization of solar selective coating technology, and advanced solar system designs. He has published two papers and has contributed in many publications. He has also worked on the development of solar steam generation system using parabolic concentrators. He is known for his acumen/skill in training engineers and professionals in solar technologies.

Kusum Lata

Kusum Lata is a Fellow, Energy–Environment Technology Division, TERI. She is responsible for initiating and coordinating projects and activities related to research and consultancy on biomass energy, which mainly includes biomass gasification. She has handled various projects, some of the prominent ones being preparation of integrated energy management master plan for Bhutan, analysis and treatment of waste water generated in biomass gasifier systems, development of biomass-gasifier-based crematorium, field testing of gasifier-based systems, formulation of biomass policy options for promotion of biomass technologies, development of clean development mechanism project design document for conversion of municipal solid waste (500 tonnes per day) into RDF (refuse-derived fuel) pellets at Okhla, New Delhi, India, and so on. She is also a member of BUN India, SESI, and Indian chapter of ICTP. She has an MSc degree in microbiology and has submitted her PhD thesis for open defence to Jiwaji University, Gwalior.

Linoj Kumar N V

Linoj Kumar N V is a Research Associate, Energy–Environmental Technology Division, TERI. He has obtained his masters degree in Environmental Technology from Cochin University of Science and Technology, Kerala. Thereafter, he has been working in the field of liquid biofuels for the last five years. He has been Principal Investigator for various projects in this area, which are mainly supported by the Department of Science and Technology, Government of India; Praj Industries Ltd, Pune; GTZ (German Technical Development Corporation); and Coir Board, Government of India. While the major focus of his laboratory studies were cellulose-based ethanol production and bio-diesel production from vegetable oil, he has also simultaneously addressed the policy issues related to the sustainability of biofuel programmes in India. He has five publications to his credit. He has recently been selected for the BOYSCAST Fellowship, one of the prestigious fellowships offered to outstanding young scientists by the Department of Science and Technology, Government of India. He is also an active member of Indo-US Network on Green Chemistry.

Stuart L Ridgway

Stuart L Ridgway did his BS mathematics from Haverford College in 1943, and obtained his PhD degree in Nuclear Physics from Princeton University in 1952. His career in politically inspired technology development began in 1957, when he led a small group at Ramo-Wooldridge for developing a lead tolerant afterburner for automobile exhaust.

He became interested in ocean thermal energy conversion in the early 1970s, when there was an Arab oil embargo in the US. He has invented two concepts: the mist lift process and the ballistic deep water pipe. He has worked at Pacific International Center High Technology Research as a Senior Mechanical Engineer during 1989/90. He has also worked as R&D associate (1973–89). He was involved in low noise scientific instrument development in Princeton Applied Research (1966–73). Besides, he was employed in the California Institute of Technology, TRW Systems, Brookhaven National Lab, Princeton University, and US Naval Research Laboratory. He has published various papers in reputed international journals.

Vidisha Salunke, Architectural Energy Analyst, Green Building Studio Inc.

An architect with a master of science from Arizona State University in building design, Vidisha Salunke has collaborated with multidisciplinary design teams of mechanical and electrical engineers, landscape architects, and construction team members on numerous projects for the past eight years. Her area of expertise is solar passive building design integrated with whole building energy analysis. A LEED-(Leadership in Energy and Environmental Design) certified professional, she has been part of LEED green building certification projects on some of the most prestigious green building projects in India and the US. Building energy simulation along with lighting and daylighting modelling and analysis are her forte. She has been involved in several outreach, training, and documentation activities funded by the European Commission, USAID (United States Agency for International Development), and ministries of the Government of India, in partnership with international subject experts and organizations.

Prashant Bhanware

Prashant Bhanware is a graduate mechanical engineer from the National Institute of Technology, Raipur, with a masters of technology in energy systems engineering from IIT Bombay. He has also cleared the Energy Auditor Examination conducted by the Bureau of Energy Efficiency. He has more than three years of experience in energy efficiency, renewable energy, and demand-side management. His areas of interests include evaluation of solar-energy-based products, assessment of renewable energy technologies, and programme development for implementation of various demand-side management technologies.

Energy and development: concerns of the current millennium

V V N Kishore, Senior Fellow, Energy–Environment Technology Division, TERI, New Delhi

The advances of human civilization in the 20th century are closely linked with the unprecedented rise in energy consumption in general, and in hydrocarbon and electricity consumption, in particular. The per capita energy consumption plotted on a time scale compatible with the human existence is shown in Figure 1.1. From about 200 W per capita, the average energy consumption has shot to about 2000 W per capita in a short span of 100 years (see Box 1 for different ways of expressing energy consumption). This can be seen as a step function or an 'energy shock', since humans have been existing since the past five million years. The 20th century was the first era dominated by fossil fuels and a 16-fold rise in their use since 1900 created the first high-energy global civilization in human history.

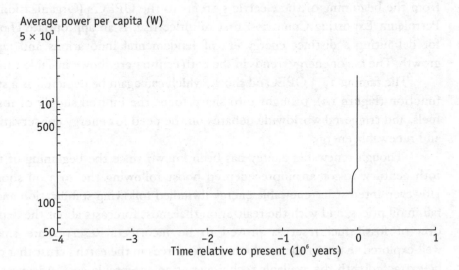

Figure 1.1 Evolution of average per capita energy consumption along the time scale of human existence
Reproduced with permission from Elsevier
Source Sorensen (1979)

Box 1 Different ways of expressing energy consumption

There are different ways of expressing energy consumption such as
EJ/year; W per capita or MTOE (million tonnes of oil equivalent).
1 EJ = 10^{18} J; 1 kg oil equivalent = 10 000 kcal

Example 1

The energy consumed by the poor, rural households in India is mainly in the form of firewood for cooking. The average consumption norm is 1 kg/person/day. Calculate the per capita power consumption in watts, assuming a calorific value of 4000 kcal/kg for firewood.

Solution

4000 kcal = $4000 \times 4.2 \times 10^3$ J

$$\text{Per capita consumption} = \frac{4000 \times 4.2 \times 10^3 \text{ J}}{1 \times 24 \times 3600 \text{ s}}$$

$$= 194.4 \text{ W, or 0.015 million Btu per capita}$$

Relative values for Germany and US are 176 million Btu per capita and 340.5 million Btu per capita, respectively.

Smil (2000) points out that the period of nine decades between 1882 and 1973, from the beginning of the electric systems to the OPEC's (Organization of Petroleum Exporting Countries) first oil price rise, is an appropriate choice for delimiting a distinct energy era of fundamental innovations and rapid growth. The major energy trends in the 20th century are shown in Table 1.1.

The famous 1973 OPEC oil shock, which can again be depicted as a step function (Figure 1.2), brought into sharp focus the limited supply of fossil fuels, and triggered worldwide debates on the need for energy conservation and renewable energy.

Though renewable energy has been known since the beginning of the 19th century, it got an unprecedented boost following the 1973 oil shock. However, interest in renewable energy dwindled following stabilization and a fall in oil prices, and with the realization that most forecasts about the depletion of fossil fuel reserves proved to be incorrect. Reserves are small, well-explored shares of the total mineral resources in the earth's crust that can be extracted with the available techniques at an acceptable cost. Advances in exploration and extraction constantly transfer fossil fuels from the broader and only poorly known resource category to the reserve pool. Resource exhaustion is thus not a matter of actual physical depletion but rather one of

Table 1.1 Energy trends in the twentieth century

Year	Population (10^9)	PEC (EJ)	Shares of PEC Coal (%)	Oil and gas (%)	Electricity generation (TWh)	Carbon intensity (tC/TJ)	World GDP (1990) (10^{12} USD)	Energy intensity (MJ/USD)
1900	1.6	22	95	5	8	24.3	2.0	11.0
1910	1.7	34	93	7	35	24.1	2.5	13.6
1920	1.8	40	88	11	85	23.3	2.7	14.8
1930	2.1	47	79	20	180	22.2	3.7	12.7
1940	2.3	57	74	25	340	22.6	4.2	13.6
1950	2.5	70	61	37	600	23.0	5.4	13.0
1960	3.0	115	52	46	2300	22.0	8.5	13.5
1970	3.7	189	34	64	5000	21.2	13.8	13.7
1980	4.4	250	31	65	8000	20.6	20.0	12.5
1990	5.3	320	30	61	11 800	18.6	27.4	11.9
2000	6.1	355	26	64	13 500	18.3	32.0	11.1

EJ – exajoules; TWh – terawatt-hour; tC – tonnes of carbon; TJ – terajoule; MJ – megajoule; PEC – primary energy consumption

Adapted from Smil (2000)

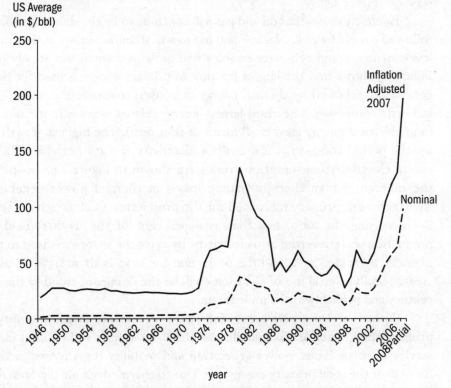

Figure 1.2 Annual average domestic crude oil prices, 1949 to present
Source <http://inflationdata.com/inflation/Inflation_Rate/Historical
_Oil_Prices_Table.asp>

eventually unacceptable costs. In spite of enormous cumulative recoveries during the past century (~250 Gt of coal, ~125 Gt of oil, and > 60 Tm^3 of gas), R/P (reserve/production) ratios of all fossil fuels in the year 2000 were substantially higher than those in 1900. The R/P ratio for coal is about 230 years; the global R/P ratio for crude oil was about 41 years in 2000 and that for natural gas was about 60 years in the late 1990s (Amoco 2000). Thus, even in the longer-term perspective of the next century of the millennium, exhaustion of fossil fuel reserves would be of primary concern. However, the highly dynamic nature of the exploration process means that although the fossil-fuelled civilization is energized by the recovery of finite, non-renewable resources, there are no serious and immediate concerns about the exhaustion of the resources. This is especially true for the industrialized countries that enjoy both economic and technological superiority. As a consequence, most projections of primary energy consumption in the world show relentless growth. One of the recent publications by IEA (2002) gives projections resource-wise (Figure 1.3), region-wise (Figure 1.4), and for energy-related services (Figure 1.5).

Figure 1.3 shows that oil and gas will continue to be the dominant sources followed closely by coal. Nuclear and hydro will stabilize almost at the current levels of usage and other renewables will grow at a small but steady pace. Figure 1.5 shows that the largest fraction of primary energy is used for power generation, followed by thermal energy in houses, commercial buildings, and industrial processes. The third-largest energy-related service is for mobility (non-electrical energy used in all forms of transport). The highest growth rate for the period 2000–30 is foreseen for electricity at 2.4% per year. The per capita electricity consumption trends are shown in Figure 1.6. In spite of the inherently high thermodynamic losses in thermal power generation, the fraction of primary fuels used for the production of electricity is steadily increasing. In 1900, less than two per cent of the world's fossil fuel production was converted into electricity. By 1950, the share increased to 10%, in 2000, it crossed 30%, and the projection for 2030 is about 45%. Thus, increased and universal use of electricity will be the dominant trend of the next century and probably of the millennium.

Figure 1.5 also shows a growing share of transportation (mobility) in primary energy use. An estimated 20% of the fossil fuels will be used for the service by 2030. Thus, power generation and mobility together will account for 65% of the total primary energy use. Due to thermodynamic dictates, more than two-third of this energy will have to be wasted so as to get the benefits of the modern energy services. This brings us to the subject of climate change, one of the main concerns of the millennium.

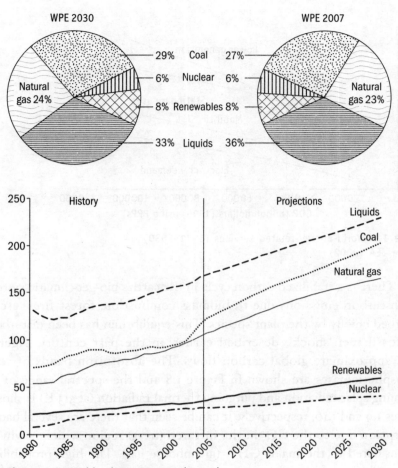

Figure 1.3 World primary energy demand
Source EIA (2007)

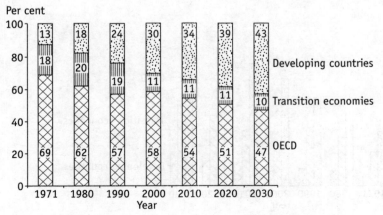

Figure 1.4 Regional share in world primary energy demand
Source IEA (2002)

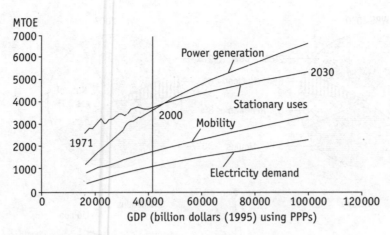

Figure 1.5 World-energy-related services (1971–2030)
Source IEA (2002)

There is a stabilized carbon cycle in the earth's bio-geoclimatic system in which carbon emissions due to biomass combustion, forest fires, etc., are absorbed chiefly by the plant species. This equilibrium has been disturbed by the fossil fuel 'shock', described earlier, in the past century. Figure 1.7 gives approximate global carbon flows. The absorption bands of several atmospheric gases are shown in Figure 1.8 and the spectral range of both incoming solar radiation and outgoing thermal radiation (at 313 K) is shown in Figures 1.9 and 1.10, respectively. It can be seen that there are several bands of absorption due to the CO_2 in the infrared (heat) region which is why CO_2 is considered as the major GHG (greenhouse gas). The high probability of

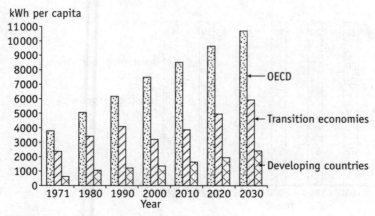

Figure 1.6 Per capita electricity consumption
Source IEA (2002)

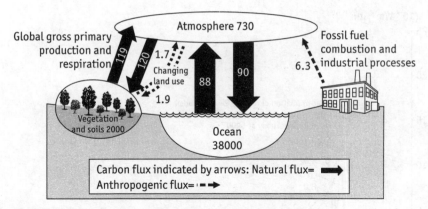

Figure 1.7 Global carbon cycle (billion metric tonnes of carbon)
Source http://www.eia.doe.gov/bookshelf/brochures/greenhouse/Chapter1.htm

relatively rapid atmospheric warning has emerged as the foremost global environment concern due to the combustion of fossil fuels. As early as in 1896, Arrhenius estimated that a geometric increase in CO_2 would result in a nearly arithmetic rise in surface temperatures, and that there would be minimum warming near the equator, maximum in the polar regions, and less warming in the southern hemisphere (Arrhenius 1896.). The current concern started in 1957 when Ravelle and Suess (1957) concluded that 'human beings are now carrying out a large-scale geophysical experiment of a kind that could not have happened in the past nor could be reproduced in future. Within a few centuries, we are returning to the atmosphere and the oceans, the concentrated organic carbon stored in the sedimentary rocks for over hundreds of millions of years.'

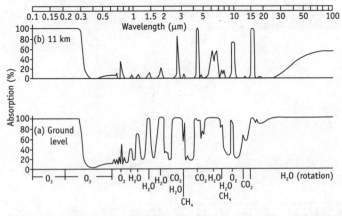

Figure 1.8 Absorption of diffuse terrestrial radiation and solar radiation by different atmospheric gases (a) at ground level (b) at a layer of atmosphere above 11 km
Source http://ceos.cnes.fr:8100/cdeom/ceos1/science/dg/dg1.htm

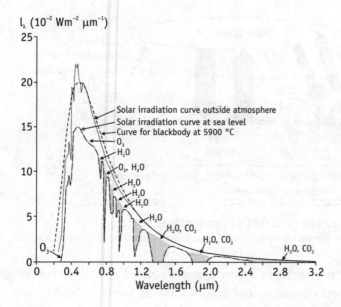

Figure 1.9 Spectral range of incoming solar radiation at the top and bottom of the earth's atmosphere, for the sun at the zenith
Source http://ceos.cnes.fr:8100/cdeom/ceos1/science/dg/dg1.htm

The first systematic measurements of the rising background CO_2 concentrations began in 1958 at two American observatories, Mauna Loa in Hawaii and the South Pole (Keeling 1998) (see Figure 1.11). Expanding computer capabilities made it possible to construct the first three-dimensional models of global climatic circulation, and their improved versions have been used to forecast changes arising from future CO_2 levels (Manabe

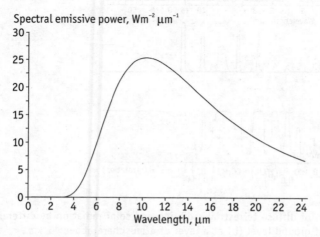

Figure 1.10 Spectral range of outgoing thermal radiation

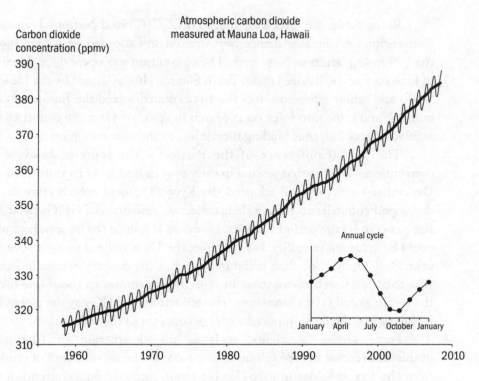

Carbon dioxide
concentration (ppmv)

Atmospheric carbon dioxide
measured at Mauna Loa, Hawaii

Figure 1.11 The 'Keeling curve', a long-term record of atmospheric CO_2 concentration measured at Mauna Loa Observatory
Source http://en.wikipedia.org/wiki/Image:Mauna_Loa_Carbon_Dioxide-en.svg

1997). During the 1980s, global climate change studies became increasingly interdisciplinary and the importance of other GHGs – CH_4, N_2O, and CFCs – was also recognized. Attempts have also been made to quantify both hydrospheric and biospheric fluxes, and the stores of carbon (Smil 1997). The key points of broad scientific consensus on all these matters are summarized in a series of reports by the IPCC (Intergovernmental Panel on Climate Change) (2001) (Houghton, Filho, Bruce, *et al.* 1996). The *Third Assessment Report* of IPCC incorporates new results for the past five years of climate change research. The global average surface temperature is projected to increase by 1.4–5.8 °C over the period 1900–2100, which will be greater than the increase over the past 10 000 years. This rise is likely to cause the mean sea level to rise by 0.09–0.88 m due to the thermal expansion of oceans, and the increased melting of glaciers and polar ice sheets. The real concern is, however, the rapid rate of temperature rise, which will give ecosystems less time to adapt, thus making them more vulnerable to the phenomenon of climate change.

Recognizing this problem, the UNFCCC (United Nations Framework Convention on Climate Change) was drafted and adopted on 9 May 1992 at the UN headquarters in New York. The convention was opened for signatures in June 1992 at the Rio de Janeiro Earth Summit. It was signed by the Heads of States and senior representatives from 154 countries (and the European Community), and came into force on 21 March 1994. As of 2003, 186 countries have ratified or acceded, thus binding themselves to the convention's terms.

The CoP (Conference of the Parties) – the supreme body of the convention – held its first session in early 1995 in Berlin. At its third session in December 1997, the CoP adopted the Kyoto Protocol, which commits the developed countries to reduce their collective emissions of GHGs by at least five per cent by the period 2008–12. However, it remains to be seen how much could be achieved in reality. For example, the US is obliged to emit seven per cent less CO_2 by 2007 than it did in 1990, but the country's emissions in the year 2000 were well above those in 1990. It contributes to about one-fifth of the entire global GHG emissions. The reduction of GHGs at the global level is thus one of the most important energy concerns of the millennium.

Some serious accidents, such as the destruction of Chernobyl's unshielded reactor in 1986 (Hohenemser 1988) or the massive spill of crude oil from the Exxon Valdez in 1989 (Keeble 1999), captured public attention with their images of horrifying damage, causing universal concern for nuclear power generation and for the excessive dependence on oil. Also, combustion of fossil fuels has been the largest source of anthropogenic (human-activity-related) emissions of SO_x and NO_x. Eventual oxidation of these compounds produces sulphates and nitrates that are responsible for the regional, and semi-continental acid deposition.

The third major energy concern of the millennium stems from the fact that about a third of the humanity does not have access to modern energy services and most of the people do not have the purchasing capacity to afford the services, even if they are made available. Figure 1.12 shows trends for societies that at a given time have the highest and the lowest average energy usage. Presently, while the poor countries use as little as 300 W per capita of energy, the rich countries use as high as 10 000 W per capita. The contradiction between rapid development and poverty has been addressed in a detailed manner in the famous Brundtland (1987) report *Our common future*. The report postulates that 'humanity has the ability to make development sustainable to ensure that it meets the needs of the present without compromising the ability of future generations to meet their own needs'. It further states that 'poverty is not only an evil in itself, but sustainable development requires meeting the basic needs of all and extending to all the

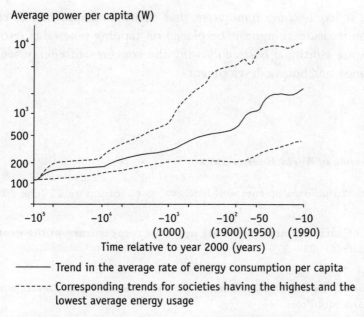

Figure 1.12 Widening gap in energy consumption trends in richer and poorer societies
Reproduced with permission from Elsevier
Source Sorensen (1979)

opportunity to fulfil their aspirations for a better life'. Consequently, modern energy services such as non-polluting cooking fuels and electricity are seen as a means to achieve 'aspirations for a better life'.

Recent global conferences have been increasingly addressing the issues of sustainable development. At the 1992 UNCED (United Nations Conference on Environment and Development), also called the Earth Summit or the Rio Summit, an ambitious strategy for sustainable development – *Agenda 21* – was launched. Though no specific chapter in Agenda 21 deals with energy, the need for new and more sustainable approaches in this key area is reflected in several chapters.

The MDGs (Millennium Development Goals) of the UN were developed in September 2000. The first of the seven goals was to halve the proportion of people living on less than one dollar a day by 2015. CSD-9, the Ninth Session of the Commission on Sustainable Development, held in April 2001, recognized the importance of energy in meeting this goal: 'To implement the goal accepted by the international community to halve the proportion of people living on less than one dollar a day by 2015, access to affordable energy services is a pre-requisite'. However, it is also realized that many developing countries find it very difficult to provide modern energy services to the poorer segments of the people at affordable prices.

Thus, it has become imperative that, apart from resorting to energy conservation measures, emphasis be placed on tapping renewable resources of energy on a continual basis, following the concerns of energy security, climate change, and human development.

References

Amoco B P. 2000
BP Statistical Review of World Energy 1999
London: BP Amoco
Details available at, <http://www.bp.com/worldenergy>, last accessed on 25 June 2005

Arrhenius S. 1896
On the influence of carbonic acid in the air upon the temperature of the ground
Philosophical Magazine 41: 237–276

Brundtland G (ed.). 1987
Our common future: The World Commission on Environment and Development
Oxford: Oxford University Press

Hohenemser C. 1988
The accident at Chernobyl: health and environmental consequences, and the implications for risk management
Annual Review Energy 13: 383–428

Houghton J J, Filho L G M, Bruce J, Lee H, Callander B A, Haites E, Harris N, Maskell K. (eds). 1996
Climate Change 1995: the science of climate change
New York: Cambridge University Press

IEA (International Energy Agency). 2002
World Energy Outlook
Paris: IEA publications. 530 pp.

IPCC (Intergovernmental Panel on Climate Change). 2001
Climate Change 2001
Cambridge: Cambridge University Press

Keeble J. 1999
Out of the Channel: the Exxon Valdez oil spill in Prince William Sound
Washington, DC: Eastern Washington University Press

Keeling C D. 1998
Reward and penalties of monitoring the earth
Annual Review of Energy Environment 23: 25–82

Manabe S. 1997
Early development in the study of greenhouse warming: the emergence of climate models
Ambio 26: 47–51

Ravelle R and Suess H E. 1957
Carbon dioxide exchange between atmosphere and ocean, and the question of an increase of atmospheric CO_2 during past decades
Tellus **9:** 18–27

Smil V. 1997
Cycles of life
New York: Scientific American Library

Smil V. 2000
Energy in the twentieth century: resources, conversions, costs, uses, and consequences
Annual Review of Energy Environment **25**: 21–51

Sorensen B. 1979
Renewable Energy
London: Academic Press. 683 pp.

2

Renewable energy utilization: desirability, feasibility, and the niches

V V N Kishore, Senior Fellow, Energy–Environment Technology Division, TERI, New Delhi

Several titles have been used in the past to characterize energy flows other than those associated with large-scale supply of modern energy services such as petrol, natural gas, and electricity. Such titles include alternative energy, non-conventional energy, ambient energy, and renewable energy, but the last term seems to have emerged as a consensus in the past decade or so. Renewable energy is an umbrella term, which includes solar energy, wind energy, hydropower, biomass energy, geothermal energy, tidal power, ocean thermal energy, wave energy, and energy from wastes, and it can be defined in several ways.

Sorensen (1979) defined renewable energy as 'energy flows that are replenished at the same rate as they are used', adding that it may include, more broadly, 'the usage of any energy storage reservoir that is being *refilled* at rates comparable to that of extraction'. The UK REAG (Renewable Energy Advisory Group) defined renewable energy as the term used to cover 'those energy flows that occur naturally and repeatedly in the environment, and can be harnessed for human benefit. The ultimate sources of most of this energy are the sun, gravity, and the earth's rotation.'

Nuclear power has been, and in some circles is still, projected by its proponents as a contending alternative to fossil fuels. However, since Chernobyl, public sentiment against nuclear power has been growing stronger in many parts of the world. In the *IEO (International Energy Outlook) 2007* reference case (<http://www.eia.doe.gov>), the nuclear share of the world's total electricity supply is projected to fall from 16% in 2004 to 12% by 2030, though in absolute terms, it will increase from 368 GW in 2004 to 481 GW in 2030. In order to give a clear identity to renewable energy, some prefer to use the term 'non-fossil, non-nuclear' energy.

Most renewable energy sources are derived from solar radiation (Figure 2.1). In recent years, forest and agro-residue wastes, municipal wastes, etc., are becoming an important source of energy which can also be termed as biomass, and hence, an indirect form of solar energy. The other renewables are

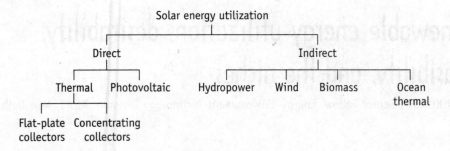

Figure 2.1 Classification of solar energy utilization

geothermal energy and tidal energy. Geothermal energy comes from the heat generated within the earth, and tidal energy results from the gravitational pull of the moon and the sun.

There have been two major lines of argument for promoting renewable energy. The first is that renewable energy is a local, natural resource and just like air and water, it is also inexhaustible. Thus, it offers enormous scope for achieving energy security that has tremendous economic and political implications for the oil-importing developing countries in the long run. The second argument is that renewable energy does not cause environmental hazards such as acid rain and greenhouse warming, and hence it is 'green'. This aspect of renewable energy is more appealing to the developed countries that are accused of being the 'primary polluters', and hence, carry the major burden of reducing greenhouse gas emissions as per the Kyoto Protocol.

Protagonists of renewable energy consistently project a higher future share of renewables in the energy mix. The exploitable resource potential is also projected to be much higher than the current use, especially for resources like solar energy, wind energy, and biomass energy. However, both the actual share and the realistic projections are lower.

For example, President Carter's target of a 20% share for renewables in the US by the year 2000 shrunk considerably under the subsequent Republican governments and the shorter-term, market-dominated policies of the 1980s. Government projections in the late 1980s estimated only a 9.5% contribution from renewables by 2000 (Jackson 1993). In fact, renewables constituted only 7% of the consumed energy supply in 2006. In 1986, the Dutch government set a target of meeting 10% of the electricity demand through renewables by the year 2000. In May 1992, the EC (European Commission) presented its proposals calling for a trebling of the percentage contribution of renewables to the electricity supply and a doubling of the contribution to the energy supply in the EC by the year 2005 (http: css.snre.umich.edu/facts, CSS03-12.pdf; mix_nl_en.pdf).

In comparison, renewable energy in the OECD (Organization for Economic Co-operation and Development) countries accounted for only 5.2% of the total primary energy supply in 2004. This share is projected to rise to 7.4% by 2020 (IEA 2001). At the global level, renewable energy accounts for about five per cent of the world's TPES (total primary energy supply), excluding bio-energy in the developing countries. If non-commercial bio-energy in the developing countries is included, the total share of renewables in TPES would be 14% (Renewables in global energy supply: an IEA fact sheet, file IEArenew_leaflet_sept2006_web.pdf).

Does renewable energy then really offer the potential to significantly displace conventional fuel supplies in the future energy mix? Also, if it does offer such potential, then why is the development taking place so slowly? Is it possible to speed up the process? Is it at all feasible to realize the Brundtland Commission's ambitious claim that renewables should be the 'foundation' for energy supply in the 21st century? Answers to such questions can emerge if one understands the constraints, both resource-based (their fluctuating nature, and technological and economic constraints) and market-based, under which renewables have to operate, compared with the elaborate organization of the modern energy systems based on fossil fuels. The rise in the prices of conventional energy could somehow always be 'contained' (despite the recent thrust on regulatory mechanisms), while the 'decline' in the renewable energy costs has not been significant (despite the thrust on increasing production volumes). It is also being realized worldwide that large- or mega-scale renewable energy projects do not take off well. Finally, it has been argued from the thermodynamic point of view that 'there are limits to the technological solutions to sustainable development (Huesemann 2003)' and that 'unless the underlying growth paradigm and its supporting values are altered, all the technical processes and manipulative cleverness in the world will not solve our problem and, in fact, will make them worse (Daly 1994)', indicating that a 100% renewables-based and yet growth-oriented society is not possible. It thus seems to be pragmatic to visualize several renewable energy 'niches' that in due course, will grow to form a significant part of the mainstream fossil-fuel-based energy systems.

As all forms of energy are used for one or the other application, satisfying one or the other need of the human society, it is appropriate to visualize an energy end-use *matrix* for renewable energy (Figure 2.2). Obviously, not all *cells* of the matrix would be relevant or feasible. For example, there would be very little sense in using expensive solar photovoltaic systems for cooking. On the other hand, the use of solar energy for domestic water heating would seem highly logical compared to burning coal in a power plant with about 30%

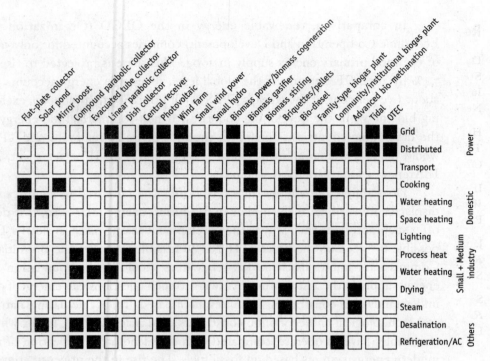

Figure 2.2 Technology–end-use matrix for renewable energy applications

efficiency to produce electricity, and then transporting it over hundreds of kilometres with the accompanying losses, merely to heat water by a few degrees. An energy end-use matrix would immediately highlight possible renewable energy options to replace, substitute, or augment the existing use of a commercial energy source. The physical, technical, and economical evaluation of the option can then be carried out to consider the future development and promotion. It is also important to understand the scientific and engineering basis for designing each resource/end-use *package*, that may require all engineering inputs that normally go into the development of any commercial product. Renewable energy product development would thus require a multi-disciplinary approach, necessitating large-scale creation of professionally trained renewable energy engineers in future.

References

Daly H. 1994
Steady-state economics
In *Ecology Key Concepts in Critical Theory*, edited by C Merchant
New Jersey: Humanities

Huesemann M H. 2003
The limits of technological solutions to sustainable development
Clean Technology Environment Policy 5: 21–34

IEA (International Energy Agency). 2001
World Energy Outlook
Paris: IEA publications. 421 pp.

Jackson T (ed.). 1993
Renewable Energy: prospects for implementation
Stockholm: Stockholm Environment Institute

Sorensen B. 1979
Renewable Energy
London: Academic Press. 683 pp.

3

Review of basic scientific and engineering principles

V V N Kishore*
Senior Fellow, Energy–Environment Technology Division, TERI, New Delhi

Renewable energy system

The renewable energy system consists of a particular energy source, a device, or a technology used to convert that source into a usable or more convenient form of energy and the end-use device. A solar homelighting system, for example, consists of sun (energy source), PV panel (conversion device), battery (storage device), and a compact fluorescent lamp (end-use device). Designers of such a system should thus be familiar with the solar radiation available, the characteristics of the PV panel, choice of batteries, and the lighting requirements of the user. It is apparent that several branches of science (meteorology, material science, electrochemistry, and illumination technology) are used in designing a proper system. Going by the variety of renewable energy sources and the range of end-use options, almost all known branches of science and engineering would be required to train a professional renewable energy engineer. This chapter attempts to provide a review of the different physical, chemical, and engineering sciences required to appreciate, use, and design renewable energy systems.

Units and dimensions

Numbers refer to quantities of known things such as distance, time, mass, money, area, volume,

*Technical assistance provided by Prashant Bhanware

speed, flow rate, density, and temperature. *Dimensions* are the basic concepts of measuring such quantities and *units* are the means of expressing the dimensions, as feet or centimetres for length, or hours or seconds for time. One should have a clear association of a particular unit with a particular quantity or dimension. For example, kWh is a unit of energy (similar to J, kcal, Btu, HP-h) but kW is a unit of power (similar to HP, kcal/h, and Btu/h), and there is often confusion between the two. If the generating capacity of a thermal power plant is stated to be 500 MW, it is not uncommon to interpret it as 500 MW per hour or as 500 MW per day, which has no meaning. A generating capacity of 500 MW (power) means that the plant has a capacity to produce 500 MWh of electricity (energy) per hour. The basic rule of handling units is quite simple: treat the units as you would treat algebraic symbols. For example, you may be able to add metres to metres or calories to calories, but you cannot add 10 m and 5 cal to get 15. One can, however, add units of the same dimension. For example, 1 HP and 2 kW can be added to get 2.746 kW, as 1 HP is equal to 0.746 kW. Conversion of units can sometimes be a vexing task but it is not difficult if one follows the basic principles.

Example 1
The speed of sound is 335 m/s. Convert this into km/h and miles/h.

Solution

$$335 \times \frac{1}{1000} \times \frac{3600}{1} = 1206 \qquad 1206 \times \frac{1}{1.61} = 749$$

$$\frac{m}{s} \times \frac{km}{m} \times \frac{s}{h} = \frac{km}{h} \qquad \frac{km}{h} \times \frac{mile}{km} = \frac{mile}{h}$$

Some basic conversions are given in Annexe 1. For online conversion of units, readers may refer to the website <www.gordonengland.co.uk/conversion>.

Properties of substances

Anything that takes up space or has mass is matter, and all known matter consists ultimately of one or more of the 118 elements listed in the periodic table (Mark 2005). Each element has an atomic structure characterized by a certain number of electrons, protons, and neutrons, and a specific atomic weight. Elements combine chemically to form compounds, the basic building blocks of which are molecules. Each compound has a well-defined molecular weight. Several naturally occurring substances are actually mixtures of several compounds. Substances exist in solid, liquid, or gaseous states at a given temperature and pressure, and can transform from one state to another upon the application of heat or pressure. Physical properties of substances, such as molecular weight, density, melting point, boiling point, and solubility are important for making engineering calculations. These, and several other properties, such as the vapour pressure data, thermodynamic properties (enthalpy, etc.), transport properties (viscosity, etc.), and kinetic data, are available in standard references, some of which are listed below.

- Perry's chemical engineers handbook
- Handbook of chemistry and physics
- Lange's handbook
- The properties of gases and liquids (Reid and Sherwood 1966)
- Internet data searches can be made at websites such as
  - http://srdata.nist.gov
  - http://www.asu.edu/lib/noble/chem/property.htm

The mole concept

The number of atoms in 12.01 g of carbon is 6.02×10^{23}. This number is termed as the Avogadro number. The atomic weight of carbon 12 is 12.01. This forms the basis of the definition of the mole. Mole can be defined as the number of molecules (or atoms) contained in a substance whose weight is exactly equal to its molecular weight. Depending on the units employed, one can define a g mole, a kg mole, or a pound mole. Thus, one g mole of carbon 12 is equal to 12.01 grams of carbon and one lb mole is equal to 12.01 lb of carbon. For the purpose of calculation, the following relations can be used.

$$g\,mole = \frac{grams}{molecular\,weight}; \; grams = (molecular\,weight) \times (g\,mole)$$

Example 2
Convert 100 kg of ethyl alcohol into g mole values.

Solution
The chemical formula of ethyl alcohol is C_2H_5OH. The molecular weight can be computed as

$$(12 \times 2) + (1 \times 5) + (16 \times 1) + (1 \times 1) = 46$$

$$g\,moles = \frac{100\,000}{46} = 2173.9$$

The importance of the mole concept lies in the fact that chemical reactions can be treated quantitatively. Thus, the chemical reaction for the combustion of carbon

$$C \text{ (carbon)} + O_2 \text{ (oxygen)} = CO_2 \text{ (carbon dioxide)}$$

can be interpreted quantitatively as '1 g mole of carbon reacts with 1 g mole of oxygen to produce 1 g mole of carbon dioxide'. If one has to calculate, for example, how much carbon dioxide is produced by the burning of 1 kg of carbon, the mole concept comes handy.

1 kg of carbon = 1000/12 = 83.33 g moles of carbon. This would produce 83.33 g mole of carbon dioxide. However, the molecular weight of carbon dioxide being 44, the mass of carbon dioxide produced would be 83.33 × 44 = 3666.7 g or about 3.67 kg.

Some conventions and definitions

Certain definitions and conventions, which are commonly used in expressing physical properties and for carrying out calculations and analysis, are as follows.

Unit prefixes

In energy and environment matters, we deal with very large and very small quantities. Thus, we have GW on one hand and μg per m³ on the other. It is important to remember the prefixes listed below.

More than one

(da)	deka	=	ten	= 10^1
(h)	hecto	=	one hundred	= 10^2
(k)	kilo	=	one thousand	= 10^3
(M)	mega	=	one million	= 10^6
(G)	giga	=	one billion	= 10^9
(T)	tera	=	one million million (trillion)	= 10^{12}
(P)	peta	=	one million billion	= 10^{15}
(E)	exa	=	one billion billion	= 10^{18}

Less than one

(d)	deci	=	one tenth	= 10^{-1}
(c)	centi	=	one hundredth	= 10^{-2}

(m) milli   = one thousandth = 10^{-3}
(μ) micro   = one millionth = 10^{-6}
(n) nano   = one billionth = 10^{-9}
(p) pico   = one trillionth = 10^{-12}
(f) femto   = 10^{-15}
(a) atto   = 10^{-18}

Mass, weight, and force

Mass is a fundamental unit, the others being distance and time. Mass is generally indestructible, except in cases involving nuclear reactions. Weight, on the other hand, is a unit of force, but in daily practice, we do not distinguish between the two. If you weigh 60 kg, you have a mass of 60 kg. However, on the moon, for example, you will weigh less, even though the mass remains 60 kg. To distinguish between the two, the terms kg_m and kg_f were used earlier. It is important to understand the difference to avoid confusion in calculations. For example, steam or air pressure is generally expressed in psi or kg/cm^2. The units lb and kg are the units of force here. A better understanding can be obtained by going back to Newton's second law of motion which states that force is proportional to the rate of change of momentum, or for constant mass, to the product of mass and acceleration.

$$F \propto ma$$
$$F = C\,ma \qquad\qquad\qquad ...(3.1)$$

where F is force, m is mass, a is acceleration (rate of change of velocity), and C is a proportionality constant, whose numerical value and units depend on those selected for F, m, and a.

In the SI system of units, the unit of force is N (Newton), which is the force required to accelerate a mass of 1 kg to a value of 1 m/s^2. Hence, if F is in N, m is 1 kg, and a is 1 m/s^2, the proportionality constant C is numerically equal to 1. Similarly, in the CGS system, the unit of force is defined as dyne, which is the force required to accelerate 1 g through 1 cm/s^2. The proportionality constant is again set to be 1.

In the American engineering system, however, the unit of force is lb_f (pound force), which is defined to be the force acting on a mass of 1 pound (lb_m) on the surface of the earth. Thus, the acceleration a is set equal to the acceleration due to gravity, which is about 32.2 ft/s^2 (9.8 m/s^2). The constant C is given a numerical value of 1/32.2 and has the units of $lb_f/[lb_m(ft/s^2)]$.

The inverse of C is denoted as g_c (gravitational constant) and has the value 32.2. Similar to lb_f, we also have kg_f (kg force) in which case C takes the value $1/9.81$ and the units of $kg_f / [kg_m (m/s^2)]$.

Volume

When dealing with quantitative terms related to mass (such as mass flow rate and heating value per unit mass), it is best to use kg as the unit (kg/s, kJ/kg, etc.). However, it is a common practice to express quantities of liquids and gases in volumetric terms. The price of petrol is thus shown in rupees per litre or cents per gallon, and the flow rate of gas is expressed in million cubic metres per day. As the liquid and gas densities are known to vary significantly with temperature, this kind of expression can result in confusion if the temperature is not specified along with the value, or if the actual density is not known. To avoid such confusion, the volumes of gases are usually expressed in Nm^3, which means that the volume has been normalized to that at 0 °C temperature and 1 atm pressure. Ideal gas laws described later govern the conversion from one set of conditions of temperature and pressure to a reference set.

Density, specific gravity, and specific volume

Density is the mass contained in a unit volume of any given substance. The common units for density are g/cm^3 or kg/m^3. It is helpful to remember that the densities of most liquids are in the vicinity of 1 g/cm^3 or 1000 kg/m^3, and that the densities of many gases at ordinary temperatures and pressures are about one thousandth of this value. Density varies with temperature and pressure, the variation being the lowest for solids and the highest for gases. For solutions and mixtures, it also varies with the concentration. Thus, the density of saline water varies from a low of 1.0 g/cm^3 for pure water to a high of 1.360 g/cm^3 for sea bitterns, and such data is crucial for construction and maintenance of salinity gradient solar ponds, which will be described in later chapters. The terms true density and bulk density are used frequently when dealing with substances having considerable porosity or air spaces. True density refers to the density of the substance if there were no void fractions, whereas bulk density includes voids. The bulk density of, say, rice husk will be considerably lower than its true density because of the large air spaces present in it.

Specific gravity is the ratio of two densities – that of a substance of that interest to that of a reference substance – and hence is dimensionless. The reference substance for liquids and solids is normally water, and for gases is frequently air. As the densities vary with temperature, it is a good practice to state the temperatures at which the densities have been measured. Thus, if the specific gravity of a salt solution is given as $d_{20}^{20} = 1.08$, it should be interpreted that the measurements of both the salt solution and the reference substance have been made at 20 °C. As mentioned above, it is common to use distilled water at 4 °C as the reference substance for which the numerical value of the density is close to 1.0 g/cm³. Thus, the numerical values of density and specific gravity (for solids and liquids) are the same in the CGS system of units. Seawater-related industries use a specific density unit °Bé called Baume, defined as follows.

$$\text{Specific gravity (60 °F/60 °F)} = \frac{145}{145 - °Bé} \qquad \qquad ...(3.2)$$

In the petroleum industry, the specific gravity of petroleum products is usually reported in terms of a hydrometer scale called API. The equation for the API scale is

$$1 API = \frac{141.5}{\text{Specific gravity (60 °F/60 °F)}} - 131.5$$

$$...(3.3)$$

The *specific volume* of any compound is the inverse of the density, that is, the volume per unit mass of the substance and is expressed in cm³/g or m³/kg.

Mole fraction and weight fraction

Mole fraction is the number of moles of a given compound in a mixture divided by the total number of moles present. Similarly, weight fraction (or more accurately the mass fraction) is nothing but the weight of the substance divided by the total weight of all substances present. These fractions are represented mathematically as follows.

Mole fraction of A = (moles of A)/(total number of moles)
Weight fraction of A = (weight of A)/(total weight)

The mole per cent or weight per cent is obtained by multiplying the above fractions by 100. For gases and liquids, the term volume fraction is also used. If the gases obey the perfect gas law, it can be shown that the volume fraction is equal to the mole fraction. For substances that usually contain moisture, the mass or mole fractions are sometimes reported on a 'dry basis', which means that the moisture or water content has been omitted while calculating the fractions. In drying calculations, it is common to use both the 'dry basis' and the 'wet basis'. Similarly, fuels containing ash in addition to moisture sometimes report fractions or heat values on an MAF (moisture- and ash-free) basis. Mole fractions can be converted into weight fraction and vice versa if the molecular weights are known, as illustrated in the following example.

Example 3
Atmospheric air contains approximately 21% oxygen and 79% nitrogen by volume. Calculate the weight fractions, assuming that air is an ideal gas mixture.

Solution
Choose a basis of 100 g mole of air. As the ideal gas law is valid, the volume fraction is same as the mole fraction. The weight of oxygen is obtained as

$$21 \text{ (g mole)} \times 32 = 672 \text{ g}$$

and the weight of nitrogen is obtained as
$$79 \text{ (g mole)} \times 28 = 2212 \text{ g}$$

The total weight is thus 2884 g. The weight fractions of oxygen and nitrogen can then be obtained as

Weight fraction of oxygen   = 672/2884   = 0.233
Weight fraction of nitrogen = 2212/2884 = 0.767

When the mole fractions or weight fractions are too small, or when solutions are involved, the term *concentration* is used. Concentration refers to the quantity of some solute per fixed amount of solvent or solution in a mixture of two or more components. It is variously expressed as g of solute/litre, g mole/litre, PPM (parts per million), mg/m^3, $\mu g/m^3$, etc. The terms molarity, molality, and normality also refer to solute concentrations.

Basis In the above example, a basis of 100 g moles was selected as a starting point for further calculation. The concept of a basis is very important for all calculations involving material and energy flows. The basis is the reference chosen for the calculations one plans to make and its proper choice makes it easier to calculate. There is no rigid rule on how to select the basis for a particular calculation, but with experience, one can make out the basis from what is given and what has to be found out.

Temperature Temperature is a measure of the thermal energy of the random motion of molecules of a substance. It is measured on four different scales: Celsius (centigrade), Fahrenheit, Kelvin, and Rankine. The first two are relative scales and the last two are absolute scales. The absolute zero on the Kelvin scale is −273.15 °C and on the Rankine scale it is −459.67 °F. In the SI system of units, Kelvin is used without the superscript. Thus, the boiling point of water will be 373 K and the melting point will be 273 K. The following formulae can be used for conversions.

$$T_K = t_C + 273.15 \qquad \qquad ...(3.4)$$
$$T_{Rank} = t_F + 459.67 \qquad \qquad ...(3.5)$$
$$t_C = (5/9)(t_F - 32) \qquad \qquad ...(3.6)$$
$$t_F = (9/5)t_C + 32 \qquad \qquad ...(3.7)$$

where subscripts K, C, Rank, and F indicate that the temperature is measured in Kelvin, Centigrade, Rankine, and Fahrenheit, respectively.

Several devices such as mercury-in-glass thermometer, a variety of thermocouples, thermistors, bimetallic strips, and pyrometers are used to measure temperature in different ranges.

Pressure and vacuum

Pressure is defined as the force per unit area. Depending upon the units employed for force, pressure is expressed as psi, kg/cm^2, N/m^2, etc. Pounds and kilograms are force terms and not mass values (for mass, kg_m can be used and for force, kg_f can be used). The pressure exerted on a unit surface area on earth due to the weight of the atmospheric air column is called the atmospheric pressure and is variously expressed as follows.

1 atm; 1.013 bar; 101.3 kPa; 14.7 psia; 760 mmHg; 34 ft of water; 1034 cm of water; 1.013×10^5 N/m^2; and 1.033 kg/cm^2

Let us consider the term psia (a stands for absolute). A pressure gauge at ambient conditions will actually show no deflection as it is not subjected to any differential pressure. To distinguish between what is shown by a pressure gauge and the actual pressure, engineers use the term psig (g stands for gauge). The conversion between the two quantities is given as follows

psia = psig + 14.7

Any system which is below atmospheric pressure is normally measured in mmHg or torr (1 torr = 1 mmHg). The study of vacuum engineering is important in renewable energy. For example, ETCs (evacuated tubular collectors) and certain kinds of absorption refrigeration systems (LiBr–H_2O) employ vacuum systems.

Chemical equations and stoichiometry

A chemical equation provides both qualitative and quantitative information about a specific chemical reaction or a set of reactions. The equation

$$C + O_2 = CO_2$$

conveys the information that
(1) carbon and oxygen combine to form carbon dioxide, and
(2) one mole of carbon combines with one mole of oxygen to form one mole of carbon dioxide. (Note that the number of carbon and oxygen atoms on both sides of the equation is same.)

Consider another equation representing the combustion of propane (a constituent of LPG).

$$C_3H_8 + O_2 = CO_2 + H_2O$$

As per this equation, propane and oxygen combine to produce carbon dioxide and water. However, the number of carbon atoms on both sides of the equation is not same, and so is the case with oxygen and hydrogen atoms. To get information on how much oxygen is needed to burn propane completely, the equation has to be balanced. This is accomplished by rewriting the equation as

$$C_3H_8 + xO_2 = yCO_2 + zH_2O$$

according to which one mole of propane combines with x moles of oxygen to yield y moles of carbon dioxide and z moles of water. The values of x, y, and z can be found out by balancing the number of atoms on both sides.

Carbon   : $3 = y$
Oxygen   : $2x = 2y + z$
Hydrogen : $8 = 2z$

It can be easily found out from the above set of equations that the values of x, y, and z are 5, 3, and 4, respectively. The balanced equation can now be written as www.tech

$$C_3H_8 + 5O_2 = 3CO_2 + 4H_2O$$

which now gives the additional quantitative information that one mole of propane has to combine with five moles of oxygen to produce three moles of carbon dioxide and four moles of water for complete combustion. One can also calculate the weights of the different compounds as shown below.

Basis 1 kg or 1000 g of propane

Number of g mol of propane	=	1000/44	=	22.73
Number of g mol of oxygen required	=	5 × 22.73	=	113.65
Weight of oxygen	=	113.65 × 32	=	3636.8 g
Number of g mol of carbon dioxide produced	=	3 × 22.73	=	68.19
Weight of CO_2	=	68.19 × 44	=	3000.3 g
Number of g mol of water produced	=	4 × 22.73	=	90.92
Weight of H_2O	=	90.92 × 18	=	1636.5 g

It can be seen that the weight of reactants on the left-hand side of the equation (1000 + 3636.8) matches the weight of the products on the right-hand side (3000.3 + 1636.5). Mass is thus conserved, and this can be used as a check to see if the equation is correctly balanced.

The amount of air required for complete combustion can also be found from the above method. As air contains about 21% of oxygen on a molar basis, each mole of oxygen is accompanied by 100/21 mol of air. Hence, 113.65 mol of oxygen required for propane combustion would require 541.2 g moles of air. Multiplying by the average molecular weight of air (28.9), we get 15 640 g or 15.64 kg of air for combustion. The air–fuel ratio is thus 15.64 for complete combustion of propane and this amount of air is referred to as *stoichiometric air*. The actual air supplied is usually higher to ensure low emissions. The excess oxygen supplied comes out in flue gases (products of combustion). Flue gas composition is routinely measured by energy engineers to get information on excess air supplied and to calculate energy losses. If oxygen (air) supply is inadequate, combustion will be incomplete and this will result in undesirable emissions such as soot and carbon monoxide.

Sometimes information on the heat of the reaction (or combustion) is also contained in the chemical equation. Heat is either released (reaction is exothermic) or absorbed (reaction is endothermic) in a process. The usual convention followed is that ΔH_R (heat of reaction) is negative for exothermic reactions and positive for endothermic reactions. With this additional information, one can calculate the heating/cooling needs of the equipment, theoretical (adiabatic) flame temperature, flue gas temperature, etc.

In engineering practice, chemical reactions are often incomplete, that is all the reactants do not get converted completely into products. To express the degree of completeness, the terms conversion or yield are used. Conversion is the fraction of the main reactant that is converted into products, whereas yield is the ratio of the weight of a particular product to the weight of a particular reactant.

Solids, liquids, vapours, and gases

Consider the following questions that frequently arise in renewable energy utilization.

1 Why cannot biogas/producer gas be stored and sold in cylinders like LPG?
2 Why instead of water, a high-temperature oil is used in some concentrating solar collectors?
3 What is the difference between an organic Rankine cycle engine and a steam engine?
4 How can one design equipment to remove tars from the producer gas in a biomass gasifier?
5 What is a suitable material other than water for storing thermal energy?
6 Why is an antifreeze material used in solar flat plate collectors in cold climates?
7 How does one select suitable materials for designing a solar adsorption cooling system?
8 Does a wind turbine perform differently in cold and hot climates?

Questions such as these require a good understanding of the properties of different materials and the knowledge of how these properties change over a range of temperatures and pressures. A huge amount of data is available on many useful compounds and mixtures, but it is not possible to have reliable experimental data on all materials. Hence, we rely on estimating or predicting properties based on certain well-known laws (such as the ideal gas law) and on empirical correlations.

The following section attempts to review some well-known principles related to properties of material.

Phases and the phase rule

Any given system is characterized by the number of pure components, phases (solid, liquid, or gas), and given conditions of temperature and pressure. The phase rule states that at *equilibrium,* these are related by an equation

$$F = C - P + 2 \qquad\qquad ...(3.8)$$

where F is the number of degrees of freedom (such as temperature and pressure, which have to be specified to determine all the intensive properties), C is the number of components, and P is the number of phases that can exist in the system. As an example, for distilled water at about room temperature, C = 1 and P = 1. Hence, the degrees of freedom would be two: these are generally temperature and pressure. This means that if temperature and pressure are specified, all other intensive properties, such as density, etc. are fixed at equilibrium. Thus, distilled water at 298 K and 1 atm can have only one density. The numerical value of the density, however, has to be decided by experimentation. There is also no way of knowing if a given substance will be in a solid, liquid, or gaseous phase at a given temperature and pressure. For this, knowledge of the so-called p–V–T (pressure–volume–temperature) properties is necessary.

p–V–T properties

For a clear understanding of the phase phenomena, the properties of the substances should be shown in three dimensions representing the three most important properties, namely pressure, volume, and temperature. As such, a representation will be somewhat complex; we can first look at two-dimensional plots, such as the p–V diagram, p–T diagram, and the T–V diagram. A typical p–V diagram is shown in Figure 3.1, in which different curves are shown for different temperatures of carbon dioxide.

Figure 3.1 shows that at a temperature of 52 °C, carbon dioxide is in a gaseous state and an increase in pressure results in a decrease in its specific volume. At 31 °C, however, there is an inflexion in the curve at about 75 atm, above which the reduction in volume is quite low even for a large increase in pressure. If the temperature is reduced to 25 °C, the shape of the curve is markedly different. At a certain pressure, there is a large change in volume, signifying condensation and formation of liquid, which takes place at constant pressure. After the liquefaction is completed, the liquid volume changes very little even for a large increase in pressure and this is reflected in

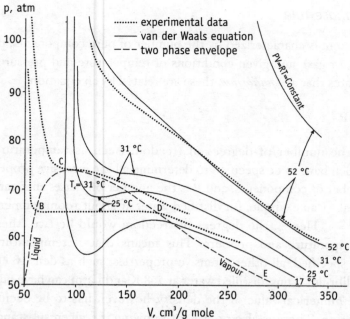

Figure 3.1 p–V–T properties of carbon dioxide

the statement that liquids are incompressible. The two-phase envelop marks the boundary between the liquid and the vapour phases, and any point within this envelop signifies the presence of both the liquid and the vapour phases in equilibrium. The point of inflexion at 31 °C is called the *critical point,* and the temperature, pressure, and volume at this point are called *critical constants.* Note that at this point, there is very little volume or density change accompanying the phase change from vapour to liquid. Also, as the pressure increases, the difference between the liquid and gas densities reduces until these are equal at the critical point. If a gas is above the critical temperature and pressure, it cannot be liquefied.

If the temperatures are reduced further, solidification will occur at some stage. The point where vapour, liquid, and solid coexist is called the triple point. When the temperatures are reduced further, there will be a region in which only the solid and the vapour phases are present. The gases (for example, iodine) go directly into the solid state in this region, a phenomenon known as sublimation.

Other representations of the p–V–T properties are pressure versus temperature and temperature versus volume. The best representation is, of course, a three-dimensional plot in which we talk about p–V–T surfaces and regions where different phases are present.

Example 4

Biogas (60% methane and 40% carbon dioxide) is a clean alternative fuel for the rural areas. Examine the feasibility of storing and transporting it in liquid form. The daily requirement of biogas for cooking is about 2 m³ for an average family. Using the following data, compute the storage volumes required for a month if storage is in the form of gas and liquid.

Density of biogas	1.128 kg/m³
Methane properties	
Boiling point	111.67 K
Critical temperature	190.7 K
Critical pressure	45.8 atm
Critical volume	99 cm³/g.mol
Density (at bp)	422.62 kg/m³
CO_2 properties	
Sublimation point	195 K
Critical temperature	304.2 K
Critical pressure	72.9 atm
Critical volume	94 cm³/g.mol
Liquid density	1032 kg/m³
(–20 °C and 19.7 atm)	

Solution

As biogas is a mixture of gases and our understanding of the phase phenomena of mixtures is not adequate, we can only have some gross conclusions. First of all, we note that the critical pressures of both the components are quite high and hence biogas can be compressed up to 45.8 atm beyond which methane is likely to exist only in a gaseous state. Also, the critical temperature of methane is 82 °C below zero at which point carbon dioxide would have sublimed (gone into the solid state directly from the gaseous state). Hence, biogas would have to be cooled much below –82 °C for liquefaction. The normal boiling point (temperature at which a liquid boils at atmospheric pressure) of methane is 161.5 °C below zero, which means that if we do not want to cool the gas to such low temperatures, we will need to compress the gas to high pressures. Hence, if we go by the critical properties of methane, biogas has to be both cooled to very low temperatures, and pressurized for storage and transportation as a liquid.

There is a method of estimating the critical constants of a mixture, according to which the mixture property is a mole-weighted average of the individual components (Kay's rule). The critical temperature of biogas can thus be estimated as

$$0.6 \times 190.7 + 0.4 \times 304.2 = 236.1 \text{ K or } -37\ ^\circ\text{C}$$

Similarly, the critical pressure can be estimated as

$$0.6 \times 45.8 + 0.4 \times 72.9 = 56.6 \text{ atm}$$

These values indicate that gas will still have to be cooled to temperatures much below zero and compressed to high pressures to achieve liquefaction. The actual phase phenomena of gas mixtures can be quite different and will have to be established through experimentation.

Calculation of volumes

The volume of gas storage can be calculated straight away as 60 m³. The calculation of liquid volume is a little more involved.

Weight of biogas to be stored	$= 60 \times 1.128$	$=$	67.7 kg
(compare this with 14.2 kg of LPG in a regular cylinder)			
Average molecular weight of biogas	$= 0.6 \times 16 + 0.4 \times 44$	$=$	27.2
kg moles of biogas to be stored	$= 67.7/27.2$	$=$	2.5

Assuming that the liquid densities given do not vary much in the range of temperatures and pressures relevant to liquefied biogas, one can proceed as follows.

Specific volume of methane on molar basis = 16/422.6 = 0.03786 m³/kg mole
Specific volume of carbon dioxide on molar basis = 44/1032 = 0.0426 m³/kg mole
Specific volume of liquefied biogas = 0.6 × 0.03786 + 0.4 × 0.0426 =
0.04 m³/kg mole
Liquid volume = 0.04 × 2.5 = 0.1 m³

If we assume that the liquid has to be stored in a cylinder with its height equal to twice the diameter, the dimensions of the tank can be calculated as d = 0.4 m and h = 0.8 m. The actual dimensions will have to be larger to accommodate some pressurized gas in the cylinder.

Vapour pressure

Gas below the critical temperature is usually called vapour as it can condense to become liquid. In the vapour–liquid region of the phase diagram, vapour and liquid coexist at equilibrium, and the pressure of the system corresponding to the equilibrium temperature is called the vapour pressure at that temperature. For a pure compound under these conditions, the number of degrees of freedom can be calculated as

$$F = C - P + 2 = 1 - 2 + 2 = 1$$

which means that only one parameter needs to be specified to describe the system. This implies that there can be only one value for vapour pressure corresponding to a given temperature. The vapour pressure increases with temperature and the vapour pressure curve is unique to a given compound. The variation of the vapour pressure of water with temperature is given in Table 3.1.

Vapour pressure increases rapidly with temperature and at 100 °C, it equals 760 mmHg, that is, the atmospheric pressure. This means that boiling takes place when vapour pressure exactly equals the atmospheric pressure. Also, if boiling or cooking is done in an open vessel in high-altitude areas where the atmospheric pressure is less than 760 mmHg, water would boil at a lower temperature. That is why longer times are needed for cooking in these areas. It is thus a good policy to promote pressure cookers in high-altitude regions to achieve higher cooking temperatures.

At a temperature of 30 °C, the vapour–liquid system at equilibrium will have a pressure of 32 mmHg, which is less than the atmospheric pressure. It means that if water is kept at 30 °C in a closed vessel and if the pressure is reduced by connecting the system to a vacuum pump, water would start boiling when pressure reaches a value of 32 mmHg. This is why an LiBr–water absorption cooling system, in which water is the refrigerant, has to operate at sub-atmospheric pressures.

The vapour pressure data for a large number of substances can be described by the Antoine equation, which is widely used for the estimation of vapour pressure.

Table 3.1 Variation of vapour pressure of water with temperature

t (°C)	0	5	10	20	30	40	50	60	70	80	90	100
p (mmHg)	4.579	6.543	9.209	17.535	31.824	55.324	92.51	149.38	233.7	355.1	525.76	760.00

$$\ln p_s = A - \frac{B}{t+C}$$

where p_s is the vapour pressure in mmHg, t is temperature in °C, and A, B, and C are constants. For water, A= 18.403, B= 3885, and C=230. Maximum error between the experimental values and those calculated using the above constants for water is about 0.5%. The Antoine equation can also be written as

$$\log_{10} (p_s) = A' - \frac{B'}{t+C}$$

where $A' = A/2.303$ and $B' = B/2.303$. The values of A', B', and C for several substances are listed in Lange's Handbook of Chemistry (Lange and Dean 1985).

The vapour pressure data is not linear over wide temperature ranges and tends to have a slight curvature. This curvature can be tackled by using a special plot known as Cox's chart. Using water as a reference substance, a special non-linear temperature scale is constructed on the y-axis, with log p_s as the x-axis. Vapour pressure data for a large number of substances will follow a straight line if plotted on this special chart. Cox's chart is extremely useful to extrapolate or interpolate vapour pressure data of any substance, if two values at different temperatures are known.

Many times, it is important to predict properties of a mixture of a pure vapour with a non-condensable gas. Practical examples are air–water vapour mixtures, water vapour–biogas mixtures, producer gas–water vapour mixtures, etc. When pure air comes in contact with water, some amount of water will evaporate into the air. If the contact is prolonged enough, an equilibrium will be attained in which the partial pressure of water vapour in air is equal to the vapour pressure of water at the given temperature. Air is then said to be saturated with water vapour. When we say that the relative humidity of air is 60% at 30 °C, it means that the partial pressure of water vapour in air is 60% of the saturation value at that temperature, which is about 32 mmHg. Thus, the partial pressure of water vapour will be 0.6 × 32 or 19.2 mmHg under this condition. It is seen that stating a value for relative humidity without mentioning the temperature has no meaning. We will learn more about air–water mixtures in the section on mass transfer operations.

Ideal gas law and real gas equations

The well-known ideal gas law equation is the result of findings by Charles, Boyle, Gay-Lussac, Dalton, and Amagat, who originally developed correlations between the temperature, pressure, and volume of gases considered ideal. The conditions for ideal behaviour are generally understood to be such under which the average distance between molecules is large enough to neglect the effect of intermolecular forces. There are no ideal gases in the real world, but deviations from the ideal gas law can be small for certain gases at certain temperatures and pressures. The ideal gas law is stated as

$$pV = nRT \qquad\qquad ...(3.9)$$

where p and T are absolute pressure and temperature, respectively, V is volume, n is number of moles, and R is universal gas constant. The numerical value of the gas constant depends upon the units employed for pressure, etc. Some common values of the gas constant are given below.

R = 8.314 J/mol K
  = 1.987 cal/mol K
  = 82.0568 cm³.atm/mol K

The relation between gas properties at different conditions can easily be obtained as

$$\frac{p_1 V_1}{T_1} = \frac{p_2 V_2}{T_2} \qquad\qquad ... (3.10)$$

This equation is generally used to *normalize* gas volumes to a set of standard conditions. The normal conditions, referred to as the NTP (normal temperature and pressure), correspond to 0 °C and 1 atm. Another set, known as STP (standard temperature and pressure), corresponds to 25 °C and 1 atm. If the volume of a gas is known at a given temperature and pressure, the corresponding volume at NTP or STP can be calculated from Equation 3.10. It is customary, for example, to mention the output of producer gas in Nm³/h. N here signifies that the normalized condition refers to NTP. The flow of natural gas is sometimes stated in standard cubic feet, meaning that the reference condition is STP. The above equation can also be used to calculate the volume of gas at any given temperature and pressure, provided that the volume is known at other conditions.

Example 5
A wood gasifier generates 2.5 Nm³ of producer gas for every kilogram of wood consumed. The exit temperature of gas is about 500 °C. Calculate the gas flow rate at the outlet of the gasifier if the wood consumption rate is 100 kg/h.

Solution

The gas flow rate at normal conditions is $100 \times 2.5 = 250$ m³/h. Assuming that the pressure at the exit of the gasifier is almost the same as the atmospheric pressure, the volumetric flow rate is calculated as $V_2 = 250 \times (773/273) = 708$ m³/h.

In order to calculate the gas density ρ, the gas law equation can be written in another form as

$$\rho = (pM / RT) \qquad \qquad \text{... (3.11)}$$

where M is the molecular weight of gas.

Ideal gas mixtures

According to Dalton's postulation of partial pressures, the total pressure of a gas mixture is equal to the sum of partial pressures exerted by individual gases. The partial pressure is defined as the pressure that would result if the same number of moles of the individual gas were to occupy the total volume of the container at the same temperature. Thus, if there are two gases with subscripts 1 and 2, we can write

$$p_1 = n_1 RT / V; \; p_2 = n_2 RT/V; \text{ and } p = p_1 + p_2$$

$$\text{Also, } \frac{p_1}{p} = \frac{n_1}{n} = \text{mole fraction} = y_1$$

Alternatively, volumes occupied by individual gases at the same total pressure and temperature are called partial volumes. Amagat's law of partial volumes, which is analogous to Dalton's law of partial pressures, states that the total volume occupied by the gas mixture is the sum of partial volumes. We can thus write

$$V_1 + V_2 = V$$

and

$$\frac{V_1}{V} = \frac{n_1}{n} = y_1$$

The above equation states that for an ideal gas law mixture, the volume fraction is equal to the mole fraction. From a practical point of view, this result is very important because the measurement of gas composition (for example by Orsat apparatus) is normally done by volumetric methods.

Real gas equations

Real gases can deviate considerably from the ideal gas law equation, especially at high pressures. Some commonly known equations to describe the real gas behaviour are the van der Waals, Berthelot, Redlich–Kwong, Beattie–Bridgeman, and Benedict–Webb–Rubin equations. The van der Waals equation, developed with some theoretical basis, is expressed as

$$\left(p+\frac{a}{v^2}\right)(v-b)=RT \qquad\qquad\qquad ...\,(3.12)$$

where v is the volume per unit mole, and a and b are known as van der Waals constants. A standard way of dealing with real gases is through the use of generalized compressibility charts that plot the compressibility factor z as a function of reduced pressure p_r and temperature T_r. The following equations describe the method.

$$z = z\,(p_r, T_r) = pv/RT$$
$$T_r = T/T_c$$
$$p_r = p/p_c$$
$$v_r = v/v_c$$

where T_c, p_c, and v_c correspond to critical conditions. Generalized compressibilitry charts are available in standard textbooks and reference books mentioned earlier.

Material and energy balances

Mass and energy balances are very important in engineering practice. In fact, these are so fundamental that they can be applied to any system that involves the flow of materials and energy. Applications range from chemical reactors and process industry to water balance for river basins and nitrogen flows in the biosphere. The basis for the mass/energy balance is the principle of conservation which states that mass/energy can neither be created nor destroyed. The material balance equation for any given system can be written as

Input – Output = Accumulation

Input represents all materials entering the system boundary, while output represents all materials flowing out of the system boundary, and accumulation refers to materials retained within or depleted from the system boundary. The basis for calculations can be one hour or one day or any convenient time interval. Material balance can be done for either continuous flows or batch systems. If a chemical reaction is involved, the above equation is modified as follows.

$$(\text{Input}) - (\text{Output}) + \begin{pmatrix} \text{Generation} \\ \text{within} \\ \text{the} \\ \text{system} \end{pmatrix} - \begin{pmatrix} \text{Consumption} \\ \text{within} \\ \text{the} \\ \text{system} \end{pmatrix} = \begin{pmatrix} \text{Accumulation} \\ \text{within} \\ \text{the} \\ \text{system} \end{pmatrix}$$

If the material is accumulated within the system, the term on the right-hand side of the equation is positive, and if it is getting depleted, the term would be negative. The following examples serve to illustrate the use of material balance equations.

Example 6
Dung slurry entering the gas pipe in a fixed-dome biogas plant

A fixed-dome biogas plant operates with variable gas pressure and variable gas volume inside the plant (Figure 3.2). Dung slurry (an equal mixture of dung and water) is fed into the digester at a rate of 100 kg per day. The gas produced in the digester pushes down the slurry level within the digester, which, in turn, causes the spent slurry to be pushed out of the digester through an outlet box. The plant is designed in a way that if 2 m^3 of gas is produced daily, the same volume of slurry is displaced inside the outlet box and hence any amount in excess (primarily equal to the volume of the slurry fed daily) goes out of the box through an overflow weir. In winter, however, gas production is less due to lower microbial activity caused by lower digester temperatures. Assuming that the gas production rate reduces by about 30% in winter, show that the slurry can ultimately enter the gas pipe. Assume further that the permanent gas space within the digester is equal to 1 m^3 and that the inlet opening of the gas pipe is at the same level as the opening for slurry outflow in the box.

Solution
It can be assumed that gas flow at about 2 kg/day is negligible compared to liquid flow (about 100 kg/day). As less amount of gas is produced, slurry level in the outlet box rises to a lower height than before and, therefore, does not reach the height of the overflow weir. Thus, there is no net outflow of slurry from the digester. Applying the material balance equation

100 (kg/day) – 0 (kg/day) = Accumulation (kg/day)

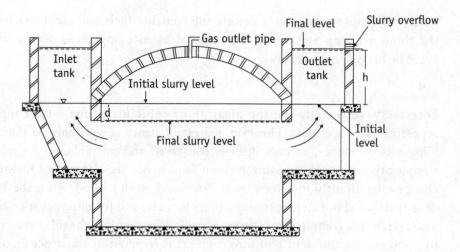

Figure 3.2 Slurry movements in fixed-dome biogas plant

As input is greater than output, accumulation term is positive. Hence, slurry accumulates in the digester at a rate of 100 kg/day or 0.1 m³/day (assuming that the density of slurry is the same as that of water). The slurry level inside the digester will rise at a rate depending on the geometry of the digester. Ten days would be needed to fill up 1 m³ of the gas space.

Note The above solution is somewhat simplistic. In reality, accumulated slurry will redistribute between the digester and the outlet tank, and hence, the initial slurry level in the outlet tank will keep on increasing. Once there is sufficient accumulation, the daily displacement of 1.4 m³ could be sufficient to push some slurry again on a daily basis through the overflow weir. If this process stabilizes, the slurry may not enter the gas pipe at all!

Example 7
Fuelwood for cooking and forest depletion

Biomass fuels, consisting of firewood, agricultural residues, and dung cakes, are the primary sources of energy for cooking in rural India. An estimate puts the annual consumption level of firewood at 252 million tonnes. India's forest area is about 64 million ha and the growing stock is estimated to be 4741 million m³ at an average of 74 m³/ha. The average annual production of firewood is 1.15 m³/ha, totalling to 72.8 million m³. Several policy-makers predicted that the unsustained fuelwood consumption in the rural areas would cause severe deforestation. However, satellite data has shown that there is not much reduction in forest area despite the fact that fuelwood continues to be the prime cooking fuel. Using the material balance approach, discuss what could be happening in reality.

Solution

Foresters traditionally use the measure of cubic metre for estimating or reporting fuelwood mass. However, material balance should only be done in kilograms or tonnes, as only mass is conserved and not volume. An added complexity is that the moisture content (and hence the density) of fuelwood changes significantly for green (wet) wood and air-dry wood. Also, the bulk density of wood is different for logs, branches, etc. and for different species of wood. Hence, a complete knowledge of bulk density values for different types of wood and for different moisture contents is required to construct material balances. Such data is not easily available, and we assume that wood has a uniform and constant bulk density of 550 kg/m³. We further assume that the system boundary consists of the forest area only. The system boundary and the various material inflows are represented schematically in Figure 3.3.

Applying the material balance equation,

0 − 252 + 40 = Accumulation

The accumulation term is negative, which means that there is a net depletion of 212 tonnes/year. This means that there is a net depletion of SB (standing biomass), which is obtained by multiplying the growing stock per unit area M by the total area A of the forests.

SB = MA   ...(3.13)

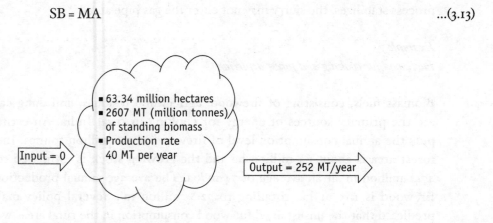

Figure 3.3 Biomass flows in a forest area

The material balance equation can be expressed mathematically as

$$\frac{d(MA)}{dt} = M\frac{dA}{dt} + A\frac{dM}{dt} = -212$$

If M is assumed to be constant, the equation reduces to

$$M\,(dA/dt) = -212,$$
$$\text{or }\ dA = (-212/M)\,dt$$

Integrating between time intervals t_1 and t_2, we get

$$A_2 - A_1 = (-212/M)\,(t_2 - t_1)$$

For a known value of M (75 m³/ha or 41.25 tonnes/ha) and for an initial value of A (63.34 million ha), the forest area remaining after a period of, say, 10 years can be calculated from the above equation as 11.95 million ha. Satellite surveys, however, have shown that the forest cover is not decreasing. Hence, if we assume that A is constant, the material balance equation can be written as

$$A\,(dM/dt) = -212$$
$$M_2 - M_1 = (-212/A)\,(t_2 - t_1)$$

The growing stock after a period of 10 years is estimated to be 7.78 tonnes/ha or 14.1 m³/ha from the above equation. This means that if the forest cover is not reducing then the forests must be thinning. If both these things are not happening, one must conclude that the material balance equations should be reformulated using new assumptions or new data. For example, any of the following can be done.

- Formulate separate material balance equations for different categories of forest land (dense forest, open forest, non-forest, etc.) with regional values for fuelwood usage
- Collect fresh information on fuelwood supply sources (private lands, road-side plantations, inter-region transport of fuelwood, availability of wild growing species such as *Prosopis juliflora*, etc.)
- Check the spread of fuelwood-saving devices such as improved stoves and biogas plants
- Check data on the increasing use of commercial fuels such as LPG

It should be emphasized that the results of material balance equations are as good as the available data. In examples such as the one cited above, good field data is often not available, and is, hence, substituted with assumptions, which may or may not be valid.

In cases where two or more components are present in material streams and chemical reactions are not occurring, material balance is applied to separate components to get the desired information. The following example on drying illustrates such a case.

Example 8

A total of 100 kg of tomatoes with an initial moisture content of 94% have been dried in a solar dryer to a final weight of 20 kg. The desired final moisture content is five per cent. Calculate the moisture content in the dried product to check if it is dried properly. Also find out the final weight if the product is dried to the desired moisture content.

Solution
Basis: 100 kg
System boundary Solar dryer

The drying process is represented in Figure 3.4.

Moisture contained in the feed material evaporates into the air stream during the drying process. Hence, the weights of the input and output streams are not equal. However, the dry matter in the feed will remain constant throughout the drying process. Hence, material balance can be made on the dry matter for which there is no accumulation.

Moisture content in fresh tomatoes = $100 \times 0.94 = 94$ kg
Weight of dry matter = 6 kg
Let the moisture content of the final product be x_2.

As the dry matter remains intact during the drying process, the material balance can be written as

Input (dry matter) = Output (dry matter)

$(1 - x_2) \times 20 = 6$; $x_2 = 0.7$ or 70%, which means that the product is under-dried.

If W_2 is the weight of the product if it was dried to five per cent moisture content, then

$(1 - 0.05) W_2 = 6$; $W_2 = 6.32$ kg

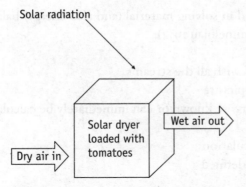

Figure 3.4 Flow diagram of a solar dryer

If chemical reactions are involved, material balance can be made on individual elements. An example of carbon balance and green house gas emissions for combustion of firewood in a stove is shown in Figure 3.5 (Smith, Khalil, Rasmussen, *et al.* 1993).

Material balances can become quite tedious if there are several streams with different components entering and leaving the system boundary, and if chemical reactions are involved. Also, recycling, bypass, and purges are common in practical applications, and these add to the complexities. It is thus

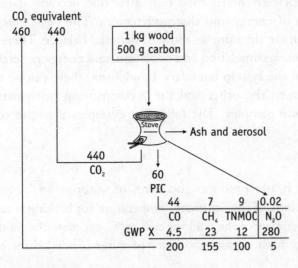

Figure 3.5 Carbon balance of a stove

TNMOC – total non-methane organic carbon; GWP – global warming potential
A figure of 23 for CH_4 implies that CH_4 is 23 times more potent than CO_2 for global warming.

best to have a strategy to be used in solving material (and also energy) balance equations as outlined below (Himmelblau 1974).

- Draw a picture of the process with all the streams
- Place all available data in the picture
- See what masses/compositions are known or can immediately be calculated for each stream
- Select a suitable basis for calculation
- Make sure the system is well-defined

As mentioned earlier, one can write

- a total material balance
- a component material balance, and
- an elemental material balance.

In addition, one can make energy balances and very often, combined material and energy balances.

Material balances can be tedious, but energy balances are complex, primarily because there are several forms of energy, and various concepts and rules are involved in converting one form to another. The complex nature of energy balances will be clearly understood only after one becomes familiar with the various forms of energy and thermodynamics. The basic rules of energy balance, however, are the same as those of material balance. There are input and output streams, accumulation and depletion, and energy generation and consumption within the system boundary. In addition, there can be conversions from one form to the other and the accompanying mathematical formulations can be quite complex. The following examples illustrate some simple energy balances.

Example 9

A domestic solar water heater produces 100 litres of water at 60 °C under normative design conditions. If the optimum temperature for bathing is taken as 39 °C and if the temperature of the cold water is 15 °C, estimate the number of persons who can take bath. Assume 30 litres of water is needed for each person.

Solution

This is a problem of mixing, and involves both material and energy balance (see Figure 3.6).

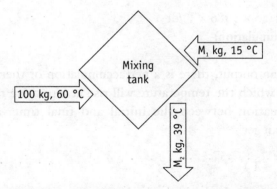

Figure 3.6 Mixing of water bodies at different temperatures

Number of people who can take bath = $M_2/30$

Hence it is required to find M_2. The energy content of each stream can be taken as equal to MC_pT, where C_p is the specific heat and T is the temperature. C_p can be taken as 1 kcal/kg °C for water. The material balance is written as

$$100 + M_1 = M_2 \qquad \qquad ...(3.14)$$

The energy balance can be written as

$$100 \times C_p \times 60 + M_1 \times C_p \times 15 = M_2 \times C_p \times 39$$
$$6000 + 15\,M_1 = 39\,M_2 \qquad \qquad ...(3.15)$$

From Equations 3.14 and 3.15, we obtain M_2 as 187.5. Hence, the number of people who can take bath is 187.5/30 = 6.25 or 6.

Example 10

A bucket containing 20 litres of water is heated by an immersion rod of capacity 1 kW. The initial temperature of water is 15 °C. Find out the time taken to heat the water to 60 °C, assuming that there are no heat losses.

Solution

This is a case of conversion of electrical energy into thermal energy. It is known that this conversion can take place with 100% efficiency. Energy is supplied at the rate of 1 kW (input) and there are no energy losses (output = 0). The thermal energy content of water can be taken as MC_pT. The energy balance can be written as

$$1 - 0 \qquad = d\,(20 \times 4.186 \times T)/dt$$
(Input) (Output) (Accumulation)

As input is greater than output, there is a net accumulation of thermal energy in water because of which the temperature will rise at a positive rate. Integrating the above equation between the initial and final time, and temperature intervals, we can get

$$t_2 - t_1 = 83.72\,(T_2 - T_1)$$

Substituting $T_2 = 60$, $T_1 = 15$, and $t_1 = 0$, we get $t_2 = 3767$ s or 62.8 minutes.

The problem would have been complicated if the heat losses were included in the output stream. Heat losses through convection and radiation would be non-linearly dependent on temperature (radiation losses vary with the fourth power of temperature), evaporation losses would depend on the difference between the vapour pressure of water (which increases exponentially with temperature) and the partial pressure of water in the surrounding atmosphere, and there would be a mass loss accompanying evaporation. The resultant differential equation would be non-linear and would require numerical techniques to solve it.

Different forms of energy

The word energy is derived from the Greek words en (in) and ergon (work). A complete and more technical treatment of energy and its conversion from one form to another will be covered in the next section. This section deals with a preliminary description of various forms of energy, listed below.
- Kinetic energy
- Gravitational potential energy
- Electrical energy
- Chemical energy
- Biochemical energy
- Electrochemical energy
- Atomic or nuclear energy

Kinetic energy

It is the energy possessed by any moving object and is given by

$$E_k = \tfrac{1}{2}\, mv^2 \qquad\qquad\qquad\qquad ...(3.16)$$

where m is mass and v is velocity of the object.

Gravitational potential energy

Energy is required to lift objects because the gravitational pull of the earth opposes the upward movement. As the object is raised (a cricket ball lifted above your head or a few thousand tonnes of water pumped into a reservoir above ground), the input energy is stored in a form called gravitational potential energy (or just potential energy). The gravitational force pulling an object towards the earth is called the weight of the object and is equal to its mass m multiplied by the acceleration due to gravity g (9.81 m/s^2). The potential energy stored in raising the object to a height h is given by

$$E_p = m \times g \times h \qquad\qquad\qquad ...(3.17)$$

Electrical energy

On a scale used to describe atomic or molecular phenomena, it is not gravity but electrical forces that are dominant.

A more familiar form of electrical energy is electricity. Electric current is the organized flow of electrons in a material, most often a metal. Metals are substances in which one or two electrons from each atom become detached and move freely through the lattice structure of the material. It is the presence of these free electrons, which allows metals to carry electrical currents. Maintaining a steady flow of electrons requires a constant input of energy because the electrons continually lose energy in collisions with the metal lattice (the resulting increase in kinetic energy of the metal atoms is the reason why wires get hot when they carry electric current).

The basic unit of charge is that carried by the electron. A much larger unit of charge and the one we use to define the electronic charge, is called the Coulomb.

1 Coulomb = 6.24273×10^{18} unit charges

or the electronic charge is defined as 1.60209×10^{-19} Coulombs. A larger unit, Faraday, is equal to 96 493 Coulomb. The rate of current flow, or ampere is defined as 1 Coulomb per second.

The potential energy per unit charge at a point in the electric field is called the potential V. The unit of potential is the volt, defined as 1 Joule (energy)/Coulomb (charge). Since a Coulomb is the quantity of charge equivalent to 6.24273×10^{18} electrons, we have

$1\,V = 1.60209 \times 10^{-19}$ J/electron

Thus, we have a new energy term called eV (electron-volt) defined as

$1\,eV = 1.60209 \times 10^{-19}$ J

In virtually every power station in the world, the generators operate on a principle discovered by Michael Faraday in 1832, that voltage is induced in a coil of wire rotating in a magnetic field B existing between the north pole N south pole S of an electromagnet, as shown in Figure 3.7.

Connecting the ends of the coil to an electric circuit through a resistance R will then allow the current (I) to flow. The electrical energy can, in turn, be transformed into heat, light, motion, or whatever, depending upon what is connected to the circuit. Thus, electricity is a convenient intermediary form of energy, used to allow energy released from one source to be converted to another quite different form.

Another, more subtle form of electrical energy is carried by electromagnetic radiation. More properly called electromagnetic energy, this is the form in which, for example, solar energy reaches the earth. It travels as a wave and can carry energy through empty space (vacuum). The wavelength (or frequency) determines its form, which includes X-rays, ultraviolet radiation, infrared radiation, microwaves, radio waves, and the small band of wavelength that our eyes can detect which we call visible light. Lower the wavelength, higher the energy. The electromagnetic spectrum is shown in Figure 3.8.

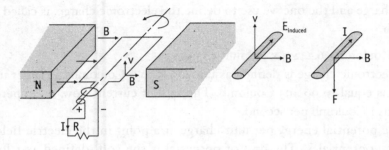

Figure 3.7 Schematic diagram of an electrical generator

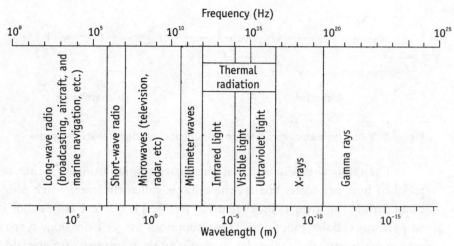

Figure 3.8 Electromagnetic spectrum

Chemical energy

Chemical energy is the energy released as a result of electron interactions in which two or more atoms and/or molecules combine to produce a more stable chemical compound. Thus, chemical energy, viewed at the atomic level, can be considered to be a form of electrical energy and it exists only as a stored energy form. If energy is released in a chemical reaction then it is called an exothermic reaction and if energy is absorbed then it is termed endothermic reaction. The energy released is commonly reported in units of calories or joules per unit mole or mass of fuel reactant. The most important source of fuel energy for the human race is the exothermic chemical reaction called *combustion*, which involves the oxidation of carbonaceous fuels.

Biochemical energy

Energy currency of the living cells is a chemical compound called ATP (adenosine triphosphate), which consists of a nitrogenous base, adenine, linked to the five-carbon sugar, ribose (Figure 3.9). A string of three phosphate molecules is linked to the sugar molecule. A phosphate group has one atom of phosphorous and three atoms of oxygen.

Most energy of the ATP molecule is in the bonds of the two phosphate groups at the end. When an ATP molecule reacts with water with the help of an enzyme, the bond between the second and the third phosphate is broken and ADP (adenosine diphosphate) is produced. The energy released can be measured as heat energy.

$$ATP + H_2O \rightarrow ADP + P_i + 30 \text{ kJ}$$

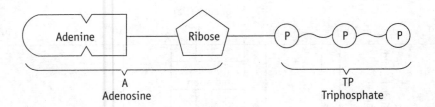

Figure 3.9 Diagrammatic representation of adenosine triphosphate molecule

The removal of the terminal phosphate residue from ATP thus releases 30 kJ of heat per mole. Hence, the bond is called the energy-rich phosphate bond. Inside the living cells, the energy released is not lost as heat but is used to perform cellular functions. These functions are cell division, synthesis of new constituents and molecules, osmotic work, transport of materials across the membrane, nervous conduction, and muscular contraction.

When a molecule of ATP gives up one energy-rich phosphate group, it transforms into ADP as mentioned above. To form a molecule of ATP again, ADP must combine with one phosphate group. The energy required for such a combination is supplied by the breakdown of organic compounds, for example glucose within the cell.

Living organisms obtain energy for synthesis of the ATP by breaking down organic compounds such as sugars, fats, and amino acids. However, the primary source of energy is the sun. Light energy from the sun is utilized by green plants through the process of photosynthesis to synthesize sugars from carbon dioxide and water. Animals, in general, depend on plants for organic compounds in the form of food, which is utilized for production of energy through the process of respiration. Although carbon dioxide, water, and oxygen are cycled between respiration and photosynthesis, energy flows in one direction in the living world (Figure 3.10).

Energy is converted from sunlight by photosynthesis and released by respiration. In the living cells, the functions of respiration and photosynthesis are performed by two special organelles, mitochondria and chloroplasts, respectively.

Mitochondria are known as the 'power houses' of the cell. These are present in all plant and animal cells, with the notable exception of bacteria and blue-green algae. A human liver cell may contain about 1000 mitochondria while a kidney cell may have 300–400.

Mitochondria perform the main functions of conversion and transfer of cellular energy, through synthesis, storage, and release of ATP for use in cellular activities, and to govern the transportation of materials and water in and

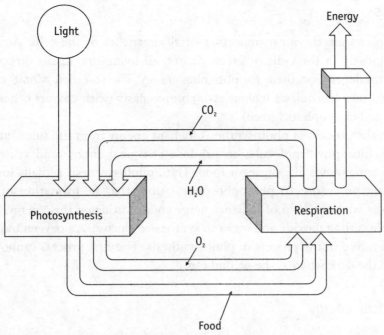

Figure 3.10 Energy flow in the living world

out of their own membranes (Figure 3.11). Thus, they are miniature biochemical factories, which produce energy-rich ATP molecules from food-stuffs, oxygen, and ADP.

The process of *phosphorylation*, that is, linking of phosphate to ADP, requires energy which is obtained by an orderly controlled release of and capture of electrons. In chemical terms, oxidation and reduction reactions involve the transfer of electrons from one molecule to another. The molecule losing the electrons is *oxidized* and the one gaining electrons is *reduced*. The energy obtained by such a transfer of electrons from one molecule to the other is utilized to attach phosphate molecule to ADP, that is to make ATP. This process is known as *oxidative phosphorylation*.

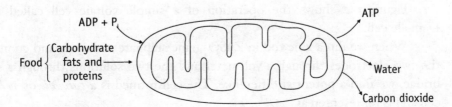

Figure 3.11 Conversion of food into energy-rich adenosine triphosphate molecule in mitochondria

Chloroplasts

Chloroplasts are the most important of all organelles of the cells. Although found mostly in the cells of green plants, all living organisms directly or indirectly depend on them for obtaining energy. The so-called *plastids* can be classified into leucoplasts (colourless), chromoplasts (with colours other than green), and chloroplasts (green).

In the process of photosynthesis, radiant energy from the sun is utilized by the chlorophyll molecules to produce chemical energy and synthesize organic compounds like sugar for food. Thus, photosynthesis actually involves two main processes: (i) photophosphorylation, that is, formation of ATP molecules with the help of radiant energy and (ii) utilizing this energy (ATP) to couple carbon dioxide and water to synthesize glucose. An oxygen molecule is also evolved in the process of photosynthesis. The first process cannot take place in the dark whereas the second can.

Electrochemical energy

Electricity is produced using an electrochemical cell, such as common battery, through the release of electrochemical energy using *redox* (reduction–oxidation) reactions.

In an electrochemical cell, current flows due to the flow of electrons in the external circuit and movement of ions through the solution in the internal circuit. The *electrochemical reaction* is one in which current is produced as a result of chemical change and which occurs due to the passage of electrical current.

The flow of current through an electrolytic solution gives electrolytic conduction. As in electrolytic conduction, the charge is carried by ions, this type of conduction will not occur unless ions of the electrolyte are free to move. Strong electrolytes (such as NaCl, HCl, and NaOH) dissociate completely into ions, whereas weak electrolytes (such as H_2CO_3, $AgCl_2$, and $ZnCl_2$) have a smaller degree of dissociation.

Figure 3.12 shows the operation of a simple voltaic cell called the Danielle cell.

When we dip a zinc rod in $ZnSO_4$ (zinc sulphate) solution, and connect the two electrodes through a voltmeter and the two solutions through a salt bridge, we find a potential difference. The cell formed is a *galvanic* or *voltaic* cell and is represented as

$$Zn(s) \ |Zn^{2+} (aq)||Cu^{2+} (aq)| \ Cu(s)$$

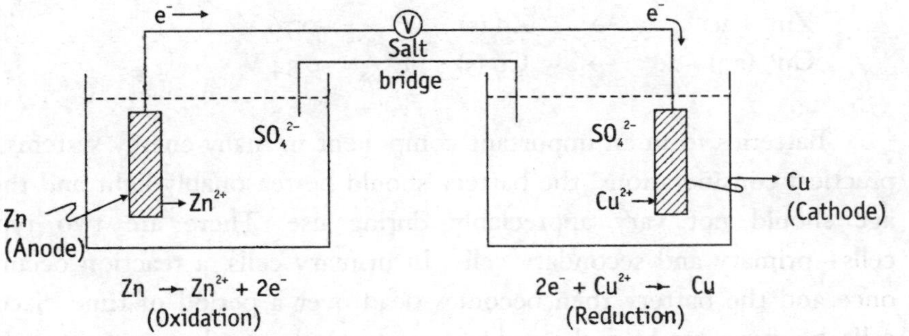

Figure 3.12 Simple voltaic cell (Danielle cell)

The potential difference is the contribution of two half cells, Zn^{2+} (aq)/ Zn (s) and Cu^{2+} (aq)/Cu (s). The zinc electrode is the anode and the copper electrode is the cathode, and oxidation and reduction reactions occur at these electrodes.

$$Zn(s) \rightarrow Zn^{2+} (aq) + 2e^- \text{ (oxidation)}$$
$$Cu^{2+} (aq) + 2e^- \rightarrow Cu (s) \text{ (reduction)}$$

Overall, Zn (s) + Cu^{2+} (aq) $\rightarrow Zn^{2+}$ (aq) + Cu (s)

The electrode potential is a measure of the tendencies of metal atoms to go into solution as metal ions. When a metal strip M is immersed in a solution of its ions M^{n+}, either of the following three can occur.

1 A metal ion M^{n+} may collide with the electrode and undergo no change.
2 A metal ion M^{n+} may collide with the electrode, gain n electrons, and be converted into a metal atom M (the ion is reduced).
3 A metal atom on the electrode M may lose n electrons to the electrode and enter the solution as M^{n+} (the metal is oxidized).

If the tendency of the metal to oxidize is especially high, there may be a slight increase in the number of electrons at the electrodes. The electrode, therefore, develops a small negative potential with respect to the solution. If this situation is established for two different metals (say copper and zinc) and if electrical contact is established between the two metal electrodes and the two solutions, electric current will flow.

Electrons pass from zinc to the copper electrode. Zinc has more negative potential than copper and has a greater tendency for oxidation.

$$Zn^{2+} + 2e^- \rightarrow Zn\ (s) \quad E^\circ_{298} = -0.76\ V$$
$$Cu^{2+}\ (aq) + 2e^- \rightarrow Cu\ (s) \quad E^\circ_{298} = +0.34\ V$$

Batteries form an important component in many energy systems. From practical considerations, the battery should be reasonably light and the voltage should not vary appreciably during use. There are two types of cells—primary and secondary cells. In primary cells, a reaction occurs only once and the battery then becomes dead over a period of time. Secondary cells, however, can be recharged by passing an electrical current through them and hence, they can be used repeatedly. Over a certain number of cycles of recharging, however, even the secondary cell has to be discarded.

Some of the well-known primary cells are the dry cell (Le clanche cell) and the mercury cell. The schematic diagram of the dry cell is shown in Figure 3.13. The electrode reactions in the dry cell are complex, but can be written approximately as follows.

Anode $Zn\ (s) \rightarrow Zn^{2+} + 2e^-$
Cathode $MnO_2 + NH_4^+ + e^- \rightarrow MnO\ (OH) + NH_3$

In the cathode reaction, manganese is reduced from the 4^+ oxidation state to the 3^+ state. Ammonia is not liberated as a gas but combines with Zn^{2+} to form $Zn\ (NH_3)_4^{2+}$ ion. Dry cells do not have an indefinite life as the acidic ammonium chloride corrodes the zinc container even when not in use.

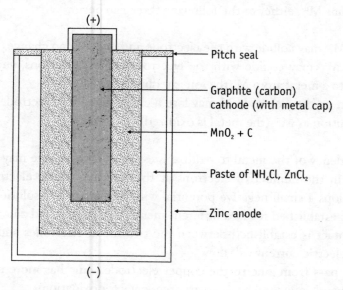

(+)

— Pitch seal

— Graphite (carbon) cathode (with metal cap)

— $MnO_2 + C$

— Paste of NH_4Cl, $ZnCl_2$

— Zinc anode

(–)

Figure 3.13 A dry cell

In the mercury cell, the zinc–mercury amalgam is the anode. A paste of HgO (mercuric oxide) and carbon is the cathode. The electrolyte is a paste of KOH (potassium hydroxide) and ZnO (zinc oxide).

The electrode reactions are

Anode $Zn \text{ (amalgam)} + 2OH^- \rightarrow ZnO \text{ (s)} + H_2O + 2e^-$

Cathode $HgO \text{ (s)} + H_2O + 2e^- \rightarrow Hg \text{ (l)} + 2OH^-$

Overall reaction $Zn \text{ (amalgam)} + HgO \text{ (s)} \rightarrow ZnO \text{ (s)} + Hg \text{ (l)}$

The cell potential is approximately 1.35 V.

Secondary cells   The most important secondary cell is the lead storage battery. It consists of a lead anode and a grid of lead packed with lead dioxide as cathode. A solution of H_2SO_4 (38% by mass) is used as an electrolyte.

Another secondary cell is the nickel–cadmium storage cell. It has a longer life compared to the lead–acid cell but is more expensive to manufacture.

Fuel cells are electrochemical cells that are designed to convert energy from the combustion of fuels such as hydrogen or methane directly into electrical energy (Figure 3.14). Fuel cells have been researched since a long time, primarily for vehicular use as they are non-polluting. Also, the conversion

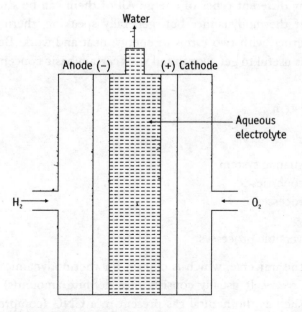

Figure 3.14 A simple fuel cell

efficiency from chemical energy to electricity is quite high and hence, they are good candidates for decentralized power generation.

Reactions occurring at the electrodes can be represented as follows.

Anode $2[H_2 (g) + 2OH^- (aq) \rightarrow 2H_2O (l) + 2e^-]$
Cathode $O_2 (g) + 2H_2O (l) + 4e^- \rightarrow 4OH^- (aq)$

Overall reaction $2H_2 (g) + O_2 (g) \rightarrow 2H_2O (l)$

Atomic or nuclear energy

The final basic form of energy is bound up in a central nuclei of atoms and is called atomic or nuclear energy. Nuclear power stations operate on much the same principles as fossil fuel plants except that the furnace in which the fuel burns is replaced by a heat-generating nuclear reaction.

Heat, work, and thermodynamics

This section aims to give the renewable energy engineer an insight into the use of basic thermodynamics for various practical applications. Thermodynamics is the study of energy and its transformation. As discussed in the previous section, there are many different types of energy. All of them can be studied using the principles of thermodynamics but, generally speaking, thermodynamics is largely concerned with two forms of energy: heat and work. Before proceeding further, it is useful to get acquainted with certain basic concepts as listed subsequently.

- Thermodynamic system
- Surroundings
- Boundary
- State of a thermodynamic system
- Thermodynamic properties
- Thermodynamic processes
- Equilibrium
- Reversible and irreversible processes

The portion of the universe, which is chosen for thermodynamic consideration, is called a *system*. It usually consists of a definite amount(s) of a specific substance(s), such as the natural gas present in a CNG (compressed natural gas) cylinder and the steam admitted into a steam turbine. A system may be homogeneous (gas or a mixture of gases, pure liquid, etc.) or heterogeneous (liquid and its vapour, two immiscible liquids, etc). The matter outside this system, or the remainder of the universe, is called the *surroundings*. The separation between the system and the surroundings is called the *boundary* of the system. The thermodynamic or macroscopic state of a given system can be defined by four observable properties (or state variables): composition, pressure, volume, and temperature. If these properties are specified, all other physical properties, such as density, viscosity, and refractive index are fixed. The thermodynamic properties thus serve to define the system completely. Properties can be intensive (independent of the mass of the system) or extensive (dependent on the mass). The thermodynamic state will not change unless there is some interaction with the surroundings. This usually takes the form of energy transfer into or out of the system.

When a thermodynamic system changes from one state to another, it happens through a process such as the addition/removal of heat/work. Some examples are isothermal (constant temperature), isobaric (constant pressure), or adiabatic (no addition/ removal of heat) processes. A system is said to be in equilibrium when its pressure, temperature, and density are uniform and unchanging. A system of compressed air in a cylinder would be in equilibrium if its pressure, temperature, etc. are uniform throughout. However, if the cylinder is heated from one side such that the temperature at one end is different from that at the other end, the system is not in equilibrium. If a thermodynamic process is not spontaneous or sufficiently slow so as to be described as a succession of equilibrium states, it is called reversible.

Temperature, heat, and zeroth law of thermodynamics

The concept of temperature is understood by human beings intuitively as something 'hotter' or 'colder' with respect to their own body temperature. Thus, while an air-conditioned room is pleasantly cold, winters in northern climates are freezing cold and steam is scalding hot. Note that in cold weather, the body is losing heat to the surroundings and in a steam sauna, the body is gaining heat. This implies that the concept of temperature is based on the energy (or heat) transfer process. It should then be possible to conclude that if two bodies at the same temperature are brought into contact, no heat will be exchanged between the two. The concept of equality of temperature can be stated as

Two bodies, each in thermal equilibrium with a third body, are in thermal equilibrium with each other.

This statement is sometimes referred to as the *zeroth law of thermodynamics*. If two bodies at different temperatures are in contact for a long enough time, there will be heat flows, which would ultimately result in the equalization of temperatures. We may thus conclude that heat is a form of energy, which flows from one body to another as a result of temperature difference. One can note the 'circular' definition: equality of temperature is associated with zero heat flow and heat flow is related to the difference in temperature (Holman 1980). As mentioned earlier, there are four different temperature scales: Celsius, Fahrenheit, Kelvin, and Rankine.

Addition or removal of heat to change the thermodynamic state of a system can be done in many different ways. For example, heat can be added at constant pressure or at constant volume. This means that the quantity of heat added/removed is dependent on the *path*, or in other words, heat is a *path function*.

Work Work W is the product of force and distance. Force F is a vector quantity, meaning that it has both magnitude and direction. Similarly, displacement is also a vector quantity. Work is expressed mathematically as

$$W = F.ds = \int_c F \cos\theta \, ds \qquad ...(3.18)$$

where the dot product indicates that work is computed by considering only the component of force in the direction of displacement (Figure 3.15).

The total work done is given by the line integral along the path. It can be seen that work is dependent on the path, that is, it is a path function. Equation 3.18 is the fundamental analytical definition of work and can always be used as a basis for calculating it. As an example, consider expansion of a gas

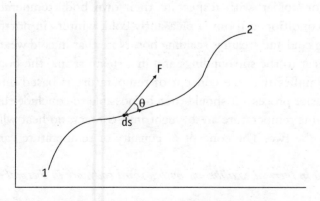

Figure 3.15 Work as a function of force and displacement

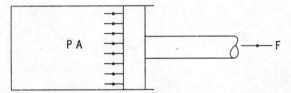

Figure 3.16 Expansion of gas in a piston–cylinder arrangement

behind a piston in a cylinder, as shown in Figure 3.16. If pressure of the gas is P and the area of the piston is A, the force exerted on the piston is

$$F = PA \qquad\qquad\qquad ...(3.19)$$

The piston moves in the direction of the applied force, and a differential displacement ds can be expressed in terms of change in the volume of gas dV as

$$ds = \frac{dV}{A} \qquad\qquad\qquad ...(3.20)$$

The differential work dW done on the piston is

$$dW = Fds$$
(θ is zero, hence cos θ is unity)

$$= PA\frac{dV}{A}$$

$$= PdV$$

Therefore,

$$W = \int_{V_1}^{V_2} PdV \qquad\qquad\qquad ...(3.21)$$

This is expressed graphically as Figure 3.17. If one considers the gas behind the piston as the thermodynamic system, there should be a differentiation between work done 'by' the system and work done 'on' the system. This is usually taken care of by selecting an appropriate sign convention, both for work and heat. The usual convention is to regard a quantity of heat as positive

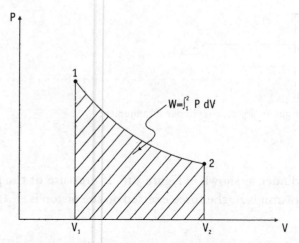

Figure 3.17 P–V diagram for expansion process

when it is transferred to the system from the surroundings. The convention for work is that it is considered positive when it is transferred from the system to the surroundings.

In the above example, work done by the system is positive. The fluid (gas) behind the piston can be considered as a closed system, or one which does not allow mass flows into or out of the system. Systems which allow mass flows are called open systems.

Example 11

Consider the gas space in a floating dome of a 2-m³ KVIC-type biogas plant.
1 Will it be classified as an open system or a closed system?
2 Calculate the work done by the system and express it as a fraction of gas energy produced per day.

Solution

(1) Gas is consumed during cooking and when all of it is used up, there is very little residual gas. As gas is produced in the digester, it accumulates in the dome. Thus, mass flows are taking place into and outside the system, and hence it can be classified as an open system.

(2) Pressure P of gas in the dome is given by the weight of dome divided by the cross-sectional area of the dome, and is about 100 mm of water column. The storage volume V of the dome is generally kept as half of the daily gas production, that is, 1 m³. As pressure is constant, work W is given by

$$W = \int PdV = P\int_{1}^{2} dV = P(V_2 - V_1)$$

$P = 100$ mm water $= 100 \times 9.8 = 980$ Pa

Therefore, $W = 980 \times 1$ (N/m^2 × m^3 or Nm or J)

The calorific value of biogas is about 20 MJ/m^3, hence the fraction of energy spent as work done is $980/40 \times 10^6 = 2.45 \times 10^{-5}$.

Equation 3.18 can be used to develop several useful relationships. Two such cases, relating to the concepts of kinetic energy and potential energy are outlined below.

If a body of mass m is acted upon by force F, its acceleration is given by Newton's second law as

$F = ma$

(a is unity in the SI system of units)

For a displacement of ds, work W is given by

$$W = \int mads = \int m \frac{du}{dt} ds = \int m \frac{du}{ds} \frac{ds}{dt} ds = \int mu.du$$

If the above equation is integrated between the limits u_1 and u_2, we get

$$W = \frac{1}{2}mu_2^2 - \frac{1}{2}mu_1^2 \qquad \qquad ...(3.22)$$

The quantity $\frac{1}{2}mu^2$ is termed as kinetic energy, which was introduced in the earlier section. Equation 3.22 shows that work done on a body in accelerating it from an initial velocity u_1 to final velocity u_2 is equal to the change in kinetic energy of the body. Conversely, if a moving body is decelerated by the action of a resisting force, work done by the body is equal to the change in kinetic energy. A range of hydro turbines and windmills operate on this principle of energy conversion.

If a body of mass m is moved from an initial elevation z_1 to a final elevation z_2, an upward force equal to the weight of the body must be exerted on it. The force acting on the body is the gravitational force, and can be expressed as mg, where g is the acceleration due to gravity. The vertical distance covered is (z_2-z_1), hence the work done is given by

$$W = mg(z_2 - z_1) \qquad \qquad ...(3.23)$$

The quantity mgz is the potential energy as described in the earlier section.

The first law of thermodynamics

Internal energy

Experiments conducted by J P Joule during 1840–78 to understand the nature of heat and work led to the concept of internal energy, and to the formulation of the first law of thermodynamics. In a series of experiments, conducted with the best of precision and accuracy possible in those times, he placed measured amounts of water in an insulated container and agitated the water with a rotating stirrer. The amount of work done by the stirrer and the temperature changes in water were measured carefully. It was found that a definite amount of work was required to raise the temperature of a unit mass of water by one degree. The original temperature of water was restored by simple contact with a cooler body. Joule was able to show conclusively that there is a definite relationship between work and heat (which established the so-called mechanical equivalent of heat, 4.186 J/cal), and hence, both are forms of energy.

In these experiments, energy was added as work, but was extracted as heat. What happened to this energy after it was added and before it was extracted? It is logical to hypothesize that this energy was stored within the water in some form, defined as the internal energy E. It pertains to the energy of molecules making up the substance. The molecules of any substance are believed to be in ceaseless motion, whether translational, rotational, or vibrational. The molecules of a solid cannot move about freely but can vibrate. Molecules of a liquid move more freely, but are still confined to the container in which the liquid is placed. The molecules of a gas, however, can move quite freely (translation), besides rotating and vibrating about specified axes (with the exception of monatomic gases like helium or neon). The addition of heat to a substance increases the molecular activity, and thus results in an increase in its internal energy. Work done on the substance has the same effect, as was shown by Joule. Thermodynamics concerns itself primarily with the changes in internal energy, and hence it is not necessary to know the absolute values of internal energy and the other quantities to be defined later. A *reference* state is usually postulated to deal with the changes. It is also postulated that the internal energy of all substances is zero at 0 K (absolute zero).

Formulation of the concept of internal energy distinguishes it from potential or kinetic energy which the substance may possess because of its position or motion as a whole.

Based on Joule's experiments, and later supported by numerous observations, the first law of thermodynamics was formulated as a universal law.

We now know that the processes involving nuclear reactions in which huge quantities of energy are produced are exceptions to the first law. The first law can be variously stated as follows.

- Energy assumes many forms but the total quantity of energy is constant. When energy disappears in one form, it appears simultaneously in other forms.
- The energy change of an isolated system (one which does not allow transport of mass and energy) is zero.
- The energy of the universe is constant.

The first law of thermodynamics applies to the system and surroundings taken together. In its most basic form, the first law can be written as

$$\Delta \text{ (Energy of the system)} + \Delta \text{ (Energy of the surroundings)} = 0$$
$$\text{or } \Delta \text{ (Energy of the system)} = \Delta \text{ (Energy of the surroundings)} \qquad ...(3.24)$$

In the above equations, Δ implies change that can be positive or negative. For a closed system (no transfer of mass across the boundary), the entire energy is transferred across the boundary as heat Q and work W. Also, there are no changes of kinetic or potential energy for a closed system; hence the only change is in the internal energy E. Equation 3.24 can thus be written as

$$\Delta E = Q - W \qquad ...(3.25)$$

The above equation is for finite changes in the system. For infinitesimal or differential changes, it can be written as

$$dE = dQ - dW$$

As stated earlier, both heat and work are path functions and hence, dQ and dW are 'non-exact' differentials, whereas internal energy is a state function, so dE is an 'exact' differential. To distinguish this subtle distinction of differentials, the above equation is usually written as

$$dE = \delta Q - \delta W \qquad ...(3.26)$$

Enthalpy

Apart from internal energy, there are several other thermodynamic functions of practical use, such as enthalpy, entropy, and free energy. Enthalpy will be defined here and other functions will be defined after the introduction of the second law. Enthalpy H is defined as

$$H = E + PV \qquad \qquad ...(3.27)$$

Note that the product PV has units of energy: pressure P has the units N/m^2 and the volume V has units of m^3, and hence, the product has the units of Nm or J. In differential form, Equation 3.27 can be written as

$$dH = dE + d(PV) \qquad \qquad ...(3.28)$$

Specific heat

The amount of heat required to raise the temperature of a unit mass (or mole) of any substance by one degree is called the specific heat. As heat supplied to a system depends on the path, it follows that the specific heat also will depend on the path. Two different kinds of specific heat quantities are generally used: specific heat at constant volume C_v and specific heat at constant pressure C_p. It can be shown that these two specific heats are related to the internal energy and enthalpy by the following relations.

$$C_v = \left(\frac{\partial E}{\partial T} \right)_v \text{ and } C_p = \left(\frac{\partial H}{\partial T} \right)_p \qquad \qquad ...(3.29)$$

The use of partial derivative signifies that differentiation is done keeping the other variable constant. Partial derivatives such as the above are used extensively in evaluating the thermodynamic properties. Hence, it may be worthwhile to recollect some basic principles.

If f is a function of two independent variables x and y

$$f = f(x, y)$$

one can write

$$df = \left(\frac{\partial f}{\partial x} \right)_y dx + \left(\frac{\partial f}{\partial y} \right)_x dy \qquad \qquad ...(3.30)$$

The second-order mixed partial derivatives are equal.

$$\frac{\partial^2 f}{\partial x \partial y} = \frac{\partial^2 f}{\partial y \partial x}$$ ...(3.31)

Another useful relation, known as the cyclical relation, is as follows.

$$\left(\frac{\partial x}{\partial y}\right)_f \left(\frac{\partial y}{\partial f}\right)_x \left(\frac{\partial f}{\partial x}\right)_y = -1$$ ...(3.32)

Thermodynamic relations for ideal gases

The important reversible non-flow processes are
- constant volume (isometric) process,
- constant pressure (isobaric) process,
- constant temperature (isothermal) process, and
- adiabatic process.

Constant volume process

Since the volume is constant, no work is done. Hence, the first law equation is written as

$$dE = dQ = C_v \, dT$$
$$\Delta E = \int dE = \int C_v dT$$ ...(3.33)

Constant pressure process

As pressure is constant, work done is given by P dV.

$$dE = dQ - P \, dV = C_p \, dT - P \, dV$$
$$dE + P \, dV = dH = C_p \, dT$$
$$\Delta H = \int dH = \int C_p dT$$ ...(3.34)

As ideal gases obey the relation PV = RT for one mole of gas, the expression for enthalpy change can be written as

$$dH = d(E + PV) = d(E + RT) = dE + RdT$$
$$C_p dT = C_v dT + R \, dT \text{ or } C_p = C_v + R$$ ...(3.35)

The above equations show that for an ideal gas, internal energy and enthalpy are functions of temperature alone.

Constant temperature process

As temperature is constant, internal energy change will be zero.

$$dE = dQ - dW = 0$$
$$Q = W$$

For an ideal gas, $P = RT/V$

Hence, $Q = W = \int PdV = \int RT\dfrac{dV}{V} = RT\ln\dfrac{V_2}{V_1}$ ...(3.36)

Adiabatic process

This is defined as one in which there is no heat addition or removal. Hence,

$$dQ = 0$$
$$dE = -PdV$$

since $dE = C_v\,dT$ for an ideal gas, we can write

$$C_v\,dT = -PdV = -\dfrac{RT}{V}dV = -\dfrac{(C_p - C_v)T}{V}dV \qquad \text{...(3.37)}$$

Dividing throughout by C_v, and rearranging, we get

$$\dfrac{dT}{T} = -\left(\dfrac{C_p}{C_v} - 1\right)\dfrac{dV}{V} = -(\gamma - 1)\dfrac{dV}{V} \qquad \text{...(3.38)}$$

where γ is the ratio of specific heats. The following relations can be derived for the adiabatic process.

$$PV^\gamma = \text{constant} \qquad \text{...(3.39)}$$

$$\dfrac{T_2}{T_1} = \left(\dfrac{V_1}{V_2}\right)^{\gamma - 1} = \left(\dfrac{P_2}{P_1}\right)^{(\gamma - 1)/\gamma} \qquad \text{...(3.40)}$$

$$W = \dfrac{P_1 V_1}{\gamma - 1}\left[1 - \left(\dfrac{P_2}{P_1}\right)^{(\gamma - 1)/\gamma}\right] \qquad \text{...(3.41)}$$

Though these equations are strictly for ideal gases, they can be applied for real gases provided the deviations from ideality are not big. The value of γ can be taken as 1.67 for monatomic gases, 1.4 for diatomic gases, and 1.3 for simple polyatomic gases.

The various processes are shown graphically on a P–V diagram in Figure 3.18. Specific heats, like other thermodynamic properties, can vary strongly with temperature and pressure. Experimental data is available for almost all known substances over a range of temperatures (Perry and Green 1997). Specific heat values for some common substances are given in Table 3.2

The temperature variation of specific heat can be captured in equations of the form

$$C_p = a + bT + cT^2 + dT^3$$

$$C_p = a + bT + cT^{-2} \qquad \qquad \dots(3.42)$$

Table 3.2 Specific heats of miscellaneous substances at 20 °C

Substance	C_p (kJ/kg K)	C_p (kcal/kg K)
Solids		
Aluminium	0.896	0.214
Copper	0.383	0.091
Iron	0.452	0.108
Brick	0.840	0.201
Glass wool	0.700	0.167
Wood	2.800	0.669
Liquids		
Water	4.180	1.000
Ammonia	4.800	1.150
Engine oil	1.900	0.450
Gases (1 atm)		
Air	1.005	0.240
Hydrogen	14.320	3.420
Carbon dioxide	0.846	0.202

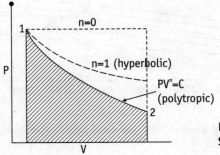

Figure 3.18 General polytropic process
Source Kadambi (1976)

Specific heat is a measure of the amount of energy that can be stored in a given substance. For solar energy applications, energy storage is an important issue. It is interesting to note that water has a large specific heat and as it is also cheaply available, it is a primary candidate for low-temperature energy storage.

To facilitate engineering calculations involving heat effects, a mean heat capacity is defined as follows.

$$\int_{T_o}^{T} C_p dT = C_{p,mean}(T - T_o) \qquad \qquad ...(3.43)$$

$$\text{or} \quad C_{p,mean} = \frac{\int_{T_o}^{T} C_p dT}{(T - T_o)} \qquad \qquad ...(3.44)$$

T_o is usually selected as 25 °C. Plots of $C_{p,mean}$ are given in Figure 3.19.

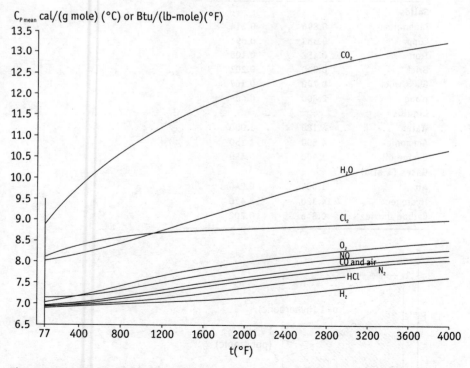

Figure 3.19 Mean molar heat capacities of gases in ideal-gas state, base temperature, 77 °F
Source Wagman (1952)

Some applications of the first law

As the first law of thermodynamics is essentially the law of conservation of energy, it can be used for formulating and solving many practical energy-related problems. Some of these are as follows.

- Application to the open (or flow) systems
- Application to chemical reactions
- Application to two-phase systems

Energy analysis of open (flow) systems

We have seen that the closed system does not allow mass transport across the boundary. An open system, on the other hand, allows flow of mass into and out of the system. For such a case, the concept of a control volume as the system can be considered, as shown in Figure 3.20.

Mass at the rate of $\dot{m}_1$ is entering the control volume and leaving at the rate $\dot{m}_2$. The velocity, density, pressure, internal energy, etc., for the entry stream are designated by the subscript 1 and those for the exit stream are designated by the subscript 2. The principle of mass conservation (referring to the section on material and energy balances) leads to the following equation.

$$\dot{m}_1 - \dot{m}_2 = \frac{dM}{dt} \qquad \qquad ...(3.45)$$

where M is the total mass within the control volume. Similarly, the energy balance can be written as

(Energy flow-in) – (Energy flow-out) = (Energy accumulated)

For a steady flow process, accumulation terms are zero, and hence, one can write

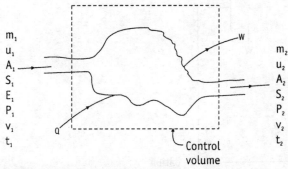

Figure 3.20 Control volume for open system

$$\overset{\bullet}{m}_1 = \overset{\bullet}{m}_2 = \overset{\bullet}{m}$$

and

Energy flow-in = Energy flow-out

Note that the energy term is now the rate at which energy is entering or leaving the control volume. Accordingly, we now speak of the rate of heat addition $\overset{\bullet}{Q}$ and the rate of work $\overset{\bullet}{W}$. The rate term can usually be identified by the dot placed on the symbol. The rate of work done is nothing but the power delivered by or to the system. The abstract picture of Figure 3.20 can be replaced with a realistic picture of a fluid flowing through various elements as shown in Figure 3.21.

The entry point 1 is at a level Z_1 from the datum, so that the fluid entering this point has a potential energy. Similarly, the kinetic energy at 1 is $\frac{1}{2}mu_1^2$. The corresponding terms for the exit stream have the subscript 2. The first law can now be written as

$$\overset{\bullet}{m}gZ_1 + \frac{1}{2}\overset{\bullet}{m}u_1^2 + \overset{\bullet}{m}E_1 + \overset{\bullet}{Q} = \overset{\bullet}{m}gZ_2 + \frac{1}{2}\overset{\bullet}{m}u_2^2 + \overset{\bullet}{m}E_2 + \overset{\bullet}{W} \qquad ...(3.46)$$

$$\text{or } \frac{\overset{\bullet}{Q}}{\overset{\bullet}{m}} - \frac{\overset{\bullet}{W}}{\overset{\bullet}{m}} = g(Z_2 - Z_1) + \frac{1}{2}(u_2^2 - u_1^2) + (E_2 - E_1) \qquad ...(3.47)$$

For negligible changes in potential and kinetic energy, the above equation can be written as

$$\overset{\bullet}{Q} - \overset{\bullet}{W} = \overset{\bullet}{m}(E_2 - E_1) \qquad ...(3.48)$$

which is similar to the first law of equation for closed systems, with the exception of flow terms present now. The work done by the system $\overset{\bullet}{W}$ is the total

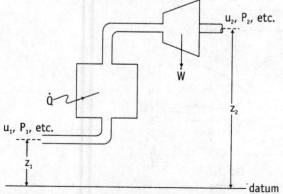

Figure 3.21 Fluid flowing through various elements

work. It is useful to distinguish between the shaft work (useful work) delivered by the system and the work accompanying P–V changes.

$$\frac{\dot{W}}{\dot{m}} = \frac{\dot{W}_s}{\dot{m}} + P_2 v_2 - P_1 v_1 \qquad \qquad ...(3.49)$$

where v is the specific volume.

Substituting in Equation 3.49 and rearranging, we get

$$\frac{\dot{Q}}{\dot{m}} - \frac{\dot{W}_s}{\dot{m}} = (E_2 + P_2 v_2) - (E_1 + P_1 v_1) = H_2 - H_1 \qquad \qquad ...(3.50)$$

If we include the potential and kinetic energy terms, we get

$$\frac{\dot{Q}}{\dot{m}} - \frac{\dot{W}_s}{\dot{m}} = g(Z_2 - Z_1) + \frac{1}{2}(u_2^2 - u_1^2) + (H_2 - H_1) \qquad \qquad ...(3.51)$$

The control volume can be a section of the pipe, the boiler or the turbine, or the entire system consisting of all these components. Equation 3.51 is the basic energy balance equation for a steady flow system.

Example 12

Saturated steam at 26.4 bars and 500 K enters a steam turbine at the rate of 1100 kg/h, and exits at 1.0815 bars and 375 K. Properties of the saturated steam are given as under. Exit velocity of steam is 5 m/s and inlet velocity can be neglected. Find out the work done by the turbine

- at 26.4 bars, 500 K, $H_1 = 2801.5$ kJ/kg and
- at 1.0815 bars, 375 K, $H_2 = 2679.1$ kJ/kg

Solution

Applying the energy balance (Equation 3.51) for the system taking heat interaction and change in potential energy as zero

$$0 - \frac{W_s}{1100/3600} = 0 + \frac{1}{2 \times 1000}\left(5^2 - 0\right) + \left(2679.1 - 2801.5\right)$$

$$W_s = 37.4 \text{ kW}$$

Joule–Thomson coefficient

When a real gas passes through a sudden contraction (for example, a porous plug) or throttling, it cools. This is referred to as the Joule–Thomson effect and is the basis for all cooling and refrigeration. As no heat is added or removed, $\dot{Q} = 0$ and as no shaft work is involved, $\dot{W}_S = 0$. If we neglect the kinetic and potential energy terms, Equation 3.51 reduces to

$$H_1 = H_2$$

If H is taken as a function of P and T, we can write

$$dH = \left(\frac{\partial H}{\partial P}\right)_T dP + \left(\frac{\partial H}{\partial T}\right)_P dT$$

$$= \left(\frac{\partial H}{\partial P}\right)_T dP + C_p \cdot dT \qquad \qquad \ldots (3.52)$$

For adiabatic throttling, dH = 0; hence, this equation can be written as

$$0 = \left(\frac{\partial H}{\partial P}\right)_T \left(\frac{\partial P}{\partial T}\right)_H + C_p \qquad \qquad \ldots (3.53)$$

The term $(dT/dp)_H$ is called the Joule–Thomson coefficient μ_J

$$\mu_J C_p = -\left(\frac{\partial H}{\partial P}\right)_T \qquad \qquad \ldots (3.54)$$

It should be noted that for an ideal gas, H is a function of temperature only and hence,

$$\left(\frac{\partial H}{\partial P}\right)_T = 0$$

It follows that for an ideal gas $\mu_J = 0$ $\qquad \qquad \ldots (3.55)$

The Joule–Thomson coefficient can be measured experimentally for any substance and can be used to calculate other thermodynamic properties using equations such as Equation 3.54.

Energy analysis of chemical reactions

Chemical reactions are accompanied by evolution or absorption of heat. As there is no work done, the energy balance equation reduces to

$$Q/m = \Delta H$$

The experimental measurement of ΔH for any given reaction can be performed. The measured heat of reaction ΔH can be reduced to a *standard heat of reaction*, which is defined as the difference between the enthalpies of products of chemical reaction and enthalpies of reactants in their standard states (reference state). A pressure of 1 atm and a temperature of 298 K are usually chosen as a standard state. The standard heat of reaction is denoted as ΔH^{o}_{298}.

The standard heat of reaction can be calculated from the *standard heat of formation*, which is defined as the heat of reaction when the compound is formed from the elements which make it up. For example, the reaction

$$C + 2H_2 \longrightarrow CH_4$$

is a formation reaction for methane as the reactants are all elements. However, the reaction

$$H_2O + SO_3 \longrightarrow H_2SO_4$$

is not a formation reaction. A formation reaction results in one mole of the compound, and hence, the standard heat of formation is expressed in units of cal/g mol. Some heats of formation and combustion for compounds are shown in Table 3.3. For a combustion reaction the products are carbon dioxide (*gas*) and water (*liquid*).

The heat of combustion values listed in Table 3.3 are also called *higher heating values* (when the water produced is in liquid form) or simply *calorific value*. Many times these are required to be converted into units of kcal/Nm³ for gases. This value can be obtained by dividing the heat of combustion by 22.4 (litre/g mol), which is the volume occupied by an ideal gas at NTP. There would be a slight difference due to the fact that standard heat of reaction is at 298 K, instead of at 273.15 K, but it is small and can be ignored.

One can estimate the standard heat of the reaction by the following formula

$$\Delta H^{o}_{298} = \sum_{products} \Delta H^{o}_{f,\,298} - \sum_{reactants} \Delta H^{o}_{f,\,298} \qquad\qquad ...(3.56)$$

which states that summation of standard heat of formation of products less summation of standard heat of formation of reactants is equal to the standard heat of reaction.

Table 3.3 Standard heat of reaction for different elements

Substance	Formula	$\Delta H^0_{f,298}$ (cal/g mol)	ΔH^0_{298} (cal/g mol of combustible material)
Methane (g)	CH_4	–17 889	–212 800
Ethane (g)	C_2H_6	–20 236	–372 820
Propane (g)	C_3H_8	–24 820	–530 600
n-Butane (g)	C_4H_{10}	–30 150	–687 640
Ethanol (g)	C_2H_5OH	–56 030	
Ethanol (l)	C_2H_5OH	–66 200	
Methanol (g)	CH_3OH	–48 050	
Methanol (l)	CH_3OH	–57 110	
Acetylene (g)	C_2H_2	54 194	–310 620
Ammonia (g)	NH_3	–11 040	
Carbon (g)	C		–94 051
Carbon dioxide (g)	CO_2	–94 051	
Carbon monoxide (g)	CO	–26 416	–67 636
Hydrogen (g)	H_2		–68 317
Hydrogen sulphide (g)	H_2S	–4815	
Sulphur dioxide (g)	SO_2	–70 960	
Water (g)	H_2O	–57 798	
Water (l)	H_2O	–68 317	

g – gas; l – liquid

Example 13

Calculate the standard heat of reaction for the *shift* reaction

$$CO_2 \text{ (gas)} + H_2 \text{ (gas)} \longrightarrow CO_2 \text{ (gas)} + H_2O \text{ (gas)}$$

which is important in gasifiers.

Solution

The heats of formation are written below each component.

$$CO_2 \text{ (gas)} + H_2 \text{ (gas)} \longrightarrow CO_2 \text{ (gas)} + H_2O \text{ (gas)}$$

$$-94051 \qquad 0 \qquad\qquad\qquad -26416 \quad -57798$$

$$\Delta H^0_{298} = [(-26416) + (-97798)] - [-94051] = 9837 \text{ cal}$$

Note that `·` value is positive, indicating that heat is absorbed in this reaction (endothermic).

The heat of reaction at a temperature T, other than 298 K, can be calculated in the following steps.

1 Bring the temperature of reactants from T to 298. The enthalpy of change for this process is

$$\Delta H_R^o = \sum_{reactan ts} \left(x \int_T^{298} C_p dT \right) \qquad ...(3.57)$$

The integration can be carried out using the temperature dependence of C_p.

2 Allow the reaction to proceed at 298 K.

3 Bring the temperature of products to T.

$$\Delta H_P^o = \sum_{products} \left(x \int_{298}^T C_p dT \right) \qquad ...(3.58)$$

then,

$$\Delta H_T^o = \Delta H_R^o + \Delta H_{298}^o + \Delta H_P^o \qquad ...(3.59)$$

Depending on whether T is higher or lower than 298, the sign of ΔH_R^o and ΔH_P^o will change. This process has been represented schematically in Figure 3.22.

Practical reactions, such as the combustion of LPG in the presence of air, are quite different from standard reactions described so far. Excess air may have to be provided (stoichiometric values are not taken), and inert gases (such as N_2) will be present. However, the first law can be used along with the knowledge of ΔH_{298}^o to analyse the effect of practical and industrial reactions.

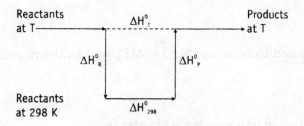

Figure 3.22 Scheme for calculating heat of reaction at temperature T

Example 14

The adiabatic flame temperature is defined as the maximum temperature attainable when a combustible gas (or liquid) is burnt with a given amount of air, and it is very important to decide the suitability of a fuel for a given application. For example, the production of lime in lime kilns requires a temperature in excess of 1000 °C, and it is desired that producer gas (which is a low calorific value gas) be used. For this, one has to calculate the adiabatic flame temperature of the producer gas.

Solution

Producer gas from a gasifier has the following composition on a moisture-free basis.

CO	21%
H_2	17%
CH_4	1%
CO_2	13%
N_2	48%

Calculate the adiabatic flame temperature with an excess air factor of 10%.

Basis One g mole of producer gas. The following steps are involved.

- Calculate stoichiometric air required for complete combustion.
- Calculate actual amount of air, considering the excess air factor.
- Calculate ΔH°_{298} for gas mixture.
- Establish the number of moles of each component in the reactant and the product stream.
- If the adiabatic flame temperature is T, the products are heated up to temperature T. The corresponding enthalpy change is

$$\Delta H^\circ_p = \sum_{products} \left(x \int_{298}^{T} C_p dT \right)$$

- If reactants are assumed to be at 298 K, $Q = \Delta H^\circ_R = 0$. As the process is adiabatic, $Q = \Delta H^\circ_T = 0$

- $\Delta H^\circ_{298} + \Delta H^\circ_p = 0$

- Using the concept of mean heat capacity, we can write

$$\left(\sum_{products} x.C_{p,mean} \right)(T - 25) = - \Delta H^\circ_{298}$$

The above equation can be solved by a trial and error method. Assume a value of T and read the C_p values from the available graphs. Calculate the left-hand side of the equation and see if it matches the right-hand side. Repeat the calculation till a close match is obtained.

The step-by-step calculations to calculate the adiabatic flame temperature are given below. Using the methods described earlier, the g.moles of different components in the reactant and product streams are obtained as listed below.

Component	g moles in reactant stream	g moles in product stream
CO	0.21	0
H_2	0.17	0
CH_4	0.01	0
CO_2	0.13	0.35
N_2	1.349	1.349
O_2	0.231	0.021
H_2O	0	0.19

ΔH°_{298} can be calculated as 25 948 cal/g mole (producer gas) by using the heat of formation data in Table 3.3. Assuming an initial guess of 1000 °C for adiabatic flame temperature, and reading the corresponding $C_{p,mean}$ values from Figure 3.19, we get

$$\sum_{\text{products}} \times C_{p,mean} = 16.16$$

$$T = 25 + \frac{25\,348}{16.16}$$

$$= 1630 \text{ °C}$$

The assumed and calculated temperatures do not match. Hence the iteration has to be repeated with a new guess. Continuing with the iteration process, T can be obtained as 1540 °C.

A similar method is used in calculating flue gas thermal energy loss in combustion devices such as stoves, furnaces, boilers, and engines. Combined with the measurement of various temperatures, this forms an important tool to understand the heat flows in various streams and to optimize the useful heat.

Heat effect accompanying phase changes

When a liquid is vaporized (boiling) or when a solid goes into a liquid state (melting), there is no change in temperature but heat is absorbed by the substance. The heat required to accomplish the phase transition is called the latent heat of vaporization or the latent heat of fusion. There is a fundamental relationship between the latent heat and the P–V–T data, the Clapeyron's equation (derived later).

$$\Delta H = T\Delta V \frac{dP_s}{dT} \qquad \qquad ...(3.60)$$

where ΔH = latent heat

ΔV = volume change accompanying phase change

$\dfrac{dP_s}{dT}$ = slope of vapour pressure curve at temperature T

Thermodynamic properties for saturated vapour–liquid system are defined using the properties of liquid and vapour separately and using the *quality* of vapour, defined as

$$x = \frac{\text{Mass of vapour}}{\text{Total mass}}$$

Refer to the P–V–T curve shown in Figure 3.23. The substance is in liquid form at A and as vaporization proceeds, it enters phase B. At any point C in between, both liquid and vapour phases are present. The quality x is proportional to BC/BA. For saturated steam (vapour), $x = 1$ and for saturated liquid x= 0. The specific volume, enthalpy, etc. are then given by

$$v = (1 - x)\,v_l + xv_v \qquad \qquad ...(3.61)$$
$$H = (1 - x)H_l + xH_v \qquad \qquad ...(3.62)$$

and so on. We will get back to the two-phase system after introduction of the second law of thermodynamics.

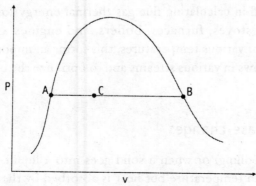

Figure 3.23 P–v–T curve for a fluid

Second law of thermodynamics

The second law of thermodynamics was formulated, just as the first law, to put certain practical observations regarding heat and work in a theoretical construct. These observations can be stated as follows.

- Heat flows from a higher temperature to a lower temperature, but never in the opposite direction, spontaneously.
- Two gases, say, nitrogen and oxygen, separated by a partition in a container, will mix spontaneously when the partition is removed. However, they will never reassemble into the individual components spontaneously.
- A resistor connected to a battery will heat up and the battery will discharge accordingly. However, it is not possible to recharge the battery by heating up the resistor.
- All mechanical work can be transformed into heat but all heat cannot be transformed into work.

    Before proceeding further into the formulation of the second law, it is useful to define certain terms.

Source and sink

A system which remains at a constant temperature, irrespective of the amount of heat withdrawn from it, is called a source or reservoir. A real-life example is the sun; any amount of solar energy used would not reduce the energy of the sun during its remaining life (probably a few billion years). Similarly, a body, which remains at the same temperature irrespective of the amount of energy added to it is defined as a sink. The closest example is the ocean.

Heat engine

A heat engine is a device or a machine, working cyclically, receiving heat Q_1 from a source at a temperature T_1 and rejecting a quantity Q_2 to a sink at temperature T_2. The difference $(Q_1 - Q_2)$ will then appear as work W. The term 'cyclical' is important and it means that the system (the working fluid inside the heat engine) is brought back to the initial thermodynamic state at the end of each cycle. As there is no change in the system at the end of a cycle, the internal energy change will be zero. According to the first law,

$$Q_1 - Q_2 = W \qquad \qquad ...(3.63)$$

Energy conversion efficiency

Energy conversion efficiency is defined as the ratio of work output to the heat received

$$\eta = \frac{W}{Q_I} \qquad\qquad ...(3.64)$$

Heat pump or refrigerator

This device will take heat from a lower-temperature reservoir and transfer it to a higher-temperature reservoir, in a cyclical operation. Work is done 'on' the system. Again by the first law

$$Q_2 + W = Q_1 \qquad\qquad ...(3.65)$$

The heat engine and the heat pump are shown schematically in Figure 3.24. COP (coefficient of performance) is used to quantify the performance of a refrigerator.

$$COP = \frac{Q_2}{W} \qquad\qquad ...(3.66)$$

The second law of thermodynamics concerns itself with η and COP. There are two ways in which the second law is stated.

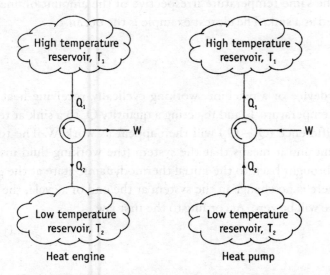

Figure 3.24 Schematic diagram of a heat engine and heat pump

Kelvin–Planck statement

It is impossible to construct a device that operates continuously in a cyclic manner, and produces no effect other than the withdrawal of energy as heat from a single reservoir and production of work. In effect, this means that it is impossible to construct a heat engine such that $Q_2 = 0$ and $\eta = 100\%$.

Clausius statement

It is impossible to construct a device which operates in a cycle and whose sole effect is to transfer heat from a cooler to a hotter body. This effectively means that it is impossible to construct a heat engine such that $W = 0$ and COP = infinity.

Reversibility

It has been stated earlier that a process, which occurs infinitesimally slowly, so that it can be reversed at any point of time, is a reversible process. It is also called a quasi-static process as it can be described as a succession of equilibrium states. All natural processes, such as the sudden expansion of gas when a valve is released or the spontaneous mixing of two fluids, are irreversible. The processes described while discussing the first law, such as quasi-static adiabatic or isothermal expansion of gas, can be called reversible. Reversibility is primarily a concept used to facilitate the theoretical formulation of thermodynamic principles.

The Carnot cycle

Consider the fictitious and reversible Carnot cycle for a heat engine as shown in Figure 3.25. The working substance at state 1 undergoes an isothermal expansion 1–2 at temperature T_1, during which an amount of heat Q_1 is absorbed. The next step is the adiabatic expansion 2–3 in which the temperature drops

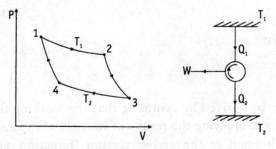

Figure 3.25 Carnot cycle on p–v diagram

to T_2. An amount of heat Q_2 is discarded at temperature T_2 during the isothermal compression process 3–4. Finally, the system returns to state 1 through the adiabatic compression process 4–1. The thermal efficiency of the cycle η is

$$\eta = \frac{W}{Q_1} = \frac{Q_1 - Q_2}{Q_1} = 1 - \frac{Q_2}{Q_1} \qquad \text{...(3.67)}$$

As the cycle consists of only reversible processes, the whole cycle is reversible, that is, it can also act as a refrigerator, absorbing heat Q_2 from a lower-temperature source and discarding an amount Q_1 to a higher-temperature sink, while absorbing work of magnitude W.

The following theorems were derived by means of well-constructed arguments involving other fictitious engines, which may be more efficient than the Carnot's engine (Holman 1980).

Theorem 1   Of all the heat engines working under the same temperature limits, Carnot engine has the highest efficiency.

Theorem 2   All reversible engines operating between the same temperature
(Corollary of 1)   limits have the same efficiency.

Theorem 3   The thermal efficiency of a Carnot cycle is independent of the working fluid, and solely depends upon the temperatures defined on a Kelvin or thermodynamic temperature scale, in which the triple point of water is assigned the value 273.15 K.

Theorem 4   The ideal gas and thermodynamic temperature scale are equivalent.

Following Theorem 3, the efficiency of the Carnot engine can be written as

$$\eta = 1 - \frac{T_2}{T_1} \qquad \text{...(3.68)}$$

It can also be shown that

$$\frac{Q_2}{Q_1} = \frac{T_2}{T_1} \qquad \text{...(3.69)}$$

Equation 3.69 can also be derived by assuming that the working fluid in the engine is an ideal gas and applying the results of isothermal expansion and adiabatic expansion obtained in the earlier section (Equation 3.67).

Theorem 4, establishing the equivalence of the ideal gas and thermodynamic scale, is thus established.

Clausius inequality and entropy

For a Carnot engine, it can also be written

$$\frac{Q_1}{T_1} + \frac{Q_2}{T_2} = 0 \qquad\qquad ...(3.70)$$

using the sign convention that Q_2 is negative as it is the heat rejected from the engine. If one considers infinitesimal quantities, Equation 3.70 can be written as

$$\frac{\delta Q_1}{T_1} + \frac{\delta Q_2}{T_2} = 0 \qquad\qquad ...(3.71)$$

Any completely reversible cycle can be broken up into a large number of small Carnot cycles as shown in Figure 3.26.

Hence, Equation 3.71 can be written as

$$\int \frac{\delta Q}{T} = 0 \qquad\qquad ...(3.72)$$

where the integration is carried out for the whole cycle. Equation 3.72 indicates that the quantity $\delta Q/T$ is the differential of a state function as it does not change when brought back to the original state after going through the cycle. The term $\delta Q/T$ is defined as the entropy change dS.

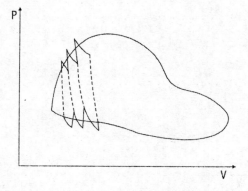

Figure 3.26 Small Carnot cycles for a complete reversible cycle

For an irreversible cycle operating between T_1 and T_2, the thermal efficiency would be lower (Theorem 1). Hence,

$$\oint \frac{\delta Q'}{T_1} < 0 \qquad\qquad ...(3.73)$$

where Q' represents the heat for an irreversible process.

Equations 3.72 and 3.73 can be combined to give the inequality

$$\oint \frac{\delta Q}{T} \leq 0 \qquad\qquad ...(3.74)$$

which is called the *Clausius inequality*.

Now consider a cycle operating between two state points 1 and 2 as shown in Figure 3.27.

R is a reversible path and I is an irreversible path.

$$\oint \frac{\delta Q}{T} = \int_1^2 \frac{\delta Q_I}{T} + \int_2^1 \frac{\delta Q_R}{T} \leq 0$$

$$\int_1^2 \frac{\delta Q_I}{T} + (S_1 - S_2) \leq 0$$

$$S_2 - S_1 \geq \int_1^2 \frac{\delta Q_I}{T}$$

or, in general,

$$\Delta S = S_2 - S_1 \geq \int_1^2 \frac{\delta Q}{T} \qquad\qquad ...(3.75)$$

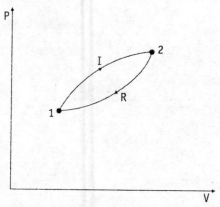

Figure 3.27 Cycle comprising of one reversible and one irreversible process

The equality sign holds for a reversible process and the inequality sign for an irreversible process. If one considers the energy changes, and hence the entropy changes of the surroundings, Equation 3.75 can be written as

$$\Delta S_{Sys} + \Delta S_{Surr} \geq 0 \qquad \qquad ...(3.76)$$

Equation 3.76 means that entropy of the universe can at the best remain constant if all processes are reversible. As all natural processes are irreversible, it leads to a hypothesis that 'the entropy of the universe is always increasing', or that the 'entropy of an isolated system can only increase or at best remain constant'. Equation 3.76 is perhaps the most useful operational statement of the second law.

The definition of entropy allows us to write

$$\delta Q_R = TdS \qquad \qquad ...(3.77)$$

which means that the heat added or removed reversibly is equal to TdS. As entropy is a thermodynamic property, it can be used as a coordinate in a diagram depicting changes in state. A useful representation is the T–S (temperature–entropy) diagram. The Carnot cycle of Figure 3.15 is shown on a T–S plot in Figure 3.28.

Entropy and statistical thermodynamics

From a microscopic point of view, an increase in the entropy can be considered as an increase in randomness of molecules. Boltzman and Gibbs (Smith and van Ness 1975) developed a mathematical formulation for microscopic treatment of entropy. They defined a quantity Ω, called the thermodynamic probability, as the number of ways the particles can be distributed among the states accessible to them. For molecules that do not possess rotational or vibrational motion, the different states can be the

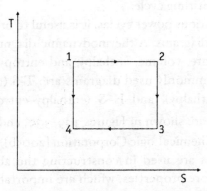

Figure 3.28 Carnot cycle on T–S diagram

different velocities of particles. The kinetic theory of gases shows that there is a *velocity distribution* for any given gas at a given temperature and that the *mean velocity* increases with an increase in temperature. One can qualitatively say that Ω is higher for higher randomness (or higher temperature). The relation postulated by Boltzman between entropy and Ω is given by the following equation.

$$S = k \ln \Omega \qquad \qquad ...(3.78)$$

$$\text{or } S_2 - S_1 = k \ln \frac{\Omega_2}{\Omega_1} \qquad \qquad ...(3.79)$$

where k is the Boltzman constant, which is equal to R/N_o (R is the universal gas constant and N_o is the Avogadro number).

Equation 3.78 allows one to define an *absolute entropy*, as compared to the entropy change dealt by classical thermodynamics. Measurements of the heat capacities at very low temperatures provide the data for calculation of entropy changes down to 0 K. This led to the postulate (originally proposed by Nernst and Plank) that 'the absolute entropy is zero for all perfect crystalline substances at absolute zero temperature'. This is now regarded as the third law of thermodynamics.

Application of second law

By far, the most profound application of the second law had been the analysis and optimization of (i) mechanical/electrical power from heat, (ii) refrigeration, and (iii) chemical thermodynamics. The thermodynamic cycle for power generation can be broadly classified into vapour power cycle, internal combustion engine cycle, and gas turbine cycle. There are also some 'external' combustion engine cycles, such as the Stirling cycle.

Before going into the details of various power cycles, it is useful to understand some practical thermodynamic diagrams. A thermodynamic diagram is one on which the temperature, pressure, volume, enthalpy, and entropy are shown by a single chart. The most commonly used diagrams are: T–S (temperature–entropy), P–H (pressure–enthalpy), and H–S (enthalpy–entropy), also called the Mollier diagram. These are shown in Figures 3.29, 3.30, and 3.31 (for more details on steam tables, see ChemicaLogic Corporation [2005b]).

Several thermodynamic relations are used in constructing the above diagrams as mentioned earlier. Two more properties, which are important in this context, are the Helmholzt function A and the Gibbs free energy G.

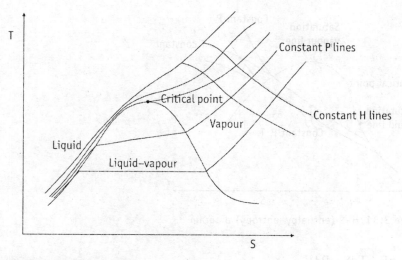

Figure 3.29 T–S diagram
Sources http://www.hss.energy.gov/NuclearSafety/techstds/standard/hdbk1012/
h1012v1.pdf, ChemicaLogic Corporation (2005a)

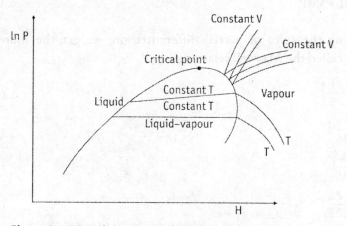

Figure 3.30 P–H (pressure–enthalpy) diagram

These are defined as

$$A = E - TS \qquad \qquad ...(3.80)$$
$$G = H - TS \qquad \qquad ...(3.81)$$

The following relations incorporating all major thermodynamic properties can easily be derived.

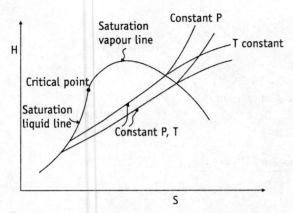

Figure 3.31 H–S (enthalpy–entropy) diagram

$$dE = TdS - PdV \qquad \qquad ...(3.82)$$
$$dH = TdS + VdP \qquad \qquad ...(3.83)$$
$$dA = -SdT - PdV \qquad \qquad ...(3.84)$$
$$dG = -SdT + VdP \qquad \qquad ...(3.85)$$

By applying the rules of partial differentiation, we get the following equations, also called the Maxwell relations.

$$\left(\frac{\partial T}{\partial V}\right)_S = -\left(\frac{\partial P}{\partial S}\right)_V \qquad \qquad ...(3.86)$$

$$\left(\frac{\partial T}{\partial P}\right)_S = \left(\frac{\partial V}{\partial S}\right)_P \qquad \qquad ...(3.87)$$

$$\left(\frac{\partial P}{\partial T}\right)_V = \left(\frac{\partial S}{\partial V}\right)_T \qquad \qquad ...(3.88)$$

$$\left(\frac{\partial V}{\partial T}\right)_P = -\left(\frac{\partial S}{\partial P}\right)_T \qquad \qquad ...(3.89)$$

The total differential for E, H, A, and G can be written as

$$dE = \left(\frac{\partial E}{\partial S}\right)_V dS + \left(\frac{\partial E}{\partial V}\right)_S dV \qquad \qquad ...(3.90)$$

$$dH = \left(\frac{\partial H}{\partial S}\right)_P dS + \left(\frac{\partial H}{\partial P}\right)_S dP \qquad \qquad ...(3.91)$$

$$dA = \left(\frac{\partial A}{\partial V}\right)_T dV + \left(\frac{\partial A}{\partial T}\right)_V dT \qquad \qquad ...(3.92)$$

$$dG = \left(\frac{\partial G}{\partial P}\right)_T dP + \left(\frac{\partial G}{\partial T}\right)_P dT \qquad \qquad ...(3.93)$$

Comparing the coefficient with those in Equations 3.82–3.85, we can write

$$T = \left(\frac{\partial E}{\partial S}\right)_V = \left(\frac{\partial H}{\partial S}\right)_P \qquad \ldots(3.94)$$

$$-P = \left(\frac{\partial E}{\partial V}\right)_S = \left(\frac{\partial A}{\partial V}\right)_T \qquad \ldots(3.95)$$

$$V = \left(\frac{\partial H}{\partial P}\right)_S = \left(\frac{\partial G}{\partial P}\right)_T \qquad \ldots(3.96)$$

$$-S = \left(\frac{\partial A}{\partial T}\right)_V = \left(\frac{\partial G}{\partial T}\right)_P \qquad \ldots(3.97)$$

The following useful equations can also be derived, relating E, H, and S to P, V, T, and C_p or C_v.

$$dE = C_v dT + \left[T\left(\frac{\partial P}{\partial T}\right)_V - P\right]dV \qquad \ldots(3.98)$$

$$dH = C_p dT + \left[V - T\left(\frac{\partial V}{\partial T}\right)_P\right]dP \qquad \ldots(3.99)$$

$$dS = C_v \frac{dT}{T} + \left(\frac{\partial P}{\partial T}\right)_V dV \qquad \ldots(3.100)$$

$$= C_p \frac{dT}{T} - \left(\frac{\partial V}{\partial T}\right)_P dP \qquad \ldots(3.101)$$

$$C_p - C_v = -T\frac{\left[\left(\partial V/\partial T\right)_P\right]^2}{\left(\partial V/\partial P\right)_T} \qquad \ldots(3.102)$$

From the above equations, all thermodynamic properties can be calculated (assuming a reference value) from the P–V–T relationships. Example of how the above equations can be used to calculate thermodynamic properties is available in standard textbooks (see Smith and van Ness 1975). Though many engineers are trained to use the available thermodynamic charts and tables such as the steam table, they seldom construct the table. However, one may be required to calculate the properties for relatively new applications such as evaluating a new working fluid (for example, cyclopentane as a CFC substitute or one of the freons for the low-temperature vapour-cycle engines), for which the above equations can be used. It should be noted, however, that the accurate calculation of thermodynamic properties for the purpose of constructing a chart is an exacting task.

We shall now discuss some common power cycles for production of mechanical power from heat.

Vapour cycles: steam power plant

The vapour power cycle is one of the most widely used thermal cycles, and accounts for most of the electrical energy generation in the world. A schematic diagram of the basic power cycle, also popularly called the Rankine cycle, is shown in Figure 3.32. The Rankine cycle on a T–S diagram is shown in Figure 3.33. (For a detailed description of thermodynamic cycles, etc. see <http://en.wikibooks.org/wiki/Applications_Engineering_Thermodynamics> and <http://en.wikibooks.org/wiki/Main_Page>.)

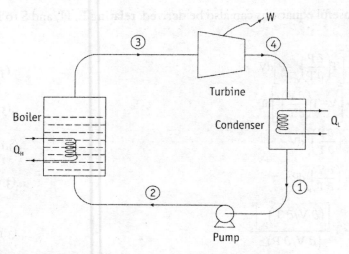

Figure 3.32 Schematic of a Rankine cycle power plant

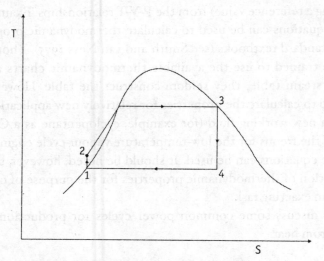

Figure 3.33 Rankine cycle on a T–S diagram

The cycle consists basically of four components: boiler, work-producing device such as a steam turbine, a condenser, and a pump. Liquid water at high pressure enters the boiler (2) where the latent heat supplied (Q_H) converts it to high-pressure vapour (3). Steam then goes to a turbine or heat engine and expands adiabatically, producing work. If this process is assumed to be reversible, it is isentropic, following the line 3–4. The low-pressure steam from the turbine is then condensed at constant pressure, by rejecting heat to a cooling medium, which is normally water. The final step is the pumping process, where work is done upon the fluid, and is represented by the isentropic line 1-2. This work is generally very small compared to the work produced by the turbine.

The cycle 1–2–3–4 is essentially theoretical and closely approaches an ideal Carnot cycle between the temperatures T_3 and T_1. The net work obtained is given by $(H_3 - H_4) - (H_2 - H_1)$ and the heat supplied is given by $(H_3 - H_2)$. Hence the efficiency of the ideal Rankine cycle is

$$\eta = \frac{\left(H_3 - H_4\right) - \left(H_2 - H_1\right)}{\left(H_3 - H_2\right)} \qquad \qquad ...(3.103)$$

The practical vapour cycle differs from the ideal cycle on several counts.

1 The liquid at point 1 is generally subcooled so that complete condensation is assumed.
2 To utilize the full potential of the high-temperature source of heat in the fuel bed and to limit condensation in the turbine (so as to reduce erosion of turbine blades), steam is generally superheated to several hundred degrees above saturation.
3 Adiabatic expansion in the turbine is never reversible; hence entropy will rise.
4 To maximize the cycle efficiency, both 'reheat' and 'regeneration' are employed. Reheat involves employing a two-stage turbine, in which steam is expanded in the first stage, returned to the boiler and reheated to the same temperature, and then expanded in the second stage. Regeneration involves preheating the liquid before it enters the boiler by transferring heat from some of the high-temperature steam.

The practical vapour cycle on a T–S diagram is shown Figure 3.34, now given by the loop 1-1'-2-3-3'-4-3"-4'-4". Each process of this cycle can be analysed by using flow equations (involving the relevant efficiencies, losses, etc.)

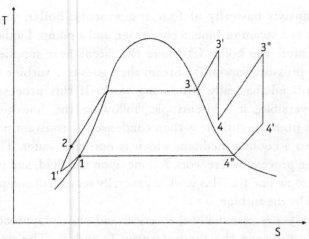

Figure 3.34 Actual Rankine cycle

and thermodynamic properties of steam (using a Mollier chart, for example) and the cycle efficiency can then be obtained. Examples of calculating cycle efficiencies can be found in Holman (1980).

Heat rate

A common term used by the power plant engineer is the heat rate defined as the heat input to boiler (in Btu) per unit electrical energy output (in kWh). In terms of overall efficiency,

$$\text{Heat rate} = \frac{3413}{\eta} \text{ Btu/kWh}$$

...(3.104)

Low-temperature organic Rankine cycle

For producing power from a low-temperature heat source such as a solar collector, solar pond, ocean thermal energy conversion, or geothermal energy, steam may not be the best working fluid. Several working fluids, such as freons, isobutane, and hexane have been used in a closed cycle Rankine engine. Such engines have to be hermetically sealed so as not to lose the fluid over long periods. In these cases, the pumping power may not be small in comparison to the work produced by the turbine, which, in turn, is low because of the lower temperatures involved. The economics of power generation in all such cases depends crucially on the efficiencies of the power cycles and on the cost the low-temperature heat. An example of solar pond power generation is given in (Kishore 1993). The organic Rankine cycles used for OTEC are described in Chapter 11.

Internal combustion engines

A steam power plant requires large heat transfer surface, both for transferring heat to the working fluid in boiler and for rejecting heat in condenser. A large heat transfer area necessitates the use of large quantities of metal, and hence involves large weights. The weight-to-power ratio of a steam power plant (in tonnes/kW) would thus be quite large. An IC (internal combustion) engine, on the other hand, employs the fuel mixture itself as the working fluid, with the result that chemical energy from the combustion process is available within the work-producing machine itself. Due to absence of large heat transfer area, the weight-to-power ratio for IC engines is much smaller, and hence they can be used as small, compact power plants, both for stationary and mobile applications. Also, as a high temperature can be attained within the combustion space, the need for high-temperature materials (required for transferring heat at high temperature to steam in a superheater for example) is minimized. The IC engine can thus operate at high efficiencies.

The IC engines can be classified as SI (spark-ignition) or CI (compression-ignition) engines. They can also be classified as four- or two-stroke engines based on the strokes per cycle. Four- or two-stroke engines are reciprocating engines in which fuel injection is intermittent. On the other hand, gas turbines employ a steady flow of fuel and air. It would be useful to consider the physical working of an IC engine (Otto cycle) before going into a cycle analysis. Consider the engine configuration shown in Figure 3.35 for the four-stroke engine.

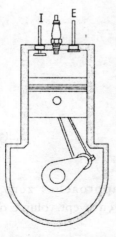

Figure 3.35 Basic arrangement for a reciprocating internal combustion four-stroke engine

When the piston is at TDC (top dead centre), the intake valve I opens and a charge of fuel and air is drawn into the cylinder as the piston moves downwards (intake stroke). At the BDC (bottom dead centre), the intake valve closes (compression stroke). At TDC, a spark is set off, initiating combustion of the fuel–air mixture and pushing the piston downwards (power stroke). At BDC, the exhaust valve E opens and the piston moves upwards to push out the combustion products (exhaust stroke). When the piston reaches TDC, the exhaust valve closes and the intake valve opens, starting the cycle again. Note that two complete revolutions of the crank shaft are involved for each power stroke.

It should be noted that while steam passes through a cycle of processes returning to its original state after each cycle, the thermodynamic cycle of the IC engine is not completely cyclical. This is because the combustion products are discharged into the atmosphere. However, a simple analysis of an IC engine can be carried out by representing the actual process with air as the working fluid. Such approximate cycles are called 'air standard' cycles. The Otto cycle described above is represented in Figure 3.36 (a) on a P–V diagram.

The air standard Otto cycle is shown in Figure 36 (b) and (c) on a P–V diagram and a T-S diagram, respectively.

It can be seen that combustion in the real Otto cycle is represented by a constant volume heat addition from an external source in an air standard Otto cycle. The thermal efficiency of the cycle is calculated by the equation

$$\eta = \frac{Q_H - Q_L}{Q_H} = 1 - \frac{Q_L}{Q_H} = 1 - \frac{mC_v\left(T_4 - T_1\right)}{mC_v\left(T_3 - T_2\right)} \qquad \ldots(3.105)$$

In the isentropic process,

$$\frac{T_2}{T_1} = \left(\frac{V_1}{V_2}\right)^{\gamma-1} = \left(\frac{V_4}{V_3}\right)^{\gamma-1} = \frac{T_3}{T_4}$$

$$\eta = 1 - \frac{T_1}{T_2} = 1 - \frac{1}{\left(V_1/V_2\right)^{\gamma-1}}$$

$$= 1 - \left(CR\right)^{1-\gamma} \qquad \ldots(3.106)$$

where CR is the compression ratio, V_1/V_2.

V_2 is the so-called dead volume. If V_2 approaches zero, CR approaches ∞. As V_2 increases, CR reduces for a given swept volume of the piston.

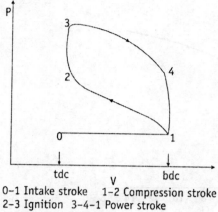

0–1 Intake stroke   1–2 Compression stroke
2–3 Ignition   3–4–1 Power stroke
1–0 Exhaust stroke

Figure 3.36(a) Otto cycle

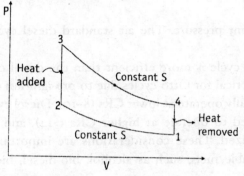

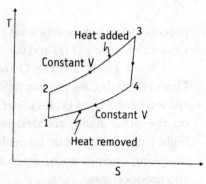

Figure 3.36(b) The air standard Otto cycle on P–V diagram

Figure 3.36(c) The air standard Otto cycle on T–S diagram

CR	1	2	4	6	8	10	12	14
η	0	0.242	0.426	0.512	0.565	0.602	0.63	0.652

A plot of η versus CR is given in Figure 3.37. It can be seen that η increases rapidly with CR at low values, and slowly at a higher CR. This theoretical result agrees with the actual tests on IC engines.

The diesel engine

In the diesel engine, the fuel–air mixture is not ignited by a spark, but by spontaneous combustion caused by a higher compression ratio. The fuel is not injected until the end of the compression process. Then it is added at such a slow rate in comparison with the rate of piston travel that the combustion

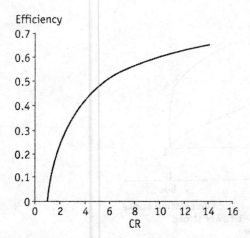

Figure 3.37 Variation of efficiency with compression ratio

process occurs at nearly constant pressure. The air standard diesel cycle is shown in Figure 3.38 (a) and (b).

For a given CR, the Otto cycle is more efficient than the diesel cycle. However, high CRs are not practical for Otto cycles due to pre-ignition difficulties, hence Otto cycles generally operate at lower CRs (8–10). Diesel cycles, on the other hand, are designed to operate at higher CRs (>14), and thus higher efficiencies can be realized. These considerations are important in designing engines using renewable fuels, such as alcohol, bio-diesel, biogas, and producer gas.

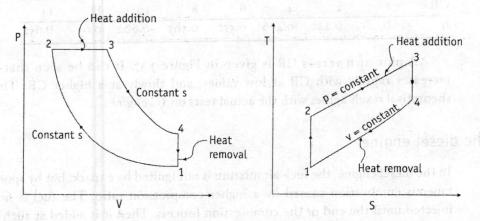

Figure 3.38 (a) The air standard diesel cycle on P-V diagram

Figure 3.38 (b) The air standard diesel cycle on T-S diagram

The thermal efficiency expression for air standard diesel cycle can be shown to be

$$\eta = 1 - (CR)^{1-\gamma} \frac{\left(R_c^{\gamma} - 1\right)}{\gamma \left(R_c - 1\right)} \qquad \qquad ...(3.107)$$

where CR is defined as for the Otto cycle, V_1/V_2 and R_c defined as the cut-off ratio equal to V_3/V_2. For the Otto cycle, V_3/V_2 is equal to 1, but for the diesel cycle it is greater than 1.

There are other cycles also such as the Wankel engine cycle and the Stirling cycle. The Stirling cycle has gained some importance in renewable energy engineering because it is an *external combustion* engine, which means that heat is added from outside. Thus, a Stirling engine can receive heat from a solar concentrator or from the combustion of solid biomass. Air, hydrogen, and helium had been used as working fluids in Stirling engines. A diagrammatic sketch of the Stirling cycle is shown in Figure 3.39 and the corresponding P–V and T–S diagrams are shown in Figure 3.40 (a) and (b). With reference to these diagrams, the following processes take place.

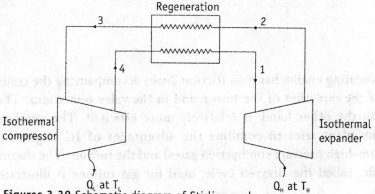

Figures 3.39 Schematic diagram of Stirling cycle

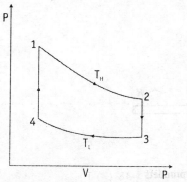

Figure 3.40 (a) Stirling cycle on p–V diagram

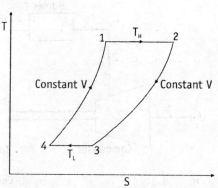

Figure 3.40 (b) Stirling cycle on T–S diagram

1–2	Constant temperature (T_H) heat addition
2–3	Constant volume heat rejection
3–4	Constant temperature (T_L) heat rejection
4–1	Constant volume heat addition

As processes 2–3 and 4–1 occur between the same temperature limits, the heat rejected in 2–3 would be equal to the heat added in 4–1. This suggests the use of a regenerator, wherein heat is transferred from step 2–3 to step 4–1 internally. If the regenerator operates ideally, then we have a reversible cycle with heat addition and removal at constant temperatures. The efficiency of such a cycle will then be

$$\eta = 1 - \frac{T_L}{T_H} \qquad \qquad ...(3.108)$$

which is the same as that for a Carnot cycle. The Stirling cycle can thus approach the highest possible efficiencies, depending upon the efficiency of the regenerator. Due to this, the Stirling engine has been explored extensively for automotive use, but so far, successful commercial designs have been elusive.

Gas turbines

The reciprocating engine has high friction losses accompanying the continual reversal of the direction of the piston and in the valve operations. The gas turbine, on the other hand, is relatively more efficient. The combustion gas turbine plant tries to combine the advantages of IC engines (high temperature–high pressure combustion gases) and the turbine. The thermodynamic cycle, called the Brayton cycle, used for gas turbine is illustrated in Figure 3.41.

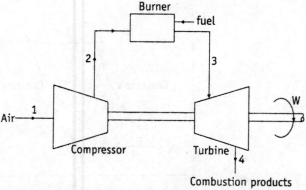

Figure 3.41 Schematic diagram of Brayton cycle

The air standard Brayton cycle is shown in Figure 3.42(a) and (b). The compressor raises the pressure of the air, and then heat is produced by burning a fuel injected into the high-pressure air stream in a burner. The high temperature–high pressure products of combustion are then expanded in the turbine to produce shaft work. Part of the turbine work is used to drive the compressor, which is mounted on the same shaft. The gas turbine cycle is a steady-flow device and work quantities, etc., can be determined with enthalpy differences.

The net work done is $(H_3 - H_4) - (H_2 - H_1)$ and the heat input is $H_3 - H_2$, hence the thermal efficiency would be

$$\eta = \frac{(H_3 - H_4) - (H_2 - H_1)}{H_3 - H_2} \qquad \qquad ...(3.109)$$

If one assumes ideal gas behaviour and isentropic conditions, the efficiency can be expressed as

$$\eta = 1 - \left(\frac{P_2}{P_1}\right)^{(1-\gamma)/\gamma} \qquad \qquad ...(3.110)$$

Most gas turbines operate on the open cycle and the exhaust gases are released to the surroundings. However, this means a loss of high-temperature gas. This loss can be recovered either by using a regenerator in which the high pressure air is pre-heated before it enters the combustion chamber, or by employing a heat recovery steam generator. Steam can be used for process heat or to drive a steam turbine, in which case it is called a combined cycle

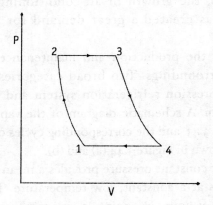

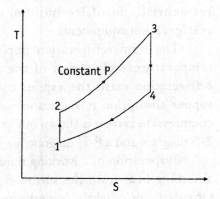

Figure 3.42(a) The air standard Brayton cycle on P-V diagram

Figure 3.42(b) The air standard Brayton cycle on T–S diagram

power plant. The overall thermal efficiency of such cycles can be quite high. It is also possible to improve the efficiency of the gas turbine cycle by using multi-stage compression with intercooling or multi-stage turbine with reheating. The design and engineering of gas turbines are quite advanced at present, and several schemes involving coal gasification and biomass gasification have been tried out using the so-called IGCC (integrated gasification combined cycle) power plants (Reddy, Kelly, Johansson, *et al.* 1993). Also, small-capacity 'micro-turbines' with a capacity rating of 70–100 kW are now available for decentralized power and heat production. It should be noted, however, that open-cycle gas turbines without heat recovery do not offer substantially higher efficiencies, compared to Otto or diesel cycles.

Gas turbine for jet propulsion

The turbojet engine, which is extensively used for aircraft propulsion, is a modified version of the open-cycle gas turbine. The hot combustion gases are expanded in the turbine only to the extent necessary to produce enough work to drive the compressor. The remaining energy is converted into high-velocity kinetic energy by using a nozzle. This generates a thrust, necessary to move the aircraft at high speeds. For further discussion on gas turbines and jet engines, one can refer to standard publications (for example Van Wylen and Sonntag 1973).

Refrigeration cycle

Refrigeration is used extensively in the preservation and transportation of food items and beverages. Also, the growth of air-conditioning in residential and office buildings has created a great demand for the refrigeration equipment.

The term refrigeration implies the production and maintenance of temperatures below that of the surroundings. Two broad categories of refrigeration exist: the vapour compression refrigeration system and the vapour absorption refrigeration system. A schematic diagram of the vapour compression system is shown in Figure 3.43, and the corresponding cycles on a T–S diagram and a P–H diagram are shown in Figure 3.44 (a) and (b).

Evaporation of a working fluid at constant pressure provides a means of absorbing heat from the surroundings at a constant, low temperature. The saturated (or slightly superheated) vapour at point 1 is compressed isentropically in the ideal cycle to point 2. Condensation produces a saturated liquid at high pressure at point 3. The high-pressure liquid is then throttled

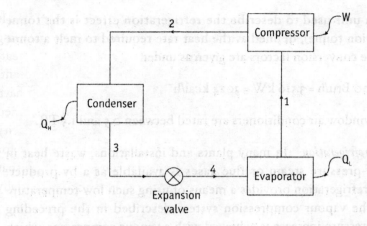

Figure 3.43 Schematic diagram of vapour compression system

to a low-pressure, low-temperature liquid at point 4 through an isenthalpic process.

Several substances have been used as refrigerants in the vapour compression system. The most common are CFCs (chlorofluorocarbons), also called freons. However, due to recent findings that the CFCs deplete the protective ozone layer in the atmosphere, these are being replaced by non-CFC refrigerants (for a complete discussion on this, refer to Dincer 2003). Ammonia is also widely used as a refrigerant in cold storages.

The coefficient of performance is defined as

$$COP = \frac{\text{Refrigeration effect}}{\text{Work done}} = \frac{Q_L}{W}$$

$$= \frac{H_1 - H_4}{H_2 - H_1} \qquad \qquad \qquad ...(3.111)$$

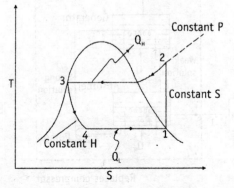

Figure 3.44(a) Vapour compression on T–S diagram

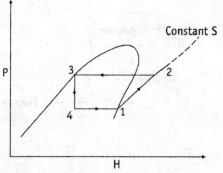

Figure 3.44(b) Vapour compression on P–H diagram

A common unit used to describe the refrigeration effect is the tonne or TR (refrigeration tonne), defined as the heat rate required to melt a tonne of ice in 24 h. The conversion factors are given as under.

1 TR = 12 000 Btu/h = 3.516 kW = 3024 kcal/h

Common window air conditioners are rated between 0.5 and 1.5 TR.

Absorption refrigeration In many plants and installations, waste heat in the form of low-pressure steam or flue gases is available as a by-product. The absorption refrigeration provides a means of using such low-temperature waste heat. In the vapour compression system described in the preceding discussion, the pressure increase is achieved with a vapour compressor, which requires a relatively large power input per unit mass of fluid. The absorption system represents a method in which the compressor is replaced by a liquid pump (which does not require high power) and a suitable heater. The basic absorption system, shown in Figure 3.45, consists of an evaporator, an absorber, a generator, and a condenser.

Absorption refrigeration involves an absorbent–refrigerant 'pair'. Common pairs are NH_3–H_2O and LiBr–H_2O. In the former, ammonia is the refrigerant and in the latter, water is the refrigerant. The ammonia vapour at point 1 is mixed with a weak solution of NH_3–H_2O in the absorber. Heat liberated in the absorption process is removed continuously. The strong solution produced is pumped to the generator (which is essentially a distillation column), where pure ammonia is generated along with a depleted (weak) solution of NH_3–H_2O by supplying waste heat. The high-pressure vapours

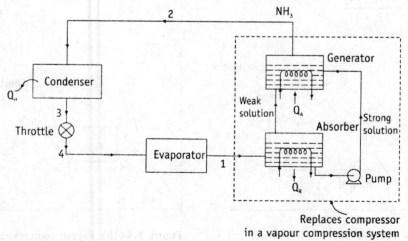

Figure 3.45 Schematic diagram of absorption refrigeration system

then go through condensation and throttling just as in the vapour compression system.

In the LiBr–H_2O system, water is used as a refrigerant and hence, vacuum has to be maintained in the evaporator to achieve a low temperature. In recent years, adsorption refrigeration systems have also been developed using the principle of adsorption–desorption to replace absorption. Water and CH_3OH (methanol) have been used as refrigerants, and silica gel, bentonite clay, etc., have been used as adsorbers. The low-temperature heat available from a solar collector and solar pond has been considered as a cheap source for absorption and adsorption refrigeration systems.

The heat pump

The term refrigeration was used because the main objective was to achieve cooling. The same device can be used to pump the heat from an ambient temperature source to a higher-temperature delivery point. Thus, one can employ the same machine for both summer cooling and winter heating. A discussion of heat pumps can be found in *Refrigeration Systems and Applications* (Dincer 2003).

Chemical thermodynamics

Chemical reactions occur with some spontaneity, characterized by a 'rate' of reaction. There are also reversible reactions, in which both the forward and backward reactions occur, but with different speeds. For example, the combustion reaction

$$C + O_2 \longrightarrow CO_2$$

is almost irreversible, meaning that the forward reaction is several times faster than the reverse reaction. However, the shift reaction

$$CO + H_2O \rightleftharpoons CO_2 + H_2$$

is reversible, with both forward and backward reactions occurring at the same time. In the former, reactants are almost completely converted into products, but this is not so in the latter reaction. A reversible reaction will ultimately come to an equilibrium, and all compounds will attain an equilibrium composition at a given pressure and temperature. Thermodynamics has been employed to throw light on the problem of chemical equilibrium by using the free energy concept. It has been possible, for example, to define the condition of equilibrium, and to show how it varies with temperature and pressure.

The free energy term introduced earlier, is defined as

$$G = H - TS \qquad \text{...(3.112)}$$

The relation that links G to other thermodynamic properties is

$$dG = VdP - SdT \qquad \text{...(3.113)}$$

At constant pressure

$$dG_p = -SdT_p \; ; \; \left(\frac{\partial G}{\partial T}\right)_p = -S \qquad \text{...(3.114)}$$

At constant temperature

$$dG_T = VdP_T; \; \left(\frac{\partial G}{\partial P}\right)_T = V \qquad \text{...(3.115)}$$

These relations give the variation of G with T and P. Substituting Equation 3.114 into Equation 3.112, we get

$$G = H + T\left(\frac{\partial G}{\partial T}\right)_p \qquad \text{...(3.116)}$$

This is referred to as the Gibbs–Helmholtz equation. Dividing by T^2 on both sides and rearranging, we can have an alternate equation

$$\left[\frac{\partial(G/T)}{\partial T}\right]_p = -\frac{H}{T^2} \qquad \text{...(3.117)}$$

In the definition of entropy, it has been shown that the entropy change dS for a reversible process is equal to $\delta Q/T$, and for an irreversible process, it is greater than $\delta Q/T$. In the equation form.

$$dS \geq \delta Q/T \qquad \text{...(3.118)}$$

Using the first law relation and the definition of free energy, one can obtain the following equation from Equation 3.113.

$$dG \leq VdP - SdT \qquad \qquad ...(3.1119)$$

At constant temperature and pressure, this becomes

$$dG_{T,P} \leq 0 \qquad \qquad ...(3.120)$$

This means that $dG_{T,P}$ will be less than zero for a spontaneous (irreversible) process, and will be equal to zero for a system at equilibrium at constant temperature and pressure. Equation 3.120 has been used successfully for calculating equilibrium conditions and to predict the direction of the reaction.

Consider the homogeneous, multicomponent system at pressure P and temperature T. The free energy for the system will then be a function of T, P, and composition.

$$G = f(T, P, n_1, n_2, ..., n_i, ...) \qquad \qquad ...(3.121)$$

where n_1, n_2, etc., are the number of moles of different constituents. Differential changes of G can then be written as

$$dG = \left(\frac{\partial G}{\partial T}\right)_{P, n_1, n_2, ...} dT + \left(\frac{\partial G}{\partial P}\right)_{T, n_1, n_2, ...} dP + \left(\frac{\partial G}{\partial n_1}\right)_{T, P, n_2, ...} dn_1 +$$

$$\left(\frac{\partial G}{\partial n_2}\right)_{T, P, n_1, n_3 ...} dn_2 + ... \qquad \qquad ...(3.122)$$

The derivative $(\delta G/\delta n_1)_{T,P,n_2,...}$ is called the partial molal free energy for the constituent n_1 and so on. This is represented as $\bar{G}_1$. For a condition of constant temperature and pressure, Equation 3.122 can be written as follows.

$$dG_{T,P} = \bar{G}_1 dn_1 + \bar{G}_2 dn_2 + ... \qquad \qquad ...(3.123)$$

The total free energy of the system is the sum of all free energies of the components. Hence, we can write

$$G_{T,P} = n_1 \bar{G}_1 + n_2 \bar{G}_2 + ... \qquad \qquad ...(3.124)$$

differentiating the above, we get

$$dG_{T,P} = \left(n_1 d\bar{G}_1 + \bar{G}_1 dn_1\right) + \left(n_2 d\bar{G}_2 + \bar{G}_2 dn_2\right) + \ldots \ldots \qquad \ldots(3.125)$$

Comparing with Equation 3.120, we can easily see that for a given temperature and pressure

$$n_1 d\bar{G}_1 + n_2 d\bar{G}_2 + \ldots \ldots \ldots = 0 \qquad \ldots(3.126)$$

The partial molar free energy is also termed as the chemical potential μ. Equation 3.126 can thus be written as

$$n_1 d\mu_1 + n_2 d\mu_2 + \ldots = 0 \qquad \ldots(3.127)$$

This is called as Gibbs–Duhem equation.
At equilibrium, Equation 3.123 can also be written as

$$dG_{T,P} = \mu_1 dn_1 + \mu_2 dn_2 + \ldots \ldots$$

$$= \sum \mu_i dn_i = 0 \qquad \ldots(3.128)$$

where the summation includes all the μdn terms for all phases constituting the complete system.

Equation 3.128 can be used to derive the phase rule, which was stated in Equation 3.8 as

$$F = C - P + 2 \qquad \ldots(3.129)$$

Refer to Glasstone (1947) for details.

Phase changes

Consider a single component system of two phases, for example liquid water and its vapour. At equilibrium, we have

$$\Delta G = 0 \qquad \ldots(3.130)$$

If $\overline{G}_A$ is the molar free energy in phase A and $\overline{G}_B$ the molar free energy in phase B, the above equation implies

$$\Delta G = \overline{G}_B - \overline{G}_A = 0$$

$$\text{or } \overline{G}_B = \overline{G}_A \qquad\qquad\qquad ...(3.131)$$

This implies that whenever two phases of the same single substance are in equilibrium at a given T and P, the molar free energy is the same in each phase. For an infinitesimal change in $\overline{G}_B$, there is a corresponding change in $\overline{G}_A$. Hence,

$$d\overline{G}_A = d\overline{G}_B \qquad\qquad\qquad ...(3.132)$$

$$V_A dP - S_A dT = V_B dP - S_B dT \qquad\qquad\qquad ...(3.133)$$

$$\frac{dP}{dT} = \frac{S_B - S_A}{V_B - V_A} = \frac{\Delta S}{\Delta V} \qquad\qquad\qquad ...(3.134)$$

where ΔS is the entropy change accompanying the phase change and ΔV is the specific volume change. If ΔH is the latent heat of vaporization, ΔS can be written as $\Delta H / T$. Equation 3.134 can thus be written as

$$\frac{dP}{dT} = \frac{\Delta H}{T \Delta V} \qquad\qquad\qquad ...(3.135)$$

This is known as the *Clayperon equation*.

If the temperature is not too near the critical point, the liquid phase specific volume is small in comparison to the vapour phase specific volume. Hence, ΔV can be approximated to V_B (assuming B to be the vapour phase). Further, assuming the vapour phase to obey the ideal gas law, one can derive

$$\frac{1}{P}\frac{dP}{dT} = \frac{\Delta H}{RT^2} \qquad\qquad\qquad ...(3.136)$$

$$\frac{d\ln P}{dT} = \frac{\Delta H}{RT^2} \qquad\qquad\qquad ...(3.137)$$

This expression is referred to as the *Clausius–Clayperon equation*. Integration gives

$$\ln\frac{P_2}{P_1} = -\frac{\Delta H}{R}\left(\frac{1}{T_2} - \frac{1}{T_1}\right) \qquad\qquad\qquad ...(3.138)$$

Chemical reactions

Consider the general chemical reaction

$$aA + bB + ... \rightleftharpoons qQ + rR + ...$$

which is in a state of equilibrium. If an infinitesimal change occurs in the composition, changes in the left side (reactants) will be accompanied by corresponding changes in the right side (products). Thus, dn_A moles of A, dn_B moles of B, etc., are consumed and dn_Q moles of Q and dn_R moles of R, etc., are formed. The free energy change accompanying the process is given by

$$dG_{T,P} = \left(\mu_Q dn_Q + \mu_R dn_R + ...\right) - \left(\mu_A dn_A + \mu_B dn_B + ...\right) \qquad ...(3.139)$$

where μs are the chemical potentials or partial molar free energies. Applying the condition of equilibrium, $dG_{T,P} = 0$, we get

$$\mu_Q dn_Q + \mu_R dn_R + ... = \mu_A dn_A + \mu_B dn_B + ... \qquad ...(3.140)$$

The various molar changes dn_A, dn_B, dn_R, etc., are not arbitrary, but are limited to each other by stoichiometry dictated by the stoichiometric parameter a, b, q, r, etc., in a proportional manner. One can thus write

$$q\mu_Q + r\mu_R + ... = a\mu_A + b\mu_B + ... \qquad ...(3.141)$$

The chemical potential μ of any component can be represented by the equation

$$\mu = \mu^\circ + RT \ln a \qquad ...(3.142)$$

where μ° is the chemical potential of a given substance in a chosen reference state and a is called the activity. It is obvious that at the reference state, $\mu = \mu^\circ$ and $a = 1$.

Substituting the above equation in Equation 3.141, we get

$$q\left(\mu_Q^\circ + RT \ln a_Q\right) + r\left(\mu_R^\circ + RT \ln a_R\right) + ... = a\left(\mu_A^\circ + RT \ln a_A\right) + b\left(\mu_B^\circ + RT \ln a_B\right) + ...$$

$$(3.143)$$

Re-arranging and simplifying, we get

$$-RT \ln\left(\frac{a_Q^q \times a_R^r \times ....}{a_A^a \times a_B^b \times ....}\right) = \left(q\mu_Q^\circ + r\mu_R^\circ + ...\right) - \left(a\mu_A^\circ + b\mu_B^\circ + ...\right)$$

$$= \Delta G°$$

$$\text{or } \Delta G_T° = - RT \ln k \qquad\qquad ...(3.144)$$

$$\text{where, } k = \frac{a_Q^q \times a_R^r \times ....}{a_A^a \times a_B^b \times ....} \qquad\qquad ...(3.145)$$

is called the equilibrium constant. The activities are related to partial pressure (in gaseous state), or mole fraction, and can be evaluated making certain assumptions, using experimental data, etc. In the gaseous state another term called 'fugacity' is defined as

$$f_i = p_i \gamma_i \qquad\qquad ...(3.146)$$

where p_i is the partial pressure and γ_i is the activity coefficient.

The term $\Delta G_T°$ is often referred to as standard free energy of the reaction at a specified temperature T and can be obtained from the heat of reaction, entropy changes, and other parameters. Standard free energies of formation at 25 °C for several compounds are available (see, for example, Smith and van Ness 1975). Equation 3.141 gives the increase in free energy when the reactants, all in their standard states, are converted into products in their standard states, and are useful both for evaluating the equilibrium condition and for predicting the path of a reaction. If $\Delta G°$ for a specified reaction is negative, the reaction can occur spontaneously. On the other hand, if it is positive, the reaction cannot occur.

The variation of equilibrium constants with pressure and temperature can also be obtained using the free energy relation, however, these will not be discussed here.

The aim of this section was to provide the renewable engineer the opportunity to appreciate the use of basic thermodynamics for various practical applications. It is so fundamental that it cannot be given a cursory treatment, but its scope is too vast to be compressed into a few pages.

Challenges to thermodynamics

It is not as if classical thermodyamic formulations have not been challenged. In his monograph entitled 'Development and thermodynamics – a search for new energy-quality markers' (Seshadri 1982), the late Prof. C V Seshadri has questioned the validity and over-riding importance given to thermodynamic formulations in the development context. As renewable energy use is intricately linked to sustainable development, any one interested in delving deeply into the philosophical aspects of thermodynamics and development can start with this monograph.

Transport phenomena

Introduction

When a fluid of mass m is moving with a velocity u, it has a momentum of mu. Hence, flow of fluids is related to transfer of momentum. When a body at a temperature T_1 is in thermal contact with another body at a lower temperature T_2, transfer of heat occurs from higher to lower temperature. Similarly, when a substance diffuses into another substance either in the same phase or in a different phase, mass transfer occurs. These three kinds of transfer, namely, momentum, heat, and mass transfer are collectively termed transport phenomena, the study of which is fundamental to all engineering applications, including renewable energy engineering. Some examples of application of transport phenomena are given below.

- Optimization of piping network for solar thermal collector fields
- Design of piping network for distribution of cooking gas (for example, biogas or producer gas) for a village community
- Calculation of pressure drop (and pumping power) in a cooling/cleaning train of a producer gas power plant
- Optimal design of blades of a wind turbine or a hydro turbine
- Fluidization of rice husk in a fluidized-bed combustion/gasification system
- Granule formation in a UASB (upflow anaerobic sludge blanket) reactor for biomethanation of organic wastes
- Analysis of biomass briquetting process
- Analysis, study, and design of a variety of solar thermal devices such as flat plate collectors, air heaters, solar ponds, solar stills, box type cookers, and solar passive components.
- Design of heat exchangers and in general, any heat transfer systems, such as furnaces, improved cook stoves, and boilers.
- Production of bio-ethanol, bio-diesel, etc., which involves distillation
- Removal of impurities, such as H_2S (in biogas) or tar (in producer gas) by scrubbing.
- Design of solar- or biomass-based drying systems
- Design and analysis of absorption/adsorption cooling systems using solar energy, biogas, and so on.

The basic equation for momentum transfer is Newton's Law of viscosity, which states that the shear stress τ between layers of fluid is proportional to the velocity gradient.

$$\tau \propto -\frac{\partial u_x}{\partial x}$$

The proportionality constant is the viscosity μ.

$$\tau = -\mu \frac{\partial u_x}{\partial x}$$

This can be written as

$$\tau = -\left(\frac{\mu}{\rho}\right)\frac{\partial}{\partial x}(\rho u_x) \qquad \qquad \ldots(3.147)$$

where ρ = density of fluid and $\mu/\rho = \upsilon$ = kinematic viscosity. Note that ρu_x is the momentum per unit volume of fluid.

The Fourier's Law of heat conduction states that the heat flow per unit area is proportional to the temperature gradient, the proportionality constant being thermal conductivity.

$$\frac{q}{A} = -k \frac{\partial T}{\partial x}$$

The negative sign implies that flow is from higher to lower temperatures and hence, the temperature gradient is negative. The above equation can also be written as

$$\frac{q}{A} = -\frac{k}{\rho C_p}\frac{\partial}{\partial x}\left(\rho C_p T\right) \qquad \qquad \ldots(3.148)$$

The quantity $(k/\rho C_p)$ is called thermal diffusivity and the quantity $(\rho C_p T)$ is the heat content per unit volume.

The rate at which a component A diffuses in a mixture of two components A and B, is given by Fick's Law of diffusion

$$N_A = -D_{AB}\frac{\partial C_A}{\partial x} \qquad \qquad \ldots(3.149)$$

where N_A is the molar rate of diffusion of A per unit area, C_A is the molar concentration in moles per unit volume, and D_{AB} is the diffusivity.

It is easy to see the similarity between Equations 3.147, 3.148, and 3.149. The proportionality constants μ/ρ, $k/\rho C_p$, and D_{AB} have the same dimensions of (length)²/time. From kinetic theory of gases, one can derive that all the three quantities are proportional to the product of mean free path λ (mean distance travelled by a gas molecule before colliding with another molecule) and the root mean square velocity of the molecule, u_m given as

$$u_m = \sqrt{8RT/\pi M}$$

Thus, there seems to be a theoretical and molecular basis for similarity between momentum, heat, and mass transfer, at least in the gaseous state. However, this is not so obvious for liquid and solid phases. All the three phenomena are presented and reviewed separately in the following sections.

Flow of fluids

In the earlier section, mass and energy balances for flowing systems have been discussed. For fluid flow, there is an additional quantity to be considered: conservation of momentum. Solutions to problems involving flow of fluids take recourse to first formulating and then solving equations of continuity (mass conservation), momentum, and energy.

Consider a fluid contained between two parallel plates placed at a distance Y. The fluid is stagnant up to a time t = 0, when the top plate is set into motion at a speed u_o. Initially, the velocity of the fluid is zero everywhere, but slowly, each layer of fluid acquires certain velocity as shown in Figure 3.46.

The layer of fluid immediately next to the top plate will have a velocity u_o, and the subsequent layer will have progressively smaller velocities until the last layer, which will have zero velocity. In other words, a velocity 'gradient' would have been established shortly after the top plate was set in motion. A certain force is required at any point on the y-axis, parallel to x-axis, to maintain the fluid motion. This force per unit area of a plane parallel to the x-axis is the shear stress exerted in the x-direction at point y. According to Newton's Law of viscosity, the shear stress is proportional to the local velocity gradient; the proportionality constant being defined as viscosity μ.

$$\tau_{yx} = -\mu \frac{du}{dy} \qquad \qquad ...(3.150)$$

The negative sign implies that as y is increasing, u is decreasing.

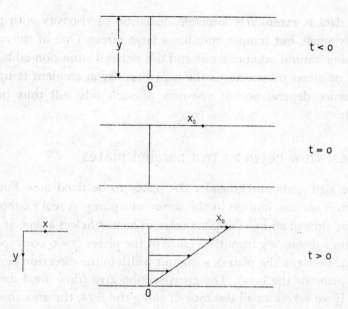

Figure 3.46 Development of a velocity gradient

Fluids obeying the above equation are called Newtonian fluids. Fluids which do not obey this law (for example, pastes, slurries, polymeric solutions, etc.), are called non-Newtonian fluids. Cattle dung, for example, or slurries made from dung would be an example of non-Newtonian fluids. Similarly, the flow of biomass through a briquetting press would also be non-Newtonian.

There are several empirical models to describe the behaviour of non-Newtonian fluids, such as the Bingham model, which is described by the equation

$$\tau_{yx} = \tau_o - \mu_o \frac{du}{dy} \qquad\qquad ...(3.151)$$

which means that the fluid behaves like a solid till a minimum stress τ_o is applied, but then onwards it behaves like a Newtonian fluid.

The power law model is described by the equation

$$\tau_{yx} = -m\left(\frac{du}{dy}\right)^n \qquad\qquad ...(3.152)$$

where m and n are constants.

Clay–water mixtures, lime solution, paper pulp, etc., can be described by the power law model.

Viscosity data is extensively available. Variation of viscosity with pressure is generally small, but temperature has a large effect. One of the major problems of using natural vegetable oils and oils derived from non-edible oil seeds as petrol or diesel substitutes is the high viscosity at ambient temperatures. Temperature dependence of viscosity of such oils will thus be an important study.

Momentum balance: flow between two parallel plates

Refer to Figure 3.46 again and consider the plates to be fixed now. But the fluid in-between is set into motion by the action of a pump. A real example of such a case is the flow of air in a flat-plate solar air heater. Select a slice of fluid contained within a distance y from the centre of the plates (y = 0 corresponds to the mid-plane between the plates), with unit width in the direction perpendicular to the plane of the book. The cross-section area (flow area) for this slice is 2y × 1. If we take a small distance dl along the flow, the area through which shear stress is acting on the top side of the slice is dl × 1. A similar stress is acting on the bottom side, on area dl × 1. If t is the shear stress acting on the top edge, the shear force is $\tau \times$ dl × 1. A similar force acts on the bottom slice. If (– dP) is the pressure difference (dP is negative, as P will decrease with l), the net force acting on the slice is (–dP) × 2y × 1 in the flow direction. The momentum balance is written as

{Rate of momentum in} – {Rate of momentum out}
+ {Sum of forces acting on the system}
= {Rate of momentum accumulation}   ...(3.153)

As rate of change of momentum is equal to force as per the equation

$$F = ma = m\frac{du}{dt} = \frac{d}{dt}(mu)$$

all force terms acting on the slice enter the momentum balance.

We will make two further assumptions.
- The flow is incompressible, meaning the density of fluid remains constant.
- The flow is steady state, meaning the accumulation term is zero.

The first assumption means that the average velocity is constant and hence 'momentum in' will be equal to 'momentum out'. The momentum balance then simplifies to a force balance, which can be written as

$$-dP \times 2y \times 1 = 2 \times \tau \times dl \times 1 \qquad \ldots(3.154)$$

According to Newton's Law, τ can be replaced by $-\mu(du/dy)$

$$-dP2y = -2\mu \frac{du(y)}{dy}dl \qquad \ldots(3.155)$$

where $u(y)$ is velocity at point y.

$$du(y) = \frac{dP}{\mu dl}ydy$$

Integration gives

$$u(y) = \frac{dP}{\mu dl}\frac{y^2}{2} + C$$

where C is the integration constant.

Near the stationary plates, the velocity is zero, giving the boundary condition as

$$u = 0 \text{ at } y = Y/2$$

Therefore,

$$0 = \frac{dP}{\mu dl}\frac{1}{2}\frac{Y^2}{4} + C$$

Substituting back we get

$$u(y) = -\frac{dP}{2\mu dl}\left(\frac{Y^2}{4} - y^2 \right) \qquad \ldots(3.156)$$

This equation implies that the velocity varies in a parabolic manner across y. The average velocity can be defined as

$$u_{av} = \frac{\int_{-Y/2}^{Y/2} u(y)dy}{\int_{-Y/2}^{Y/2} dy} \qquad \ldots(3.157)$$

Performing the integration, we get

$$u_{av} = -\frac{-dP}{12\mu dl}\left(Y^2\right) \qquad \qquad ...(3.158)$$

The volume flow rate Q per unit width is given by

$$Q = (Y \times 1)u_{av}$$

$$Q = -\frac{-dP}{12\mu dl}\left(Y^3\right) \qquad \qquad ...(3.159)$$

If B is the total width, the total volumetric flow rate is

$$Q = -\frac{(-dP)}{12\mu dl}BY^3 \qquad \qquad ...(3.160)$$

The maximum velocity can be obtained by substituting y = 0 in Equation 3.156.

$$u_{max} = \frac{(-dP)}{8\mu dl}Y^2 = 1.5u_{av} \qquad \qquad ...(3.161)$$

Let R be the shear stress on the fluid element per unit of wetted surface area. This is balanced by the pressure drop dP. A force balance gives

$$R (1 \times 2dl) = (-dP) \times Y \times 1 \qquad \qquad ...(3.162)$$
$$R = (-dP)(Y/2dl)$$

The fanning friction factor f is defined as

$$f = \frac{R}{-(1/2\,\rho u_{av}^2)} \qquad \qquad ...(3.163)$$

Substituting from Equations 3.162 and 3.158, and simplifying, we get

$$f = \frac{12}{Re} \qquad \qquad ...(3.164)$$

where Re is defined as the Reynolds number $\rho u_{av} Y/\mu$.

Similar equations can be derived for any flow geometry: circular pipes, flow in the annulus, flow around spheres, etc. For flow in a circular pipe of diameter d, one can derive the following equations starting from momentum balance on a differential slice and integrating across the cross-section.

$$(-dP) = \frac{32\mu dl u_{av}}{d^2} \qquad \qquad ...(3.165)$$

$$Q = \frac{(-dP)\pi d^4}{128\mu dl} \qquad\qquad ...(3.166)$$

$$u_{max} = 2u_{av} \qquad\qquad ...(3.167)$$

$$\text{and } f = \frac{16}{Re} \qquad\qquad ...(3.168)$$

$$\text{with } Re = \frac{\rho ud}{\mu} \qquad\qquad ...(3.169)$$

Equation 3.166 is the well-known Hagen–Poiseuille equation, derived experimentally by Hagen in 1839 and independently by Poiseuille in 1840.

Equation 3.168 shows that the friction factor is dependent only on the Reynolds number. This is true only for laminar flow condition, which is found to prevail for $Re < 2.1 \times 10^3$. Beyond this, Equation 3.168 is not applicable. The friction factor, however, depends on Re, and also on the surface roughness ε, represented by a roughness factor ε/d. The value of ε for different materials is given in Table 3.4.

The friction factor for a given Re and roughness can be estimated graphically by Moody's Diagram (see Perry and Green 1997). For turbulent flow in smooth pipes, f can be estimated by the Blasius equation

$$f = \frac{0.079}{Re^{0.25}}\left(4000 < Re < 10^5\right) \qquad\qquad ...(3.170)$$

For non-circular cross-sections, a hydraulic diameter d_h can be used in place of d. This is defined as

$$d_h = \frac{4 \times \text{cross-sectional area}}{\text{wetted perimeter}} \qquad\qquad ...(3.171)$$

Table 3.4 Surface roughness for different materials

Material	ε (mm)
Drawn tubing (brass, lead, glass, etc.)	0.00152
Wrought iron	0.0457
Galvanized iron	0.152
Cast iron	0.259
Concrete	0.305–3.05
Riveted steel	0.014–0.14

For an annulus of outer dia d_2 and inner dia d_1, $d_h = d_2 - d_1$, and for a rectangular duct of dimensions a and b, $d_h = 2\,ab/(a + b)$.

The hydraulic diameter method should not be used for laminar flow conditions, as it would lead to significant errors.

Equation (3.162) for a circular pipe of length L can be written as

$$R\pi dL = (-\Delta P)\frac{\pi}{4}d^2 \qquad \qquad ...(3.172)$$

From the definition of f (Equation 3.163), the above equation can be re-arranged as

$$\left(\frac{\Delta P}{\rho}\right) = \frac{2fu^2 L}{d} \qquad \qquad ...(3.173)$$

This equation is used for calculating pressure drops in pipelines of straight sections.

In addition to frictional losses, there are losses due to bends, sudden contraction, expansion, etc. In the engineering practice, such losses are expressed as a product of the term $\rho u^2/2$, and a head loss coefficient K. Values of K for some types of fittings or valves are given in Table 3.5.

Energy of a fluid in motion

The total energy per unit mass of a fluid in motion consists of the internal energy E, pressure energy (P/ρ), potential energy (gz), and kinetic energy $(\tfrac{1}{2}u^2)$. Application of the first law of thermodynamics between any two sections of fluid flow leads to the equation

$$E_2 + \frac{P_2}{\rho_2} + gz_2 + \frac{1}{2}u_2^2 = E_1 + \frac{P_1}{\rho_1} + gz_1 + \frac{1}{2}u_1^2 + Q - W \qquad ...(3.174)$$

Table 3.5 Friction losses through fittings and valves

Type of fitting or valve	K
45° elbow	0.35
90° elbow	0.75
Union	0.04
Gate valve (open)	0.17
Gate valve (half open)	4.50
Foot valve	70.00
Sudden contraction and entrance from area A_1 to A_2	$0.50\,[1- (A_2/A_1)]$

where Q is the net heat absorbed from the surroundings and W is the net work done by the fluid.

u_2 and u_1 are the mean velocities at sections 1 and 2. As the velocity of the fluid at different points along the cross-section is likely to be different, the terms $1/2\,(u_2^2)$ and $1/2\,(u_1^2)$ do not represent the actual kinetic energy of a slice of fluid. Thus, a correction factor α is introduced to modify the kinetic energy terms as $u_2^2/2\alpha_2$ and $u_1^2/2\alpha_1$. If the velocity is same throughout the cross-section, $\alpha = 1$. For laminar flow in pipes of circular cross-section, $\alpha = \frac{1}{2}$; α is taken as 1 for turbulent flow.

For isothermal systems, $E_1 = E_2$. Making the energy balance over a differential length and then integrating, Equation 3.174 can be written as

$$\frac{u_2^2}{2\alpha_2} + gz_2 + \int_{P_1}^{P_2} \frac{dP}{\rho} = \frac{u_1^2}{2\alpha_1} + gz_1 + W_s - \frac{Q_f}{\dot{m}} \qquad \qquad ...(3.175)$$

where W_s is the shaft work done 'on' the fluid and $Q_f/\dot{m}$ is the energy dissipated due to friction 'to' the surroundings.

For incompressible fluids (ρ is constant, $dP/\rho = (P_2 - P_1)/\rho$), and for the case of uniform velocity ($\alpha = 1$) and frictionless conditions, Equation 3.175 can be simplified as

$$\frac{P_1}{\rho} + \frac{u_1^2}{2} + gz_1 = \frac{P_2}{\rho} + \frac{u_2^2}{2} + gz_2 \qquad \qquad ...(3.176)$$

This equation is known as the Bernoulli's equation, and Equation 3.175 is referred to as engineering Bernoulli's equation.

Example 15

A tapered pipe, having a cross-sectional area of 0.5 m² at entry and 0.25 m² at exit, carries water at 70 °C. The exit point is 4 m above the entry. The velocity of water at entry is 2 m/s and pressure is 140 kPa. Determine the pressure at the exit neglecting the frictional losses.

Solution

Applying mass balance

$(\rho AV)_{inlet} = (\rho AV)_{exit}$

Density (at 70 °C, ρ_{water} = 977.7 kg/m³) will remain same.

Therefore,

$$0.5 \times 2 = 0.25 \times V_{exit}$$

$$V_{exit} = 4 \text{ m/s}$$

Applying Bernoulli's Law (Equation 3.176)

$$\frac{140 \times 1000}{977.7} + \frac{2^2}{2} = \frac{P_{exit}}{977.7} + \frac{4^2}{2} + 9.81(5-0)$$

$$P_{exit} = 86.2 \text{ kPa}$$

Flow across beds of solids

A few parameters are defined first to facilitate further discussion.

The porosity ε is defined as the ratio of the volume of voids in the bed to the total volume (voids plus solid) of the bed.

$$V = \varepsilon.V_o \qquad \qquad ...(3.177)$$

where V is the volume of voids and V_o is the total volume of the bed. If the bed is uniform throughout the length, the cross-sectional area is defined in a similar way.

$$A = \varepsilon A_o \qquad \qquad ...(3.178)$$

where A is the cross-sectional area for flow and A_o is the empty bed cross-sectional area.

If u is the average velocity of flow, and u_o is the velocity based on the empty cross-section A_o

$$Au = A_o u_o$$

Therefore, $u = \dfrac{u_o}{\varepsilon}$ $\qquad \qquad ...(3.179)$

The equivalent diameter d_p of a non-spherical particle is defined as the diameter of a sphere having the same volume as the particle. The sphericity ϕ_s is defined as the ratio of the surface area of this sphere to the actual surface area of the particle. If a_p is the surface area of the particle and V_p is the volume, one can write

$$V_p = \frac{1}{6}\pi d_p^3 \qquad \qquad ...(3.180)$$

$$\phi_s = \frac{\pi d_p^2}{a_p} \qquad \qquad ...(3.181)$$

Re-arranging, we get

$$\frac{a_p}{V_p} = \frac{6}{\phi_s d_p} \qquad \qquad ...(3.182)$$

The sphericity of some materials is given in Table 3.6.

The modified Reynolds number Re_p is defined as

$$Re_p = \frac{d_p G}{\mu} \qquad \qquad ...(3.183)$$

where G is the mass velocity. The mass velocity is defined as the mass flow rate divided by the cross-section of the empty tower. This parameter is widely used in distillation columns, packed beds, absorption towers, etc.

$$G = \frac{\rho A_o u_o}{A_o} = \rho u_o \qquad \qquad ...(3.184)$$

The friction factor f_p for a packed bed is defined as

$$f_p = \frac{(-\Delta P)\phi_s d_p \varepsilon^3}{\rho u_o^2 L(1-\varepsilon)} \qquad \qquad ...(3.185)$$

where L is the depth of the bed.

The correlation between f_p and Re_p was developed by Ergun based on experimental data. The Ergun equation can be written as

$$f_p = \frac{150(1-\varepsilon)}{Re_p} + 1.75 \qquad \qquad ...(3.186)$$

Table 3.6 Sphericity of miscellaneous materials

Material	ϕ_s
Spheres, cubes, short cylinders (L = d_p)	1.00
Round sand	0.83
Coal dust	0.73
Crushed glass	0.65
Raschig rings	0.30

More accurate, and somewhat modified correlations can be found in standard handbooks (Perry and Green 1997). Equations 3.185 and 3.186 can be used to estimate pressure drops in packed beds. Iron filings packed in a column to remove H_2S from biogas, or wood chips in a fixed-bed gasifier are typical examples in renewable energy field for fluid flow in packed beds.

A liquid or gas flowing at low velocities through a porous bed of solid particles does not cause the particles to move. If the fluid velocity is steadily increased, a point is eventually reached when the particles start lifting. This phenomenon is called fluidization. The gas velocity $u_{o,m}$ at which fluidization begins is called the minimum fluidization velocity and is dependent on d_p, ρ_p, ρ, and μ. This can be found experimentally, but can also be estimated from semi-empirical equations.

When the fluid velocity becomes large enough, all particles are entrained in the fluid and carried along with it. This fact is used in pneumatic conveying of solids. When the solids have to be recovered, the gas or air is passed through cyclones where solid particles get separated.

The separation of solids from gas–solid systems is an important subject. Producer gas from a biomass gasifier, for example, contains particulate matter, among others, as an impurity, which has to be removed for applications such as power generation. It thus becomes important to design equipment for removal of solid particulate matter from the gas stream, which involves using the principles of fluid mechanics.

There are other topics of fluid flow, such as flow past an aerofoil to calculate the lift and drag on the blade of a wind mill; measurement of fluid flow rates; and selection of pumps, blowers, and fans, which are important for renewable energy application, just as in any branch of engineering. But covering all such aspects is beyond the scope of this book.

Heat transfer

Fourier's Law of heat conduction has already been introduced in Equation 3.148. The thermal conductivities of miscellaneous materials are given in Table 3.7.

Thermal conductivities of solids can either increase or decrease with temperature. Most solids show a linear variation with temperature. Conductivities of most liquids decrease with temperature though water is an exception. Conductivity of gases increases with temperature.

Apart from conduction, heat can be transferred by other mechanisms, namely, convection, radiation, and evaporation (or condensation). Consider

Table 3.7 Thermal conductivities of selected materials

Material	Thermal conductivity (W/m K)
Copper	385
Aluminium	211
Steel	47.6
Ice	2.26
Concrete walls	1.73
Glass	1.05
Brick	0.242
Water (20 °C)	0.596
Asbestos sheet	0.319
Gypsum plaster	0.170
Wood	0.13–0.16
Mineral wool, glass wool, polystyrene	0.034
Air	0.026

the simple practical case of water boiling in a kettle on a gas stove as shown in Figure 3.47.

The LPG burning at the ports of the burner produces flue gases at high temperatures. The walls of the vessel (both bottom and sides) receive this heat

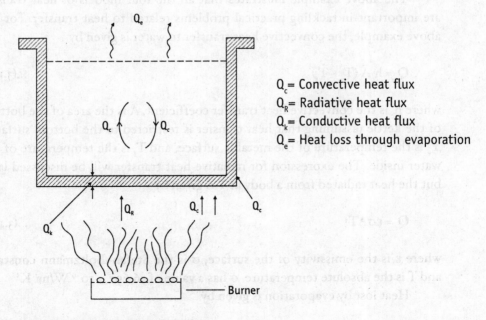

Q_c = Convective heat flux
Q_R = Radiative heat flux
Q_k = Conductive heat flux
Q_e = Heat loss through evaporation

Figure 3.47 Modes of heat transfer

through convection. Also, the flame is generally at a high temperature. The bottom of the kettle also receives radiant energy from the flame. Heat received by the walls of the kettle by both convection and radiation quickly raises the temperature of the vessel material and hence, a temperature gradient is established between the outer surface of the walls and inner surface that is in contact with water. The temperature gradient causes heat flow through conduction. As the kettle is generally made of metals like copper, aluminium, or steel, and as the wall thickness is small, heat transfer rates are very large. Heat received by the bottom layers of water in the kettle causes the temperature to rise, and because the density of water reduces at higher temperatures, much of the hot water tends to float towards the top layer of the kettle. At lower temperature, the top layer tends to sink to the bottom, thus establishing a natural circulation. This is called convection. Heat transfer by convection is very efficient in liquids. As the temperature of water increases, the vapour pressure also increases, and water starts evaporating into the surrounding air. As stated in earlier sections, the rate of evaporation depends on the difference between the vapour pressure of water and partial pressure of water vapour in the atmosphere (which is the basis for the definition of relative humidity or absolute humidity). The amount of heat transferred to the atmosphere by evaporation is equal to the mass of water evaporating multiplied by the latent heat of vaporization.

The above example illustrates that all the four models of heat transfer are important in tackling practical problems related to heat transfer. For the above example, the convective heat transfer to water is given by

$$Q_c = h_c A (T_v - T_f) \qquad\qquad ...(3.187)$$

where h_c is the convective heat transfer coefficient, A is the area of the bottom of the kettle (assuming that heat transfer is restricted to the bottom surface), T_v is the temperature of the metallic surface, and T_f is the temperature of the water inside. The expression for radiative heat transfer will be discussed later but the heat radiated from a body at T is given by

$$Q_r = \varepsilon \sigma A T^4 \qquad\qquad ...(3.188)$$

where ε is the emissivity of the surface, σ is the Stefan–Boltzmann constant, and T is the absolute temperature. σ has a value of 56.697×10^{-9} W/m^2 K^4.

Heat lost by evaporation is given by

$$Q_e = A \frac{h_c \lambda}{1.6 C_s P_t}(p_s - p_w)$$ ...(3.189)

where h_c is the convective heat transfer coefficient, λ is the latent heat of vaporization, p_s is the vapour pressure of water at the given temperature, p_w is the partial pressure of water vapour in air, P_t is the atmospheric pressure, and C_s is the humid heat capacity of air. The above equation is valid only for air–water mixtures. The different modes of heat transfer will now be discussed in detail.

Steady conduction in a slab

Consider a slab of area A and thickness Y as shown in Figure 3.48.

Application of Fourier's Law of heat conduction for an element dy gives

$$Q = -kA \frac{dT}{dy}$$

or $Qdy = -kAdT$

For steady-state conditions, Q is constant. Integrating the above equation

$$Q \int_0^Y dy = -kA \int_{T_1}^{T_2} dT$$

$$QY = -kA(T_2 - T_1)$$

or $$Q = -\frac{kA}{Y}(T_2 - T_1)$$

$$= \frac{kA}{Y}(T_1 - T_2)$$

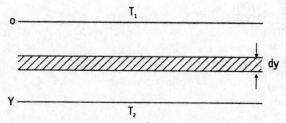

Figure 3.48 Heat conduction in a slab

The above equation can also be written as

$$Q = \frac{\Delta T}{R}$$

...(3.190)

where, R = (Y/kA)

R is the thermal resistance for heat flow. The term kA/Y is also termed conductance. Equation 3.190 is analogous to the flow of electricity.

$$i = \frac{\Delta V}{R}$$

where i is the current, ΔV is the potential difference, and R is the electrical resistance. The similarities between flow of electricity and flow of heat, as shown in Figure 3.49, can be used to solve a variety of heat transfer problems using the electrical network theory.

Compound slab of different materials

Consider a slab consisting of different materials as shown in Figure 3.50. Material 1 with thermal conductivity k_1 has a thickness of Y_1 and so on. A practical example is a brick wall with cement plaster on both sides and a thermocole sheet as insulation. Thus (1) and (3) can be cement plaster, (2) can be brick, and (4) can be thermocole.

Temperatures at the interface are as shown. Application of Equation 3.190 to different layers gives

$$Q_1 = \frac{T_1 - T_2}{R_1} \qquad Q_2 = \frac{T_2 - T_3}{R_2} \qquad Q_3 = \frac{T_3 - T_4}{R_3}$$

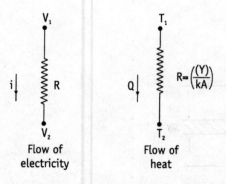

Flow of electricity   Flow of heat

Figure 3.49 Heat conduction and electrical conduction

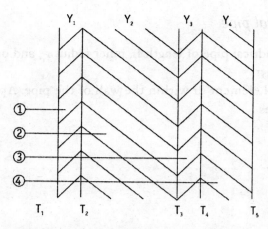

Figure 3.50 Conduction in a compound slab

$$\text{and } Q_4 = \frac{T_4 - T_5}{R_4}$$

where $R_1 = (Y_1/K_1 A)$, etc.

At steady-state conditions, there is no accumulation or depletion of heat, hence, $Q_1 = Q_2 = Q_3 = Q_4 = Q$. One can thus write,

$$QR_1 = T_1 - T_2$$
$$QR_2 = T_2 - T_3$$
$$QR_3 = T_3 - T_4$$
$$QR_4 = T_4 - T_5$$

Adding the above equations, we get

$$Q(R_1 + R_2 + R_3 + R_4) = T_1 - T_5$$

$$\text{Therefore, } Q = \frac{T_1 - T_5}{R}$$

where R is the summation of all resistance. This example is for all resistances in series. Just as in electrical networks, one can have resistances in parallel, series or in series–parallel combination. An example of tackling multiple heat transfer resistances is the analysis of various heat fluxes in a flat plate solar collector, which would be dealt in detail in Chapter 5.

Steady heat flow in a cylindrical pipe

Consider a thick-walled cylindrical pipe of length L, inner radius r_1, and outer radius r_2 as shown in Figure 3.51.

Consider a differential element dr within the wall of the pipe. Application of Fourier's Law gives

$$Q = -k.2\pi r.L \frac{dT}{dr}$$

or

$$\frac{dr}{r} = -\frac{k2\pi L}{Q} dT$$

Integrating the above equation, we get

$$\int_{r_1}^{r_2} \frac{dr}{r} = -\frac{k2\pi L}{Q} \int_{T_1}^{T_2} dT$$

Q is constant for steady heat flow and hence, could be taken out of the integral.

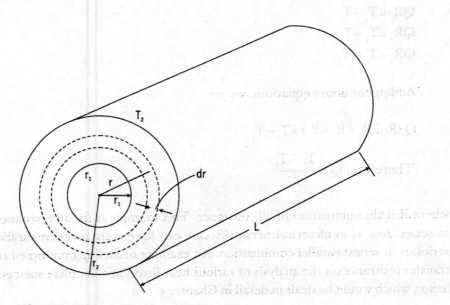

Figure 3.51 Conduction in a cylindrical pipe

$$\ln r_2 - \ln r_1 = -\frac{k2\pi L}{Q}(T_2 - T_1)$$

$$\ln\left(\frac{r_2}{r_1}\right) = \frac{k2\pi L}{Q}(T_1 - T_2)$$

$$\text{or } Q = \frac{k2\pi L}{\ln(r_2/r_1)}(T_1 - T_2)$$

Multiplying and dividing by $(r_2 - r_1)$ on the right-hand side of the above equation, we get

$$Q = k\frac{2\pi(r_2 - r_1)L}{\ln(r_2/r_1)}\frac{(T_1 - T_2)}{(r_2 - r_1)}$$

$$= kA_{lm}\frac{(T_1 - T_2)}{(r_2 - r_1)} \qquad \qquad \qquad ...(3.191)$$

where A_{lm} is the log-mean area given by

$$A_{lm} = \frac{A_2 - A_1}{\ln A_2 - \ln A_1} = \frac{2\pi(r_2 - r_1)L}{\ln(r_2/r_1)}$$

Equation 3.191 can now be written as

$$Q = \frac{(T_1 - T_2)}{R}$$

where R is the resistance given by $R = \dfrac{(r_2 - r_1)}{kA_{lm}} = \dfrac{\ln(r_2/r_1)}{2\pi kL}$

Composite cylindrical wall

The above treatment can be extended to a cylindrical tube made of two or three layers of different materials with different k values (Figure 3.52).

Following the same treatment as the composite slab, we can write

$$Q = \frac{T_1 - T_2}{R_1} \qquad \qquad \qquad R_1 = \frac{\ln(r_2/r_1)}{2\pi k_1 L}$$

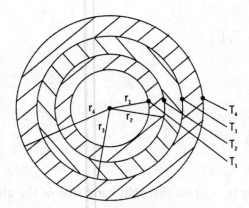

Figure 3.52 Conduction in a composite cylinder

$$= \frac{T_2 - T_3}{R_2} \qquad\qquad R_2 = \frac{\ln(r_3 / r_2)}{2\pi k_2 L}$$

$$= \frac{T_3 - T_4}{R_3} \qquad\qquad R_3 = \frac{\ln(r_4 / r_3)}{2\pi k_3 L}$$

$$Q = \frac{T_1 - T_4}{R} \qquad\qquad R = R_1 + R_2 + R_3$$

Example 16

The high-temperature zone of a biomass gasifier consists of a steel cylinder of wall thickness 3 mm, with an inner ceramic layer of 50 mm thickness, and an outer mineral wool layer. The innermost diameter is 30 cm. The reactor temperature is 1000 °C and the length of cylinder is 50 cm. The outside convective heat transfer coefficient is 20 W/m²K and the ambient temperature is 30 °C. Calculate the thickness of the mineral wool layer if the outer surface temperature has to be kept at 60 °C.

Solution

For the given system configuration

$r_1 = 0.15$ m, $r_2 = 0.2$ m, $r_3 = 0.203$ m

$k_1 = 1.2116$ W/mK, $k_2 = 45$ W/mK

$k_3 = 0.0415$ W/mK

$T_1 = 1000$ °C, $T_4 = 60$ °C, $T_o = 30$ °C

$h_c = 20$ W/m²K, $l = 0.5$ m

$r_4 = ?$

Now,

$$Q = \left(h_c A \Delta T\right)_{conv} = \left(\frac{\Delta T}{R}\right)_{cond}$$

$$20 \times 2\pi r_4 l(60-30) = \frac{1000-60}{R}$$

$$R = \frac{47}{2\pi l \times 30 \times r_4}$$

$$R = \frac{1}{2\pi l}\left[\frac{\ln(r_2/r_1)}{k_1} + \frac{\ln(r_3/r_2)}{k_2} + \frac{\ln(r_4/r_3)}{k_3}\right]$$

$$= \frac{1}{2\pi l}\left[\frac{\ln(0.2/0.15)}{1.2116} + \frac{\ln(0.203/0.2)}{45} + \frac{\ln(r_4/0.203)}{0.0415}\right]$$

From the above two equations for R, we get

$$\frac{47}{30\,r_4} = 0.2378 + \frac{\ln\left(r_4/0.203\right)}{0.0415}$$

By trial and error, one can get

$$r_4 = 0.2585 \text{ m}$$

Hence, thickness of mineral wool layer = 55.5 mm.

Heat transfer in extended surface

Extended surfaces or fins are used to increase the area available for heat transfer. An example is the radiator of a diesel engine. A fin and a tube arrangement is widely used as the basic configuration of a solar flat-plate collector. The heat conduction equation can be used to calculate the efficiency of such fins. A simple rectangular fin is shown in Figure 3.53.

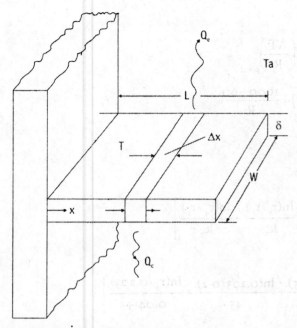

Figure 3.53 Heat transfer in a fin

A heat balance on the differential element Δx gives

$$Q\mid_x - Q\mid_{x+\Delta x} = h_c(2W\Delta x)(T-T_a)$$

Dividing throughout by Δx and taking the limit as $\Delta x \to 0$, we get

$$-\frac{dQ}{dx} = 2Wh_c(T-T_a)$$

Now, $Q = -k(W\delta)\dfrac{dT}{dx}$ by Fourier's law.

Thus, we can write

$$kW\delta\frac{d^2T}{dx^2} = 2Wh_c(T-T_a)$$

$$\text{or } \frac{d^2T}{dx^2} = \frac{h_c}{k(\delta/2)}(T-T_a) \qquad \qquad ...(3.192)$$

The boundary conditions for solving this equation are

$T = T_w$ at $x = 0$, where T_w is the temperature at $x = 0$

$$\frac{dT}{dx} = 0 \text{ at } x = L$$

The second boundary condition is obtained by assuming that the fin is insulated at the free end. Equation 3.192 can be solved as

$$\frac{(T - T_a)}{(T_w - T_a)} = \cosh N(x/L) - (\tanh N)\sinh N(x/L) \qquad ...(3.193)$$

where $N = L\sqrt{\dfrac{h_c}{k(\delta/2)}}$

The efficiency of the fin can be defined as

$$\eta = \frac{\text{Actual heat dissipated by the fin surface}}{\text{Heat dissipated if the entire fin surface is kept at } T_w}$$

From Equation 3.193 it can be derived that

$$\eta = \frac{\tanh N}{N}$$

A similar equation will be encountered while calculating the efficiency of a solar flat-plate collector.

Unsteady conduction

Unsteady conditions are characterized by variation of temperatures and heat fluxes with time. The generalized heat conduction equation with heat source (or sink) in rectangular coordinates, with constant thermal conductivity (Gupta and Prakash 1976) can be written as

$$\frac{\partial T}{\partial t} = \left(\frac{k}{\rho C}\right)\left(\frac{\partial^2 T}{\partial x^2} + \frac{\partial^2 T}{\partial y^2} + \frac{\partial^2 T}{\partial z^2}\right) + \frac{Q'}{\rho C} \qquad ...(3.194)$$

where, T = Temperature
t = time
Q' = heat production per unit volume
x,y,z = cartilinear co-ordinates
ρ = density
C = specific heat, and
$k/\rho c$ = thermal diffusivity which was introduced earlier.

Equation 3.194 can be used for solving many types of problems, but only a few cases will be considered here.

i)   One dimensional heat conduction with heat generation

Equation 3.194 takes the form

$$\frac{\partial T}{\partial t} = \alpha \frac{\partial^2 T}{\partial x^2} + \frac{Q'}{\rho C}$$

For a steady or pseudo-steady condition (temperature changing slowly with time), the equation can be written as

$$k\frac{d^2 T}{dx^2} + Q' = 0 \qquad\qquad ...(3.195)$$

These equations have been used to predict the performance of non-convecting salinity gradient solar ponds (Rao, Kishore, and Vaja 1990).

ii)   Two-dimensional conduction with heat generation

Equation 3.194 takes the form

$$\frac{\partial^2 T}{\partial x^2} + \frac{\partial^2 T}{\partial y^2} + \frac{Q'}{k} = 0$$

This can be used to analyse the performance of a box-type solar cooker (Rao, Das, and Karmakar 1994[a, b]).

In many practical cases, heat transfer is also accompanied by fluid flow. Simple conduction equations such as those discussed above will no longer be adequate. One has to consider three different sets of equations.

- Continuity equation (mass conservation)
- Momentum balance (Navier-Stokes equation)
- Energy balance

As such equations become too complex, numerical methods are often used for solving problems. Techniques of the CFD (computational fluid dynamics) are generally used for such situations. One example of such an approach is the analysis of biomass burning stove (Ravi, Kohli, and Ray 2002).

Convective heat transfer

Many engineering heat transfer problems involve heat exchange between a solid surface and a fluid in contact with it. Apart from temperature differences, heat transfer rates also depend on fluid properties and velocity. There is a fine interplay between fluid dynamics and heat transfer in all such situations. Thus, factors such as whether the fluid is liquid or gas, whether the flow is laminar or turbulent, whether the solid surface is rough or smooth, whether condensation or evaporation occurs, etc., all decide the rate of heat exchange. If the fluid movement is caused by density differences, as shown in Figure 3.47, heat exchange is called free convection or natural convection. On the other hand, fluid can be forced into motion through the action of a pump or a blower. Heat transfer in such a situation is called forced convection. The equation for heat flow (Equation 3.187) can also be written as

$$Q = \frac{\Delta T}{\left(1/h_c A \right)}$$

The term $(1/h_c A)$ would then be the convective heat transfer resistance, which can be added to the other resistances to get an overall resistance. Such a situation is shown in Figure 3.54.

Figure 3.54 represents a simple case of a double-pipe heat exchanger, in which a hot fluid flows in the inner pipe, and a cold fluid flows in the annulus between the outer pipe and the inner pipe. The hot fluid cools down and cold fluid heats up during the heat exchange process. Under steady conditions, a temperature profile as shown in Figure 3.54 develops. There will be a small temperature gradient in the bulk of fluids on both sides, but for practical purposes, the fluids can be assumed to be at average temperatures of $T_{h,av}$ and $T_{c,av}$. A large temperature drop (or gain) occurs in a thin layer adjacent to the wall. Most of the resistance for heat flow comes from this film. Heat transfer across

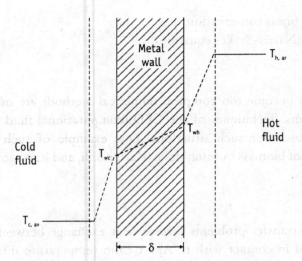

Figure 3.54 Heat transfer in heat exchanger tube

the wall is through conduction. The following equations can be written for heat fluxes.

$$Q = h_{c,h} A_h \left(T_{h,av} - T_{wh} \right) \qquad \ldots(3.196)$$

$$= \frac{k_w A_{lm}}{\delta} \left(T_{wh} - T_{wc} \right) \qquad \ldots(3.197)$$

$$= h_{c,c} A_c \left(T_{wc} - T_{c,av} \right) \qquad \ldots(3.198)$$

where A_h, A_c, and A_{lm} are the heat transfer areas of hot side, cold side, and log mean of these two; $h_{c,h}$ is the convective heat transfer coefficient on the hot side; $h_{c,c}$ is the same corresponding to the cold side; and k_w is the thermal conductivity of the wall. The above equations can be combined as

$$Q \left[\frac{1}{h_{c,h} A_h} + \frac{\delta}{k_w A_{lm}} + \frac{1}{h_{c,c} A_c} \right] = \left(T_{h,av} - T_{c,av} \right) \qquad \ldots(3.199)$$

It is useful to define an overall heat transfer coefficient U, based on either A_h or A_c. U is often based on the outside surface area of the hot pipe, that is, A_c. We can then write

$$Q = U_o A_c \left(T_{h,av} - T_{c,av} \right) \qquad \ldots(3.200)$$

where the subscript o refers to the outside area. Comparing Equations 3.199 and 3.200, we have

$$\frac{1}{U_o.A_c} = \frac{1}{h_{c,h}.A_h} + \frac{\delta}{k_w A_{lm}} + \frac{1}{h_{c,c}.A_c} \qquad ...(3.201)$$

For a pipe of inside radius r_i and outside radius r_o, the overall heat transfer coefficient U_o can be written as

$$U_o = \frac{1}{(r_o/r_i)(1/h_i) + (r_o/k_w)(\ln[r_o/r_i] + (1/h_o)} \qquad ...(3.202)$$

Note that the heat transfer coefficients are changed to h_i and h_o to represent the inside and outside. Equation 3.202 is widely used in heat exchanger designs.

The individual heat transfer coefficients h_i and h_o cannot be computed easily for most situations. The designer will have to depend on experimental data, which is quite voluminous. To reduce the available experimental data to usable forms, recourse is taken to what is known as 'dimensional analysis'. This is not covered in this chapter, and the reader can refer to standard texts (such as Coulson and Richardson 1999). Application of dimensional analysis to convective heat transfer gives the following equation between several dimensionless groups.

$$\left(\frac{hl}{k}\right) \propto \left(\frac{lu\rho}{\mu}\right)^{n_1} \left(\frac{C_p.\mu}{k}\right)^{n_2} \left(\frac{\beta g \Delta T l^3 \rho^2}{\mu^2}\right)^{n_3} \qquad ...(3.203)$$

where

$\left(\dfrac{hl}{k}\right)$ = Nusselt number

$\dfrac{lu\rho}{\mu}$ = Reynold's number

$\dfrac{C_p \mu}{k}$ = Prandtl number

$\dfrac{\beta g \Delta T l^3 \rho^2}{\mu^2}$ = Grashof number

l = characteristic length, which depends upon the geometry.

The characteristic length for a pipe is equal to D, for two parallel plates it is the width between the plates, etc. n_1, n_2, and n_3 can be either positive or negative numbers. β is the volumetric coefficient of expansion (= 1/T for ideal gases). For natural convection, Nu is a function of Pr and Gr and for forced convection it is a function of Re and Pr.

In problems of natural convection, sometimes the Rayleigh number (Ra) is used, which is a product of Gr and Pr.

$$Ra = \frac{g\beta\Delta Tl^3}{\upsilon\alpha}$$

where υ is the kinematics viscosity μ/ρ and α is the thermal diffusivity $k/\rho c$.

The rate of heat transfer between two parallel plates inclined at some angle to the horizontal is of importance to flat plate collectors. The relationship between Nu and Ra for such a situation is given by Duffie and Beckman (1991).

$$Nu = 1 + 1.44\left[1 - \frac{1708(\sin 1.8\theta)^{1.6}}{Ra\cos\theta}\right]\left[1 - \frac{1708}{Ra\cos\theta}\right]^{+} + \left[\left(\frac{Ra\cos\theta}{5830}\right)^{1/3} - 1\right]^{+}$$

...(3.204)

where $0 < \theta < 75°$ is the tilt from horizontal and + sign in the exponent indicates that only positive values of the term in the square brackets are to be used (use zero if the term is negative). The 75° correlation can be used for vertical surfaces also with little loss of accuracy.

Convection suppression devices, such as honeycomb structures, have been tried out extensively for reducing convective losses in solar flat-plate collectors. Equations for calculating Nu for some such geometries are given in Duffie and Beckman (1991). For fully developed turbulent flow inside tubes (Re > 2100), the Petukhov Equation (Duffie and Beckman 1991) can be used.

$$Nu = \frac{(f/8)RePr}{1.07 + 12.7\sqrt{(f/8)}\left(Pr^{2/3} - 1\right)}\left(\frac{\mu}{\mu_w}\right)^n$$

...(3.205)

with $f = (0.79\ln Re - 1.64)^{-2}$

...(3.206)

(μ/μ_w) is the ratio of fluid viscosities at bulk temperature and at wall temperature. n = 0.11 for heating and 0.25 for cooling. The hydraulic diameter (4 × flow area/ wetted perimeter) can be used for non-circular cross-sections. Numerous correlations exist for several specific cases.

For solar air heaters used in drying applications, the following equation can be used for fully developed turbulent flow with one side heated and the other side insulated.

$$Nu = 0.0158 Re^{0.8} \qquad ...(3.207)$$

The wind convection coefficient for collectors exposed to a wind speed u is obtained by the equation (Duffie and Beckman 1991).

$$h_w = 2.8 + 3.0u \qquad ...(3.208)$$

Nucleate or pool boiling

Heat transfer to a boiling liquid is a necessary step in operations such as steam generation, and distillation. Examples for renewable energy applications are boiling of freon in solar collectors in solar water pumps, boiling of an organic fluid in the so-called ORC (Organic Rankine Cycle) engines, which have been the prime movers in power generation through solar ponds, geothermal power, and OTEC (ocean thermal energy conversion), etc. In such applications, the fluid temperature remains constant corresponding to the saturation pressure, but the wall temperatures of the vessel can reach very high values. The relation between Q and ΔT initially follows a linear equation such as Equation 3.187, but when ΔT increases, it deviates significantly, as shown in Figure 3.55.

In the free convection zone, the heat flux increases, more or less linearly with ΔT. But in the nucleate boiling zone, the increase is more steep, and is represented by the equation

$$\frac{Q}{A} = a\left(\Delta T\right)^n \qquad ...(3.209)$$

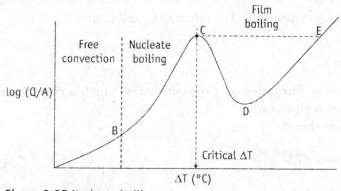

Figure 3.55 Nucleate boiling

where a is a proportionality constant and n lies between 3 and 4. At the end of this zone, the heat flux reaches a peak. The temperature difference corresponding to this is called the critical ΔT. After this, the heat flux decreases with ΔT, reaches a minimum at D, and again starts increasing. At very high ΔTs (of the order of a few thousand degrees), heat flux values go beyond the peak value corresponding to point C. Within the range of C-D-E, the boiling process is unstable and heat cannot be removed at the rate corresponding to ΔT.

If heat supply is not suddenly reduced, the metal surface temperatures might reach values that can damage or even melt the wall, leading to accidents. Such situations can occur in nuclear reactors. A theoretical and experimental study of solar flat-plate collector with boiling organic liquids can be found in Kishore, Gandhi, Marquis, *et al.* (1984).

Radiative heat transfer

It has been mentioned that the flux emitted by a black body (emissivity = 1) at a temperature T is given by

$$\frac{Q}{A} = \sigma T^4 \qquad \qquad ...(3.210)$$

where σ is the Stefan–Boltzmann constant. Radiation is emitted over the entire electromagnetic spectrum. Energy emitted at a given wavelength λ can be obtained by Plancks Law (Duffie and Beckman 1991).

$$E_{\lambda,b} = \frac{2\pi h C_o^2}{\lambda^5 [\exp(hC_o / \lambda kT) - 1]} \qquad \qquad ...(3.211)$$

where h is Planck's constant, k is the Boltzmann constant, and C_o is the velocity of light. $(2\pi h C_o^2)$ and (hC_o/k) are called Planck's First and Second Constants C_1 and C_2, respectively. The values of C_1 and C_2 are

$C_1 = 3.7405 \times 10^{-16}$ m^{2}W
$C_2 = 0.0143879$ m K

Equation 3.210 can be integrated over the entire wavelength region to get the total flux at a given temperature.

One can show that

$$\frac{Q_b}{A} = \int_o^\infty E_{\lambda,b} d\lambda = \sigma T^4 \qquad \qquad ...(3.212)$$

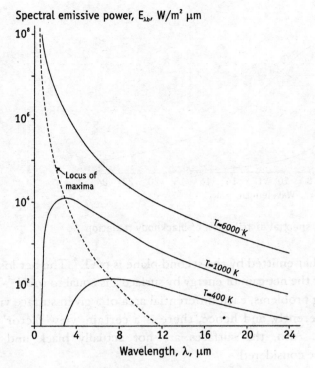

Spectral emissive power, $E_{\lambda b}$, W/m² µm

Locus of maxima

T=6000 K

T=1000 K

T=400 K

Wavelength, λ, µm

Figure 3.56 Spectral distribution of blackbody radiation

Equation 3.212 can thus be actually derived from the quantum mechanical principles.

The spectral distribution of Equation 3.211 is shown graphically in Figure 3.56. It can be seen that each curve (at a given T) goes through a peak at a particular value of λ, denoted by λ_{max}.

The relation between λ_{max} and T is given by Wien's displacement law

$$\lambda_{max} T = 2897.8 \ (\mu m)K \qquad \qquad ...(3.213)$$

The normalized spectral distribution $E_{\lambda,b}/(E_{\lambda,b})_{max}$ is shown in Figure 3.57 for different temperatures.

Radiation exchange between two surfaces

The simplest type of radiation exchange between two surfaces is where each surface can see only the other, that is, where the surfaces are very large parallel planes, and where both surfaces are black. The energy emitted by the first

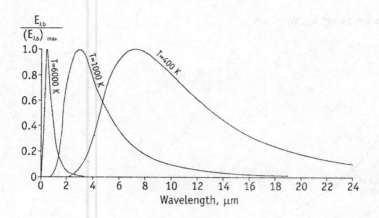

Figure 3.57 Normalized spectral distribution of blackbody radiation

plane is $\sigma A T_1^4$, and that emitted by the second plane is $\sigma A T_2^4$. The net loss of energy by surface 1 or the net gain of energy by surface 2 is equal to $\sigma A(T_1^4 - T_2^4)$. In actual engineering problems, each differential area of a given surface views the other areas differently, and hence, there is a certain 'view factor' between two surfaces. Also, the surfaces are not actually black and the emissivities have to be considered.

To understand the view factors, consider Figure 3.58 depicting how a differential area dA_1 of surface 1 sees a differential area dA_2 in another surface 2. 1–2 is the line connecting the centres of these elements separated by a distance r. ϕ_1 is the angle between the normal of the surface dA_1 and line 1-2; and ϕ_2 is the similar angle for dA_2.

The radiation from dA_1 which impinges on dA_2 (assuming both the surfaces are black), is given by McCabe and Smith (2000).

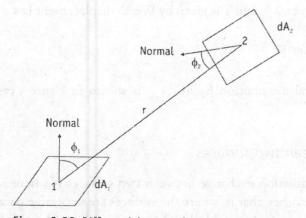

Figure 3.58 Differential areas of two surface

$$dq_{dA_1 \to dA_2} = \frac{\sigma T_1^{\,4}}{\pi r^2} \cos \phi_1 \cos \phi_2 dA_1 dA_2 \qquad \qquad ...(3.214)$$

Similarly,

$$dq_{dA2 \to dA_1} = \frac{\sigma T_2^{\,4}}{\pi r^2} \cos \phi_1 \cos \phi_2 dA_1 dA_2$$

The net rate of transfer is given by

$$dq_{12} = \sigma \frac{\cos \phi_1 \cos \phi_2 dA_1 dA_2}{\pi r^2} \left(T_1^{\,4} - T_2^{\,4} \right) \qquad \qquad ...(3.215)$$

The integration of Equation 3.215 for a given combination of surfaces can be quite complicated depending upon the geometry but the result can be expressed as

$$q_{1-2} = \sigma A F \left(T_1^{\,4} - T_2^{\,4} \right)$$

where A is either of the surfaces (A_1 or A_2) on which F is based. F is called view factor or geometric factor.

If surface A_1 is chosen for A

$$q_{12} = \sigma A_1 F_{12} \left(T_1^{\,4} - T_2^{\,4} \right)$$

If A_2 is chosen for A

$$q_{12} = \sigma A_2 F_{21} \left(T_1^{\,4} - T_2^{\,4} \right)$$

Thus, $A_1 F_{12} = A_2 F_{21}$

If there are N surfaces, the net heat transfer to a surface is given by

$$Q_i = \sum_{j=1}^{N} \varepsilon_i \varepsilon_j A_i \bar{F}_{ij} \sigma \left(T_1^{\,4} - T_2^{\,4} \right) \qquad \qquad ...(3.216)$$

The factor $\overline{F}_{ij}$ is the total exchange factor between surfaces i and j (Duffie and Beckman 1991). ε_i and ε_j are the emissivities of different surfaces.

Heat transfer problem in solar energy applications largely involve two surfaces. For N = 2, Equation 3.216 can be written as

$$Q_1 = \frac{\sigma\left(T_2^{\,4} - T_1^{\,4}\right)}{[(1-\varepsilon_1)/\varepsilon_1 A_1] + (1/A_1 F_{12}) + [(1-\varepsilon_2)/\varepsilon_2 A_2]} \qquad \text{...(3.217)}$$

For the case of two infinite parallel plates (as in flat-plate collectors), the areas A_1 and A_2 are equal and the view factor F_{12} is unity. Equation 3.217 can then be written as

$$Q = \frac{\sigma A\left(T_2^{\,4} - T_1^{\,4}\right)}{[(1/\varepsilon_1) + (1/\varepsilon_2) - 1]} \qquad \text{...(3.218)}$$

For a small surface surrounded by a large enclosure, such as a flat plate radiating to the sky, A_1/A_2 approaches zero, and F_{12} is equal to 1. Equation 3.217 then becomes

$$Q_1 = \varepsilon_1 A_1 \sigma(T_2^{\,4} - T_1^{\,4})$$

If the second surface is sky, it would become important to know the value of T_{sky}. The sky can be considered a black body at some equivalent temperature T_{sky}. The sky radiation can also be measured and is sometimes called the downward atmospheric radiation. It becomes important in deciding the performance of evaporative and nocturnal cooling (passive cooling techniques). See Kishore, Ramana, and Rao (1979) for a case study on nocturnal cooling.

The atmosphere is transparent in the wavelength region 8–14 μm, which is called the atmospheric window. Outside this window, the atmosphere has absorbing bands covering the infrared spectrum, chiefly contributed by absorption bands in molecules such as water, and CO_2 present in air. The sky temperature would thus, be dependent on the relative humidity besides the ambient temperature. Berdahl and Martin (1984) used extensive data to relate the effective sky temperature to the dew point temperature, dry bulb temperature, and hour from midnight t by the following equation.

$$T_{sky} = T_a \left[0.711 + 0.0056 T_{dew} + 0.000073 T_{dew}^2 + 0.013 \cos(15t) \right]^{1/4} \qquad \text{...(3.219)}$$

The other equations generally used are

$$T_{sky} = T_a - 6$$

$$\left(T_{sky} + 273\right) = \left(T_a + 273\right)\left(0.55 + 0.061\sqrt{p_a}\right)^{0.25}$$

where p_a is the partial pressure of water vapour in the atmosphere in mmHg.

Radiation heat transfer coefficient

It is often convenient to define a radiation heat transfer coefficient h_r in line with the convective heat transfer coefficient h_c

Noting that

$$T_2^4 - T_1^4 = (T_2 - T_1)(T_2 + T_1)(T_2^2 + T_1^2)$$

one can write Equation 3.217 as

$$Q = h_r A_1 (T_2 - T_1) \qquad \qquad ...(3.220)$$

where

$$h_r = \frac{\sigma\left(T_2^2 + T_1^2\right)(T_2 + T_1)}{[(1-\varepsilon_1)/\varepsilon_1] + (1/F_{12}) + [(1-\varepsilon_2)/\varepsilon_2](A_1/A_2)} \qquad ...(3.221)$$

When T_1 and T_2 are close together, the average temperature $\bar{T}$ can be used to evaluate h_r. One can write

$$(T_2 + T_1)(T_2^2 + T_1^2) = T_2^3 + T_2 T_1^2 + T_1 T_2^2 + T_1^3$$

If $T_1 \simeq T_2 \simeq \bar{T}$, then the RHS of the above equation becomes $4\bar{T}^3$. The numerator of Equation 3.221 becomes $4\sigma\bar{T}^3.\bar{T}$ can be estimated without actually knowing both T_1 and T_2, and the equation of radiation heat transfer is reduced to linear equation. This methodology is used in calculating heat losses from the cover system of solar flat-plate collector (Duffie and Beckman 1991).

Heat exchanger calculations

It becomes necessary to design heat exchangers in renewable energy applications, just as in any engineering practice. Some examples are hot gas heat recovery in producer gas application, gas-to-air heat exchangers for drying applications, hot brine-to-water heat exchangers for solar pond applications, etc. A method of using effectiveness – NTU (number of transfer units) – is reviewed here for the case of a counter-current heat exchanger. The notations for the hot and cold streams are shown in Figure 3.59.

Heat gained by cold fluid = heat lost by hot fluid = heat exchanged between the two fluids.

$$Q = \dot{m}_h C_{p,h} \left(T_{h,i} - T_{h,o} \right) = \dot{m}_c C_{p,c} \left(T_{c,o} - T_{c,i} \right) = UA\Delta T_{lm}$$

where ΔT_{lm} is the log mean temperature difference

$$\Delta T_{lm} = \frac{\left(T_{h,o} - T_{c,i} \right) - \left(T_{h,i} - T_{c,o} \right)}{\ln\left(T_{h,o} - T_{c,i} \right) - \ln\left(T_{h,i} - T_{c,o} \right)} \qquad \qquad ...(3.222)$$

The maximum possible heat transfer would occur when the outlet temperature of the cold fluid equals the inlet temperature of the hot fluid, and is given by

$$Q_{max} = \left(\dot{m}C_p \right)_{min} \left(T_{h,i} - T_{c,i} \right)$$

where $\left(\dot{m}C_p \right)_{min}$ is the minimum of the two streams. The effectiveness ε of the heat exchanger would then be

$$\varepsilon = \frac{Q}{Q_{max}} = \frac{\dot{m}_h C_{p,h} \left(T_{h,i} - T_{h,o} \right)}{\left(\dot{m}C_p \right)_{min} \left(T_{h,i} - T_{c,i} \right)}$$

$$Q = \varepsilon Q_{max} = \varepsilon \left(\dot{m}C_p \right)_{min} \left(T_{h,i} - T_{c,i} \right) \qquad ...(3.223)$$

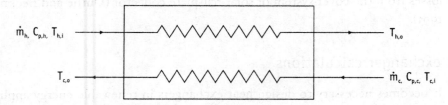

Figure 3.59 Counter-current heat exchanger

ε is a function of the NTU and the capacity rate ratio C*. These are defined as

$$NTU = \frac{UA}{\left(\dot{m}.C_p\right)_{min}}$$

$$C^* = \frac{\left(\dot{m}C_p\right)_{min}}{\left(\dot{m}C_p\right)_{max}}$$

Curves of ε versus NTU for C* as a variable parameter are given for a number of heat exchanger configurations in Kays and London (1964). The heat exchanger surface area A approaches infinity as ε approaches unity. An optimal value for A can be found by trial and error process using the ε- NTU correlation.

Mass transfer operations

Mass transfer operations are characterized by transfer of a substance from one phase to another or within the phase itself on a molecular scale. For example, when water vapour evaporates from a surface of water into a flowing air stream, water molecules diffuse into air and are then carried away. This process is different from that of, say, pumping water from one place to another. Transfer of mass in the former case occurs through the concentration gradient, which is the essential driving force. The various mass transfer operations can be classified as follows.

There are several other processes in which the mass transfer does not occur through the concentration gradient but by gravitational or other methods. Typical examples are as follows.

- Separation of fine particles in a gas stream through gravity settling or cyclone or sand beds
- Removal of liquid through filtration, sedimentation, etc.
- Size reduction through pulverization
- Agglomeration (pelletizing, briquetting)

In the following sections, some basic theoretical treatment of various mass transfer operations, along with empirical correlations will be presented.

Category	Sub-category	Example
Gas–liquid operations	Distillation	• Separation of ethanol from fermented molasses • Recovery of methanol catalyst in bio-diesel manufacture
	Absorption (desorption, stripping)	• Removal of tar, etc., from producer gas • Removal of H_2S from biogas • Separation of NH_3–H_2O solution (as in absorption refrigeration)
	Humidification/dehumidification	• Air cooling (for example, desert cooler)
	Evaporation	• Solar stills
Gas-solid operations	Adsorption/desorption	• Solar biomass adsorption cooling system (for example, CH_3OH^- Silica gel) • Removal of tars from producer gas through charcoal beds, etc.
Vapour–liquid–solid	Drying	• Drying of a variety of agro-products using solar energy, biomass heat, etc.
Liquid–liquid operations	Extraction	• Solvent extraction of vegetable oils

Molecular diffusion

If a mixture of two components A and B in a liquid or gaseous phase is uniform in concentration everywhere, no changes occur. But if the concentration is not uniform, the solution is brought to uniformity spontaneously by diffusion, the components moving from higher concentration to lower concentration.

Consider the mixture of fluids consisting of components A and B in a container with a partition as shown in Figure 3.60. The concentration of substance A to the left of point P is higher than that on the right, and the concentration of substance B to the right of point P is higher than that on the left. When the partition is removed, molecules of A will diffuse to the right (from higher to lower concentration), whereas molecules of B will move towards the left. If u_A and ρ_A are the velocity and density of A, then the mass flux of A across P is area$\times u_A \times \rho_A$. As area is taken as unity, the mass flow rate is $u_A\rho_A$. As the volume on either side of the section at P is constant, the volume rate of A should be equal to that of B, but should be in the opposite direction.

$$u_B = -u_A$$

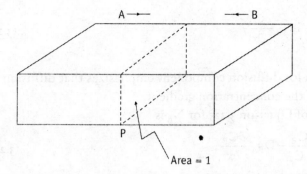

Figure 3.60 Molecular diffusion in a box

The number of moles will be equal to the mass divided by the molecular weight. Hence, it can be written that

$$N_A = u_A \rho_A / M_A = u_A C_A$$
$$N_B = u_B \rho_B / M_B = u_B C_B$$

where N_A is the rate at which moles of A pass an observer at P and C_A is the molar concentration (moles per unit volume) of A. N_B and C_B are defined similarly. The net rate of movement of moles past the observer and can be defined as

$$N_A + N_B = u_A C_A + u_B C_B$$

The molar average velocity can be defined as

$$u_M = \frac{u_A C_A + u_B C_B}{C_A + C_B} = \frac{u_A C_A + u_B C_B}{C} = \frac{N_A + N_B}{C}$$

where C is the total concentration $C_A + C_B$.

Now the flux N_A of moles moving past the fixed position P would be a combination of two fluxes: one which happens due to random motion of molecules of A type and the second, which is due to diffusion. The flux due to motion is equal to the mean volume rate multiplied by the molar concentration of A. The diffusion flux, given by Fick's Law, is stated in the beginning of this section. Thus, one can write

$$N_A = u_M C_A - D_{AB} \frac{\partial C_A}{\partial x} \qquad \qquad ...(3.224)$$

$$= \left(N_A + N_B\right)\frac{C_A}{C} - D_{AB}\frac{\partial C_A}{\partial x} \qquad \qquad ...(3.225)$$

The negative sign for diffusion component emphasizes that diffusion occurs in the direction of the concentration gradient.

The counterpart of Equation 3.225 for N_B is

$$N_B = \left(N_A + N_B\right)\frac{C_B}{C} - D_{BA}\frac{\partial C_B}{\partial x} \qquad \qquad ...(3.226)$$

Adding Equations 3.225 and 3.226, we get

$$-D_{BA}\frac{\partial C_A}{\partial x} = D_{BA}\frac{\partial C_B}{\partial x}$$

If $C_A + C_B$ is constant, it follows that

$$D_{AB} = D_{BA}$$

The above formulation considered diffusion in only one direction, but equations can be written for all three directions in Cartesian coordinates and in other coordinate system. Equations can also be formulated for multi-component systems. In some solids, D_{AB} can be direction-sensitive, but for liquids and gases it is not. Practical diffusion problems involve formulation and solution of continuity equation, momentum equation, and energy equation along with the relevant diffusion equations, and can be extremely tedious in mathematical terms. Both steady-state and unsteady-state cases can be analysed.

In case of steady-state conditions for diffusion in one direction, Equation 3.225 can be integrated to yield

$$\int_{C_{A_1}}^{C_{A_2}} \frac{-dC_A}{N_A C - C_A\left(N_A + N_B\right)} = \frac{1}{C D_{AB}} \int_{x_1}^{x_2} dx \qquad ...(3.227)$$

where 1 is the beginning of diffusion path (high C_A) and 2 represents the end of diffusion path (low C_A).

If $x_2 - x_1$ is written as Δx, the above equation can be written as

$$N_A = \frac{N_A}{N_A + N_B} \cdot \frac{C.D_{AB}}{\Delta x} \cdot \ln\frac{\left\{N_A/\left(N_A + N_B\right) - \left(C_{A_2}/C\right)\right\}}{\left\{N_A/\left(N_A + N_B\right) - \left(C_{A_1}/C\right)\right\}} \qquad ...(3.228)$$

For diffusion in gases, and where the ideal gas law can be applied, concentrations can be replaced by partial pressures and mole fractions.

$$\frac{C_A}{C} = \frac{p_A}{P_T} = X_A$$

...(3.229)

where p_A is the partial pressure of component A, P_T is the total pressure, and X_A is the mole fraction of A.

There are two situations, which occur frequently, as discussed below.

1 *Steady-state diffusion of A through non-diffusing B*

This can occur, for example, when H_2S is removed from biogas (a mixture of methane and carbon dioxide) through a reaction with iron filings. As only H_2S is taken in by the iron filings, it will develop a concentration gradient in the gas phase, especially close to the surface of the iron filings. If A represents H_2S and the rest of biogas is represented by B, we have

$$N_A = const$$
$$N_B = 0$$
$$and\ N_A/(N_A + N_B) = 1$$

Equation 3.228 can then be written as

$$N_A = \frac{D_{AB} P_T}{RT(\Delta x)} \ln \frac{P_T - p_{A_2}}{P_T - p_{A_1}}$$

In practice, the concentration of H_2S is generally small and the term $(P_T - p_{A_2})/(P_T - p_{A_1})$ is nearly equal to 1. The above equation can be simplified assuming $(P_T - p_{A_2}) / (P_T - p_{A_1}) = 1 + \varepsilon$ (where ε is a small quantity) and $\ln(1 + \varepsilon) = \varepsilon$. It can then be written as

$$N_A = \frac{D_{AB} P_T}{RT(\Delta x)} \frac{p_{A_1} - p_{A_2}}{p_{B,av}}$$

...(3.230)

where $p_{B,av} = p_{B1} = p_{B2}$.

2 *Steady-state equimolal counter diffusion*

This situation frequently occurs in distillation operations. For this, $N_A = -N_B =$ constant, and Equation 3.228 becomes indeterminate. But one can start with the differential equation 3.225, which can be written for gases as

$$N_A = (N_A + N_B)\frac{p_A}{P_T} - \frac{D_{AB}}{RT}\frac{dp_A}{dx}$$

Since, $N_A = -N_B$; $N_A + N_B = 0$

$$N_A = -\frac{D_{AB}}{RT}\frac{dp_A}{dx}$$

Integrating between the limits p_{A_1} and p_{A_2}, we get

$$N_A = -\frac{D_{AB}}{RT(\Delta x)}\left(p_{A_1} - p_{A_2}\right)$$

Diffusivities of some pairs of substances are given in Table 3.8.

Diffusivities of gases can be estimated from theoretical consideration also. Equations for prediction of diffusivities are given in Bird, Stewart, and Lightfoot (2001). Diffusivities in liquids can be estimated by similar equations.

Mass transfer coefficient

In practical applications of mass transfer operations, fluids are always in motion and often turbulent conditions prevail. Hence, equations such as Equation 3.228 are seldom used. For practical situations, it is customary to describe the mass transfer flux in terms of the mass transfer coefficients. The flux is then written as

$$N_A = k_G \, (p_{A_1} - p_{A_2}) \quad \text{gas phase}$$

$$N_A = k_L \, (C_{A_1} - C_{A_2}) \quad \text{liquid phase}$$

Table 3.8 Diffusivities of selected pairs of substances

Substance pair A–B	Temperature °C	D_{AB} (cm²/s)
Gases		
Water–air	0	0.220
Water–CO_2	0	0.138
CO_2–air	0	0.138
CO–O_2	0	0.185
H_2–air	0	0.611
Benzene–air	0	0.77
Liquids		
Ethanol–water	25	1.28×10^{-5}
Methanol–water	25	1.6×10^{-5}
Ammonia–water	25	2.0×10^{-5}
Glycerol–water	25	0.94×10^{-5}
Sucrose–water	25	0.56×10^{-5}

Just as in heat transfer, there are dimensionless numbers for correlating mass transfer coefficients. Some of these are defined below.

Sherwood number: $Sh = \dfrac{k_L \delta}{D_{AB}}$ (δ = characteristic length)

Schmidt number: $Sc = \dfrac{\mu}{\rho D_{AB}}$

Peclet number:　　$Pe = Re \times Sc$

Stanton number:　$St = Sh/Pe$

Correlations for estimating mass transfer coefficients exist and are based both on theoretical consideration (for example, Danckwerts' surface renewal theory) and experimental data. The reader may referred to standard textbooks on mass transfer such as Treybal (1981) for further understanding.

Interphase mass transfer

So far, we considered only diffusion of substances within a single phase. In many mass transfer operations, transfer of substances occurs across the phases. For example, when the tar ladden producing gas (gas phase) is contacted with water (liquid phase), water-soluble tars move from gas phase to liquid phase. When ethanol–water mixtures are fed into a distillation column for separating ethanol, they goes from vapour phase to liquid phase. The rate of diffusion of a given component within each phase is dependent upon the respective concentration gradients. But what is the rate of transfer between phases? Before answering this question, we should study the equilibrium characteristics between phases.

Consider the boiling point diagram of two completely miscible liquids, for example, benzene and toluene or ethanol–water, as shown in Figure 3.61.

The boiling point of substance A at a given pressure is T_A, and that of B is T_B at the same pressure. For mixtures in between, concentration in the liquid phase and vapour phase will be as shown. For example, for a temperature T between T_A and T_B, the concentration of A in liquid phase will be x_e and that in the vapour phase y_e. If the system is in equilibrium at T, there will be a definite y_e corresponding to x_e. One can plot y_e versus x_e to get the equilibrium diagram as shown in Figure 3.62.

There are substance pairs that do not obey the simple form of Figure 3.61, which are called azeotropes for which the mixture boiling points can be

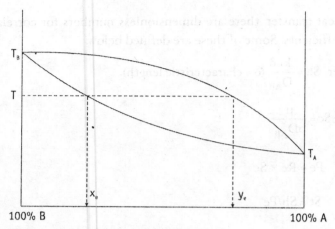

Figure 3.61 Two-phase equilibrium

higher than those of either A or B. Refer to standard texts (McCabe and Smith 2000) for further information. The important thing to note is that there exists an equilibrium diagram relating the equilibrium concentration of a substance A definitively in both phases.

The concentration profiles across the phases, including the interface (for example, liquid–gas interphase) would thus look as shown in Figure 3.63.

The concentration of A in the bulk of the gas phase is y_{AG}, which falls to y_{Ai} at the interphase. In the liquid, the concentration falls from x_{Ai} at the interface to x_{AL} in the bulk liquid. In what was called a two-film or two-resistance theory, Lewis and Whitman (1924) suggested that there is no resistance to mass transfer across the interface separating the phases and that

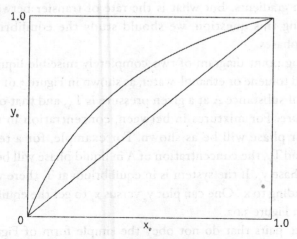

Figure 3.62 Vapour–liquid equilibrium curve

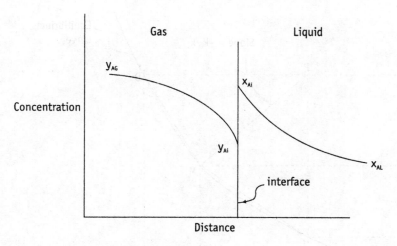

Figure 3.63 Inter-phase mass transfer

y_{Ai} and x_{Ai} correspond to the equilibrium values. Thus, the only resistances for mass transfer are those present in the fluids themselves. Except for a few special cases, this theory had been found to be valid.

Various concentrations may be shown graphically in Figure 3.64. The point P represents the bulk phase concentrations. For steady-state operation, the mass transfer rate in the gas phase could be equal to that in the liquid phase. If k_y and k_x are the corresponding mass transfer coefficients, one can write

$$N_A = k_y \left(y_{AG} - y_{Ai}\right) = k_x \left(x_{Ai} - x_{AL}\right)$$

Re-arranging, we get

$$\frac{\left(y_{AG} - y_{Ai}\right)}{\left(x_{AL} - x_{Ai}\right)} = -\frac{k_x}{k_y}$$

This is the slope of line PM. The overall driving force for mass transfer is the difference $\left(y_{AG} - x_{AL}\right)$. However, the equilibrium gas phase composition corresponding to x_{AL} is y_A^*, which is as good a measure as x_{AL} itself. An overall mass transfer coefficient can now be defined as

$$N_A = K_y \left(y_{AG} - y_A^*\right)$$

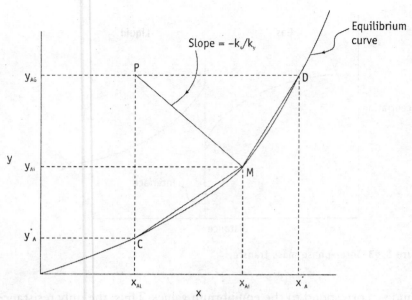

Figure 3.64 Steady-state inter-phase mass transfer

Referring to Figure 3.64, one can write

$$y_{AG} - y^*_A = (y_{AG} - y_{Ai}) + m'(x_{Ai} - x_{AL})$$ ...(3.231)

where m' is the slope of the line CM. Substituting from previous equations, we get

$$\frac{N_A}{K_y} = \frac{N_A}{k_y} + \frac{m'N_A}{k_x}$$

or

$$\frac{1}{K_y} = \frac{1}{k_y} + \frac{m'}{k_x}$$ ...(3.232)

This relation is similar to the overall heat transfer coefficient defined in the earlier section on heat transfer.

Material balances and operating lines

When a phase G containing material A is contacted with another phase L with the purpose of transferring part of A into the L phase, it is generally done in a mass transfer equipment. A distillation column, an absorption tower, etc. are examples of such equipment. Both G stream and L stream will be flowing through the equipment, either in a co-current, counter-current manner, or cross-current manner. If both phases contain a non-diffusing component (such as air in air–water mixtures or water in NH_3–water mixtures), the mass flow rates of the non-diffusing components remain constant throughout the contacting equipment. The mole fractions x and y, based on total molar stream, can be converted to represent moles of A per unit mole of the non-diffusing component X. It can be shown that x and X are related as follows.

$$X = \frac{x}{1-x}$$

or

$$Y = \frac{y}{1-y}$$

If the flow rates of the non-diffusing components are expressed as G_S and L_S, and if X_1 and Y_1 are the concentrations of A at the inlet, the material balance at any intermediate point (X, Y) gives

$$G_s (Y_1 - Y) = L_s (X - X_1) \qquad \qquad ...(3.233)$$

which suggests that the amount of A removed from the G stream is equal to the amount of A gained by the L system. The above equation represents a straight line connecting the point (X, Y) to (X_1, Y_1) in the X–Y coordinates, and is called the 'operating line'. The operating line and the equilibrium line can be used in a graphical construct to determine the number of ideal 'stages' required to obtain the desired outlet concentration of A. An illustration of this graphical constant is shown in Figure 3.65 for a tray-type absorber.

A stage is defined as a contacting device into which both the streams (G-stream and L-stream, for example) enter, brought in thorough contact during which mass transfer occurs across the phases, and both streams get separated and leave the device. An example is a bottle containing CO_2, into which water is poured, thoroughly shaken, kept for some time to settle the phases, and then the remaining CO_2 is removed, leaving soda in the bottle. An 'ideal' or theoretical stage is one in which both the streams achieve their equilibrium concentration. A continuous contacting process, such as a co-current or counter-current process, can be thought of as a series of stages. Further discussion on stages and methods of designing mass transfer equipment is beyond the scope of this book and the reader is referred to standard texts on the subject (such as McCabe and Smith 2000).

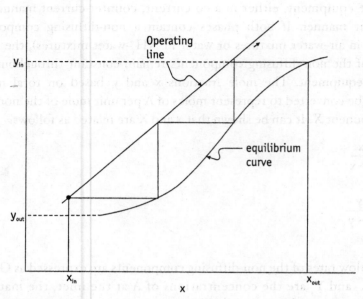

Figure 3.65 Graphical construction to obtain number of equilibrium stages

Raoult's Law and Henry's Law

It is desirable to have certain characteristics of solutions, and these are called ideal solutions. Most practical systems are non-ideal and approach ideality only as a limit. Mixtures of molecules, which are similar, such as benzene and toluene, would represent near-ideal solutions.

Raoult's Law states that when a gas mixture in equilibrium with an ideal solution follows the ideal gas law, the partial pressure of the component A in gas phase equals the product of its vapour pressure at the same temperature and its mole fraction $x_{L,e}$ in the liquid solution.

$$P_A = P_s x_{L,e} \qquad \qquad ...(3.234)$$

For non-ideal solutions which deviate from Raoult's Law, experimental data suggests that the equilibrium gas phase composition varies linearly with the liquid phase composition.

$$y_{G,e} = \frac{P_A}{P_T} = m.x_{L,e} \qquad \qquad ...(3.235)$$

where m is a constant. This is called the Henry's Law. Both these laws are widely used in mass transfer.

There is a large variety of mass transfer operations such as distillation, gas absorption, desorption (or stripping), humidification, dehumidification, drying, absorption, liquid–liquid extraction, crystallization, and leaching. Although many of these could be important to renewable energy applications, only a few will be discussed briefly here.

Humidification operations

These refer to the operation in which a gas is brought into contact with a pure liquid in which it is insoluble. The best and most practical example is the air–water system, which is used in the evaporative cooling system. As mass transfer in such cases is accompanied by a simultaneous heat transfer also, some analysis involving enthalpy characteristics would be necessary.

Variation of the vapour pressure of the pure liquid with temperature is important, and can be obtained by the Clausius-Clapeyron equation mentioned in an earlier section.

$$\frac{dP_s}{dT} = \frac{\lambda'}{T(v_G - v_L)}$$

where λ' is the molal latent heat, and v_G and v_L are the molar specific volumes of vapour and liquid phases. $v_L \ll v_G$, and in the above equation can be integrated to yield

$$\ln P_s = -\frac{\lambda'}{RT} + \text{Const} \qquad \qquad ...(3.236)$$

It is useful to plot vapour pressures in relation to a reference substance. Equation 3.236 can be written as

$$\log P_s = \frac{M\lambda}{M_r \lambda_r} \log P_{s,r} + \text{Const} \qquad \qquad ...(3.237)$$

where $P_{s,r}$, M_r, and λ_r correspond to a reference substance. M is the molecular weight and λ is the latent heat per unit mass. Equation 3.237 suggests that P_s plotted against $P_{s,r}$ at the same temperature on a log–log graph yields a straight line. Such a line can be used for interpolation between temperatures. For gas–liquid mixtures, some liquid will evaporate into the gas stream and hence the gas stream will contain some amount of liquid substance (we can call this A). The exact amount will depend upon the temperature and the conditions of saturation. If the gas stream is saturated, the partial pressure will be equal to the vapour pressure of the substance at the given temperature; otherwise it will be lesser. One can define an 'absolute humidity' for the gas phase as

$$Y = \frac{\text{mass of A}}{\text{mass of B}} = \frac{p_A}{P_T - p_A} \frac{M_A}{M_B}$$

where B refers to the pure gas stream.

When the gas phase is saturated $Y = Y_s$, one can then define a relative humidity as the ratio of the partial pressure of A to the saturation vapour pressure.

$$RH = 100 \times \frac{p_A}{P_S}$$

At saturation, $p_A = P_S$ and RH will be 100%. We can then have curves for different RH as shown in the psychrometric chart of Figure 3.66.

The dew point is the temperature at which the gas phase becomes saturated when cooled at constant total pressure out of contact with the liquid. The dew point corresponding to point F is shown in Figure 3.66.

The humid volume, humid heat capacity, and enthalpy are defined for vapour–gas mixtures based on the unit mass of dry gas.

Adiabatic saturation curves

Consider an operation in which a certain quantity of vapour–gas mixture at point F, as shown in Figure 3.66, is brought into intimate contact with an

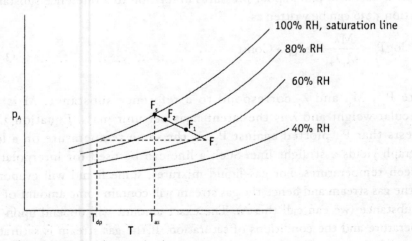

Figure 3.66 Psychrometric chart

adequate quantity of liquid (in the form of a spray, for example) so that entire water evaporates and the vapour–gas mixture leaves at a different condition of humidity and temperature. As no heat is gained from or lost to the atmosphere, the process is adiabatic. Also, as the liquid has evaporated, heat for evaporation would have come from the vapour gas stream, and so, the outlet stream would be cooler. Depending on the relative proportions of liquid and gas quantities, the end conditions would be different. The locus of all such end points is called the adiabatic saturation curve which is shown as $F–F_1–F_2$ in Figure 3.66. This line will ultimately meet the saturation curve at point F_s, at which the vapour–gas mixture is saturated. The temperature corresponding to F is T_{as}, and is called the adiabatic saturation temperature. For any vapour–gas mixture, there exists an adiabatic saturation temperature T_{as}, such that if contacted with liquid at T_{as}, the gas will become humidified and cooled. Sensible heat given up by the gas in cooling equals the latent heat required to evaporate the added vapour.

Wet bulb temperature

The wet bulb temperature is the steady-state, non-equilibrium temperature reached by a small mass of liquid exposed to a continuous stream of gas under adiabatic conditions. A theory of wet bulb thermometry can be developed by applying the relevant heat and mass transfer equations for a drop of liquid evaporating into a gas stream. The wet bulb depression $(T_G – T_w)$, where T_G is the gas temperature and T_w is the wet bulb temperature is obtained as (Treybal 1981).

$$T_G – T_w = \frac{\lambda_w \left(Y'_w – Y' \right)}{\left(h_G / k_Y \right)}$$

...(3.238)

where λ_w = latent heat of evaporation at T_w

         Y'_w = absolute humidity at T_w (kg vapour/kg gas)

         Y' = absolute humidity in gas stream

         h_G = heat transfer coefficient

and k_Y = mass transfer coefficient based on humidities

Experimental correlations showed that

$$\frac{h_G}{k_Y} = C_s \left(\frac{Sc}{Pr} \right)^m$$

...(3.239)

where m is a constant and C_s is the humid heat capacity. Sc and Pr are Schmidt and Prandtl numbers defined earlier. For combining with Equation 3.238, we can write

$$\frac{Y' - Y'_w}{T_G - T_w} = -\frac{(h_G/k_Y)}{\lambda_w} = -\frac{C_s}{\lambda_w}\left(\frac{Sc}{Pr}\right)^m$$

This is the equation of a straight line drawn on the humidity chart with a slope of $-[(h_G/k_Y)/\lambda_w]$ and intersecting the saturation line at T_w. This line is called a psychrometric line. There is a deviation in slopes between the adiabatic saturation line and the psychrometric line, the difference in slopes depending on the relative magnitudes of C_s and h_G/K_Y. For the air–water system

$$\frac{h_G}{k_Y C_s} = 1 \qquad \qquad \qquad ...(3.240)$$

which is called the Lewis relation. For this condition, the adiabatic saturation line and the psychometric line become essentially the same.

The above relations, along with basic mass and energy balance equations, are useful in analysing situations where direct contact of a gas with pure liquid is done. There are several practical situations for such operations, as listed below.

- For cooling a hot gas, direct contact with water may provide a very economical and effective means; eliminating the need for a heat exchanger
- A hot liquid can be cooled by a colder gas by transfer of sensible heat and by evaporation
- Humidifying air for controlling the moisture
- Dehumidifying air in air-conditioning operations

Gas absorption

Gas absorption is an operation in which a gas-containing vapour comes in contact with a liquid with the aim of preferentially dissolving one or more components of the gas into the liquid stream. If the solute is valuable, it can be recovered by distilling the liquid later. An example of renewable energy application, as cited before, is the removal of tarry vapour from producer gas by scrubbing with water. This operation is presently done rather crudely because there is not much data on the types and properties of water-soluble tars. Some systems use empty columns, some use columns with some packing, and others use venturi or centrifugal scrubbers to remove impurities. Using the techniques employed in designing gas absorption systems, producer gas cleaning equipment can be designed more efficiently and optimally. When mass transfer occurs in the opposite direction (from the liquid to the gas), the operation is called desorption or stripping.

Solubilities of different gases or vapours in a solvent depend both on the partial pressure of the gas or vapour in the gas stream and on temperature. Equilibrium solubility data for several solute–solvent pairs (for example, NH_3–H_2O, SO_2–H_2O, etc.) is available. Such data is the starting point for design of absorption columns. For ideal gases and ideal solutions, the Raoult's law is applicable. For non-ideal solution, Henry's Law is applicable at least in the low-concentration region. There are technical and economic criteria for choosing appropriate solvents. In the gasifier cases, for example, a suitable organic solvent such as anisole can be used for removing tars, but it is expensive and has to be recovered by distillation, which is cumbersome for a small system.

Use of material balance equations and equilibrium data can provide information on ideal stages, plates, or transfer units. Knowledge of mass transfer rates and interfacial areas can be used to obtain efficiencies and dimensions of absorption columns in a realistic way. Terms like HTU (height of a transfer unit) or NTU are defined to obtain overall heights of columns. Graphical integration methods are used to obtain the NTU.

Adsorption

Certain solids have properties to preferentially concentrate specific substances from solutions into their surface. Some well-known adsorbents are Fuller's earth (natural clays), activated clay, charcoal, activated charcoal, alumina, silica gel, molecular sieves, etc. The amount of solute adsorbed at equilibrium (kg/kg) is dependent on the partial pressure of the solute in the gas stream (or concentration is a liquid stream). Such equilibrium curves are called adsorption isotherms.

Adsorption is used for a variety of operations such as removal of colour, objectionable odours, and impurities. Adsorption is also the basis for operation of gas chromatography. The adsorbed gases can be released back by heating, which is called desorption. There is some kind of hysterisis in the adsorption–desorption equilibrium curves. Adsorption is an exothermic process, and the concentration of the adsorbed gas decreases with increased temperature. Some typical equilibrium curves are shown in Figure 3.67. The phenomenon of hysterisis is shown in Figure 3.68.

A well-known example of adsorption in renewable energy application is solar cooling. Water-geolite (Ramos, Espinoza, and Horn 2003) and methanol-silica gel (Mande, Ghosh, Kishore, *et al.* 1997) have been successfully used to

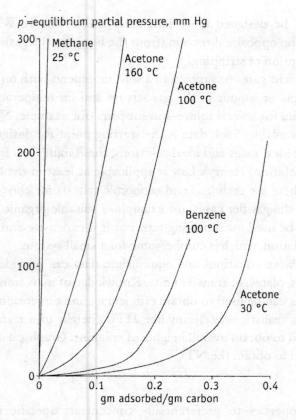

Figure 3.67 Some adsorption isotherms

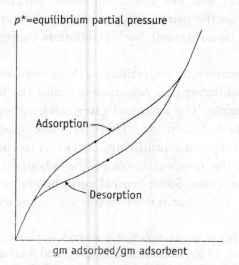

Figure 3.68 Adsorption-desorption hysterisis

design adsorption cooling systems employing solar heat or waste heat from engines. For a complete understanding of adsorption, standard books on mass transfer can be referred.

Drying

It is useful to review some of the basic definitions in the context of drying.

Moisture content, wet basis (x)   Amount of moisture (water) present (in kg) per kg of wet material. It is also called weight fraction, and is always less than 1.0. Percentage moisture is expressed as 100x.

Moisture content, dry basis (X)   Amount of moisture present (in kg) per kg of dry material. This quantity can be more than 1.0. Percentage is expressed as 100X. The relation between x and X is as follows.

$$X = x / (1 - x) \text{ or } x = X / (1 + X)$$

Equilibrium moisture content (X*)   The moisture contained in a wet solid or liquid solution exerts a vapour pressure to an extent depending upon the nature of moisture, nature of the solid, and the temperature. If a wet solid is exposed to a continuous supply of fresh gas containing a fixed partial pressure of the vapour p_w, the solid will either lose moisture by evaporation or gain moisture from the gas until the vapour pressure of the moisture of the solid equals p_w. The solid and gas are then said to be in equilibrium, and the moisture content of the solid is termed its equilibrium moisture content at the prevailing conditions. The equilibrium moisture content of some materials is shown in Figure 3.69.

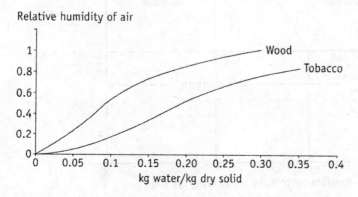

Figure 3.69 Equilibrium moisture contents of some materials at 25 °C

Bound moisture, unbound moisture content, and free moisture   With reference to Figure 3.69, consider, for example, wood with a high initial moisture of 50% on wet basis, which translates to 1.0 kg water per kg dry solid. If it is exposed to a continuous supply of air with a relative humidity of, say, 60%, the wood will lose moisture by evaporation till its equilibrium moisture is about 0.12 kg per kg of dry solid. Water lost till this point is called *free moisture* and any more loss is not possible as long as the relative humidity of the air remains at 60%. Only free moisture can be evaporated, and the free moisture content of a solid depends on the vapour content of the gas. The moisture contained in the wood up to a point corresponding to 100% relative humidity of air is called *bound water* (referring to Figure 3.69 this value for wood is about 0.3 kg per kg dry solid) and the moisture in excess of this is called *unbound water.* In the bound moisture regime (for example, in the moisture range of 0–0.3 kg/kg dry solid for wood), the moisture contained within a substance exerts an equilibrium vapour pressure less than that of pure water at the same temperature. Beyond this regime, the moisture exerts a vapour pressure equal to that of pure water at the same temperature. The various definitions are shown graphically in Figure 3.70.

Vapour pressure, partial pressure, and relative humidity   Pure liquid water, in equilibrium with its vapour, has a definite pressure corresponding to a

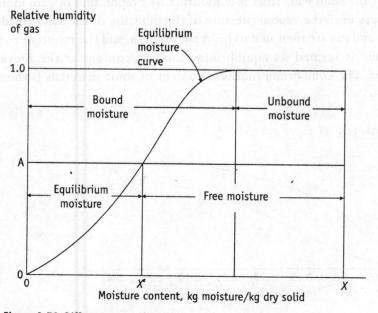

Figure 3.70 Different types of moisture

given temperature. This is called the *vapour pressure* ($p_s(T)$) of water at the given temperature T. The vapour pressure of water at 30 °C, for example, is 31.8 mmHg. When water evaporates into air, it will be in the form of vapour, as a component with a certain concentration along with other known gases like CO_2. Its concentration is expressed as *partial pressure* (p_w). If air is completely saturated with water vapour at 30 °C, the partial pressure will be equal to the vapour pressure at 30 °C and cannot be increased arbitrarily. The *relative humidity* of air, defined earlier, will then be 100%. If air has a relative humidity of, say, 60%, the partial pressure will be 0.6 × 31.8, or 19.1 mmHg. The absolute humidity is expressed as kg of water per kg dry air, which can be calculated from the following equation.

Absolute humidity = $(18/29) \times p_w/(760 - p_w)$

The relations between temperature, absolute humidity, relative humidity etc., are represented graphically in the psychrometric chart introduced earlier and shown in Figure 3.71.

Understanding and possessing a working knowledge of psychrometric charts greatly helps in designing suitable driers for different weather conditions. For example, relative humidities of air in coastal areas and during rainy season are quite high (80%–90%), which means that open sun drying alone cannot reduce moisture contents to the extent desired. It will thus be necessary to heat the air. The heating process can be represented by a horizontal line on the psychrometric chart. For example, when air with 90%

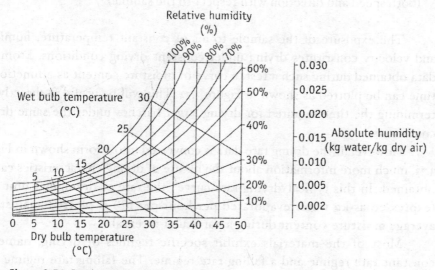

Figure 3.71 Psychrometric chart with details

relative humidity at 30 °C is heated to 50 °C, the relative humidity decreases to about 42%, which means that the air now has more capacity to 'hold' moisture and hence when contacted with the material to be dried, can cause faster drying. The drying rate, however, is unique to each material, as discussed below.

Drying kinetics

In order to set up drying schedules and to determine the size of the equipment, it is necessary to know the time required to dry a substance from one moisture content to another under specified conditions. One may also wish to estimate the influence of different drying conditions on drying times. Unfortunately, the present knowledge of the mechanisms of drying is incomplete and hence, it is necessary to rely upon some experiments. Measurements of the rate of batch drying are relatively simple to make and provide much information for design.

The rate of drying can be determined for a sample of substance by suspending it in a cabinet or duct, in a stream of air, from a balance. Also, the following conditions should be as close as possible to actual conditions in the contemplated drier.

- The sample should be similarly supported in a tray or frame
- It should have the same ratio of drying to non-drying surface
- It should be subjected to similar conditions of radiant heat transfer
- The air should have the same temperature, humidity, and velocity (both speed and direction with respect to the sample)

The exposure of the sample to air of constant temperature, humidity, and velocity constitutes drying under constant drying conditions. From the data obtained during such a test, a curve of moisture content as a function of time can be plotted as shown in Figure 3.72. This will be useful in directly determining the time required for drying larger batches under the same drying conditions.

However, if the drying rate data is converted into a form shown in Figure 3.73, much more information about the nature of drying characteristics can be obtained. In this typical 'drying rate curve', the rate of drying per unit area (expressed as kg water evaporated per hour per m²) is plotted against the average moisture content during a particular interval.

Most of the materials exhibit specific regimes of drying, namely a constant rate regime and a falling rate regime. The falling rate regime may

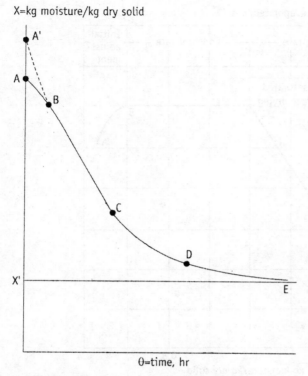

X=kg moisture/kg dry solid

θ=time, hr

Figure 3.72 Moisture loss as a function of time

further be subdivided into different regions depending on the shapes of the curves. In the constant rate regime, it is assumed that the moisture is unbound, and hence, the evaporation takes place as if it is occurring from a thin film of water. The drying surface adjusts quickly to a temperature depending on the conditions of the air. The rate of evaporation is then proportional to $(p_s - p_w)$, where p_s refers to the vapour pressure at the surface temperature. Since evaporation of moisture absorbs latent heat, the liquid surface will tend to cool, but the fall in temperature is compensated by convective heat gain from the air stream. The liquid surface will thus come to an equilibrium temperature such that the rate of heat gain from the surroundings to the surface exactly equals heat loss due to evaporation. As conditions are constant, the rate of drying will also be constant.

When the average moisture content has reached a critical value X_c, the surface water film has shrunk in area so that dry spots start appearing on the surface. The net effective surface available for evaporation keeps reducing, and so does the drying rate. On further drying, the rate at which moisture can be transported to the surface from the inner layers becomes the controlling step. This is the general explanation given to the occurrence of the falling rate

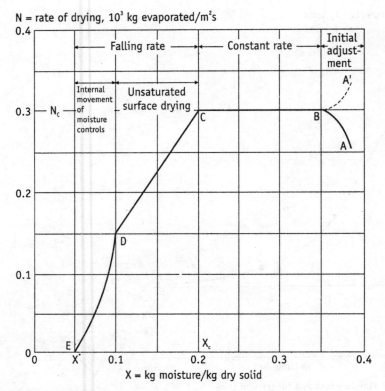

N = rate of drying, 10³ kg evaporated/m²s

Figure 3.73 Drying rate curve

regime. Drying stops ultimately when the moisture content of the solid reaches a value equal to the equilibrium moisture content given by curves such as those shown in Figure 3.69. Drying time can be determined from drying rate curve by mathematical integration methods.

It is possible to estimate drying rates in the constant-rate regime from a knowledge of air conditions (temperature, humidity, and velocity), radiation received, tray geometry, etc. Based on mathematical analysis of the heat and mass transfer processes, the following general conclusions are drawn regarding the effect of various parameters on drying rates.

Effect of gas velocity If radiation and conduction through the solid are neglected, drying rate is proportional to $G^{0.71}$ for parallel flow and to $G^{0.37}$ for perpendicular flow, where G is the mass velocity (mass flow rate divided by cross-sectional area) of air. If radiation and conduction are present, the effect of air velocity is less important. Air velocities of 3–4 m/s are recommended in tray driers in order to obtain uniform conditions of drying. Values such as these can be used to design tray or cabinet driers.

Effect of gas temperature   Increased air temperature increases drying rate. In the absence of radiation, the rate is directly proportional to the difference of dry bulb and surface temperatures. Surface temperatures or adiabatic saturation temperatures of air leaving the drier can be estimated by heat flow and psychrometric calculations. If radiation and conduction are not significant, the surface temperature will be equal to the wet bulb temperature of the air.

Effect of air humidity   Drying rate varies directly as the difference of humidity levels of water surface and air, and hence, increasing air humidity reduces drying rate.

Effect of thickness of drying solid   If heat conduction through solid occurs, drying rate decreases with increased thickness.

For fixed-bed, through-circulation driers such as those used in many traditional methods, the following theoretical treatment applies.

Consider a case where bed of solids has an appreciable thickness with respect to the size of the particles, as in Figure 3.74.

The evaporation of unbound moisture into the gas occurs in a relatively narrow zone, which moves slowly through the bed. The gas leaving this zone is saturated at the adiabatic saturation temperature of the entering gas, which is also the surface temperature of the wet particles. The rate of drying is constant as long as the zone is entirely within the bed. When the zone first

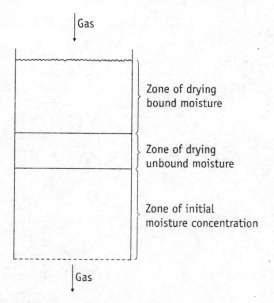

Gas

Zone of drying
bound moisture

Zone of drying
unbound moisture

Zone of initial
moisture concentration

Gas

Figure 3.74 Drying zone

reaches the end of the bed, the rate of drying begins to fall because the gas no longer leaves in a saturated condition. In the case of shallow beds composed of large particles (fish smoking or cardamom drying are close examples), the gas leaves the bed unsaturated from the beginning, but as long as each particle surface remains fully wet, there will still be a constant-rate period. The falling rate begins when the surface moisture is depleted. The rate of drying can be calculated by the equation

$$N/N_{max} = 1 - \exp(-NTG)$$

Here N is the rate of drying, N_{max} is the maximum rate, and NTG is the number of 'gas transfer units' in the bed. NTG depends on parameters such as gas phase mass transfer coefficient, interfacial surface per unit volume of the bed, and the thickness of the bed. Empirical equations to estimate NTG are given in ref [1].

Theoretical considerations outlined as above can help in deciding on parameters such as bed height to achieve optimum drying conditions. However, in practice, the bed is often disturbed by overturning to obtain uniform drying, in which case any attempt to model the system can become extremely complicated.

Annexe 1: Units and conversions

Quantity	Governing equation	SI unit	Other common units	Some conversions
Length (D)		m	inch, ft, mile, km mile = 1.61 km	1 inch = 2.54 cm 1 ft = 30.48 cm
Area (A)	• Area of a rectangle of sides a and b = a × b • Area of a circle of radius r = πr^2	m^2	ft^2, km^2, acre, ha	1 ha = 10 000 m^2 = 2.471 acres
Volume (V)	• Volume of a cube of side a = a^3 • Volume of a cylinder of radius r and height h = $\pi r^2 h$	m^3	gal litre, ft^3	1 US gal. = 3.785 litres 1 m^3 = 35.3 ft^3
Time (t)		s	min, h, y	1 h = 3600 s 1 yr = 8760 h
Mass (m)		kg	lb	1 kg = 2.205 Lbs
Speed, velocity	u = D/t	m/s	ft/s, km/h	1 m/s = 3.28 ft/s = 3.6 km/h
Acceleration	a = du/dt	m/s^2	ft/s^2	1 m/s^2 = 3.28 ft/s^2
Density	ρ = m/V	kg/m^3	kg/litre, lb/ft^3	1 kg/m^3 = 1000 kg/ litre = 77.87 lb/ft^3
Flow rate	Q = V/t	m^3/s	litre/h, gal/h	1 m^3/s = 0.2778 litre/h = 1.0514 gal/h
Momentum	mu	kg.m/s	lb.ft/s	1 kg.m/s = 7.2343 lb.ft/s
Force	F = ma	N	kg_f, dyne	1 N = 10^5 dyne = 0.1020 kg_f
Work, energy	W = FD	J	kg_f.cm, calorie, Wh, Btu	1 J = 0.000948 Btu = 0.239 cal = 3600 Wh = 10.197 kg_f.cm
Pressure	p = F/A	Pa or N/m^2	bar, atm, kg_f/cm^2	1 Pa = 10^{-5} bar = 1.0132×10^{-5} atm = 1019.7 kg_f/cm^2
Power	P = W/t	W	Btu/h, hp, calorie/s	1 W = 3.4128 Btu/h = 1/736 hp = 0.239 calorie/s
Viscosity (dynamic)	$\tau = \mu \dfrac{du}{dx}$	$N.s/m^2$	P (poise)	1 $N.s/m^2$ = 10 P
Surface tension		N/m	kg_f/cm, lbf/in	1 N/m = 0.00102 kg_f/cm = 0.00225 lbf/in
Temperature (T)		K	°C	1 K = °C + 273 ° °F = °C (1.8) + 32 °C = (°F – 32)/1.8
Heat flux	$q = -k \dfrac{dT}{dx}$	W/m^2	Calorie/h.cm^2	1 W/m^2 = 0.086 cal/h.cm^2 = 0.3171 Btu/h.ft^2
Thermal conductivity	$q = -k \dfrac{dT}{dx}$	W/m.K	Btu.ft/h.ft^2.°F	1 W/m.K = 1.8724 Btu.ft/h.ft^2.°F

Continued

Annexe 1 Continued

Quantity	Governing equation	SI unit	Other common units	Some conversions
Heat transfer coefficient	$\dot{Q} = hA\Delta T$	$W/m^2.K$	$Btu/h.ft^2. °F$	$1\ W/m^2.K = 0.1761\ Btu/h.ft^2.°F$
Specific heat	$Q = m.C_p.\Delta T$	$J/kg.K$	$Btu/lb.\ °F$ $Cal/g\ °C$	$1\ J/kg.K = 0.000774\ Btu/lb.°F$

SI – Systeme international d unites; ft - foot; km – kilometre; ha – hectare; gal – gallon; cm – centimetre; s – second; min – minute; h – hour; y – year; kg – kilogram; lb – pound; N – Newton; kg_f – kilogram force; dyn – dyne; J – joule; cal – calorie; Pa – Pascal; W – watt; atm – standard atmosphere; Btu – British thermal unit; hp – horsepower; K – Kelvin; °C – Centigrade; °F – Farenheit

References

Berdahl P and Martin M. 1984
Emissivity of clear skies
Solar Energy 32(5): 663–664

Bird R B, Stewart W E, and Lightfoot E N. 2001
Transport Phenomenon
New York: Wiley. 912 pp.

ChemicaLogic Corporation. 2005a
Pressure–enthalpy diagram for water and steam: based on the IAPWS-97 formulation for general and scientific use
Details available at, <http://www.chemicalogic.com/download/ mollier_chart_iapws97.pdf>, last accessed on 28 September 2005

ChemicaLogic Corporation. 2005b
SteamTab™ product information
Details available at, <http://www.chemicalogic.com/steamtab/ product_information.htm>, last accessed on 28 September 2005

Coulson J and Richardson J F. 1999
Coulson and Richardson's Chemical Engineering: fluid flow, heat transfer and mass transfer, (vol. 1)
Oxford: Butterworth-Heinemann. 928 pp.

Dincer I. 2003
Refrigeration Systems and Applications
Sussex: John Wiley & Sons Ltd. 598 pp.

Duffie J A and Beckman W A. 1991
Solar Engineering of Thermal Processes
New York: John Wiley & Sons. 919 pp.

Glasstone S. 1947
Thermodynamics for Chemists
New York: D Van Nostrand Company, Inc. 522 pp.

Gupta C P and Prakash R. 1976
Engineering Heat Transfer
Roorkee: Nem Chand and Brothers

Himmelblau D M. 1974
Basic Principles and Calculations in Chemical Engineering
New Jersey: Prentice-Hall Inc. 542 pp.

Holman J P. 1980
Thermodynamics
Auckland: McGraw-Hill. 770 pp.

Kadambi V. 1976
Introduction to Energy Conversion: basic thermodynamics (volume 1)
New Delhi: Wiley Eastern. 239 pp.

Kays W M and London A L. 1964
Compact Heat Exchangers
New York: McGraw-Hill. 271 pp.

Kishore V V N. 1993
Economics of solar pond generation
In *Renewable Energy Utilization: scope, economics, and perspectives*, pp. 53–68,
edited by V V N Kishore
New Delhi: Tata Energy Research Institute

Kishore V V N, Gandhi M R, Marquis Ch, Rao K S. 1984
Analysis of flat plate collectors charged with phase changing fluids
Applied Energy 17(2): 133–149

Kishore V V N, Ramana M V, and Rao D P. 1979
An experimental and theoretical study of a natural water cooler
Proc. Nat. Sol. En. Convention held at Indian Institute of Technology Bombay,
pp. 482–488

Lange N A and Dean J A. 1985
Lange's Handbook of Chemistry
New York: McGraw-Hill. 1792 pp.

Lewis W K and Whitman W G. 1924
Principles of gas absorption
Industrial and Engineering Chemistry 16:1215–1220

Mande S, Ghosh P, Kishore V V N, Oertel K, Sperengel U. 1997
**Development of an advanced solar-hybrid adsorption cooling system for
decentralized storage of agricultural products in India**
In *Proceedings of CLIMA 2000 International Congress*
Brussels: ATIC, the Belgian HVAC Engineers Association

Mark W. 2005
WebElementsTM Periodic Table (professional edn)
Details available at, <www.webelements.com>, last accessed on 17 July 2005

McCabe W L and Smith J. 2000
Unit Operations of Chemical Engineering
McGraw-Hill Education

Perry R H and Green D W. 1997
Perry's Chemical Engineers' Handbook, 2640 pp.
McGraw-Hill

Ramos M, Espinoza R L, and Horn M J. 2003
Evaluation of a zeolite–water solar adsorption refrigerator
[ISES Solar World Congress 2003, Sweden, June 14–19, p. 5]

Rao D P, Das T C Thulasi, and Karmakar S. 1994a
Solar box-cooker: Part I–modeling
Solar Energy 52(3): 265–272

Rao D P, Das T C Thulasi, and Karmakar S. 1994b
Solar box-cooker: Part II–analysis and simulation
Solar Energy 52 (3): 273–282

Rao K S, Kishore V V N, and Vaja D (eds). 1990
Solar Pond: scope and utilization
Vadodara: Gujarat Energy Development Agency

Ravi M R, Kohli S, and Ray A. 2002
Use of CFD simulation as a design tool for biomass stoves
Energy for Sustainable Development IV(2): 20–27

Reddy A K N, Kelly H, Johansson T B, Williams R.H. 1993
Renewable Energy: sources for fuels and electricity, 460 pp.
London: Earthscan Publications Ltd.

Reid R C and Sherwood T K. 1966
The Properties of Gases and Liquids
New York: McGraw-Hill

Seshadri C V. 1982
Development and Thermodynamics: a search for new energy-quality markers
Chennai: Sri AMM Murugappa Chettiar Research Centre

Smith J M and van Ness H C. 1975
Introduction to Chemical Engineering Thermodynamics, 632 pp
Auckland: McGraw-Hill.

Smith K R, Khalil M A K, Rasmussen R A, Thorneloe S A, Manegdeg F, Apte M. 1993
Greenhouse gases from biomass and fossil fuel stoves in developing countries: a Manila pilot study
Chemosphere 26(1–4): 479–505

Treybal R E. 1981
Mass-Transfer Operations, 784 pp.
Auckland: McGraw-Hill

Van Wylen G J and Sonntag R E. 1973
Fundamentals of Classical Thermodynamics, 752 pp.
New York: John Wiley

Wagman D D (ed.). 1952
Selected values of chemical thermodynamics properties
[National Bureau of Standards, Circular 500]

Bibliography

Kuhn G. 2004
Environmental and Hygiene Engineering, 2004
Details available at, <www.gkehe.8m.com/data.htm>, last accessed on 21 July 2005

Ricci J E. 1951
The Phase Rule and Heterogeneous Equilibrium
New York: Van Nostrand Reinhold

Stanley H E. 1971
Introduction to Phase Transitions and Critical Phenomena, 308 pp.
London: Oxford University Press

Taylor C F and Taylor E S. 1962
The Internal Combustion Engine, 668 pp.
Scranton: International Textbook Co.

US Department of Energy. 1992
DOE fundamentals handbook: thermodynamics, heat transfer, and fluid flow (vol. 1)
Details available at, <http://www.eh.doe.gov/techstds/standard/hdbk1012/h1012v1.pdf>, last accessed on 14 August 2005

Weast Robert C. 1983
CRC Handbook of Chemistry and Physics: a ready reference book of chemical and physical data. 176 pp.
Florida: CRC Press

Zucrow M. J. 1957
Jet Propulsion and Gas Turbines
New York: John Wiley

4

The solar energy resource

Amit Kumar, Director and Mahesh C Vipradas, Fellow
Energy–Environment Technology Division, T E R I, New Delhi

Introduction

The sun is a body of intensely hot gaseous matter and the temperature in its interior is estimated to be between 8 and 40 million Kelvin. The sun is practically a nuclear fusion reactor in which hydrogen (four protons) combines to form helium (that is, one helium nucleus) and the difference in the mass lost is released as heat. The rate of energy emitted from the sun is 3.8×10^{23} kW, of which the earth intercepts only a tiny fraction, approximately 1.7×10^{14} kW. However, even this tiny amount is many thousand times greater than the present consumption rate of energy on the earth. Thus, the sun can be considered as a source of inexhaustible energy. However, solar radiation is a dilute source of energy with a maximum flux density of 1.3 kW/m², which is a low value for technological utilization. Also, due to the geometry of the sun–earth movements, there are large variations in the amount of solar radiation received at any given location; the largest variation being at the poles. The variations are both diurnal (during the day) and seasonal. The presence of clouds, dust, etc., in the atmosphere further reduces the availability of solar energy. Thus, precise information about the quality and quantity of solar resources at a specific location is essential for optimal design of equipment, which would convert solar energy into the desired form for practical applications. Such information is also useful in assessing the potential of utilization of solar energy, and in the comparative performance evaluation of different solar collectors.

Almost 99% of the energy of solar radiation is contained in the wavelength band of 0.15–4.0 μm, comprising the ultraviolet, visible, and near infrared regions. About 40% of the solar radiation received on the earth's surface on clear days is visible radiation (0.4–0.7 μm), while 51% is infrared radiation (0.7–4.0 μm) (Figure 4.1).

Solar radiation received on the earth's surface, called global radiation, has two components, namely (i) beam radiation that comprises direct radiation from the sun and (ii) diffuse radiation, the reflected radiation from the atmosphere. Solar energy can be utilized in a number of ways. For example, in the photosynthesis process, the absorbed solar energy is converted into chemical forms of energy. PV systems based on semiconductor materials can convert solar radiation directly into electrical current. The solar thermal energy systems first convert the incident solar energy into heat through a variety of solar collectors, and then the heat is used for a variety of applications such as water heating, air heating, cooking, process heating, and power generation. Solar energy can also be used intelligently in buildings to reduce the energy consumption for heating and/or cooling. For optimum design of such systems, accurate solar radiation data – global, direct, and diffuse – is essential. Table 4.1 lists the data requirement of different technologies/applications.

It is always best to use measured solar radiation data for the design of solar energy systems or for performance evaluation, at the location where the solar systems are to be installed. However, such data is not available for many places and, therefore, some calculations, approximations, and interpolations

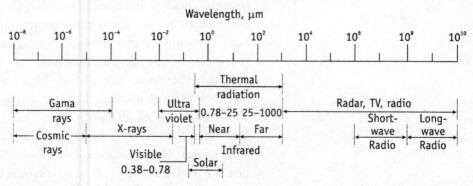

Figure 4.1 Spectrum of electromagnetic radiation
Reproduced with permission from John Wiley & Sons (Asia) Pvt. Ltd
Source Duffie and Beckman (1991)

Table 4.1 Solar radiation data requirements for different solar energy applications

Technology/application	Data required (on hourly basis)
Solar heating and cooling (below 100 °C) (using flat plate collectors)	Global and diffuse solar radiation
Passive architecture	Global solar radiation
Medium/high-temperature industrial process heating (using concentrating collectors)	Direct normal (or beam) radiation
Solar thermal power generation (using concentrating collectors)	Direct normal (or beam) radiation
Photovoltaic applications	Global solar radiation

have to be made. In this context, it is useful to have an understanding of the nature of extraterrestrial radiation, atmospheric attenuation, and the effects of the orientation of the receiving surface.

Solar radiation outside the earth's atmosphere

The sun: structure and characteristics

The sun is a plasmatic, almost spherical, body, with a diameter of 1.392×10^9 m and is at an average distance of 1.5×10^{11} m (150 million km) from the earth. On account of the periodic changes in its diameter, its brightness also changes periodically. The sun is composed of about 80% hydrogen, 19% helium, and small traces of nearly all the known elements. It is made up of many layers of gases, and these layers get progressively hotter towards the centre of the sun. The temperature of the outermost layer is 5770 K, which is nearly the same as the effective black body temperature of the sun (temperature of a black body radiating the same amount of energy as the sun). This is the layer that radiates energy into the solar system. The rate of energy emitted from the sun is 3×10^{23} kW.

The solar constant

As the earth's orbit is elliptical, the distance between the sun and the earth varies by 1.7%. At a distance of 1.495×10^{11} m (the mean earth–sun distance), the sun subtends an angle of 32°. The solar constant, G_{SC}, is the energy received from the sun, per unit time, on a unit area of surface perpendicular to the direction of propagation of the radiation at the mean earth–sun distance. There were several measurements of the solar constant, but the value, 1367 W/m², adopted by the World Radiation Centre, is now commonly used.

Spectral distribution

The extra-atmospheric distribution of energy in the solar radiation spectrum can be found either by direct measurements or by extrapolating beyond the atmosphere of the surface spectrometric measurement data.

Figure 4.2 shows one possible spectral distribution of solar energy according to sea-level observations and the distribution obtained by extrapolating beyond the earth's atmosphere. The shaded areas show the effects of absorption by atmospheric gases. Solar radiation values for different wavelength intervals can be taken directly from those prepared by Smith and Goettlieb (WMO 1981).

Sun–earth geometric relationship

Geometry of the earth–sun–time system

The orbit of earth around the sun is elliptical. Thus, the distance between the earth and the sun varies from time to time. It is minimum (147.1×10^6 km) at winter solstice on 21 December, and the point in the orbit is known as perihelion. It is maximum (152.1×10^6 km) at summer solstice on 21 June, and the point in the orbit is known as aphelion. The difference between the maximum and minimum distances is only 3.4% (Figure 4.3). The angles at

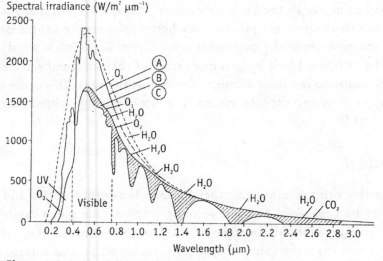

Figure 4.2 Spectral distribution of solar radiation outside the atmosphere and at sea level. (A) Spectral distribution of radiation from a black body at a temperature of 6000 K. (B) Spectral distribution of solar radiation outside the atmosphere. (C) Spectral distribution at sea level
Source WMO (1981)

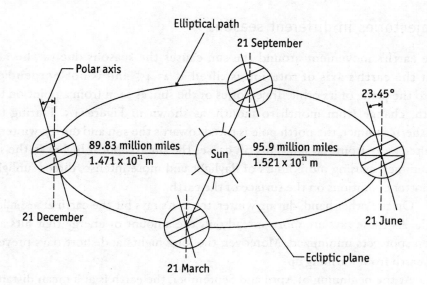

Figure 4.3 Motion of the earth around the sun
Source Goswami, Kreith, and Kreider (2000)

which sun's rays impinge on earth in different seasons and the location of tropics is shown in Figure 4.4.

The sun is so far away from the earth that all its rays may be considered as parallel to one another when they reach the earth.

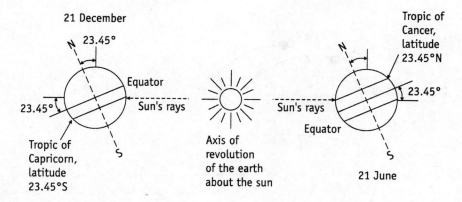

Figure 4.4 Location of tropics
Source Goswami, Kreith, and Kreider (2000)

Sun's trajectories in different seasons

The earth's movement around the sun causes the seasons due to the fact that the earth's axis of rotation is tilted at 23.45° and is not perpendicular to the plane of its orbit. Trajectories of the sun, as seen from a point on the earth, change from month to month, as shown in Figure 4.5. During the northern summer, the north pole is tilted towards the sun and during winter, it is tilted away from the sun. This tilt causes the sun to appear higher in the sky in summer, causing more hours of daylight and more intense, direct sunlight, or hotter conditions on the surface of the earth.

On the other hand, during winter, the sun's rays hit the earth at a shallow angle. As these rays are more spread out, the amount of energy that hits any given spot gets minimized. Moreover, the long nights and short days prevent the earth from warming up.

At the beginning of April and September, the earth is at a mean distance of 149.6×10^6 km from the sun. The sun's visible diameter also changes as the earth moves round its orbit. In January, the angle subtended by the diameter of the disc is at its maximum at 32'36", and in July, it is minimum at 31'32". When the earth is at its mean distance from the sun, this angle is about 32' (Figure 4.6). The changes in the distance throughout the year lead to changes in the radiation flux reaching the earth from the sun; the radiation flux varies inversely with the square of the distance. The variation of the extraterrestrial radiation with the time of the year can be estimated using Equation 4.1.

$$G_{on} = G_{sc}\left(1 + 0.033\cos\frac{360n}{365}\right) \qquad ...(4.1)$$

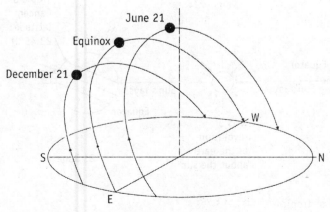

Figure 4.5 Sun trajectories

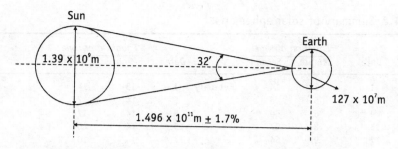

Figure 4.6 Relative geometry of the sun and the earth
Reproduced with permission from John Wiley & Sons (Asia) Pvt. Ltd
Source Duffie and Beckman (1991)

where G_{on} is the extraterrestrial radiation on the nth day of the year (n = 1 for 1 January) and G_{SC} is the solar constant.

Some definitions

It will be useful to introduce some basic definitions at this stage before proceeding further.

Zenith

The point directly overhead.

Solar time

This is the time based on the apparent angular motion of the sun across the sky and does not coincide with the local clock time. The solar noon, for example, is the time during which the sun crosses the meridian of the observer, and need not be 12 noon according to local time. All solar calculations are done in solar time. The solar time can be calculated from local time for any location by the following equation

$$\text{Solar time} = \text{LST} + \text{ET} + 4\,(l_{st} - l_{local}) \qquad \qquad ...(4.2)$$

where LST is local standard time, ET is the equation of time, l_{st} is the standard time meridian, and l_{local} is the local longitude. Both l_{st} and l_{local} recokned positive, *west* of the GM (Greenwich Meridian) and negative *east* of the GM. The time correction in Equation 4.2 is in minutes. ET may be calculated from the following equation or estimated from Table 4.2.

$$\text{ET (in minutes)} = 9.87 \sin 2B - 7.53 \cos B - 1.5 \sin B$$

where $B = (n-81)\,360/364$ and n = day of the year.

Table 4.2 Summary of solar ephemeris*

Date		Equation of time		Date		Equation of time	
		Minute	Second			Minute	Second
January	1	-3	14	February	1	-13	34
	5	5	6		5	14	2
	9	6	50		9	14	17
	13	8	27		13	14	20
	17	9	54		17	14	10
	21	11	10		21	13	50
	25	12	14		25	13	19
	29	123	5				
March	1	-12	38	April	1	-4	12
	5	11	48		5	3	1
	9	10	51		9	1	52
	13	9	49		13	-0	47
	17	8	42		17	0	13
	21	7	32		21	1	6
	25	6	20		25	1	53
	29	5	7		29	2	33
May	1	2	50	June	1	2	27
	5	34	17		5	1	49
	9	3	35		9	1	6
	13	3	44		13	-0	18
	17	3	44		17	0	33
	21	3	24		21	1	25
	25	3	16		25	2	17
	29	2	51		29	3	7
July	1	-3	31	August	1	-6	17
	5	4	16		5	5	59
	9	4	56		9	5	33
	13	5	30		13	4	57
	17	5	57		17	4	12
	21	6	15		21	3	19
	25	6	24		25	2	18
	29	6	23		29	1	10
September	1	-0	15	October	1	10	1
	5	1	2		5	11	17
	9	2	22		9	12	27
	13	3	45		13	13	30
	17	5	10		17	14	25
	21	6	35		21	15	10
	25	8	0		25	15	46
	29	9	22		29	16	10
November	1	16	21	December	1	11	16
	5	16	23		5	9	43
	9	16	12		9	8	1
	13	15	47		13	6	12
	17	15	10		17	4	47
	21	14	18		21	2	19
	25	13	15		25	0	20
	29	11	59		29	-1	39

*Since each year comprises 365.25 days, the precise value of declination varies from year to year. *The American Ephemeris and Nautical Almanac*, published each year by the US Government Printing Office, contains precise values for each day of each year.
Source Goswami, Kreider, and Kreith (2000)

Air mass

Air mass m is the path length of radiation through the atmosphere, considering the vertical path at sea level as unity. At sea level, m = 1, when the sun is at the zenith (that is, directly overhead) and m = 2 for a zenith angle θ_z of 60°.

$$m = \frac{1}{\cos \theta_z} \qquad\qquad ...(4.3)$$

for zenith angles θ_z between 0° and 70° at sea level.

Since at a higher zenith angle, the effect of the earth's curvature becomes quite significant, it must be taken into consideration.

Beam radiation

The part of solar radiation that reaches the earth without any change in direction is called beam radiation. It is also known as direct radiation.

Diffuse radiation

The solar radiation received by the earth after its direction gets changed because of scattering in the atmosphere is known as diffuse radiation.

Total solar radiation

The sum of the beam and the diffuse components of solar radiation is called total solar radiation. Total solar radiation on a horizontal surface is commonly known as global radiation.

Irradiance

The solar irradiance G is the rate at which the radiant energy is incident on a unit area of a surface. Denoted in terms of W/m^2, it is used for beam, diffuse, or spectral radiation with suitable subscripts.

Insolation

The incident solar energy radiation (or irradiation) is also termed as insolation. While H is insolation for the day, I is the insolation for a specific time period, usually one hour. H and I are expressed in $W\text{-}hr/m^2/day$ and $W\text{-}hr/m^2/hr$, respectively. When the values are measured on an hourly basis, I is numerically equal to G. Again, H and I can (1) represent beam, diffuse, or total radiation and (2) be on any orientation.

Example 1

Calculate the solar time corresponding to 12:00 (IST or Indian Standard Time) at Pondicherry (11.92° N, 79.92° E) on 17 July. The standard meridian for IST is 82.5 °E (Mirzapur).

Solution

For 17 July

$$n = 31 + 28 + 31 + 30 + 31 + 30 + 17 = 198$$

For India

Standard time longitude, $l_{st} = 82.5°$

$$B = \frac{360\,(198-81)}{364}$$

$$= 115.7°$$

$$ET = 9.87 \sin(2 \times 115.7) - 7.53 \cos(115.7) - 1.5 \sin(115.7)$$
$$= -5.8 \text{ min}$$

$$4(l_{st} - l_{local}) = 4\,(-82.5 - (-79.92))$$
$$= -10.32 \text{ min}$$

$$\text{Solar time} = LST + ET + 4(l_{st} - l_{local})$$
$$= 12:00 - 16.12 \text{ min}$$
$$\simeq 11:44 \text{ h}$$

Solar radiation on different surfaces

Different angles: explanation and relations

The geometric relationships between a plane of any particular orientation relative to the earth at any time (whether the plane is fixed or moving relative to the earth) and the incoming beam radiation, that is, the position of the sun relative to that plane, can be described in terms of several angles as follows.

Latitude (ϕ) It is the angle made by the radial line joining the location to the centre of the earth with the projection of the line on the equatorial plane. By convention, latitude is measured as positive for the northern hemisphere. It varies as $-90° \leq \phi \leq 90°$.

Solar declination Since the earth's axis of rotation is inclined at an angle of 23.45° to the axis of its orbit around the sun, this tilt causes seasonal variations in available solar radiation at any location. The angle between the earth–sun line (through their centres) and the plane through the equator is called the solar declination. It varies between −23.45° on 21 December and +23.45° on 21 June. Declinations towards the north of the equator are positives whereas those to the south are negatives. Solar declination can be estimated using the following expression.

$$\delta = 23.45 \sin\left[\frac{360}{365}(284 + n)\right] \qquad \qquad ...(4.4)$$

where n is the day number during a year (1 January being n = 1).

Table 4.3 can be used to find the day of the year n and corresponding declination.

Slope (β)   As shown in Figure 4.7, it is the angle between the plane of the surface concerned and the horizontal. It varies as $0 \le \beta \le 180°$. (β > 90° means that the surface has a downward-facing component.)

Surface azimuth angle (γ)   It is the angle made in the horizontal plane between the line due south and the projection of the normal to the surface on the horizontal plane. As per convention, due south is taken as zero, east of south as positive, and west of south as negative. Hence, it varies as $-180° \le \gamma \le 180°$.

Hour angle (ω)   It is the angular displacement of the sun, east or west of the local meridian, due to rotation of the earth on its axis at 15° per hour, morning negative and afternoon positive.

Table 4.3 Recommended average days for months and value of n by months

Month	n for ith day of the month	For average day of the month		
		Date	n, day of year	δ, declination
January	i	17	17	- 20.9
February	31 + i	16	47	- 13.0
March	59 + i	16	75	- 2.4
April	90 + i	15	105	9.4
May	120 + i	15	135	18.8
June	151 + i	11	162	23.1
July	181 + i	17	198	21.2
August	212 + i	16	228	13.5
September	243 + i	15	258	2.2
October	273 + i	15	288	- 9.6
November	304 + i	14	318	18.9
December	334 + i	10	344	- 23.0

Source Duffie and Beckman (1991)

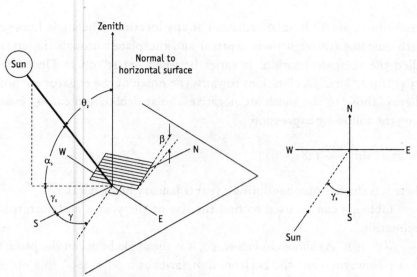

Figure 4.7 Different angles for a tilted surface
Reproduced with permission from John Wiley & Sons (Asia) Pvt. Ltd
Source Duffie and Beckman (1991)

Angle of incidence (θ)   It is the angle between the beam radiation on a surface and the normal to that surface.

Zenith angle (θ_z)   It is the angle between the vertical and the line to the sun, that is, the angle of incidence of the beam radiation on a horizontal surface.

Solar altitude angle (α_s)   It is the angle between the horizontal and the line of the sun, that is, the complement of the zenith angle.

Solar azimuth angle (γ_s)   It is the angular displacement from the south of the projection of the beam radiation on the horizontal plane. Displacements east of south are negative and west of south are positive.

The relationship between the angle of incidence of beam radiation (θ) and other angles is expressed as follows.

$$\cos\theta = \sin\delta\sin\phi\cos\beta - \sin\delta\cos\phi\sin\beta\cos\gamma + \cos\delta\cos\phi\cos\beta\cos\omega$$
$$+ \cos\delta\sin\phi\sin\beta\cos\gamma\cos\omega + \cos\delta\sin\beta\sin\gamma\sin\omega$$

and   ...(4.5)

$$\cos\theta = \cos\theta_z\cos\beta + \sin\theta_z\sin\beta\cos(\gamma_s - \gamma) \qquad \text{...(4.6)}$$

For more details see Faiman, 2003.

Example 2

Calculate the angle of incidence on a surface located at Mumbai (19°07′ N, 72°51′ E) at 11:00 a.m. (solar time) on 17 July, if the surface is tilted at an angle of 30°, with the horizontal and is pointing 10° east of south.

Solution

$n = 198$

$\delta = 21.2°$ (from Table 4.3)

$\phi = 19.11°$

$\omega = -15°$

$\beta = 30°$

$\gamma = 10$

$$\cos\theta = \sin\delta \sin\phi \cos\beta - \sin\delta \cos\phi \sin\beta \cos\gamma + \cos\delta\cos\phi \cos\delta \cos\omega +$$
$$\cos\delta \sin\phi \sin\beta \cos\gamma \cos\omega + \cos\delta \sin\beta \sin\gamma \sin\omega$$

$= 0.36 \times 0.32 \times 0.86 - (0.36 \times 0.94 \times 0.5 \times 0.98) + 0.93 \times 0.94 \times 0.86 \times$
$0.96 + 0.93 \times .032 \times 0.5 \times 0.98 \times 0.96 + 0.93 \times 0.5 \times 0.17 \times 0.25$

$= 0.099 - 0.165 + 0.72 + 0.139 + 0.019$

$\cos\theta = 0.812$

Thus,

$\theta = 35.7°$

Extraterrestrial radiation on a horizontal surface

The extraterrestrial solar irradiance G_o on a horizontal surface is given by the expression

$$G_o = G_{sc}\left(1 + 0.033 \cos\frac{360n}{365}\right)\cos\theta_z \qquad \dots(4.7)$$

where G_{sc} is the solar constant and n is the day of the year. G_o for a horizontal surface, anytime between sunrise and sunset, may be expressed as

$$G_o = G_{sc}\left(1 + 0.033 \cos\frac{360n}{365}\right)(\cos\phi\cos\delta\cos\omega + \sin\phi\sin\delta) \qquad \dots(4.8)$$

The extraterrestrial radiation on a horizontal surface during a specific time period (between starting hour angle ω_1 to ending hour angle ω_2 can be estimated by integrating Equation 4.8 from ω_1 to ω_2.

$$I_o = \frac{12 \times 3600}{\pi}G_{sc}\left(1 + 0.033\cos\frac{360n}{365}\right)$$
$$\times\left[\cos\phi\cos\delta\left(\sin\omega_2 - \sin\omega_1\right) + \frac{\pi(\omega_2 - \omega_1)}{180}\sin\phi\sin\delta\right] \qquad \dots(4.9)$$

The daily extraterrestrial radiation H_o on a horizontal surface is obtained by integrating Equation 4.8 over the period from sunrise to sunset, as shown below.

$$H_o = \frac{24 \times 3600 G_{sc}}{\pi} \left(1 + 0.033 \cos \frac{360n}{365} \right) \times \left(\cos\phi \cos\delta \sin\omega_s + \frac{\pi\omega_s}{180} \sin\phi \sin\delta \right)$$

...(4.10)

where ω_s is the sunset hour angle in degrees, given by the following expression.

$$\cos\omega_s = \frac{-\sin\phi \sin\delta}{\cos\phi \cos\delta} = -\tan\phi \tan\delta$$

...(4.11)

Example 3
Estimate the daily total extraterrestrial radiation on a horizontal surface at Mumbai (19°07' N, 72°51' E) on 17 July

$$n = 198$$
$$\delta = 21.2$$
$$\phi = 19.11$$

Solution
The sunset hour angle $\cos\omega_s = -\tan\phi \tan\delta$

$$H_o = \frac{24 \times 3600 \times 1367}{\pi} \left(1 + 0.033 \cos\left(\frac{360 \times 198}{365}\right) \right)$$

$$\times \left(\cos 19.11 \cos 21.2 \sin 82.29 + \frac{\pi \times 82.29}{180} \sin 19.11 \sin 21.2 \right)$$

$$= 37595200 \times 0.968 \times 1.042 \text{ J/m}^2$$

$$= 37.92 \text{ MJ/m}^2$$

Solar radiation available on the earth's surface

Attenuation of solar radiation by the atmosphere

All atmospheric components contribute to the attenuation of direct solar radiation on its path to the earth's surface. The attenuation of the direct flux of radiation takes place due to absorption and scattering, which simultaneously affect all parts of the solar spectrum. In the upper layers of the atmosphere, the main processes are the absorption in the x-ray and ultra-violet regions of the solar spectrum and scattering in the violet and blue ranges. As the radiation penetrates to the lower layers of the atmosphere, the attenuation affects the longer wavelength portions of the solar radiation.

Solar radiation is scattered not only by gas molecules and water vapour, but also by solid aerosol particles, liquid drop components (cloud particles and fog), and ice crystals. The scattering process in the atmosphere results in the production of scattered radiation, part of which goes back to the space. In the upper layers of the atmosphere, the absorption of solar radiation is caused by oxygen, ozone, and nitrogen oxides and in the lower stratosphere and troposphere, by water vapour, carbon dioxide, aerosols, and other minor components.

The scattering theory implies that large particles produce intensive scattering of the light falling through them. Thus, clouds, which consist of a large number of water drops or ice crystals, produce a marked attenuation of the direct solar radiation passing through them.

Albedo

Albedo is the fraction of the global radiation that is reflected by the receiving surface. Albedo characterizes the reflective property of a surface, object, or a whole system to which it is related. Thus, one can speak of the albedo of a desert, forest, glacier, cloud, atmosphere, sea, continent, hemisphere, or the planet as a whole.

The reflectivity of any natural underlying surface can be measured for the global solar radiation flux or for its individual spectral regions. In the latter case, one measures the spectral albedo. The albedo of global solar radiation is determined from the following relationship.

$$\rho = \frac{\int_0^\infty \rho_{\lambda,m} \left(G_{b,\lambda,m} + G_{d,\lambda,m} \right) d\lambda}{\left(G_b + G_d \right)_m} \qquad \qquad ...(4.12)$$

where $\rho_{\lambda,m}$ is the spectral albedo; $(G_b + G_d)_m$ is the global radiation flux (the sum of the direct radiation on a horizontal surface, G_b, plus the diffuse radiation, G_d) falling on a horizontal surface through an atmospheric mass, m. The upward radiation flux contains reflected and diffuse components. For each type of surface, there is a characteristic relationship between these components. The quantity of albedo produced by the solar radiation flux depends, in differing degrees, on the height of the sun at the time of measurement.

Beam and diffuse components of hourly and daily radiation

Clearness index

The clearness index is the ratio of average radiation on a horizontal surface to the average extraterrestrial radiation. The monthly average clearness index is the ratio of the monthly average daily radiation on a horizontal surface to extraterrestrial radiation.

$$\overline{K}_T = \frac{\overline{H}}{\overline{H}_o}$$

The daily clearness index K_T is defined as ratio of daily radiation to the extraterrestrial radiation.

$$K_T = \frac{H}{H_o}$$

Similarly, the hourly clearness index is defined as

$$k_T = \frac{I}{I_o}$$

H_o and I_o are given by Equations 4.9 and 4.10.

The clearness index is location-specific. However, it has been observed by Liu and Jordan that locations with same value of $\overline{K}_T$ have similar values of K_T.

Beam and diffuse components of hourly radiation

Calculation of total radiation incident on differently oriented surfaces using data on a horizontal surface requires that the beam and diffuse components of the radiation be treated separately. Furthermore, estimates of beam radiation are essential for evaluating the performance of concentrating collectors.

The diffuse fraction I_d/I, the fraction of the hourly radiation on a horizontal plane that is diffuse, is correlated with the hourly clearness index k_T. Available data shows that the correlation of (I_d/I) with k_T follows the curve shown in Figure 4.8. This curve can be represented by correlations such as the one shown below. Some more correlations will be given in a later section.

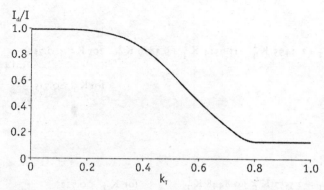

Figure 4.8 The ratio I_d/I as a function of the hourly clearness index k_T.
Reproduced with permission from John Wiley & Sons (Asia) Pvt. Ltd
Source Duffie and Beckman (1991)

$$\frac{I_d}{I} = \begin{bmatrix} 1.0 - 0.09k_T & \text{for } k_T \leq 0.22 & ...(4.13a) \\ 0.9511 - 0.1604\,k_T + 4.388\,k_T^2 - 16.638\,k_T^3 + 12.336\,k_T^4 & \text{for } 0.22 < k_T \leq 0.80 & ...(4.13b) \\ 0.165 & \text{for } k_T > 0.80 & ...(4.13c) \end{bmatrix}$$

Beam and diffuse components of daily radiation

Many studies have established that the average diffuse fraction H_d/H is a function of the day's clearness index K_T.

This correlation for different seasons (expressed in terms of the sunset hour angle ω_s) is as follows (Figure 4.9).

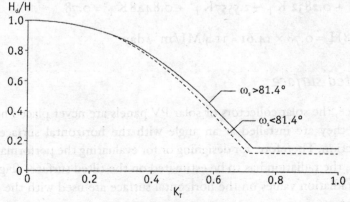

Figure 4.9 Suggested correlation of daily diffuse fraction with K_T
Reproduced with permission from John Wiley & Sons (Asia) Pvt. Ltd
Source Duffie and Beckman (1991)

For $\omega_s < 81.4°$

$$\frac{H_d}{H} = \begin{cases} 1.0 - 0.2727\, K_T + 2.4495\, K_T^2 - 11.9514\, K_T^3 + 9.3879\, K_T^4 & \text{for } K_T < 0.715 \\ 0.143 & \text{for } K_T \geq 0.715 \end{cases}$$ ...(4.14a)

For $\omega_s > 81.4°$

$$\frac{H_d}{H} = \begin{cases} 1.0 + 0.2832\, K_T - 2.5557\, K_T^2 + 0.8448\, K_T^3 & \text{for } K_T < 0.722 \\ 0.175 & \text{for } K_T \geq 0.722 \end{cases}$$ ...(4.14b)

Example 4
Estimate the diffuse radiation on a horizontal surface on 17 July at Mumbai, the global radiation being 14.61 MJ/m²/day.

Solution
From Example 3, the daily total extraterrestrial radiation
$H_o = 37.92$ MJ/m²
Sunset hour angle $\omega_s = 82.29°$

Thus, $K_T = \dfrac{H}{H_o} = \dfrac{14.61}{37.92} = 0.38$

using Equation 4.14b.

$$\frac{H_d}{H} = 1.0 + 0.2832\, K_T - 2.5557\, K_T^2 + 0.8448\, K_T^3 = 0.78$$

$$H_d = 0.78\,H = 0.78 \times 14.61 = 11.4 \text{ MJ/m}^2/\text{day}$$

Radiation on a tilted surface

In actual practice, the solar collectors or solar PV panels are never placed horizontally, rather they are installed at an angle with the horizontal surface to maximize collection. Therefore, for designing or for evaluating the performance of solar systems, the radiation has to be estimated on the tilted surfaces (Figure 4.10). For this, radiation values on the horizontal surface are used with the following expression.

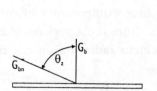 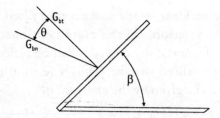

Figure 4.10 Beam radiation on horizontal and tilted surfaces
Reproduced with permission from John Wiley & Sons (Asia) Pvt. Ltd
Source Duffie and Beckman (1991)

$$R_b = \frac{G_{b,t}}{G_b} = \frac{G_{b,n} \cos\theta}{G_{b,n} \cos\theta_z} = \frac{\cos\theta}{\cos\theta_z} \qquad \text{...(4.15)}$$

where $\cos\theta$ and $\cos\theta_z$ are determined from Equation 4.5

In the northern hemisphere, the optimum azimuth angle γ for flat plate collectors is usually 0° (and in the northern hemisphere $\gamma = 180°$). Thus, in the northern hemisphere, R_b can be calculated from the following expression

$$R_b = \frac{\cos(\phi-\beta)\cos\delta\cos\omega + \sin(\phi-\beta)\sin\delta}{\cos\phi\cos\delta\cos\omega + \sin\phi\sin\delta} \qquad \text{...(4.16)}$$

and at solar noon, noon $R_{b,\,noon}$ as

$$R_{b,noon} = \frac{\cos|(\phi-\beta)-\delta|}{\cos|\phi-\delta|} \qquad \text{...(4.17)}$$

For a plane rotated continuously about a horizontal east–west axis (to maximize the incident beam radiation) the ratio of beam radiation on the plane to that on a horizontal surface is given as

$$R_b = \frac{\left(1 - \cos^2\delta\sin^2\omega\right)^{1/2}}{\cos\phi\cos\delta\cos\omega + \sin\phi\sin\delta} \qquad \text{...(4.18)}$$

In order to calculate radiation on a tilted surface when only the total radiation on a horizontal surface is known, one needs directions from where the beam and diffuse components reach the tilted surface. The direction of the incoming diffuse radiation (or its distribution over the sky dome) depends upon the cloudiness and atmospheric clarity. Essentially, the diffuse radiation is composed of the following three parts.

1 An isotropic part, received uniformly from all of the sky dome.
2 A circumsolar diffuse part that is a result of the forward scattering of solar radiation; it is concentrated in the part of the sky around the sun.
3 Horizon brightening, which is most evident in clear skies and is concentrated near the horizon.

The incident solar radiation on a tilted surface, therefore, is made up of (1) beam radiation, (2) the aforementioned three components of diffuse radiation from the sky, and (3) radiation reflected from those surfaces that are seen by the tilted surface. Thus, the total incident radiation on the tilted surface can be given by the expression

$$I_T = I_{T,b} + I_{T,d,iso} + I_{T,d,cs} + I_{T,d,hz} + I_{T,refl} \quad \quad ...(4.19)$$

where iso, cs, hz, and refl denote isotropic, circumsolar, horizon, and reflected radiation, respectively. The relation of total radiation on the tilted surface to that on the horizontal surface R is therefore

$$R = \frac{\text{Total radiation on the tilted surface}}{\text{Total radiation on a horizontal surface}} = \frac{I_T}{I} \quad \quad ...(4.20)$$

Example 5

Calculate the ratio of beam radiation on a horizontal surface to the titled surface at Mumbai at 11:00 a.m. (solar time) on 17 July; if the surface is tilted at an angle of 30° to the horizontal and is pointing 10° east of south.

Solution

From Example 2

$\cos \theta = 0.812$, $\phi = 19.11°$, $\delta = 21.2°$, and $\omega = -15°$

Since $\cos \theta_z = \cos \phi \cos \delta \cos \omega + \sin \phi \sin \delta$

$= 0.94 \times 0.93 \times 0.96 + 0.32 \times 0.36$

$= 0.851 + 0.118 = 0.969$

Thus, $R = \dfrac{0.812}{0.969} = 0.838$

Isotropic diffuse model

In this model, derived by Liu and Jordan, it is assumed that the total radiation on the tilted surface comprises (1) beam radiation, (2) isotropic diffuse radiation, and (3) diffusively reflected radiation from the ground. If a surface is at an angle β from the horizontal, then

$$I_T = I_b R_b + I_d \left(\frac{1+\cos\beta}{2} \right) + I\rho_g \left(\frac{1-\cos\beta}{2} \right) \quad \quad ...(4.21)$$

$$\text{and } R = \frac{I_b}{I} R_b + \frac{I_d}{I} \left(\frac{1+\cos\beta}{2} \right) + \rho_g \left(\frac{1-\cos\beta}{2} \right) \quad \quad ...(4.22)$$

where ρ_g is the ground reflectance, taken as 0.2 when there is no snow and 0.7 when there is a snow cover.

Anisotropic diffuse model

While the isotropic diffuse model is easy to use, it tends to underestimate I_T. Therefore, improved models were developed that also include the circumsolar diffuse and horizon-brightening components on a tilted surface (Figure 4.11).

As per the HDKR (Hay, Davies, Klucher, and Reindl) model, the total radiation on the tilted surface is expressed as

$$I_T = \left(I_b + I_d A_i\right) R_b + I_d \left(1 - A_i\right)\left(\frac{1+\cos\beta}{2}\right)\left[1 + f\sin^3\left(\frac{\beta}{2}\right)\right] + I\rho_g\left(\frac{1-\cos\beta}{2}\right) \quad ...(4.23)$$

Example 6

Estimate beam, diffuse, and reflected components of solar radiation on a surface with tilt angle 30° facing south at New Delhi (28° N), for the hour starting 11:00 a.m. (solar time) on 17 April using the isotropic model. The following values are given.

I = 3.06 MJ/m²

ρ_g = 0.20

ϕ = 28

δ = 10.14

n = 107

Solution

$$I_o = \frac{12 \times 3600}{\pi} 1367\left(1 + 0.033\cos(105.5)\right)$$

$$\left[\cos 28 \cos 10.14\left(\sin(0) - \sin(-15)\right) + \frac{\pi(0 - (-15))}{180}\sin 28 \sin 10.14\right]$$

$$I_o = 18797599\,(0.991) \times [0.866\,(0.25) + 0.26 \times 0.082]$$

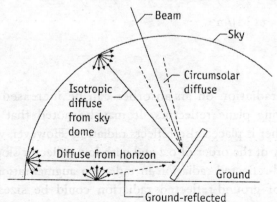

Figure 4.11 Beam, diffuse, and ground-reflected radiation on a tilted surface
Reproduced with permission from John Wiley & Sons (Asia) Pvt. Ltd
Source Duffie and Beckman (1991)

$$= 4.30 \text{ MJ/m}^2$$

$$k_T = \frac{I}{I_o} = \frac{3.06}{4.30} = 0.71$$

From Equation 4.13b

$$\frac{I_d}{I} = 0.9511 - 0.1604 \, k_T + 4.388 \, k_T^2 - 16.638 \, k_T^3 + 12.336 \, k_T^4 = 0.23$$

$$I_d = 0.23 \times 3.06 = 0.70$$
$$I_b = 0.77 \times 3.06 = 2.35$$

From Equation 4.16

$$R_b = \frac{\cos(\varphi - \beta)\cos\delta\cos\omega + \sin(\varphi - \beta)\sin\delta}{\cos\varphi\cos\delta\cos\omega + \sin\varphi\sin\delta}$$

$$= \frac{\cos(-2)\cos10.14\cos(-7.5) + \sin(-2)\sin10.14}{\cos28\cos10.14\cos(-7.5) + \sin28\sin10.14}$$

$$= \frac{0.97 - 0.0061}{0.86 + 0.08}$$

$$= \frac{0.964}{0.94} = 1.025$$

From Equation 4.21

$$I_T = I_b R_b + I_d \left(\frac{1 + \cos\beta}{2}\right) + I\rho_g \left(\frac{1 - \cos\beta}{2}\right)$$

$$= 2.35 \times 1.025 + 0.70 \left(\frac{1 + \cos30}{2}\right) + 3.06 \times 0.2 \left(\frac{1 - \cos30}{2}\right)$$

$$= 2.41 + 0.65 + 0.04 = 3.1 \text{ MJ/m}^2$$

Radiation augmentation

The amount of incident radiation on an absorber can be increased (or supplemented) by employing plane reflectors. It may be noted that the ground on which the absorber is placed also reflects radiation. However, with ground reflectance normally of the order of 0.2 and with low collector slopes, the contribution of ground-reflected radiation in radiation augmentation is small. The contribution of ground-reflected radiation could be sizeable when it is snow-covered or collector slopes are high. In such cases, ground reflectance is of the order of 0.6–0.7.

For a more general case, consider the geometry as shown in Figure 4.12. It essentially consists of two intersecting planes, namely, (1) the receiving surface C (for example, a solar collector) and (2) a diffuse reflector R. The angle between these planes is Ψ. If the reflector is horizontal, angle Ψ equals 180 − β where β is slope of the tilted surface from the horizontal.

The incident radiation on surface C at anytime will be the sum of the (a) beam component of the radiation, (b) diffuse component from the sky-portion seen by the surface, (c) diffusively reflected radiation from surface R, and (d) ground-reflected radiation from the ground outside R.

The radiation reflected onto surface C from surface R is the product of the total radiation incident on R, its reflectance, and view factor F_{R-C}. Figure 4.13 shows values of F_{R-C} as a function of c/b and r/b for Ψ = 90°, 120°, and 150°.

If the surfaces C and R are very long (as in the case of a long array of collectors), the view factor is given by Hottel's method.

$$F_{R-C} = \frac{c + r - s}{2r} \qquad \qquad ...(4.24)$$

where s is the distance from the upper edge of the collector to the outer edge of the reflector. This can be calculated from the following equation.

$$s = \left[c^2 + r^2 - 2cr\cos\psi \right]^{1/2} \qquad \qquad ...(4.25)$$

It is essential to know the radiation incident on the reflector. The beam component is calculated by using R_b for the orientation of the reflector surface. The diffuse component is estimated from the view factor F_{R-S}.

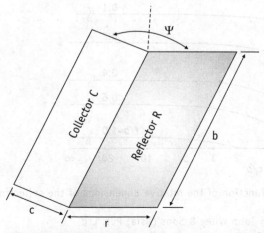

Figure 4.12 Geometric relationship of receiving and reflecting surfaces

For any orientation of surface R

$$F_{R-S} + F_{R-C} + F_{R-G} = 1 \qquad \qquad ...(4.26)$$

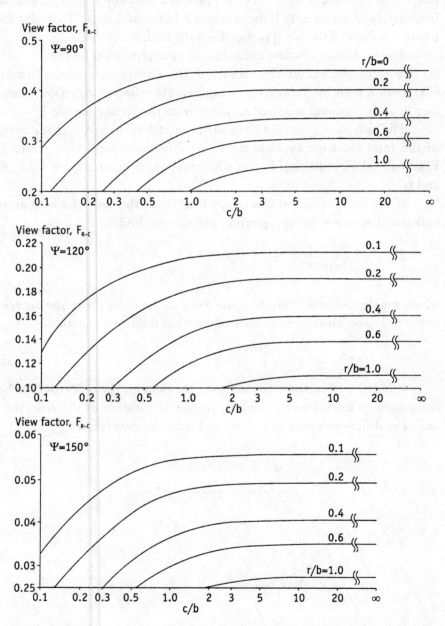

Figure 4.13 Values of F_{R-C} as a function of the relative dimensions of the collecting and reflecting surfaces

Reproduced with permission from John Wiley & Sons (Asia) Pvt. Ltd

Source Duffie and Beckman (1991)

where F_{R-S} is the view factor from surface R to sky, F_{R-C} is the view factor from surface R to the collector surface C, and F_{R-G} is view factor from surface R to the ground.

For a horizontal reflector, F_{R-G} is zero whereas for a long array of collectors, it will be small. Thus, Equation 4.26 can be re-written as

$$F_{R-S} = (1 - F_{R-C})$$

...(4.27)

where F_{R-C} is estimated either from Figure 4.13 or from Equation 4.24.

The average angle of incidence of reflected radiation may be taken as that of the radiation from the midpoint of surface R to the midpoint of surface C as shown in Figure 4.14.

The angle of incidence θ_r of radiation reflected from surface R onto surface C is given by the following equation.

$$\sin\theta_r = \frac{r \sin\psi}{s}$$

...(4.28)

The total reflected radiation from surface R (having an area A_R) to surface C (having an area A_C), assuming that R has a diffuse reflectance of ρ is given by

$$A_C\, I_r = [I_b R_b + (1 - F_{R-C})I_d]\,\rho A_R F_{R-C}$$

...(4.29)

where I_b is beam radiation, I_d is diffuse radiation, and I_r is reflected radiation.

Radiation on moving surfaces

Angles for moving surfaces

In case of some solar collectors, particularly the concentrating ones, the collector surface moves or 'tracks' the sun so as to maximize the quantum of

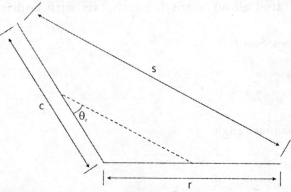

Figure 4.14 Approximate angle of incidence of reflected radiation θ_r

incident beam radiation. This is achieved by minimizing the angle of incidence of beam radiation on the collector surface.

The tracking can be of the following two types.

1 Rotation about a single axis (that is, horizontal east–west; horizontal north–south; vertical or parallel to the earth's axis)

2 Rotation about the two axes

(a) For a plane rotated about a horizontal east–west axis, with a single adjustment daily to ensure that the beam radiation is normal to the surface at noon, the angle of incidence is given by

$$\cos \theta = \sin^2 \delta + \cos^2 \delta \cos \omega \qquad \qquad ...(4.30a)$$

The slope of this surface will be

$$\beta = |\phi - \delta| \qquad \qquad ...(4.30b)$$

and the surface azimuth angle

if $(\phi - \delta) > 0, \gamma = 0^\circ$

if $(\phi - \delta) < 0, \gamma = 180^\circ$ $\qquad \qquad ...(4.30c)$

(b) For a plane rotated about a horizontal east–west axis with continuous adjustment

$$\cos \theta = (1 - \cos^2 \delta \sin^2 \omega)^{1/2} \qquad \qquad ...(4.31a)$$

and slope of the surface will be

$$\tan \beta = \tan \theta_z |\cos \gamma_s| \qquad \qquad ...(4.31b)$$

and

if $|\gamma_s| < 90, \gamma = 0^\circ$

if $|\gamma_s| > 90, \gamma = 180^\circ$. $\qquad \qquad ...(4.31c)$

(c) For a plane rotated about a north–south axis with continuous adjustment

$$\cos \theta = (\cos^2 \theta_z + \cos^2 \delta \sin^2 \omega)^{1/2} \qquad \qquad ...(4.32a)$$

The slope will be

$$\tan \beta = \tan \theta_z |\cos (\gamma - \gamma_s)| \qquad \qquad ...(4.32b)$$

and the surface azimuth angle

if $\gamma_s > 0, \gamma = 90^\circ$

if $\gamma_s < 0, \gamma = -90^\circ$ $\qquad \qquad ...(4.32c)$

(d) For a plane with a fixed slope rotated about a vertical axis, the angle of incidence

$$\cos \theta = \cos \theta_z \cos \beta + \sin \theta_z \sin \beta \qquad ...(4.33a)$$

The slope is fixed, that is

$$\beta = constant \qquad ...(4.33b)$$

The surface azimuth angle

$$\gamma = \gamma_s \qquad ...(4.33c)$$

(e) For a plane rotated about a north–south axis parallel to the earth's axis with continuous adjustment

$$\cos \theta = \cos \delta \qquad ...(4.34a)$$

and the slope varies continuously, that is

$$\tan\beta = \frac{\tan\phi}{\cos\gamma} \qquad ...(4.34b)$$

The surface azimuth angle

$$\gamma = \tan^{-1}\frac{\sin\theta_z \sin\gamma_s}{\cos\theta' \sin\phi} + 180C_1C_2 \qquad ...(4.34c)$$

where $\cos\theta' = \cos\theta_z \cos\phi + \sin\theta_z \sin\phi \qquad ...(4.34d)$

$$C_1 = \begin{bmatrix} 0 \ if \left(\tan^{-1}\frac{\sin\theta_z \sin\gamma_s}{\cos\theta'\sin\phi} \right) + \gamma_s = 0 \\ \\ 1 \ otherwise \end{bmatrix} \qquad ...(4.34e)$$

$$C_2 = \begin{bmatrix} 1 \ if \ \gamma_s \geq 0 \\ \\ -1 \ if \ \gamma_s < 0 \end{bmatrix} \qquad ...(4.34f)$$

(f) For a plane continuously tracked about the two axes

$$\cos \theta = 1 \qquad ...(4.35a)$$
$$\beta = \alpha \qquad ...(4.35b)$$
$$\gamma = \gamma_s \qquad ...(4.35c)$$

Beam radiation on concentrators

Most concentrating collectors, which track the sun, utilize only beam radiation. At any given time, the beam radiation on a surface is given by

$$G_{bT} = G_{bn} \cos \theta \qquad ...(4.36)$$

where G_{bn} is beam radiation on a plane normal to the incident radiation. The angle cos θ can be calculated using Equations 4.30a, 4.31a, 4.32a, 4.33a, 4.34a, or 4.35a, as appropriate.

It may be noted that the uncertainties associated in the estimation of beam radiation are greater as compared to the estimation of total radiation. Therefore, it is preferable to use measured data from a pyrheliometer as far as possible.

Utilizability

The solar radiation availability varies, as explained in previous sections, due to various reasons. In case of solar collectors, to collect useful energy, the solar radiation received should be higher than critical radiation. Critical radiation is defined as the radiation received by an absorber (after all optical losses) which just balances the heat lost by it at a given operating temperature. Thus, if the incident radiation is more than the critical radiation I_{TC}, only then there will be net energy generation from the collector. For a given hour, utilizable radiation would be $(I_T - I_{TC})$. The utilizability thus could be estimated as

$$\phi_h = \frac{(I_T - I_{TC})^+}{I_T} \qquad \qquad ...(4.37)$$

The + sign indicates that only the positive values are to be used, and explained graphically in Figure 4.15, which shows the radiation available on a tilted surface during the day. The ratio of the shaded area to the total area under the curve for a given hour (11:00 a.m.–12 noon in this case) gives the hourly utilizability ϕ_h. I_T is the total area under the curve between 11:00 a.m.–12 noon and I_{TC} is the shaded area (or the area above G_{TC}, the critical radiation level).

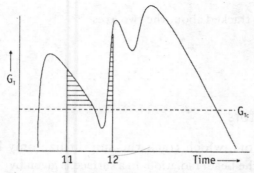

Figure 4.15 Radiation on a tilted surface during the day

The monthly average utilizability can be estimated using Equation 4.37 as

$$\phi = \frac{1}{N}\sum_{}^{N}\frac{(I_T - I_{TC})^+}{\bar{I}_T} \qquad\qquad ...(4.38)$$

For a given collector, the monthly average utilizable energy for that hour can be estimated from the monthly utilizability as $\bar{I}_T\phi$, if the critical radiation level is known. With the availability of monthly average radiation data, the utilizability ϕ can be estimated as explained below.

A plot of $I_T/\bar{I}_T$ for New Delhi, for the month of April and for the hour ending at 11:00 a.m. (solar time) is shown in Figure 4.16. The x-axis is fraction of time f for which radiation is less than I_T.

It can be inferred from Figure 4.16 that for the month of April, the hourly radiations for the hour ending 11:00 a.m. are quite identical and have very less variation.

The critical radiation ratio is defined as the ratio of critical radiation to monthly average radiation.

$$X_c = \frac{(I_{TC})}{\bar{I}_T}$$

In Figure 4.16, for $X_c = 0.8$, only for 7.5% of the time the radiation is below I_{TC}. The shaded area gives the utilizability for that location, collector, and hour in that month.

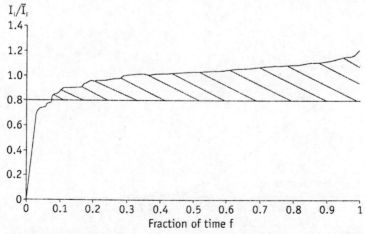

Figure 4.16 Distribution of X_c on a south-facing surface in New Delhi

The critical radiation is a function of collector properties. Thus, a collector with low heat loss, that is with lower level of critical radiation, will collect higher utilizable energy.

Thus, as the critical radiation ratio varies, the utilizability varies. Figure 4.17 is derived from Figure 4.16 by changing the value of $I_T/\bar{I}_T$.

It has been established that the utilizability ϕ, even though derived for one hour during the month, is same for all the hours.

Figure 4.17 also indicates that the days are quite identical in the month of April in Delhi, as the cumulative distribution curve is very close to the identical days line.

Measurement of solar radiation

Instruments
Pyranometer

The pyranometer is used to measure the total or global solar radiation (Figure 4.18) and when provided with a shading ring, measures the diffuse radiation (Figure 4.19). The instrument generally measures the solar radiation in the wavelength range of about 0.13–3 μm and consists of the following basic components

- Detector or sensor protected by glass dome(s)
- Instrument body with a spirit level, adjustable levelling screws, and a desiccant chamber

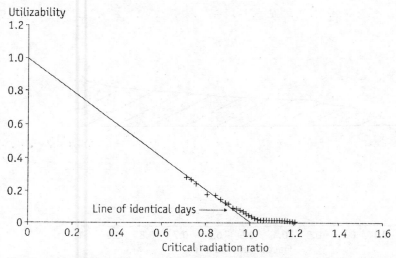

Figure 4.17 Utilizability curve for New Delhi (estimated from Figure 4.16)

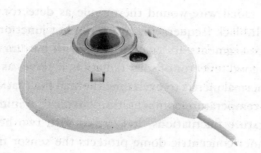

Figure 4.18 Pyranometer
Published with permission from Kipp and Zonen B.V. <www.kippzonen.com>

- Radiation shield to protect the instrument case from direct sunlight
- Electrical connector for the output signal

There are two main types of solar radiation sensors (1) photo-sensors and (2) thermal sensors. Photo-sensors, represented by silicon PV cells, are the most popular and easy-to-use devices. However, since these sensors respond selectively with respect to the wavelength of incident radiation, their use is limited to the measurement of special wavelengths or short spectral bands. Therefore, for precise measurements over the entire range of wavelengths, thermal sensors (that convert the heating effect of solar radiation into electrical signals) are used. Thermal sensors include thermopiles and thermistor-bolometers.

The commonly used pyranometer works on the thermoelectric principle, where thermocouples are arranged in series (thermopile). The black

Figure 4.19 Pyranometer with shading ring
Published with permission from Kipp and Zonen B.V. <www.kippzonen.com>

and white pyranometer has a radial wire-wound thermopile as detector. Its three segments are coated with black lacquer; this forms the hot junction of the thermopile. Its other three segments are coated white with barium sulphate. As the white colour is a good reflector of solar radiation, it serves as the cold junction and has a large thermal inertia to overcome thermal fluctuations. The instrument has a built-in temperature compensation circuit to minimize the effect of ambient temperature fluctuations. It is fitted with two hemispherical domes, where the inner concentric dome protects the sensor from transient convection effects. The case is painted white with barium sulphate. The factors to be considered to obtain reliable measurements include sensitivity (5 mV/cal.cm^{-2}min^{-1}), response changes due to ambient temperature fluctuations (±0.5% from –20 to 40 °C), and response time (1 second). Pyranometers must be calibrated periodically by using either a pyrheliometer or a calibrated pyranometer acting as a reference.

There are bimetallic pyranometers, also known as pyranographs, that work on the thermo-mechanical principle. In this type of pyranometer, the movement of the free end of the bimetallic strip is magnified by mechanical linkages that are recorded by means of a moving arm and pen on a drum. Since the drum is powered by a spring, the pyranograph is suitable as a stand-alone instrument in remote sites. While bimetallic pyranometers are popular on account of their simplicity, they are sensitive to the level of solar flux as well as the ambient temperature.

Shaded pyranometer

The shaded pyranometer is used to measure diffuse solar radiation. It needs a shading device so that the sensor cannot 'see' the sun. This ensures that the radiation coming directly from the sun is not measured.

Pyranometer with shading ring

This is an inexpensive device for diffuse solar radiation measurement. However, the measurements require correction because the shading ring partially obstructs the passage of diffuse radiation (Figure 4.19).

Pyranometer with shading disc

In the case of a pyranometer with a shading disc (for diffuse solar radiation measurement), the measured data does not require correction because the disc obstructs only the path of direct radiation. On the other hand, it requires a sun-tracking device that can be of one or two axes.

Figure 4.20 Pyranometer with shading disc
Published with permission from Kipp and Zonen B.V. <www.kippzonen.com>

Figure 4.20 shows a pyranometer with a shading disc.

Pyrheliometer

This equipment is used to measure direct (or beam) solar radiation. Pyrheliometers must follow the sun to measure only direct sunlight and to avoid the diffuse component. This is done by using a collimator tube over the sun and tracking the sun continuously. The Angstrom electrical compensation pyrheliometer is one of the most accurate and convenient instruments used in the measurement of direct solar radiation. It consists of two strips of manganin foil coated on one side with Parsons black lacquer. The strips are mounted side by side and a thermo-junction is attached to the back side of each strip. This is placed in a holder, which is mounted at the base of a cylindrical metal tube (collimator). The collimating tube, blackened from inside, has a 5°43'30" field of view. The tube is filled with dry air at atmospheric pressure and its viewing end is sealed by a removable, 1 mm thick, crystal quartz window. An azimuth elevation mechanism is provided to help in directing the cylindrical tube at the sun. A reversible shutter at the front end of the tube permits one of the strips to be shaded from the sun, while the other is exposed to direct solar radiation (Figure 4.21).

During operation, the exposed strip is heated by the sun's direct radiation and its temperature rises. The shaded strip is heated electrically to the same temperature as the exposed strip so that the rate of energy absorbed by the exposed strip is thermoelectrically compensated by the rate

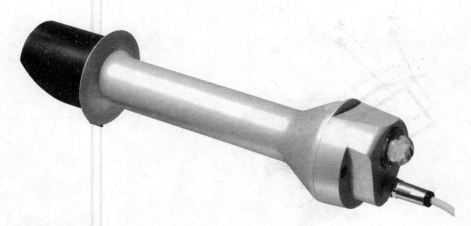

Figure 4.21 Pyrheliometer
Published with permission from Kipp and Zonen B.V. <www.kippzonen.com>

of electrical energy supplied to the shaded strip. The equivalence of the temperatures of the strips is determined by the thermo-junctions attached to the reverse sides of the strips by means of a sensitive null detector.

Pyrheliometers are classified by the WMO (World Meteorological Organization) according to their accuracy (as well as the accuracy of their auxiliary equipment). The accuracy is judged against the criteria of sensitivity, stability of the correction factor, maximum error due to variations in ambient temperature, errors due to spectral response of the receiver, non-linearity of response, opening angle, time constant of the system, and effect of auxiliary equipment. Thus, pyrheliometers are classified in the following manner.

- Absolute pyrheliometer
- Reference-standard pyrheliometer
- First-class pyrheliometer
- Second-class pyrheliometer

The World Radiation Centre, Davos, maintains absolute pyrheliometers against which reference standard pyrheliometers are calibrated. These reference standard pyrheliometers are then used to calibrate first-class and second-class pyrheliometers (Iqbal 1983).

Campbell-Stokes sunshine recorder

This equipment only measures the number of insolation hours. The sunshine duration is defined as the amount of time the sun's disc is not obscured by clouds and the quantity measured by a sunshine recorder is the amount of time, in which direct solar radiation has sufficient intensity to activate the recorder (Figure 4.22).

Figure 4.22 Sunshine recorder

The Campbell-Stokes sunshine recorder uses the heat of the direct solar radiation to activate the instrument. It consists of a polished solid glass sphere about 10 cm in diameter with the axis mounted parallel to the earth's north–south axis. The sphere is supported in a spherical bowl so that the sun's image is brought to focus on a chemically treated thin card held in a groove inside the bowl. The bowl is part of the spherical shell and surrounds the lower half of the focusing sphere. The sphere acts as a lens and the focused image moves along the card bearing a time scale, and the bright sunshine burns a path along this paper. The bowl is cut in such a manner that at all times of the day, the sun's image can fall on some part of the inner surface containing the card. The card, printed with hour lines and placed in an east–west direction, allows the duration and exact time of sunshine to be measured. To accommodate the sun's movement north and south (effect of declination), three types of cards are used: long curved for summer, short for winter, and straight for spring and autumn.

Solar radiation passing through the glass sphere is focused on the card, and a brown line is burnt on the paper whenever the incident radiation is above a certain minimum threshold level. If the radiation is below this level, the burnt line is interrupted by an unburnt length. The number of hours of the bright sunshine during a day is then determined by measuring the total length of the burnt segments.

Careful adjustments must be made with reference to the levelling of the base and the north–south orientation of the recorder. The recorder is not sensitive to weak sunlight when the sun is within 5° of the horizon (early mornings and late evenings) and corrections have to be made. Similarly, humidity affects the card's ability to establish a fixed threshold level. In moist conditions, the burning may not commence until about 280 W/m², whereas in a dry climate, it may begin at 70 W/m². However, the current standard for sunshine duration is the period of time during which a threshold of 120 W/m² is exceeded.

Albedometer

The albedometer is a device that measures both the global solar radiation as well as the reflected radiation. Albedo is a measurement that quantifies how much the medium (vegetation, buildings, mountains, soil, snow, etc.) reflects the solar radiation falling onto it. Its value is calculated by dividing the total solar radiation reflected by the medium, by the global solar radiation incident at the location. To ensure more precise measurement, the albedometer is generally installed several metres above the surface (Figure 4.23).

Accuracy of meteorological measurements

Meteorological measurements made at a site are mainly intended to determine the climatic conditions likely to affect the operation of a solar system to be set up at a later date on that site. Its exact position is, of course, not known and even within a well-determined microclimate, there may be some variability in the meteorological conditions, both spatially and temporally.

In sum, the process usually adopted is as follows.

1 Measurements are taken for a sufficiently long period, preferably several years.
2 These measurements are corrected with the corresponding climatic data of the nearest meteorological stations for which a long series of observations are available.
3 The most probable values of the usual statistics to be taken for the site under study (averages, extremes, standard deviations, etc.) are deduced.

For all meteorological parameters, the WMO specifications and the instruments as per Meteorological Department Standards should be employed as far as possible. Besides, arrangements for regular checks and

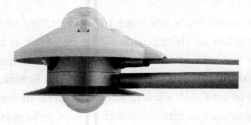

Figure 4.23 Albedometer
Published with permission from Kipp and Zonen B.V. <www.kippzonen.com>

calibrations, and also maintenance, should be made in accordance with the operational instructions.

Quality control of measurements

In order to maintain a high standard in the quality of measurements, checks should be made for all elements of the measuring process. In the most usual case of measuring global radiation, it is thus appropriate to carry out the following checks.

- Pyranometer
  - Adjustments and initial calibration by the national centre
  - Setting up the instruments in accordance with clearly defined standards
  - Day-to-day maintenance in accordance with instructions (daily cleaning, etc.)
  - Annual check of the calibration

- Associated recorder or integrator
  - Periodic checks of proper installation of electrical wiring
  - Regular checks of the calibration of the integrator or recorder
  - Continuous check of the associated electric power supply
  - Day-to-day maintenance of the instruments.

- Data collection
  - Using the calibration factor of the measuring unit used
  - Applying corrections, as necessary, to each measurement
  - Checking totals and means (10-day, monthly) and extreme values

Prediction of available solar radiation

Different models

Model for estimating surface solar radiation

As explained in the earlier section, while crossing the atmosphere, solar radiation undergoes interactions of spreading and absorption by the atmospheric components and the earth's surface. Figure 4.24 represents, in a very simplified fashion, the main interaction processes of solar and thermal radiation within the atmosphere–earth system. The values shown refer to the mean global effect of all processes for each component of the radioactive balance of the planet, and these may vary considerably from one region to the other and from one season to the next.

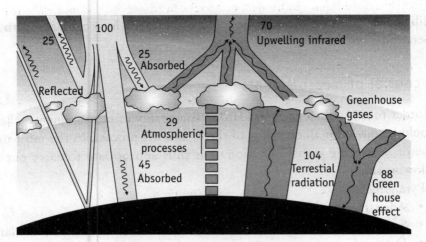

Figure 4.24 Flow diagram of the interaction processes of solar radiation with the earth's atmosphere.
Source Pereira, Banda, Martins, *et al.* (2002)

Clouds, atmospheric gases and particles, and the surface reflect about 30% of the incident solar radiation. The remaining 70% is absorbed, thereby heating the system and causing water evaporation or convection. The energy absorbed by the earth–atmosphere system rebounds into the infrared spectrum of electromagnetic radiation, of which six per cent comes from the surface and 64% comes from clouds and atmospheric components.

The radiation models are broadly classified into two categories.

- *Parametric models* These require detailed information about atmospheric conditions like cloud cover and turbidity coefficient, etc. The Iqbal model and the ASHRAE model fall under this category.
- *Decomposition models* These use information on global radiation to predict the beam and sky components. Some examples are the Angstrom; Hay; Liu and Jordan; and Orgill and Hollands models.

Methods used for estimating solar radiation

Three methods are used to derive solar radiation when no measurements are available.

- Estimating extraterrestrial radiation, allowing for its depletion by absorption and scattering by atmospheric gases, dust, aerosols, and clouds. This is a theoretical method and requires some approximation of the absorbing and scattering properties of the atmosphere.
- Assessing on the basis of other meteorological elements such as the duration of bright sunshine and cloudiness using regression techniques.

This is an empirical method in which the cloud field is defined by either the total cloud amount or the measured number of sunshine hours.

- Making use of geostationary satellites to estimate the solar radiation incident on the earth's surface from visible brightness measurements.

Estimation of global radiation

For calculating the global radiation on a horizontal surface, the model given by Angstrom in 1924 is used. The equation relates monthly average daily radiation to clear-day radiation at the location in question and the average fraction of possible sunshine hours.

$$\frac{\overline{H}}{\overline{H}_o} = a + b\frac{\overline{n}}{\overline{N}}$$

...(4.39)

where $\overline{H}$ is the monthly average daily radiation on a horizontal surface, $\overline{H}_o$ is the extraterrestrial radiation, $\overline{n}$ is the monthly average daily hours of bright sunshine $\overline{N}$ is the maximum daily hours of bright sunshine for the same period, and a, b are empirical constants.

Angstrom gave the original values of these empirical constants for different stations. Later on, various scientists worked on these values and gave more accurate and contemporary values.

For New Delhi, the values given by Angstrom are a = 0.256 and b = 0.454 and those given by Garg are a = 0.341 and b = 0.446.

Estimation of diffuse components

Monthly diffuse components

The monthly average radiation for the diffuse components is calculated using the models by Collares–Pereira–Rabl and Erbs.

The curves given by Collares–Pereira–Rabl are based on the summations of daily total and diffuse radiation. They found a seasonal dependence of the relationship which they expressed in terms of the sunset hour angle of the mean day of the month.

$$\frac{\overline{H}_d}{\overline{H}} = 0.775 + 0.00606(\overline{\omega}_s - 90) - [0.505 + 0.00455(\overline{\omega}_s - 90)]\cos(115\overline{K}_T - 103)$$

...(4.40)

Erbs developed a monthly average diffuse fraction correlation from the daily diffuse correlations. The equations are given as follows.

For $\bar{\omega}_s \leq 81.4$
and
$0.3 \leq \bar{K}_T \leq 0.8$

$$\frac{\bar{H}_d}{\bar{H}} = 1.391 - 3.560\bar{K}_T + 4.189\bar{K}_T^2 - 2.137\bar{K}_T^3 \qquad \qquad ...(4.41)$$

and for $\bar{\omega}_s > 81.4$
and
$0.3 \leq \bar{K}_T \leq 0.8$

$$\frac{\bar{H}_d}{\bar{H}} = 1.311 - 3.022\bar{K}_T + 3.427\bar{K}_T^2 - 1.821\bar{K}_T^3 \qquad \qquad ...(4.42)$$

Hourly diffuse components

The diffuse components of hourly radiation are calculated using the models given by Orgill and Hollands, Erbs, and Reindl.

The diffuse fraction was estimated by Orgill and Hollands using the clearness index k_T as the only variable. The correlation between the hourly diffuse fraction I_d/I and the clearness index k_T is given by

$$\frac{I_d}{I} = \begin{bmatrix} 1 - 0.249k_T & \text{for } k_T < 0.35 \\ 1.577 - 1.84k_T & \text{for } 0.35 < k_T < 0.75 \\ 0.177 & \text{for } k_T > 0.75 \end{bmatrix} \qquad ...(4.43)$$

Erbs *et al.* studied the same kind of correlation with data from five stations in US. The data was of short duration, ranging from one to four years. In each station, hourly values of normal direct irradiance and global irradiance on a horizontal surface were registered. Diffuse irradiance was obtained as the difference of these quantities. The relationships are given as follows.

$$\frac{I_d}{I} = \begin{bmatrix} 1 - 0.094k_T & \text{for } k_T < 0.22 \\ 0.9511 - 0.1604k_T + 4.388k_T^2 + 16.638k_T^3 + 12.336k_T^4 & \text{for } 0.22 < k_T < 0.8 \\ 0.165 & \text{for } k_T > 0.8 \end{bmatrix}$$

$$...(4.44)$$

which is same as Equation 4.13.

Reindl *et al.* estimated the diffuse fraction using two different models developed with measurements of global and diffuse irradiance on a horizontal surface registered at five locations in the US and the Europe. The model estimates the diffuse fraction using the clearness index as the input data.

$$\frac{I_d}{I} = \begin{cases} 1.02 - 0.248k_T & \text{for } k_T < 0.3 \\ 1.45 - 1.67k_T & \text{for } 0.3 < k_T < 0.78 \\ 0.147 & \text{for } k_T > 0.78 \end{cases} \quad \ldots(4.45)$$

where I_d is the hourly diffuse radiation, I is the hourly total global radiation, and k_T is the hourly clearness index.

Solar mapping using satellite data

Potential investments in the area of solar energy and its large-scale applications require reliable information about the availability of solar resources. The mapping of solar resources from satellite images lowers costs because of reduced dependence on ground stations. Besides, it ensures fewer uncertainties than would be the case when interpolating data from distantly located ground-measuring stations. Nonetheless, some ground stations are necessary to provide quality-measured data for (1) validation of these models and (2) improvement of the estimating models using satellite images.

Artificial satellites

The major obstacle in applying physical models is the collection of atmospheric data with the desired levels of precision and reliability. The development of the remote sensing technology through artificial satellites led to substantial progress in the development of computational models. Indeed, satellite images are tools of great value in obtaining the cloud coverage and other atmospheric parameters required for modelling radiative processes in the atmosphere. The initial studies for the utilization of data obtained from satellites – in estimating the solar radiation reaching the surface – used orbital satellite data. However, the cloud cover evaluation was very inaccurate because of their small time resolution (one image per day). Stationary satellites started being utilized at the beginning of the 1980s. The satellite data obtained with good time and space resolution (30 minutes and around 4 km at nadir) allowed a better evaluation of atmospheric parameters and consequently, a more accurate estimate of incident solar radiation.

Various satellites in orbit around the earth – namely polar satellites and geostationary satellites – are used for different purposes.

Polar satellites

Polar orbit satellites are located in a limited orbit (about 800 km) around the planet. They survey the earth's surface from one pole to the other. As the periods of their orbits are from one to two hours, these satellites are able to observe the earth at points that have the same geometry of illumination from the sun. Each satellite passes through the same observation point over the earth's surface once every 12 hours (once during the day and again during the night).

Geostationary satellites

These are positioned at a height of approximately 36 000 km above the earth. They rotate around the planet at the same speed at which the earth rotates about its axis, thus remaining stationary over a fixed point on the earth (normally over the equator). These satellites are very useful because they are able to observe almost one half of the planet from the same vantage point. They are thus ideal for communication and in remote sensing as they can observe the same points on earth repeatedly. In general, they are capable of providing images at a half-hourly resolution, making it possible to observe cloud movements and meso-scale phenomena such as hurricanes over the planet with great reliability. These satellites have sensors at various wavelengths (channels), allowing the detection of different characteristics of the atmosphere and the surface of the earth. The field of view of a sensor is narrow, to be able to observe the surface at resolutions better than 1 km. A representation of the satellite scanning is shown in Figure 4.25.

Several countries around the world make use of geostationary satellites, so that the whole planet may be observed at all times.

- Two US satellites cover the Eastern Pacific and the American continent— GOES-W over the Pacific and GOES-E over the Amazon basin.
- Two European Union satellites, at 0° latitude and 0° longitude, permit observation of the Atlantic Ocean, Africa, and Europe and cover the Indian Ocean.
- One Indian satellite covers the Indian and Pacific oceans.
- One Russian Federation satellite covers the Indian Ocean.
- One Japanese satellite covers the Eastern Pacific.

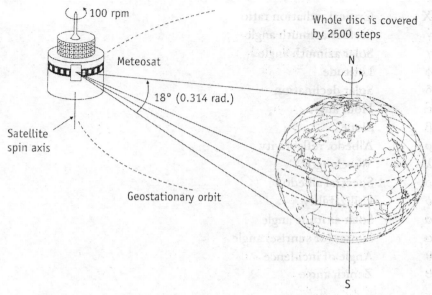

Figure 4.25 Representation of a satellite scanning system

Nomenclature

ET	Equation of time
G_o	Extraterrestrial solar irradiance in a horizontal surface, W/m²
G_{on}	Extraterrestrial radiation on the nth day, W/m²
G_{sc}	Solar constant, W/m²
H	Daily irradiance, MJ/m²
H_o	Daily extra terrestrial radiation on a horizontal surface, MJ/m²
I	Hourly irradiance, MJ/m²
K_T	Daily clearness index
k_T	Hourly clearness index
$\overline{K}_T$	Monthly average daily clearness index
$\overline{k}_T$	Monthly average hourly clearness index
l_{Local}	Local longitude
LST	Local standard time
l_{st}	Standard time meridian
m	Air mass
n	Day of the year
R	Ratio of total radiation on tilted plane to that on the plane of measurement
R_b	Ratio of beam radiation on a tilted plane to that on the plane of measurement

X_c	Critical radiation ratio
γ	Surface azimuth angle
γ_s	Solar azimuth angle
ϕ	Latitude
δ	Solar declination
ω	Hour angle
β	Slope
ρ	Albedo, reflectivity
λ	Wavelength
$\rho_{\lambda,m}$	Spectral albedo
ϕ_n	Utilizability
α_s	Solar altitude angle
ω_s	Sunset (or sunrise) angle
θ	Angle of incidence
θ_z	Zenith angle

References

Duffie J A and Beckman W A. 1991
Solar Engineering of Thermal Processes
New York: John Wiley and Sons

Faiman D. 2003
<http://physweb.bgu.ac.il/COURSES/EnvPhys/SolPhysLect2(2003).pdf
last accessed on 20 March 2007

Goswami D Y, Kreith F, and Kreider J F. 2000
Principles of Solar Engineering
Philadelphia: Taylor and Francis

Iqbal M. 1983
An Introduction to Solar Radiation
New York: Academic Press

Mani A and Rangarajan S. 1982
Solar Radiation Over India
New Delhi: Allied Publishers Pvt. Ltd

Pereira E B, Banda F Z, Martins F R, Abreu S L, Mantelli Netto S L. 2002
Short Training Course in Solar Radiation Assessment of SWERA
[Short Training Course. INPE, Recife - Brazil on 25 May 2002]

WMO (World Meteorological Organization). 1981
Meteorological Aspects of the Utilization of Solar Radiation as Energy Source
[Technical Note No. 172 (No. 557)]
Geneva: WMO

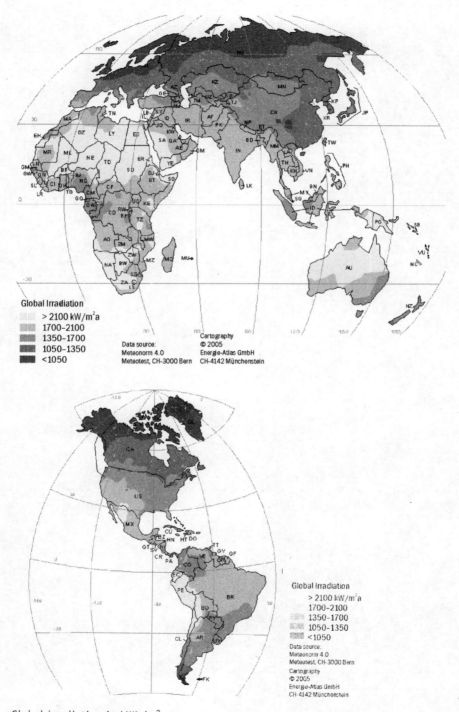

Global irradiation in kWh/m² a

Source Reproduced with permission of Energie-Atlas GmbH, Switzerland
<www.energie-atlas.ch>

5

Solar photovoltaic technology

Parimita Mohanty, Associate Fellow
Energy–Environment Technology Division, TERI, New Delhi

The solar PV (photovoltaic) technology is primarily a semiconductor-based technology used to convert solar radiation into direct electricity via the photo electric effect. A basic PV system comprises PV modules/solar cells and the BoS (balance of system) that includes support structure, wiring, storage, power electronics, etc.

Solar cell, or the PV cell is one of the most important components of PV systems and this chapter will focus on solar cell technology. The performance of various solar cells depends upon parameters such as type and structure of material, band gap energy of material, doping strength, absorption coefficient of material, etc. Hence the basic characteristic properties of different materials should be known in order to understand the behaviour of solar cells.

Physics of semiconductors

The following section will provide a brief description of semiconductors and their properties. However, any standard handbook of solid state physics and electronics should be referred to for a detailed description.

Types of solids

Solids can be differentiated on the basis of size of the individual crystal grains in the material and can be broadly classified as crystalline and

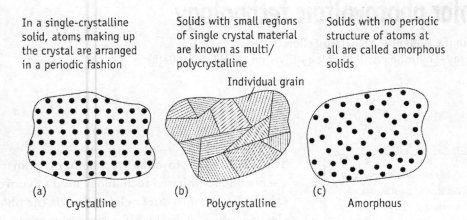

In a single-crystalline solid, atoms making up the crystal are arranged in a periodic fashion

Solids with small regions of single crystal material are known as multi/ polycrystalline

Solids with no periodic structure of atoms at all are called amorphous solids

Individual grain

(a) Crystalline   (b) Polycrystalline   (c) Amorphous

Figure 5.1 Different types of solids

amorphous. Crystalline solids can be further categorized into single crystalline and polycrystalline solids. The fundamental difference between single crystal, polycrystalline (multicrystalline), and amorphous solids is the periodicity of the atomic arrangement (Figure 5.1).

In *single crystalline solids*, the constituent atoms, ions, or molecules are arranged in a regular and periodic pattern as shown in Figure 5.1(a). A single pattern is followed throughout the crystal. A good example is sodium chloride.

A *polycrystalline solid* or *polycrystal* comprises many individual *grains* or *crystallites* as shown in Figure 5.1(b). Each grain can be thought of as a single crystal, within which the constituent atoms, ions, or molecules are arranged in a periodic pattern. In case of a polycrystalline solid, the grain dimension is of the order of 1µm. Almost all common metals and many ceramics are polycrystalline.

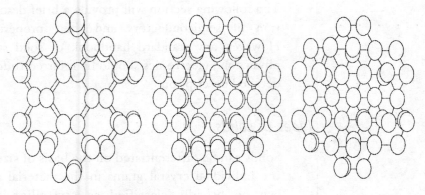

Figure 5.2 Crystalline structures
Source Pillai (1999)

Amorphous materials have no long-range order, or a periodic structure. Carbon is an example for amorphous solids.

An *ideal crystal* has an atomic structure that is repeated periodically across its whole volume (Figure 5.2). Even at infinite length scales, each atom is related to every other equivalent atom in the structure by translational symmetry.

Semiconductors

Characteristic properties such as mechanical, thermal, optical, magnetic, and electronic behaviour differ for different solids. With reference to the electrical conductivity σ, defined as the current density J per unit applied electric field E (that is, σ = J/E) of the material, solids can be basically divided into three types: conductors, insulators, and semiconductors. σ is expressed in units of mho/cm. Good electrical conductors such as copper, iron, and aluminium have conductivities of the order of 10^6 mho/cm. On the other hand, the electrical conductivities of insulators such as quartz (SiO_2) are very low of the order of 10^{-16} mho/cm.

The electrical conductivity of semiconductors is less than that of conductors and more than that of insulators. Semiconductors can be elements such as Si (silicon) and Ge (germanium) or compounds such as CdS (cadmium sulphide), CdTe (cadmium telluride), GaAs (gallium arsenide), or alloys such as $Ga_x Al_{1-x} As$, where x can take any value between zero and one. Many organic compounds such as anthracene exhibit semiconducting properties.

Characteristic properties of semiconductors: band theory

Characteristic properties of different solids can be explained by the band theory. In an isolated atom, there are certain definite discrete energy levels but in a crystal, as the atoms are not isolated and are within each other's electric field, the energy level of each atom splits into a number of energy levels equal to the number of atoms present in the crystal as shown in Figure 5.3a. As a large number of atoms are present in a crystal, the number of energy levels is also very large. These energy levels are so close to each other that they are assumed to be continuous and form an energy band (Figure 5.3b). Spacing between the atoms (that is, interatomic distance d) varies from one crystalline solid to another and is fixed for a particular solid. Hence, the width of a band depends on the type of the solid crystal and is greater for the solid for which the interatomic distance is less.

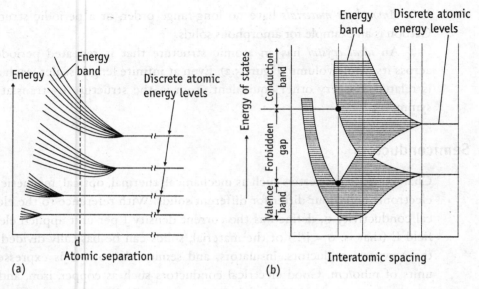

Figure 5.3 Band diagram

As electrons are present at lower energy levels, the lower energy bands are usually completely filled or occupied by electrons. The band corresponding to these energy levels is called VB (*valence band*).

However, the band corresponding to higher energy levels may be completely empty or partially filled and is called CB (*conduction band*). It may be noted that Pauli's exclusion principle restricts the number of electrons that can be accommodated in a band.

Just as electrons in an isolated atom cannot reside anywhere between the two allowable energy levels, electrons in a crystal cannot reside anywhere between VB and CB. The gap between these two energy bands is called the *forbidden energy gap* or *band gap*, or *energy gap*. Band gap is the energy required to move an outer shell electron from VB to CB. Width of the band gap is generally denoted as E_g (= $E_c - E_v$), where E_c is the lowest energy level of CB and E_v is the highest energy level of VB.

Depending upon the band gap, any material can be divided into three different categories: (a) conductor, (b) semiconductor, and (c) insulator. The band gap in the case of a conductor is very less, as either VB overlaps CB or there is a very less gap between these two bands (Figure 5.4a). In case of an insulator, the band gap is very high, of the order of several eV (Figure 5.4c).

In case of a semiconductor, VB is completely filled and CB is empty but the band gap is quite small, about 1 eV (Figure 5.4b). Even at room temperature, an electron in VB can acquire some thermal energy, which is greater than

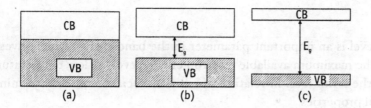

Figure 5.4 Band structures of conductor, semiconductor, and insulator

the band gap, and jump from VB to CB. These electrons are free to move under the influence of even small electric fields. A hole (that is, empty state due to the absence of an electron) is created in VB when an electron jumps from VB to CB. The number of electrons in CB and number of holes in VB create EHP (electron–hole pairs). Electrons and holes are known as charge carriers in semiconductors.

The electrical conductivity of a substance depends upon two factors:

1 The status of CB; whether it is partially filled or completely empty.
2 The width of the band gap (E_g).

Energy density of allowed state

To calculate how electrons and holes distribute themselves within a solid, the number of available states per unit volume per unit energy should be known. The available state is first calculated in k-space (momentum space), and then the E–P (energy–momentum) relation ($E = 1/2\ mv^2 = P^2/2m$, where m is the mass and v the velocity) is used in parabolic bands to give the density of the states in terms of energy. The number of available states N (E) per unit volume with energy E is given as

- N (E) = 0   within forbidden gap   ...(5.1)
- N_c (E) = const* $(E - E_c)^{1/2}$ in CB   ...(5.2)
- N_v (E) = const* $(E_v - E)^{1/2}$ in VB   ...(5.3)

where E_c is the lowest energy of CB and E_v is the highest energy in VB. N_c and N_v correspond to conduction band and valence band respectively. That means there is no available energy state in the forbidden gap.

Fermi level

The fermi level is an important parameter in the band theory, which gives an idea about the maximum available electron energy levels at low temperatures. Position of the fermi level in relation to CB is a crucial factor in determining the electrical properties.

The probability of occupation of an allowed state of any given energy E can be calculated from statistical considerations, taking into account constraints imposed by Pauli's exclusion principle. The result is the Fermi–Dirac distribution function f (E), given by

$$f(E) = \frac{1}{1 + e^{(E - E_f)/kT}} \qquad \qquad ...(5.4)$$

where f (E) is the probability of occupancy of a given state with energy E, k is the Boltzmann's constant, T is the absolute temperature, and E_f is the fermi level. The function is shown in Figure 5.5.

At finite temperatures, the number of electrons which reach CB and contribute to the current can be modelled by the fermi function f(E). This function gives the probability that an available energy state at E will be occupied by an electron at temperature T (Figure 5.5).

It is seen from Equation 5.5 that at temperature 0 K, the exponential term is zero for $E < E_f$ and infinite for $E > E_f$. When $E < E_f$, f(E) = 1, which means the probability that all the quantum states below E_f may be occupied by the electron is unity. Therefore, all these states will be filled by electrons. On

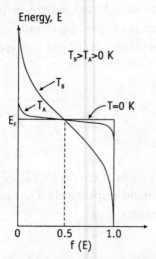

Figure 5.5 Fermi-Dirac distribution functions
Source Chattopadhyay, Rakshit, Saha, *et al.* (1988)

the other hand, when $E > E_f$, $f(E) = 0$. Therefore, all quantum states above E_f will be unfilled. Hence, fermi level is defined as the maximum energy level that can be occupied by the electrons at 0 K.

Intrinsic semiconductor

When conductivity of the semiconductor is determined solely by the thermally generated carriers, the semiconductor is termed an intrinsic semiconductor. It is a semiconductor in an extremely pure form. However, in an intrinsic semiconductor, the number of electrons in CB and number of holes in VB are less and so is the current-carrying capacity of the material. As shown in Figure 5.6, the fermi level in case of the intrinsic semiconductor lies between VB and CB. In such a case, the density of free electrons in CB is equal to the density of free holes in VB. Many intrinsic semiconductors are available but very few have a practical application in electronics. Ge and Si are the two most frequently used intrinsic semiconductors.

A schematic diagram of the energy bands for germanium at 0 K and at room temperature (at 300 K) are shown in Figure 5.7. At 0 K, as there is no

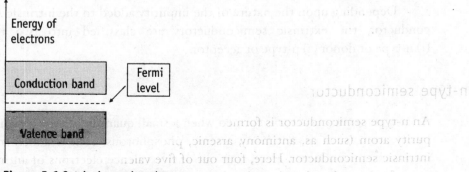

Figure 5.6 Intrinsic semiconductor

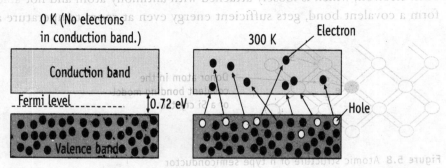

Figure 5.7 Energy band for germanium

thermal vibration, electrons are not present in CB. However, at room temperature (300 K), some electrons in VB acquire thermal energy and jump from VB to CB and take part in the conduction of electricity. Hence, in an intrinsic semiconductor, electron hole pairs (EHPs) are created even at room temperature due to thermal vibration.

Under the influence of an electric field, conduction through the semiconductor is both by free electrons and holes.

Extrinsic semiconductor

At room temperature, very few EHPs are generated due to thermal agitation. Thus, the current conducting capability of intrinsic semiconductors is very less.

When a small quantity of impurity is mixed in a pure semiconductor (one atom of impurity in 10^7 atoms of pure semiconductor), conductivity of the semiconductor increases substantially. Such an impure semiconductor is called an *extrinsic semiconductor*. The process of deliberately adding on a desirable impurity atom to a pure semiconductor to modify its property in a controlled manner is called *doping*.

Depending upon the nature of the impurity added to the intrinsic semiconductor, the extrinsic semiconductors are classified into two types: (1) n-type or donor (2) p-type or acceptor.

n-type semiconductor

An n-type semiconductor is formed when a small quantity of pentavalent impurity atom (such as, antimony, arsenic, phosphorous, etc.) is added to the intrinsic semiconductor. Here, four out of five valence electrons of antimony atom form covalent bonds with valence electron of silicon atoms (Figure 5.8). Fifth electron, which is loosely attached with antimony atom and not able to form a covalent bond, gets sufficient energy even at room temperature and

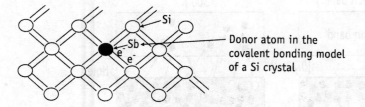

Figure 5.8 Atomic structure of n type semiconductor
Source Gupta and Kumar (2000)

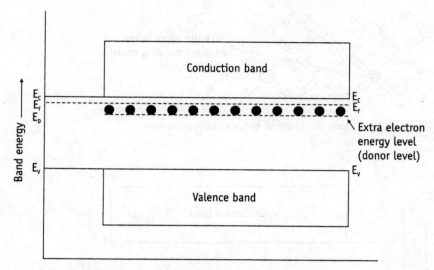

Figure 5.9 Energy bands of a n-type semiconductor

detaches completely and moves freely within the crystal. So n-type semiconductor has extra free electrons, which are known as majority carrier. However, holes generated due to thermal vibration do not contribute much to the current flow and are called minority carriers. As the impurity atom donates electrons to the crystal conductivity, it is called donor atom. An n-type semiconductor has (1) electron e^- as majority carrier and (2) thermally generated holes as minority carrier.

In case of the n-type semiconductor, the fermi level lies just below CB. An additional energy level (donor level) lies below CB, which represents the ground state of the additional electron of the impurity atom (Figure 5.9).

p-type semiconductor

A p-type semiconductor is formed when a small quantity of a trivalent impurity atom, that is, boron, aluminium, or indium, etc., is mixed with the intrinsic semiconductor (Figure 5.10). There is deficiency of one electron and this deficiency is called a hole. A p-type semiconductor has a large number of holes (majority carrier). A p-type semiconductor has (1) holes as majority and (2) thermally generated electrons as minority carrier.

In case of p-type semiconductor, the fermi level lies just above VB. The extra energy level above VB is due to extra holes present in the material (Figure 5.11).

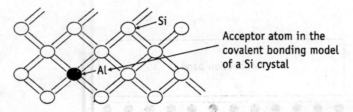

Figure 5.10 Atomic structure of p-type semiconductor

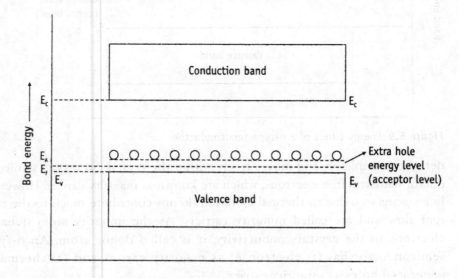

Figure 5.11 Energy bands of a p-type semiconductor

Equilibrium carrier concentration in semiconductor

The density of electrons in a semiconductor is related to the density of available states and probability that each of these states is occupied. The density of occupied states per unit volume and energy is simply the product of the *density of states* $N(E)$ as given in Equation 5.2 and the *fermi function* $f(E)$ given by Equation 5.4. The electron density per unit energy equals

$$dn_o = f(E) * N_c(E) dE \qquad ...(5.5)$$

Since holes correspond to empty states in VB, probability of having a hole equals the probability that a particular state is not filled. So the hole density per unit energy equals

$$dp_o = [1 - f(E)] N_V(E) dE \qquad ...(5.6)$$

The density of carriers is then obtained by integrating the density of carriers per unit energy over all the possible energies within a band.

Therefore, electron concentration at equilibrium can be obtained by integration as

$$n_o = N_V e^{(E_f - E_c)/kT} \qquad \qquad ...(5.7)$$

and the hole concentration at equilibrium will be

$$p_o = \left(1 - \frac{1}{1 + e^{(E_v - E_f)/kT}}\right) N_V \simeq [1 - (1 - e^{(E_v - E_f)/kT})] N_V$$
$$= N_V e^{(E_v - E_f)/kT} \qquad \qquad ...(5.8)$$

Concept of carrier transport in semiconductors

This section discusses the process involved in transferring carriers from one location to the other in a semiconductor. In a semiconductor, charge carriers can move from one location to another by two processes: drift and diffusion. In case of drift, carriers move due to the application of an external electric field whereas in case of diffusion, carriers move due to concentration gradient.

Drift of carriers

Drift is the motion of a charge particles in response to an applied electric field. There is no net motion of carriers in absence of an electric field (Figure 5.12a). However, a carrier drifts in a particular direction when an electric field is applied (Figure 5.12b). When an electric field E is applied, the carrier drifts with a certain velocity, which after a short period of time reaches a constant velocity called drift velocity v_d for the given electric field. The drift velocity is proportional to the electric field strength E and is different for different semiconductors.

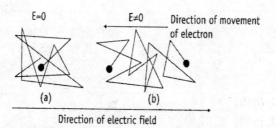

Figure 5.12 Net motion of carriers in presence of electric field

So, $v_d \propto E$

$$\Rightarrow v_d = \mu E \qquad \text{...(5.9)}$$

where v_d is expressed in cm/s, E is expressed in volt/cm, and mobility μ is defined as the ease with which the carrier flows. The higher the impurity and temperature, the lower the mobility.

Current density due to drift of carriers

Current density is defined as the electrical charge e (electron or hole) crossing unit surface (normal to the flow) per unit time due to the drift under influence of an applied electric field (Figure 5.13). If n is the density of electrons in the conduction band of semiconductor then net charge per unit volume for conduction of electric current will be equal to ne. So, electron current density J_e is expressed as

$$J_e = nev_e = ne\mu_e E \qquad \text{...(5.10)}$$

where e is the charge of the electron expressed in Coulomb, n is the concentration of electrons expressed in electrons/cm^3, v_e is the drift velocity of the electron, E is the electric field applied, and μ_e is the mobility of the electron.

Similarly, hole current density J_h is expressed as

$$J_h = epv_h = ep\mu_p E \qquad \text{...(5.11)}$$

where p is the concentration of hole expressed in holes/cm^3 and μ_p is the mobility of the hole. Then, total current density will be the summation of electron current density and hole current density expressed as

$$J_{total} = J_e + J_h$$

$$J_{total} = e\,(n\mu_e + p\mu_p)\,E \qquad \text{...(5.12)}$$

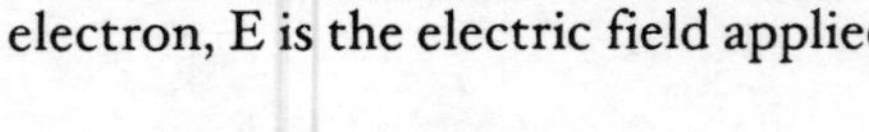

E(x) ──────→   ──────────→ J_p (drift)

　　　　　　　　 ←────────── J_n (drift)

Figure 5.13 Direction of the currents

Conductivity

According to Ohm's law

$$\text{Conductivity}(\sigma) = \frac{J_{total}}{E}$$

Therefore, Conductivity $(\sigma) = e(n\mu_e + p\mu_p)$ ...(5.13)

and,

$$\text{Resistivity } (\rho) = \frac{1}{e(n\mu_e + p\mu_p)} \qquad \text{...(5.14)}$$

ρ is a strong function of doping and, therefore, the doping level of a substrate is to be specified.

For n-type conductor, $n \gg p$ and hence

$$\rho_n \approx \frac{1}{e\mu_e n}$$

For n-type semiconductor, n will be approximately equal to the number of donor atoms (N_D). Hence

$$\rho_n \approx 1/e\mu_e N_D \qquad \text{...(5.15)}$$

Similarly, for p-type semiconductor

$$\rho_p \approx \frac{1}{e\mu_p N_A} \qquad \text{...(5.16)}$$

where N_A is the number of acceptor atoms.

Diffusion of carrier

The diffusion of charge particles occurs due to the concentration gradient. Carriers move from a higher concentration to a lower one due to their random physical motion (Figure 5.14).

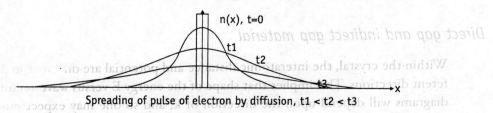

Spreading of pulse of electron by diffusion, t1 < t2 < t3

Figure 5.14 Diffusion of carriers at different time t

Current flow due to diffusion of carrier

Current flow due to diffusion is given by

$$J'_n = eD_n \frac{dn}{dx} \qquad \qquad ...(5.17)$$

$$J'_p = -eD_p \frac{dp}{dx} \qquad \qquad ...(5.18)$$

where, D_n and D_p are the diffusion coefficients, J'_n is the current flow due to diffusion of electron, and J'_p is the current flow due to diffusion of hole.

Total current in semiconductor

Total current in a semiconductor is the sum of the drift current and the diffusion current produced by the charge carriers, and is defined as J.

J = electron current (drift, diffusion) + hole current (drift, diffusion)

$$J = J_p + J_n$$

where $J_n = J_e + J'_n = en\mu_e E + eD_n \frac{dn}{dx}$

$$J_p = J_h + J'_p$$

$$J_p = ep\mu_p E - eD_p \frac{dp}{dx} \qquad \qquad ...(5.19)$$

J_n is the sum of drift current and diffusion current produced by electron, and J_p is the sum of drift current and diffusion current produced by hole.

Absorption coefficient (α) The absorption coefficient of a material indicates how far light having a specific wavelength can penetrate the material before being absorbed. A small absorption coefficient means light is not readily absorbed by the material. A good light absorbing material should have a high absorption coefficient so that approximately 99% of the incident light is absorbed by a layer with a thickness of only 1 μm.

Diffusion length It is the distance travelled by a charge carrier before recombining with an opposite charge carrier.

Direct gap and indirect gap material

Within the crystal, the interatomic distance and potential are different in different directions. This implies that shape of the energy E versus wave vector k diagrams will depend upon the direction of k, and so one may expect much more complicated band shapes. There will, in general, be a number of minima in CB. However, the maximum in VB is always at k = 0. Based on these considerations, two general categories of semiconductors may be defined: direct band gap semiconductors and indirect band gap semiconductors.

Direct band gap semiconductors

For direct band gap semiconductors, the minimum of the lowest conduction band occurs at k = 0 (Figure 5.15). These semiconductors are good light absorbers and absorb light in a layer few micrometers thick. GaAs, InSb, InP, and CdS are examples of direct band gap semiconductors.

Indirect band gap semiconductors

For indirect band gap semiconductors, the minimum of lowest conduction band does not occur at k = 0 (Figure 5.16). Si and Ge are few examples of indirect band gap semiconductors. Several hundred micro metres of indirect band gap semiconductors are necessary to absorb all light.

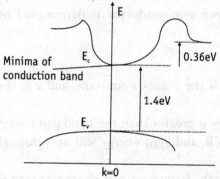

Figure 5.15 Direct band gap semiconductor (GaAs)

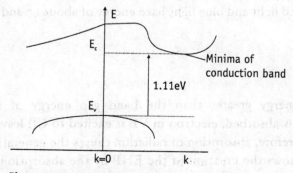

Figure 5.16 Indirect band gap semiconductor (Si)
Source Pillai (1999)

In case of direct band gap semiconductor, electron transition between the bands occurs in a single event involving only energy change. In case of indirect band gap semiconductors, electron transition between the top of VB and the minimum of CB involves both energy and momentum changes.

Generation and recombination of charge carriers in semiconductors

When photons (electromagnetic radiation can also be thought of as consisting a large number of particles called photons) of electromagnetic radiation strike a semiconducting material, the photon with energy higher than the band gap energy of the semiconducting material is able to free the electrons. These free electrons then contribute to the electric current. If the photon energy is less than the band gap energy, then it is not possible to remove electrons. In that case, either radiation passes through the material or is absorbed as heat. So, in order to move certain electrons from VB to CB, a certain frequency of electromagnetic radiation is needed. No matter how intense the radiation is (or how large the number of photons), it should have some minimum energy to move electrons from VB to CB. The minimum energy to produce an EHP for a given semiconductor is determined by the equation

$$E_g = h\nu$$

where E_g is the band gap energy, h is the Planck's constant, and ν is the frequency of radiation.

Unless and until photon energy is greater than the band gap energy, the electron cannot be removed from VB, and light energy will pass through the material (Figure 5.17).

Figure 5.18 shows the wavelength, frequency, and photon energy of the solar radiation. The entire spectrum of sunlight covers about 0.5–2.9 eV (from infrared to ultraviolet). Red light and blue light have energy of about 1.7 and 2.7 eV, respectively.

Generation of carriers

When radiation with energy greater than the band gap energy of the semiconducting material is absorbed, electron in VB is excited to CB leaving behind a hole in VB. Therefore, absorption of radiation causes the generation of an EHP. Figure 5.19 shows the creation of the EHP by the absorption of radiation.

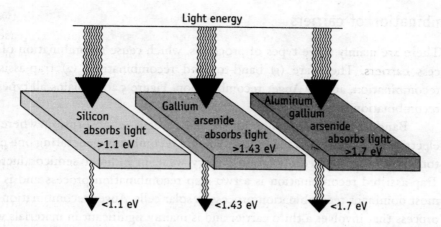

Figure 5.17 Interaction of light energy with different semiconductors

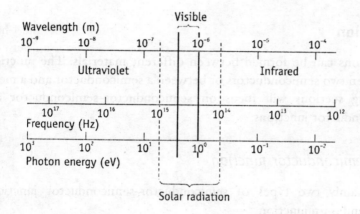

Figure 5.18 Solar radiation in the electromagnetic spectrum

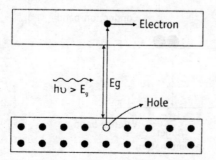

Figure 5.19 Creation of electron–hole pair

Recombination of carriers

There are mainly three types of processes, which cause recombination of excess carriers. These are (1) band-to-band recombination, (2) trap-assisted recombination, and (3) Auger recombination. Figure 5.20 explains all types of recombination processes.

Band-to-band recombination is the reverse of absorption where an electron in CB falls into the hole of VB and recombines, generating one photon. This type of recombination is very weak in indirect semiconductors. Trap-assisted recombination is a two-step recombination process and is the most dominant recombination process in solar cells. Auger recombination is a process that involves a third carrier and is mainly significant in materials with a high doping concentration.

In general, a large number of recombination centres exist typically at the surface of the material and affect the performance of the device.

Junction formation

Several junctions can be formed between different materials. The junctions can be between two semiconductors or between a semiconductor and a metal. The following sections will focus on semiconductor–semiconductor and metal–semiconductor junctions.

Semiconductor–semiconductor junction

There are mainly two types of semiconductor–semiconductor junctions: p-n junction and p-i-n junction.

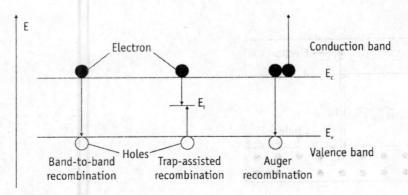

Figure 5.20 Recombination of carriers

p-n junction

The flow of current in any one direction would not be possible without some asymmetry, that is, one side of the device should be different from the other. Therefore, all practical semiconductor devices have built-in asymmetry. When the asymmetry is built by a p-type semiconductor on one side and an n-type semiconductor on the other, the structure is known as a p-n junction. A p-n junction is also known as p-n junction diode. A p-n junction is classified into unbiased p-n junction and biased p-n junction.

Unbiased p-n junction

When p-type and n-type semiconductors are suitably treated to form a p-n junction, the following happens.

- On account of the difference in the concentration of the charge carriers in the two sides of the p–n junction, initially, electrons from the n-region diffuse through the junction into the p-region and combine with the holes in there.
- Due to this process, the immobile ions (N_D and N_A) in the neighbourhood of the junction will prevent further diffusion of charge carriers and an equilibrium condition is reached (Figure 5.21).

In other words, a barrier is created against further movement of the charge carriers, that is, holes and electrons. This barrier is called *potential barrier* or *junction barrier* and is denoted as V_o. The potential barrier is of the order of 0.1–0.3 V. The potential barrier gives rise to an electric field that is directed from the n- to the p-region. However, outside this barrier, on each side of the junction, the material remains neutral.

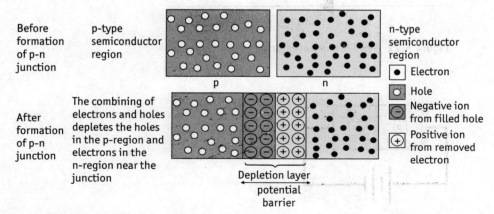

Figure 5.21 Unbiased p-n junction formation

Due to the existence of the potential barrier across the junction, transfer of an electron from the n- to the p-side requires expenditure of a certain amount of energy termed as the barrier energy E_B and is given by $E_B = eV_o$, where e is the charge of the electron and V_o is the potential difference across the two junctions.

Figure 5.22 is the symbolic representation of a p-n junction. The direction of arrow is from the p-side to the n-side, which signifies the direction of the current.

Biased p-n junction

In case of a biased p-n junction, a potential difference is applied across the p-n junction. There are two methods of biasing a p-n junction: *forward biasing* and *reverse biasing*.

Forward biasing

In forward biasing, the positive terminal of the battery is connected to the p-side and the negative terminal to the n-side of the junction and potential difference of V is applied across the junction (Figure 5.23).

Here, the forward-bias voltage (V) opposes the potential barrier (V_o), thus reducing the depletion region. The potential barrier is reduced and at some forward voltage (0.1–0.3V), it is eliminated altogether, and when $V > V_o$, current starts flowing in a forward direction.

In forward-biased condition, the junction offers low resistance to the current flow.

Figure 5.22 Symbolic representation of p-n junction

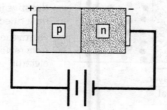

Figure 5.23 Forward biasing

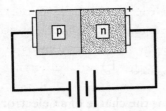

Figure 5.24 Reverse biasing

Reverse biasing

In reverse biasing, the positive terminal of the battery is connected to the n-side and the negative terminal to the p-side of the junction (Figure 5.24).

In the reverse-biased condition, a transient current flows and both the carriers are pulled away from the junction. The increase in barrier energy is given by eV, where V is the voltage applied across the junction. When the potential formed by the widened depletion layer equals the applied voltage, the current ceases except for the small thermal current generated by the minority carriers. This current I_o is called reverse saturation current, which is independent of the applied reverse voltage (V) and increases with the temperature of the diode. The p-n junction conducts the current easily when forward biased and practically no current flows when it is reverse biased.

The forward current is typically of the order of a few mA, whereas the reverse saturation current is of the order of μA or even less.

V-I characteristic of a p-n junction

The V-I characteristic of a p-n junction is a curve, which shows the variation of the voltage across the junction with the current.

Under equilibrium condition, V = 0 and no current flows.

In forward-biased condition, current flow is mainly due to diffusion of carriers and it increases exponentially. In reverse-bias condition, drift current is zero and both electron and hole diffusion components are negligible due to large barriers. However, the reverse saturation current I_o is independent of the voltage.

In forward-biased condition, the probability that a carrier can diffuse across the junction is proportional to exp (eV/kT). Therefore, the diffusion current in forward bias is I_o exp (eV/kT). As discussed earlier, the total current crossing the diode is the sum of the diffusion and drift currents.

The total current I flowing through a p-n junction due to application of voltage V across the junction is given by

$$I = I_{diffusion} + I_{drift} - I_{generation}$$

$$= I_o \left[\exp\left(\frac{eV}{\eta kT} \right) - 1 \right] \qquad \ldots(5.20)$$

(as $I_{diffusion} = I_o \exp [eV/\eta kT]$, $I_{drift} = 0$, $I_{generation} = I_o$)

where I_o is the reverse saturation current, e is the charge of an electron (1.6×10^{-19} C), k is the Boltzmann's constant (1.38×10^{-23} J/K), T is temperature in K, and η is a numerical constant depending upon the material of the diode (for Ge, $\eta = 1$ and for Si, $\eta = 2$).

At room temperature (T = 300 K), Equation 5.20 becomes

$$I = I_o \left[\exp\left(\frac{39V}{\eta} \right) - 1 \right] \qquad \ldots(5.21)$$

A typical V-I characteristic, as shown in Figure 5.25, can be drawn by using Equation 5.20.

Example 1

The saturation current density J_o of p-n junction Ge diode is 250 mA/m² at a temperature T of 300 K. Calculate the voltage that would have to be applied across the junction to cause a forward current density J of 10⁵ A/m² to flow.

Solution

Given J_o = 250 mA/m²; T = 300 K; and J = 10⁵A/m²

$$I = I_o \left[\exp(eV/kT) - 1 \right]$$

(as $\eta = 1$ for Ge)

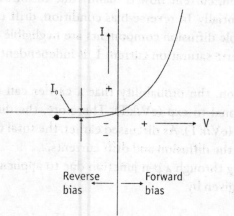

Figure 5.25 A typical current–voltage characteristic of a p-n junction diode

Dividing both the sides of Equation 5.20 by the area of the diode (A) to get the saturation current, we have

$$\frac{I}{A} = \frac{I_o}{A}\left[\exp\left(\frac{eV}{kT}\right) - 1\right]$$

$$\Rightarrow J = J_o\left[\exp\left(\frac{eV}{kT}\right) - 1\right]$$

$$\Rightarrow \frac{J}{J_o} = \exp\left(\frac{eV}{kT}\right) - 1$$

$$\Rightarrow \exp\left(\frac{eV}{kT}\right) - 1 = \frac{10^5}{250 \times 10^{-3}}$$

$$\Rightarrow \exp\left(\frac{eV}{kT}\right) - 1 = 4 \times 10^5$$

$$\Rightarrow \exp\left(\frac{eV}{kT}\right) = 4 \times 10^5 \text{ (neglecting 1)}$$

$$\left(\frac{eV}{kT}\right) = 12.9$$

$$V = \frac{12.9 \times 1.38 \times 10^{-23} \times 300}{1.6 \times 10^{-19}} = .33 \text{ V}$$

Load line of a diode circuit

Load line of a diode circuit is used to determine the operating point of the circuit. Figure 5.26 shows a basic diode circuit consisting of a diode connected in series with a load resistor R_L and an input battery.

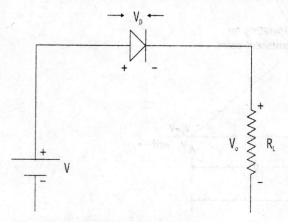

Figure 5.26 Basic diode circuit
Source Gupta and Kumar (2000)

Applying Kirchoff's law to the basic diode circuit

$$V = V_D + V_o = V_D + (I \times R_L) \qquad \qquad ...(5.22)$$

When I = 0, Equation 5.23 reduces to $V_D = V$

This gives a point A (V_D, 0) on the x-axis as shown in Figure 5.27. The point B corresponds to $V_D = 0$ and $I = V/R_L$.

A line passing through A and B gives all possible combinations of V_D and I that satisfy Equation 5.22. This line is called the *load line*. Intersection of the load line with the static characteristic curve of the diode provides a point Q known as the *quiescent point* or *operating point*.

From Equation 5.22

$$I = \frac{V}{R_L} - \frac{V_D}{R_L} \qquad \qquad ...(5.23)$$

If Equation 5.23 is compared with the equation for a straight line (that is, $Y = mX + C$, where m is the slope of the line) then

$$Y = I, X = V_D, \text{ and } m = -(1/R_L)$$

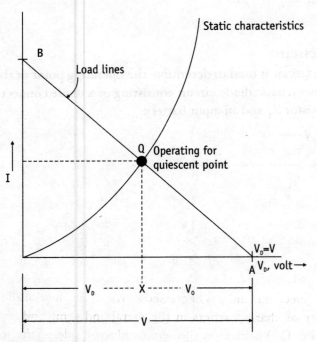

Figure 5.27 Load line curve of a diode

Thus, for a given input voltage, effect of a higher load will be a steeper load line and hence, a higher current at the operating point Q.

p-i-n junction

The p-i-n diode has heavily doped p-type and n-type regions separated by an intrinsic region. It acts like an almost constant capacitance at reverse bias and behaves as a variable resistor when forward biased. The forward resistance of the intrinsic region decreases with an increase in current. As its forward resistance can be changed by varying the bias, it can be used as a modulating device for AC (alternating current) signals. It is used in microwave switching applications.

Metal–semiconductor junction

A semiconductor–semiconductor junction is not the only type of junction that is important in electronics. Metals are necessary to connect the semiconductor to batteries. Also, metal strips, which act as the interconnections, must be able to carry adequate current and make good contact with the devices. Therefore, metal has to form a junction with semiconductor. The region where the metal forms a junction with the semiconductor is characterized by the *Schottky barrier* or *Ohmic contact*.

Schottky barrier diode

The Schottky diode has characteristics that are similar to those of the p-n junction diode. The Schottky barrier is created at the junction of a metal and a semiconductor. The Schottky barrier height for n- or p-type semiconductors depends upon the metal and semiconductor properties. However, it is found experimentally that the Schottky barrier height is almost independent of the metal employed.

When a metal and a semiconductor are brought into contact, due to the difference in their work function, a potential drop occurs in the interfacial region. (Work function is defined as the energy required by an electron at the fermi level in order to be liberated from the metal. It is the work done in liberating an electron at the fermi level from the metal. If ϕ is the voltage equivalent of the work function then $E_w = \phi e$, where E_w is the work function, e is the charge of the electron, and ϕ is expressed in volts.) Because of the differences in availability of charge carriers in the metal and semiconductor, this entire potential drop (V_o) occurs on the semiconductor side of the junction

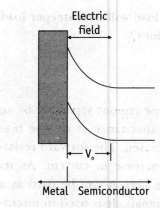

Figure 5.28 Band diagram of a Schottky barrier

(Figure 5.28). To produce a Schottky barrier on an n-type semiconductor, a metal with a work function larger than that of the semiconductor has to be chosen. Similarly, for a p-type semiconductor, metal with a work function smaller than that of the semiconductor has to be chosen. If the barrier height is small, the saturation current will be very high. Therefore, in order to limit large saturation current, a Schottky barrier with a large barrier height has to be chosen as most of the minority carrier will be able to cross the barrier. However, for many narrow band gap semiconductors, it is difficult to obtain large Schottky barriers.

Current flow in Schottky diode

The V-I characteristics of a Schottky diode show the rectifying character similar to those of the p-n junction diode. Figure 5.29 shows the rectifying characteristic of Schottky diode. Schottky diode is a unipolar device as it has electrons as majority carriers on both sides of the junction. When the applied voltage changes, electrons from the semiconductor experience variable potential barriers, whereas flow of electrons from the metal remains unchanged.

When the applied voltage is zero, current flow from the metal to the semiconductor must balance the current flow from the semiconductor to the metal. Therefore, the net current is zero.

Forward bias allows electrons to flow from the semiconductor to the metal, increasing the current (Figure 5.29), whereas reverse bias suppresses the electron flow from the semiconductor while flow from the metal is unaffected. The saturation current in the Schottky barrier turns out to be much higher than that in the p-n diode with similar built-in voltage.

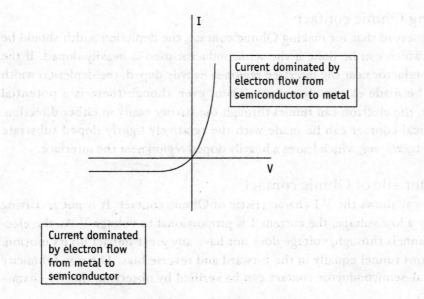

Figure 5.29 V-I characteristic of a Schottky diode

Ohmic contact

From the previous section (on Schottky barrier), it is observed that there is a barrier that restricts the flow of electrons. However, it is possible to create metal–semiconductor junctions, which have a linear non-rectifying V-I characteristic. Such junctions or contacts are called Ohmic contacts. Here the forbidden region can become so thin that carriers can pass through the region by quantum mechanical tunnelling (Figure 5.30).

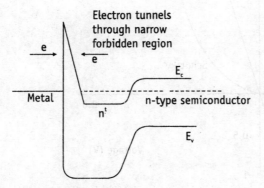

Figure 5.30 Ohmic contact

Establishing Ohmic contact

It is observed that for making Ohmic contact, the depletion width should be small, which can be done if the semiconductor used is heavily doped. If the semiconductor near the interface region is heavily doped, the depletion width could be made extremely narrow. Then, even though there is a potential barrier, the electron can tunnel through the barrier easily in either direction. Electrical contact can be made with the relatively lightly doped substrate region by *alloying*, which leaves a heavily doped region near the interface.

V-I characteristic of Ohmic contact

Figure 5.31 shows the V-I characteristic of Ohmic contact. It is not rectifying and for a low voltage, the current I is proportional to voltage V. As the electron tunnels through, voltage does not have any great impact on its motion. Electrons tunnel equally in the forward and reverse bias. The non-Ohmicity in metal–semiconductor contact can be verified by observing the V-I characteristic.

Homo-junction and hetero-junction

When a junction is made between two similar materials (the same band gap), it is called homo-junction (Figure 5.32a) and when a junction is formed between two different materials (of different band gaps), it is called hetero-junction (Figure 5.32b).

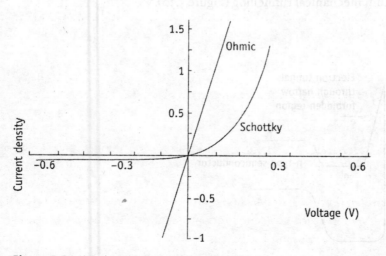

Figure 5.31 V-I characteristic of Ohmic contact

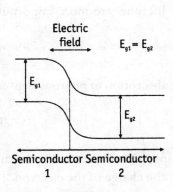

Figure 5.32 (a) Homo-junction

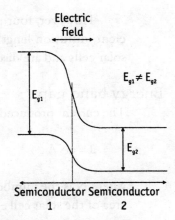

Figure 5.32 (b) Hetero-junction

Introduction to photovoltaic technology

A majority of PV cells are semiconductor-based junction devices, which convert solar radiation into direct electricity. PV cells are also known as solar cells which exhibit the PV effect. PV effect is a phenomenon by virtue of which a voltage difference is created across a p-n junction as a result of absorption of photons. Three processes are necessary for the PV action.

1 Solar radiation falling on the solar cell should be capable of producing EHPs.
2 Excess charge carriers of opposite sign (electrons and holes) must be separated at some electrostatic inhomogeneity (for example, a p-n junction or metal–semiconductor contact).
3 These photogenerated carriers in the material must be able to move across the region to the junction.

Requirements for a material to form solar cells

One key requirement of an efficient solar cell is the capacity to convert as much sunlight into electricity as possible. Of the three requirements explained above, the first and third are associated with the current generation aspects of the PV effect whereas second is necessary for voltage generation.

There are certain parameters, which determine the operation of a solar cell. These parameters are semiconductor band gap, the absorption coefficient, the diffusion length, lifetime of the minority carriers, doping, recombination, and density of states in the energy gap.

However, four parameters, namely, energy band gap, absorption coefficient, diffusion length, and minority carrier lifetime, are more important for solar cells and are discussed hereafter.

Energy band gap

The current produced by a solar cell upon the absorption of photons is given as

$$I = eNA \qquad\qquad ...(5.24)$$

where N is the number of photons incident upon the solar cell, A is the surface area of the solar cell exposed to light, and e is the charge of the electron.

Thus, materials with lower band gap energies can generate a large number of charge carriers and thus the current. Hence, one can assume that a material with the lowest band gap would make the best PV cell but this is not true.

The band gap energy also influences the strength of the electric field, which determines the maximum voltage the cell can produce ($V = E_g/e$). Higher the band gap energy of the material, higher the generated open circuit voltage.

In conclusion, low-band gap cells have high current but low voltage, while high-band gap cells have high voltage and low current. A compromise is necessary in the design of solar cells to set the maximum power. It is observed that cells made of materials with band gaps between 1 and 1.8 eV can be used efficiently in PV devices (Table 5.1).

Absorption coefficient (α) Photon absorption is an important property and the absorber materials of the solar cell should have a large absorption coefficient. Large absorption coefficient of a material means photons are absorbed fast before travelling a large distance. As the necessary absorber thickness is approximated by $1/\alpha$, large α permits the absorber layer to be thin (that is, photons are absorbed near the surface). As a result, less material is needed,

Table 5.1 Potential materials to be used in manufacture of a solar cell

Material	Band gap energy (eV)
Crystalline silicon	1.12
Amorphous silicon	1.75
Copper indium diselenide	1.05
Cadmium telluride	1.45
Gallium arsenide	1.42
Indium phosphite	1.34

thus reducing the material cost. In general, the direct band gap semiconductors have a higher α ($>10^4$/cm) as compared to indirect band gap semiconductors and thus are desirable.

Diffusion length   The photogenerated carriers in absorbing material must be able to move across the region to the junction. Carriers that recombine before arriving at the junction are lost and cannot contribute to the output current. In general, the diffusion length of the minority carrier in a solar cell should be large.

Minority carrier lifetime   This is the most fundamental determination of the effectiveness of a semiconductor as a PV candidate. The minority carrier lifetime in a solar cell should be large.

Working of solar cells

When light is incident upon a solar cell, the following phenomena occur.

1 Photons whose energy is comparable to the band gap of the semiconductor get absorbed.
2 A few photons are reflected back from the front surface of the cell.
3 Some photons are lost as they are blocked from reaching the crystal by the current collector (metal grid) that covers part of the front surface.
4 Some photons whose energy is less than the band gap pass right through the solar cell.

When a photon of light with an appropriate amount of energy penetrates a silicon solar cell near the p-n junction and encounters one of the silicon atoms (Figure 5.33), the following events happen.

- It dislodges one of the electrons, which leaves behind a hole (as shown in Figure 5.33).
- The generated carriers tend to move to the layer where it acts as majority carrier, that is, the electron tends to migrate towards the layer of n-type silicon, whereas the hole tends to move towards the layer of p-type silicon.
- The electron can be collected by the current collector on the front surface of the cell to generate an electric current in the external circuit and then re-appear in the layer of p-type silicon.
- It recombines with the hole in the layer of p-type silicon.

The other components shown in Figure 5.33, viz, the anti-reflection coating, cover glass, back contact, etc., are discussed later.

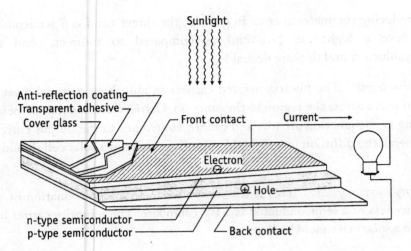

Figure 5.33 Working of a solar cell
Source http://www1.eere.energy.gov/solar/solar_cell_materials.html

Characteristic curve of a solar cell (I-V curve)

The resultant equivalent circuit is shown in Figure 5.34.

When a solar cell is not irradiated, it is just like a semiconductor diode (p-n junction diode) and follows the ideal diode equation. When a solar cell is irradiated, photogenerated current (I_L) flows through the circuit exactly opposite to that of the diode current (I_D). The output current I is then equal to the difference between photogenerated (light-generated) current I_L and the diode current I_D (Figure 5.34).

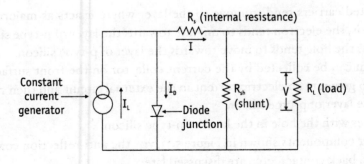

Figure 5.34 Equivalent circuit of solar cell
Source Garg and Prakash (1997)

$$I = I_L - I_D - I_{sh}$$

$$= I_L - I_O \left[\exp \left\{ \frac{e(V + R_s I)}{\eta kT} \right\} - I \right] - \left(\frac{V + R_s I}{R_{sh}} \right) \qquad ...(5.25)$$

where I_L is the light-generated current, I_D the diode current, I_o the saturation current, R_s the series resistance of the solar cell, R_{sh} the shunt resistance, and T is the junction temperature. Saturation current is the current of minority carriers by thermal agitation and accelerated within the built-in field of the p-n junction.

Additional resistance arises from the bulk n- and p-layers and the contacts, giving rise to a series resistance (R_s). Losses occur from junction leakage providing a parallel or shunt resistance (R_{sh}) across the diode.

I-V curve

The I-V curve of a solar cell can be plotted by varying voltage applied across the solar cell and measuring the corresponding current flowing through it. The I-V curve of a solar cell is shown in Figure 5.35.

The I-V curve for a solar cell passes through three significant points, that is short-circuit current I_{sc}, open-circuit voltage V_{oc}, and MPP (maximum power point).

Short-circuit condition

A circuit is called a short circuit when V = 0 (Figure 5.36). In such a case, $I = I_{sc}$, where I_{sc} is the short-circuit current. Hence Equation 5.25 becomes

$$I_{sc} = I_L - I_o \left[\exp \left(e \, R_s \, I_{sc} / \eta kT \right) - I \right] - \left(R_s \, I_{sc} / R_{sh} \right) \qquad ...(5.26)$$

The series resistance is very small and thus under short-circuit condition, it can be neglected $(R_s \rightarrow 0)$. Therefore, Equation 5.26 reduces to

$$I_{sc} = I_L$$

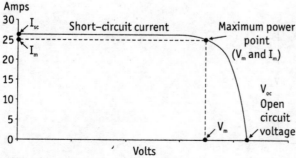

Figure 5.35 I-V curve of solar cell

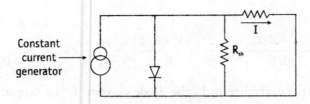

Figure 5.36 Short-circuit condition

Hence, the short-circuit current is equivalent to the light-generated current, which is proportional to irradiance.

Short-circuit current (I_{sc}) depends on the following factors.

- *Area of the solar cell:* larger the area of the solar cell, larger will be I_{sc}
- *Number of photons* striking solar cells (solar radiation intensity)
- *Spectrum of the incident light:* for most solar cell measurements, the spectrum is standardized to the AM 1.5 spectrum
- *Optical properties* (absorption and reflection) of solar cell
- *Collection probability* of solar cell, which depends on the surface passivation and minority carrier lifetime in the base material

Open-circuit condition

The circuit is in the open-circuit condition when current flowing through the circuit is zero, that is, I = 0. In such a case, $V = V_{oc}$ and the shunt resistance R_{sh} tends to ∞ (Figure 5.37).

Therefore, Equation 5.25 reduces to

$$V_{oc} = \frac{\eta kT}{e} \left[\ln \left((I_L + I_o)/I_o \right) \right] \qquad \qquad ...(5.27)$$

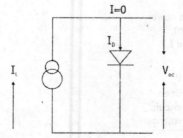

Figure 5.37 Open-circuit condition

The open-circuit voltage increases logarithmically with an increase in irradiance level. The open-circuit voltage depends upon the following.

- Saturation current of the solar cell
- Light-generated current

The following requirements have to be met for obtaining high open-circuit voltage.

- High carrier diffusion length
- Low recombination
- High doping concentration
- Low crystal volume

Maximum-power operation

Power obtained from the solar cell in a day varies with the irradiance level. If the cell's terminals are connected to a variable resistance R (Figure 5.38b), the maximum power from the solar cell is obtained at a particular operating point called MPP. The voltage and current at MPP are known as V_m and I_m, respectively (Figure 5.35).

The operating point will be determined by intersection of the I-V curve of solar cell and load line. For resistive load, the loadline is a straight line with a slope I/R. The maximum power is obtained for R = R_{opt} (Figure 5.38), at a point A.

$$\text{where } R_{opt} = \frac{V_m}{I_m}$$

It is always desirable to connect the load in a way that it will always work under the MPP.

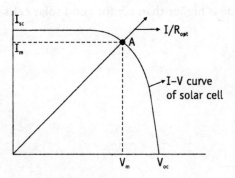

Figure 5.38a Locating MPP

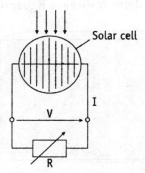

Figure 5.38b

Power delivered by solar cell

Electrical power that can be extracted from a solar cell is proportional to its area and to the intensity of sunlight that strikes the area. It is measured in W_p. Under STCs (standard test conditions), a solar cell of 1 W_p will supply power of 1 W. The STC is defined as: global irradiation of 1000 W/m^2, AM 1.5, and COT (cell operating temperature) 25 °C. A typical 4-inch (10.2 cm) diameter crystalline Si solar cell or a 10 cm × 10 cm multicrystalline cell should provide a power of 1–1.5 W under STCs. However, STCs are rarely found in real-world operating conditions. In real-field conditions, on a bright sunny day, global irradiance on ground is usually 700–800 W/m^2. Again, when the ambient temperature is 45 °C, COTs go as high as 60–65 °C. In reality, a 1 W_p cell never produces power of 1 W, rather on a bright sunny day, it produces 60%–70% of that, that is, 0.6–0.7 W. COT is discussed in subsequent paragraphs.

Fill factor

The FF (fill factor) is defined as the ratio of maximum power (peak power) from the solar cell to the product of the open-circuit voltage and short-circuit current. It is expressed as

$$FF = \frac{V_m \times I_m}{V_{oc} \times I_{sc}} = \frac{P_m}{V_{oc} \times I_{sc}} \qquad \qquad ...(5.28)$$

Graphically (Figure 5.39), FF is the measure of squareness of the I-V characteristic of a solar cell and is also the area of the largest rectangle, which will fit in the I-V curve.

A higher voltage will give a higher possible FF. For a perfect cell, FF = 1. However, parasitic resistance reduces FF and in real devices 0.7 < FF < 0.95. Maximum FF can be seen for smallest series resistance R_s and largest shunt resistance R_{sh}, and its value is higher than 0.7 for good solar cells.

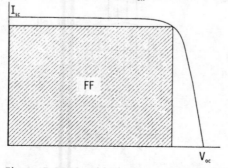

Figure 5.39 Graphical representation of fill factor

Conversion efficiency

Conversion efficiency η is the ratio of the maximum power output (P_{opt}) delivered by the PV cell to the input power (P_{in}) received at a given cell operating temperature T.

$$\eta = P_{opt}/P_{in} = P_{opt}/AG_T$$

where $P_{in} = AG_T$; G_T = solar irradiance in W/m², A = cell area (in m²)
P_{opt} is expressed in W

$$\eta = \frac{V_{oc} \times I_{sc} \times FF}{P_{in}} = \frac{V_{oc} \times I_{sc} \times FF}{AG_T} \qquad \qquad ...(5.29)$$

as $P_{opt} = P_m = I_m V_m$

Solar cell technologies

Performance of a solar cell depends upon three key factors.

1 Type of absorbing material used (such as different semiconductors), which absorbs light (photons) and converts it into EHPs.
2 Type of junction (semiconductor junction) formed, which separates the photogenerated carriers (electrons and holes).
3 Contacts on the front and back of the cell that allow the current to flow to the external circuit.

Several approaches are used to classify solar cells. One approach is the type and structure of light absorbing material used, such as single crystal, polycrystalline, or amorphous. Devices can also be categorized with respect to the number of junctions used in the cell: single junction and multi-junctions or tandem arrangements.

The efficiency regime of the cell can also be used to classify it intermediate efficiency (>10%), high efficiency (>20%), and very high efficiency (>30%). Irradiance can be used to differentiate between cell designs; flat plate or concentrator.

The cell can also be classified in terms of how light enters the device (front wall or back wall), the support material configuration (substrate or superstrate), the semiconductor composition (elemental or compound), or end use (terrestrial or space).

Depending upon the type of light-absorbing material used, solar cell technology is broadly classified into silicon-based technologies and compound semiconductor-based technologies. In recent years, other

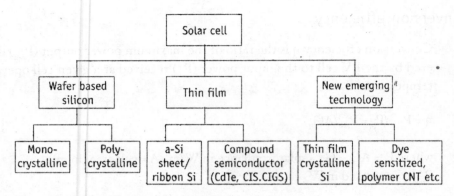

Figure 5.40 Classification of solar cell technologies

emerging technologies such as organic solar cells have come up. Figure 5.40 summarizes the solar cell technologies. However, of all solar cell technologies, wafer-based crystalline silicon is one of the most reliable and most developed technology. Wafer-based crystalline silicon requires a considerable thickness of material (several hundred microns) as it is a poor absorber of light. On the other hand, thin film solar cell technologies require a smaller amount of material (1–3 μm) for the device to function. One of the most important criteria for a thin-film solar cell is that the material to be employed is a direct band gap material with a high optical absorption coefficient. Silicon solar cell technologies can thus be divided into wafer-based crystalline silicon solar cell technology and thin-film solar cell technology.

Wafer-based crystalline silicon solar cell technology

The wafer-based crystalline silicon solar cell manufacturing process involves three stages (Figure 5.41).

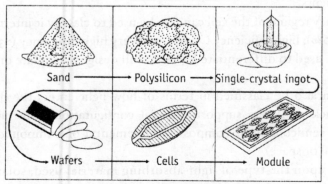

Figure 5.41 Wafer-based solar cell manufacturing process
Source Lasnier and Ang (1990)

1 Material preparation and shaping
2 Cell processing
3 Cell interconnection and encapsulation

Material preparation and shaping

Figure 5.42 illustrates the steps involved in material preparation and shaping, and the stepwise approach for the same is given in subsequent sections.

Quartzite to metallurgical-grade silicon

Quartzite rocks and special strata of quartz sand are the most suitable raw materials for the production of elemental silicon. The commercial extraction process uses the crystalline form of silicon dioxide, quartzite, with 90% or more silica, rather than sand, the main constituent of which is also silica. Sand, with its high amount of impurities, is too expensive to process and so is economically unviable.

Quartzite is reduced in large arc furnaces, by carbon, to produce elemental silicon. The chemical reaction involved in this process can be written as:

$$SiO_2 \text{ (s)} + 2C \text{ (s)} \rightarrow Si \text{ (l)} + 2CO \text{ (g)}$$

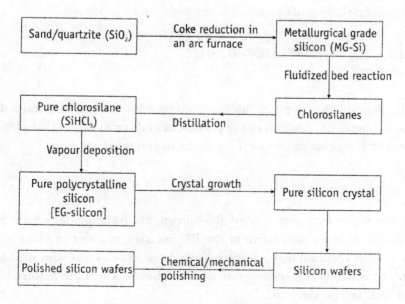

Figure 5.42 Flow diagram for material preparation

In this process, MG (metallurgical grade) silicon is produced with a purity of 98%–99%, which is low for electronic application. Therefore, MG silicon has to be purified further.

Metallurgical-grade silicon to pure polycrystalline silicon (electronic-grade silicon)

The standard method for purifying MG-silicon to EG (electronic grade)-silicon is the Siemens C process. Silicon produced by this method is 99.99% pure and is polycrystalline in form. In this process, $SiHCl_3$ (trichlorosilane) is first obtained, and then reduced to EG-Si.

A bed of fine MG-silicon particles is fluidized and chlorinated with HCl in the presence of a copper catalyst form $SiHCl_3$. $SiHCl_3$ thus formed has to be purified through subsequent fractional distillation (to reduce impurities by a factor of 10^4–10^5) to produce EG (pure) $SiHCl_3$.

$$MG- Si + 3HCl \longrightarrow SiHCL_3 + H_2$$

The CVD (chemical vapour deposition) method is subsequently used to isolate pure silicon from the EG-$SiHCl_3$. Vaporized $SiHCl_3$ is decomposed and reduced with hydrogen at about 1000 °C, resulting in the deposition of silicon on the hot surface of a bridge in the shape of an inverted U. The bridge is heated in a reactor by passing an electric current through it. Through this process, polycrystalline silicon (EG-silicon) rods can be produced.

$$\begin{array}{l} SiHCl_3 + H_2 \longrightarrow EG - Si + 3HCl \\ \text{(vaporized)} \end{array}$$

The deposition process is characterized by a low throughput and a discontinuous operation. Analysis of the present module costs shows that 50% of the manufacturing cost can be ascribed to the silicon substrate alone.

Solar grade silicon concept

Due to the high processing cost of EG-silicon, SG (solar-grade) silicon has been considered as an alternative in the PV industry, in order to make solar cells the silicon required may not necessarily be as pure as EG-silicon. It has not been considered as a second-rate material but as a specific material for efficient solar cell production.

Pure polycrystalline silicon to pure silicon crystal (crystal growth)

The step after production of polycrystalline silicon is crystallization. The crystal growth could be of single crystal Si growth or polycrystal Si growth. The crystal growth method is broadly divided into

- bulk growth leading to ingots, which have to be cut into wafers, and
- ribbon pulling.

Figure 5.43 summarizes the different processes involved in crystal growth. In the ribbon pulling technique, silicon consumption is four to eight times less than in forming ingots, and there is no need for slicing the ingots into wafers. Though there is an economic advantage in ribbon growth as compared to forming ingots it is not widely adopted in the industry, as the techniques are less mature.

Single crystal silicon growth

There are two bulk growth technologies for single crystal silicon growth.

1 Ingot pulling
2 Ingot casting

The CZ (Czochralski) and FZ (floatzone) techniques use the ingot-pulling approach for growing ingot from molten silicon. The ingot grown is then

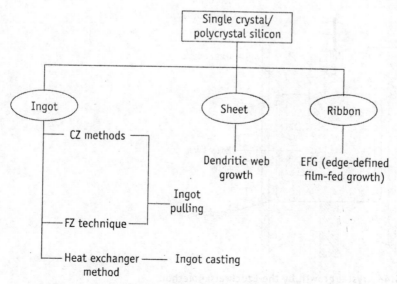

Figure 5.43 Different processes involved in crystal growth

sawn into wafers. However, newer processes, such as dendritic web growth and EFG (edge-defined film-fed growth) are used for growing ribbons, which can then directly be used as wafers.

Ingot-pulling technology
Czochralski method

It is the major commercial method of manufacturing ingots of single crystal silicon. Figures 5.44 and 5.45 show how a crystal is grown.

To start the crystallization process, a small silicon crystal is properly cooled and used as a seed in molten silicon. When the seed is pulled out slowly, silicon solidifies at the interface with the melt and due to controlled pulling, silicon atoms arrange themselves according to the crystallographic structure of the seed. This yields an ingot of single-crystal silicon. The cylindrical ingots are typically 1 m long and 7.5 cm in diameter. A p-type crystal is manufactured directly by adding boron into the melt. Crystals with diameters as large as 15 cm and with length of 1.5 m can be grown using the continuous process. The rechargeable process also allows the crucible to be reused several times, thus reducing the cost of replacement.

In another recent development, a magnetic field is applied to reduce interaction between molten silicon and crucible, thus reducing the usual contamination by carbon and oxygen.

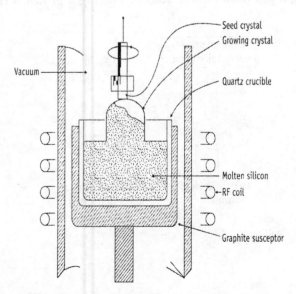

Figure 5.44 Crystal growth by the Czochralski method
Source Garg and Prakash (1997)

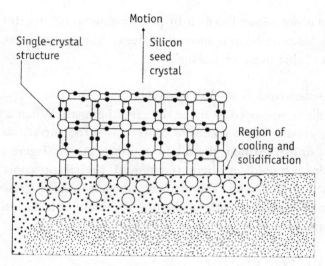

Figure 5.45 How a crystal is grown

Floating zone technique

Here, the polycrystalline ingot is placed over the single-crystal seed and the interface between the two is melted using a movable heating coil (Figure 5.46). As the moving coils are moved up the stationary ingot, the material below the molten interface gets solidified into a single-crystal silicon. The process is called floating zone process (Figure 5.46) as the molten silicon is supported by its own surface tension.

Ingot casting technology
Heat exchanger method

In HEM (heat exchanger method), the single-crystal seed is placed in a crucible (where molten silicon is kept), which is placed on a high temperature heat

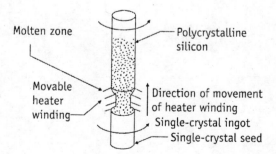

Figure 5.46 Float zone method
Source Lasnier and Ang (1990)

exchanger. The seed is not allowed to melt by passing gaseous helium through the heat exchanger. Molten silicon is allowed to freeze from a seed crystal to form a nearly perfect cubic single-crystal ingot.

Ribbon growth: edge-defined film-fed growth

A graphite-shaped die is employed to shape the crystal growth. When a seed crystal of a a single crystal silicon comes in contact with the surface of the melt, the melt gets drawn through the die by capillary action (Figure 5.47). As the crystal is slowly raised (at about 15 cm/min), the silicon cools and freezes into a single-crystal ribbon. The main advantage of ribbon growth is the simultaneous multi-ribbon growth in one operation and efficient utilization of silicon material with no loss involved in slicing.

Sheet growth: dendritic web growth

A silicon ribbon forms between a pair of dendrites, which provide stability to the edge for width control. It is a slow process without any requirement of die.

Polycrystalline silicon growth

Polycrystalline silicon is made of many grains of single-crystal silicon, the size and quality of which determine the performance of the solar cell. The grains should be as large as possible so that losses such as recombination losses will be minimum at the grain boundaries. The grain size has a major detrimental effect on polycrystalline silicon solar cell characteristics and efficiencies. Figure 5.48 shows the production of multicrystalline/polycrystalline Si ribbon.

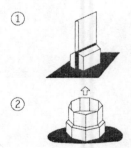

Figure 5.47 Methods of producing Si ribbons
1 EFG (the edge-defined film-fed growth process)
2 RSG (Ribbon growth on substrate)

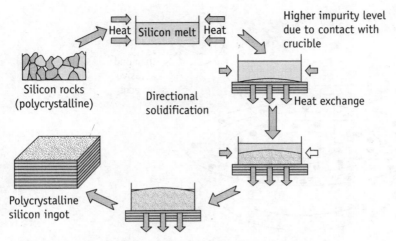

Figure 5.48 Polycrystalline silicon growth

Almost all the methods explained above are used for manufacturing polycrystalline silicon solar cell under less controlled conditions. However, ingot pulling is used mainly in single crystal growth and ingot casting is used for polycrystalline silicon growth.

Silicon crystal ingot to silicon wafer

There are several ways of slicing an ingot into wafers. As the major constraint to crystalline silicon solar cell is its high manufacturing cost, it is understood that low-cost production of crystalline solar cell is possible if efficient cutting and wafering techniques are incorporated (as 50% of the silicon is wasted by the wafering operation itself). A common method is by using an inner diameter saw, as shown in Figure 5.49.

Here, the output is very less because of slow slicing rate (about 30 wafers/h). In addition to a slow rate, the main disadvantage of this process is the amount of silicon wasted. Almost half the silicon gets wasted as *kerf loss*.

However, a multi-blade slurry saw can decrease the *kerf loss* to a large extent. In this case, a thinner wafer of 200 mm can be produced as compared to 450 mm in the case of inner diameter saws. Figure 5.50 shows the production process from multicrystalline ingot to wafer. There are other material removal processes and Table 5.2 shows the comparison of different silicon slicing methods.

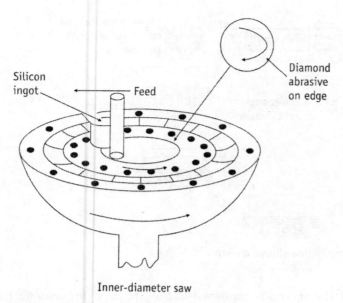

Figure 5.49 Inner-diameter saw

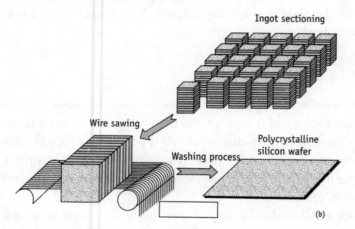

Figure 5.50 Multiple slurry saw machine; Production of polycrystalline Si wafer
Source Lasnier and Ang (1990)

Table 5.2 Comparison of different silicon slicing methods

Slicing method	Mode of application	Production rate wafers/h	Yield (%)	Surface quality	Technical reliability
ID sawing	Conventional	Average	High	Average	High
ID sawing	Advanced	High	Very high	Good	Very high
MBS	Conventional	Low	Average	Average	Good
MBS	Advanced	Average	High	Good	Fair

ID – internal diameter; MBS – multi-blade slurry sawing
Source Lasnier and Ang (1990)

Silicon wafers to polished silicon wafers

After ingot sawing, about 30-mm thick surface layer is damaged. This damaged layer is removed by acid etching (mixture of $1HF$ [48%] + $1CH_3COOH$ [99.8%] + $5HNO_3$ [65%] at a rate of 1 mm/min at 20 °C the surface is also polished with alumina powder.

After polishing, it is cleaned using one of the following procedures.

- Simple cleaning (hot chlorethylene + hot acetone + hot methanol + rinsing in distilled water)
- RCA cleaning (10 minutes in each of $1H_2O_2$ [30%] + $4H_2SO_4$ [96%], $1H_2O_2$ [30%] + NH_4OH (30%) + $5H_2O$, $1H_2O_2$ [30%] + $1HCl$ [37%] + $5H_2O$ and then HF+1/2 distilled water).

Once the wafer is produced, the next step is cell processing, that is, conversion of wafer into solar cell. Cell processing involves junction formation, contact formation, etc., which can be done through a variety of deposition techniques. The general deposition techniques are discussed in the section below.

Deposition techniques

There are basically three types of deposition techniques which are involved in the fabrication of solar cells. These are (1) PVD (physical vapour deposition), (2) CVD (chemical vapour deposition), and (3) liquid phase deposition.

Physical vapour deposition

PVD includes three processes.

1 Vacuum evaporation
2 Sputtering
3 Epitaxial layer deposition

Vacuum evaporation

In the vacuum evaporation process, the metal (target) which is to be deposited over a substrate is placed in a crucible and heated in vacuum to a very high temperature so as to cause melting and evaporation. The metallic vapour is then deposited on the substrate. Vacuum evaporation is one of the most reliable deposition techniques but is very expensive.

Sputtering

In the sputtering process, the coating material or the constituent of the coating material is vaporized from the solid target by high-energy particle bombardment and is then deposited as thin film on a suitable substrate (Figure 5.51). When the high-energy particle bombardment is done by the RF (radio-frequency) potential, it is called RF sputtering.

The RF sputtering comprises two units: (1) power supply and (2) vacuum system. A chamber is used for sputtering which is to be evacuated. However, the chamber is not fully evacuated because discharge will not take place. For this reason, Ar (argon) gas, which is inert in nature, is used. Power supply in the RF sputtering system contains an electronic circuit, which can generate an RF of 13.56 MHz. The RF potential can be applied between the target and the substrate, and substrate is perfectly grounded.

The target is placed inside the evacuated chamber and is made up of a material which is to be evaporated and deposited over the substrate. The substrate is placed just over the target (Figure 5.51).

Due to this high field, electric breakdown occurs, which will ionize the gas (Ar) present in between the target and the substrate. When the target is negatively charged, it will attract the Ar⁺ ion. This high-velocity ion will bombard the target. Due to bombardment of the high energetic ion, the target becomes very hot and the metal becomes vaporized. The metal vapours thus generated spread over the substrate and condense on the surface of the substrate. As target is very hot, some cooling arrangement is made around the target. The sputtered atoms have a very high energy (5–10 eV) compared to thermally evaporated atoms. Sometimes magnetic fields are applied to increase deposition rate and confine the sputtered atom to the substrate region.

Epitaxial layer deposition

Epitaxial layer deposition is a process by which a thin layer of single-crystal material is deposited on single-crystal substrate; epitaxial growth occurs in

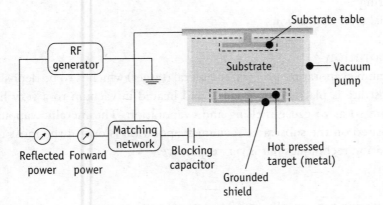

Figure 5.51 Radio-frequency sputtering technique

such a way that the crystallographic structure of the substrate is reproduced in the growing material. Also, crystalline defects of the substrate are reproduced in the growing material.

Chemical deposition

Chemical deposition includes
1 CVD
2 Hotwire CVD
3 Plasma-enhanced CVD
4 Electro-deposition
5 Spray pyrolysis

Chemical vapour deposition

CVD is a chemical process in which the target which is to be deposited over the substrate, is carried by a gas or a liquid, usually at atmospheric pressure. The chemical, which is to be deposited, is heated by using some heating arrangement. Some carrier gas such as oxygen is bubbled through the target solution and the vapour of the target along with the carrier gas moves to the chamber where substrate is placed. In this case, substrate wafers can be placed in the chamber. When vapour enters into the chamber, it sprays over the substrate and a film is deposited over the substrate uniformly (Figure 5.52).

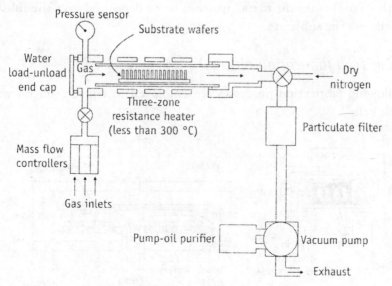

Figure 5.52 Chemical vapour deposition process
Source Lasnier and Ang (1990)

The deposition rate is controlled by concentration of the solution, carrier gas flow rate, substrate temperature, reaction rate at the substrate, and diffusion within the carrier gas.

Plasma-enhanced CVD In case of plasma-enhanced CVD, the reaction is driven by plasma (Figure 5.53). Rest of the process of this deposition technique is similar to the CVD method.

Electro-deposition

In the electro-deposition process, an anode and a cathode are immersed in a suitable electrolyte that contains metallic solution (that is, solution of those metals, which are to be deposited over the substrate), complexing ion, and preservative. Complexing ion prevents the formation of any oxide or hydroxy ion. When electricity is supplied from outside, metal ions are emitted from the anode and deposited on the cathode. Along with the metal ion, some other ions are also emitted from the electrode, which could degrade the quality of deposition. Preservative is used to take care of such unwanted ions.

Spray pyrolysis

In the CVD technique, the compound that constitutes the solid film impinges upon the substrate in a true vapour phase so that reaction can start as soon as it reaches the substrate. In spray pyrolysis process, a liquid droplet of the compound that constitutes the metal, which is to be deposited over the substrate is sprayed over the substrate.

Cell processing (cell fabrication process)

The following fabrication processes are employed to convert silicon wafers into solar cells.

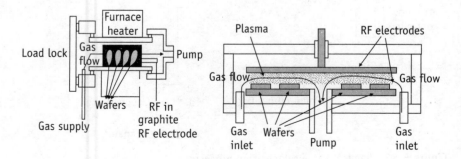

Figure 5.53 Plasma-enhanced chemical vapour deposition process

- p-n junction formation
- Possible back p⁺ region formation
- Electrical contact formation (front and back metal contact)
- Anti-reflection layer deposition

p-n junction formation

The wafer formed through the process described earlier is either p- or n-type. However, to make one solar cell, one p-n junction has to be created to separate the photogenerated carriers. A p-n junction can be created either by a high temperature diffusion process or by an ion implantation process, which is a cold junction process. The advantages of cold-junction processing over high-temperature processing are lower energy consumption, full automation, and lower overall cost. In both the processes, diffusion can be from a solid or a vapour phase.

High temperature diffusion

Standard solar cell technologies produce a base layer material with dopant during crystal growth. The p- or n-type silicon must then be formed upon the base layer depending on the impurity type of the initial wafer. To form a p-type layer on an n-type silicon, boron is commonly used as the impurity from sources such as B_2O_3 (solid), BCl_3 (gas), or BBr_3 (liquid). To form an n-type layer on a p-type silicon, phosphorus is used as impurity from sources such as P_2O_5 (solid) or $POCl_3$ (liquid). Figures 5.54 and 5.55 show the diffusion process for solid, liquid, and gaseous sources.

For a liquid source such as $POCl_3$, a carrier gas (a mixture of N_2 and O_2 in the ratio of 3:1) is bubbled through it at about 950 °C. After a diffusion time of 10 minutes, a diffused layer of about 0.25 mm is formed.

The formation of a p-junction follows Fick's law of diffusion, according to which, if a base layer of doped silicon is coated with phosphorus and then heated, the phosphorus atoms will diffuse into the solid silicon from the regions of high concentration to the regions of low concentration. Diffusion is arrested by lowering the temperatures of an n-type collector transition region, in which the dominant dopant switches from phosphorus to boron.

Cold junction processing

Cold junction processing could be an alternative to a slow, energy-intensive, high-temperature diffusion process. Here, the dopants are ionized and

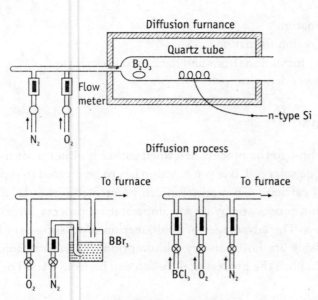

Diffusion process

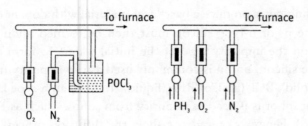

Figure 5.54 High temperature diffusion process to form p-type layer
Source Lasnier and Ang (1990)

Figure 5.55 High temperature diffusion process to form n-type layer
Source Lasnier and Ang (1990)

accelerated to penetrate the target through their kinetic energy. Due to their collision, surface of the base material gets damaged, which can be rectified through annealing.

Advantages of cold junction processing over thermal annealing are less usage of energy, and greater control over dopant concentration and junction depth.

Possible back p⁺ region formation

A back p⁺ region may be formed to improve the cell performance by creating a back surface field, which reduces the carrier recombination loss at the back surface. This can be done by heavy doping of semiconductor.

Electrical contact formation

Once the photogenerated carriers are separated, they have to be collected to flow in an electrical circuit. Therefore, an electrical contact has to be established. The front surface electrical contact (through which light passes) should be made in a way that it can collect as many carriers as possible and at the same time does not shade more than 10% of the cell's surface. The back surface contact can simply be metal sheets. However, good ohmic contact has to be established between the metal and semiconductor interface in order to reduce losses. Aluminium and silver are generally used on the back side. The following processes are used for electrical contact formation.

- Vacuum evaporation
- RF sputtering
- Electroplating
- Electro-less plating
- Screen printing

Of all the processes, vacuum evaporation, RF sputtering, and screen printing are popularly used.

Vacuum evaporation

Metal, which is to be deposited over the cell, is placed in a crucible and heated in vacuum to a very high temperature to cause melting and evaporation.

Metal contact at the back surface

The metallic contact is created at the back surface of the p-type silicon. Metal such as aluminium is deposited by vacuum evaporation on the entire back surface. Aluminium is heated in a vacuum so that it melts, vaporizes, and then deposits on the cooler side of the solar cell. Later a heat treatment at 500–800 °C for about 20 minutes is given to the wafer so that a part of aluminium diffuses into the p-layer, thus lowering the contact resistance.

Metal contact at the front surface

The front contact is also made by vacuum evaporation. After deciding the grid shape and pattern layers, Ti (titanium), Pd (palladium), and Ag (silver) are deposited one over the other by vacuum evaporation. First, a thin layer of Ti is deposited on the n-type silicon, which provides a good adherence to silicon. After this, a layer of Pd is deposited and then finally, at the top, a layer of Ag is deposited. The process is quite capital intensive.

Screen printing

Electrical contacts are usually formed by screen printing. This technology is simple and inexpensive. Here, metal is painted on the entire front surface of the cell and subsequently etched away from the unnecessary parts using a photographic technique called photolithography (Figure 5.56). Photolithography is an optical means for transferring patterns onto a substrate. It is essentially the same process that is used in lithographic printing. In screen-printing process, the photoresist is coated on the metal coating of the substrate. The chrome on glass photo mask is placed over the substrate wafer (Figure 5.57). When uniform UV radiation is passed through the glass photo mask, a latent image would be created in photoresist (a).

There are three separate layers of metal: a thin layer of Ti as a bottom layer for good adherence, a top layer of Ag for low resistance and solderability, and a middle Pd layer for restricting the reaction between Ti and Ag. For front contacts, the screened paste normally consists of Ag powder combined with frit (low melting point glass composites) and organic binders. For rear surface, aluminium is often added. After screening each side, pastes are dried by heating to 350–400 °C. Then the organic solvents evaporate and the metal powder becomes a conducting path for electrical current. Glass present in the paste provides good adherence to the silicon surface. Contacts must be fired at above 700 °C to give reasonable metal resistivity. The metal finger lines cannot be made very thin by screen printing.

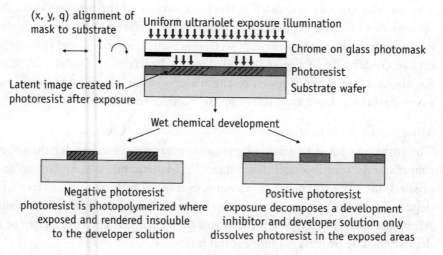

Figure 5.56 Screen printing process through photolithography
Source <http://www.ee.washington.edu/research/microtech/cam/PROCESSES/PDF%20FILES/Photolithography.pdf>

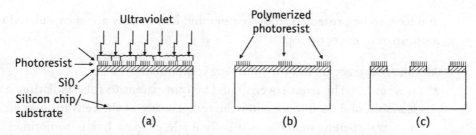

Figure 5.57 Screen printing process

Anti-reflection coating

Untreated silicon reflects more than 30% of incident light. However, a thin layer of transparent material can be deposited over the cell before and after the metal contact, which can reduce the reflectivity of the surface and absorb as much incident light as possible. A single layer of anti-reflection coating will reduce the cell surface reflection to 10% whereas a double coating will reduce the surface reflection to three per cent. SiO_2 (silicon dioxide), SiN_x (silicon nitride), TiO_2 (titanium dioxide), and Ta_2O_5 (tantalum pentoxide) are generally treated as anti-reflection coating. The main advantage of SiN_x is that in addition to acting as an anti-reflection coating, it also passivates the emitter surface.

Cell encapsulation and interconnection (module fabrication)

It was mentioned earlier that under STCs, a typical 4-inch (10.2 cm) diameter crystalline silicon solar cell, or a 10 cm × 10 cm polycrystalline cell will provide power of 1–1.5 W at a voltage of 0.5–0.6 V. Such a low voltage is not sufficient to run most appliances. So, to increase the voltage, solar cells are connected in series to make one module. However, during the manufacturing process, there is some variation in the quality of cells and it is very difficult to maintain the precise quality for all cells. In general, the current produced by commercial cells has a higher degree of dispersion than the voltage. Therefore, solar cells with a given output are first sorted and similar solar cells are connected in series to form a module.

Crystalline silicon modules consist of individual cells connected and encapsulated between a transparent front, usually glass, and a backing material. Encapsulation is required for avoiding degradation, as oxygen present in air will corrode the surface.

Each and every cell is not encapsulated rather all cells are interconnected and then the whole system is encapsulated. The PV cells are thin and brittle and have to be protected from weather and hence, they are encapsulated with a transparent cover on top.

Module construction has the following components.

- *Front cover* The front cover should be transparent to solar radiation, easily cleanable, and should not allow the temperature of the cells to go high. The front transparent material is usually made of glass but is sometimes also made of a UV-stabilized plastic sheet.
- *Encapsulant* The encapsulation not only provides mechanical strength to the cells but also protects them from birds, dirt, dust, moisture penetration, etc. It also saves interconnections and metallic contacts from the weather and wind. The material used for encapsulation should be UV-stabilized, and able to withstand temperature extremes and thermal shock. The most widely used encapsulant is EVA (ethylene vinyl acetate).
- *Solar cells and the metal interconnect.*
- *Back cover* The back of the module is covered with a layer of tedlar or mylar. A frame of aluminium or other composite material provides the necessary mechanical stability to the module (Figure 5.58).

Figures 5.58 (a) and (b) show a module made of single-crystalline silicon whereas Figure 5.58 (c) shows polycrystalline silicon module.

Maximum and achieved efficiencies

Figure 5.59 shows the maximum efficiencies that could be obtained with different materials. From Figure 5.59, it can be seen that the best confirmed efficiency of C-Si is near the maximum of the curve. It is also observed that CIS (copper indium diselendie) and a-Si:H (hydrogenated amorphous silicon) are potential candidates for the thin film solar cell application. Though GaAs is the material of choice, it is very expensive and used only for space application.

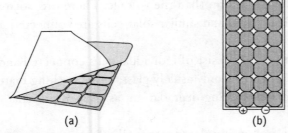

(a)   (b)   (c)

Figure 5.58 Module structure before lamination and as it is viewed from the rear
Source Solanki (2004)

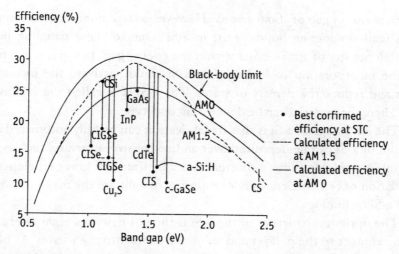

Figure 5.59 Maximum efficiency of different materials

Thin-film solar cell technology

In thin-film solar cell technology, a very thin film of material is used which is sufficient to absorb light and give the desired result. Here, less material is used with a relatively easy manufacturing process. Though the module efficiency of thin-film solar cell is lesser (about 8%–10%) compared to that of wafer-based crystalline silicon solar cells, potentially it is cheaper to manufacture. So, this technology has the potential of reducing the module cost through lower material and lower energy consumption. In case of thin film solar cells, integrally connected modules are produced directly without the interconnection of cells.

Some of the emerging thin-film solar cells are as follows.

- a-Si
- CdTe
- CIS, CIGS (copper indium gallium diselenide)
- Thin-film crystalline silicon

Studies show that if cells are made with thin-film crystalline Si such that they trap enough light, cell efficiencies greater than 20% can be obtained even with just a few micron thick layer.

Amorphous silicon solar cell

The structure of a-Si differs from that of crystalline silicon. The randomness in the atomic arrangement of a-Si material makes it a direct-gap material with an optical energy gap of about 1.75 eV. However, a large number of incomplete bonds (called dangling bonds) exist in a-Si. Due to these dangling bonds, very high density of states exists within the energy gap. In 1969, it was found that the incorporation of hydrogen in a-Si could passivate the incomplete bonds and reduce the density of states. The a-Si:H is called the intrinsic (i) layer. Then, the material can be doped as n- or p-type.

The advantage of a-Si is that its fabrication can be fully automated and a thin film of a-Si can be deposited over an inexpensive substrate, such as glass or stainless steel. However, a shorter carrier lifetime and lower efficiency and degradation of cell efficiency under exposure to light are the major drawbacks of the a-Si technology.

The optimum structure of the a-Si is the p⁺-i-n, with a transparent conducting contact to the p⁺ layer and an ohmic contact to the n-layer. As both n and p materials have poor transport properties, the i-layer (a-Si : H) is incorporated to improve transport properties.

Production of amorphous silicon photovoltaic modules

Figure 5.60 shows the process followed for the production of a-Si solar cells/module.

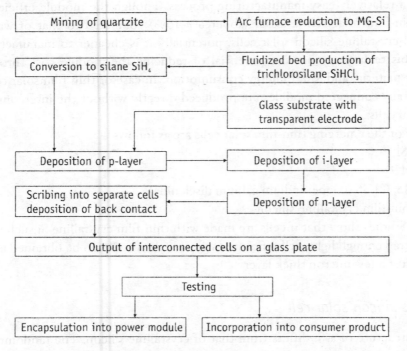

Figure 5.60 Flow diagram for the production of a-Si solar cells/module

Generally, SiH_4 is decomposed in an RF glow discharge reactor near a heated substrate. The exposure time, discharge power, and flow rate all control the film thickness.

Amorphous silicon fabrication process

The different fabrication processes for a-Si modules can be classified as follows.
- Glow-discharge method
- Sputtering method
- CVD method

Glow-discharge method

A multi-chamber deposition system by the RF glow discharge is the most common approach for fabrication of high-quality a-Si solar cells. It consists of four reactors in a large chamber. The first reaction chamber is for hydrogen plasma etching and the other three chambers are for deposition of p^+-i-n layers (Figure 5.61). First, the substrate is placed in the p-chamber, where SiH_4 and B_2H_6 are present. When discharge occurs, B_2H_6 breaks down and boron is deposited over the substrate. The thickness of the p-type layer is 10 nm. When p-type doping is over, the second chamber is opened where only SiH_4 is applied. Therefore, a layer of intrinsic a-Si:H layer is deposited. Its thickness is 0.1–0.2 mm. Then the substrate goes to the third chamber, where n-layer (thickness of 20 nm) is formed when discharge occurs. Then, the metal plate of aluminium (~200 nm) is deposited for metal contact.

The main disadvantage of a-Si solar cell is degradation in the performance of the silicon solar cell when exposed to light. It is because the light absorption by the i-layer creates additional defects, thus increasing the density

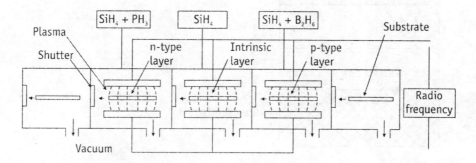

Figure 5.61 Multi-chamber deposition system by radio frequency glow discharge

of trapping and scattering states. This effect is called the *Staebler–Wronski effect*. The main decrease in efficiency occurs in the first 100 hours of exposure after which the cell parameters remain fairly constant.

Multi-junction solar cell (tandem solar cells)

Multi-junction cells are the combination of cells where more than one p-n junction solar cells of different bandage are stacked together. In amorphous solar cell as a Staebler-Wronski effect depends primarily on the thickness of the i-layer, it can be reduced by creating a multi-junction structure (tandem structure) where absorption of light can be split into two or three separate layers. Such cells are more stable and have a higher efficiency than the single-junction solar cells (Figure 5.62).

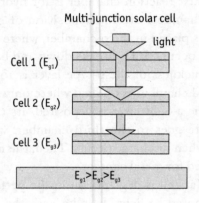

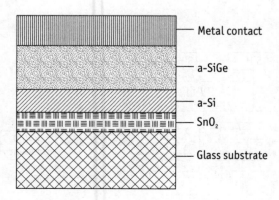

Figure 5.62 Multi-junction structure based on a a-Si:H and a-Ge:H

Copper indium diselenide cells

CIS (copper indium diselenide) is a direct gap material and can be either p- or n-type. The electronic property of CIS depends on the copper/indium ratio. Though a p-n homo-junction of CIS is possible, it is unstable and inefficient. However, it is observed that a hetero-junction of CIS with CdS (cadmium sulphide) would be the best possible option. Again, since CdS is only grown as n-type, CIS has to be of p-type in order to create the p-n junction. Figure 5.63 shows the flow chart for production of CIS solar cells.

Fabricating of CIGS (copper indium gallium diselenide) solar cell

Figure 5.64 shows the structure of the CIGS solar cell. Initially, a molybdenum-based electrode is sputtered and deposited over the 7059 corning glass substrate. Such a type of glass is used because it can resist higher temperatures for a longer time. Molybdenum is used for making ohmic contact with the CIS layer once the back contact is framed.

A CIS layer is deposited over the molybdenum-plated glass by three source vacuum evaporation or RF sputtering process. In the three-source evaporation process in an evacuated chamber, copper indium and selenium are placed at the vertices of one isosceles triangle. CdS, window layer, and top contacts can be deposited using various deposition techniques as discussed earlier.

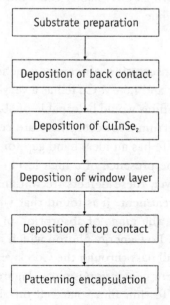

Figure 5.63 Flow chart of the production of copper indium diselenide module

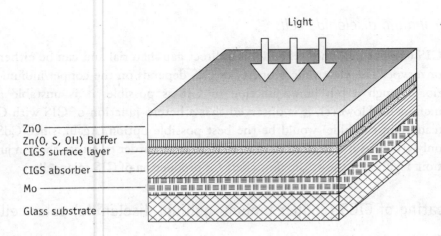

Figure 5.64 Structure of the copper indium gallium diselenide solar cell

For efficient use, CIS has to be heavily doped at the metal contact side and lightly doped near the junction. The optimum device used is a very thin layer of cadmium sulphide with a wide gap window layer of highly conductive material. ZnO (zinc oxide) is basically used as a window layer. The function of the window layer is to minimize the optical reflection and absorption losses and transport the photogenerated majority carrier current to the outside circuit with minimal electrical resistance losses. It is also found that incorporation of gallium into the CIS (known as CIGS solar cells) improves the fill factor and requires fewer cells per module.

Cadmium telluride cells

It is observed that highest solar energy conservation efficiency is achieved with semiconductor of band gap between 1.5 and 1.6 eV. CdTe has a very ideal band gap of 1.4 eV. In addition to this, CdTe films would be capable of absorbing all the useful energy from the solar spectrum. However, there are certain problems associated with CdTe. Though CdTe has an ideal band gap for solar absorption, its performance is limited by the presence of a high level of defects at the middle of the band gap. However, these defects (recombination states) can be reduced in post-deposition treatment. It is found that when a hetero-junction of Cds and CdTe is formed, the range of absorption of light will be increased due to different band gaps. Low-energy photon which is less than the band gap of CdS (that is, 2.4 eV) will pass through the CdS layer and may be absorbed in CdTe. In such case, current generation is more. Figure 5.65 shows the typical structure of the CdS/CdTe thin film solar cell on glass a substrate.

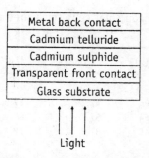

Figure 5.65 Typical structure of a cadmium telluride/cadmium sulphide solar cell on a glass substrate

Fabrication of cadmium telluride solar cell

First, the glass substrate is cleaned by an organic or inorganic cleaner. Generally, trichloro ethylene vapour is used as an organic cleaner. Once the glass is cleaned, ITO (indium tin oxide) is deposited as a transparent front contact. The function of the transparent layer is to allow maximum light to pass through it and at the same time make good electrical contact. The ITO-coated glass is put in a vacuum chamber for CdS deposition. In the deposition process, 0.5-mm thick CdS layer is coated over the glass substrate following one of the deposition techniques. The cadmium sulphite coated substrate is put in a chamber for CdTe deposition. The deposition of CdTe is prepared basically in three ways: close space sublimation, vacuum evaporation, and electro-deposition.

Cell parameters of silicon solar cell

The theoretical power conversion efficiency for a solar cell based on a single semiconducting material is limited to 31%. One of the fundamental limitations of solar cell efficiency is the band gap of the semiconductor from which the cell is made. The band gap of the material cannot perfectly match the broad range of solar radiation, which has usable energy in the photon range of 0.4–4 eV. Light with energy below the band gap of the semiconductor will not be absorbed and thus, not be converted; light with energy above the band gap will be absorbed but the excess energy will be lost in the form of heat. Decades of research in developing single-material solar cells has led to cell efficiencies, which lie close to the theoretical limit. However, different layers could be stacked to receive photons at all energies, reaching efficiencies better than 70%. However, in that case, there may be many other problems such as cell stability, etc. Table 5.3 shows the cell parameters of different solar cells.

Table 5.3 Confirmed cell parameters of selected silicon solar cells (measured under the global AMI-5 spectrum (1000 W/m^2 at 25 °C)

Classsification	Efficiency η (%)	V_{oc} (V)	I_{sc} (mA/cm^2)	FF (%)
Silicon cells				
Si (crystalline)	24.7±0.5	0.706	42.2	82.8
Si (multicrystalline)	20.3±0.5	0.664	37.7	80.9
Si (thin film transfer)	16.6±0.4	0.645	32.8	78.2
III–V cell				
GaAs (crysrlline)	25.1±0.8	1.022	28.2	87.1
GaAs (thin film)	23.3±0.7	1.011	27.6	83.8
GaAs (multicrystalline)	18.2±0.5	0.994	23.0	79.7
InP (crystalline)	21.9±0.7	0.878	29.3	85.4
Polycrystalline thin film cells				
CIGS (cell)	18.4±0.5	0.669	77.0	—
CIGS (submodule)	16.6±0.4	2.643	75.1	—
CdTe (cell)	16.5±0.5	0.845	75.5	—
Amorphous/Nanocrystalline Si cells				
Si (nanocrystalline)	10.1±0.2	0.539	24.4	76.6
Multijunction devices				
a-Si/CIGS (thin film)[a]	14.6±0.7			
a-Si/uc-Si (thin submodule)	**11.7±0.2**	**5.462**	**2.99**	**71.3**

V_{oc} – open-circuit voltage; I_{sc} – short-circuit current;   FF – fill factor;   [a] – unstabilized results

Compiled from Progress in *Photovoltaic Res Appl.* 2005, **13**: pp. 49–54, John Wiley & Sons Ltd

Emerging technologies

There are several emerging technologies for solar cell/module production. Some of these technologies are thin film monocrystalline silicon solar cell, micromorph tandem solar cell, organic solar cells, etc.

Thin film monocrystalline silicon solar cell

The process involved in manufacturing thin film monocrystalline silicon solar cell allows the transfer of epitaxially deposited (mono) crystalline silicon thin films of high electronic quality from a host wafer onto a foreign substrate.

The epitaxial film is grown on top of a monocrystalline silicon wafer, which was treated to allow the consecutive separation of the film. The treatment of the wafer is rather simple: the wafer is anodically etched in an HF (hydrogen fluoride) containing solution. Etching is a two-step process with the etching current switching from low to high current density. This process leads to a layer structure with a low-porosity layer of several 100 nm thickness on top of a thin buried layer of high porosity. When the wafer is heated to high temperatures of about 1100 °C, both the porous layers recrystallize, leading to a monocrystalline layer on top containing large voids, which is called QMS (quasi-monocrystalline silicon). The buried layer with high porosity, instead, forms a layer with only small amounts of silicon. This layer is mechanically weak and allows the detachment of the QMS layer from the original wafer. During recrystallization, pores of the QMS film close completely near the surface. Therefore, one obtains an undisturbed surface, which allows for high-quality epitaxial growth. The QMS technique enables one to repeatedly use a silicon wafer for epitaxy. After device fabrication, the layer is detached from the wafer and transferred to a substrate suitable for a desired application. Figure 5.66 shows the photograph of such a transferred film. The films are flexible even though they have a monocrystalline structure.

The possibility of reusing an individual silicon wafer several times as a host wafer for the epitaxial deposition and the subsequent formation of the solar cell opens the way for a cost-effective thin-film technology while preserving the high material quality of monocrystalline silicon.

Micromorph tandem solar cell

Considerable progress has been made so far in solar cell research, in terms of device design and material quality. One of the most important inventions is the micromorph tandem solar cell, combining a layer made of microcrystalline silicon with an amorphous silicon layer (mc-Si/a-Si:H).

Figure 5.66 Photograph of a 14-µm-thick epitaxial silicon film on top of a 1.5-µm-thick quasi-monocrystalline silicon film
Source http://solar-club.web.cern.ch/solar-club/SolPV/autres/organics.html

Organic solar cell

Organic and nano solar cell production uses low-cost materials and involves a simpler and less energy-intensive manufacturing process compared to conventional solar cells. In addition, the organic solar cells' ultra-thin flexible material can be applied to large surfaces by printing or spraying the material onto a roll of plastic, making it flexible, versatile, and easy to install.

Working of organic and nano solar cells

Many materials and approaches are used in making organic and nano solar cells. However, the basic principle is the same, that is, using light to create a separation of charge inside a device. One of the most advanced organic solar cells is DSSC (dye-sensitized solar cell), invented in 1990 by Prof. Michael Graetzel of the Swiss Federal Institute of Technology. It works on a principle similar to photosynthesis. The DSSC cell uses an organic dye (photosensitizer) to absorb sunlight and create pairs of charged particles. These charged particles are transported through nanoporous materials to collector electrodes, creating electrical current.

Figure 5.67 shows the structure of DSSC. DSSC is created using layers of electron-acceptor and electron-donor organic PV materials between the two electrodes. These organic PV materials release electrons when light shines on them. Electron 'donor' material conducts negatively charged particles (electrons) to the negative electrode, while the electron 'acceptor' material conducts positively charged particles (holes) to the positive electrode. The separation of charge created by this effect flows through the circuit as electricity (Figure 5.68).

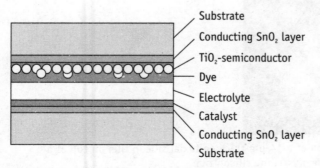

Figure 5.67 Structure of dye-sensitized solar cell

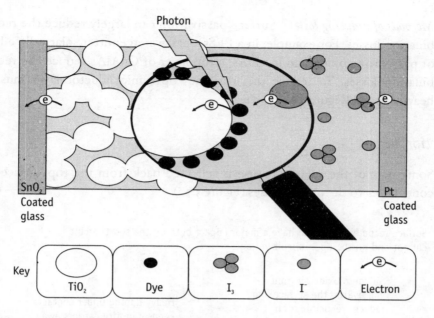

Figure 5.68 Working of dye-sensitized solar cell

Different losses in solar cells

The power of a solar cell and, therefore, its efficiency is limited by the following factors.

- Fundamental losses such as recombination and reflection losses
- Unabsorbed radiation and excessively strong radiation
- Current loss due to collection inefficiency
- Shading effect
- Losses due to series and shunt resistances

All these losses along with the possible methods of reduction are discussed in the subsequent sections.

Losses due to recombination

Recombination (process by which an electron combines with a hole and the pair is annihilated) occurs mostly at the semiconductor surface, at the lattice defect, or at the impurities site. It also occurs at the Ohmic metal contacts to the semiconductor.

Methods of reducing losses Surface passivation can largely reduce the recombination losses. For example, in a PESC (passivated emitter solar cell), a layer of passivating oxide and in GaAs, a top layer of GaAlAs reduce the recombination losses. To reduce the loss, the top semiconductor layer must be heavily doped (Figure 5.69).

Reflection losses

Some part of the solar radiation is reflected back from the top surface and contributes to the overall losses (Figure 5.70).

Surface recombination can have a major impact both on the short-circuit current and on the open-circuit voltage.

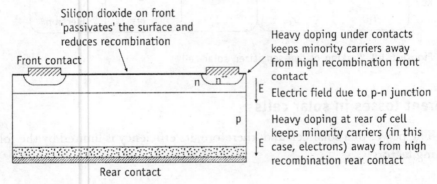

Silicon dioxide on front 'passivates' the surface and reduces recombination

Front contact

Heavy doping under contacts keeps minority carriers away from high recombination front contact

E Electric field due to p-n junction

Heavy doping at rear of cell keeps minority carriers (in this case, electrons) away from high recombination rear contact

Rear contact

Figure 5.69 Surface passivation

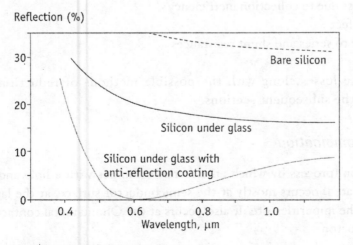

Reflection (%)

Bare silicon

Silicon under glass

Silicon under glass with anti-reflection coating

Wavelength, μm

Figure 5.70 Variation of reflection losses

Methods of reducing losses By texturing, or by applying ARC (anti-reflection coating) (Figure 5.70), reflection losses can be reduced from about 30% to 4%. A thin layer of dielectric material with a specially chosen thickness can act as an ARC. A single crystalline substrate can be textured by etching along the faces of the crystal planes whereas texturing of polycrystalline wafers is more difficult due to different orientation (Figure 5.71).

Unabsorbed radiation and excessively strong radiation

A considerable amount of solar radiation is not absorbed by the semiconductor because of its inability to absorb the lower band gap light. Again, when radiation with energy higher than the band gap of the solar cell is absorbed, in addition to EHPs, it also generates heat. These are the two major factors contributing to losses from a solar cell.

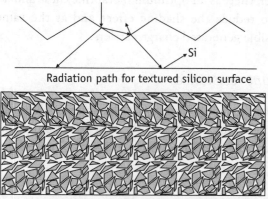

Radiation path for textured silicon surface

- A single crystalline substrate can be textured by etching along the faces of the crystal planes
- Etching in alkaline or acidic solution
- Texturing of multi crystalline wafers are more difficult due to different orientations

Figure 5.71 Surface texturing to minimize reflection

Methods of reducing the losses A tandem structure can reduce the losses as it consists of a stack of several cells of different band gap, and photons of different energy levels are absorbed by different layers. The top cell must be made of a larger band gap so that light of shorter wavelength can be absorbed first. The transmitted light is then converted by the bottom cell.

Current losses due to collection inefficiency

All the generated carriers are not able to reach the junction. Some of them recombine before reaching the junction and so do not contribute to the current. The ratio of charge carriers that are able to reach the p-n layer to the total number of charge carriers is known as the *collection efficiency*. In a crystalline silicon solar cell, carrier transport is good because of the good transport property of the material. However, in amorphous and polycrystalline thin films, electric fields are required to pull carriers. Therefore, collection efficiency can be influenced during the manufacturing process itself.

Shading effect

Shading effect is observed due to the top contact design and the metal grid pattern. The metal grid pattern obstructs some of the radiation from entering into the material. However, there is an optimum metal thickness and a height-to-width ratio in order to reduce the shading effect and at the same time, collect the maximum possible generated charge carriers.

Losses due to series and shunt resistances

Series and shunt resistances reduce efficiency of the solar cell by dissipating power in the resistances. Series resistance R_s depends mainly upon the resistivity of the grid contact and the surface layer (Figure 5.72). The

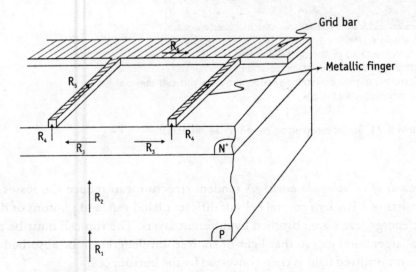

Figure 5.72 Series resistance in a solar cell

series resistance is the sum of

- resistance R_1 of the back Ohmic contact with the p-type semiconductor
- the p-type base material resistance R_2
- resistance R_3 due to the sheet resistance of the n or n$^+$ surface layer (n$^+$ means the n layer is heavily doped)
- the contact resistance R_4 between the top Ohmic contact and the n or n$^+$ layer
- resistance R_5 along the metallic fingers and the resistance R_6 along the grid bar

Shunt resistance is due to current leakage at the junction and depends on the method of junction construction.

Photovoltaic system engineering

This section will give some aspects of the PV system engineering. However, detailed discussion about the PV systems and their designing is beyond the scope of this book.

A PV system consists of the following components.

- PV generator
- Battery
- Charge controller
- Inverter/converter
- Mounting structure and tracking device
- Interconnections and other devices

BoS (Balance of System) includes all components mentioned above except the PV generator. In other words, a PV system consists of a PV generator and a BoS. However, components of a PV system depend upon the type of configuration, which, in turn, depends upon the application.

Photovoltaic generator

The heart of the PV system is the PV generator or PV array. Solar cells are connected in series to form a module, and modules are connected in series or parallel to form a PV array (Figure 5.73). The number of cells in a module is governed by the voltage of the module. The nominal operating voltage of the system has to be matched with the nominal voltage of the storage subsystem. Most module manufacturers, therefore, have standard configurations, which can work with a 12-V storage system (batteries). Allowing for some over-voltage to charge the battery and to compensate for a lower output under less than

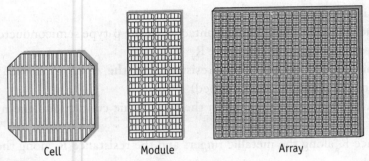

Cell   Module   Array

Figure 5.73 Hierarchy of photovoltaic generators

perfect conditions, it is found that 33–36 solar cells connected in series form one module ensuring reliable operation.

Parameters influencing photovoltaic system operation

Temperature and irradiance are the major parameters influencing the PV system operation.

Effect of temperature on voltage of photovoltaic module

The open-circuit voltage V_{oc} of an individual silicon solar cell reduces by 2.3 mV for every one-degree rise in temperature T of the solar cell. Therefore, the voltage coefficient is negative and is expressed as

$$dV_{oc}/dT = -2.3 \text{ mV}/°C$$

As a module consists of large number of cells in series, the voltage coefficient can be expressed as

$$dV_{oc}/dT = -2.3 \times n_c \text{ mV}/°C$$

where n_c is the number of cells connected in series.

The above equation can be integrated to obtain

$$V_{oc}(T_c) = V_{oc}(\text{at } 1 \text{ kW/m}^2) - 0.0023.n_c (T_c - T_{STC}) \quad\quad ...(5.30)$$

As mentioned, it is important to note that the voltage is determined by the operating temperature of the solar cells, which is not the same as the ambient temperature (T_a).

Hence, the characterization of the PV module is completed by measuring the NOCT (normal operating cell temperature), which is defined as the cell temperature when the module operates under the following conditions in open circuit.

Irradiance = 0.8 kW/m²

Spectral distribution = AM 1.5

Ambient temperature = 20 °C

Wind speed > 1 m/s

NOCT is used to determine the cell temperature T_c for other conditions of module operation by using the following equation.

$$T_c = T_a + ([(NOCT - 20) / 0.8] \times G) \qquad ...(5.31)$$

where G is the global irradiance expressed in kW/m².

Effect of temperature and irradiance on current and voltage of photovoltaic module

The short-circuit current I_{sc} of a module is proportional to the irradiance and at a given irradiance is given by

$$I_{sc} \text{ (at 1 kW/m²) G (in kW/m²)} \qquad ...(5.32)$$

Therefore, short-circuit current varies continuously in a day. Voltage is a logarithmic function of the current, which varies linearly with irradiance. Hence voltage also varies logarithmically with irradiance. Therefore, in a day, voltage varies less than the current with irradiance. Figure 5.74 shows the effect of irradiance and temperature on current and voltage of PV module. Figure 5.75 gives a flow chart for the calculation of module parameters under operating conditions. Thus, operating parameters of the PV modules vary with irradiance and temperature.

The manufacturer specifies the following parameters.

- Short-circuit current I_{sc}
- Open-circuit voltage V_{oc}

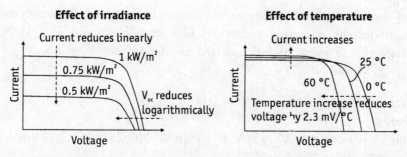

Figure 5.74 Effect of irradiance and temperature on cell/module performance

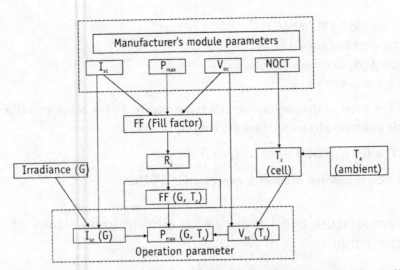

Figure 5.75 Flow chart showing module parameters under operating conditions

- Maximum power P_{max}
- NOCT

Based on these, the operation parameters can be calculated as follows.
- Short-circuit current at a given irradiance is calculated from Equation 5.32.
- The solar cell temperature T_c is obtained from Equation 5.31.
- Once T_c is known, open-circuit voltage V_{oc} can be calculated by using Equation 5.30.
- From the manufacturer's module parameter, FF can be calculated. If the sheet resistance of the module is known, which is independent of the module condition, then FF under a particular temperature and irradiance can be calculated. The FF can be assumed to be constant.
- Once short-circuit current, open-circuit voltage, and FF under operating conditions are known, the maximum power of the module can be calculated.

Example 2
Find out the parameters of a module formed by 33 solar cells connected in series, under the operating conditions $G = 700$ W/m² and $T_a = 34$ °C. The manufacturer's value under standard conditions are $I_{sc} = 3A$, $V_{oc} = 21$ V, $P_{max} = 46$ W, and NOCT $= 45$ °C.

Solution
Given
I_{sc} (at 1 kW/m²) = 3 A; $V_{oc} = 21$ V; $P_{max} = 46$ W; NOCT $= 45$ °C; $G = 700$ W/m²; $T_a = 34$ °C; $n_c = 33$; $T_{STC} = 25$ °C

1   Short-circuit current (I_{sc}) $= I_{sc}$ (at 1 kW/m²) × G (in kW/m²)

$= 3 \times 0.7$ (kW/m²) = 2.1 A

2   Solar cell temperature (T_c) $= T_a + ([(NOCT - 20)/0.8] \times G$ [kW/m²])

$= 34 + (0.7 \times [45 - 20]/0.8)$

$= 55.8\,°C$

3   Open-circuit voltage (V_{oc}) $= V_{oc}$ (at 1 kW/m²) $- 0.0023 \times n_c \times (T_c - T_{STC})$
at 55.8 °C

$= 21 - 2.27$

$= 18.73$ V

4    $FF = \dfrac{P_{max}}{V_{oc} \times I_{sc}}$ $= 46/(3 \times 21) = 0.71$

5    $P_{max}(G, T_c) = FF \times V_{oc} \times I_{sc}$ $= 2.1 \times 18.73 \times 0.71$

$= 27.92$ W

It can be noted that the manufacturer's P_{max} is given as 46 W, whereas with the given field condition, P_{max} is 27.92. That means the module is operating at about 60% of its nominal rating.

Interconnection of photovoltaic modules

Modules are interconnected to constitute PV array/PV generator. These interconnections have suitable bypass and blocking diodes. These diodes protect the modules and prevent the PV generator to act as a load when not irradiated. The modules are connected in series or in parallel depending upon the voltage and current requirement at the output.

The number of PV modules to be connected in series N_s is determined by the selected DC voltage and the number of modules in parallel N_p can be determined by the current required from the generator (Figure 5.76).

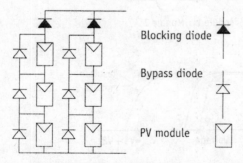

Figure 5.76 Photovoltaic array with blocking and bypass diodes

In Figure 5.76, N_s = 3 and N_p = 2, therefore, the output voltage of the PV generator will be three times that of the module, and the generator current will be twice that of the module. This is when the modules are exactly identical but in practice this is not so because of the following reasons.

- During the manufacturing process, there is some variation in the quality of modules and it is difficult to maintain the exact quality for all modules. In general, the current produced by commercial modules suffers a higher degree of dispersion than the voltage.
- Different operating conditions may exist for different parts of the same PV generator, for example, shading of one region of a module compared to the other.

Due to this variation in operating conditions and operating parameters of different modules, mismatch losses occur. When two or more modules with different electrical properties or under different conditions (mismatch modules) are connected, the output power is determined by the module with the lowest output power. Mostly, mismatches occur due to differences in either the short-circuit current or open-circuit voltage.

Mismatch in modules connected in series

Mismatch in modules connected in series could be due to mismatch in short-circuit current or in open-circuit voltage.

Mismatch in open-circuit voltage of modules connected in series

Figure 5.77 gives the I-V characteristic of the resultant modules. The figure shows that as modules are connected in series and there is no mismatch in current rating, the same current will flow through all modules, and the total voltage is the summation of the voltages of all modules.

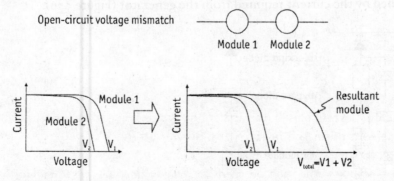

Figure 5.77 I-V characteristic of the resultant modules of different V_{oc}

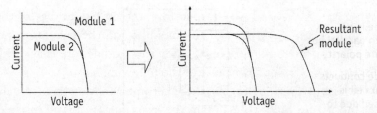

Figure 5.78 I–V characteristic of the resultant modules of different I_{sc}

Mismatch in short-circuit current

Figure 5.78 shows the I-V characteristic of the resultant modules. When there is a mismatch in short-circuit current, the current flowing through the circuit is due to the module with the least current (current of poor module). The extra power generated by the good module is dissipated in the poor module. Such mismatch could result in irreversible damage.

In certain circumstances, a cell/module can operate as 'load' for other cells/modules acting as 'generators'. Consequently, the cell/module, which acts as a load, dissipates energy and its temperature increases. If the cell temperature rises above a certain limit (85–100 °C), the encapsulating materials can be damaged, and this will degrade the performance of the entire module/string. This is called hot spot formation. Hot spot heating occurs where there is one low current value solar cell in a string of several high current solar cells.

Mismatch in modules connected in parallel

Mismatch in parallel connection is not as severe as in series. The voltage across the cell/module combination is always the same and the total current from the combination is the sum of the currents in the individual cells/modules.

Role of bypass diode and blocking diode in a photovoltaic array

The destructive effect of hot spot heating can be circumvented with the use of a bypass diode. The bypass diode is connected in parallel but with opposite polarity (Figure 5.79).

As it is reverse biased, it starts conducting when a solar cell is reverse biased due to a mismatch in the short-circuit current. Normally, for a 36-cell module, two bypass diodes are used.

A blocking diode may be used for minimizing the mismatch losses. It blocks the current from flowing into the shaded module from the parallel

Bypass diode is connected in parallel but with opposite polarity

Bypass diode conducts when a solar cell is reverse biased due to mismatch in short-circuit current

The maximum reverse bias across the poor cell is reduced to about a single diode drop

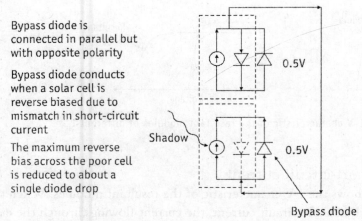

Figure 5.79 Role of bypass diode

module. It also prevents the flow of current from the battery to the module during night.

Battery storage

As solar modules can generate power only when exposed to sunlight, there is a need for batteries to store energy collected during high sunshine hours and make it available during night and periods of low sunlight. There is a demand for frequent charging and discharging of battery in the PV system. So, the characteristics of the battery used here are not the same as car battery which is not suitable for deep discharge. The different types of batteries available in the market for PV applications are lead acid battery (most common), nickel cadmium, and NiMH (nickel metal hydride). A typical solar system battery lasts for three to five years.

The performance of the battery can be characterized by three parameters.

1 Ah (ampere hour) capacity
2 SoC (state of charge)
3 DoD (depth of discharge)

Ah capacity The capacity of the battery can be defined by Ah. A battery capacity of 60 Ah means either 1 A current can be drawn from the battery for 60 hours or that 60 A can be drawn for 1 hour.

SoC The SoC at a particular instant indicates the amount of charge available with the battery at that instant. For example, for a 40 Ah battery, at a particular instant, if SoC is 40% that means that capacity of the battery is

16 Ah (40% of 40 Ah). For a 2V battery, the electrolyte composition (and density) varies from about 40% by weight of H_2SO_4 at full charge of 2.15 V at 25 °C to above 16% by weight of H_2SO_4 when fully discharged, having voltage of 1.98 V. The change in electrolyte's specific gravity provides a convenient method of determining the state of charge of a cell.

DoD The DoD is a measure of the depth or extent of discharge for a longer life of battery.

The life span of a battery depends upon the following .

- *Charging/discharging cycles* The more a battery is charged, the shorter its lifetime. Also, the more a battery is discharged, the shorter its lifetime.

The battery capacity (in Ah) depends on the following.

- *Discharging current* The higher the discharging current, the lower the capacity and vice versa.
- *Temperature* The capacity of a battery depends upon temperature and for temperatures below 25 °C, it is reduced by about 0.6%/°C.
- *Age of battery* The capacity of a battery also depends upon its age.

Charge controller/regulator

During daytime, when sunshine is in abundance, the battery can get overcharged and at night because of the heavy load and its usage, the battery can become discharged. Both these states – overcharging and deep discharging – drastically reduce the life of a battery. The PV charge controller protects the battery from such extreme states. A charge controller must be compatible with the type of battery used. Many charge controllers have LED (light emitting diode) on the front face to show the status but it is not always advisable to follow the LED status.

A PV system charge controller prevents the battery from charging once it reaches a fully charged state by automatically disconnecting the module(s) from the battery bank. It also prevents the drawing of additional energy from a battery when a dangerously low battery level is reached by disconnecting the supply of power to the load.

The different types of charge controllers/regulators available in the market are (a) switch on/off regulators, (b) PWM (pulse width modulation) charge regulators, which charge the battery with constant voltage or constant current (they are the most commonly used regulators in the PV systems), and (c) the most complex MPPT (maximum power point tracking) charge regulators.

Inverter

The current produced by the PV system is of the DC type but most of the electrical appliances available in the market are AC-powered. An inverter is used to convert DC into AC for the AC load to run and is basically a power-conditioning unit.

According to the form of output voltage, an inverter output can be shaped like a rectangle, trapezoid, or a sine wave. The most expensive and best quality inverters are those where the output voltage is a sine wave. For large systems, three-phase inverters are available in the market. Inverters connecting a PV system and the public grid are designed purposefully to allow energy transfers to and from the public grid. There are three types of inverters: central inverters that are used for a wide power range from 1 to 100 kW or even more (used in large applications), and string inverters and module inverters that are used in small PV systems.

DC and AC loads

Appliances, motors, and equipment powered by DC and AC are known as DC and AC loads, respectively.

Photovoltaic system design guidelines and methodology

Sizing of PV systems is done as per the specific application requirement, such as stand-alone operation or grid-interactive operation. Since a huge capital cost is involved in the installation of the PV system, oversizing the system will lead to high price of the electricity generated. On the other hand, undersizing the system reduces the reliability. Hence an optimum sizing has to be arrived at. The array peak power and battery capacity of a PV system for any specific application would depend on the location of the system. The following section will discuss the common methodology followed for sizing the system. It will also give an overall idea about the sizing of PV system. However, the detailed sizing can be done as per the specific application requirement.

Planning and sizing of the solar photovoltaic system

While designing a PV system, one has to make a correct assessment of the system specifications, including the possible losses, connected loads and its time variation, and the likely degradation of the system components due to ageing.

Pre-sizing of the photovoltaic system

The solar PV system design for any site requires some essential input parameters such as incident solar radiation, load profile of the appliances, ambient temperature, wind velocity, etc., which are likely to vary widely from site to site. The ambient temperature significantly affects the COT and hence the output energy. So it is necessary to have access to the data and information on all these components in order to carry out an optimum PV system design exercise.

Site specificity

The output of the PV array is site-specific/dependent. For the generation of maximum power, it is important to choose the best site for the array. Maximum power will be obtained if the module area is exposed to direct sunlight for a longer period of the day.

In the northern hemisphere, the module should be kept in a south-facing location. Again the site should be chosen in such a way that the module is exposed to direct sunshine without the obstruction of any shadow. Shadow not only affects the generation but also contributes to hot spot formation in the module.

In general, the array is tilted at a certain angle to the horizontal in order to receive the maximum global radiation. The tilt angle is generally kept as equal to the latitude of the site for an yearly average optimum performance.

Minimum possible array capacity

The electrical energy generated by the PV array must at least be equal to the energy consumed. Any capacity less than this minimum will necessarily result in system failure, possibly after some time.

Minimum possible battery capacity

The minimum possible battery capacity is specified by the daily energy requirement. While designing the battery capacity, the days of autonomy have to be taken into consideration. One- to three-days autonomy is generally taken depending upon the geographical location and climatic conditions, and type and need of the application. Days of autonomy specify the number of consecutive days where the stand-alone system will meet a definite load without the solar energy input.

Sizing a solar photovoltaic system

The following steps are taken for sizing a stand-alone solar PV system.

Step 1: *Obtaining the electric load data and energy demand*

Load data includes information on appliances to be powered, their number, nominal power and voltage of each appliance. The number of hours of operation of each appliance during a 24-hour period is also required. The power rating of selected appliances is given in Table 5.4.

Table 5.4 Nominal power rating of selected appliances

Appliance	Nominal power (W)
Tube light with ballast	55
Compact fluorescent lamp (CFL)	7/9/11/18
Ceiling fan	60
Colour TV	150
Refrigerator	150
Radio	20

From the load data and hours of use, a load curve can be constructed for a typical day. As an example, let us consider the case of sizing a stand alone PV system for a village with 50 households. Each household requires AC load of 2 CFLs (each of 11W) and a radio/tape recorder with 25W rating. These will be used for 4 hours in the evening from 6–10 pm. Hence the load of each household is 47W, say 50W and the total load of the village would then be 2500W or 2.5kW. The curve would be as shown in Figure 5.80.

Load curves for most practical situations will not be as simple as the case of Figure 5.80. There will be some loads in the day time, such as flour mill, etc., and irrigation loads which are seasonal. The area under the load curve gives the energy demand. For the case of Figure 5.80, the connected load is also peak load of 2.5 kW and the energy demand is 2.5 × 4 or 10 kWh per day.

If there are DC loads, one should obtain the DC and AC loads and energy demands separately.

In order to calculate the electric load, demand profile is to be known. Information on the demand profile can be obtained from the following.

- Working hours of the system (round the year, only during the day or/ also at night)
- Energy consumption in a given time period (days in a week/month/year)

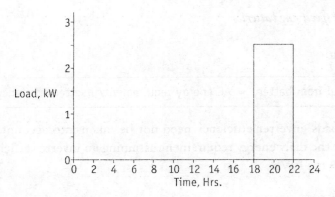

Figure 5.80 Load curve for a typical day for a remote village

- Energy consumption in various uses

Once demand profile is known, calculate the following.

- Total connected load in watts (AC and/or DC)
- Number of hours of operation per day

Multiply the total connected load with hours of operation per day to get the average daily energy requirement in watt-hour.

Average daily AC energy required	= total connected AC load [W] × hours of operation per day (h)

It is important to find out the DC watt-hour requirement and AC watt-hour requirement separately.

Let the total connected AC load (AC watt) be 2.5 kW and the hours of operation per day be 4 hours, then the daily AC energy requirement would be 10 kWh.

Step 2: *Specifying an inverter*

Calculate the total connected AC load. Capacity of the inverter should be the nearest higher value. For example, here the total connected AC load is 2.5 kW. So the rated capacity of the inverter would be 2.5 kW. The load on the inverter should be as close as possible to its rated capacity in order to get better efficiency. However, there is no need of an inverter for DC load.

Step 3: *Sizing and specifying batteries*

In case of AC load

> Energy required from battery = AC energy requirement/inverter efficiency

In case of DC loads, inverter efficiency need not be taken into account. For the present case, the daily energy requirement assuming an inverter efficiency of 90% is 10/0.9 = 11.1 kWh.

- As there will be days without sunshine (as in monsoon), one has to decide the days of autonomy, during which time the load will be met only through the batteries. The battery capacity in Ah is then calculated by multiplying the daily energy requirement with number of days of autonomy and dividing by the rated battery voltage. Multiply 'average Ah per day required' with the desired 'days of autonomy' to determine the 'required battery bank capacity'.

> Required battery bank capacity in Ah = (average Wh required per day × days of autonomy)/battery voltage

For the present case, if days of autonomy is 2, then the required battery bank capacity in Ah would be 1110 × 2 /48 ≃ 460 Ah.

- The required battery capacity in Ah has to be divided by the maximum DoD to determine the final battery capacity in Ah, which must be purchased. If the desired Ah battery is not available, then the battery of nearest higher value should be purchased.

> Final battery capacity in Ah = required battery capacity in Ah/DoD

For the present case, assuming a DoD of 80%, the final battery capacity would be equal to 460/0.8 = 515 Ah. Therefore, the recommended battery bank capacity would be 48 V, 600 Ah.

Once the battery bank capacity is found out, the nominal voltage and Ah of each battery can be decided from the manufacturer's specifications.

- Divide the battery bank voltage with the nominal voltage of the battery to find out the number of batteries connected in series. In this case, if each battery is of 2 V, (normal voltage of the battery is 2 V)

> The number of batteries connected in series = 48/2 = 24

- Similarly, divide 'final battery bank capacity in Ah' with 'Ah of each battery' to find out the number of batteries connected in parallel.

> The number of batteries connected in parallel = 600/600 = 1

Step 4: *Sizing and specifying an array*

The array sizing should be such that it enables meeting the average Ah per day needed with the nominal operating voltage.

The average Ah per day that has to be supplied by the array to the battery is calculated by dividing the daily average Ah taken from the battery to the battery efficiency.

> Average Ah per day to be supplied by the array = daily average Ah taken from the battery / efficiency of the battery

(Battery efficiency is in the range of 80% to 85%)

For the present case assuming a battery efficiency of 85%,

> Daily average Ah per day to be supplied by the array = 230/0.85 = 270

- The number obtained is divided by the 'peak sunshine hours per day' available after considering the system's mounting scheme. The result gives the desired 'array peak amps'. The value of peak sunshine per day varies for different places.

> Array peak amps = average Ah per day to be supplied by the array/ peak sunshine hours per day

In this case, the array peak amps would be 270 Ah/5 h = 54 A, where peak sunshine hour per day is taken as 5 h.

- Divide the array peak ampere with the peak ampere per module to find the number of modules in parallel. 'Peak ampere per module' under operating condition should be found based on manufacturer's module specification (refer to Figure 5.75). It should account for the temperature and insolation correction.

> So, modules in parallel = array peak ampere/peak ampere per module

- Let the peak ampere per 75 W module be 4 A.

> So, modules in parallel = 54/4 = 13.5 ≃ 14

- Number of modules in series can be obtained by dividing the battery bank voltage with the nominal module voltage.

> So, modules in series = battery bank voltage/nominal module voltage

- Let the nominal module voltage be 12 V.

> So modules in series = 48/12 = 4

Step 5: *Specifying a controller*

- Multiply the module short-circuit current (which should be listed in the manufacturer's specification) with the number of modules in parallel to obtain 'maximum array amps' the controller would encounter under a short-circuit condition. Compare this to the 'controller array amps' or charging current, which should be listed in the manufacturer's specification.
- Enter the DC total connected watts or the maximum DC watts that are to be allowed through the controller at any given time. Divide this with the DC system voltage to calculate the maximum DC load ampere the controller will encounter. Compare this to the manufacturer's specifications for load ampere and list under controller load amps.

Step 6: *Sizing system wiring*

- Divide the AC total connected watts by the DC system voltage to obtain the maximum DC amperes continuous. This value will be used to size the wire from the battery to the inverter.

Problem 1

Find the forward current flow of a p-n junction germanium diode if the temperature is 27 °C and reverse saturation current is 1 mA for an applied forward bias of 0.2 V.

Problem 2

Determine the DC and AC resistance for a p-n junction at 150 mV forward bias if the reverse saturation current for the junction is 1 mA at 300 K.
(Hint: r_{dc} = V/I whereas r_{ac} = dV/dI)

Problem 3

A module has 36 solar cells connected in series. The manufacturer's values under standard test conditions are short-circuit current I_{sc} = 3 A; V_{oc} = 20.4 V; P_{max} = 46 W; NOCT = 44 °C. Find out the parameters of the module under the operating conditions G = 600 W/m² and T_g = 34 °C.

Problem 4

Design a stand-alone PV system, which is to be installed in Delhi for supplying power to a house having following AC load.
   Two fans of 65 W
   Three 9 W CFLs
   One 150 W refrigerator
Refrigerator and fans are to be operated for three hours every day and all CFLs for five hours a day. The system should be so designed that it can run smoothly for three consecutive foggy days in Delhi. The standard manufacturing specification can be taken.

Nomenclature

D	Inter-atomic distance (cm)
D_n	Electron diffusion coefficient
D_p	Hole diffusion coefficient
E	Energy (J)
Ec	Lowest energy of the conduction band (eV)
E_f	Fermi level (eV)
E_g	Forbidden energy gap (eV)/band gap valence

E	Applied electric field
E_v	Highest energy of the band (eV)
$f(E)$	Fermi-Dirai distribution function
FF	Fill factor
G	Global irradiance (kW/m^2)
h	Planck's constant
I_m	Current at maximum power point (A)
I_o	Reverse saturation current (A; mA)
I_m	Current at maximum power point (A)
J	Current density (A/m^2)
J_e	Electron current density (A/m^2)
J_h	Hole current density (A/m^2)
J_{total}	Total current density (A/m^2)
k	Boltzmanns constant
M	Mass (g)
ν	Frequency of radiation (cm^{-1})
N_e	Electron density of states
N_h	Hole density of states
e	Carrier charge (c)
s	Electrical conductivity (mho/cm)
T	Temperature (°C)
V	Voltage applied
V_d	Drift velocity (m/s^2)
V_m	Voltage at maximum power point (V)
α	Absorption coefficient (cm^{-1})

References

Chattopadhyay D, Rakshit P C, Saha B, Purkait M N. 1988
Foundation of Electronics
New Delhi: New Age International (P) Ltd

Garg H P and Prakash J. 1997
Solar Energy Fundamentals and Applications
New Delhi: Tata McGraw-Hill Publishing Company Ltd

Gupta C L and Kumar V. 2000
Handbook of Electronics, 36th edn.
Meerut: Pragati Prakashan

Lasnier F and Ang T G. 1990
Photovoltaic Engineering Handbook
Bristol: IOP Publishing

Pillai S O. 1999
Solid State Physics, Third edition
New Delhi: New Age International (P) Ltd

Bibliography

Brendel R. 2003
Thinfilm Crystalline Silicon Solar Cells: physics and technology
Berlin: Wiley–VCH Weinheim

Chopra K L and Das S B. 1983
Thin Film Solar Cells
New York: Plenum, USA

Das B K, Singh S N, Lasnier F, Ang T G. 1990
Photovoltaic Engineering Handbook
Bristol: IOP Publishing

Fahrenbruch A L and Bube R H. 1983.
Fundamentals of Solar Cells
New York: Academic Press

Fonash S J. 1981
Fundamentals of Solar Cells
New York: Academic Press

Green M A. 1982
Solar Cells: operating principles, technology and system applications
Englewood Cliffs, NJ: Prentice-Hall Inc.

Hill R (ed.). 1989
Applications of Photovoltaics
Bristol: Adam Hilger Publishing

Hohmeyer O. 1988.
Social Costs of Energy Consumption
Berlin: Springer-Verlag

Koltun M M. 1988.
Solar Cells: their optics and metrology
New York: Alterton Press, Inc.

Luque A and C L (eds). 1990
Physical Limitations to Photovoltaic Energy Conversion
Bristol: Adam Hilger Publishing

Mehta V K. 1980
Principles of Electronics
New Delhi: S Chand & Co

Mehta V K. 1991
Principles of Electronics
New Delhi: S Chand & Co

Moller H J. 1993
Semiconductors for solar cells
Massachusetts: Artech House Inc.

Neville R C. 1978
Solar Energy Conversion: the solar cell
Amsterdam: Elsevier

Streetman Ben G. 1995
Solid State Electronic Devices
New Delhi: Prentice-Hall of India Pvt. Ltd

Takahashi K and Konagai M. 1986
Amorphus Silicon Solar Cells
New York: Wiley-Interscience

Van Overstraeten and Mertens H P. 1986
Physics, Technology and Use of Photovoltaics
Bristol: Adam Hilger Publishing

Willaims A F (ed.). 1986
The Handbook of Photovoltaic Applications
Atlanta: Fairmont Press

Zweobel K. 1990
Harnessing Solar Power: the photovoltaics challenge
New York: Plenum Publishing Corporation

Solar thermal engineering

Shirish S Garud, Fellow, and V V N Kishore, Senior Fellow
Energy–Environment Technology Division, TERI, New Delhi

6

Introduction

As discussed in Chapter 4, solar energy is received on the earth's surface in the form of radiation comprising mostly the visible and infrared wavelength ranges of the electromagnetic spectrum. A black body (or any absorbing surface) intercepting the solar radiation can convert solar energy into heat, which can then be transferred to a heat transfer fluid. The absorbing surface, due to the high temperature gained from solar radiation, will also lose heat through various mechanisms, described in detail in Chapter 3. When the area of heat loss is more or less same as the area of collection, the ratio of heat lost to energy received is rather high, as the solar radiation fluxes are not too large. For example, solar radiation intensities on a horizontal surface vary between 300 and 800 W/m² on a clear sunny day for a horizontal surface, and for a plate with a temperature that exceeds ambient temperature by about 40 °C and with a heat loss coefficient of 8 W/m² °C, the heat loss is 320 W/m². Thus, the losses are of the same order of magnitude as the energy received. Consequently, such solar thermal systems attain maximum temperatures of the order of 100 °C or so and are called low-temperature solar thermal devices. These are, nevertheless, very important because a large amount of heat is required at low temperatures in a variety of applications such as residential heating, domestic water heating, drying of several agro-products, and cooking. This results in the saving of high-grade energy such as electricity or LPG (liquid petroleum gas). This has been, and continues to be the rationale for adopting low-temperature solar thermal devices. Some of the solar thermal devices in this category are solar water heating collectors, collecting-cum-storage systems, shallow solar ponds, salinity gradient solar ponds, active or passive systems for residential or commercial buildings, swimming pool heaters, solar stills, and solar cookers.

As discussed in Chapter 4, solar radiation received on the earth's surface is of two types: beam and diffuse. The beam or direct radiation can be focused onto a line or a point using either reflectors (mirrors) or refractors (lenses). Thus, radiation collected over large areas can be directed to receivers of much smaller areas, leading to a reduction in the area of heat loss as compared to the area of collection. The solar flux received is several times higher than the heat losses, which makes it possible for the receiver to attain very high temperatures. As the position of the sun changes continuously with respect to the earth, the collector will have to 'track' the sun in order to keep receiving high fluxes at the receiver throughout the day. These collectors are called concentrating collectors and are of two types: line-focusing and point-focusing. The temperature of the receivers can range between 250 °C and 1500 °C, depending on the concentration ratio, which is defined as the ratio of the collecting area to the receiving area and can vary from 10 to 10 000. Concentrating collectors are used primarily for power generation applications, though recent applications include industrial process heating and institutional cooking.

Due to the diurnal and seasonal variations of the solar radiation, the heat collected will have to be stored, as the demand for heat seldom matches the supply of energy over long periods of time. Heat can be stored as sensible heat, latent heat, or chemical energy. Various heat storage systems will be discussed later in this chapter.

Low-temperature solar systems: flat plate collector

The flat plate collector has been at the centre of considerable commercial as well as research activities in solar thermal engineering for the past 50 years or so. It is simple to construct, does not need tracking, and requires little maintenance.

To understand the working of a flat plate collector, consider a black surface insulated at the bottom and exposed to solar radiation. As the black surface absorbs radiation in the complete wavelength region of the solar spectrum, it gains heat, and starts losing heat to the surroundings by conduction, convection, and radiation. Using the equations derived in Chapter 3, the equilibrium temperature of a black surface can be calculated. Wind convection is one of the main causes of heat loss. For certain assumed conditions of radiation intensity, surface absorption, etc., the plate temperature as a function of wind velocity is plotted as curve A in Figure 6.1.

It can be seen that the difference between the plate temperature and the ambient temperature drops off rapidly as the wind velocity increases. Now consider enclosing the black surface in a glass enclosure. Glass has the property of transmitting most of the visible part of the radiation and part of the near-infrared radiation. However, it is opaque to the infrared wavelength range in which the hot plate emits radiation. Thus, it acts like a greenhouse besides isolating the hot surface from the wind convection currents to a large extent. Using the equations for heat losses, to be derived later, one can calculate the plate temperatures for a black surface with a glass enclosure, shown as curve B in Figure 6.1. It can be seen that the plate temperatures are much higher, even for high wind velocities. The black surface can now be a part of the heat exchanger that delivers heat to the colder fluid passing through it. The simplest way to extract the heat is to pass the fluid below the hot surface (Figure 6.2).

An analysis of a simple flat plate water heater depicted in Figure 6.2 is presented by Grossman, Shitzer, and Zvirin (1977). If instead of water, air is

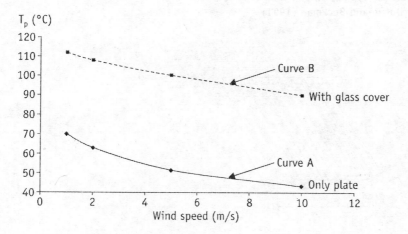

Figure 6.1 Temperature profiles of the hot absorber surface exposed to sunlight under covered and non-covered conditions

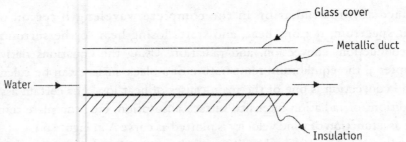

Figure 6.2 A simple flat plate collector for heating water

passed through this water heater, we get the classic solar air heater, analysed in detail by Whillier (1963), and presented along with other types of air heaters by Sukhatme (1984e). The most standard and accepted method of heat extraction in a liquid flat plate collector, however, is the tube fin arrangement (Figures 6.3[a] and 6.3[b]).

The useful heat that can be obtained from a collector would be equal to the net solar energy falling on the plate minus the net heat loss to the

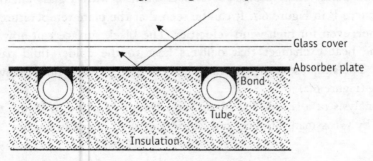

Figure 6.3(a) Typical fin and tube-type solar collector
Source Duffie and Beckman (1991)

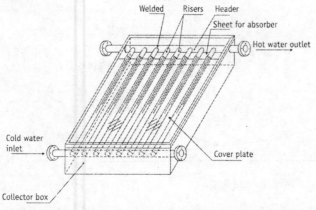

Figure 6.3(b) Flat plate collector assembly showing header, footer, and distribution pipes
Source IS 12933 (2003)

surroundings. A bare plate would receive maximum radiation but the introduction of a cover (or a 'cover system' consisting of two or more glazings) reduces the energy received, the reduction being due to reflection, transmission, and absorption of radiation by the cover system. The heat losses, on the other hand, are primarily through convection and radiation from the top (called top losses) and through conduction from the bottom and the side insulation (called bottom and side losses). It is important to study these losses for estimating the efficiency or performance of the collector.

Transmission, reflection, and absorption of radiation are dependent on the properties of the medium of transmission, wavelength of radiation, and the angle of incident radiation. We assume that all properties are independent of wavelength, which is true for glass with low iron content, but not for other cover materials such as PVC (polyvinyl chloride) or polyethylene. Solar radiation is only slightly polarized and has perpendicular as well as parallel components. Fresnel has derived expressions for the reflection of unpolarized light passing from medium 1 to medium 2, which can be written as

$$r_{pp} = \frac{\sin^2(\theta_2 - \theta_1)}{\sin^2(\theta_2 + \theta_1)} \qquad \qquad ...(6.1)$$

$$r_{pa} = \frac{\tan^2(\theta_2 - \theta_1)}{\tan^2(\theta_2 + \theta_1)} \qquad \qquad ...(6.2)$$

$$r = \frac{I_r}{I_i} = (r_{pp} + r_{pa})/2 \qquad \qquad ...(6.3)$$

where θ_1 and θ_2 are the angles of incidence and refraction, respectively, as shown in Figure 6.4, r is reflectance, and r_{pp} and r_{pa} are the perpendicular and parallel components of the unpolarized light.

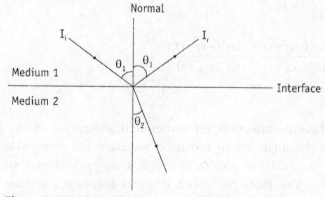

Figure 6.4 Angles of incidence and reflection

For the present discussion, medium 1 is air and medium 2 is glass. For applications such as solar pond, medium 2 can be water. The angles θ_1 and θ_2 are related to the refractive indices of the media n_1 and n_2 by Snell's law.

$$n_1 \sin \theta_1 = n_2 \sin \theta_2 \qquad \qquad ...(6.4)$$

The refractive index of air is approximately 1.0 and the average refractive index of glass is 1.526.

Example 1
Calculate the reflectance of glass at normal incidence and at an incidence angle of 75°.

Solution
For normal incidence, both θ_1 and θ_2 are zero and it can be shown that

$$r(0) = \left(\frac{n_1 - n_2}{n_1 + n_2} \right)^2 \qquad \qquad ...(6.5)$$

$$r(0) = \left(\frac{n - 1}{n + 1} \right)^2 \qquad \qquad ...(6.6)$$

For n = 1.526

$$r(0) = \left(\frac{0.526}{2.526} \right)^2 = 0.0434$$

For $\theta = 75°$, the refraction angle is obtained as

$$\theta_2 = \sin^{-1} \left(\frac{\sin 75}{1.526} \right) = 39.27$$

$$r(75) = \frac{1}{2} \left(\frac{\sin^2(-35.73)}{\sin^2(114.27)} + \frac{\tan^2(-35.73)}{\tan^2(114.27)} \right)$$

$$= 0.2577$$

The losses through reflection, refraction, and absorption in the cover system determine the quantum of radiation available for conversion into useful heat. When radiation passes through a cover material such as glass, there are two interfaces per cover, whereas Figure 6.4 depicts only one interface. The radiation reflected and transmitted is different for each

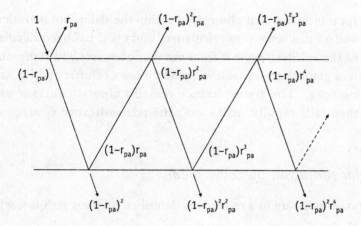

Figure 6.5 Transmission through a single cover considering only reflectance and transmission
Source Duffie and Beckman (1991)

component of polarization, hence it is necessary to treat the perpendicular and parallel components separately. The ray tracing method is generally used to estimate the transmittance (Figure 6.5).

Neglecting the absorption in the cover material and considering only the parallel component, the transmittance τ_{pa}, is obtained by summing up all the transmission terms for the parallel component.

$$\tau_{pa} = (1 - r_{pa})^2 + (1 - r_{pa})^2 \, r^2_{pa} + (1 - r_{pa})^2 \, r^4_{pa} + \ldots\ldots$$

$$\tau_{pa} = (1 - r_{pa})^2 \sum_{n=0}^{\infty} r^{2n}_{pa}$$

$$\tau_{pa} = (1 - r_{pa})^2 \, \frac{1}{(1 - r^2_{pa})}$$

$$\tau_{pa} = \frac{1 - r_{pa}}{1 + r_{pa}} \qquad \qquad \ldots(6.7)$$

Similarly, transmittance for the perpendicular component τ_{pp} is

$$\tau_{pp} = \frac{1 - r_{pp}}{1 + r_{pp}} \qquad \qquad \ldots(6.8)$$

The components τ_{pa} and τ_{pp} are not equal at normal incidence and the average transmittance is given by

$$\tau^\circ = \frac{1}{2}\left(\frac{1 - r_{pa}}{1 + r_{pa}} + \frac{1 - r_{pp}}{1 + r_{pp}} \right) \qquad \qquad \ldots(6.9)$$

Superscript o indicates that absorption within the slab is not considered. Equations 6.1 and 6.2 can be used to calculate r_{pa} and r_{pp}. τ^o has been calculated as a function of the incident angle θ for a number of covers and represented graphically. These graphs are available in standard texts (Duffie and Beckman 1991; Sukhatme 1984). The transmittance remains almost constant up to $\theta = 40°$, and then falls rapidly. At $\theta = 90°$, the transmittance is zero, as is expected.

Absorption of solar radiation by cover media

The absorption of radiation in a medium is described by Bouguer's law, which can be stated as

$$\frac{dI}{ds} = -KI \qquad \qquad ...(6.10)$$

where dI is the amount of radiation absorbed in the width ds of the path length and K is extinction coefficient of the medium. The negative sign on the right-hand side of Equation 6.10 indicates that I decreases with increase in s. Consider a differential element dx as shown in Figure 6.6.

$$\cos\theta_2 = \frac{dx}{ds}; \qquad ds = \frac{dx}{\cos\theta_2}$$

Substituting in Equation 6.10, we have

$$\frac{dI}{I} = -K\frac{dx}{\cos\theta_2} \qquad \qquad ...(6.11)$$

The extinction coefficient K varies from approximately 4 m⁻¹ for good-quality glass to about 32 m⁻¹ for poor-quality glass. It is dependent on the iron content of the glass and can generally be determined by observing the edge of the glass. A greenish edge indicates high iron content and a high value of K. Low-iron glasses are generally preferred for solar applications. Typically,

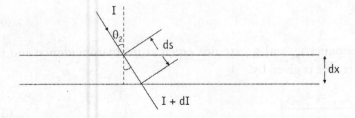

Figure 6.6 Absorption of the incident ray in transparent cover

the best transmittance of high-iron containing glass is about 88%, whereas low-iron glasses have transmittance of 92% or above. One can observe that transmittance plays an important role since the radiation transmitted through good-quality glass is higher by five per cent or more than normal glass.

Integrating Equation 6.11 between the limits I_1 to I_2, and 0 to L, where L is the cover plate thickness, we get

$$\tau^a = \frac{I_2}{I_1} = \exp\left(-\frac{KL}{\cos\theta_2}\right) \qquad \qquad ...(6.12)$$

Superscript a indicates transmittance considering absorption in the cover plate.

The transmittance τ, reflectance ρ, and absorptance α of a single cover can be derived using the ray tracing techniques similar to the ones described earlier. The parallel and perpendicular components of the polarization are obtained and the average of the two is used for application. For practical collector applications, however, the transmittance of a single cover can be approximated by

$$\tau = \tau^o \tau^a \qquad \qquad ...(6.13)$$

where t^o and t^a are given by Equations 6.9 and 6.12. The absorptance of a single cover can be approximated by

$$\alpha = 1 - \tau^a \qquad \qquad ...(6.14)$$
As $\rho = 1 - \tau - \alpha$
$$\rho = \tau^o - \tau \qquad \qquad ...(6.15)$$

Example 2
Calculate α, τ, and ρ for a glass cover with following attributes:
Thickness = 4 mm, K = 25, and angle of incidence = 75°.

Solution
For angle of incidence of 75° from Example 1, the angle θ_2 = 39.27°. Now using Equation 6.1

$$r_{pp} = \frac{\sin^2(\theta_2 - \theta_1)}{\sin^2(\theta_2 + \theta_1)}$$

Thus, $r_{pp} = 0.410$
Similarly, $r_{pa} = 0.105$

Substituting in Equation 6.9, we get
$\tau^\circ = 0.614$

Similarly, $\tau^a = 0.88$
Now $\tau = \tau^\circ \, \tau^a = 0.54$
$\alpha = 1 - \tau^a = 0.12$, and
$\rho = \tau^\circ \, \tau = 0.614 - 0.54 = 0.074$

The values of transmittance as a function of the angle of incidence for a single cover for different values of KL are shown in Figure 6.7.

Transmittance–absorptance product

After passing through the cover system, the solar radiation is mostly absorbed by the collector plate, and a small amount is reflected off the plate. The reflected radiation is not entirely lost, but is re-reflected and re-absorbed (Figure 6.8).

If τ is the transmittance of the cover system and α is the absorptance of the plate at a given angle of incidence, an amount $\tau\alpha$ is absorbed by the plate

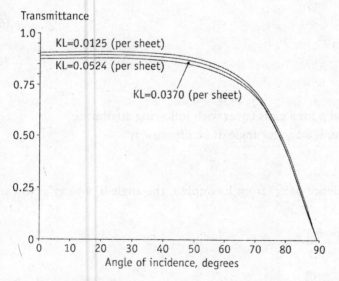

Figure 6.7 Transmittance of a single cover for three types of glass
Source Duffie and Beckman (1991)

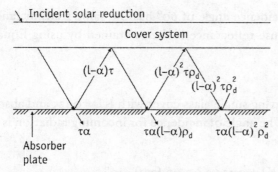

Figure 6.8 Transmittance–absorptance product of cover plate absorber system
Source Duffie and Beckman (1991)

and $\tau(1-\alpha)$ is reflected back to the cover system. The absorptance of the plate also depends on the angle of incidence. Table 6.1 gives the values of absorptance for various angles of incidences for a plain black surface (Goswami, Kreith, and Kreider 2000).

The reflection from the absorber plate is assumed to be diffuse. The amount reflected back to the plate from the cover system is $\tau(1-\alpha)\rho_d$, where ρ_d is called diffuse reflectance. The reflection continues as shown in Figure 6.8, and the summation of all the absorption terms is

$$(\tau\alpha) = \tau\alpha \sum_{n=0}^{\infty}\left[(1-\alpha)\rho_d\right]^n \qquad \ldots(6.16)$$

$$= \frac{\tau\alpha}{1-(1-\alpha)\rho_d} \qquad \ldots(6.17)$$

The symbol $(\tau\alpha)$ is not the product of τ and α, but should be considered as a property of the cover–absorber system. Studies show that transmittance or reflectance of the diffuse radiation can be obtained by defining an equivalent angle of beam radiation that gives the same transmittance or reflectance as diffuse radiation. For solar applications, this angle is 60°, which means that

Table 6.1 Angular variation of absorptance of lamp black paint

Angle of incidence (°)	Absorptance (α)
0–30	0.96
30–40	0.95
40–50	0.93
50–60	0.91
60–70	0.88
70–80	0.81
80–90	0.66

Source Goswami, Kreith and Kreider (2000)

beam radiation with an incidence angle of 60° has the same transmittance as the diffuse radiation. Diffuse reflectance is then obtained by using Equation 6.15 for τ evaluated at 60°.

Example 3
Find (τα) for a collector having single glass cover with KL = 0.125 and absorber having absorptance 0.90. The angle of incidence for incoming radiation is 40°.

Solution
For single glass cover with KL = 0.125, τ = 0.90 for θ = 40°.

Similarly, τ at 60° (the effective angle of incidence for reflected radiation) = 0.82.

Now, $\theta_2 = \sin^{-1}(\sin 40/1.526) = 24.91°$
$\tau_a = \exp(-KL/\cos \theta_2)$
$= 0.986$
$\rho_d = \tau_a - \tau = 0.086$

Hence, $(\tau\alpha) = \tau\alpha/[1 - (1 - \tau\alpha)\rho_d]$
$= 0.90 \times 0.90/[1 - (1 - 0.90) \times 0.086]$
$= 0.817$
Alternatively, it is possible to estimate $\rho_d = 1 - \tau_r$.
τ_r can be found from Figure 6.5.

For most practical application, it is found that
$(\tau\alpha) = 1.01 \tau\alpha$ ...(6.18)
As τ and α are dependent on the incident angle, the property (τα) also varies. Relationship between $(\tau\alpha)/(\tau\alpha)_n$ and the incident angle θ is shown in Figure 6.9. The graphs are not sensitive to KL and hence can be applied to all materials having a refractive index close to that of glass (Table 6.2).

Figure 6.9 provides the theoretical basis for evaluating the 'incident angle modifier' of a given collector during collector testing, which will be described later.

Table 6.2 Average refractive index of some glazing materials in the solar spectrum

Material	Average refractive index
Glass	1.526
Polycarbonate	1.600
Polyvinyl fluoride	1.450
Polyfluorinated ethylene propylene	1.340

Source Duffie and Beckman (1991)

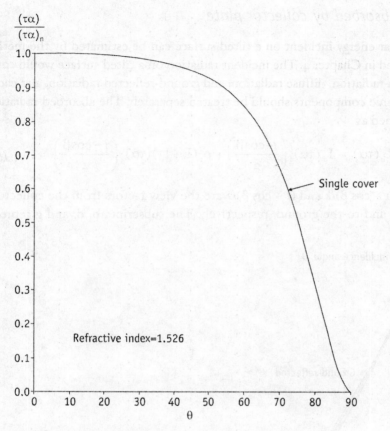

Figure 6.9 Typical $(\tau\alpha)/(\tau\alpha)_n$ curve
Source Duffie and Beckman (1991)

It has been assumed earlier that transmission, etc., are independent of wavelength, in case of low-iron glass, but not for glass with higher iron content and for many plastic films. Transmittance as a function of wavelength is available for many common cover materials (Duffie and Beckman 1991; Goswami, Kreith, and Kreider 2000). The transmittance over the entire wavelength region can be obtained by integration.

If a film of low refractive index is deposited onto a transparent slab up to an optical thickness of $\lambda/4$, radiation of wavelength λ reflected from the upper and lower surfaces of the film will have a phase difference of 180° and will cancel each other.

The reflectance will decrease, and the transmittance will increase correspondingly. Such coatings are used in camera lenses, binoculars, etc., but are expensive. Some methods such as etching are also available for reducing the reflectance of glass used in solar collectors.

Radiation absorbed by collector plate

The solar energy incident on a tilted surface can be estimated by the methods described in Chapter 4. The incident radiation on a tilted surface would consist of beam radiation, diffuse radiation, and ground-reflected radiation, and each of these three components should be treated separately. The absorbed radiation S is obtained as

$$S = I_b R_b (\tau\alpha)_b + I_d (\tau\alpha)_d \left(\frac{1 + \cos\beta}{2} \right) + \rho_g (I_b + I_d)(\tau\alpha)_g \left(\frac{1 - \cos\beta}{2} \right) \qquad ...(6.19)$$

where $(1 + \cos\beta)/2$ and $(1 - \cos\beta)/2$ are the view factors from the collector to the sky and to the ground, respectively. The subscripts b, d, and g represent

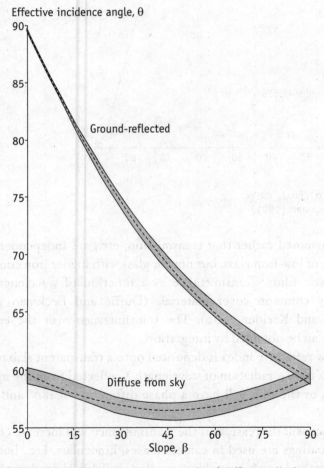

Figure 6.10 Effective incidence angle of diffuse radiation
Source Duffie and Beckman (1991)

beam, diffuse, and ground. For a given collector tilt β, one can define an 'effective incidence angle for transmittance of diffuse radiation' (Figure 6.10).

Figure 6.10 has been obtained by assuming that diffuse radiation from the sky, as well as from the ground, is isotropic and by integrating the beam transmittance over an appropriate incidence angle (Brandemuehl and Beckman 1980).

Sometimes, it is convenient to define an average value for $(\tau\alpha)$ as

$$S = (\tau\alpha)_{av} I_t \qquad \qquad \qquad ...(6.20)$$

When the beam radiation is high, an approximate value of $(\tau\alpha)$ can be obtained as follows.

$$(\tau\alpha)_{av} = 0.96(\tau\alpha)_b \qquad \qquad \qquad ...(6.21)$$

The values of I_t are measured directly sometimes. For example, a pyranometer is mounted on the plane of the collector during testing. Equations 6.20 and 6.21 can then be used to obtain S.

The monthly average value can be obtained using the average values for $\overline{H_b}, \overline{R_b}$, etc. (Duffie and Beckman 1991).

Selective surfaces

The energy absorbed by the collector plate is dependent on the absorptance α. Dull black paints are used to get high values for α. However, a black surface will also have high emittance, as absorbance and emittance are related. For an opaque surface

$$\rho_\lambda + \varepsilon_\lambda = \rho_\lambda + \alpha_\lambda = 1 \qquad \qquad ...(6.22)$$

where ρ_λ = reflectance, ε_λ = emittance, and α_λ = absorptance

and

$$\rho + \alpha = 1 \qquad \qquad \qquad ...(6.23)$$

Hence

$$\varepsilon_\lambda = \alpha_\lambda \qquad \qquad \qquad ...(6.24)$$

The black surface will lose energy through convection and radiation, and a surface with higher ε will have higher losses also. An ideal surface should thus have a high absorbance (or low reflectance) in the solar spectrum ($\lambda < 3 \ \mu m$) and low emissivity (high reflectance) in the infrared spectrum (Figure 6.11).

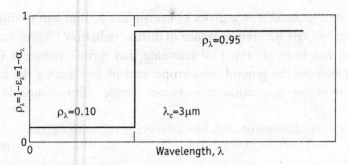

Figure 6.11 Characteristics of a selective surface

In reality, the selectivity is achieved by polishing a metal surface and coating it with a thin surface layer, which is transparent to long-wave radiation but has high absorptance for shorter wavelength solar radiation. Surface layers made of nickel black or black chrome have been found to be quite suitable as selective surfaces. These are deposited onto a polished metal plate by chemical treatment, vacuum deposition, or by electroplating. The thickness of the layer is only about 0.1 μm. The spectral reflectance of black chrome as a function of wavelength is shown in Figure 6.12.

The examples of calculating emissivities of such surfaces at any given temperature by suitable integration are given in standard books (Duffie and Beckman 1991).

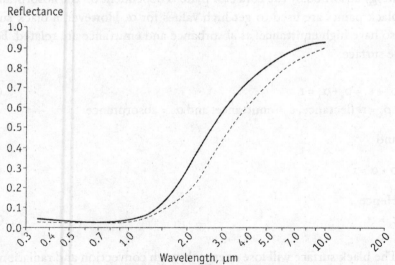

—— Black chrome on dull nickel after 30-day humidity test; $\alpha = 0.961$, $\varepsilon = 0.088$
---- Black chrome on dull nickel before humidity test; $\alpha = 0.964$, $\varepsilon = 0.116$

Figure 6.12 Spectral reflectance of black chrome surface
Source Duffie and Beckman (1991)

Having obtained a value for absorbed radiation by using Equation 6.19, we can now derive an energy balance equation for the flat plate collector.

If A_c is the area of the collector, the energy input would be equal to A_cS. If T_p is the mean plate temperature of the absorber, the heat lost at any given time can be represented by

$$Q_L = A_c U_L (T_p - T_a) \qquad \ldots(6.25)$$

where U_L is an overall heat loss coefficient and T_a is the ambient temperature. The 'useful' heat, mentioned in the beginning of this chapter, can thus be written as

$$Q_u = A_c [S - U_L (T_p - T_a) \qquad \ldots(6.26)$$

If no heat is extracted, $Q_u = 0$, and Equation 6.26 can be written as

$$T_p^S = T_a + (S/U_L) \qquad \ldots(6.27)$$

where T_p^S is the stagnation temperature for a given radiation level. It can be seen that the stagnation temperature can be used as a measure of the heat loss coefficient U_L and has actually been used in the proposed standard for solar cookers for India. U_L can also be obtained by calculating various heat transfer resistances represented in the electrical network analogy shown in Figure 6.13.

The collector plate at temperature T_p loses heat to the cover both by convection and radiation with the heat loss q_{p-c} given by

$$q_{p-c} = h_{c1}(T_p - T_c) + \frac{\sigma\left(T_p^4 - T_c^4\right)}{(1/\varepsilon_p) + (1/\varepsilon_c) - 1} \qquad \ldots(6.28)$$

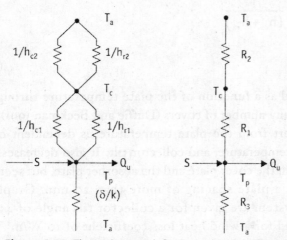

Figure 6.13 Thermal network for a single cover flat plate collector

The radiation loss can be expressed as h_{r1} $(T_p - T_c)$, where

$$h_{r1} = \frac{\sigma(T_p + T_c)(T_p^2 + T_c^2)}{(1/\varepsilon_p) + (1/\varepsilon_c) - 1} \qquad ...(6.29)$$

Therefore,

$$\begin{aligned}
q_{p-c} &= h_{c1}(T_p - T_c) + h_{r1}(T_p - T_c) \\
&= (h_{c1} + h_{r1})(T_p - T_c) \\
&= \frac{(T_p - T_c)}{\left[\dfrac{1}{(h_{c1} + h_{r1})}\right]}
\end{aligned} \qquad ...(6.30)$$

From Equation 6.30, it can be deduced that resistance R_1, shown in Figure 6.13, is given as

$$R_1 = \frac{1}{h_{c1} + h_{r1}} \qquad ...(6.31)$$

The heat transfer resistance from the cover to the ambient temperature can similarly be obtained as

$$R_2 = \frac{1}{h_{c2} + h_{r2}} \qquad ...(6.32)$$

The wind convection coefficient h_{c2} can be obtained from Equation 3.208 of Chapter 3. The top loss coefficient U_t is given as

$$\frac{1}{U_t} = \frac{1}{(h_{c1} + h_{r1})} + \frac{1}{(h_{c2} + h_{r2})} \qquad ...(6.33)$$

$$U_t = \frac{R_1 R_2}{R_1 + R_2} \qquad ...(6.34)$$

U_t can be obtained as a function of the plate temperature through an iterative procedure, for any number of covers (Duffie and Beckman 1991). The top loss coefficient, apart from the plate temperature, is dependent on ε_p, wind velocity, ambient temperature, and collector tilt. It also decreases with the plate spacing between the cover plate and the absorber plate, but seems to be nearly constant for a plate spacing of more than 20 mm. Graphs of U_t for a single cover system are given for a collector tilt angle of 45° for $\varepsilon_p = 0.95$ and $\varepsilon_p = 0.1$ and for a wind heat loss coefficient of 10 W/m² °C in Figure 6.14.

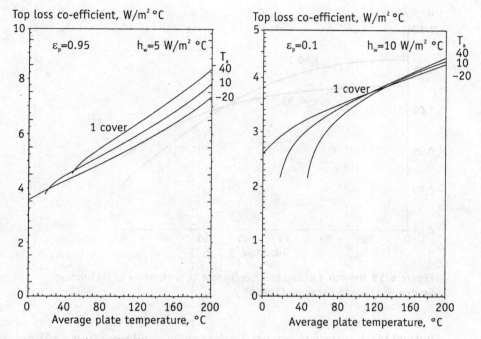

Figure 6.14 Dependence of U_t on T_p and ε_p
Source Duffie and Beckman (1991)

An empirical equation for U_t was developed by Klein, and is described in Duffie and Beckman (1991). The variation of $U_t(\beta)/U_t(45°)$ is shown in Figure 6.15.

The heat loss from the bottom of the collector is through the insulation, and is given by

$$U_b = \frac{k}{\delta_i} \qquad \qquad ...(6.35)$$

where k is the thermal conductivity and δ_i is the thickness of the insulating material. The edge losses can also be included in U_b. The overall heat loss coefficient U_L is given as

$$U_L = U_t + U_b \qquad \qquad ...(6.36)$$

The reduction in the values of the top loss coefficient (and hence that of U_L) for selectively coated surfaces is apparent from Figure 6.14. For example, the value of U_t (45°) at a plate temperature of 80 °C, ambient temperature of 40 °C, and wind loss coefficient of 10 W/m² °C is about 6.5 W/m² °C for a

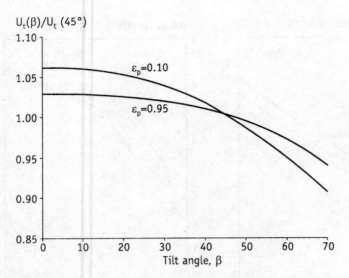

Figure 6.15 Variation of top loss coefficient U_t with angle of inclination
Source Duffie and Beckman (1991d)

normal black surface ($\varepsilon_p = 0.95$). Under similar conditions, it is 3.5 W/m² °C for selectively coated surfaces ($\varepsilon_p = 0.1$).

Several efforts have been made to reduce the value of top loss coefficient by other means. Equation 6.33 shows that a reduction in h_{c1} will also reduce U_t. The convective heat transfer coefficient can be reduced by evacuating the space between the plate and the cover, or by using convection suppression devices. Convection suppression has been achieved in practice by using a transparent honeycomb-like structure between the plate and the cover. Recent advances in convection suppression and heat loss reduction include the development of TIM (transparent insulation material) technology. TIM represents a new class of insulation materials wherein air gaps and/or evacuated spaces are used to reduce heat loss. Unlike conventional insulation materials, these materials are transparent to solar radiation and hence can be used to prevent convective heat loss from the top surface of the collector absorber.

Typically, transparent insulation consists of a transparent cellular (honeycomb) array where air pockets are formed in an air layer. Kaushika and Sumathy (2003) have presented a detailed review of TIM.

Plate fin collectors for water heating have been made in a variety of ways, with the twin aims of improving heat transfer rate from the plate to the flowing water and of reducing the cost.

Presently, however, selectively coated plate fins have become more or less standardized. The makers of solar thermal systems procure the fin tube assemblies and assemble the complete collectors in smaller factories. A single copper plate fin (with a single copper tube) typically has the dimensions shown in Figure 6.16.

One-dimensional conductive heat transfer in metallic fins has been described in detail in Chapter 3. The equations derived in that chapter can now be applied to solar collectors.

Referring to Figure 6.17, heat balance on the element is given as

$$k \, \delta \, b \left. \frac{dT}{dx} \right|_x - k \, \delta \, b \left. \frac{dT}{dx} \right|_{x + \Delta x} = S \, \Delta x \, b - U_L \, \Delta x \, b \, (T - T_a) \qquad \text{...(6.37)}$$

where k is the thermal conductivity of the material, δ is the fin thickness, b is a small width in direction perpendicular to x, S is the solar radiation

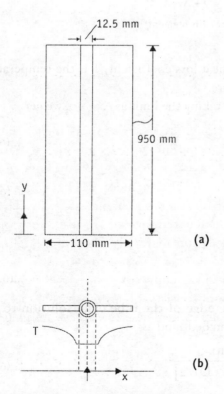

(a)

(b)

Figure 6.16 Typical dimensions (a) and temperature (b) profile of single copper plate fin with selective coating

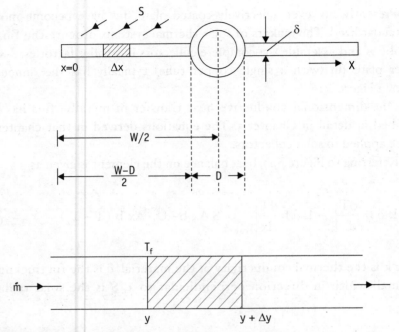

Figure 6.17 Heat balance on solar collector fin element

absorbed by the fin, U_L is the total heat loss coefficient, T is the temperature at x, and T_a is the ambient temperature.

Dividing throughout by Δx and taking the limit as $\Delta x \rightarrow 0$, we get

$$-k\delta \frac{d^2T}{dx^2} = S - U_L (T - T_a) \qquad \qquad ...(6.38)$$

or

$$\frac{d^2T}{dx^2} = \frac{U_L}{k\delta}(T - T_a - \frac{S}{U_L}) \qquad \qquad ...(6.39)$$

The boundary conditions are

$$\frac{dT}{dx}\bigg|_{x=0} = 0; \qquad T\big|_{x=(W-D)/2} = T_b \qquad \qquad ...(6.40)$$

where T_b is the temperature at the edge of the tube. The solution to the second-order differential equation can be found as

$$\frac{T - T_a - S/U_L}{T_b - T_a - S/U_L} = \frac{\cosh mx}{\cosh[m(W-D)/2]} \qquad \qquad ...(6.41)$$

where

$$m = \sqrt{\frac{U_L}{k\delta}} \qquad \qquad ...(6.42)$$

The heat conducted into the tube can be obtained by

$$q_{fin} = -k\,\delta\,b\,\frac{dT}{dx}\bigg|_{x=(W-D)/2}$$

$$= \frac{k\delta b m}{U_L}\left[S - U_L(T_b - T_a)\right]\tanh\left[\frac{m(W-D)}{2}\right] \qquad ...(6.43)$$

Equation 6.43 is for heat conducted from one side of the tube and hence the total heat conducted is twice this amount. Taking b as unity (for unit width in the y-direction), the equation can be written as

$$q_{fin} = (W-D)F\left[S - U_L(T_b - T_a)\right] \qquad ...(6.44)$$

where F is the fin efficiency, given as

$$F = \frac{\tanh[m(W-D)/2]}{[m(W-D)/2]} \qquad ...(6.45)$$

This is the standard fin efficiency equation, in which F decreases steadily with decrease in $\sqrt{(U_L/k\delta)\left[(W-D)/2\right]}$.

It can be seen that F would be same for different metals, provided $k\delta$ is kept constant. If δ_c is the thickness of a copper plate, the equivalent thickness of, say, an aluminium plate will be $(k_{Cu}/k_{Al})\delta_c$ or $1.8\delta_c$. The choice would ultimately depend upon other factors such as long-term durability and the cost of the material.

The energy collected above the tube region will also contribute to the useful heat and is given by

$$D[S - U_L(T_b - T_a)].$$

The total useful energy thus obtained is

$$q_u = \left[(W-D)F + D\right]\left[S - U_L(T_b - T_a)\right] \qquad ...(6.46)$$

For a quasi-steady state, useful heat must be ultimately transferred to the fluid flowing inside the pipe. The resistance to heat flow from T_b to T_f, the fluid temperature, consists of the contact resistance of the bond between the plate and the tube and the convective heat transfer resistance of the fluid. Combining these two resistances

$$q_u = \frac{(T_b - T_f)}{\left[\dfrac{1}{\pi D_i h_{fi}} + \dfrac{1}{C_b}\right]} \qquad \ldots(6.47)$$

where D_i is the inside tube diameter and h_{fi} is the liquid heat transfer coefficient. The bond conductance C_b can be estimated if thermal conductivity, bond width, and thickness are known, but is best measured by experiments. Whillier and Saluja (1965) showed that the bond conductance should be greater than 30 W/m² °C in order to avoid loss of performance.

Combining Equations 6.46 and 6.47 one can eliminate T_b to obtain

$$q_u = W F'\left[S - U_L (T_f - T_a)\right] \qquad \ldots(6.48)$$

where the collector efficiency factor F′ is given by

$$F' = \frac{(1/U_L)}{W\left[\dfrac{1}{U_L [D + (W - D)F]} + \dfrac{1}{C_b} + \dfrac{1}{\pi D_i h_{fi}}\right]} \qquad \ldots(6.49)$$

The collector efficiency factor F′ is essentially a design parameter and is constant for a given collector. Variation of F′ with other parameters such as tube spacing, $k\delta$, and U_L can be plotted using Equation 6.49. Such plots are available in standard textbooks (Duffie and Beckman 1991). Equation 6.48 gives the useful heat gain as a function of fluid temperature T_f. However, T_f will increase as the fluid is flowing through the collector. The temperature distribution in the flow direction will depend on the flow rate of the fluid with reference to the collector area, and can be obtained by simple heat balance.

Bonding methods

In the early 1970s, mechanical clamping and rivetting were common bonding technologies. These methods were commonly used for mild steel fins and galvanized fin tube combination. However, the conductivity of such bonds was low due to improper thermal contact between the tube and the fin metal.

Subsequently, soldering technology was developed, which was cheaper, but the solders generally tin based, which melt at about 100 °C, causing loss of performance at stagnation temperature and the bonding is also not uniform.

Later on, a copper eutectic alloy containing a high percentage of silver was developed. This alloy melts at temperatures above 200 °C and also has conductivity close to that of copper. It comes in the form of paste and hence is easy to apply and gives uniform contact throughout the fin length. The high cost of the alloy is its major disadvantage. Another method of forming channels between two aluminium sheets, called roll bonding, was developed. This method is commonly used in refrigerator evaporator modules. For solar applications, however, aluminium tubes or channels cannot be used due to corrosion problems, hence some companies tried to develop roll bond sheets with a copper tube inserted in the channel. These types of fins were common for black-painted collectors. However, the recent trend is to use selective coated fins of copper. Recently, some companies have introduced high-frequency welding of copper to copper. This method gives a direct joint between the tube and the fin and hence the bond conductivity is as good as that of copper. However, the width of the contact area is very small (approximately 1 mm), which restricts flow of heat to the tube.

Heat balance of fluid

As Equation 6.48 has been developed for unit length in the y-direction (direction of fluid flow), one can make a heat balance on the section Δy for the one tube shown in Figures 6.16 and 6.17.

$$\dot{m}C_p \left.T_f\right|_{y+\Delta y} - \dot{m}C_p \left.T_f\right|_y = \Delta y\, q_u \qquad \qquad ...(6.50)$$

where $\dot{m}$ is the fluid flow rate for a single tube.

Dividing throughout by Δy, substituting for q_u from Equation 6.48, and taking the limit as $\Delta y \to 0$, we get

$$niC_p \frac{dT_f}{dy} = WF'\left[S - U_L(T_f - T_a)\right] \qquad \qquad ...(6.51)$$

Integrating between the limits $T_{f,i}$ and $T_{f,o}$ and between y = 0 to L, we get

$$\frac{T_{f,o} - T_a - S/U_L}{T_{f,i} - T_a - S/U_L} = \exp(-WL\, U_L F'/\dot{m}C_p) \qquad \qquad ...(6.52)$$

where $T_{f,i}$ is the inlet fluid temperature, $T_{f,o}$ is the outlet fluid temperature, and L is the length of the collector plate. We have made an implicit assumption that F' and U_L are independent of temperature. Noting that WL is the area of the single tube fin arrangement, Equation 6.52 can be written as

$$\frac{T_{f,o} - T_a - S/U_L}{T_{f,i} - T_a - S/U_L} = \exp(-A_c U_L F'/\dot{m}C_p) \qquad ...(6.53)$$

If the collector consists of several tube fins, as is usually the case, $\dot{m}$ will be the total flow rate of the fluid entering the collector, and A_c is the collector area.

One can see that the right-hand side of Equation 6.53 approaches 1 as $\dot{m}$ increases, making $T_{f,o} \to T_{f,i}$. It is, thus, clear that the maximum heat that could have been transferred, from the plate to the fluid would be when the entire collector is at $T_{f,i}$. It is convenient to define a quantity, which is the ratio of the actual useful energy gained by the fluid to the maximum energy that could have been transferred.

This ratio can be written as

$$F_R = \frac{\dot{m}C_p(T_{f,o} - T_{f,i})}{A_c\left[S - U_L(T_{f,i} - T_a)\right]} \qquad ...(6.54)$$

It can be shown that

$$F_R = \frac{\dot{m}C_p}{A_c U_L}\left[1 - \exp\left(-\frac{A_c U_L F'}{\dot{m}C_p}\right)\right] \qquad ...(6.55)$$

Dividing both sides by F', we get

$$\frac{F_R}{F'} = \frac{\dot{m}C_p}{A_c U_L F'}\left[1 - \exp\left(-\frac{A_c U_L F'}{\dot{m}C_p}\right)\right] \qquad ...(6.56)$$

The quantity F_R/F' is, thus, a function of a single dimensionless variable $\left(\dot{m}C_p/A_c U_L F'\right)$ and is plotted in Figure 6.18. This graph can be used to calculate F_R when other parameters are known.

Equation 6.54 can be written as

$$Q_u = \dot{m}C_p(T_{f,o} - T_{f,i}) = A_c F_R\left[S - U_L(T_{f,i} - T_a)\right] \qquad ...(6.57)$$

Dividing by $(A_c G_T)$ throughout, we get an expression for instantaneous collector efficiency.

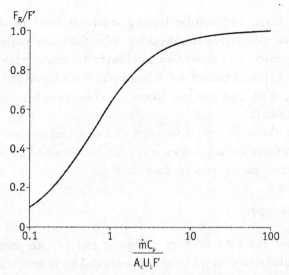

Figure 6.18 Relationship between F_r/F' and $\dot{m}C_p/A_cU_LF'$
Source Duffie and Beckman (1991d)

$$\eta = \frac{Q_n}{A_cG_T} = F_R\left[(\tau\alpha)_{avg} - U_L\frac{T_{f,i} - T_a}{G_T}\right]$$ ...(6.58)

Note that G_T is numerically equal to $I_T/\Delta t$, where Δt is the time interval for measuring solar radiation, which is usually 1 hour. G_T has the unit of W/m² whereas I_T is expressed in Wh/m². For hourly measurements, G_T and I_T are numerically equal. Equation 6.58 is often referred to as HWB (Hottel–Whillier–Bliss) equation.

Equation 6.58 shows that if the instantaneous collector efficiency is plotted against $(T_{f,i} - T_a)/G_T$, a straight line should result with the intercept equal to $F_R(\tau\alpha)_{avg}$ and the slope equal to $-F_RU_L$. If tests are conducted when most of the radiation is beam radiation and during noon time when incident angle is almost normal, the two parameters obtained, that is, $F_R(\tau\alpha)_n$ and F_RU_L, describe the collector functioning quite unambiguously. A third parameter, the incident angle modifier $K_{\tau\alpha}$, is defined based on the variation of $(\tau\alpha)/(\tau\alpha)_n$ with the angle of incidence θ (refer to Figure 6.9). The following functional form has been suggested for $K_{\tau\alpha}$.

$$K_{\tau\alpha} = 1 + b_o\left(\frac{1}{\cos\theta} - 1\right)$$ ...(6.59)

where b_o is termed as the 'incident angle modifier coefficient'. Different testing standards are followed in different countries. The ASHRAE 93–77 standard is followed in USA, and similar testing methods exist in other countries. In India, IS 12933 (2003) is followed for solar flat plate collector testing. Gupta and Garg (1967) have described a collector testing method for solar air heaters in India. A typical collector effciency plot for a liquid flat collector is shown in Figure 6.19, and the test loops for collection testing are shown in Figures 6.20 (a) and (b).

It is apparent that the collector efficiency changes throughout the day. The collector performance for a given location can be obtained by using Equation 6.58, if the collector parameters are known.

Other types of collector geometries

The analysis employed for the tube-fin type collector can be extended to other types of collectors also. For example, the conventional air heater, shown in Figure 6.21, has been analysed using the same methodology.

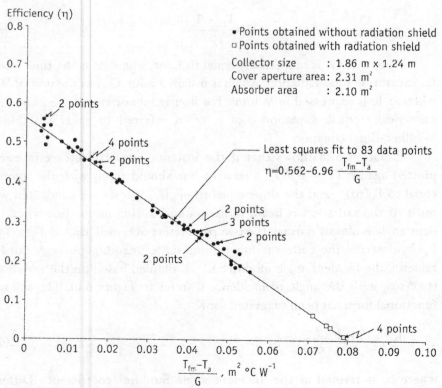

Figure 6.19 Solar collector test
Source Kishore and Katam (1987)

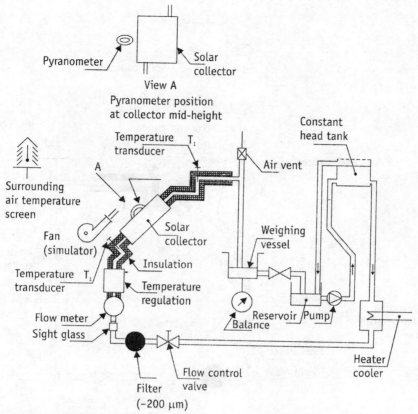

Figure 6.20(a) Test loop for solar collector test set-up (open loop)
Source IS 12933 (2003)

Heat balance equations for the length Δy and for width b in the x-direction can be written as follows.

For the collector plate

$$S\,b\,\Delta y - U_t\,b\,\Delta y\,(T_p - T_a) = h_{pf}\,b\,\Delta y\,(T_p - T_f) + h_{r,pb}\,(T_p - T_b)\,b\,\Delta y$$

$$\ldots(6.60)$$

where S, U_t, and T_p have the same meaning as those for liquid collectors, T_f is the air temperature at y, h_{pf} is the convective heat transfer coefficient between the plate and the fluid, $h_{r,pb}$ is the radiative heat transfer coefficient between the collector plate and the insulated bottom plate, and T_b is the temperature of the bottom plate. $h_{r,pb}$ is given by

$$h_{r,pb} = \frac{\sigma(T_p^2 + T_b^2)(T_p + T_b)}{\left(1/\varepsilon_p + 1/\varepsilon_b - 1\right)} \qquad\ldots(6.61)$$

which is similar to Equation 6.29.

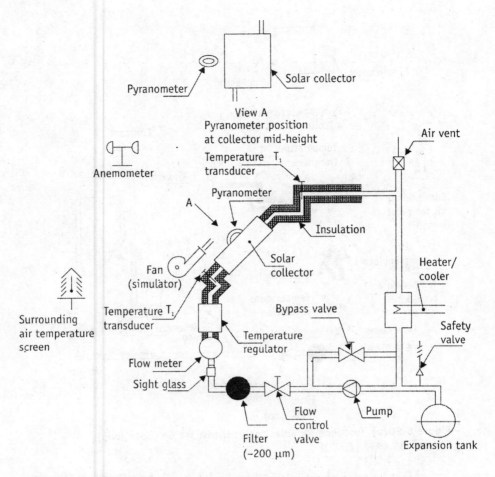

Figure 6.20(b) Test loop for solar collector test (closed loop)
Source IS 12933 (2003)

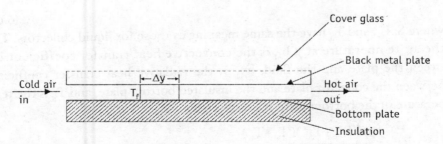

Figure 6.21 Cross-section of a conventional solar air heater

The term $b\Delta y$ cancels throughout, and we have

$$S - U_t\left(T_p - T_a\right) = h_{pf}\left(T_p - T_f\right) + h_{r,pb}\left(T_p - T_b\right) \qquad \text{...(6.62)}$$

A similar equation is obtained for the bottom plate

$$h_{r,pb}\left(T_p - T_b\right) = h_{bf}\left(T_b - T_f\right) + U_b\left(T_b - T_a\right) \qquad \text{...(6.63)}$$

where h_{bf} is the convective heat transfer coefficient from the bottom plate to the air, U_b is the bottom loss coefficient, and T_a is the ambient temperature. The useful heat gain by the fluid is given by

$$\dot{m}C_p\,\Delta T_f = b\Delta y\,h_{pf}\left(T_p - T_f\right) + b\Delta y\,h_{bf}\left(T_b - T_f\right) \qquad \text{...(6.64)}$$

or

$$\frac{\dot{m}C_p}{b}\frac{dT_f}{dy} = h_{pf}\left(T_p - T_f\right) + h_{bf}\left(T_b - T_f\right)$$

With some simplifying assumptions, it can be shown that (Dickinson, Clark, and Iantuore 1976)

$$\frac{\dot{m}C_p}{b}\frac{dT_f}{dy} = F'\left[S - U_L\left(T_f - T_a\right)\right] \qquad \text{...(6.65)}$$

where

$$F' = \left(1 + \frac{U_L}{h_e}\right)^{-1} \qquad \text{...(6.66)}$$

$$U_L = U_t + U_b \qquad \text{...(6.67)}$$

and

$$h_e = \left[h_{pf} + \frac{h_{r,pb}\,h_{bf}}{\left(h_{r,pb} + h_{bf}\right)}\right] \qquad \text{...(6.68)}$$

The useful heat gain is given by

$$Q_u = A_c\,F_R\left[S - U_L\left(T_{f,i} - T_a\right)\right] \qquad \text{...(6.69)}$$

$$F_R = \frac{\dot{m}C_p}{A_c\,U_L}\left[1 - \exp\left(-\frac{F'A_c\,U_L}{\dot{m}C_p}\right)\right] \qquad \text{...(6.70)}$$

where A_c is the collector area.

The heat transfer coefficients h_{pf} and h_{bf} can be taken as equal and can be obtained as follows.

$$Nu = \frac{0.01344 (Re)^{0.75}}{1 - 1.586 (Re)^{-0.125}} \qquad \ldots (6.71)$$

where Nu is the Nusselt number hD_e/k, Re is the Reynold's number $\rho V D_e/\mu$, and D_e is the equivalent diameter.

$$D_e = \frac{4 \times \text{Cross-sectional area of the duct}}{\text{Wetted perimeter}} \qquad \ldots (6.72)$$

The above treatment assumes that the flow is fully developed (which is the case if $L/D_e \geq 30$) and is fully turbulent. The pressure drop across the collector can be obtained by the Blasius equation

$$f = 0.079 \, Re^{-0.25} \qquad (6.73)$$

It is also common to have solar air heaters where the air flows between the cover and the plate (Figure 6.22[a]).

The useful heat gain for the type of collector is given by

$$Q_u = A_c F_R \left[S - U_L \left(T_{f,i} - T_a \right) \right] \qquad \ldots (6.74)$$

where

$$U_L = \frac{\left(U_t + U_b \right) \left(h_1 h_2 + h_1 h_{r,pc} + h_2 h_{r,pc} \right) + U_t U_b \left(h_1 + h_2 \right)}{h_1 h_{r,pc} + h_2 U_t + h_2 h_{r,pc} + h_1 h_2} \qquad \ldots (6.75)$$

and

$$F' = \frac{h_{r,pc} h_1 + h_2 U_t + h_2 h_{r,pc} + h_1 h_2}{\left(U_t + h_{r,pc} + h_1 \right) + \left(U_b + h_2 + h_{r,pc} \right)) - h_{r,pc}^2} \qquad \ldots (6.76)$$

where h_1 and h_2 are the convective heat transfer coefficients as shown in thermal network (Figure 6.22[b]).

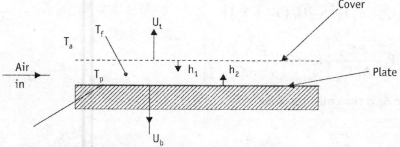

Figure 6.22(a) Solar air collector with air flow between the cover plate and the absorber

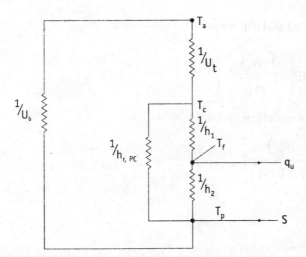

Figure 6.22(b) Thermal network of solar air heater

Example 4

Most solar dryers operate as a one-pass system, that is, ambient air enters the solar collector and exits as hot air, which is fed into the drying cabinet. Drying of several food crops requires the hot air to be at a specific temperature. This would mean that the flow rate of air also be regulated as the radiation intensity changes in order to keep a constant output temperature.

1 Derive an expression giving the flow rate as a function of radiation intensity.
2 Calculate the mass velocity ($\dot{m}/A_c$) under the following conditions

$U_L = 8.0$ W/m²

$T_{f,o} - T_{f,i} = 30$ °C

$\tau\alpha = 0.85$

$I_t = 900$ W/m²

$F' = 0.74$

$C_p = 1.0$ kJ/kg °C

Solution

1 For $T_{f,i} = T_a$, Equation 6.69 can be written as

$$Q_u = A_c F_R S$$

$$= A_c S \frac{\dot{m}C_p}{A_c U_L}\left[1 - \exp\left(-\frac{F'A_c U_L}{\dot{m}C_p}\right)\right] \qquad \text{(a)}$$

For a given rise in temperature T, q_u can also be written as

$$Q_u = \dot{m}C_p\left(T_{f,o} - T_{f,i}\right) = \dot{m}\,C_p\,\Delta T \qquad \text{(b)}$$

From Equations (a) and (b), we get

$$\frac{U_L \Delta T}{S} = \left[1 - \exp\left(-\frac{F' A_c U_L}{\dot{m} C_p} \right) \right]$$

Re-arranging the equation yields

$$\frac{\dot{m}}{A_c} = \frac{\left(F' U_L / C_p \right)}{\ln\left[1 - \left(U_L \Delta T / S \right) \right]^{-1}} \tag{c}$$

2    $\dfrac{U_L \Delta T}{(\tau\alpha) I_t} = \dfrac{8 \times 30}{0.85 \times 900} = 0.235$

$F' U_L / C_p = \dfrac{0.740 \times 8}{1} = 5.92 \, \text{g/m}^2.\text{s}$

$\dfrac{1}{\ln\left[1 - \left(U_L \Delta T / S \right) \right]^{-1}} = 3.73$

Therefore $\dot{m}/A_c = 22 \, \text{g/m}^2 \cdot \text{s}$

or $79.2 \, \text{kg/m}^2 . \text{h}$

Trickle type collectors

In these collectors, water at a small flow rate passes down a corrugated absorber sheet. It is distributed uniformly by a header pipe at the top and collected in a channel at the bottom. A theoretical and experimental investigation of such a roof collector has been done by Maru, Kishore, and Gomkale (1986).

Collector with phase-changing fluids

Solar collectors charged with low boiling point organic fluids such as freon-11, n-hexane, acetone, n-pentane, etc. have been employed for applications such as solar water pumping (Kishore, et al. 1986; Pytlinski 1978; Rao and Rao 1976). The fluid entering the collector is first heated to boiling temperature and is subsequently maintained at that temperature. The fluid then exits the collector as vapour at high pressure. This high pressure vapour can then be used for developing mechanical power for applications such as pumping. Analysis and testing of collectors charged with phase-changing fluids have been reported by Kishore, Gandhi, and Rao (1984) and Kishore, Gandhi, Marquis, *et al.* (1984).

Thermo-siphoning in solar energy systems

As flat plate collectors have to be kept inclined to receive maximum radiation for a given latitude, the heated water rises to the top due to its lower density. This principle can be used for circulating water within the solar system. Similarly, heated air rises to the top of the collector in natural circulation dryers. The natural convection flow caused due to density differences has been used to induce ventilation through solar chimneys (Figure 6.23).

Thermo-siphoning in solar water heating collectors

The analysis of thermo-siphon solar water heating systems (sometimes also called natural circulation systems) was done by Close (1962), Gupta and Garg (1968), and Ong (1974). Schematic diagram of thermo-siphon solar water heating system along with temperature distribution at various points is shown in Figure 6.24.

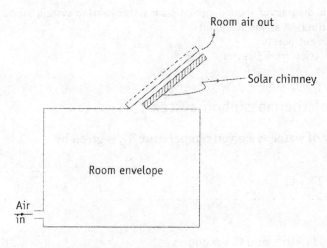

Figure 6.23 Solar chimney

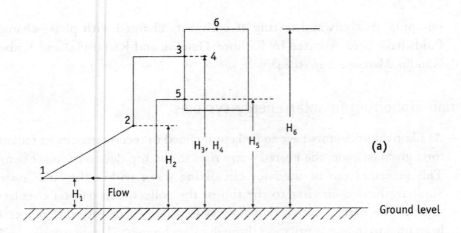

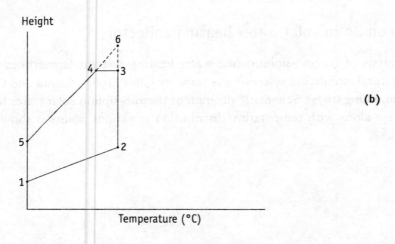

Figure 6.24 Schematic diagram of thermo-siphon solar water heating system along with temperature distribution at various points
Note 1–6 are measurement points
Reproduced with permission from Elsevier
Source Ong (1974)

Theoretical analysis of thermo-siphon system

The specific gravity of water at a given temperature T_m is given by

$$SG = A\,T_m^2 + B\,T_m + C$$

where,
$A = -1.25 \times 10^{-6}$, $B = 5.83 \times 10^{-5}$, and $C = 0.99967$ ...(6.77)

The thermo-siphon head causing flow is equivalent to the area under the height–temperature curve of Figure 6.24, and is given by Ong (1974) as

$$H_f = 0.5(T_5 - T_3)(2\,AT_m + B_3)f(H) \qquad \qquad ...(6.78)$$

where

$$f(H) = 2(H_3 - H_1) - (H_2 - H_1) - (H_3 - H_5)^2/(H_6 - H_5) \qquad ...(6.79)$$

The thermo-siphon head is balanced by the pressure drop due to friction and other losses in pipes, which is represented as follows.

$$H_f'\left[f'(l/d) + k\right]u^2/2g \qquad \qquad ...(6.80)$$

where f' is the friction factor, l is length, d is diameter, k is the equivalent number of velocity heads lost by the flow in passing through bends, etc., and u is the velocity. As shown in Chapter 3, the friction factor f' for laminar flow is given by

$$f' = 64/Re.$$

Natural circulation flows are generally small, and hence laminar flow can be assumed. The total pressure drop through N pipes in the collector panel, headers, and piping outside the collector can be written as

$$H_{f,e} = \left[f_e\left(l_1/d_1\right) + k_e\right]\left(\frac{8\dot{m}^2}{\pi^2 g \rho^2 d_1^4 N^2}\right) \qquad ...(6.81)$$

where f_e and k_e are the equivalent friction factors and the head loss given by

$$f_e = f_1 + f_2 N^2 \frac{l_2}{l_1}\left(\frac{d_1}{d_2}\right)^5 + f_3 N^2\left(\frac{l_3}{l_1}\right)\left(\frac{d_1}{d_3}\right)^5 \qquad ...(6.82)$$

and

$$k_e = k_1 + k_2 N^2\left(\frac{d_1}{d_2}\right)^4 + k_3 N^2\left(\frac{d_1}{d_3}\right)^4 \qquad ...(6.83)$$

1, 2, and 3 represent a bank of N parallel collector tubes, each of length l_1 and diameter d_1; two headers of total length l_2 and diameter d_2; and two lengths of connecting pipes of total length l_3 and diameter d_3.

A mean temperature T_m for the system is defined as

$$T_m = \frac{(T_1 + T_2)}{2} = \frac{(T_5 + T_3)}{2}$$

By writing instantaneous (hourly) energy balance for the system and by employing a finite difference method, T_m and $\dot{m}$ can be found by employing a finite difference method (Ong 1974).

Solar thermo-siphon systems have become popular due to their intrinsic advantage of not requiring external energy for operating the system.

Cheaper alternatives to conventional solar water heaters

As conventional solar water heaters have higher initial costs as compared to commercial fuel-based water heaters (for example, electric geysers), several attempts have been made to design cheaper systems. Some of these are discussed below.

Collector-cum-storage water heater

This is a rectangular or cylindrical metal box (usually galvanized iron or mild steel to keep the cost low), one side of which is painted black. The metal box is kept inside a wooden enclosure (or a similar low-cost enclosure) with one side open for fixing a cover glass (Figure 6.25).

If the height of the metal box is 50 mm, it can hold 50 litres/m² of collector area. On the basis of a rough calculation for a 24-hour average radiation level of 200 W/m² and an average collector efficiency of 25%, a temperature rise of about 20 °C can be obtained for such a box-type collector. If the initial temperature of water is 30 °C, the final water temperature would be 50 °C, which is quite adequate for bathing, etc. Such

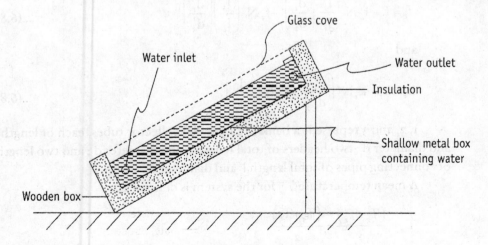

Figure 6.25 Simple collector-cum-storage solar water heater

box-type water heaters have been investigated in India at CAZRI (Central Arid Zone Research Institute), Jodhpur, and at TERI.

Shallow solar ponds

SSPs (shallow solar ponds) have long been considered potential alternatives for conventional flat plate collectors. One of the earlier applications of SSPs was in desalination (Hodge, Thompson, Groh, *et al.* 1966). An SSP prototype facility was built and operated to supply hot water to the Sohio Uranium Mill near Grants, New Mexico (Dickinson, Clark, and Iantuore 1976). A compact SSP for hot water preparation for military and recreation purposes was reported by Kudish and Wolf (1979).

SSP consists of a shallow bed of water contained within two plastic layers – black plastic layer at the bottom and transparent layer at the top – with suitable insulation and container box, and another glazing to reduce heat losses (Figure 6.26).

The temperature build-up over the day can be obtained by solving the equation

$$(MC)_p \frac{dT_p}{dt} = A_c (\tau\alpha)G - A_c U_L (T_p - T_a) \qquad \qquad ...(6.84)$$

where $(MC)_p$ is the mass-specific heat product of water in the pond, T_p is the temperature of water at a given time t, A_c is the area of the pond exposed to sunlight, and $(\tau\alpha)$ and U_L are same as those in flat plate collectors. As fins and

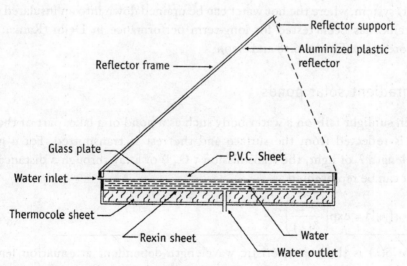

Figure 6.26 Sectional view of shallow solar pond

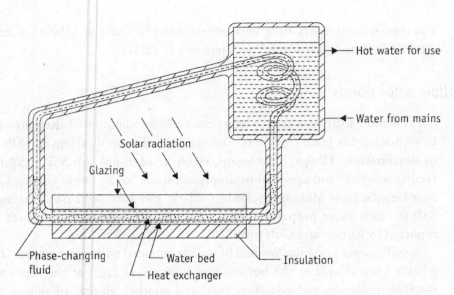

Figure 6.27 Shallow solar pond based domestic solar water heater

fluid flow are not involved, F′ and F_R are unity. A theoretical and experimental investigation of SSPs with continuous heat extraction has been proposed (Kishore, Gandhi, and Rao 1986). A domestic solar water heater based on SSP with a heat pipe heat exchanger, shown in Figure 6.27, has been studied experimentally (Gandhi and Kishore 1983).

A portable SSP water heater has also been proposed (Kishore, Ranga Rao, and Raman 1987). Temperature increments of up to 30 °C over the day have been reported. An SSP-DHW (shallow solar pond based domestic hot water) system, where the hot water can be drained down into an insulated storage tank, has been tested for long-term performance in Delhi (Raman and Kishore 1992) (Figures 6.28 and 6.29).

Salinity gradient solar ponds

When sunlight falls on a water body such as a pond or a lake, part of the energy is reflected from the surface and the rest is transmitted. For a given wavelength λ of light, the transmission τ (λ, l) of a ray through a distance l in water can be represented as

$$\tau(\lambda,l) = \exp\left(\frac{-l}{\delta(\lambda)}\right) \qquad \qquad ...(6.8_5)$$

where δ(λ) is the characteristic wavelength-dependent attenuation length. Attenuation of light results from absorption, molecular scattering, and scatter-

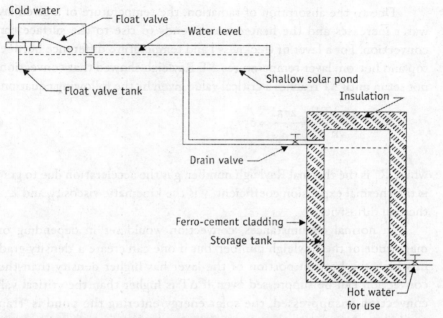

Figure 6.28 Schematic diagram of a shallow solar pond based domestic solar water heater
Source Raman and Kishore (1992)

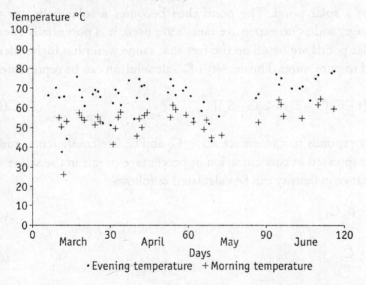

Figure 6.29 Storage tank temperatures of the SSPDHW system

ing from suspended particles. The higher the magnitude of $\delta(\lambda)$, the lower is the penetration of radiation. Experimental values for distilled water show that water is relatively transparent to the visible wavelength (0.4–0.7 μm), which includes much of the solar spectrum (Hull, Nielsen, and Golding 1989).

Due to the absorption of radiation, the temperature of a given layer of water increases and the heated water tends to rise to the surface through convection. For a layer of thickness l and temperature difference between the top and bottom layer temperatures ΔT, Rayleigh showed that convection does not set in until ΔT reaches a critical value given by the following equation

$$R_T = \frac{g\alpha\Delta T l^3}{vk_T} = \frac{27\pi^4}{4} = 657.5 \qquad ...(6.86)$$

where R_T is the thermal Rayleigh number, g is the acceleration due to gravity, α is the thermal expansion coefficient, v is the kinematic viscosity, and k_T is the thermal diffusivity.

In normal circumstances, convection would set in depending on the magnitude of the Rayleigh number, but if one can create a 'density gradient', in which the bottom portion of the layer has higher density than the top, convection can be suppressed even if ΔT is higher than the critical value. If convection is suppressed, the solar energy entering the pond is 'trapped', resulting in higher temperatures of the lower layers of water in the pond, from which heat can be extracted for useful purposes. This is the principle of operation of a solar pond. The pond thus becomes a solar collector with built-in storage, and as no expensive metals are used, it is potentially cheaper. Practical solar ponds are based on the fact that saline water has higher density as compared to pure water. The density of a salt solution can be represented as

$$\rho = \rho_o [1 - C_T(T - T_o) + C_s(S - S_o)] \qquad ...(6.87)$$

where ρ_o corresponds to a reference state; C_s and C_T are coefficients; and S is the salinity expressed as concentration or percentage of salt in the saline solution. The change in density can be calculated as follows.

$$\frac{\partial\rho}{\partial T} = -\rho_o C_T \qquad ...(6.88)$$

$$\frac{\partial\rho}{\partial S} = \rho_o C_s \qquad ...(6.89)$$

and

$$\frac{\partial\rho}{\partial x} = \frac{\partial\rho}{\partial S}\frac{\partial S}{\partial x} + \frac{\partial\rho}{\partial T}\frac{\partial T}{\partial x} \qquad ...(6.90)$$

where x is the vertical co-ordinate increasing downward. For densities to remain stable, it is necessary that

$$\frac{\partial \rho}{\partial x} > 0 \qquad\qquad\qquad ...(6.91)$$

Substituting from Equations 6.88 and 6.89 and re-arranging, we get

$$R_\rho = \frac{C_s\left(\partial S/\partial x\right)}{C_T\left(\partial T/\partial x\right)} > 1 \qquad\qquad ...(6.92)$$

This is the criterion for static stability and provides the salinity gradient values for a given temperature gradient. In real ponds, there is another criterion called dynamic stability criteria, in which both thermal and mass diffusivity are considered in the double-diffusive system (Hull, Nielsen, and Golding 1989).

In real solar ponds, there are three distinct zones: UCZ (upper convective zone) or the surface zone, NCZ (non-convective zone) or the gradient zone, and LCZ (lower convective zone) or the storage zone. A schematic diagram of the solar pond is shown in Figure 6.30 (a).

The UCZ is formed due to wind effects, evaporation, etc., and can be maintained at a thickness of about 30–50 cm. The NCZ has a thickness of 1–1.5 m and the LCZ has a thickness of about 1.5 m. The density gradient can be created artificially using a diffuser method (Kishore and Kumar 1996). Temperatures in LCZ and NCZ build up rapidly once the salinity gradient is established in a clear pond. The rise of temperature in the LCZ for the 6000 m² solar pond of Bhuj is shown in Figure 6.30 (b).

With convection suppressed, NCZ can be treated as a transparent conducting solid with a heat generating source (solar radiation absorbed).

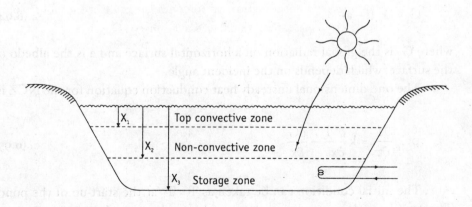

Figure 6.30a Schematic diagram of the solar pond

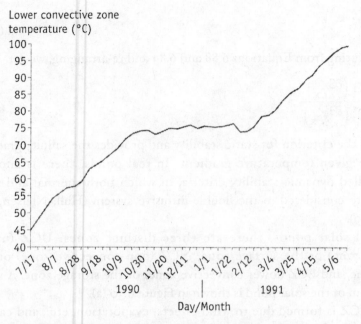

Figure 6.30(b) Temperature history of the storage zone for the 6000 m² solar pond in Bhuj
Source Kishore and Kumar (1996)

Choosing a co-ordinate system with x = 0, corresponding to the surface of the pond, the solar radiation at depth x is given by

$$G(x) = G_s\, g(x) \qquad\qquad ...(6.93)$$

where G_s is the radiation immediately below the surface and is given by

$$G_s = G_o\,(1-a) \qquad\qquad ...(6.94)$$

where G_o is the global radiation on a horizontal surface and a is the albedo of the surface, which depends on the incident angle.

The one-dimensional unsteady heat conduction equation for the NCZ is

$$\rho C_p\, \frac{\partial T}{\partial t} = k\, \frac{\partial^2 T}{\partial x^2} - \frac{\partial G}{\partial x} \qquad\qquad ...(6.95)$$

The initial condition can be taken as $T = T_o$ at the start-up of the pond. The two boundary conditions required to solve the above equation are obtained by heat balance on UCZ and NCZ. With a suitable functional form for G(x), Equation 6.95 can be solved using numerical techniques. One general method of solving it is by applying the Crank–Nicolson method (Joshi and Kishore 1985a; Joshi, Kishore, and Rao 1984).

A useful expression for obtaining the efficiency of the solar pond can be derived by assuming pseudo-steady state conditions in which $\delta T/\delta t = 0$. Equation 6.95 can then be written as

$$k\frac{d^2T}{dx^2} = \frac{d}{dx}(G(x)) \qquad \qquad ...(6.96)$$

An energy balance for UCZ gives

$$Q_s = G_s[1 - g(x_1)] + k\frac{dT}{dx}\bigg|_{x=x_1} \qquad \qquad ...(6.97)$$

where Q_s is the sum of heat losses (convective, radiative, and evaporative) from the surface and x_1 is the depth of UCZ. A similar equation for LCZ can be written as

$$Q_u = G_s g(x_2) - k\frac{dT}{dx}\bigg|_{x=x_2} - Q_b \qquad \qquad ...(6.98)$$

where Q_u is the useful heat extracted, Q_b is the bottom loss to the ground, and x_2 corresponds to the interface between NCZ and LCZ. Equation 6.96 can be solved using the boundary conditions of Equations 6.96 and 6.98 (Kishore and Joshi 1984; Kooi 1979).

$$Q_u = G_s(\tau\alpha) - U_t(T_b - T_s) - Q_b \qquad \qquad ...(6.99)$$

where

$$(\tau\alpha) = \frac{\int_{x_1}^{x_2} g(x)dx}{x_2 - x_1} \qquad \qquad ...(6.100)$$

and

$$U_t = \frac{k}{x_2 - x_1} \qquad \qquad ...(6.101)$$

Q_s and Q_b can be related to the ambient and ground conditions, respectively. Taking assumed or measured profiles for g(x), thermal efficiencies of solar ponds can be obtained. However, such results are applicable only for yearly average performance (Joshi and Kishore 1986).

The attenuation function g(x) had been fitted to various functional forms. The Rabl–Nielsen model is expressed as

$$g(x) = \sum_{i=1}^{4} r_i \exp\left(-\mu_i\, x \sec\theta_r\right) \qquad\qquad ...(6.102)$$

where r_i and μ_i are the constants for a particular seawater and θ_r is the angle of refraction.

Bryant and Colbeck proposed a simple two-parameter model

$$g(x) = a - b \ln\left(x \sec\theta_r\right) \qquad\qquad ...(6.103)$$

The one-parameter model proposed by Hawlader and Brinkworth is expressed as

$$g(x) = (1-F)\exp[-\mu(x-\delta)\sec\theta_r] \qquad\qquad ...(6.104)$$

where F is taken as 0.4 and δ as 0.06 m. The effect of using different attenuation models on performance predictions has been studied by Joshi and Kishore (1985b).

Considerable work on solar ponds has been done worldwide, including India (Rao, Kishore, and Vaja 1990). The largest solar pond in Asia, the 6000 m² solar pond at Bhuj, India, operated in an industrial environment and supplied process hot water to the Kuchch dairy for more than two years (Kishore and Kumar 1996). The solar pond at Pondicherry is producing electricity since 2004. A very large number of applications, including desalination, bromine recovery, manufacture of magnesium chloride, improved salt production, and so on have been identified for coastal areas in India.

Evacuated tube collectors

Evacuated or vacuum tube collectors are fast becoming popular in the world market. Emmett first proposed the concept of an evacuated tube collector in 1909. With recent advances in vacuum and sealing technology and the development of selective coating on glass surfaces, the evacuated tube collectors are now mass-produced in various countries.

Essentially, these are based on the Dewar vacuum flask concept, wherein the convective losses from the collector surface are reduced by providing vacuum around the absorber. There are two major design configurations in evacuated tube collectors

- Single-glass tube
- Double-glass tube

Single-glass tube evacuated collectors

In single-glass tube collectors, either a heat pipe is used to extract heat from the collector or a simple U-tube with fin is provided to circulate the fluid (Figure 6.31 a, b). The metal tube or heat pipe tube and the glass tube covering it are hermetically sealed to form a vacuum tight joint. The air between the gap is extracted from the other end using a vacuum pump and then the end is sealed. Activated barium getter is provided to absorb the gases, which can diffuse through the glass tube. Sometimes a small ripple reflector is provided below the collector to improve the concentration of the solar radiation from below. The sealing of the glass to metal joint is the most important area in these collectors. These types of collectors have a few advantages as listed below.

- Higher heat transfer efficiencies.
- No fluid present inside the glass collectors.
- Easy to use as an indirect heating element, especially when the outside conditions are freezing or hard water is to be heated.

Double-glass tube evacuated collectors

Double-glass tube collectors (Figure 6.31c) are easy to manufacture but are less efficient than single-glass tube collectors. They have two glass tubes

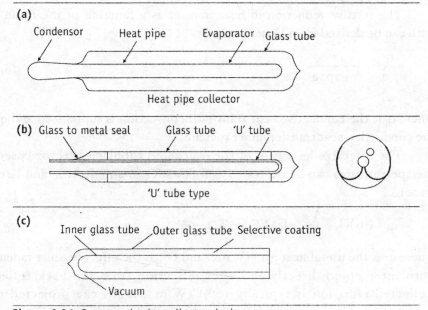

Figure 6.31 Evacuated tube collector designs
Source Goswami, Kreith, and Kreider (2000)

attached to each other at one end while the other end of both the tubes is closed. The space between them is evacuated and a selective absorbing coating is applied on the outer surface of the inner tube. They can be used directly to heat water stored in the inner tube and are commonly used in domestic water heating systems. Domestic solar water heating systems based on double-glass tube collectors are now commonly available worldwide.

Evacuated tube collector thermal analysis

Conductive heat transfer between two surfaces having low-pressure gas in the interim space is given by the following equation (Goswami 2006).

$$q_l = k\Delta t/(g + 2p) \qquad \qquad ...(6.105)$$

where q_l is the heat loss, k is the constant, Δt is the temperature gradient, g is the gap between surfaces, and p is the mean free path of molecules.

For air, the mean free path at atmospheric temperature and pressure is about 70 μm. If 99% air is removed from a tubular collector, the mean free path increases to 7 mm, and conduction heat transfer is almost unaffected. However, the mean free path increases to 7 cm at 10^{-7} torr, which is substantially greater than the heat transfer path length (gap between the glass tubes), which is of the order of 20 mm. This reduces the conductive heat transfer substantially.

The relative reduction in heat transfer as a function of the mean free path can be derived from Equation 6.105

$$\frac{q_{vac}}{q_l} = \frac{1}{1 + 2p/g} \qquad \qquad ...(6.106)$$

where q_l is the conductive heat transfer if convection is suppressed and q_{vac} is the conductive heat transfer under vacuum.

The effective heat gain of the evacuated tubular collector based on the aperture area can be expressed as follows (Goswami, Kreith, and Kreider 2000b).

$$q_u = (\tau\alpha)G_{eff}\frac{A_{tb}}{A_{cl}} - U_L\left(T_{abs} - T_a\right)\frac{A_{abs}}{A_{cl}} \qquad \qquad ...(6.107)$$

where q_u is the useful heat gain (W/m²) and G_{eff} is the effective solar radiation, both intercepted directly and after reflection from the back reflector (reflected radiation is typically 10%) (W/m²); A_{tb} is the projected tube area (m²), A_{cl} is the total collector area (m²), U_L is the overall heat loss coefficient (W/m²K), T_{abs} is the absorber temperature (°C), T_a is the ambient temperature (°C), and A_{abs} is the projected area of the absorber (m²).

Bekey and Mather have shown that a tube spacing of one diameter apart maximizes the energy output (Goswami, Kreith, and Kreider 2000).

Thermal storage systems

Thermal storage of energy is essential in most solar thermal systems due to a mismatch between the need and availability of energy. In this section, we will study various types of thermal storage systems used in solar systems.

Types of thermal energy storage

Sensible heat storage

Sensible heat storage is the most commonly used storage method in solar heating systems. Sensible heat is stored in a material by raising its temperature.

The important characteristics of good sensible heat storage media are high value of heat storage density, chemical stability, and an ability to withstand repeated heating and cooling cycles (Table 6.3).

Water is the most commonly used media for sensible storage in low- and medium- temperature systems. It is stored in hot water tanks duly insulated with 50–100 mm thick glass wool/mineral wool insulations.

For large systems, such as district heating systems, large underground water storage systems have been used. The earth itself can be used as an insulating material. In the next section, we will analyse the thermal performance of a well-mixed water storage tank (Figure 6.32).

Thermal analysis of a well-mixed tank

A well-mixed tank receives energy from the collectors and supplies energy to the load when required. The main assumption here is that water in the tank is at uniform temperature T_{tank}, which changes over a period of time. In this analysis, we will study the change in temperature over a period of time.

Hot water from the collector at a temperature $T_{c,\,out}$ enters the tank and relatively low-temperature water at a temperature T_{tank} enters the solar system from the tank.

The hot water at temperature T_{tank} is supplied to the load. The make-up water is supplied at temperature T_i to the tank.

Table 6.3 Sensible heat storage media and their properties

Storage medium	Temperature range (°C)	Density (kg/m³)	Specific heat (J/kg K)	Energy density (kWh/m³K)	Thermal Conductivity (W/mK)
Water	0–100	1000	4190	1.16	0.63 at 38 °C
Water (10 bar)	0–180	881	4190	1.03	—
50% ethylene glycol–50% water	0–100	1075	3480	0.98	—
Dowtherm A® (Dow Chemical Co.)	120–260	867	220	0.53	0.122 at 260 °C
Therminol 66® (Monsanto Co.)	9–343	750	2100	0.44	0.106 at 343 °C
Rock	—	1600	880	0.39	—
Molten salt	142–540	1680	1560	0.72	0.61
Aluminium	m.p. 660	2700	920	0.69	200
Draw salt (50NaNO₃–50KNO₃)[a]	220–540	173	1550	0.75	0.57
Cast iron	m.p. 1150–1300	7200	540	1.08	42.0
Taconite	—	3200	800	0.71	—
Liquid sodium	100–760	750	1260	0.26	67.5
Fireclay	—	2100–2600	1000	0.65	1.0–1.5

m.p. – melting point
[a] Composition in per cent by weight
Source Goswami, Kreith, and Kreider (2000)

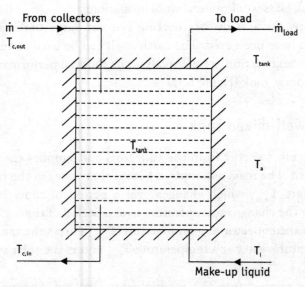

Figure 6.32 Well-mixed water storage tank
Source Sukhatme (1984)

Now the thermal heat balance equation of the tank can be written as

$$\left[\left(\rho V C_p\right)_{liq} + \left(\rho V C_p\right)_{tank}\right]\frac{dT_{tank}}{dt} = q_u - q_{load} - (UA)_{tank}(T_{tank} - T_a) \quad \text{...(6.108)}$$

Following the analysis given by Sukhatme (1984), if q_{load} is the useful energy supplied to the load from the storage and q_u is the useful energy gained from the solar thermal system, we get

$$\frac{q_u - q_{load} - (UA)_{tank}(T_{tank} - T_a)}{q_u - q_{load} - (UA)_{tank}(T_i - T_a)} = \exp\left[-\frac{(UA)_{tank}\, t}{(\rho V C_p)_e}\right] \quad \text{...(6.109)}$$

where $(UA)_{tank}$ is the product of the overall heat transfer coefficient and the surface area of the tank (W/K), q_u is the useful heat gain in collector (W), T_{tank} is the tank water temperature (°C), T_a is the ambient temperature (°C), and $(\rho V C_p)_e$ is the combined heat capacity of the tank material and water (J/K). t is the time (s) and T_i is the tempeature at $t = 0$.

Alternatively,

$$q_u = \dot{m}\, C_p(T_{c,out} - T_{c,in}) = \dot{m}\, C_p((T_{c,out} - T_{tank})$$

$$q_{load} = \dot{m}_{load} C_p(T_{tank} - T_i)$$

The equation then becomes

$$\left(\rho V C_p\right)_e \frac{dT_{tank}}{dt} = \dot{m} C_p\left(T_{c,out} - T_{tank}\right) - \dot{m}_{load} C_p\left(T_{tank} - T_i\right) -$$

$$(UA)_{tank}\left(T_{tank} - T_a\right)$$

$$\text{...(6.110)}$$

Depending on the operating conditions, q_u or q_{load} may be zero at any particular time. Equation 6.110 can be used to calculate tank temperatures. The assumption that q_u, q_{load} and T_a are constants holds good if the time interval of integration is kept small.

Latent heat storage

Latent heat storage system stores heat energy in the form of latent heat. Heat is stored when the solid storage media absorbs heat and melts. The heat is released when required and during the process the medium cools down and solidifies. Hence, latent heat storages are also known as PCM (phase change material) storages and are used for medium and high temperature storages, especially for applications like solar thermal power generation.

A good latent heat storage medium should have

- melting point in the required range
- high value of the latent heat of fusion
- small volume change during the phase change
- negligible supercooling or superheating for the phase change
- high thermal conductivity in both phases

In the past few decades, a number of such materials have been investigated (Table 6.4).

Most of these materials are inorganic salt compounds or complex organic chemicals used singly or in mixture (Sukhatme 1984). The application of a particular substance depends on the melting point. For example, $CaCl_2.6H_2O$ or paraffin wax having melting points between 30 °C and 50 °C are used for space heating applications, while sodium hydroxide or lithium fluoride having melting points above 300 °C can be considered for high-temperature applications.

Table 6.4 Physical properties of latent heat storage materials

Storage medium	Melting point (°C)	Latent heat (°C)	Specific heat (kJ/kg °C)		Density (kg/m³)		Energy density (kWh/m³K)	Thermal conductivity (W/mK)
			Solid	Liquid	Solid	Liquid		
$LiClO_3$–$3H_2O$	8.1	253	—	—	1720	1530	108.0	—
Na_2SO_4–$10H_2O$ (Glauber's Salt)	32.4	251	1.76	3.320	1460	1330	92.7	2.250
$Na_2S_2O_3.5H_2O$	48.0	200	1.47	2.390	1730	1665	92.5	0.570
$NaCH_3COO.3H_2O$	58.0	180	1.90	2.500	1450	1280	64.0	0.500
$Ba(OH)_2.8H_2O$	78.0	301	0.67	1.260	2070	1937	162.0	0.653
$Mg(NO_3).6H_2O$	90.0	163	1.56	3.680	1636	1550	70.0	0.611
$LiNO_3$	252.0	530	2.02	2.041	2310	1776	261.0	1.350
$LiCO_3/K_2CO_3 (35:65)^a$	505.0	345	1.34	1.760	2265	1960	188.0	—
$LiCO_3/K_2CO_3/Na_2CO_3$ (32:35:33)^a	397.0	277	1.68	1.630	2300	2140	165.0	—
η–Tetradecane	5.5	228	—	—	825	771	48.0	—
η–Octadecane	28.0	244	2.16	—	814	774	52.5	0.150
HDPE (cross-linked)	126.0	180	2.88	2.510	960	900	45.0	0.361
Steric acid	70.0	203	—	2.350	941	347	48.0	0.172

^{a} Composition in per cent by weight
l – liquid; HDPE – high density polyethylene
Source Goswami, Kreith, and Kreider (2000)

Latent heat storages are normally more compact than sensible heat storages. However, the design of the heat exchanger system is complicated because the material changes phase during heat extraction.

Thermochemical storage

In thermochemical storages, the energy is stored in the form of heat of chemical reactions. Thermochemical storages were first studied to meet the requirement of high-density and high-temperature storage media.

Thermochemical storage is a two-step process (Garg, Mullick, and Bhargava 1985).

1 *Storage or endothermic mode* In this step, energy is absorbed, thereby chemical bonds are either broken or re-arranged, in the process producing more energetic materials, which are stored.

2 *Discharge or exothermic mode* In this step, the reaction is reversed, producing thermal energy, and while doing so the original material is regenerated.

Table 6.5 (Garg, Mullick, and Bhargava 1985) shows the properties of various chemical heat storage systems.

Solar box cooker

The solar box cooker is a very popular cooker in various developing countries. Mullick, Kandpal, and Saxena (1987) have studied the performance of this solar cooker in detail.

Table 6.5 Properties of various chemical storage systems

Reaction	Conditions of reaction Pressure (kPa)	Temperature (°C)	Component (phase)	Pressure (kPa)	Temperature °C	Density (kg/m^3)	Energy density (kWh/m^3)
$MgCO_3(s) + 1200$ kJ/kg = $MgO + CO_2$ (g)	100	427–327	$MgCO_3(s)$ CO_2 (l)	100 7400	20 31	1500 465	187
$Ca(OH)_2$ (s) + 1415 kJ/kg = $CaO(s) + H_2O$ (g)	100	572–402	$Ca(OH)_2$ (s) H_2O (l)	100	20	1115	345
$SO_3(g) + 1235$ kJ/kg = SO_2 (g) + $1/2O_2(g)$	100	520–960	SO_3 (l) SO_2 (l) O_2 (g)	100 630 10 000	45 40 20	1900 1320 130	280

s – solid, l – liquid, g – gas
Source Garg, Goswami et al. (2000)

Solar box cooker consists of an outer box made of aluminium or fibre glass. The inner tray is blackened to absorb the solar radiation. A mirror is used for augmenting the solar radiation falling on cooker aperture.

Cooker pots are made of steel or aluminium. These are blackened on the top and sides. Glass wool is used as an insulation between the inner tray and the outer box (Figure 6.33).

Analysis of solar box cooker performance

The performance of the solar box cooker is measured using two figures of merit.

The first figure of merit, $F_1 = \eta_o/U_L$, is determined from no-load, stagnation temperature test as

$$F_1 = \frac{(T_{ps} - T_a)}{I} \qquad \qquad ...(6.111)$$

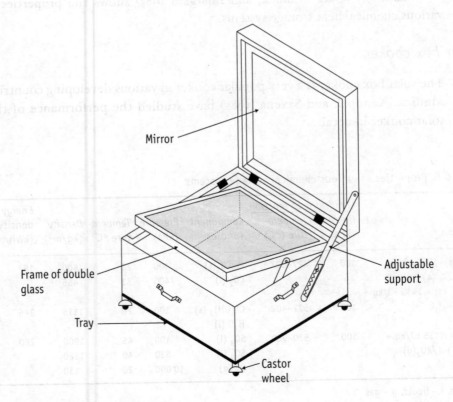

Figure 6.33 Box-type solar cooker

where T_{ps} is the tray (plate) stagnation temperature, I is the solar radiation on horizontal surface, T_a is the ambient temperature at the time when stagnation temperature is reached, η_o is the optical efficiency of the cooker, and U_L is the heat loss factor.

The second figure of merit, F_2 is obtained from full-load water heating test[1] given by the equation

$$F_2 = F' \eta_0 C_r = \frac{F_r (MC_w)}{A \Delta t} \ln \frac{(1 - 1/F_r)[(T_{w1} - \overline{T_a})/I]}{(1 - (1/F_r)[(T_{w2} - \overline{T_a})/I]} \qquad \ldots(6.112)$$

where F' is the heat exchanger efficiency factor, A is the aperture area of the cooker, T_{w1} is the initial water temperature, T_{w2} is the final water temperature, Δt is the time interval required to heat water from T_{w1} to T_{w2}, and C_r is the ratio $(MC)_w/(MC)'_w$. $(MC)'_w$ is the product of heat mass and specific heat (heat capacity) of water, cooker pots, and interior of the cooker.

The Bureau of Indian Standards has adopted this method of determining efficiency of the cooker. During testing, the temperature of water is kept below the boiling temperature. The box-type cookers can attain temperatures up to 120–140 °C.

Concentrating collector system

As mentioned in the beginning of this chapter, the solar flat plate collector has a limitation of operating in a relatively low-temperature region, that is, below 100 °C. This is because the area of the collection of solar radiation is the same as the area of the receiver (black plate), the thermal losses from which are proportional to the area. One can re-write Equation 6.26 as

$$Q_c = A(\tau\alpha) I - A U_L (T_p - T_a) \qquad \ldots(6.113)$$

The highest temperature attained under given conditions can be reached when there is no heat extraction, that is, $Q_c = 0$. The stagnation temperature T_p^s, for flat plate collectors, as given in Equation 6.27, would be in the range of 150–200 °C. Now consider the case in which the collection area A_c is higher than the receiving area A_r. Equation 6.113 can now be written as

$$Q_c = A_c(\tau\alpha) I - A_r U_L (T_r - T_a) \qquad \ldots(6.114)$$

[1.] Full-load is as defined by the manufacturer.

where T_r is the receiver temperature. The stagnation temperature is obtained when $Q_c = 0$

$$T_r^s = T_a + \left(\frac{A_c}{A_r}\right)\frac{(\tau\alpha)}{U_L} \qquad \qquad ...(6.115)$$

It is apparent from the above equation that receiver temperatures can be increased to any desired level by increasing the area ratio (A_c/A_r). A variety of methods for increasing the collection area for a given receiver area have been tried, starting from simple reflection to more complicated geometries (Figure 6.34).

The parabola is the most commonly used reflecting surface, as it has the property of focusing (or imaging) all radiation falling on the surface in the direction parallel to the axis. A parabola is the set of all points in a plane equidistant from a given line L (diretrix) and a given point F (focus) (Figure 6.35). The equation of a parabola is given as

$$y = \left(\frac{x^2}{4a}\right) \qquad \qquad ...(6.116)$$

A parabolic trough is generated by dragging the parabola in the direction perpendicular to the plane of paper. The focal point thus becomes a focal line. Such a collector is called a line focussing collector.

On the other hand, if the parabolic curve is rotated around the y-axis, one gets a three-dimensional surface of revolution called the paraboloid or the

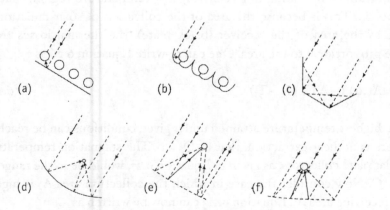

Figure 6.34 Possible concentrating collector configurations: (a) tubular absorbers with diffuse back reflector; (b) tubular absorbers with specular cusp reflectors; (c) plane receiver with plane reflectors; (d) parabolic concentrator; (e) Fresnel reflector; (f) array of heliostats with central receiver

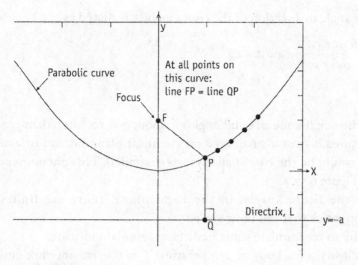

Figure 6.35 Parabola

parabolic dish. The focal point remains at the same place, but it now receives radiation from a cylindrical zone parallel to y-axis.

It is not possible to obtain indefinitely large temperatures at the focal point. Though the sun rays falling on the earth seem to be parallel, they are not. This is because of the finite sun–earth geometry, shown schematically in Figure 6.36.

Any point on the surface of the earth subtends a solid angle Ω_{S-E} with the solar disc. This angle can be calculated using the definition of solid angle.

$$\Omega_{S-E} \text{ (in steradian)} = \frac{\pi R_{sun}^2}{D_{SE}^2} \qquad \qquad \text{...(6.117)}$$

The radius of the sun is 1.39×10^9 m and the mean sun–earth distance is 1.496×10^{11} m. The solid angle Ω_{S-E} can thus be calculated as 6.8×10^{-5} steradians.

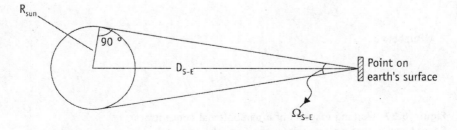

Figure 6.36 Sun–earth geometry

The half angle in a section of the cone can be calculated as

$$\sin\theta = \frac{6.95 \times 10^8}{1.496 \times 10^{11}} = 4.652 \times 10^{-3}$$

$$\theta = 0.266°$$

When the small cone of solid angle of about 7×10^{-5} steradian reaches the focus, it spreads over a small area on the focal plane instead of a single point (which would be the case if all rays were parallel). This phenomenon is illustrated in Figure 6.37.

Due to the finite spread in the focal plane, there are limits for concentration ratios that can be achieved.

It is useful to recapitulate some basic facts on solar radiation.

The 'radiosity' of a body at temperature T is the radiant flux emitted from a unit surface area. For a black body, it is given as σT^4 (refer to Chapter 3). If we assume the sun to be a black body radiation at T_{sun}, the radiosity G_{sun} will be σT^4_{sun}. The radiant flux ϕ is defined as the radiosity for a given area. While G has the unit W/m², ϕ has the unit W. Referring to Figure 6.36, the value for ϕ for sun on the entire spherical surface is given by

$$\Phi_{sun} = 4\pi R^2_{sun} G_{sun} \qquad ...(6.118)$$

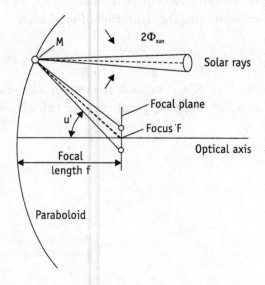

Figure 6.37 Working diagram of a paraboloidal concentrator
Reproduced with permission from the Gujarat Energy Development Agency
Source Zakhidov, Umarov, and Weiner (1992)

If one draws another sphere with diameter as D_{S-E}, the radiant flux passing through the surface of this sphere would be equal to Φ_{sun}, assuming that there is no loss of energy. Now, if $G_{s,c}$ is the flux density on this outer sphere, we can write

$$\Phi_{sun} = 4\pi R_{sun}^2 \, G_{sun} = 4\pi D_{S-E}^2 \, G_{s,c} \qquad \text{...(6.119)}$$

As stated in Chapter 4, $G_{s,c}$ is the solar constant at the mean sun–earth distance. From the measured value of $G_{s,c}$, 1367 W/m², one can calculate G_{Sun} as 63.2 MW/m² and T_s as 5777 K. In reality, the sun is not a black body. The brightness at the centre of the solar disc is 1.2 times the average brightness. This is called limb darkening.

The radiant flux Φ can be treated like a vector quantity with a magnitude and a direction. In a radiation field, Φ depends on the location, size, shape, and orientation of the area element dA through which the radiation is passing. Consider the solid angle $d\Omega$ in direction Ω with angle θ to the normal of dA (Figure 6.38).

Then, the irradiance I is the derivative with respect to area

$$I = \frac{d\Phi}{dA} \quad \text{or} \quad \int I \, dA = \Phi$$

Irradiance has the same meaning as radiosity, the difference being that the radiosity is the flux 'emerging' from a radiation and irradiance is the flux 'incident' on a surface. The radiance B is the derivative of I with respect to the solid angle projected onto the axis normal to dA.

$$B = \frac{d^2\Phi}{d\Omega \, \cos\theta \, dA} \qquad \text{...(6.120)}$$

$$I = \int B \cos\theta \, d\Omega \qquad \text{...(6.121)}$$

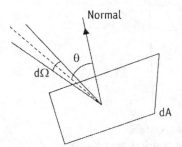

Figure 6.38 Sketch showing incident angle

For the case where radial symmetry exists and all radiation comes from a cone of angular aperture around the normal to the receiver area (Figure 6.39a), it can be shown that

$$I = \pi B \sin^2 \theta_a \qquad \qquad ...(6.122)$$

Now consider a concentrator system shown in Figure 6.39(b), in which radiant flux Φ_1 enters as input into area A_1 and goes out as output flux Φ_2 through area A_2. Assuming no losses in the concentration, radiant energy is conserved. One can then write

$$\Phi_1 = \Phi_2 \quad \text{or} \quad A_1 E_1 = A_2 E_2 \qquad \qquad ...(6.123)$$

E_1 and E_2 are averaged over A_1 and A_2, respectively.

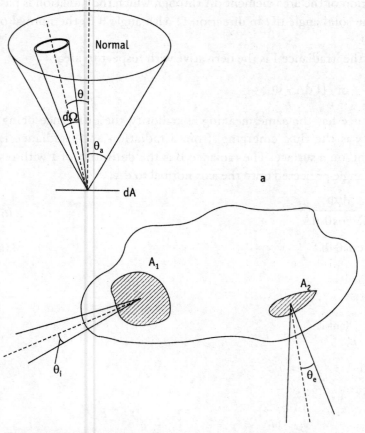

Figure 6.39b Conceptualization of a concentration system
Reproduced with permission from Springer Science and Business Media
Source Winter, Sizmann, and Vant-Hull (1991)

From thermodynamic considerations, it can be shown that the quantity $A\sin^2\theta$, also called 'Etendue', is conserved in a concentration system. Equation 6.123 can be used to define the concentration ratio C as

$$C = \frac{A_1}{A_2} = \frac{E_2}{E_1} \qquad \qquad ...(6.124)$$

The term (A_1/A_2) is called the area concentration ratio and (E_2/E_1) is called the flux concentration ratio. As the quantity $A_1\sin^2\theta$ will be equal to $A_1\sin^2\theta_e$, one can write

$$C = \frac{A_1}{A_2} = \frac{\sin^2\theta_e}{\sin^2\theta_i} \qquad \qquad ...(6.125)$$

The maximum concentration ratio is obtained when $\theta_e = 90°$ (hemispherical angular distribution) and when θ_i is equated to the solar aperture angle of $0.267°$. Substituting the values, one can write

$$C_{max} = 46\ 200 \qquad \qquad ...(6.126)$$

For line focusing concentration (parabolic trough), it can be shown that

$$C_{max} = \frac{1}{\sin\theta} = 215 \qquad \qquad ...(6.127)$$

In deriving the above, it has been assumed that entrance and exit of radiation are in the same media, with equal indices of refraction ($n_1 = n_2$). If these happen to be different, one can show

$$C_{max} = \frac{n_2^{\ 2}}{\sin^2\theta_i} \qquad \qquad \text{for point focus}$$

$$C_{max} = \frac{n_2}{\sin\theta_i} \qquad \qquad \text{for line focus}$$

Note that $n_1 = 1$ corresponds to vacuum. By varying values of n_2, θ_i, and θ_e, one can get a range of values of concentration ratio for different cases, including primary concentration (only one reflection) and secondary (secondary-level reflection) stage concentration (Winter, Sizmann, and Vant–Hull 1991).

Real concentrators, of course, attain lesser concentration and flux densities. The spatial flux distribution on the plane of the receiver, especially, is not flat but has a gaussian shape. The two-dimensional function for the so-called 'degraded' sun can be expressed as

$$D(x,y) = \frac{I}{2\pi\sigma_x\sigma_y} \exp\left[-\frac{I}{2}\left(\frac{x^2}{\sigma_x^2} + \frac{y^2}{\sigma_y^2}\right)\right] \sum_{i,j=I}^{\infty} C_{ij}\, H_i\left(\frac{x}{\sigma_x}\right)\, H_j\left(\frac{x}{\sigma_y}\right)$$

The reader is referred to Winter, Sizmann, and Vant–Hull (1991) for a complete understanding. The different non-idealities of a concentrating system can be classified as follows.

- Specularity error
- Surface or contour error
- Mechanical error
- Tracking error
- Receiver error

Each of these errors will contribute individually to the gaussian shape. The sum of the errors can be represented by an effective σ_{beam} given by

$$\sigma_{beam}^2 = \sigma_{specular}^2 + \sigma_{contour}^2 + \sigma_{mechanical}^2 + \sigma_{tracking}^2 + \sigma_{receiver}^2$$

Some calculated distributions of flux densities along with an experimental distribution are shown in Figure 6.40 (Zakhidov, Umarov, and Weiner 1992).

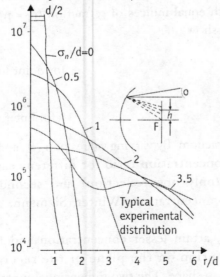

Figure 6.40 Distribution of radiance at the focal plane according to model (91)
Reproduced with permission from the Gujarat Energy Development Agency
Source Zakhidov, Umarov, and Weiner (1992)

The technology of solar concentration has been evolving for many years. In recent years, the so-called Schaeffler collector has become popular for institutional cooking in India. Also, a concentrating system has been tried for process heat in industry by IIT, Bombay (Figure 6.41). A 50-kW solar thermal power plant based on parabolic trough was established in the late 1980s at SEC (Solar Energy Centre), Gual Pahari, Haryana, and efforts are being made at present to convert it into a biomass hybrid. Hybridization with biomass gasifiers is believed to reduce costs of power generation. A Solar Dish–Stirling system for small-scale power generation is also being tried out at Vellore Institute of Technology, Vellore, Tamil Nadu.

With the prices of petroleum fuels rising consistently, solar concentration systems for high temperature process heat seems to hold considerable promise. However, huge research and development efforts would be needed to realize industrial-level prototypes in India.

Figure 6.41 Scheffler concentrator

Nomenclature

a	Albedo of the surface
A	Aperture area of the cooker (m^2)
A_c	Area of the collector (m^2)
A_r	Receiver area (m^2)
A_{ct}	Total collector area (m^2)
A_{tb}	Projected tube area (m^2)
A_{abs}	Projected area of the absorber (m^2)
A_1/A_2	Area ratio
b	Width (m)
b_o	Incident angle modifier coefficient
B	Radiance (W)
C	Concentration ratio
C_b	Bond conductance (W/m)
C_r	Ratio of $(MC)_w/(MC)'_w$
d	Diameter (m)
D_e	Equivalent diameter (m)
D_i	Inside tube diameter (m)
E_2/E_1	Flux concentration ratio
f'	Factor
F	Fin efficiency
F'	Collector efficiency factor
g	Gravitational acceleration (m/s^2)
G_o	Global radiation on a horizontal surface (W/m^2)
G_s	Radiation immediately below the surface (W/m^2)
$G_{s,c}$	Solar constant
G_{eff}	Effective solar radiation (W/m^2)
G_{sun}	Radiosity
h_1, h_2	Convective heat transfer coefficient (W/m^2K)
h_{bf}	Convective heat transfer coefficient from the bottom plate to the air (W/m^2K)
h_{fi}	Fluid heat transfer coefficient (W/m^2K)
h_{pf}	Convective heat transfer coefficient between the plate and the fluid (W/m^2K)
$h_{r,pb}$	Radiative heat transfer coefficient between the collector plate and the bottom plate (W/m^2K) in air heater
I	Irradiance, W/m^2
k	Equivalent number of velocity heads lost by the flow in passing through bends, thermal conductivity (W/mK)

k_T	Thermal diffusivity (m²/s)
K	Extinction coefficient of the medium
K_{ta}	Incident angle modifier
l	Thickness of water layer (m), length, m
L	Length of the collector plate (m)
L	Cover plate thickness (m)
$\dot{m}$	Fluid flow rate for a single tube (kg/s)
MC	Mass-specific heat product of water in the pond (J/K)
$\dot{n}_1, n_2$	Refractive indices of the media
Nu	Nusselt number
p	Mean free path molecule (m)
q_t	Conductive heat transfer if convection is suppressed (W)
q_u	Useful heat gain (W)
$\dot{q}_{p-c}$	Heat loss, W
$\dot{q}_{load}$	Useful energy supplied to the load from the storage (W)
q_{vac}	Conductive heat transfer under vacuum (W)
Q_b	Bottom loss to the ground (W)
Q_L	Heat lost (W)
Q_s	Sum of heat losses (W)
Q_u	Useful heat (W)
r	Reflectance of unpolarized light
r_{pa}	Parallel component of the unpolarized light
r_{pp}	Perpendicular component of the unpolarized light
Re	Reynolds number
R_T	Thermal Rayleigh number
S	Salinity (kg/m³)
S	Solar radiation absorbed by the fin (W/m²)
t	Time (s)
T	Temperature
u	Velocity (m/s)
U_b	Heat loss coefficient from the bottom of the collector (W/m²K)
U_L	Overall heat loss coefficient (W/m²K)
U_t	Top loss coefficient (W/m²K)
$(UA)_{tank}$	Product of the overall heat transfer coefficient and the surface area of the tank (W/K)
x	x-coordinate
y	y-coordinate
Φ	Radiant flux (W)
θ	Angle

τ	Transmittance
τ_{pa}	Transmittance for the parallel component
τ_{pp}	Transmittance for the perpendicular component
ρ	Reflectance
ρ_d	Diffuse reflectance
α	Absorptance, thermal expansion coefficient
γ	Wavelength (m)
β	Collector tilt
δ	Fin thickness (m)
δ_i	Thickness of the insulating material (m)
δ_c	Thickness of a copper plate (m)
η	Efficiency
η_o	Optical efficiency
ν	Kinematic viscosity (m²/s)
$\delta(\lambda)$	Characteristic wavelength-dependent attenuation length (m)
Ω	Solid angle

References

Brandemuehl M J and Beckman W A. 1980
Transmission of diffuse radiation through CPC and flat plate collector glazing
Solar Energy 24(511)

Close D J. 1962
The performance of solar water heaters with natural circulation
Solar Energy 6(33)

Dickinson W C, Clark A F, and Iantuore A. 1976
in *Proceedings of the International Solar Energy Society*
Solar Energy Conference 5(117)

Duffie J and Beckman W A. 1991
Solar Engineering of Thermal Processes
New York: John Wiley and Sons, Inc.

Gandhi M R and Kishore V V N. 1983
Experimental performance of shallow solar pond of novel design
In *Proceedings of the Sixth Miami International Conference on Alternative Energy Sources* [Sixth Miami International Conference on Alternative Energy Sources, Florida]

Garg H P, Mullick S C, and Bhargava A K. 1985
Solar Thermal Energy Storage
Reidel Publishing Company

Goswami Y D, Kreith F, and Kreider J N. 2000
Principles of Solar Thermal Engineering, Second edition
Philadelphia: Taylor and Francis

Grossman G, Shitzer A, and Zvirin Y. 1977
Heat transfer analysis of a flat-plate solar energy collector
Solar Energy 19: 493–502

Gupta C L and Garg H P. 1967
Performance studies of solar air heaters
Solar Energy II(25)

Gupta C L and Garg H P. 1968
System design in solar water heaters with natural circulation
Solar Energy 12(163)

Hodge C N, Thompson T L, Groh J E, Frieling D H. 1966
Progress Report 194
USA: Office of Saline Water Research and Development

Hull J R, Nielsen C E, and Golding P. 1989
Salinity-gradient Solar Ponds
Boca Raton, Florida: CRC Press

IS 12933. 2003
Solar flat plate collector-specifications
New Delhi: Bureau of Indian Standards

Joshi V and Kishore V V N. 1985a
Computer simulation of the performance of the solar pond
Reg. J Energy Heat Mass Transfer 7(2): 97–106

Joshi V and Kishore V V N. 1985b
A numerical study of the effects of solar attenuation modelling on the performance of solar ponds
Solar Energy 35(4): 377–380

Joshi V and Kishore V V N. 1986
Applicability of steady state equations for solar pond thermal performance predictions
Energy II(8): 821–827

Joshi V, Kishore V V N, and Rao K S. 1984
A digital simulation of non-convecting solar pond for Indian conditions
Renewable Energy Sources: International Progress edited by T. Nejat Veziroglu
Elsevier, pp. 207–220

Kaushika N D and Sumathy K. 2003
Solar transparent insulation materials: a review
Renewable and Sustainable Energy Reviews 7: 317–351

Kishore V V N and Joshi V. 1984
A practical collector efficiency equation for non-convecting solar ponds
Solar Energy **33**(5): 391–395

Kishore V V N and Katam S. 1987
Flat Plate Collector Testing—an overview
SESI Journal **1**: 11–17

Kishore V V N and Kumar A. 1996
Solar pond: an exercise in development of indigenous technology at Kutch
Energy for Sustainable Development **III** (1)

Kishore V V N, Gandhi M R, and Rao K S. 1984
Analysis of flat-plate collectors charged with phase-changing fluids
Applied Energy **17**: 133–149

Kishore V V N, Gandhi M R, and Rao K S. 1986
**Experimental and analytical studies of shallow solar pond system with
continuous heat extraction**
Solar Energy **36**(3): 245–256

Kishore V V N, Ranga Rao V V, and Raman P. 1987
A portable shallow solar pond water heater
Solar and Wind Technology **4**(2): 201–204

Kishore V V N, Gandhi M R, Marquis Ch, Rao K S. 1984
Testing flat plate collectors charged with phase-changing fluids
Applied Energy **17**: 155–168

Kishore V V N, *et al.* 1986
**Development of solar (thermal) water pump prototype – an Indo Swiss
experience**
Solar Energy **36**(3): 257–265

Kooi C F. 1979
The steady state salt gradient solar pond
Solar Energy **23**: 37–45

Kudish A I and Wolf D. 1979
A compact shallow solar pond hot water heater
Solar Energy **21**(317)

Maru L V, Kishore V V N, and Gomkale S D. 1986
A roof collector for industrial hot water production
Energy **II**(7): 651–657

Mullick S C, Kandpal T C, and Saxena A K. 1987
Thermal test procedure for box-type solar cooker
Solar Energy **39**: 353–360

Ong K S. 1974
A finite-difference method to evaluate the thermal performance of a solar water heater
Solar Energy 16: 137–147

Pytlinski J T. 1978
Solar energy installations for pumping irrigation water
Solar Energy 21: 255–262

Raman P and Kishore V V N. 1992
Performance of a shallow solar pond based domestic hot water system (100 LPD), in North Indian Climate
[Paper presented in the National Seminar on Urban–Rural Alternative Energy Management, Pondicherry University, 7–8 February 1992]

Rao D P and Rao K S. 1976
A solar water pump for lift irrigation
Solar Energy 18: 405–411

Rao K S, Kishore V V N, and Vaja D (eds). 1990
Solar pond: scope and utilisation
Vadodara: Gujarat Energy Development Agency

Sukhatme S P. 1984
Thermal Energy Storage. Solar Energy – principles of thermal collection and storage
New Delhi: Tata McGraw-Hill Publishing Company Ltd

Whillier A. 1963
Black painted solar air heaters of conventional design
Solar Energy 8(31)

Whillier A and Saluja G. 1965
Effects of materials and of construction details on the thermal performance of solar water heaters
Solar Energy 9(21)

Winter C J, Sizmann R L, and Vant-Hull L L (eds.). 1991
Solar Power Plants: fundamentals, technology, systems, economics
New York: Springer-Verlag

Zakhidov R A, Umarov G Y A, and Weiner A A. 1992
Theory and calculation of applied solar energy concentrating systems
Vadodara: Gujarat Energy Development Agency

Elements of passive solar architecture

Vidisha Salunke, Research Associate, and V V N Kishore, Senior Fellow
Energy–Environment Technology Division, TERI, New Delhi

Introduction

Ancient man shielded himself from the extremes of weather by living in caves and making wood fires. The cave is probably the earliest example of a passive structure for dwelling, and modern passive concepts like earth berming, earth-air tunnel, etc., evolved out of it. Wood fires continue to provide comfort during the cold seasons in remote locations and in developing countries such as Bhutan. A vast variety of passive architectural designs suitable for the given geo-climatic conditions have been developed throughout the world (see, for example, Bahadori 1978). With the advent of industrialization and availability of cheap energy, modern buildings have become energy guzzlers. Current global energy consumption figures suggest that buildings use the largest percentage (45%) for providing heating, cooling, and lighting. A further five per cent is used in building construction (Krishnan, Baker, Yannas, *et al.* 2001). Mounting concerns such as energy security, climate change, etc., have become a driving force in finding 'green' or passive solutions in recent times. The romantic idea of a 'zero energy' house gained popularity in the 1970s and 1980s, but such an idea is either too expensive or limited to a narrow bandwidth of climatic parameters. The present thinking is that buildings can and should use as little fossil fuel-based energy as dictated by the geo-climatic conditions. In other words, one can make a building envelope 'greener than the greenest' in the given climatic conditions.

There are several excellent books on passive and low energy architecture (Athienitis and Santamouris 2000; Cook 1989; Givoni 1994, Krishnan, Baker, Yannas, *et al.* 2001; Santamouris 2006; Szokolay 1985), which a serious student of passive architecture must consult. This chapter provides a review of basic concepts and design principles used in energy-efficient buildings.

Climate and human thermal comfort

Weather is defined as the state of atmosphere with respect to variables such as temperature, moisture, wind velocity, sunshine, precipitation, barometric pressure, etc. It varies widely with time for a given location/region. Climate can be defined as the average weather over a period of several years,[1] and is characterized by the same parameters of temperature, moisture, etc. Based on monthly average data from weather stations, countries can be divided into different climatic zones. India for example is divided into six zones, as follows (Bansal and Minke 1998).

1 Hot and dry
2 Warm and humid
3 Moderate
4 Cold and cloudy
5 Cold and sunny
6 Composite

The criteria for classification are given in Table 7.1 and Figure 7.1.[2]

Table 7.1 Criteria for climatic zones in India

Climatic zone	Mean monthly temperature (°C)	Mean relative humidity (%)	Precipitation (mm)	Number of clear days
Hot and dry	>30	<55	<5	>20
Warm and humid	>30	>55	>5	<20
Moderate	25–30	<75	<5	<20
Cold and cloudy	<25	>55	>5	<20
Cold and sunny	<25	<55	<5	>20
Composite	Six months or more do not fall into any of the above categories			

[1] Worldwide, climatic zones are also classified as equatorial, tropical, sub-tropical, desert, grassland, and temperate.
[2] Different countries have different classifications. For example, Australia has the following zones: hot-humid, warm-humid, hot-dry summer with warm winter, hot-dry summer with cold winter, temperate, and cool temperate.

Figure 7.1 Climate map of earth
Source <http://www.metoffice.gov.uk/education/data/climate/index.html>

It is logical to state that habitats in different climatic zones require different methods of achieving comfort. Buildings in cold climates would need heating, those in hot and dry climates would need simple cooling, while those in hot and humid climates would need cooling as well as humidity control. Before assessing the requirements for heating, cooling, and humidity control, it is necessary to understand the effect of different parameters on human thermal comfort.

A large number of experimental studies have been conducted in the past to understand the relation between comfort and external parameters, but Fanger (1970) was probably the first to establish such a relationship from theoretical considerations. He found that for a given activity level, the skin temperature T_S and the heat lost by evaporation of sweat from the surface of the skin Q_{SW} are the basic physiological variables that determine the feeling of comfort in a human body. The thermoregulatory system of human beings is quite effective and will create a heat balance for the body within a wide range of environmental variables, even if comfort does not exist. For instance, a person subjected to extreme cold will start shivering and one exposed to hot and humid weather will start sweating profusely. Both shivering and sweating are the body's mechanisms directed at achieving a state of comfort. Both parameters, T_S and Q_{SW}, are functions of the activity level, which, in turn, determines the heat production within the body. Energy released by the oxidation process in the human body, called the metabolic rate, is converted into external power (for example, if someone is carrying a load or digging a pit) and internal body heat. If η is the efficiency of conversion of metabolic heat to the work done, one can represent it as

$$Q_M = Q_H + W \qquad\qquad ...(7.1)$$
$$\eta = W/Q_M \qquad\qquad ...(7.2)$$
$$Q_H = Q_M (1 - \eta) \qquad\qquad ...(7.3)$$

where Q_M is the metabolic rate, Q_H is the internal heat production, and W is the work done, all expressed in kcal/h or in kW. Dividing both sides by the surface area of the body A_S (also called the Du Bois area), we get the rates per unit surface area.

$$Q_H/A_S = (Q_M/A_S)(1 - \eta) \qquad\qquad ...(7.4)$$

The values of Q_M/A_s and η for some activities are given in Table 7.2. Refer to Fanger (1970) and ASHRAE (2005) for an extensive list.

There is no significant heat storage within the body, and in order to maintain an essentially constant internal body temperature, internal heat production should be equal to the heat dissipated through processes such as respiration, sweat evaporation, etc. There are four routes for dissipation of heat.

1   Heat loss by water vapour diffusion through skin (Q_d)
2   Heat loss due to evaporation of sweat (Q_{sw})
3   Heat loss by evaporation of respiration (Q_{re})
4   Dry respiration heat loss (Q'_{re})

Three of the above can be expressed by the following equations.

Table 7.2 Metabolic rates for different activities

Activity	Q_M/A_S (kcalh^{-1}m^{-2})	η
Sleeping	35	0
Seated	50	0
Walking, at the rate of 3 km/h	100	0
Office/lab work	50–80	0
Driving	50–100	0
Playing tennis	230	0.0–0.1
Pick and shovel work	200–240	0.1–0.2
Digging pits	300	0.2

$$Q_d = 0.35\,A_s\,(1.92\,T_s - 25.3 - p_a) \hspace{3cm} ...(7.5)$$
$$Q'_{re} = 0.0023\,Q_M\,(44 - p_a) \hspace{3cm} ...(7.6)$$
$$Q_{re} = 0.0014\,Q_M\,(34 - T_a) \hspace{3cm} ...(7.7)$$

where p_a is the partial pressure of water vapour in ambient air (mm Hg) and T_a is the ambient temperature (°C). The net heat lost to the surroundings, which for the body is the clothing envelope, is given by

$$Q_H - Q_{sw} - Q_d - Q_{re} - Q'_{re}$$

The dry heat transfer from the skin to the outer surface of the clothing can be expressed as

$$Q_{cl} = A_s\,(T_s - T_{cl})\,/\,R_{cl} \hspace{3cm} ...(7.8)$$

where T_{cl} is the clothing temperature and R_{cl} is the total heat transfer resistance from the skin to the outer clothing surface. Transfer of heat between the skin and the outer surface is quite complicated, involving convection and radiation processes in the intervening air spaces and conduction through clothing material. To simplify calculations, a dimensionless clothing parameter I_{cl}, also called the clo-unit, has been introduced as

$$R_{cl} = 0.18\,I_{cl} \hspace{3cm} ...(7.9)$$

Equation 7.8 can thus be written as

$$Q_{cl} = A_s\,(T_s - T_{cl})\,/\,0.18\,I_{cl} \hspace{3cm} ...(7.10)$$

Also, a factor f_{cl} has been defined as the ratio of the surface area of the clothed body to that of the nude body. Clo-values and f_{cl} values for some clothing ensembles are given in Table 7.3.

Table 7.3 Clo-values for some clothing ensembles

Clothing	I_{cl} clo	f_{cl}
Nude	0	1
Light summer clothing	0.5	1.1
Typical business suit	1	1.15
Heavy traditional European business suit	1.5	1.15–1.2
Polar weather suit	3.0–4.0	1.3–1.5

Heat transfer from the clothing surface to the surroundings occurs through two parallel routes – radiation and convection – expressed in Equations 7.11 and 7.12, respectively.

$$Q_R = 3.4 \times 10^{-8} A_S f_{cl} [(T_{cl} + 273)^4 - (T_{MR} + 273)^4] \qquad \ldots(7.11)$$

$$Q_C = A_S f_{cl} h_c (T_{cl} - T_a) \qquad \ldots(7.12)$$

where T_{MR} is the mean radiant temperature and h_c is the convective heat transfer coefficient.

The heat balance between the clothing and the surroundings can thus be written as

$$Q_{cl} = Q_R + Q_C \qquad \ldots(7.13)$$

Heat transfer from the body to the clothing envelope and then to the surroundings can be expressed as a double heat balance equation given below.

$$Q_H - Q_{SW} - Q_d - Q_{re} - Q'_{re} = Q_{cl} = Q_R + Q_C \qquad \ldots(7.14)$$

Substituting all the terms for the heat transfer and dividing by A_S, we get

$$(Q_M/A_S)(1-\eta) - (Q_{SW}/A_S) - 0.35[1.92 T_S - 25.3 - p_a] -$$
$$0.0023(Q_M/A_S)(44 - p_a) - 0.0014(Q_M/A_S)(34 - T_a)$$

$$= (T_s - T_{cl})/0.18 I_{cl}$$

$$= 3.4 \times 10^{-8} f_{cl} [(T_{cl} + 273)^4 - (T_{MR} + 273)^4] + f_{cl} h_c (T_{cl} - T_a) \qquad \ldots(7.15)$$

Experiments have shown that both the skin temperature T_S and the sweat secretion Q_{SW} are functions of (Q_H/A_S).

Regression analysis of the available data gives

$$\overline{T}_s = 35.7 - 0.032(Q_H/A_S) \qquad \ldots(7.16)$$

$$\overline{Q}_{SW} = 0.42 A_S \{(Q_H/A_S) - 50\} \qquad \ldots(7.17)$$

It can be seen from the above equations that for constant comfort, $\overline{T}_S$ decreases while sweat secretion increases with an increase in the activity level. The above are for average values, but in reality, both T_S and Q_{SW} must lie within narrow limits of comfort. Substituting Equations 7.16 and 7.17 in equation 7.15 results in the formulation of the comfort equation

$$(Q_M/A_S)(1-\eta) - 0.42\,[(Q_M/A_S)(1-\eta) - 50] -$$
$$0.35\,[43 - 0.061\,(Q_M/A_S)(1-\eta) - p_a] - 0.0023\,(Q_M/A_S)(44 - p_a) -$$
$$0.0014\,(Q_M/A_S)(34 - T_a)$$

$$= 3.4 \times 10^{-8}\,f_{cl}\,[(T_{cl} + 273)^4 - (T_{MR} + 273)^4] + f_{cl}\,h_c\,(T_{cl} - T_a) \qquad \ldots(7.18)$$

with T_{cl} given by the following equation

$$T_{cl} = 35.7 - 0.032\,(Q_M/A_S)(1-\eta) - 0.18\,I_{cl}\,[(Q_M/A_S)(1-\eta) -$$
$$0.42\,\{(Q_M/A_S)(1-\eta) - 50\} - 0.35\,\{43 - 0.061\,(Q_M/A_S)(1-\eta) - p_a\} -$$
$$0.0023\,(Q_M/A_S)(44 - p_a) - 0.0014\,(Q_M/A_S)(34 - T_a)] \qquad \ldots(7.19)$$

h_c is a function of air speed and is given as

$$h_c = 2.05\,(T_{cl} - T_a)^{0.25} \quad \text{for } V < 0.1 \text{ m/s}$$
$$h_c = 10.4\,\sqrt{V} \qquad\qquad \text{for } V > 0.1 \text{ m/s} \qquad \ldots(7.20)$$

The above set of equations imply that for a given activity level (Q_M/A_S) and for a given clothing envelope (I_{cl}, f_{cl}), it is possible to calculate several combinations of T_a, p_a, T_{MR}, and V, which will create optimum thermal comfort for persons under steady-state conditions. Several 'constant comfort' lines can thus be drawn on a psychrometric chart to indicate conditions of optimum comfort. However, as T_S and Q_{SW} lie within the lower and upper bound for practical comfort, one actually has 'comfort zones' with boundaries as comfort lines (ASHRAE 2005; Fanger 1970).

The sensation of comfort or discomfort primarily depends on the variables discussed above. Thermal preferences are, however, influenced by a number of subjective factors given below.

- *Acclimatization* Exposed to a new set of climatic conditions, the human body will adapt in about 30 days and by that time, thermal preferences of that individual will change. Long-term, endocrine adjustments constitute the acclimatization process. These may involve a change in the basal metabolic heat production, an increase in the quantity of blood (to produce and maintain a constant vasodilation), and an increase in sweat rate.
- *Age and gender* Thermal preferences may be influenced by age and sex. Older people usually prefer higher temperatures as their metabolism is slower. Women too have slightly slower metabolism, and hence, prefer, on an average, 1 °C higher temperature than men.

- *Body shape* The surface-to-volume ratio also has an influence. A thin person has a much greater body surface than a short, stout person of the same weight. He would prefer and tolerate a higher temperature so as to dissipate more heat than a plump person.
- *Subcutaneous fat* Presence of fat under the skin is an excellent thermal insulator. A fat person needs cooler air to dissipate the same amount of heat.
- *State of health* During an illness, the metabolic rate may rise but as proper functioning of the regulatory mechanism is generally impaired, the range of tolerable temperatures could be narrower.

Of the four parameters – ambient temperature, humidity, mean radiant temperature, and wind speed – the first two seem to be more important for indoor comfort. Hence, the comfort zone can be represented quite well on the psychrometric chart. The comfort zone range as per European norms was based on experiments carried out on Caucasians, and it is generally known that the ranges for T_s and Q_{sw} are different for subjects of different races. With mild air movements, the comfort zone can be expanded as shown in Figure 7.2.

It can be seen from Equations 7.18 and 7.19 that the partial pressure pa of water is a more fundamental parameter than relative humidity, which is commonly reported. The pa value of water can be calculated for a given set of values of Ta and RH from the following equations.

$$p_a = (p_s)_{T_a} RH \qquad \qquad ...(7.21)$$

where RH is the relative humidity and $(p_s)_{T_a}$ is the saturation vapour pressure of water at ambient temperature T_a.

p_s can also be obtained from the Antoine equation, which can be written (in a slightly modified form) as

$$In\ (p_s)_{T_a} / (p_s)_{ref} = [3885/(T_{ref} + 230)] - [3885/(T_a + 230)] \qquad ...(7.22)$$

where T_{ref} is the reference temperature (100 °C) and $(p_s)_{ref}$ is the reference vapour pressure (757.7 mm Hg).

When p_a is plotted against T_a, daily variations generally fall into a narrow band for a given location. Typical summer ranges for different Indian cities are shown in Figure 7.2 along with the comfort zone (Kishore 1988).

Representing hourly average weather parameters in the manner shown in Figure 7.2 can provide insights into the relative ease with which comfort can be provided. For moderate climates, many weather points could come quite close to the comfort zone, and hence, simple measures like providing fans, night ventilation, etc., may suffice. For cold climates, weather points would lie

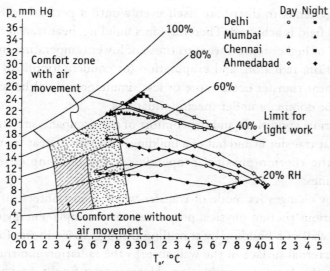

Figure 7.2 A graphic representation of outdoor conditions for the hottest month of the year for different Indian cities. Values for international cities would be interesting.
Source Kishore (1988)

towards the left of the comfort zone and hence, heating would be the primary requirement. Weather points will be to the right of the comfort zone in hot, dry climates, and simple cooling (represented by a horizontal line on the psychrometric chart), such as evaporative cooling or nocturnal cooling, can provide adequate comfort. Hot, humid climates are represented by weather points crowding the top right side of the comfort zone (such as Chennai in Figure 7.2), which indicates that humidity control would be essential to feel comfortable. Composite climates would have a wide fluctuation, with the weather points lying to the left and right sides of the comfort zone, and it is not an easy task to find passive design solutions for such climates.

Indoor values of temperature, humidity, etc., would be different from the outdoor parameters, as the building envelope experiences both mass and energy exchanges with the surroundings. Various streams of thermal exchange would constitute the heating and cooling 'loads' if the building is to be maintained within a narrow range of temperature and humidity conditions.

Heat exchange in buildings

It is difficult to consider the thermal behaviour of buildings and to design solar passive heating and cooling systems without understanding the nature of heat and how it flows from one place to another.

Heat energy tends to distribute itself evenly until a perfectly diffused, uniform thermal field is achieved. Therefore, in a building, heat transfer takes place from areas of higher temperature to those of lower temperature by conduction, convection, radiation, and evaporation (or condensation). All these mechanisms of heat transfer occur more or less simultaneously, although one or another may be dominant under specific situations.

Indoor thermal conditions and the comfort of the occupants are affected by the rate of heat transfer in and out of a building. This rate of heat transfer is dependent on the thermo-physical properties of the building envelope, besides other things.

Heat energy changes its mode of transfer when it encounters different substances of various thermo-physical properties. For example, energy of the sun reaches the wall of a building by travelling through space in the form of radiation. The external surface of the wall absorbs the radiation and transfers heat inwards in the form of conduction. Heat energy finally reaches indoor air by convection. There is, however, a delay in the transfer of heat into/out of the building, depending on the material properties and thickness. If the delay is more, the building is better isolated from the surroundings. Hence, the choice of building materials assumes significance. The thermo-physical properties of some common building materials are given in Table 7.4.

Thermal properties of materials

Heat flow by conduction forms a large fraction of the total heat gains/losses in a building envelope. Conduction heat transfer is discussed briefly in Chapter 3. The conductivities of some building materials are listed in Table 7.5.

Table 7.4 Thermo-physical properties of common building materials

Material	Density (kg/m^3)	Thermal conductivity (W/mK)	Specific heat capacity (Wh/kgK)
Aluminium	2700	165	0.25
Construction steel	7850	60	0.13
Normal concrete	2400	2.1	0.24
Foam concrete	400–800	0.14–0.27	0.24
Solid brick	1200–2000	0.50–0.96	0.24
Lime/sand stone	1000–2200	0.5–1.3	0.24
Glass	2480	0.8	0.19–0.26
Earth	1450–2040	0.5–2.6	0.24
Wood	600–800	0.13–0.20	0.66
Polystyrene hard foam	10–30	0.03–0.04	0.42
Mineral wool	8–500	0.03–0.05	0.24
Water	1000	0.60	1.2

Table 7.5 Conductivity and resistivity of some materials

Material	Conductivity k (W/m °C)	Resistivity 1/k (m °C/W)
Asbestos		
Loose	0.034	29.40
Sprayed	0.046	21.75
Asbestos cement sheet		
Light	0.216	4.63
Average	0.360	2.78
Dense	0.576	1.74
Asphalt	0.576	1.74
Brickwork commons		
Light	0.806	1.24
Average	1.210	0.83
Dense	1.470	0.68
In light-weight bricks	0.374	2.68
In engineering bricks	1.150	0.87
Concrete		
Ordinary, dense	1.440	0.69
Clinker aggregate	0.403	2.48
Expanded clay aggregate	0.345	2.90
Cork slab		
Natural	0.245	4.08
Regranulated baked	0.043	23.20
Eel grass blanket	0.039	25.60
Glass wool		
Quilt	0.043	23.20
Blanket	0.034	29.40
Mineral wool		
Felt	0.042	23.80
Rigid slab	0.037	27.00
Onozote (expanded ebonite)	0.049	20.40
Plasterboard, gypsum	0.029	34.50
Plastering		
Gypsum	0.159	6.33
Vermiculite	0.461	2.17
Plywood	0.201	4.98
Polystyrene foam slab	0.138	7.25
Rendering, sand-cement	0.033	30.30
Stone		
Granite	0.532	1.88
Limestone	2.920	0.34
Sandstone	1.530	0.65
Slate	2.09	0.478
Strawboard	1.295	0.77
Timber		
Softwood	0.093	10.75
Hardwood	0.138	7.25
Wood chipboard	0.160	6.25
Wood fibre softboard	0.108	9.26
Wood wool slab		
Light	0.065	15.38
Dense	0.082	12.20

Resistivity is the reciprocal of conductivity ($1/k$) and is measured in m °C/W. Higher the resistivity value, better is an insulator. While conductivity is the property of a material, the corresponding property of a body of a given thickness is described as conductance C.

If a body consists of several layers of different materials, its total resistance will be the sum of the resistances of individual layers. Conductance of such a multi-layer body can be found by determining its total resistance and taking its reciprocal. While resistances are additive, conductance is not.

There are a few other properties of materials that affect the rate of heat transfer into and out of a building.

- Surface characteristics with respect to radiation—absorptivity, reflectivity, and emissivity
- Surface coefficient
- Heat capacity
- Transparency to radiation of different wavelengths

The actual heat flow across a building element depends not only on the thermal conductivity but also on the thickness d of the element. The greater the thickness, lower will be the rate of heat flow.

Referring back to Chapter 3, the flow of heat under steady-state conditions across a wall element of surface area A and thickness d, built of a material having a given value of k and subjected to a temperature gradient of $T_2 - T_1$, is given by the formula

$$Q_c = \frac{k}{d} A(T_2 - T_1) \qquad \qquad ...(7.23)$$

where Q_c is the rate of heat flow from the warmer to the colder surface in kcal/h.

While calculating the rate of heat flow between indoor and outdoor air, thermal resistance of air layers adjacent to the surfaces must be taken into account. A film of still air forms on any surface, its thickness decreasing as the velocity of the adjacent air increases. As the thermal conductivity of air is very low – and consequently its resistivity is high – the air film attached to a surface gives considerable resistance to the heat flow across that surface. The reciprocal of resistance of the air film is known as the surface coefficient, denoted by h_i for internal surface and h_o for external surface.

Thus, the overall resistance of a single layer wall to the heat flow R between the air on either side is given by

$$R = 1/h_i + d/k + 1/h_o \qquad \qquad ...(7.24)$$

The rate of heat flow (q) per unit area from internal to external air, under steady-state conditions, can be calculated from the formula

$$q = \frac{(T_i - T_o)}{R} \qquad \qquad ...(7.25)$$

where T_i and T_o are the indoor and outdoor temperatures respectively. The reciprocal of thermal resistance ($1/R$) is termed as thermal transmittance K.

When T_o is greater than T_i, heat transfer occurs in the opposite direction, from outside to inside. However, the skin temperature of the wall will depend not only on T_o, but also on the surface properties and radiation falling on the surface.

Example 1

Calculate the thermal transmittance of a concrete wall (thickness 20 cm) plastered on both sides (internal plaster 0.1 cm, external plaster 0.2 cm). Thermal conductivity of concrete is 1.2, that of external plaster 0.8, and of internal plaster 0.6 kcal/h m²°C. Internal and external surface coefficients are 7 and 18, respectively.

Solution

R = 1/7 + 0.01/0.6 + 0.2/1.2 + 0.02/0.8 + 1/18 = 0.407
K = 1/R = 1/0.407 = 2.46 kcal/h m²°C

Surface characteristics with respect to radiation: absorptivity, reflectivity, and emissivity

The external surface of any opaque material has three properties determining its behaviour with respect to radiant heat exchange—its absorptivity, reflectivity, and emissivity.

Radiation impinging on an opaque surface may be absorbed or reflected—it is fully absorbed by a perfectly black surface and fully reflected by a perfect reflector. However, most surfaces absorb part of the radiation and reflect the rest.

If the absorptivity is denoted by a and reflectivity by r then

$$r = 1 - a \qquad \qquad ...(7.26)$$

The emissivity e is the relative power of a material to emit radiant energy. For any specific wavelength, absorptivity and emissivity are numerically equal, that is, a = e. Emissivity of a perfectly black surface is 1.0; for other surfaces, values range from 0.05 for some highly polished metals to about 0.95 for ordinary building materials.

Radiation is absorbed selectively, according to the wavelengths incident on the surface. Thus, a fresh whitewash has an absorptivity of about 0.12 for short-wave solar radiation but absorptivity for long-wave radiations from other surfaces at ordinary temperatures is about 0.95. Consequently, this surface also has an emissivity of 0.95 for long wavelengths, and it is a good radiator as it readily loses heat to colder surfaces; but at the same time is a good reflector of solar radiation. On the other hand, a polished surface has a very low absorptivity and emissivity for both short- and long-wave radiation. Thus, while being a good reflector of radiation, it is a poor radiator and can hardly lose its own heat by radiative cooling.

The colour of a surface is a good indicator of its absorptivity for solar radiation. Absorptivity decreases and reflectivity increases with lightness in colour. However, colour does not indicate the behaviour of a surface with respect to longer-wave radiations. Thus, black and white paints have very different absorptivities for solar radiation. A black surface becomes much more heated on exposure to the sun. But long-wave emissivities of the two colours are equal, which are therefore cooled equally at night by radiation to the sky. The absorptivities and emissivities of some selected materials are given in Table 7.6.

The effect of incident solar radiation on a building surface is to increase the heat gain by a factor dependent on magnitude of the radiation, outside convective heat transfer coefficient, and the absorptivity of the surface. The concept of 'sol-air' temperature had been developed taking this factor into account. The sol-air temperature is defined as that temperature of air which

Table 7.6 Absorptivities and emissivities of various surfaces

Material or colour	Short-wave absorptivity	Long-wave emissivity
Aluminium foil (bright)	0.05	0.05
Aluminium foil (oxidized)	0.15	0.12
Galvanized steel (bright)	0.25	0.25
Aluminium paint	0.50	0.50
Whitewash (new)	0.12	0.90
White oil paint	0.20	0.90
Grey colour (light)	0.40	0.90
Grey colour (dark)	0.70	0.90
Green colour (light)	0.40	0.90
Green colour (dark)	0.70	0.90
Ordinary black colour	0.85	0.90

will result in the same heat flux into building, taking into account the effects of incident radiation. It is defined as

$$T_{sa} = T_o + (I \times a)/h_o \qquad \qquad ...(7.27)$$

where T_{sa} = sol-air temperature

T_o = outside air temperature

I = radiation falling on the plane of the surface (W/m²)

a = absorptivity of the surface, and

h_o = outside heat transfer coefficient (W/m²K)

Heat flows into and out of the building envelope are never at a steady state. Building surfaces, especially walls, have 'thermal mass', which is the product of heat capacity and total mass of a given element. As ambient conditions such as temperature, air velocity, incident radiation, etc., keep changing throughout in a periodic manner, heat flows into and out of the building are also periodic. The situation is represented in Figure 7.3, which depicts the response of inside temperature to variations outside the building. Two things have to be noted: (i) the temperature fluctuations inside the building are generally dampened compared to the magnitude of variations outside, and (ii) there is a time lag between the occurrences of maximum temperature outside and inside the buildings. Both the dampening and time lag are dependent on the material properties and thickness. The hourly heat fluxes are thus quite different from those calculated assuming steady-state conditions, using

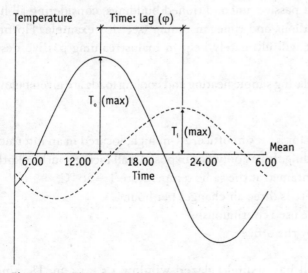

Figure 7.3 Time-lag and decrement factor
Reproduced with permission from Orient Longman Pvt. Ltd
Source Koenigsberger, Ingersoll, Mayhew, *et al.* (1973)

Equation 7.23, for example. Calculation of unsteady state heat fluxes and temperatures involves solving of the general, three-dimensional unsteady-state heat condition equation, and can be quite tedious. Methods have been developed to compute the cooling and heating 'loads' for conditioned buildings, where the indoor comfort conditions are maintained. However, there are no simple methods for similar computations for unconditioned or passive buildings.

Heat gains of a building occur through the following means.

1 Conductive heat flows through walls, roof, windows, doors, ground, etc.
2 Convective flows through air exchange with surroundings
3 Direct gain of radiation through glass or other transparent building materials
4 Heat generated through appliances such as lights, fans, computers, etc.
5 Heat generated through metabolic activity of people

The cooling loads are in general higher than heating loads because heat generation from items 3–5 alone reduce the heating load in cold climates.

Calculation of the various heating and/or cooling loads is beyond the scope of this book, and the reader is referred to standard books (ASHRAE 1997). Understanding of quantitative heat flows, however, is extremely important in designing passive buildings. Many architects base their designs largely on qualitative arguments and, of course, on past experience. Climatic performance simulation of passive, unconditioned buildings, considering all heat fluxes, can be quite tedious and timeconsuming (see, for example, Hoffman and Feldman 1981) but will ultimately help in mainstreaming passive design concepts.

Examples of calculating simple heating and cooling loads are given below.

Example 2
Calculate the heat load for a 5 × 5 m office, 2.5-m high, located in an intermediate floor of a large building. It has only one exposed wall facing south, all other walls adjoin rooms maintained at the same temperature T_i = 20 °C.

The ventilation rate is three air changes per hour.

3 × 40 W lights are used continuously.

Four clerks occupy the office.

The exposed wall has a single glazed window 1.5 × 1.5 m, U-value = 4.48 W/m² °C.

The exposed wall is made of concrete blocks, 200-mm thick, plastered, U-value = 1.35 W/m² °C.

The design outdoor temperature $T_o = -1\ °C$

Solution

Temperature difference $\Delta T = 20\ °C - (-1\ °C) = 21\ °C$

Area of window $= 2.25\ m^2$

Area of wall $= (5 \times 2.5) - 2.25 = 10.25\ m^2$

Applying $Q_c = U \times A \times \Delta T$

$Q_c = [(2.25 \times 4.48) + (10.25 \times 1.35)]21 = 502\ W$

Volume of the room $= 5 \times 5 \times 2.5 = 62.5\ m^3$

Ventilation rate $= 62.5 \times 3\ /60 \times 60 = 0.052\ m^3/s$

Applying $Q_v = 1300 \times V \times \Delta T$

$Q_v = 1300 \times 62.5 \times 21 = 1420\ W$

The internal generation is given by

$Q_i = (3 \times 40) + (4 \times 50) = 320\ W$

Net heat lost $= 502 + 1420 - 320 = 1602\ W$

The heating system should provide heat at this rate to compensate for the net losses.

Example 3

Calculate the cooling load for the office described in Example 1. Additional details are as given below, all other parameters remaining same as in Example 2.

Design outdoor temperature $T_o = 26\ °C$

Incident solar radiation $= 580\ W/m^2$

Absorbance of wall surface $a = 0.4$

Surface conductance $h_o = 10\ W/m^2\ °C$

Solar heat gain factor for window $= 0.75$

Solution

Temperature difference $\Delta T = 26\ °C - 20\ °C = 6\ °C$ for conduction and for ventilation through the window.

For the opaque area, the sol-air temperature is

$T_{sa} = 26 + (580 \times 0.4)/10 = 49.2\ °C$

Temperature difference $(\Delta T) = 49\ °C - 20\ °C = 29\ °C$ for opaque wall area

Area of window $= 2.25\ m^2$

Area of wall $= 10.25\ m^2$

Applying $Q_c = U \times A \times \Delta T$

$Q_c = (2.25 \times 4.48)\ 6 + (10.25 \times 1.35)\ 29 = 462\ W$

Volume of room $= 5 \times 5 \times 2.5 = 62.5\ m^3$

Ventilation rate is $62.5 \times 3\ /\ 60 \times 60 = 0.052\ m^3/s$

Applying $Q_v = 1300 \times V \times \Delta T$

$Q_v = 1300 \times 62.5 \times 6 = 405\ W$

The internal gain is same as in previous example at 320 W

The solar direct gain is given by $Q_s = A \times I \times \theta$

$= 2.25 \times 580 \times 0.75 = 979\ W$

The net heat gain is the sum of all gains

$Q_{cooling} = 462 + 405 + 320 + 979 = 2166\ W$

The cooling system should remove heat at this rate.

Some known concepts of passive heating in cold climates, passive cooling in hot climates, and combination of the two for composite climates are discussed below.

Concepts of passive heating

Traditional methods of keeping a building warm and comfortable in cold weather include the fireplace and stove (*bukhari*). Passive techniques rely on minimizing heat losses from the building envelope and allowing solar radiation to enter.

Minimization of surface to volume ratio

A larger surface area of the building envelope leads to greater heat loss for a given volume of the building. This occurs due to higher conduction losses from the building components. For instance, for a given living area of a dwelling unit, the transmission loss will be higher for a bungalow than for a unit in a multi-storeyed building.

Similarly, for the same volume, various building forms could have different surface areas, therefore, a different surface-to-volume ratio and hence, different heat losses. The surface-to-volume ratios for some geometries are given in Figure 7.4.

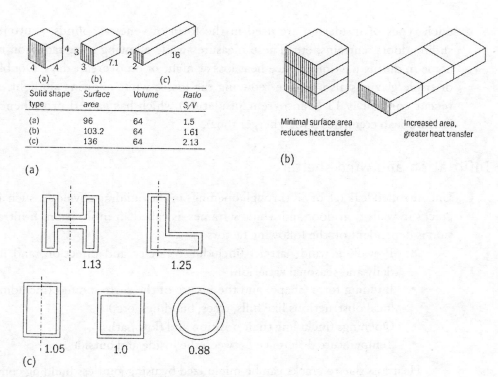

Solid shape type	Surface area	Volume	Ratio S/V
(a)	96	64	1.5
(b)	103.2	64	1.61
(c)	136	64	2.13

Figure 7.4 Surface-to-volume ratios of some geometries

Insulation

The amount of heat flowing through a building component depends upon the thermal conductivity of materials and the thickness of the individual layers of which the component is made of and is represented by its U-value. The smaller the value, the lesser the amount of heat flowing through it.

Thermal insulation plays an important role in increasing the thermal resistance of the building envelope. Insulation can be placed either on the inner side of the building element (wall/roof /floor) or on the outer surface. Some of the commonly used pre-formed insulation materials are:

- Expanded polystyrene slabs
- Extruded polystyrene slab
- Polyurethane/polyisocyanurate slabs
- Perlite boards

Another type of insulation is in situ, that is, sprayed in the form of polyurethane foam. Hollow bricks also help in increasing insulation.

Windows are particularly vulnerable to heat loss. Hence, it is quite beneficial to use movable insulation for periods when solar radiation is absent.

Such types of insulation are used in the form of venetian blinds, shutters, sliding doors, curtains, etc. These measures while allowing solar radiation to come in during daytime reduce heat loss at night or on overcast days. Double or triple glazings also help in reducing the losses while allowing sunlight. A recent concept is TI (transparent insulation) which has gained prominence in European countries (Goetzberger 1992).

Infiltration and wind shelter

Uncontrolled leakage of air through openings in the building envelope, such as cracks in walls near door and window frames, is called air infiltration. Infiltration is dependent on the following factors.

- Prevailing wind pattern (including velocity and direction, and its daily and seasonal variation)
- Building form (shape) and the nature of the surroundings (including local obstructions like hills, trees, buildings, etc.)
- Openings (including their position and flow paths)
- Temperature difference between the inside and outside

Heat loss due to cracks can be minimized by using jointless building components and also employing sealants, profile seals, joint bands, foils, etc. Cold winds in winter are a source of large heat loss from buildings due to convection as a result of increased air infiltration. This infiltration can be reduced by providing a wind shelter with appropriate site and landscape planning.

An analysis of the building site should be made to determine if there exist wind-protected areas. The designer should avoid open areas, hilltops, and valley floors that are directly exposed to the prevailing winter winds. The existing tree stands, hillocks, and hedges may also be available as windshields.

Solar heat gain and storage

Solar heating systems fall into two groups: direct and indirect. A direct system relies wholly on the form and construction of the building itself or on its parts to capture and then utilize the sun's rays to provide the building with warmth. Its effectiveness is, therefore, entirely dependent on a building's orientation and planning. An indirect system, on the other hand, does not rely much on the actual building. To a large extent, it consists of mechanical components, which can either be attached to a building or stand independent of it and even then, such systems need to be optimized by careful planning. Indirect heating systems are generally more applicable where the building is already constructed but direct systems, being integral to the building design, are easier to incorporate in new construction. Other power sources, such as electricity, gas, oil, or coal, do not in themselves affect the form that a plan takes.

The principle of passive solar energy use primarily involves the following so as to achieve the usual degree of thermal comfort expected inside a building.

- Collection of the external radiative energy
- Storage of the collected energy
- Controlled use of the stored energy

It will be clear that for anything other than the simplest of these, a decision to make use of passive solar heating systems has significant implications for the planning of the building concerned. Hence, without a doubt, such a decision has to be part of the initial brief.

Passive solar heating systems are of two kinds: those that allow the heat gained from the sun to be stored for later use and those that do not.

Passive solar heating systems without storage

In solar heating systems, the window is at the lowest level of complexity. Properly managed, a south-facing window (in the Northern Hemisphere) can be a source of annual net heat gain, rather than net heat loss. The detailed design of the window, its components, and their control mechanism are of key importance. This includes the area of glazing, its type, its location/orientation, frame characteristics, and shading.

Window-box collector

This is a clever device to increase the collection area of a window without increasing the window's liability to the heat loss at night. The collector is simply a flat, multi-layered panel, sealed on three of its four edges and fitted to the outside of the building so as to allow the passage of air between itself (via the unsealed fourth edge) and the building's interior and vice versa (Figure 7.5). The collector is usually fitted immediately below a window from which it slopes down and away. Starting from the top, the panel's layers are as follows: glass, upper airspace, black metal plate, lower airspace, insulation, and base. The two airspaces interconnect at the lower edge of the collector, across its full width. Sunlight passes through the glass and strikes the black metal plate, which largely absorbs it. The plate then warms the air that surrounds it, which expands, becomes less dense, and rises through the upper airspace to enter the building. In the process, cooler air is drawn into the collector from

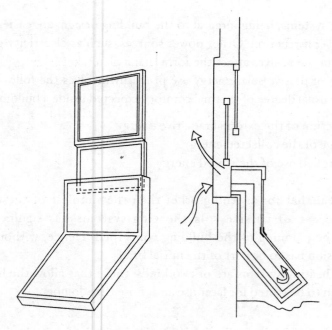

Figure 7.5 The window-box collector

the building by way of the lower airspace, is itself heated, and then discharged back into the building via the upper airspace. This cycle continues as long as there is sufficient solar energy available to make it possible. When the cycle stops, colder, denser air accumulates inside the bottom of the collector and, for as long as the building's interior temperature remains higher than the outside, prevents re-entry of warm air into the collector and the heat loss that would otherwise occur.

Passive solar heating systems with storage

The advantage of systems with a storage capacity is that the heat can be made available when the user requires it and not only when the system is providing it, as happens with the window-box collector. Heat can be stored in any dense material. The principle is simple: heat the dense material, insulate it so that heat cannot escape unintentionally but have a method of releasing this heat on demand. Ideally, the dense material should be cheap, readily available, easy to store, and have a high thermal capacity. In practice, this can be concrete, masonry, or water.

Direct solar gain

Direct solar gain is the simplest of all storage systems and may be suitable for south-facing spaces (Northern Hemisphere) where

- there is considerable glazing on the south side,
- windows are equipped with efficient heat conservation features to prevent heat loss at night,
- the structure enclosing the space consists of materials with high thermal capacity and/or elements within the space may have that quality, and
- the north, east, and west sides are well insulated.

Sunlight is allowed to enter the space through the glazed windows and to fall upon areas of high thermal capacity, for example, a ceramic tiled concrete floor slab or unlined masonry walls (Figure 7.6). Heat is absorbed and stored in them during the day and is then gradually released after sunset as the air temperature begins to fall. The space that benefits is usually the one receiving sunlight, as the stored heat is released mainly by radiation. It is, at times, possible to arrive at configurations that permit some of the stored heat to be radiated to spaces other than those facing south.

It is important for the success of the system to have daytime or summer shading devices in front of the windows to control excessive solar gain and to have insulating devices at night time to prevent thermal loss.

A slightly more sophisticated version of this system employs skylights or roof monitors instead of, or in addition to, windows. The monitors are positioned so as to direct sunlight onto storage walls well inside the building's perimeter, thereby allowing a greater area of the floor plan to be addressed by passive solar heating.

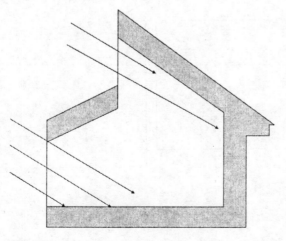

Figure 7.6 Direct gain system for passive heating

Top-lit atrium

The top-lit atrium is a more advanced way of capitalizing on direct solar gain. An area in the centre of the building acts as a collector, absorbing heat into its floor and walls. Depending upon the circumstances, the roof of the atrium can be wholly glazed or else equipped with monitors that point in various directions. Either way, heat can be collected throughout the day. It is still necessary, however, to provide the means of combating heat loss at night and overheating during the day.

The Trombe wall

The Trombe wall (named after its French inventor, Felix Trombe) is actually a large-scale version of the window-box collector described earlier. In this case, the entire face of the building becomes the collector. This consists of a south-facing, massive, dark-coloured wall (such as, concrete painted black or a dark natural stone) placed 100 mm behind an outer wall that is entirely glazed (Figure 7.7). At the top and bottom of the heavy wall, there are ventilation slots through to the interior spaces of the building. Air heated between the stone and the glass rises and enters the building via the upper slot. Meanwhile, cooler air is drawn out of the building through the lower slot to be heated and to continue the cycle. Thus, during the day, the Trombe wall creates a convection loop that supplies warm air into the building.

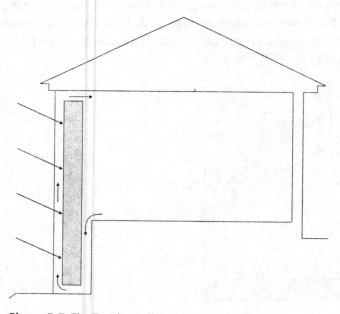

Figure 7.7 The Trombe wall

However, this is only part of the wall's advantage as it also capitalizes on direct solar gain. It absorbs heat into the masonry or concrete during the day and radiates that heat into the interior during the night. The effectiveness of this system depends not only on having a total mass sufficient to store the amount of heat required, but also on the thermal lag (time taken for heat to pass from outside to inside of wall) that ensures peak radiation internally when it is actually required. One of the architectural disadvantages of the Trombe wall may be reduced opportunities for an external view on the south.

The greenhouse

Greenhouses can be used as collectors of solar energy for the purpose of heating buildings. Air in the greenhouse is heated by the sun and is then vented by way of openings into the building it serves, to be replaced by cooler air from inside the building (Figure 7.8). At night, access to the greenhouse is shut off. Although this is the simplest use of a greenhouse, it requires careful management. The greenhouse may also be required to maintain its own night-time temperature during winter. This may be achieved through direct solar gain with the help of a masonry rear wall, a well-insulated concrete floor slab, or water drums exposed to the sun during the day.

To make optimal use of the greenhouse, it has to be designed as an integral part of new buildings, and not as an afterthought. Then the idea of conduction loop can be exploited to a far greater degree to encompass the entire building, rather than merely those spaces immediately adjacent to the greenhouse. Warm air from the greenhouse can be ducted first to the top of the building, then down the north side, eventually to return to the greenhouse via ducts beneath the floor.

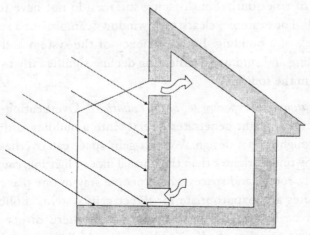

Figure 7.8 The greenhouse for passive heating

The integration of a greenhouse for heating a building clearly has a major impact on the entire plan. It affects not just the siting of the greenhouse but also its area, volume, and cross-sectional shape, all of which have to be determined in relationship to the configuration of the building as a whole.

Guidelines for the design of passive solar heating systems

Analyse building thermal load patterns   An important concept of passive solar design is to match the time when the sun can provide daylighting and heat to a building with the time when the building needs them. This will determine which passive solar design strategies are most effective. Commercial buildings have complicated demands for heating, cooling, and lighting, therefore, their design would require a detailed computer analysis by an architect or an engineer.

Integrate passive solar heating with daylighting design   A passive solar building that makes use of sunlight as a heating source should also be designed to take advantage of sunlight as a lighting source. However, each use has different design requirements that need to be addressed. In general, passive solar heating benefits from 'beam sunlight', directly striking dark-coloured surfaces. Daylighting, on the other hand, benefits from gentle diffusion of sunlight over large areas of light-coloured surfaces. Integrating the two approaches requires an understanding and coordination of daylighting, passive design, electric lighting, and mechanical heating systems and controls.

Design the building's floor plan to optimize passive solar heating   Orient the solar collection surfaces, for example, appropriate glazings in windows and doors, preferably within 15° of true south. South-facing surfaces do not have to be all along the same wall. For example, clearstorey windows can project a south sun deep into the back of a building. Both efficiency of the system and the ability to control shading and summer overheating decline significantly as the surface shifts away from the south.

Identify appropriate locations for exposure to 'beam radiation'   Overheating and glare can occur whenever sunlight penetrates directly into a building and this must be addressed through proper design. A 'direct-gain' space can overheat in full sunlight and is many times brighter than the normal indoor lighting, causing intense glare. Generally, rooms and spaces where people stay in one place for more than a few minutes are inappropriate for direct gain systems. Lobbies, atria, or lounges can be located along the south wall where direct sun penetrates. Glazings that optimize the desired heat gain, daylighting, and yet avoid burdening the cooling load, should be chosen.

Avoid glare from low-sun angles   In late morning and early afternoon, the sun enters through south-facing windows. The low angle allows the sun beam to penetrate deep into the building beyond the normal direct-gain area. If the building and occupied spaces are not designed to control the impact of the sun's penetration, the occupants will experience discomfort from glare. Careful sun-angle analysis and design strategies will ensure that these low-sun angles are identified and addressed. For instance, light shelves can intercept the sun and diffuse daylight.

Locate thermal mass to be hit by a low-angle winter sun   Building design should incorporate sufficient amount of correctly located thermal mass to effectively contribute to harvesting of heat as well as its uniform re-radiation to internal space.

Passive cooling concepts

Some of the well-known concepts of passive cooling include
- manipulating building form and orientation
- shading
- vegetation
- insulation
- ventilation
- evaporative and nocturnal cooling

Building form and orientation

The optimum shape of a building is widely considered to be the one that loses the least amount of heat in winter and accepts the least amount of heat in summer. Olgyay (1963) dispelled the myth that the square form was the optimum for most climates. An elongated form oriented along the east–west axis can perform better.

The maximum amount of solar radiation is interrupted by the horizontal surface of the roof followed by east and west walls and then by the south and north wall during the summer for latitudes below the Tropic of Cancer in the Northern Hemisphere. It is, therefore, desirable that the building is oriented with the longest wall facing north and south, and short walls facing east and west. The intention is to minimize the wall area exposed to the intense morning and evening sun. It is observed that significant heat gains occur between 06:30 a.m. and 10.00 a.m. on the east walls and between 2:00 p.m. and 5:30 p.m. on the west walls, when the sun is low in the sky. Consequently, windows on these walls are sources of high heat gain and should either be eliminated or reduced in size.

The volume of a building is approximately related to its thermal capacity, while the exposed surface area is related to the rate at which it gains or loses heat. A summary of requirements for conditions of optimum heat gain and heat loss for a simple rectangular building form in different climates is presented in Table 7.7 (Santamouris and Asimakopoulos 1996).

It is also advisable to place service spaces that are unconditioned on the east and west sides. Garages, staircases, toilets, and service shafts can act as buffer spaces to minimize the ingress of heat into the habitable areas of the building.

Orientation should also take into consideration prevailing local wind patterns; hot winds should be blocked in hot and dry climates and captured and channelled to improve ventilation in warm and humid climates.

Table 7.7 Requirements of building form for different climate types

Climate	Elements and requirements	Purpose
Warm humid	Minimize building depth	For ventilation
	Minimize west-facing wall	To reduce heat gain
	Maximize south and north walls	To reduce heat gain
	Maximize surface area	For night cooling
	Maximize window wall	For ventilation
Composite	Controlled building depth	For thermal capacity
	Minimize west wall	To reduce heat gain
	Limited south wall	For ventilation and some winter heating
	Medium area of window wall	For controlled ventilation
Hot dry	Minimize south and west walls	To reduce heat gain
	Minimize surface area	To reduce heat gain and loss
	Maximize building depth	To increase thermal capacity
	Minimize window wall	To control ventilation heat gain and light
Mediterranean	Minimize west wall	To reduce heat gain (summer)
	Moderate area of south wall	To allow (winter) heat gain
	Moderate surface area	To control heat gain
	Small to moderate window	To reduce heat gain but allow winter light
Cool temperate	Minimize surface area	To reduce heat loss
	Moderate area of north and west walls	To receive heat gain
	Minimize roof area	To reduce heat loss
	Large window wall	For heat gain and light
Equatorial upland	Maximize north and south walls	To reduce heat gain
	Maximize west-facing walls	To reduce heat gain
	Medium building depth	To increase thermal capacity
	Minimize surface area	To reduce heat loss and gain

Shading

Application of shading must take into consideration the interaction of several factors.

- Obstructing the solar heat gains from reaching the envelope and the interiors of the building
- Non-interference with winter solar gains
- Control of intense daylight by diffusing it in a uniform manner into the space
- Unobstructed view from the windows
- Admission and regulation of the ventilation of adjacent spaces

These design objectives and the corresponding shading techniques differ according to the latitude, location, type of building, its schedule of operation, occupant activities, internal heat gains, and expected comfort conditions.

Good shading strategies can save 10%–20% of the energy required for cooling. A group of buildings in a cluster can be so spaced as to mutually shade each other. However, effectiveness of the shading depends on the configuration and layout of the cluster, which the architect may not have much control over. Internal and external shading devices can be used extremely effectively for solar control.

Internal and external shading devices

The main aim of shading devices is to protect openings from direct solar radiation, while their secondary intent is to protect openings from diffused and reflected radiation. There are two primary classifications of shading devices.

- *Fixed elements* Fixed elements are mainly external, and include horizontal overhangs, vertical fins, combination of horizontal and vertical elements, and balconies. Internal elements include light shelves and louvres.
- *Adjustable elements* Adjustable elements can be external shading elements in the form of tents, awnings, pergolas, or internal elements such as curtains, venetian blinds, rollers, and window shutters (Figure 7.9). Adjustable devices can be lifted, rolled, and drawn back from the window either manually or automatically in response to, the changing radiation and daylighting levels.

External devices are more effective as they obstruct solar radiation even before it reaches the interior of the building. Internal devices stop from the radiation that has already penetrated inside only the portion that can be reflected by their surfaces, while the remainder is absorbed, convected, and radiated to the room. Thus, effectiveness of shading devices is mainly determined by their reflectivity. Additionally, these may conflict with daylighting

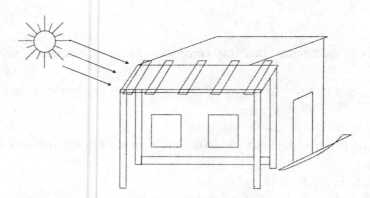

Figure 7.9 Pergolas for shading of south facades
Source Santamouris and Asimakopoulos (1996)

and ventilation requirements as they block the openings in most cases. External shading devices are generally considered about 35% more efficient than the internal ones.

Olgyay and Olgyay (1957) have suggested a method by which the designing of shading devices is carried out in four steps.

1 The times when shading is needed (overheated period) is determined. According to this methodology, provisions for shading are required at any time when the outdoor air temperature exceeds 70 °F (21 °C) at a latitude of approximately 40°. For every 5° latitude change towards the equator, the limiting temperature is elevated by 0.75 °F (0.42 °C).

2 The position of the sun when the shading is needed is determined by using a sun-path diagram. The overheated period is marked on the sun-path. This is done by constructing a table of average temperature for every hour in every month, thus obtaining the times of the overheated period. The boundary lines of the overheated period are then transferred to the sun-path diagram.

3 The type and position of the shading device are determined. The 'shading mask' of a given shading device is plotted on a protractor having the same scale as the sun-path diagram.

4 The dimensions of the shading device are determined so as to interrupt the sunlight during the overheated period and to let it in during the underheated period.

Maximum solar radiation is incident on the roof. Hence, it is helpful if the roof surface is protected from the sun as far as possible during the day. This can be designed as an integral part of the building or it can be a separate cover.

Shading provided by external means should not interfere with night-time cooling. A cover of concrete or galvanized iron sheets over the roof not only provides protection from direct radiation but also prevents radiation from escaping into the cool night sky. An alternative method is to provide a cover of deciduous plants or creepers. Evaporation from the leaf surfaces lowers the daytime temperature of the cover and at night time, it may even be lower than the sky temperature.

Another shading device in some traditional buildings is developed by covering the entire roof surface by small, closely packed inverted earthen pots. In addition to shading, this system provides a layer of still air over the roof, which acts as insulation, impeding heat flow into the building during the day. This arrangement also permits an upward heat flow and loss to the night sky. Although this technique is thermally efficient, there are practical difficulties of maintenance and lack of usability of the rooftop.

An effective roof shading system is a removable canvas roof, which can be mounted close to the roof during the day and can be rolled away during night. The upper side of the cover should be white or light to minimize the amount of radiation absorbed and the consequent heat gain through it. The various methods of roof shading are shown in Figure 7.10.

Reflective surfaces

Certain colours reflect solar radiation more (and absorb less) than others. If external surfaces are painted in such a manner that their emission in the long-wave region is high, then heat flux transmitted into the building is reduced considerably (Table 7.8).

It is seen that whitewash has lower reflectivity than aluminium but stays cooler when exposed to solar radiation because of its high emissivity at low temperatures.

External colour and finish influence light-weight structures more due to their low resistance to heat flow and low thermal capacity.

Vegetation

Rational use of vegetation around the building can offer significant shading. Vegetation can reduce the impinging solar radiation and modify air tempera-tures, outdoors and indoors, to effectively reduce the cooling loads of buildings. The choice of trees should be very carefully based on the shape and character of the plant, both during summer and winter, and on the shadow shape they provide. Their position should be chosen strategically to provide shade at the most critical hours of the day. Trees on the east–south-east and

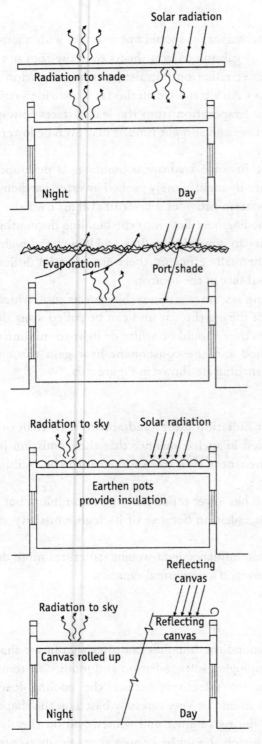

Figure 7.10 Some methods of shading the roof
Reproduced with permission from Elsevier
Source Bansal, Minke, and Hauser (1994)

Table 7.8 Reflectivity of the surfaces for solar radiation and their emission in the long-wave region

Material	Reflectivity (solar radiation)	Emissivity (low temperature)
Aluminium foil (bright)	0.95	0.05
Aluminium foil (oxidized)	0.85	0.12
Polished aluminium	0.80	0.05
Aluminium paint	0.50	0.50
Galvanized steel bright	0.75	0.25
Whitewash new	0.88	0.90
White oil paint	0.80	0.90
Grey colour (light)	0.60	0.90
Grey colour (dark)	0.30	0.90
Green colour (light)	0.60	0.90
Green colour (dark)	0.30	0.90
Red brick	0.40	0.90
Glass	0.08	0.90

west–south-west give the best results as the sun is at a low altitude in the morning and evening, and casts long shadows (Figure 7.11). Bushes can act as vertical wingwalls to shade windows on the east and west (Figure 7.12). Vertical trellis covered by vines or creepers are very useful to shade facades.

A study quoted by Bansal, Minke, and Hauser (1994) has shown that the ambient air temperature under a tree adjacent to a wall is about 2 to 2.5 °C cooler than the unshaded areas. This cooler microclimate immediately adjacent to a building results in less heat gain through conduction by walls and windows.

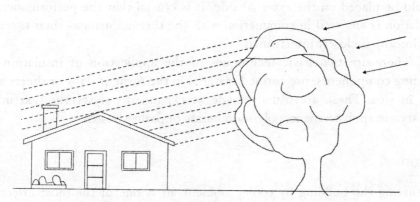

Figure 7.11 Trees at east and west sides for shading in summer
Source Santamouris and Asimakopoulos (1996)

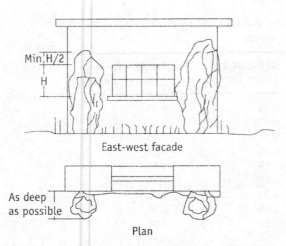

Figure 7.12 Bushes for shading of east and west windows
Source Santamouris and Asimakopoulos (1996)

Insulation

The external surface of the building gets heated due to solar radiation that is absorbed and due to convective heat gain from hotter ambient air. This heat is then transmitted inwards through the building envelope. One important factor that determines the quantity of heat that reaches the interiors of the building is insulation in the walls and the roof.

The use of insulation is a very important aspect of energy efficiency in buildings. In hot climatic conditions, thermal mass of the building envelope should be strongly coupled with the interiors in order to absorb internal heat gains easily. This means that the layer of insulation in the building element should be placed on the external side. It is crucial that the performance of insulation is analysed in conjunction with the thermal mass as their thermal relationship is closely intertwined.

There are two basic techniques of the application of insulation to building components: pre-formed material in the form of slabs or sheets and cast in situ. These are usually made of expanded polystyrene, extruded polystyrene, polyurethane, polyisocyanurate or perlite boards.

Ventilation

Ventilating the building by cooler ambient air is one of the most effective means of cooling the indoors. Convective cooling can take place by natural means in which internal air is replaced by the ambient air.

Air flow in a building is either due to pressure differences caused by temperature differences (stack effect) or due to the pressure difference of the wind.

Stack ventilation depends upon the buoyancy of warmer air to rise and leave the building through an outlet located at a higher level while heavier cooler air is pulled in through a lower inlet (Figure 7.13). The rate of flow induced by thermal force is given by the formula

$$M_v = 0.117\,A\,[H(T_i - T_a)]^{1/2} \qquad\qquad ...(7.28)$$

where M_v is the ventilation rate in m^3/s, A is the unobstructed area of inlet in m^2, H is the vertical distance between inlet and outlet in m, T_i is the average temperature of indoor air in °C, T_a is the average temperature of ambient air in °C.

However, by conventional design, it is usually not possible to achieve the required ventilation rates for occupant comfort.

In order to maximize the advantage of ventilation, designers manipulate building components. Some elements typically used to facilitate the ventilation of indoor spaces are discussed below.

Various parameters that affect wind forces entering through windows are:

- Climate
- Wind speed and direction
- Size and location of inlet and outlet openings
- Volume of the space
- Shading devices
- Internal partitions

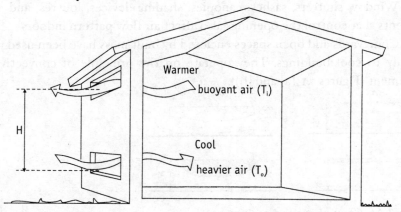

Figure 7.13 Stack effect
Source Bansal, Minke, and Hauser (1994)

Wind movement inside a building is governed by aeromotive forces and the temperature differential between the inside and the outside. Air moves from a high-pressure zone to a low-pressure zone through openings in the building envelope.

To assist sensible air movement, it is essential to provide cross ventilation. In the absence of an outlet opening, there is no effective air movement through a building even in case of strong winds. An opening only on the windward side leads to a pressure build-up indoors and increases discomfort. Similarly, an opening only on the leeward side can lead to oscillating pressures which are uncomfortable for occupants.

For effective convective cooling, air movement should be directed at the surface of the human body. Therefore, the inlets and outlets must be positioned such that movement is created up to 2 m from the floor level (Figure 7.14).

The sizes of the inlet and outlet openings need to be carefully calculated to make the most of the available wind outside. For a given area and a total wind force (area × pressure), the largest air velocity is attained through a small inlet opening with a large outlet. This is due to the air being forced through a small aperture and also due to the Venturi effect, that is, in the imaginary funnel connecting the inlet and the outlet, a sideways expansion of the air jet further accelerates the particles. Such an arrangement is useful when the direction of wind is fairly constant. However, when that is not the case, a large inlet is preferable, which enables the total volume of air that passes through to be larger. The best arrangement is large, full wall openings on both sides with adjustable shutters that can channel the air flow in the required direction, in keeping with the varying direction of the wind.

Window shutters, sashes, canopies, shading devices, louvres, and other elements that control the openings also affect air flow pattern indoors.

Courtyards and open spaces enclosed by built mass have been used traditionally to cool buildings. They operate on the principle of convective air movement (Figures 7.15[a] and [b]).

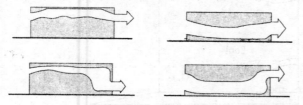

Figure 7.14 Effect of opening positions
Reproduced with permission from Orient Longman Pvt. Ltd
Source Koenigsberger, Ingersoll, Mayhew, *et al.* (1973)

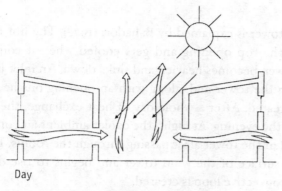

Figure 7.15(a) Courtyard effect (day)
Reproduced with permission from Elsevier
Source Bansal, Minke, and Hauser (1994)

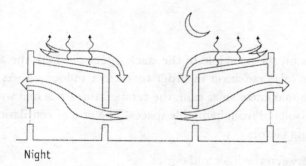

Figure 7.15(b) Courtyard effect (night)
Reproduced with permission from Elsevier
Source Bansal, Minke, and Hauser (1994)

During the day, solar radiation incident to the courtyard heats up the air that rises to escape the enclosure. To replace it, cool air from the ground level flows through the openings of the room, thus setting up the air movement.

During the night, the process is reversed. As radiation and convection cool the warm roof surface, a stage is reached when its surface temperature equals the dry bulb temperature of ambient air. On further cooling by radiation, condensation may also occur. If the roof surface is sloping inwards, this cool air sinks into the court and enters habitable spaces through low level openings. To make this system work efficiently, a parapet wall should be raised around the perimeter of the roof to prevent mixing of air.

This technique works well in warm and humid climate. A temperature drop of 4–7 °C below ambient is possible.

Wind tower

The principle of a wind tower is explained by Bahadori (1978). The hot ambient air enters through the top opening and gets cooled when it comes in contact with the cool tower, becomes heavier, and sinks down. An inlet is provided to the rooms from the tower and, along with an opening on the outer side, a cool draught is created. After a whole day of heat exchange, the wind tower becomes warm by the evening. At night, the cooler ambient air comes in contact with the bottom of the tower after passing through the rooms. It gets heated up by the warm surface of the wind tower and begins to rise due to buoyancy, and a reverse convective loop is created.

This system works well in hot and dry regions, where the diurnal range of temperatures is very high. For a wind tower approximately 4-m high, a temperature drop of about 4 °C is observed with an indoor air velocity of 1 m/s (Figure 7.16).

Solar chimney

A solar chimney works on the principle of the stack effect but here the air is deliberately heated by solar radiation in order to extract indoor air. As the chimney is designed to maximize solar gain, the temperatures reached within are very high, and it is isolated from habitable spaces. The rate of ventilation is affected by the following factors.

- Vertical distance between the inlet and outlet
- Cross-sectional area of the inlet and outlet
- Geometrical construction of the solar absorbing plate
- Inclination angle

Solar chimneys can be used effectively for ventilation in regions of low wind speed.

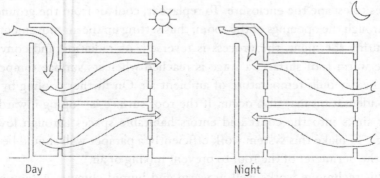

Figure 7.16 Daytime and night-time operations of a wind tower
Reproduced with permission from Elsevier
Source Bansal, Minke, and Hauser (1994)

Evaporative and nocturnal cooling

The principles of evaporation are discussed in the section on mass transfer in Chapter 3. Briefly, evaporation of water from a film or a pond or a drop occurs if the difference between the vapour pressure at the given water temperature and the partial pressure of water in the atmosphere is positive. A water body exposed to the atmosphere loses/gains heat through conduction, convection, radiation, and evaporation. The temperature attained by such water bodies can be calculated by a heat balance method (Kishore, Ramana, and Rao 1979). Experiments have shown that water ponds exposed to the night sky can attain temperatures below the wet bulb temperature in favourable conditions. A roof sack-cloth cooling technique has been developed in India (Jain and Rao 1974; Kumar and Jain 1981) and is found to be effective in hot arid conditions. The clear night sky can act as a sink with low temperatures. The sky temperatures can be obtained by the equation

$$T_{sky} = T_a (0.55 + 0.061\sqrt{p_a})^{1/4} \qquad\qquad ...(7.29)$$

where T_{sky} and T_a are measured in Kelvin, and p_a is the partial pressure of water vapour in the atmosphere in mm Hg. An ingenious method of cooling a building by nocturnal and evaporative cooling in summer, and heating by solar radiation in winter was developed by Hay and Yellot (1969), as described in the following section.

Passive concepts for composite climates

As mentioned earlier in this chapter, buildings in composite climate require both heating and cooling, which makes passive design more challenging and complex. A few systems based on passive concepts are as follows.

- Earth–air tunnel
- Sky-therm system
- Solar chimney based hybrid systems

Earth–air tunnel

It is well known that ground temperatures at a depth of about 4-5 m below the surface remain constant throughout the year at a value close to the annual average. Hence, a tunnel of suitable length in which ambient air enters at one end would provide heated/cooled air at the other end. Earth–air tunnels are studied quite well (Sodha, Bansal, Kumar, et al. 1986). A schematic diagram of the earth–air tunnel installed at the Gual Pahari campus of TERI is shown in Figure 7.17. A parametric prediction of the cooling potential has also been carried out (Mihalakakou, Santamouris, Asimakopoulos, *et al.* 1995).

The sky-therm system

The sky-therm system, originally conceived by H Hay in the 1960s and developed further in collaboration with J I Yellot (Hay and Yellot 1969; Yellot and Hay 1969), makes use of roof ponds with movable insulation for collection, storage, and dissipation of heat. In this simple but very effective system, black plastic bags filled with water are placed on a metallic roof. An insulation panel can be slid in or out so as to either expose the roof to the sky or to shield it (Figure 7.18).

In winter, the black plastic bags are exposed to sunlight and they act as collection-cum-storage devices of solar energy. The bags are insulated in the night by sliding the insulation panel. The metallic roof transfers the stored heat into the room by conduction and radiation. In summer, and under dry conditions, the black plastic bags are exposed to night sky and they lose heat by nocturnal radiation to sky. The roof is insulated during the daytime, so that room heat is transferred to the colder bags. During hot and humid conditions, the roof deck is flooded with water, so that heat loss at night is enhanced due to evaporation. The occupants reported that heating and cooling are even and comfortable. Sky-therm systems are useful in hot/arid zones and for single dwellings. Making a sturdy and leak-proof metallic roof and providing for sliding insulation pose some problems, which have been adequately addressed in practical buildings. A system similar in concept to the roof pond was experimented with at IIT Kanpur in the early 1980s under the sponsorship of TERI.

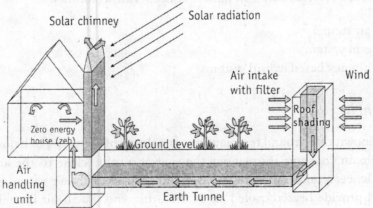

Figure 7.17 Schematic diagram of earth–air tunnel system

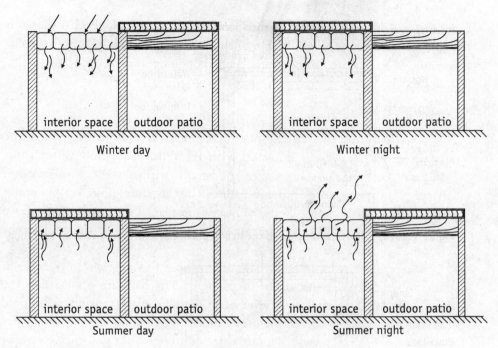

Figure 7.18 Operating principles of sky-therm system

Solar chimney-based hybrid system

The solar chimney-based hybrid system, designed by TERI, was constructed and studied at the Gual Pahari campus of TERI under a project funded by the Ministry of Non-conventional Energy Sources. The design consists of the following elements (Raman, Mande, and Kishore 2001).

- Reduction of heat loads is achieved by insulating the outside surfaces, using double glazed windows and providing an ante-room at the entrance.
- The entire south wall is converted into a solar collector by painting it black and fixing a single glazing. Bottom and top vents into the room and a closable exhaust at the top edge of the wall are provided. The collector wall acts like a solar chimney in summer and like a Trombe wall in winter.
- Evaporative cooling of the roof is done by the sack-cloth system with a provision for air passage space below the cooling surface. Ambient air enters this space, gets cooled, and enters the room from a roof vent. The solar chimney aids in creating a natural draught. A schematic diagram of the system for summer and winter operations is shown in Figures 7.19 (a) and (b) and a photograph of the system is shown in Figure 7.20. Typical day per-

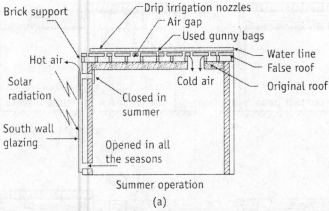

Figure 7.19(a) Schematic diagram of the hybrid passive system during summer operation

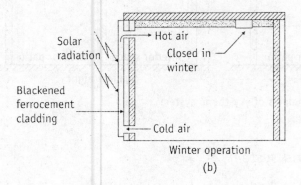

Figure 7.19(b) Schematic diagram of the passive system during summer operation

Figure 7.20 Overall view of the heating-cum-cooling design developed at TERI

formance of the system for summer and winter is shown in Figure 7.21. It can be seen that when the ambient temperature varied from 26 °C to 42 °C in summer, the room could be maintained around 32 °C. For a winter day when the temperature varied from 2 °C to 17 °C, the room was kept at about 16 °C. The design of the system is such that existing structures can be modified to incorporate passive features.

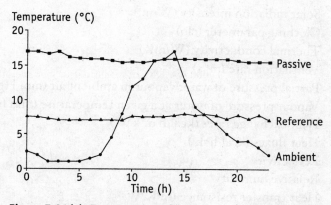

Figure 7.21(a) Temperature profile of the passive building for a typical winter day operation

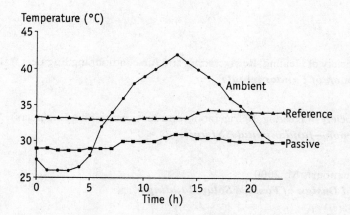

Figure 7.21(b) Temperature profile of the passive building for a typical summer day operation

Nomenclature

A_s	Surface area of the body, Du Bois area (m²)
a	Absorptivity of a surface
A	Unobstructed inlet area of a building opening (m²)
d	Thickness of a conducting element (m)
f_{cl}	Clothing surface ratio
h_c, h_i, h_o	Convecting heat transfer coefficient (W/m²k)
I	Solar radiation intensity (W/m²)
I_{cl}	Clothing parameter (clo)
k	Thermal conductivity (W/mK)
M_v	Ventilation rate (m³/s)
p_a	Partial pressure of water vapour in ambient air (mm Hg)
p_s	Vapour pressure of water at a given temperature (mm Hg)
Q_M, Q_H, Q_{sw}	Heat exchange rates (kcal/h or kW)
q_c, q	Heat fluxes (kcal/h m²)
r	Reflectivity
RH	Relative humidity
R_{cl}, R	Heat transfer resistance (m²K/W)
T_a, T_s, T_{cl}	Temperature (°C or K)
V	Air speed (m/s)
W	Work done per unit time (kcal/h or kW)
η	Efficiency of conversion

References

ASHRAE (American Society of Heating, Refrigerating, and Air-conditioning Engineers). 1997
1997 ASHRAE Handbook of Fundamentals
Atlanta: ASHRAE

ASHRAE (American Society of Heating, Refrigerating, and Air-conditioning Engineers). 2005
2005 ASHRAE Handbook—fundamentals: SI units
Atlanta: ASHRAE

Athienitis A K and Santamouris M. 2000
Thermal Analysis and Design of Passive Solar Buildings
UK: Stylus Publishing

Bahadori M N.1978
Passive cooling systems in Iranian architecture
Scientific American **238**(2): 144–154

Bansal N K and Minke G (eds). 1998
Climatic Zones and Rural Housing in India
Julich: German–Indian Cooperation in Scientific Research and Technological Development/
KFA

Bansal N K, Minke G, and Hauser G. 1994
Passive Building Design
Elsevier Science B V.

Cook J (ed.). 1989
Passive Cooling
The MIT Press

Fanger P O. 1970
Thermal Comfort: analysis and applications in environmental engineering
New York: McGraw-Hill

Givoni B. 1994
Passive and Low Energy Cooling of Buildings
van Nostrand Reinhold

Goetzberger A (ed.). 1992
Special issue on transparent insulation
Solar Energy 49(5)

Hay H R and Yellot J I. 1969
Natural air conditioning with roof pond and movable insulation
ASHRAE Trans. 75(1): 165–177

Hoffman M E and Feldman M. 1981
Calculation of the thermal response of buildings by the total thermal time constant method
Building and Environment 16(2): 71–85

Jain S P and Rao K R. 1974
Experimental study on the effect of roof-spraying cooling on unconditioned and conditioned building
Building Science 9: 9–16

Kishore V V N. 1988
Assessment of natural cooling potential for buildings in different climatic conditions
Building and Environment 23(3): 215–223

Kishore V V N, Ramana M V, and Rao D P. 1979
An experimental and theoretical study of a natural water cooler, pp. 482–488
[Proceedings of National Solar Energy Convention, IIT Bombay, Organized by Solar Energy Society of India]

Koenigsberger O H, Ingersoll T G, Mayhew A, Szokolay S V. (1973)
Manual of Tropical Housing and Building, Part 1
Orient Longman Pvt. Ltd

Krishnan A, Baker N, Yannas S, Szokolay S V (eds). 2001
Climate Responsive Architecture
New Delhi: Tata McGraw-Hill

Kumar V and Jain S P. 1981
Automatic water actuated switch for cooling building by roof surface evaporation
Energy Management 5: 127–129

Mihalakakou G, Santamouris M, Asimakopoulos D, Tselepidaki I. 1995
Parametric prediction of the buried pipes cooling potential for passive cooling applications
Solar Energy 55(3): 163–173

Olgyay V. 1963
Design with Climate
Princeton: Princeton University Press

Olgyay V and Olgyay A. 1957
Solar Control and Shading Devices
Princeton: Princeton University Press

Raman P, Mande S, and Kishore V V N. 2001
A passive solar system for thermal comfort conditioning of buildings in composite climates
Solar Energy 70(4): 319–329

Santamouris M. 2006
Environmental Design of Urban Buildings: an integrated approach
UK: Earthscan Publications Ltd

Santamouris M and Asimakopoulos D (eds). 1996
Passive Cooling of Buildings
James & James (Science Publishers) Ltd

Sodha M S, Bansal N K, Kumar A, Bansal P K, Malik M A S. 1986
Solar Passive Building: science and design
New York: Pergamon Press

Szokolay S V. 1985
Thermal comfort and passive design
In *Advances in Solar Energy*, vol. 2, edited by K Boer and J Duffie
New York: Plenum Press

Yellot J I and Hay H R. 1969
Thermal analysis of a building with natural air conditioning
ASHRAE Trans. 75(1), 179–190

8

Wind energy resources

Jami Hossain, Adviser (Technical)
Indian Wind Energy Association, New Delhi

Overview of wind energy developments

Current scenario

Over the past two decades or so, wind has emerged as a significant source of energy meeting some of the electricity requirements in many parts of the world. The modern wind turbine technology has evolved as a highly reliable and rugged system that can function unattended or with minimal maintenance support for the next 20 years or more to generate grid-quality electricity.

In view of the global climate change concerns and the fact that generation of electricity from wind is essentially pollution-free, there is a world-wide acceptance and support for this technology. At the end of 2007, nearly 94 000 MW of wind-power-generating capacity was installed across the world. Wind power industry is also one of the fastest growing industries in the world. Between 1990 and 2005, worldwide, the wind industry grew at an average annual growth rate of nearly 24% (World Wind Energy, press realese, 2008). Table 8.1 gives the wind-power-generating capacity in different countries. India with nearly 8000 MW of installed capacity as of March 2007 ranks fourth in the world and is currently the third-largest market in the world. The largest wind turbine in India was a prototype rated at 2 MW and was commissioned in November 2004. It had been designed, manufactured, and installed near Kanyakumari by the

Table 8.1 Total global wind-energy-generating capacity in megawatts (ending December 2005)

Ranking total 2007	Country/ region	Total capacity installed 2007 [MW]	Additional Capacity 2007 (Difference 2007–2006) [MW]	Rate of growth 2007 [%]	Ranking total 2006	Total capacity installed end 2006 [MW]	Total capacity installed end 2005 [MW]
1	Germany	22.247,4	1.625,4	7,9	1	20.622,0	18.427,5
2	USA	16.818,8	5.215,8	45,0	3	11.603,0	9.149,0
3	Spain	15.145,1	3.515,1	30,2	2	11.630,0	10.027,9
4	India	7.850,0	1.580,0	25,2	4	6.270,0	4.430,0
5	China	5.912,0	3.313,0	127,5	6	2.599,0	1.266,0
6	Denmark	3.125,0	−11,0	−0,4	5	3.136,0	3.128,0
7	Italy	2.726,1	602,7	28,4	7	2.123,4	1.718,3
8	France	2.455,0	888,0	56,7	10	1.567,0	757,2
9	United Kingdom	2.389,0	426,2	21,7	8	1.962,9	1.353,0
10	Portugal	2.130,0	414,0	24,1	9	1.716,0	1.022,0
11	Canada	1.846,0	386,0	26,4	12	1.460,0	683,0
12	The Netherlands	1.747,0	188,0	12,1	11	1.559,0	1.224,0
13	Japan	1.538,0	229,0	17,5	13	1.309,0	1.040,0
14	Austria	981,5	17,0	1,8	14	964,5	819,0
15	Greece	873,3	115,7	15,3	16	757,6	573,3
16	Australia	817,3	0,0	0,0	15	817,3	579,0
17	Ireland	805,0	59,0	7,9	17	746,0	495,2
18	Sweden	788,7	217,5	38,1	18	571,2	518,0
19	Norway	333,0	7,9	2,4	19	325,0	268,0
20	New Zealand	322,0	151,0	88,3	25	171,0	168,2
21	Egypt	310,0	80,0	34,8	21	230,0	145,0
22	Belgium	286,9	92,6	47,7	22	194,3	167,4
23	Taiwan (China)	280,0	92,3	49,2	23	187,7	103,7
24	Poland	276,0	123,0	80,4	26	153,0	73,0
25	Brazil	247,1	10,2	4,3	20	236,9	28,6
26	Turkey	206,8	142,2	220,0	31	64,6	20,1
27	Korea (South)	191,3	15,0	8,5	24	176,3	119,1
28	Czech Republic	116,0	59,5	105,3	34	56,5	29,5
29	Finland	110,0	24,0	27,9	28	86,0	82,0
30	Ukraine	89,0	3,4	4,0	29	85,6	77,3
31	Mexico	86,5	0,0	0,0	27	86,5	2,2
32	Costa Rica	74,0	0,0	0,0	30	74,0	71,0
33	Bulgaria	70,0	34,0	94,4	37	36,0	14,0
34	Iran	66,5	19,1	40,4	36	47,4	31,6
35	Hungary	65,0	4,1	6,8	33	60,9	17,5
36	Morocco	64,0	0,0	0,0	32	64,0	64,0
37	Estonia	58,1	25,1	76,1	39	33,0	33,0
38	Lithuania	52,3	−2,7	−4,8	35	55,0	7,0

Continued

Table 8.1 *Continued*

Ranking total 2007	Country/ region	Total capacity installed 2007 [MW]	Additional Capacity 2007 (Difference 2007–2006) [MW]	Rate of growth 2007 [%]	Ranking total 2006	Total capacity installed end 2006 [MW]	Total capacity installed end 2005 [MW]
39	Luxembourg	35,3	0,0	0,0	38	35,3	35,3
40	Argentina	29,8	2,0	7,2	40	27,8	26,8
41	Latvia	27,4	0,0	0,0	41	27,4	27,4
42	Philippines	25,2	0,0	0,0	42	25,2	25,2
43	Jamaica	20,7	0,0	0,0	43	20,7	20,7
44	Guadeloupe	20,5	0,0	0,0	44	20,5	20,5
45	Tunisia	20,0	0,0	0,0	45	20,0	20,0
46	Chile	20,0	18,0	900,0	59	2,0	2,0
47	Colombia	19,5	0,0	0,0	46	19,5	19,5
48	Croatia	17,8	0,6	3,5	47	17,2	6,0
49	South Africa	16,6	0,0	0,0	48	16,6	16,6
50	Russia	16,5	1,0	6,5	49	15,5	14,0
51	Guyana	13,5	0,0	0,0	50	13,5	13,5
52	Curaçao	12,0	0,0	0,0	51	12,0	12,0
53	Switzerland	11,6	0,0	0,0	52	11,6	11,6
54	Romania	9,0	6,2	226,1	57	2,8	0,9
55	Israel	7,0	0,0	0,0	53	7,0	7,0
56	Slovakia	5,0	0,0	0,0	54	5,0	5,0
57	Faroe Island	4,1	0,0	0,0	55	4,1	4,1
58	Ecuador	3,1	3,1		73	0,0	0,0
59	Cape Verde	2,8	0,0	0,0	56	2,8	2,8
60	Nigeria	2,2	0,0	0,0	58	2,2	2,2
61	Cuba	2,1	1,7	366,7	67	0,5	0,5
62	Jordan	1,5	0,0	0,0	60	1,5	1,5
63	Martinique	1,1	0,0	0,0	61	1,1	1,1
64	Belarus	1,1	0,0	0,0	62	1,1	1,1
65	Indonesia	1,0	0,2	25,0	63	0,8	0,8
66	Eritrea	0,75	0,00	0,0	64	0,75	0,75
67	Peru	0,70	0,00	0,0	65	0,70	0,70
68	Uruguay	0,60	0,45	300,0	70	0,15	0,15
69	Kazakhstan	0,50	0,00	0,0	66	0,50	0,50
70	Namibia	0,47	0,22	88,0	69	0,25	0,25
71	Netherl. Antilles	0,33	0,33		74	0,00	0,00
72	Syria	0,30	0,00	0,0	68	0,30	0,30
73	North Korea	0,01	0,00	0,0	71	0,01	0,01
74	Bolivia	0,01	0,00	0,0	72	0,01	0,00
	Total	**93.849,1**	**19.695,8**	**26,6**		**74.153,3**	**59.033,0**

Source World Wind Energy Association, Bonn

Figure 8.1 Modern wind turbines

Pune-based Suzlon Energy Ltd. World-wide, the largest wind turbine set up so far is a 6 MW machine installed in Germany by General Wind, a wind farm development company.

Wind resource assessment

At any location, the pre-condition for harnessing wind energy is that the winds should blow sufficiently strongly to economically justify the investment in a wind power plant. Therefore, prior knowledge of the wind resource at a given site is necessary to plan for a wind energy project.

Today, when thousands of megawatts of wind power projects are either being planned or projected to come up in many parts of the world, wind resource knowledge has become an aspect of tremendous importance. In fact, it has evolved into a specialized discipline and is needed for different reasons at different stages of the initiatives with varying levels of detailing.

Regional

Knowledge of wind speed is needed at various regional levels—a continent, a country, or a large widespread area. Referred to as 'areas of interest', these regions are being looked upon by the industries, governments, and other stakeholders to promote wind energy projects. At the regional level, the main parameter of importance is the mean annual wind speed at a given height above the ground level.

Based on this information, one is able to assess the potential for harnessing wind energy in an economically viable manner. These estimates are not expected to be very precise but provide sufficient information to the decision-makers to evolve strategies for wind power development in that region. These strategies may include government policies to encourage investments in the wind power projects, launching of programmes for detailed wind monitoring exercises at specified locations or to study the availability of grid infrastructure, etc. The estimates can also be used as indicators of the potential by private sector players, such as the suppliers of instruments to measure wind speeds, and suppliers of wind turbine equipment.

While wind resource assessment at the regional level gives a broad idea of wind speeds in the region, the information is not very precise. For example, on the basis of its long coastal line and the wind measurements carried out for meteorological purposes, one may conclude that Tamil Nadu has a high potential for wind energy projects, or on the basis of the geology of the Leh valley, it may appear that it is a potentially windy area. However, these assessments cannot indicate the precise location of the projects.

The different ways of assessing the average wind speed in a region vary from very simple observations of biological and geomorphologic indicators to employing very sophisticated techniques using wind flow models, satellite images, digitized terrain, and the GIS (Geographical Information System).

Site assessment

More detailed and precise information is needed for the selection of the actual sites to set up wind energy projects. Wind speeds have to be measured in accordance with internationally acceptable norms at relevant heights.[1] The measurements are made by installing a monitoring mast and setting up wind-measuring instruments, and are carried out for at least one year to obtain information on wind speeds in all the seasons. The Centre for Wind Energy Technology, Chennai, recommends at least 3 year record of data in order to factor in year-to-year variations in wind speeds.

Often, data collected from a site is correlated with long-term data from the nearest meteorological site to study year-to-year variations in the

---

[1] Relevant heights are the heights at which commercially available wind turbines are set up. These are also known as hub heights.

wind speeds. Wind speeds measured at a given monitoring mast have to be extrapolated over a larger area for planning wind energy projects.

Micrositing

Wind turbines are installed in clusters or arrays known as wind farms. Planning a wind farm requires identification of precise locations of individual wind turbines in an array (Figure 8.2). The precise locations are to be determined for optimal generation from the wind farm, which requires the consideration of many factors, such as the following.

- Terrain features of sectors in different directions[2]
- Surface characteristics of sectors in different directions
- Obstructions, buildings, and trees
- Wind speed distribution of sectors in different directions
- Rotor diameter
- Spacing between wind turbines

Wind speeds measured in various sectors located in different directions at the monitoring mast are converted into wind speeds from the sectors located in the same direction at any other location on the site using the wind-speed extrapolation techniques presented in this chapter. This

Figure 8.2 Aerial view of a 201-megawatt wind farm at Satara in Maharashtra, India

---

[2] The 360° of the horizontal plane is normally divided into 12 equal sectors to record or assess the prevalent wind direction.

information is used to optimally plan the wind farm layout optimally. Sometimes, additional measurements are needed to validate the results. Software packages that help in designing the wind farm layout are also available.

Global wind system

Sun is the ultimate primary source of energy. A part of the total solar energy received on the earth manifests itself in the form of wind, which is the motion of air resulting from the conversion of the potential energy of the atmosphere into kinetic energy. In the macro-meteorological sense, the winds are movements of air masses in the atmosphere (Freris 1990).

Since long, the earth's mean surface and atmospheric temperatures have remained unchanged; there has been a net balance of radiant energy received by the earth and emitted back into the space. Although there is a total radiative balance, it does not hold for the individual latitude zones from the equator to the poles. The incoming radiation in the tropics is in excess of the outgoing, while the reverse holds true for higher latitudes and the polar regions. This, along with the rotation of the earth results in the creation and maintenance of large-scale wind systems of the globe (Figure 8.3). This, in turn, results in the general circulation of the atmosphere, which transports heat and air mass from the excess to the deficit regions. Air tends to move from regions of high pressure to those of low pressure. Therefore, the force and speed of

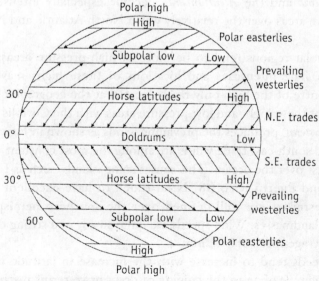

Figure 8.3 Global wind systems

the winds generated are proportional to the gradient of pressure. However, in many large areas and in some seasons, the actual pattern can be vastly different from the picture presented here. These variations are primarily due to the irregular heating of the earth's surface in time and space.

As a result of the average overheating of the earth in the equatorial region, we have a low-pressure belt, also known as the *doldrums*. Here, the warm air rises in a strong convective flow. Late afternoon showers are common from the resulting adiabatic cooling, which is most pronounced at the time of the highest *diurnal* temperature.

Ideally, there are two belts of high pressure and relatively light winds which occur symmetrically around the equator at 30° N and 30° S latitudes. These are called the *subtropical calms, subtropical highs,* or *horse latitudes*. The high-pressure pattern is maintained by the vertically descending air inside the pattern. This air is warmed adiabatically and, therefore, develops a low relative humidity with clear skies. The dryness of the descending air is responsible for the bulk of the world's great deserts that lie in the horse latitudes.

There are two more belts of low pressure, which occur at 60° S latitude and 60° N latitude, called the *subpolar lows*. In the Southern Hemisphere, due to the oceans, the belt is fairly stable and does not change much from summer to winter. In the Northern Hemisphere, however, there are large landmasses and strong temperature differences between the land and water. These differences cause the lows to reverse and become highs over land in winter (the *Canadian* and *Siberian highs*). At the same time, the lows over the oceans, called the *Iceland low* and the *Aleutian low*, become especially intense and stormy low-pressure areas over the relatively warm North Atlantic and North Pacific Oceans.

Finally, the polar regions tend to be more of high-pressure areas than low-pressure areas. The intensities and locations of these highs may vary widely, with the centre of the high only rarely located at the geographic pole. The combination of these high- and low-pressure areas with the Coriolis force (see section *Wind systems*) produces the prevailing winds as shown in Figure 8.3. The north-east and south-east trade winds are among the most constant winds on the earth, at least over the oceans. This causes some islands, such as Hawaii (20° N latitude) and Puerto Rico (18° N latitude), to have excellent wind resources. The westerlies are well-defined over the southern hemisphere because of lack of landmasses. Wind speeds are quite steady and strong during the year, with an average speed of 8–14 m/s.

The wind speeds tend to increase with the increase in latitude in the Southern Hemisphere, leading to the coining of descriptive terms *roaring for-*

ties, furious fifties, and *screaming sixties.* This means that islands and parts of the southern continents in these latitudes should prove to be good wind resource regions. Tasmania, an island in the south of Australia, is a good example.

In the Northern Hemisphere, the westerlies are quite variable and may be masked or completely reversed by more prominent circulation around moving low- and high-pressure areas. This is particularly true over the large landmasses of the continents.

Physics of wind

Wind is the movement of air relative to the surface of the earth. Also known as surface winds, as opposed to the upper atmospheric movements, they result from changes in the climatic parameters. One of the basic principles is that the air in the atmosphere flows from high-pressure zones to low-pressure zones or from low-temperature zones to high-temperature zones. Such a circulation of wind is called the *wind system.*

Wind systems

Surface winds are a result of uneven heating of the earth's surface by solar radiation. The solar radiation absorbed by the earth's surface varies with latitude—it is maximum at the equator and minimum at the poles. This creates pressure gradients in the atmosphere and the resulting *pressure gradient force* causes air movement from high-pressure zones to low-pressure zones. The vertical pressure gradients are cancelled by the force of gravity, while the horizontal gradients are the main cause of air movement near the surface of the earth that we know as the wind system. Ideally, in the lower layers of the atmosphere, closer to the earth's surface, the horizontal pressure gradients should establish a wind system between the poles and the equator, as the air rises at the equator and sinks at the poles. However, the wind system is not that simple. Other factors related to the earth's rotation and its surface characteristics, seasons, continents, oceans, and mountains result in a fairly complex system of surface winds. Some of the other wind systems and factors that influence winds due to the local topographical or climatic reasons are described in the subsequent sections.

Geostrophic winds

The earth rotates at a speed of about 166 m/s at the equator, which tapers down to zero at the poles. A 19th-century French engineer–mathematician,

Gustave-Gaspard Coriolis, introduced the concept of a force in rotating co-ordinate systems as an apparent deflection of the path of an object that moves within a rotating co-ordinate system. The object appears to deviate from its path because of the motion of the co-ordinate systems (Manwell, McGrowan, and Rogers 2002). Named after him, the *Coriolis force* – in the context of the earth's rotation – is an inertial force that results in an eastward force component for the surface winds moving from the equator towards the poles and a westward force component for the winds moving from the poles to the equator.

The winds balanced by the Coriolis and the pressure gradient forces are known as *geostrophic winds*. An air parcel initially at rest moves from high pressure to low pressure because of the pressure gradient force. However, as the air parcel begins to move, it is deflected by the Coriolis force to the right in the Northern Hemisphere (to the left in the Southern Hemisphere). As the wind gains speed, deflection increases until the Coriolis force equals the pressure gradient force (Figure 8.4). At this point, the wind blows parallel to the isobars and is referred to as *geostrophic wind*.

Frictional force

The surface of the earth exerts a frictional drag on the air blowing just above it. This friction can act to change the wind's direction and to slow it down, and keep it from blowing as swiftly as the wind above the boundary layer (see section *Boundary layer wind* below). Actually, the differences in the terrain directly affects the friction exerted. For example, as the surface of a calm ocean is quite smooth, the wind blowing over it does not move up, down, or around any features. In contrast, features such as trees and forests, hills and undulating terrain, and buildings force the wind to slow down and/or change direction frequently.

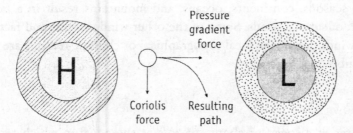

Figure 8.4 The resulting path due to the Coriolis and the pressure gradient forces

Boundary layer wind

Frictional contact with the features on the ground affects the speed and direction of the wind. As the wind altitude above the ground increases, this effect diminishes till it is negligible. At about 1–2 km above the ground, no effect can be seen. The depth of the atmosphere up to which friction plays a role in atmospheric motion is referred to as the *boundary layer*. Within the boundary layer, this friction keeps the wind from being *geostrophic*.

Monsoon system

Monsoons are large-scale regional wind systems. The seasonal temperature gradients and the temperature gradients between the oceans surfaces, and the landmasses of the continents result in large-scale seasonal movement of winds known as the monsoons. Monsoons are active for approximately six months from the north-east and six months from the south-west, principally in the southern Asia and parts of Africa. The origin of the word 'monsoon' can be traced to the Persian word *mausam* meaning weather.

During the summer months, the Afro-Asian continent is strongly heated, resulting in the generation of an extensive heat low-pressure zone, stretching from Eastern Sahara to East China. Under its influence, cool and humid air from the Indian Ocean begins to move from early June towards the trough of low pressure as the south-west monsoon. With the southward movement of the sun, the cooling landmass gives rise to a high-pressure area with its centre over eastern Siberia.

Summer monsoons have a dominant westerly component and a strong tendency to converge, rise, and produce rain. Winter monsoons have a dominant easterly component and a strong tendency to diverge, subside, and cause drought. Both are the results of differences in annual temperature trends over land and sea. The Indian monsoon is a perfect example of the monsoon system.

Sea and land system

The temperature gradient between land and sea or large water bodies such as lakes and landmasses occurs almost on a daily basis. The winds that commonly blow in all coastal regions are called the *land and sea breeze*. Land areas warm up and cool down more rapidly than the water bodies. For this reason, cooler, denser air often blows from water to land during the day, and from land to water at night. Because the temperature contrast is usually greater during the day in summer, the sea breeze is the strongest.

Figures 8.5 (a) and (b) illustrate the sea and land breeze system. During the night, the land cools down faster than the sea. This creates a relatively high pressure zone over the land, and the cool and denser air on the land moves towards the warmer low-pressure zone over the sea. Similarly, during the day, while the sea remains cool, the land surface warms up faster, resulting in a low-pressure zone over the land. Cool air from the sea moves to fill the low-pressure zone over the land.

The sea breeze normally starts some time in the noon and continues well into the night.

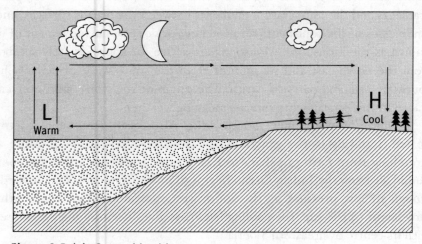

Figure 8.5 (a) Sea and land breeze

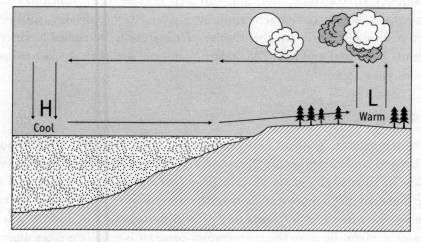

Figure 8.5 (b) Sea and land breeze

Mountain valley system

Mountain wind is induced by differential heating or cooling along the mountain slopes. During the day, the sun heats the mountain slopes causing the air from the valley to move upslope. This is known as the mountain breeze or anabatic wind. At night, as the slopes cool, the motion is reversed and the air moves downslope to the valley. This is known as the katabatic wind. Depending on the topographic configuration, such winds may be relatively gentle or may blow in strong gusts (Figure 8.6).

Mountain winds occur across the world. Adiabatic temperature changes induced by air flows over a mountain also result in a type of mountain wind.[5]

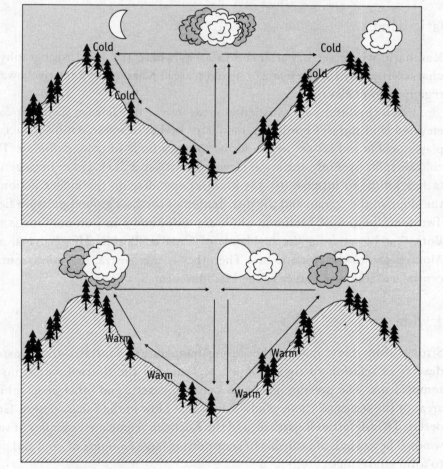

Figure 8.6 Mountain wind system

[5] Adiabatic temperature changes are those that occur without the addition or subtraction of heat; they occur in the atmosphere when bundles of air are moved vertically. When air is lifted, it enters a region of lower pressure and expands. This expression is accompanied by a reduction in temperature (adiabatic cooling). When air subsides, it contracts and experiences adiabatic warming.

As air ascends on the windward side of the mountain, its cooling rate may be moderated by the heat that is released during the process of precipitation. However, having lost much of its moisture, the descending air on the leeward side warms up adiabatically faster than it is cooled on the windward ascent. Thus, the effect of this wind, if it reaches the surface, is to produce warm, dry conditions. Usually, such winds are gentle and result in slow warming, but at times may result in strong currents with a rapid increase in temperature. We have such winds on the slopes of the Himalayas, the 'waav' in Kashmir, the German Föhn in the Alps, Chinook in the North American Rockies, Ghibli in Libya, and Zonda in the Andes of Argentina.

Katabatic wind

Katabatic winds are experienced in areas where the local topography is characterized by the presence of a cold plateau adjacent to a relatively warm region of lower elevation.

Such conditions are satisfied in areas in which major ice sheets or cold, elevated land surfaces border warmer, large bodies of water. Air over the cold plateau cools and results in the formation of a body of cold, dense air. This cold air spills over into the lower elevations with speeds that vary from gentle (a few km/h) to intense (93–185 km/h), depending on the incline action of the slope of the terrain and the distribution of the background pressure field. Two special types of katabatic winds are well-known in Europe. One is the Bora that blows from the highlands of Croatia, Bosnia, Herzegovina, and Montenegro to the Adriatic Sea. The other is the Mistral that blows out of central and southern France to the Mediterranean Sea.

Desert wind

Strong winds blow due to the strong temperature contrasts that exist in deserts in relation to the adjoining regions and the diurnal variations in temperature. The pressure gradients between deserts and other geographical areas or the sea cause these winds. Winds that blow in the Sahara desert, Gobi desert, Thar desert in Rajasthan, and the Kachchh region are examples of such winds. These winds are capable of transporting large amounts of sand and dust to form sand dunes.

Mechanics of wind

A simple approach to studying the mechanics of the wind system is to model it with the following.

- Pressure force
- Coriolis force
- Inertial force
- Frictional force

The force causing the air movement is the difference in air pressure between the two regions.

The pressure force on the air (per unit mass) F_p is given by (Manwell, McGowan, and Rogers 2002)

$$F_p = (-1/\rho)(\partial p/\partial n) \qquad \qquad ...(8.1)$$

where ρ is the density of air and n is the direction normal to the lines of constant pressure. The pressure gradient normal to the lines of constant pressure is given by $\partial p/\partial n$. The above equation can be derived from the boundary layer equations, assuming laminar flow (see general textbooks on heat and mass transfer, for example, Burmeister [1983] or Eckert and Drake [1972]).

As soon as the wind motion is established, a deflective force, the Coriolis force, is produced, which alters the direction of the motion. The Coriolis force (per unit mass) F_c is given by

$$F_c = fU \qquad \qquad ...(8.2)$$

where U is the wind speed and f the Coriolis parameter [$f = 2\omega\sin(\phi)$]. ϕ represents the latitude and ω represents the angular rotation of the earth. The Coriolis force, therefore, is a function of wind speed and latitude.

If no other forces are present, the Coriolis force will exactly balance the pressure gradient force and the wind will flow parallel to the isobars. For straight or slightly curved isobars, the resultant wind is called the geostrophic wind (U_g).

We have $F_p = F_c$
Using Equations 8.1 and 8.2, we have

$$(-1/\rho)(\partial p/\partial n) = fU_g$$

Thus, the geostrophic wind, U_g, a resultant of the pressure gradient force causing the wind and the Coriolis force is given by

$$U_g = (-1/f\rho)(\partial p/\partial n) \qquad \text{...(8.3)}$$

The geostrophic wind is still a simplification of forces at work. Normally, the isobars are curved, which further imposes a centrifugal force. The resultant is the gradient wind U_{gr} (Figure 8.7).

The gradient wind is also parallel to the isobars and is a result of the balance of the forces given by Manwell, McGowan, and Rogers (2002).

$$U_{gr}^2/R = -\,fU_{gr} - (1/\rho)(\partial p/\partial n) \qquad \text{...(8.4)}$$

where R is the radius of curvature of the path of the air particles, and

$$U_{gr} = U_g - U_{gr}^2/fR \qquad \text{...(8.5)}$$

Finally, the gradient wind is subjected to friction near the earth's surface. The forces of friction reduce the wind speeds near the surface but progressively become negligible as one approaches the boundary layer.

Wind power density

The energy content in the wind can be estimated as follows. Imagine a ring of area A (as shown in Figure 8.8) through which the wind flows. From the continuity equation, the mass flow rate $\dot{m}$ is given by

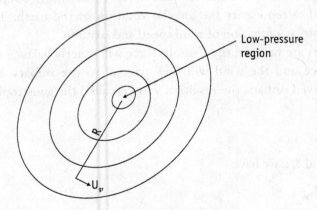

Figure 8.7 Gradient wind U_{gr} for radius of curvature R

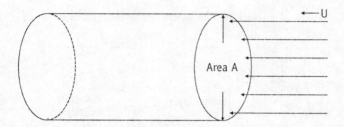

Figure 8.8 Wind flowing through a ring of area A

$$\dot{m} = \rho A U \qquad \qquad \qquad ...(8.6)$$

The kinetic energy per unit time, or the power of the flow is given by

$$P = (½)(\dot{m})U^2 \qquad \qquad ...(8.7)$$

Using Equation 8.6, the power contained in the wind per unit area is

$$P/A = (½)\rho U^3 \qquad \qquad ...(8.8)$$

Atmospheric boundary layer

Wind speeds are known to vary with an increase in height. On the surface of the earth, friction acts as a drag, and close to the surface, the wind speed is nearly zero. Wind speed increases as the height above the ground level increases up to a point, after which there is no further increase in wind speeds.

The transition zone between the surface of the earth and the free atmosphere, where the air motion is practically unaffected by the friction and turbulence induced by the surface irregularities, is called the atmospheric boundary layer. In the lowest 100 m or so of the boundary layer, the effect of surface friction is most pronounced and is of significance for wind power generation. On the other hand, at a height of about 1–3 km above the ground level, the wind is hardly influenced by the surface characteristics and features of the earth. The boundary layer is normally about 1 km thick, but under the conditions of large static stability, it may be as thin as 30 m, and in highly convective conditions, it could be up to 3 km. The boundary layer over the sea surface is thinner than that over the land surface because of the smoother sea surface and lesser friction. The situation would be somewhat complex in the case of high mountain ranges such as the Himalayas.

The boundary layer generally comprises (i) surface layer or the Prandtl layer, and (ii) transition layer, also known as the spiral layer or the Ekman

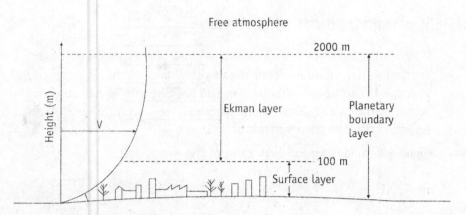

Figure 8.9 Demarcation within the boundary layers

layer. In Figure 8.9, we can see the surface layer; the dominant force is the frictional drag, decreasing with height. It is about 50–100 m thick and is higher over land than over sea. Prandtl (1931) has shown that the variation of wind immediately above the ground is in accordance with the logarithmic law for wind speeds from a given direction.

The transition layer or Ekman layer extends from the top of the Prandtl layer to about 1 km. In this region, wind is influenced by surface friction, the pressure gradient, and the Coriolis force. The variation of wind speed in the boundary layer with elevation is called the vertical profile of the wind speed or vertical wind shear.

All wind turbines are set up within the boundary layer and the modern wind turbines that are rated at nearly 5 MW can have their rotor blade tips going up to 150 m above the ground level. Knowledge of the vertical profile of the wind speed characteristics is immensely important for wind project developers, not only to ascertain the energy capture they can hope to have at the height they set up the wind turbines, but also for design purposes, and for understanding dynamic structural loads on the wind turbine structure.

There are two areas of interest.
1   Instantaneous variation in wind speed as a function of height over time scales of a second.
2   Seasonal or monthly variation in wind speed with height.

Different methodologies apply to these two problems. Assessment of instantaneous wind speed variations is beyond the scope of this chapter.

Height extrapolation of wind speeds

There are several studies on the variation in wind speed with height. Golding (1976) has described in detail the efforts, and surveys undertaken in Europe and the US to assess variations in wind speed with height and altitude. The wind profile in the atmospheric surface layer is commonly described by the equation (Burton, Sharpe, Jenkins, *et al.* 2001).

$$U(z) = (U_*/k)[\ln(z/z_o) - \Psi] \qquad \qquad ...(8.9)$$

It describes the wind speed U at height z by means of the friction velocity U_* and the surface roughness length z_o. k is the von Kármán constant which is taken to be 0.4, and Ψ is a function which depends on stability.

According to Hassan and Sykes in Freris (1990), the variation of the mean wind speed with the height in the surface layer can be adequately represented by the Prandtl logarithmic law model.

The surface roughness length z_o depends on both the size and the spacing of the roughness elements such as grass, crops, and buildings. The typical values of z_o are about 0.01 cm for water or snow surfaces, 1 cm for short grass, 25 cm for tall grass or crops, and 1–4 m for the forest and city. In practice, it is difficult to precisely determine z_o from the appearance of a site. The same is true for U, which is a function of surface friction, air density, and Ψ. As these are not linear equations, the three equations have to be solved for the three unknowns U_*, z_o, and Ψ. Moreover, one needs the surface roughness by the direction sector and for different seasons, which further complicates the problem. This approach is generally not preferred in wind prospecting and engineering. The standard values for the terrain surface roughness are given in Table 8.2 and can be used where we have wind speed measurements at only one height.

These values can be used in the standard formula for the logarithmic profile of wind sheer.

$$U(z)/U(z_r) = \ln (z/z_o)/\ln(z_r/z_o) \qquad \qquad ...(8.10)$$

where U(z) is the wind speed at height z and $U(z_r)$ is the wind speed at reference height z_r. Reference height is the height at which wind speed measurements have been carried out.

The need to find simpler approaches has led many people to look for simpler, empirical expressions that yield satisfactory results. The most common of these simpler expressions is the power law, expressed as

Table 8.2 Terrain surface roughness z_0

Roughness class	Roughness length (m)	Landscape type
0	0.002	Water surface
0.5	0.0024	Completely open terrain with a smooth surface, for example concrete runways in airports, mowed grass, etc.
1	0.03	Open agricultural area without fences and hedgerows and very scattered buildings. Only softly rounded hills
1.5	0.055	Agricultural land with some houses and 8-m tall sheltering hedgerows at a distance of approximately 1250 m
2	0.1	Agricultural land with some houses and 8-m tall sheltering hedgerows at a distance of approximately 500 m
2.5	0.2	Agricultural land with many houses, shrubs, and plants, or 8-m tall sheltering hedgerows at a distance of approximately 250 m
3	0.4	Villages, small towns, and agricultural land with many tall sheltering hedgerows, forests, and very rough and uneven terrain
3.5	0.8	Larger cities with tall buildings
4	1.6	Very large cities with tall buildings and skyscrapers

Source <http://www.windpower.org/en/stat/unitsw.htm#roughness>, last accessed in January 2005

$$U_2/U_1 = (z_2/z_1)^\alpha \qquad \qquad ...(8.11)$$

where U_2 and U_1 are the mean wind speeds at heights z_2 and z_1, respectively, and α is the power law index.

$$\alpha = (\log U_2 - \log U_1)/(\log z_2 - \log z_1) \qquad \qquad ...(8.12)$$

The power law index can be used to extrapolate the mean wind speed, frequency distributions, and even direction-wise distribution of wind speed.

If wind speeds are not known at two heights, terrain surface roughness can be used in various empirical relations to compute α. One of the empirical relations given by Counihan (1975) is

$$\alpha = 0.096 (\log_{10} z_0) + 0.0016(\log_{10} z_0) + 0.24 \qquad \qquad ...(8.13)$$

Across the world, at many wind measurement sites, the average value of α has been determined and is found to be closely approximated by 1/7. This has led to the common reference to Equation 8.11 as the 1/7 power law equation. In view of the large number of possible values that α can assume, the approximation should be used only if site-specific data is not available.

Example 1

The monthly average wind speeds recorded at 10-m height at a site Surajbari, Gujarat, are given in the table below. The site is in an open agricultural field with very scattered low building structures. Estimate the mean monthly wind speeds at 20-m height by the extrapolation techniques described in this chapter and plot them against the actual wind speeds given in the table below. Try to ascertain if there is anything exceptional about this site.

Month	Jan.	Feb.	Mar.	Apr.	May	Jun.	Jul.	Aug.	Sep.	Oct.	Nov.	Dec.
Speed: 10 m (kmph)	8.83	9.92	11.56	16.51	23.36	24.65	23.32	18.68	14.48	9.55	8.59	8.33
Actual speed at 20 m (kmph)	13.10	14.41	15.69	20.46	26.83	28.47	27.94	22.89	18.20	12.83	13.31	13.18

Solution

The corresponding wind speeds at 20-m height are calculated separately using Equations 8.10 and 8.13, and the 1/7 power law assuming $z_0 = 0.03$ for an open agricultural area without fences and hedgerows, and very scattered buildings (Table 8.2). The calculated results are compared against the actual wind speeds at 20-m height in Figure 8.10. It can be seen that all three methods closely follow the actual wind speeds and that there is better approximation in summer months. In case of average annual wind speeds recorded at 10 m and 20 m heights at this site, using Equation 8.12 we find that the power law index is 0.39. This is an exceptionally high figure and one can conclude that perhaps it is an exceptional site.

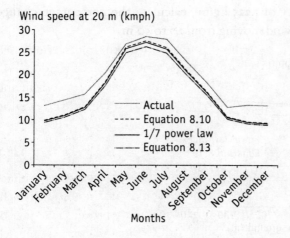

Figure 8.10 Height extrapolation of average monthly wind speed at Surajbari from 10 to 20 m

Example 2

At a site, Kethanur in Tamil Nadu, wind speeds at heights 10 and 20 m have been recorded as 19.90 and 21.91 kmph (Mani 1994, page 349, table 47). The terrain is open agricultural field similar to that in Example 1. Compare the vertical profile of wind speeds till 150 m above the ground level, using all the approaches given in this chapter by plotting them.

Solution

Using Equation 8.12 at this site, we find that the power law index is 0.14. In the wind speed values plotted till 150 m above the ground level, shown in Figure 8.11, it can be seen that the power law index, 1/7 power law, and the logarithmic profile (Equation 8.10) approaches are in agreement, while the empirical approach of Equation 8.13 to calculate z_0 leads to somewhat different results.

Example 3

At a site, Okha Madhi in Gujarat, the mean annual wind speed measured at a height of 10 m above ground level is 16.89 kmph. The site is located in the midst of a salt lake, which stretches around the site for several miles.

1 Use Equation 8.13 to estimate the mean annual wind speed at 20 m above the ground level on this site.
2 If the actual measurements show a mean annual wind speed of 19.05 kmph at 20 m height, calculate the power law index and use it to estimate the mean annual wind speed at 50 m height above the ground level.
3 Assuming an air density of 1.225 kg/m³, calculate the percentage change in energy content of the wind moving from 20 to 50 m.

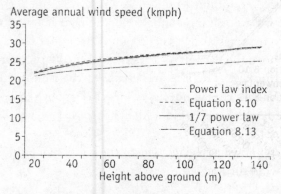

Figure 8.11 Extrapolation of mean annual wind speeds till 150 m above ground at Kethanur, Tamil Nadu

Solution

1 From Table 8.2, we take surface roughness z_0 for a calm open sea as applicable to this site. Therefore, we have $z_0 = 0.20$. Using Equation 8.13, we compute the power law index α at the site as 0.18.

We have $u_1 = 16.89$ kmph, $z_1 = 10$, $z_2 = 20$

From Equation 8.11, we have $z_2/z_1 = 2$
Therefore, $u_2 = 2^{0.18} \cdot u_1$
or $u_2 = 1.1334 * 16.89 = 19.14$ kmph

2 We apply Equation 8.12 to obtain $\alpha = 0.1907$ and apply Equation 8.11 as in 1.
We have $u_1 = 19.05$ kmph, $z_1 = 20$, $z_2 = 50$
or $u_2 = 2.5^{0.1907} * 19.05 = 22.68$ kmph

3 Energy content is given by Equation 8.7. For a unit area
$E = \frac{1}{2}\rho u^3$
At 20–m height, $E = 19.05$ kmph $= 5.29$ m/s. Therefore, $E_{20} = \frac{1}{2} * 1.225 * (5.29)^3$
or $E_{20} = 90.75$ W/m². Similarly, at 50-m height, $E_{50} = 153.34$ W/m²

Therefore, the increase in the energy content of the wind in moving from 20-m height to 50-m height is 68.9%.

Flow over hills and mountains

Hills, small mountains, ridges, and cliffs obstruct the flow of the wind. In the discussion on wind systems, we have seen that the very process of creation of winds is altered by large mountains and mountain ranges such as the Himalayas. However, here we are talking about topographical features on a smaller scale, that is, hilly features above the ground level that are a few hundred metres in height above the datum. As a result of this obstruction, the air is channelled over or around it. The size and shape of the hill obviously affect the flow pattern. When the wind flows over a smooth hill, the streamlines[4] are compressed, which results in a higher intensity of wind near the top of the hill (Figure 8.12).

This phenomenon, also known as the *speed-up* of the wind, makes hilltops a particularly good site for wind turbines.

---

[4] Parallel lines along the direction of fluid flow are known as streamlines that never intersect.

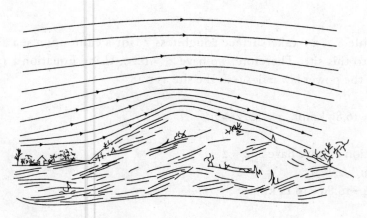

Figure 8.12 Streamlines over a hill

According to Hassan and Sykes in Freris (1990), the turbulence levels over a hill are generally lower than those over the ground level if the hill is not so steep as to cause the flow to separate. Hills with slopes less than 1:4 are considered suitable. Hassan and Sykes report of several analytical studies that have established an understanding of the wind flow over small rounded hills.

Figure 8.13 shows the flow field over a hill of height H and length L, which is defined as the distance from the crest to the point where the height of the ground is half the height of the hill.

The mean wind speed profile upstream of the hill is assumed to be defined by the log-law model. For mean wind speeds over the summit of the hill

$$U(z) = U_o(z) + u(z) \qquad ...(8.14)$$

where $U(z)$ is the mean wind speed at height z above the hill, $U_o(z)$ is the upstream mean wind speed at the same height above the ground, and $u(z)$ is

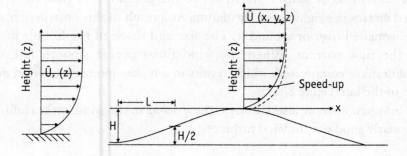

Figure 8.13 Wind flow speed-up over a hill

the speed-up of the mean wind speed. According to Hassan and Sykes, the speed-up has been found to depend linearly on the slope of the hill and can be used to define a non-dimensional speed-up Δu by

$$\Delta u(z) = [U(z) - U_o(z)]/[(H/L)U_o(L)] \qquad \qquad ...(8.15)$$

Hassan and Sykes cite studies undertaken by Jackson (1979), Mason (1979), and Taylor (1979) to present a simplified expression.

$$\Delta u(z) = \frac{1}{[1 + z/L]^2} \qquad \qquad ...(8.16)$$

Assuming that roughness on the hill is the same as for the area downwind, if the wind moves from a rough plain to a smooth hill, there are two factors accelerating the flow.

- Change in the roughness factor
- Speed-up on the slope

The two effects are approximately additive, which would make such a location an ideal site for wind power generation.

In nature, the terrain normally tends to be complex and ideal smooth hills are rather uncommon. Many high-wind areas lie in mountainous regions as the hills and ridges speed up the air flow. While selecting sites for wind turbines in a complex terrain, it is important to take into account the fact that the turbulence around hill sites tends to be greater. Moreover, the vegetation on hill slopes is often rougher and the speed-up due to the hill, to some extent, could get cancelled out by the deceleration due to the roughness change.

Steep hills with slopes greater than 30°, or less if very rough, behave as bluff bodies and cause flow separation at the leading edge or the peak, depending on the actual shape of the hill. The hill imposes an additional drag on the wind flow, which leads to increased turbulence and perhaps flow separation. This causes very low wind speeds near the separation region. As hills are seldom symmetrical, some hills can have good flow characteristics for certain sectors of the wind directions, but very poor characteristics for other sectors.

The air flow at the hill is also influenced by the terrain features upwind. Additional turbulence and shelter effects may be important. There are many numerical models, which try to model the effect of orography, but the complexity of most hilly or mountainous areas makes this difficult. The effect of atmospheric stability can also often lead to air flows that are difficult to predict.

Wind measurements, instrumentation, and data characteristics

Knowledge of the long-term wind resource characteristics at a given location is a pre-requisite to any initiative or effort to harness wind energy. There are various approaches to assessing the wind resource characteristics. As the basic purpose of such assessments is to identify areas that experience sufficiently strong wind speeds that can be considered for harnessing energy or electricity from the wind, it is important to arrive at assessments that will hold for a long period of time or provide a good representation of the occurrence of wind speeds and fluctuations in them. Therefore, one is looking for data that will provide information on the strength and variations in wind speeds over the long term.

In an exercise on wind resource assessment, though ultimately well-calibrated instruments are required to be set up to measure wind speeds, this is done only after going through a process of initial screening of possible locations in the region.

A wind resource assessment exercise, therefore, involves two distinct activities.

1   Screening of probable sites in the region based on general observation of terrain, surface characteristics, vegetation, and local knowledge
2   Measurements over extended periods of time (sometimes two to three years) using instruments

Screening of probable sites

Screening of sites based on geomorphological and biological indicators involves gaining insights into the wind resource at different locations in the region of interest. It helps the site engineer in the following.

- Evaluating alternative sites for relative merits and suitability
- Identifying the actual location where wind speeds should be measured
- Assessing various locations in relation, and with respect, to the site where wind speeds are measured
- Identifying the locations for wind energy projects
- Designing a wind farm layout

Various indicators that can be observed to reach an understanding of the wind flow are described below. Some of these indicators are based on the experience gained in identifying wind resource sites in India (Hossain and Kishore 1989).

Topography

A topographical map of the region should be examined for highlands, plateaus, ridges, hills, and cliffs. Elevated locations are naturally prone to higher wind speeds due to the effect of the boundary layer and speed-ups associated with the slopes. Topographical maps also indicate the complexity of the terrain. In a highly complex terrain, wind flow is also likely to be complex and perhaps turbulent. On the other hand, plain terrain and smooth surfaces give rise to more uniform wind flow. Elevated areas and slopes should be marked on the topographical map.

Local knowledge

Though it would be difficult for a scientist or a technician to have knowledge of wind-prone areas in a region without carrying out measurements, local people generally have this information. They know of the seasons, the locations, and even a particular time-of-day when wind speeds are high. For example, where the region of interest includes water bodies and areas like sea shores and/or large lakes, the fishermen always know exactly when the wind speeds would be maximum. Similarly, in the case of a mountainous terrain, where it is very difficult to identify good wind sites, local knowledge can provide insights that would otherwise not be possible.

Judgement

Wind speeds and their directions can be assessed within ±10% merely by judgement. Golding (1955) presented a compilation devised by Admiral Sir Francis Beaufort of British Navy in 1805. This is now very common and is known as Beaufort scale. The Beaufort scale is shown in Table 8.3 and can be useful in arriving at some level of quantification of information gathered in this manner.

Eolian landforms

On the surface of the earth, there are often forms and features that have been shaped by strong and persistent winds. These are called the *Eolian features* or *Eolian landforms*. Observation and study of these forms provide us with insights into the strength of the winds, their flow patterns, and sometimes the prevalent direction. Figure 8.14 shows the wind-swept smooth and rounded landforms, highlands, and ridges at Dhank in Gujarat, an area within 100 km

Table 8.3 Wind speed Beaufort scale

Wind speed at 10-m height			
m/s	kmph	Beaufort scale	Wind
0.0–0.4	0.0–1.5	0	Calm
0.4–1.8	1.5–6.5	1	Light air
1.8–3.6	6.5–13	2	Light breeze
3.6–5.8	13–21	3	Gentle breeze
5.8–8.5	21–40	4	Moderate breeze
8.5–11	40–51	5	Fresh breeze
11–14	51–61	6	Strong breeze
14–17	61–76	7	Moderate gale
17–21	76–90	8	Fresh gale
21–25	90–104	9	Strong gale
25–29	104–122	10	Whole gale

Figure 8.14 Wind-swept ridges of Dhank, Gujarat

of the Arabian Sea. Wind speeds at this site are the highest in Gujarat. Some of the Eolian features are discussed in the following sections.

Sand dunes

Sand dunes are one of the common Eolian features found in arid regions. Dunes tend to be elongated parallel to the dominant wind flow. The wind picks up the finer materials from the site where wind speeds are higher and deposits them at the site where wind speeds are lower. The size distribution of sand at a given site thus gives an indication of the average wind speed, with coarser sand indicating higher wind speed. The movement of a sand dune over a period of several years is proportional to the average wind speed. This movement can be recorded by a satellite or aerial photographs. The following are some of the different types of dunes (Figure 8.15).

- *Barchan* Crescent-shaped, solitary, tips point downwind, indicate constant wind direction
- *Linear dunes* Indicate the possibility of a ridge
- *Transverse dunes* Wind direction goes over the top of the crest and not parallel along the ridge as seen with linear dunes
- *Parabolic* Crescent-shaped, but ends point upwind
- *Star dunes* Show lack of directional predominance in wind

Playa lake

The wind scours out a depression in the ground that fills with water after rains. When the water evaporates, the wind scours out the sediments at the bottom. These lakes go through a maturing process and their stage of maturity gives a relative measure of the strength of the wind.

Sediment plumes

Other Eolian features include sediment plumes from dry lakes and streams, and wind scour, where air-borne materials gouge out streaks in exposed rock surfaces.

Figure 8.15 Different types of dunes

Vertical cliffs

Cliffs often carry on them marks created by the particles in the wind. The face of vertical cliffs exposed to the winds carries marks of sand particles that have been blown into the cliff.

Wave-like landforms

Often, one comes across wave-like forms or ripples on the surface, particularly in wind-swept valleys and deserts. These too are indicators of strong winds.

It is difficult to get a precise estimate of the average wind speeds through Eolian landforms. Further research can hone this area as a more effective tool for wind prospecting. The distance by which the sand dunes are moved, the size of dust or soil particles, average distances, and wave-like ripples in landforms can be calibrated with average wind speeds in a region. Eolian features, however, in the absence of any other data, dramatically increase the level of confidence in selecting sites or in comparing different sites. An experienced meteorologist or a wind-prospecting engineer can use the Eolian indicators very effectively.

Biological indicators

Winds also leave tell-tale marks on plants. While Eolian indicators may be more useful in areas that lack vegetative cover, in areas that have some trees, shrubs, etc., one can examine the biological indicators that offer more precise information than the Eolian features.

Strong winds deform trees and shrubs so that they indicate an integrated record of the local wind speeds during their lives. Different species of trees show up this effect differently. Therefore, in a site-prospecting exercise, to the extent possible, one should examine only one species of trees. Often one can corroborate the findings by examining both biological indicators and Eolian indicators to arrive at a better understanding of the wind flow in a region.

Tree deformation caused by winds shows up best on coniferous evergreens because their exposure to the wind is relatively constant during the year. On the other hand, species like deciduous trees that shed their leaves seasonally change the exposed area tremendously. However, tree trunks may still carry on them marks of the long-term impact of wind speeds.

According to Katzhan (1989), winds blowing below an average speed of 4 m/s have little effect on the vegetation, and trees cannot generally survive winds averaging speed above 12 m/s. Among the significant biological indicators of unidirectional wind speed are foliar deformation, girth asymmetry, and growth ring asymmetry.

Putnam (1948) lists five types of foliar deformation in trees: *brushing, flagging, throwing, wind clipping,* and *tree carpets.*

Brushing

A tree is said to be *brushed* when the branches are bent to leeward (downwind). It is usually seen only on the deciduous trees and other trees that are not evergreen and occurs due to the light prevailing winds. It is only useful in ascertaining the seasonal wind direction (Figure 8.16).

Flagging

A tree is said to be *flagged* when the wind causes its branches to stretch out leeward, leaving the windward side bare, so that the tree appears like a flagpole carrying a banner in the breeze. This is an easily observable and measurable effect that occurs over a range of wind speeds important to wind power applications (Figure 8.17).

Windthrowing

A tree is said to be *windthrown* when the main trunk as well as the branches are deformed such that they lean away from the prevailing wind. This effect is

Figure 8.16 Brushing

Figure 8.17 Flagging
Source <http://www.marietta.edu/~biol/biomes/images/alpine/flagging_5496.jpg>

produced by the same mechanism that causes flagging, except that the wind in this case is strong enough to modify the growth of the upright leaders of the tree as well as the branches (Figure 8.18).

Clipping

Trees are said to be *wind clipped* when the wind has been sufficiently severe to suppress the leaders and hold the tree tops to a common, abnormally low level. Every twig that rises above that level is promptly killed so that the upper surface is as smooth as a well-kept hedge.

Carpeting

Tree carpets are the extreme case of clipping in which a tree may grow only a few centimetres tall before being clipped. The branches will grow out along the surface of the ground, appearing like a carpet because of the clipping

Figure 8.18 Windthrowing

action. The result may be a tree 10-cm tall but extending 30 m to the leeward of the sheltering rock where the tree sprouted (Figure 8.19).

A rating scale for tree deformation has been proposed (Figure 8.20). In this scale, Class 0 corresponds to no wind damage; Classes I–IV to various degrees of flagging; Class V to flagging plus clipping; and Class VI to throwing. Class VII is a flagged tree with the flagging caused by factors other than a strong prevailing wind, such as salt spray from the ocean or mechanical damage from a short, intense storm. In coastal areas, particularly in India, care has to be taken to account for Class VII type of flagging.

Crown deformation

A common kind of wind deformation of trees is a pronounced asymmetry in the crown of the tree so that it grows more in the direction away from the prevalent direction of the wind. This deformation would represent the prevalent seasonal wind direction rather than the strength of the wind speed, though in high-wind-speed regions, the deformation may be more pronounced, leading to flagging or carpeting. In areas where sufficiently strong winds are experienced from different directions, crown deformation may not be a good indicator. For a given species of trees, an index of deformation can be computed by drawing an imaginary vertical line from the tip of the crown to the ground and computing the ratio of the distances from the vertical line to the outer periphery at the widest plane level of the crown. Again, only one species of trees should be considered for comparison (Figure 8.21).

Figure 8.19 Near carpeting of a tree

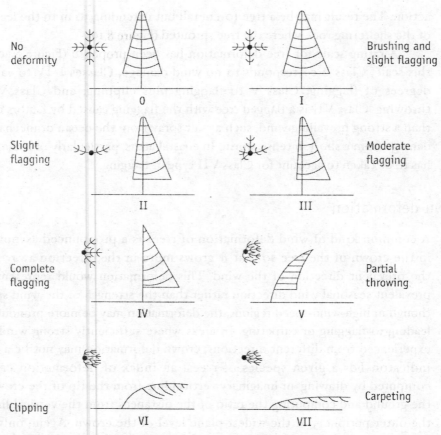

No deformity — 0

Brushing and slight flagging — I

Slight flagging — II

Moderate flagging — III

Complete flagging — IV

Partial throwing — V

Clipping — VI

Carpeting — VII

Figure 8.20 Deformation indices

Ring asymmetry

Long-term effect of the winds can also be seen in the asymmetry of the tree rings. Again, trees at locations that experience wind speeds uniformly from all directions may not show a high level of asymmetry in the tree rings, though the average wind speeds at these sites will be reasonably high (Figure 8.22).

The average wind speed at which these effects occur varies from one species to another. Therefore, for a given species, calibration is necessary.

In a complex terrain, deformations also show the location of high wind speed zones caused by the reflection or deflection of wind from certain terrain features. It is difficult to gain such insights, merely by theoretical or analytical approaches. Tables 8.4 and 8.5 summarize all the indicators of windy sites that are generally useful in wind prospecting.

Figure 8.21 Crown deformation
Source <http://www.dcstechnical.com.au/oz/Earlier%20Photos/Earlier%20Photos.htm>

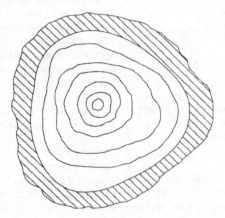

Figure 8.22 Ring assymetry

Instrumentation

A wind-measuring system tracks, records, and displays wind speeds and/or wind direction. The instrument used for recording wind speed is known as an anemometer while the direction is recorded by a wind vane. The wind-measuring system consists of sensors, signal conditioning units, and displays, and/or recorders. The devices can be read out magnetically, optically, electrically, or mechanically with virtually no error.

Table 8.4 Indicators of windy sites

Gentle slopes Gentle slopes of up to 30° in the prevalent wind direction result in the speed-up in approaching winds

Highlands, ridges, and cliffs Generally wind speeds are higher on elevated terrain from 20 m to more than 500 m

Temperature gradients Proximity of the site to the sea, mountains, or deserts indicates that it may be subjected to temperature gradients that result in high winds

Soil Lack of fine particles in the soil is an indicator of the fact that the winds experienced are strong enough to carry away the fine particles (Figure 8.23)

Rounded terrain features These features are an indicator of the fact that the site is wind swept and that over a period of time, the sharp edges on rocks, etc., have worn away

Biological indicators Different types of biological indicators discussed in the text reveal some characteristics of windy sites

Table 8.5 Natural indicators of wind speed and direction

Type of observation	Observation	Quantifiable entity	Output
Topography	Terrain features	Slopes and elevation	Identification of potential sites such as highlands, plateaus, ridges, hill tops, etc. Also identification of shadow areas where the wind speeds are likely to be low
Local knowledge	Information	—	Identification of locations that are known to experience high wind speeds, time-of-the-day of high wind speeds, and seasonal variation in wind speeds
Eolian landforms	Sand dunes	Seasonal movement of sand dunes, size of grains sites	Locations with high wind speeds/relative evaluation of
	Playa lake	—	Greater insights
	Sediment plumes	—	Greater insights
	Vertical cliffs	Predominant direction	Insight into wind flow
Biological indicators	Brushing	Strength and direction	Relative evaluation of sites
	Flagging	Strength and direction	Relative evaluation of sites
	Windthrowing	Strength and direction	Relative evaluation of sites
	Clipping	Strength and direction	Identification of high wind sites/relative evaluation of sites
	Carpeting	Strength and direction	Identification of high wind sites/relative evaluation of sites
	Crown deformation	Deformation ratio	Identification of high wind sites/relative evaluation of sites
	Ring asymmetry	Asymmetry in annual rings	Identification of high wind sites/relative evaluation of sites

Figure 8.23 Lack of fine particles on the surface; Dhank in Gujarat

Wind vane

Wind vanes are known to have been used extensively in the past. A wind vane is a simple device that consists of a flat asymmetrical plate (vane) mounted on a vertical shaft on which it is free to revolve. The wind vane revolves with the movement of the wind and comes to a position that offers least resistance to the wind flow (Figure 8.24). The direction of the wind is observed by noting the position of the pointer against the fixed rods mounted beneath the vane showing the four directions. Indication at a distance can be provided using contact closures, potentiometers, or servo-mechanisms. In the electrical selsyn wind vane, a self-synchronous motor connected to the vane spindle transmits the position of the vane to a receiver selsyn where the direction is indicated on a meter graduated from 0° to 360°. In photoelectric wind vanes, a circular disc programmed in grey code is mounted on the vane shaft. An optocoupler straddling the disc provides digital signals corresponding to the movement of the vane. The direction can be recorded using either purely mechanical recorders, such as the twin-pen pattern used with pressure tube anemometers, or electrical recorders.

Figure 8.24 A photoelectric wind vane

Wind speed measurement

Various types of anemometers have been designed to measure wind speeds. These include the propeller, cup, pressure plate, pressure tube, hot wire, Doppler acoustic radar, and laser anemometers. The propeller and the cup anemometers depend on the rotation of a small turbine for their output while the others have basically no moving parts (Figure 8.25).

The rotating cup anemometer is the most commonly used wind instrument. It has three or four cups mounted on a vertical axis on which the cups can rotate freely. Various methods are used to measure the speed of the rotation of cups. In the cup counter anemometer, a mechanical counter indicates the 'run of wind' and gives the mean wind speed over specified intervals of time. In the cup generator anemometer, the cup wheel drives a small electrical generator whose current output is indicated on a meter, and which is graduated in kmph or m/s. Normally, the generator requires no external power source and is conveniently coupled to a simple DC voltmeter for visual readout or to an analog-to-digital converter for digital use. The major disadvantages are that the brushes required for the generator must be maintained periodically, and there is a susceptibility to noise in a recording system dependent on the voltage level. Modern photoelectric anemometers use a light chopper fixed to the cup wheel to measure the speed of the rotation of the cups.

The propeller-type anemometer has a propeller-like device mounted on a horizontal axis. Again the rotational speed of the propeller has to be translated into the wind speed. The digital anemometer uses a slotted disk, an LED (light-emitting diode), and a phototransistor to obtain a train of constant amplitude pulses with frequency proportional to the anemometer angular

Figure 8.25 A photoelectric anemometer

velocity. Wind speed can be determined either by counting pulses in a fixed time period to get frequency or by measuring the duration of a single pulse. In either case, the noise immunity of the digital system is much better than an analog system.

Disadvantages of the digital anemometer are its relative complexity and the power consumption of the LED in battery-powered applications. One LED may easily draw more current than the remainder of a data acquisition system. This, however, would not be a consideration where a power source is available.

Yet another type of anemometer is the pressure plate or normal plate anemometer. The hot wire anemometer depends on the ability of the air to carry away heat. The resistance of a wire varies with temperature. Thus, as the wind blows across a hot wire, it tends to cool off with a corresponding decrease in resistance.

Sonic anemometers or Doppler acoustic radars, as they are often called, use sound waves reflecting off small blobs or parcels of air to determine the wind speed.

Data presentation

Most wind-measuring systems are equipped with modern data acquisition systems that record instantaneous or average wind speeds over time scales of one second or less, while the analysis of wind resource requires information over time scales of one hour to one year (8760 hours). The data acquisition system should have the capability to record average wind speeds over time scales of a few minutes. It is normal to maintain average hourly wind speed records. However, for design purposes and to study the turbulence intensity, a time series even in seconds may be used. Maximum gusts, even if they are of short duration, also need to be recorded for the purpose of design.

For the purpose of wind resource assessment, annual frequency distribution of average hourly wind speeds is adequate. Normally, this data is recorded in bins of 1 m/s or 1 kmph width in each of the 12 direction sectors. In addition to frequency distribution, the average monthly wind speeds, average monthly power law index, and air density should also be computed.

Wind distribution

Wind speeds at any given location vary widely throughout the year, but nevertheless follow a statistical distribution. It has been shown (Lysen 1982, Manwell, McGowan, and Rogers 2002) that a good representation of wind speed frequency distribution is the Weibull distribution function, charact⸴rised by the shape parameter k and scale parameter c. The cumulative probability distribution function F(U) is given by

$$F(U) = 1 - \exp\left(-\frac{U}{c}\right)^k \qquad \qquad ...(8.17)$$

where U is the wind speed. The probability density function f(U) is obtained by differentiating F(U)

$$f(U) = \frac{dF}{dU} = \left(\frac{k}{c}\right)\left(\frac{U}{c}\right)^{k-1} e^{-(U/c)^k} \qquad ...(8.18)$$

The mean velocity is obtained by integrating (8.18)

$$\bar{U} = \int_0^\infty f(U).dU$$

It can be shown that

$$\bar{U} = c.\Gamma\left(1 + \frac{1}{k}\right) \qquad \qquad ...(8.19)$$

where Γ is the gamma function* defined as

$$\Gamma(n) = \int_0^\infty e^{-x}.x^{n-1}.dx \qquad \qquad ...(8.20)$$

In terms of the gamma function, F(U) and f(U) are given by

$$F(U) = 1 - \exp\left[-\Gamma^k(1+1/k)(U/\bar{U})^k\right] \qquad ...(8.21)$$

$$f(U) = (k/U)(U/\bar{U})^k . \Gamma^k(1+1/k) . \exp\left[-\Gamma^k(1+1/k)(U/\bar{U})^k\right] \qquad ...(8.22)$$

Manwell, McGowan, and Rogers (2002) and Lysen (1982) have presented empirical expressions that can be used as good approximations of gamma function, as given below.

$$\Gamma^k(1+1/k) = 0.568 + \frac{0.434}{k} \qquad \qquad ...(8.23)$$

$$c/\bar{U} = (0.568 + 0.434/k)^{-1/k}$$

$$k = (\sigma_U/\bar{U})^{-1.086}$$

where σ_U is the standard deviation
or

$$\Gamma(n) = (\sqrt{2\pi n})\, n^{n-1}.e^{-n}\left[1 + \frac{1}{12n} + \frac{1}{288n^2} + \frac{139}{51840n^3}\right] \qquad ...(8.24)$$

* <http://mathworld.wolfram.com/GammaFunction.html>, last accessed in February 2007.

when k = 2 the Weibull distribution reduces to a Rayleigh distribution, which is often used for making assessments on wind energy output at a given site in the absence of any knowledge on frequency distribution. For Rayleigh distribution,

$$\sigma_U / \overline{U} = 0.523$$

A typical wind speed distribution is shown in Figure 8.26.

There are different approaches to height extrapolation of parameters c and k, which are based on empirical relations. Poje and Civindini (1988) mention three approaches that are described below for relating shape parameters k_1 and k_2 and scale parameters C_1 and C_2 at heights z_1 and z_2 respectively.

Method A is based on the work of Mikhail and Justus (1981)

$$c_2 = c_1 (z_2/z_1)^{\alpha_A} \qquad \qquad ...(8.25)$$

$$\alpha_A = a + b \ln c_1 \qquad \qquad ...(8.26)$$

$$a = \frac{1}{\ln(\overline{z}/z_o)} \qquad \qquad ...(8.27)$$

$$b = \left(\frac{0.0881}{1 - 0.0881 \ln(z_1/10)} \right) \qquad \qquad ...(8.28)$$

$$k_2 = k_1 \left(\frac{1 - 0.0881 \ln (z_1/10)}{1 - 0.0881 \ln(z_2/10)} \right) \qquad \qquad ...(8.29)$$

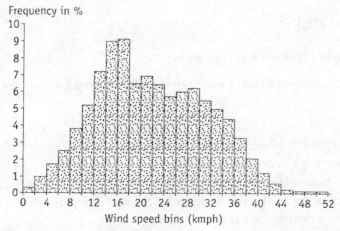

Figure 8.26 Relative frequency distribution at Arasampalayam, Tamil Nadu

Method B is based on the work of Justus (1978)

$$c_2 = c_1 (z_2/z_1)^{\alpha_B} \qquad \qquad ...(8.30)$$

$$\alpha_B = \frac{1}{\ln(\overline{z}/z_o)} - 0.0881 \ln(c_1/b) \qquad \qquad ...(8.31)$$

$$\overline{k}_2 = \overline{k}_1 /[1 - 0.0881 \ln (z_2/1_1)] \qquad \qquad ...(8.32)$$

Method C is based on the work of Spera and Richards (1979)

$$c_2 = c_1 (z_2/z_1)^{\alpha_C} \qquad \qquad ...(8.33)$$

$$\alpha_C = \alpha_o \frac{1 - \log c_1 / \log U_h}{1 - \alpha_o \log(z_1/z_r)/\log U_h} \qquad \qquad ...(8.34)$$

where $\alpha_o = (z_o/z_r)^{0.2}$

U_h is the homogeneous wind speed, and z_r is the reference height (Equation 8.10).

$$k_2 = k_1 \frac{1 - \alpha_o \log(z_1/z_r)/\log U_h}{1 - \alpha_o \log(z_2/z_r)/\log U_h} \qquad \qquad ...(8.35)$$

Of the above methods, Poje and Civindini, based on the work they carried out for a large number of sites, recommend Method B. As in Method B, extrapolation of c requires computation of α_B, which is to be further computed from a relation (Equation 8.31) that requires knowledge of surface roughness z_o, an easier approach is to first calculate k_2 using Equation 8.32 and then to use the relation between k and c, given by Lysen, to estimate c_2.

$$\overline{U} = c * \Gamma\left(1 + 1/k\right) \qquad \qquad ...(8.36)$$

An approximation of Equation 8.36 is given by

$$U/c = -0.0083k^6 + 0.1165k^5 - 0.6374k^4 + 1.6982k^3 - 2.1758k^2 + 1.0032k + 1 \qquad ...(8.37)$$

Example 4

The average wind speed at a height of 20 m at Narsimhakonda is 20.08 kmph.
1 Assuming the Rayleigh distribution, calculate the scale factor c.
2 Given that the power law index at the site is 0.27, develop the per cent frequency distribution of the site at 50-m height and compare it with the per cent frequency distribution at 20-m height.

Solution

1 Value of the shape factor k for Rayleigh distribution = 2
   Using Equation 8.22, we have
   $G = \Gamma^k(1+1/k) = 0.79$
   Now $\Gamma (1+1/k) = G^{1/k}$
   In this case since k = 2, we have
   $\Gamma (1+1/k) = \sqrt{G} = 0.886$
   From Equation 8.17, we have
   $c = \bar{U}/\Gamma (1+1/k)$
   Therefore, c = 20.08/0.886 = 22.7 kmph

2 Power law index or $\alpha = 0.27$
   Assuming k = 2 at low height
   We calculate k_{50} by using Equation 8.32, which is 2.17
   Now using Equation 8.36, we have $\bar{U}/c_{50} = 0.889514$ or $c_{50} = 25.7/0.889514 =$
   28.89
   $c_{50} = 25.7/0.886 = 29$ kmph

From Equation 8.19, we have the probability of occurrence of a certain wind speed given by

$$f(U_i) = k/c(U_i/c)^{k-1} \exp [- (U_i/c)^k] \qquad \ldots(8.38)$$

The per cent frequency corresponding to a given bin of wind speeds can be found from Equation 8.38. We apply this equation to arrive at the frequency distribution. It can be seen in Figure 8.27 that the effect of height extrapolation on wind distribution is to flatten it out, that is, the occurrence of wind speeds is over a wider range of wind speed values.

Estimation of Weibull parameters

The Weibull distribution is used in wind resource assessment to apply the results from the records obtained at one location over a larger area. There are different approaches to estimating the Weibull parameters.
The expression for standard deviation is given by

$$\sigma^2 = \int_0^\infty (U - \bar{U})^2 f(U) dU \qquad \ldots(8.39)$$

It can be shown that
$$\sigma = c \sqrt{[\Gamma (1 + 2/k) - \Gamma^2(1 + 1/k)]}$$

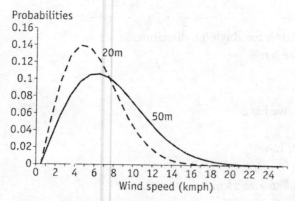

Figure 8.27 Frequency distributions at 20 and 50 m heights at Narsimhakonda

Using Equation 8.19, we get

$$\sigma/\overline{U} = \sqrt{[\,\Gamma\,(1 + 2/k) - \Gamma^2\,(1 + 1/k)]/\,\Gamma\,(1 + 1/k)} \qquad ...(8.40)$$

Lysen (1982) has plotted this function, which is shown in Figure 8.28. Therefore, if standard deviation is calculated as

$$sd = \sqrt{\sigma^2} = \sqrt{\left[\frac{\Sigma U_n^2}{N} - \left(\frac{\Sigma U_n}{N}\right)^2\right]}$$

then the corresponding k value can be found from Figure 8.28. Alternatively, k can also be estimated using the maximum likelihood estimation.[5]

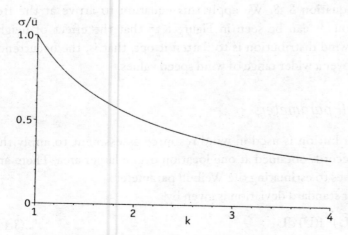

Figure 8.28 Weibull parameter: standard deviation plot

Wind rose

A wind rose is a graphic representation of the information on direction-wise distribution of wind speeds.

In a wind rose 360 degrees of a circle representing all the directions is divided into 12 sectors of 30 degrees each as shown in Figure 8.29. (A wind rose may also be drawn for 8 or 16 sectors, but 12 sectors tend to be the standard set by European Wind Atlas.)

The radius of the boundary in each sector presents the relative frequency of winds from that sector as a part of total wind from all the sectors. The second inner boundary gives the same information, but multiplied by the average wind speed in each particular direction. The result is then normalized to add up to 100 per cent. This tells how each sector contributes to the average wind speed at a particular location.

The innermost boundary gives the same information as the first, but multiplied by the cube of the wind speed in each particular location. The result is then normalized to add up to 100 per cent. This shows how much each sector contributes to the energy content of the wind at a particular location.

Wind roses vary from one location to the next. They actually are a form of meteorological fingerprint.

A look at the wind rose is extremely useful for siting wind turbines. If a large share of the energy in the wind comes from a particular direction, then one should have as few obstacles as possible, and as smooth a terrain as possible in that direction.

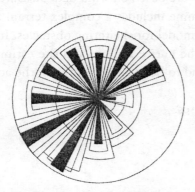

Figure 8.29 A typical wind rose

Spatial wind resource assessment tools: GIS and satellite data

All over the world, the installed wind power capacity is more than 75 000 MW. This capacity has been growing at an average rate of 20% over the past few years. To keep up with the pace, there is a need to identify regions rich in wind resource. Wind measurements at a given site or location have to eventually be extrapolated over a wide area and there is a need for tools, techniques, and approaches that can assess the wind resource in a given region with an acceptable level of accuracy, even in absence of wind speed measurements at those exact locations.

In many countries in the North, particularly Europe, more and more offshore wind projects are coming up. In the offshore areas, traditionally, wind monitoring activities have remained limited and generally, it is impractical to set up monitoring masts in these areas. Here, again, there is a need for wind resource assessment at a spatial level rather than at a point level.

Many tools have emerged over the past few years to carry out spatial assessment of wind resource. These are discussed in the following sections.

Wind atlas analysis and application programme

There has been a need to develop models that would enable the project developers and planners to extrapolate the wind speed from a monitoring mast records to a region in general, or given the terrain and surface characteristics, to arrive at a good understanding of the wind resource in a region. Perhaps, the earliest model to have been developed and applied on a large scale is WASP (Wind Atlas Analysis and Application Programme).[6]

Developed by the Riso National Laboratories, Denmark, WASP is a PC program for predicting wind climates and power productions from wind turbines and wind farms. The predictions are based on wind data measured at stations in the same region. The programme includes a complex terrain flow model, a roughness change model, and a model for sheltering obstacles. It has been used to develop a wind atlas for the entire Europe and the estimates arrived at through WASP have been found to be in agreement with the actual measurements.

WASP is used for the following purposes.
- Wind farm production
- Wind farm efficiency
- Micro-siting of wind turbines

---

[6] Details available at, <http://www.wasp.dk>, last accessed in Janaury 2005.

- Power production calculations
- Wind resource mapping
- Wind climate estimation
- Wind atlas generation
- Wind data analysis

The central point in the wind transformation model of WASP – the so-called wind atlas methodology – is the concept of a *regional* or *generalized wind climate,* or *wind atlas.* The RWC (regional wind climate) is the hypothetical wind climate for an ideal, completely flat terrain with uniform surface roughness, assuming the same overall atmospheric conditions as those of the measuring position.

To deduce the wind climate at a location of interest from the RWC and to introduce the effect of different terrain features such as terrain height variations, surface roughness, and sheltering obstacles, a flow model shown in Figure 8.30 is used. To deduce the regional wind climate from the measured wind in the actual terrain, the same flow model is used, but in a reverse manner, to remove the local terrain effects.

Satelite data and GIS-based techniques

The GIS (Geographical Information System) techniques are also being developed and used for spatial analysis of wind speeds. The technology itself has evolved tremendously over the past few years and numerous GIS platforms are available commercially. The GIS enables superimposition of a vast number of layers, which in case of wind could be contours, surface roughness, vegetation, forest area, crown cover, habitation, solar radiation, cloud cover, etc.

In context of India, for example, superimposition of layers of roads and electricity transmission networks along with wind resource maps can help identify areas where wind resources can be harnessed most economically. GIS is being increasingly used in India not only to identify suitable sites but also to plan the entire wind farm with access roads, positioning of each wind turbine and layout of power evacuation system.

SAT–WIND, a research project currently under way at the Riso National Laboratory, Denmark, is developing techniques of offshore wind resource assessment using offshore wind observations from satellites with the help of various technologies such as passive microwave, altimeter, scatterometer, and imaging SAR (synthetic aperture radar). Data from such technologies has

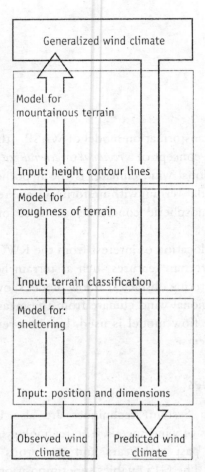

Figure 8.30 Flow diagram of a Wind Atlas Analysis and Application Program model

been available for more than a decade. Figure 8.31 shows the radar image of Horns Rev Windfarm in Denmark.

Though this project is specifically for offshore assessments, the technologies and tools can also be applied to large onshore projects. Such techniques, which are still under development, can be used for selection of suitable areas or regions for large-scale wind projects or for implementing large-scale wind energy programmes. Though the level of accuracy from such techniques would not be comparable with the exact measurements at a location, it is felt that where large regions are under investigation, a lower absolute accuracy on the wind estimate may be acceptable. In some cases, SAT-WIND or GIS type of analysis can be carried out prior to setting up of costly meteorological masts.

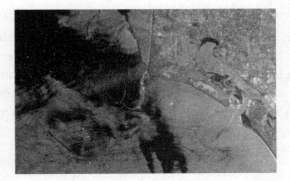

Figure 8.31 European Remote Sensing Satellite-2 synthetic aperture radar image of Horns Rev wind farm in Denmark

Exercises

Question 1

A parcel of air from the Sahara desert under pressure is moving towards the north. In which direction will its path get modified due to the Coriolis force?

Question 2

At Jaisalmer, the wind speed at 10-m height is 14.92 kmph. Using three different approaches, calculate the mean annual wind speeds at 20-m height.

- If actual measurements at 20-m height indicate a mean annual wind speed of 17.37 kmph, calculate the power law index.
- Which of the three methods used above are in close agreement with the power law index approach?

Question 3

A wind project developer is evaluating different sites to arrive at a site with maximum mean annual wind speed at 50 m above the ground level. From the sites given in the following table, which of the sites will be selected by him?

| Site | Mean annual wind speed (kmph) | |
	10 m	20 m
Alagiyapandiyapuram	18.13	20.88
Ayikudi	18.16	21.35
Talayuthu	18.84	20.51

Question 4

At a site Lamba in Gujarat, the mean annual wind speed at 20-m height is 20 kmph and the power law index is 0.13. A wind prospecting engineer is evaluating a site Lamba-2 in a very similar terrain about 10 km from Lamba. At this site, no measurements have been carried out so far. The engineer estimates the surface roughness to be 0.025. Estimate the mean annual wind speed at 30-m height above the ground level at Lamba-2.

Question 5

Average wind speed at 20-m height at a site MPR dam is 19.85 kmph.

- Assuming Rayleigh distribution, calculate scale factor c.
- Given that the power law index at the site is 0.08, develop the per cent frequency distribution of the site at 50-m height and compare it with the per cent frequency distribution at 20-m height.
- If the actual shape parameter at 20-m height is 1.9 and the scale parameter is 21.7, plot the frequency distribution at 50-m height.

Nomenclature

A	Area (m²)
c	Scale parameter
f	Coriolis parameter
F_p	Pressure force on the air (per unit mass)
F_c	Coriolis forcer (per unit mass)
H	Height of hill
k	Shape parameter
L	Monin-Obukhov length
m	Mass flow rate (kg/s)
n	Direction normal to the lines of constant pressure
P	Power (watt)
R	Radius of curvature of the path of air particles
U	Wind speed
U_*	Friction velocity
U_g	Geostrophic wind
U_{gr}	Gradient wind
U_h	Homogenous wind speed
$U_o(z)$	Upstream mean wind speed at the same height above the ground
Δu	Non-dimensional speed-up
z	Height

z_o	Surface roughness length
z_r	Reference height
α	Power law index
Ψ	Constant
Γ	Gamma function
ρ	Density
$\partial p/\partial n$	Pressure gradient normal to the lines of constant pressure
ω	Angular rotation of the earth

References

Burmeister L C. 1983
Convective Heat Transfer
New York: Wiley Interscience. 790 pp.

Burton T, Sharpe D, Jenkins N, Bossanyi E. 2001
Wind Energy Handbook
London: John Wiley & Sons Ltd

C-WET. 2001
Wind Energy Resource Survey in India–VI
C-WET. 696 pp.

Eckert E R G and Drake R M. 1972
Analysis of Heat and Mass Transfer
Tokyo: McGraw-Hill Kogakusha Ltd

Freris L L. 1990
Wind Energy Conversion Systems
New Jersey: Prentice Hall International. 388 pp.

Golding E W. 1976
The Generation of Electricity by Wind Power
Trowbridge: Redwood Burn Ltd

Hossain J, Thukral K, Kishore V V N, Ramesh S, Pachauri R K. 1988
Wind Electric Generation: some economic and policy implications
Urja, **xxiii**: 303–305

Hossain J and Kishore V V N. 1989
Siting for Stand Alone Wind Electric Generators in Complex Terrain of Ladakh
In *Renewable Energy for Rural Development*
New Delhi: Tata McGraw-Hill

Justus C G. 1978
Wind and Wind System Performance
Philadelphia, PA: Franklin Institute Press
[Cited by Poje and Cividini]

Katzhan M T. 1989
Assessing regional wind energy resources with biological indicators: a decision-analytic approach
Solar Energy 42(1): 15–25

Lysen E H. 1982
Introduction to Wind Energy
The Netherlands: Publication SWD 82-1 Steering Committee Wind Energy Developing Countries. 310 pp.

Mahmmud F, Watson S, Woods J, Halliday J, Hossain J. 1996
The economic potential for renewable energy sources in Karnataka, India
[Proceedings of the EuroSun'96 Conference, Freiburg, Germany, September 1996]

Mahmmud F, Woods J, Watson S, Halliday J, Hossain J. 1996
A GIS tool for the economic assessment of renewable technologies
[Proceedings of the 18th British Wind Energy Association Conference, Exeter, 1996]

Mani A. 1990
Wind Energy Resource Survey for India–I
New Delhi: Allied Publishers Ltd. 347 pp.

Mani A. 1992
Wind Energy Resource Survey for India–II
New Delhi: Allied Publishers Ltd

Mani A. 1994
Wind Energy Resource Survey for India–III
New Delhi: Allied Publishers Ltd

Mani A and Mooley D A. 1983
Wind Energy Data for India
New Delhi: Allied Publishers Ltd

Manwell J F, McGowan J G, and Rogers A L. 2002
Wind Energy Explained: theory, design, and application, 577 pp.
London: John Wiley & Sons Ltd.

Mikhail A S and Justus C G. 1981
Comparison of height extrapolation models and sensitivity analysis
Wind Engineering 5(2) [cited by Poje and Cividini]

Poje D and Civindini B. 1988
Assessment of wind energy potential in Croatia
Solar Energy 41(6): 543–554

Spera D A and Richards T R. 1979
Modified power low equations for vertical wind profiles
[Conference and Workshop on Wind Energy Characteristics and Wind Energy Siting, 19–21 June 1979]

Introduction to wind turbine technology

9

Jami Hossain, Adviser (Technical)
Indian Wind Energy Association, New Delhi

History and evolution of wind turbines

Wind has been used for its force and energy ever since the ancient times, the use of sails in shipping being one such example. Interestingly, the windmill itself happens to be one of the earliest inventions of mankind and its use can be traced almost 4000 years back. Hammurabi, the Babylonian emperor in the 17th century BC, is believed to have used windmills in an irrigation scheme.

In Persia, windmills were used for grinding grain as early as 200 BC (Cheremisinoff 1978). According to Gipe (1995), windmills were found in Persia till AD 900. The earliest findings of windmill structures date back to the seventh century AD in Persia and the adjoining areas, including Iraq and the Arab region (Figure 9.1). The first known documented design is also that of a Persian windmill—one with vertical sails made of bundles of reeds or wood, which were attached to the central vertical shaft by horizontal struts (Figure 9.2). These were the drag-type vertical axis windmills (described later in the chapter).

The Persian windmills were housed in a building that had funnel-shaped openings in the walls. These openings directed the wind onto the cloth 'sails' that turned the shaft. Grinding stones, one attached to the shaft and another kept stationary, were used for grinding grains.

Figure 9.1 Ruins of a Persian windmill

Figure 9.2 Sketch of a typical Persian windmill
Source <http://www.catpress.com/bplanet9/eolica.htm>
last accessed on 20 August 2008

It is believed that the concept of windmills reached Europe either through Morocco or through the crusaders. Windmills first appeared in England in 1137 AD. They had horizontal shafts, and were used by the Dutch to pump water off the sea-reclaimed land.

The windmill type, that we know as traditional European, emerged in the medieval times and was fundamentally different in design from the vertical axis windmill of Persia. These mills usually had a four-blade horizontal axis rotor. More than 100 000 windmills of this type are believed to have been functional at a given point of time in medieval Europe. These windmills were mainly used for grinding grain, sawing wood, and sometimes for pumping water (Figure 9.3).

Windmills for pumping water are also known to have been widely used in the United States in the 18th and 19th centuries. These were multi-bladed fan-like horizontal axis windmills with a rather advanced regulatory mechanism compared to the traditional windmills of Europe (Figure 9.4).

In 1854, Daniel Halladay, a New England machinist, obtained the first American windmill patent (Pat No. 11 629). His windmill, designed to pump water, had four wooden blades that pivoted and self-adjusted according to wind speed. It had a tail, which caused it to turn in the wind (Figure 9.5).

Modern wind energy technology is a story of the challenges faced in overcoming a wide range of engineering problems, which include breakthroughs in aerodynamics, aerofoil design, structural engineering, materials, and power electronics. A wind turbine is a highly sophisticated machinery capable of competing with well-entrenched conventional technologies, such as gas turbines, steam turbines, hydro turbines, and modern combustion systems deployed for power generation. Apart from advances in design, engineering, and manufacture of modern wind turbines, there have also been advances in the understanding of wind as a resource, its fluctuation, and its behaviour in different kinds of terrains.

In wind energy technology, as in other electricity-generation technologies utilizing steam, hydro, or gas, there are transmission and

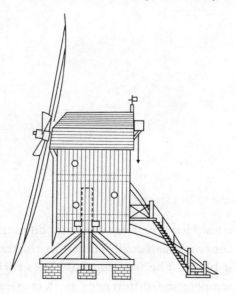

Figure 9.3 Typical traditional horizontal axis windmill

Figure 9.4 American windmill for pumping water

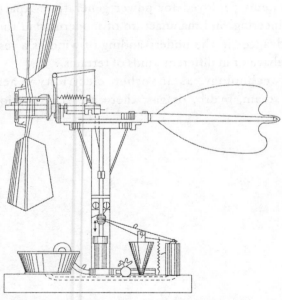

Figure 9.5 Daniel's windmill (the first one to be patented in the US)

generation systems involving turbines. However, a fundamental difference is that the working fluid – wind – is not controlled at all. Therefore, the turbine must face the ambient wind as it blows. The working fluid goes through almost negligible pressure and temperature differences as it crosses the turbine, unlike steam, gas, or hydro turbines where the working fluid is either controlled by or is subject to significantly higher temperature and pressure gradients. Also, because of low air density of ambient air, the sweep area of the turbine per unit of energy conversion is much higher. Therefore, even the modern wind turbine happens to be an unusually large structure.

In the past, different approaches to designing wind turbines have been attempted. The rotor has always been the most crucial part of a wind turbine. Different types of wind turbines evolved with various designs and constructions of the rotor (Manwell, McGowan, and Rogers 2002). Two types of wind turbine designs have been developed depending on the orientation of the axis of the rotor: horizontal axis wind turbines and vertical axis wind turbines (Figures 9.6 [a] and 9.6 [b]).

It can be seen from the figures that apart from the main differences in the axis of rotation, there are many other differences in design.

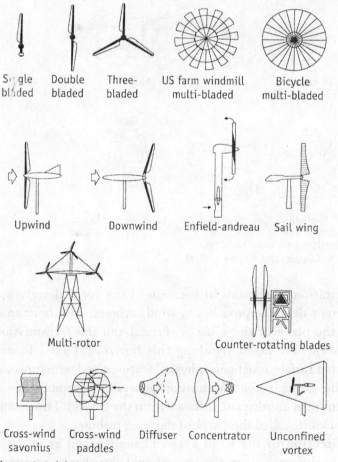

Single bladed   Double bladed   Three-bladed   US farm windmill multi-bladed   Bicycle multi-bladed

Upwind   Downwind   Enfield-andreau   Sail wing

Multi-rotor   Counter-rotating blades

Cross-wind savonius   Cross-wind paddles   Diffuser   Concentrator   Unconfined vortex

Figure 9.6 (a) Horizontal axis wind turbines
Source Manwell, McGowan, and Rogers (2002)

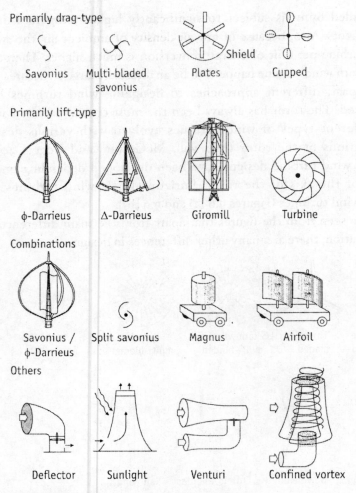

Figure 9.6 (b) Vertical axis wind turbines
Source Manwell, McGowan, and Rogers (2002)

The orientation of the axis of rotation of the rotor determines two distinctly different design approaches in wind turbines. In a horizontal-axis wind turbine, the plane of the rotor is vertical and the transmission and generation systems are normally along this horizontal axis. To prevent interference from surface roughness, physical features, and structures close to the ground, the entire rotor–transmission–generator assembly has to be mounted on a tower at an adequate distance from the ground. This assembly is housed in an enclosure called the *nacelle* of the wind turbine.

As the direction of the wind keeps changing, and as it has to be perpendicular to the rotor plane, the rotor should also change its orientation every time there is a change in wind direction. Also, as the entire transmission and generator assembly has to be on the axis of the rotor, *yaw mechanism* is applied to rotate the entire nacelle on a vertical axis. All modern wind turbines have wind vanes on top of the nacelle or at a certain location on the tower to sense the change in wind direction. The control system of the wind turbine communicates with the *yaw motor* to operate the yaw mechanism. Although this appears to be a rather cumbersome approach to designing wind turbines, the horizontal-axis wind turbine has emerged as the most predominant commercial wind turbine type all over the world. Figure 9.7 shows a typical three-bladed commercial horizontal axis wind turbine along with the main components.

The second type of wind turbine is the vertical axis one that despite having many aesthetic and structural advantages over the horizontal axis type of wind turbine, has not emerged as an acceptable commercial model. In a vertical axis wind turbine, the axis of rotation of the rotor is obviously vertical, which allows the transmission and generation components to be

Figure 9.7 A 2-MW horizontal axis wind turbine
(The size of this wind turbine can be judged from the people standing at its bottom.)

Figure 9.8 A typical vertical axis wind turbine
Source <http://www.ecopowerusa.com/png/F19B.jpg>, last accessed on 1 April 2007

placed below the rotor. Depending on the type of rotor design, it looks like an egg beater, or drums stacked one over the other or vertical aerofoil, as in case of the Darrieus rotor (Figure 9.8).

Rotor aerodynamics

The horizontal-axis wind turbine, normally with three-bladed rotors, has emerged as a universally acceptable commercial wind turbine design type. This chapter will focus mainly on this type of wind turbine.

Axial momentum theory

As the rotor rotates, kinetic energy of the wind is converted into utilizable rotational energy at the shaft. A simple model based on the linear momentum theory developed more than 100 years back to predict the performance of ship propellers can be used to determine the power from an ideal turbine rotor and the effect of the rotor operation on the local wind field. According to Manwell, McGowan, and Rogers (2002), the application of the linear momentum theory to wind turbine rotors is attributed to Betz (1926).

The analysis assumes a control volume as shown in Figure 9.9. The turbine, represented by a uniform 'actuator disk', creates a discontinuity of pressure in the stream tube of air flowing through it. The only flow is across the ends of the stream tube.

Following are the assumptions with regard to air flowing through the stream tube.

- The static pressure far upstream and far downstream of the rotor is equal to the undisturbed ambient static pressure.
- It is a steady-state time-invariant fluid flow.
- The fluid is homogenous and incompressible.
- There is no frictional drag.
- The thrust over the disk is uniform.

From the conservation of linear momentum, we have thrust T on the rotor equal and opposite to the change in the momentum of the air stream.

$$T = U_1(\rho A U_1) - U_4(\rho A U_4) \qquad \qquad ...(9.1)$$

where ρ is the air density, A is the cross-sectional area of the actuator disk, and U is the air velocity as shown in Figure 9.9(a).

Under steady-state flow

$$(\rho A U_1) = (\rho A U_4) = \dot{m}$$

where m is the mass flow rate. Therefore,

$$T = \dot{m}(U_1 - U_4) \qquad \qquad ...(9.2)$$

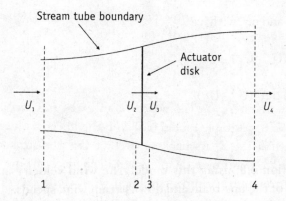

Figure 9.9a Stream tube

As the velocity U_4 behind the rotor is less than the free stream velocity U_1, the thrust is positive.

As no work is done on either side of the actuator disk, the Bernouli function can be applied to the control volumes on either side. In the stream tube upstream of the disk

$$p_1 + \tfrac{1}{2}\,\rho\,U_1^2 = p_2 + \tfrac{1}{2}\,\rho\,U_2^2 \qquad \text{...(9.3)}$$

In the stream tube downstream of the disk, we have

$$p_3 + \tfrac{1}{2}\,\rho\,U_3^2 = p_4 + \tfrac{1}{2}\,\rho\,U_4^2 \qquad \text{...(9.4)}$$

We have assumed that the far upstream and the far downstream pressures are equal, that is, $p_1 = p_4$. The condition of the steady-state fluid flow requires that velocity across the actuator disk remains the same, that is, $U_2 = U_3$.

Now, thrust T can also be expressed as the force resulting from the pressure difference over the actuator disk, that is,

$$T = A_2(p_2 - p_3) \qquad \text{...(9.5)}$$

From Equations 9.3 and 9.4, we have

$$p_2 - p_3 = \tfrac{1}{2}\,\rho\,(U_1^2 - U_4^2) \qquad \text{...(9.6)}$$

Therefore, the thrust is

$$T = \tfrac{1}{2}\,\rho\,A_2\,(U_1^2 - U_4^2) \qquad \text{...(9.7)}$$

From Equations 9.2 and 9.7, we have

$$\dot{m}(U_1 - U_4) = \tfrac{1}{2}\,\rho\,A_2\,(U_1^2 - U_4^2)$$
or
$$\rho\,A_2\,U_2\,(U_1 - U_4) = \tfrac{1}{2}\,\rho\,A_2\,(U_1^2 - U_4^2)$$
or
$$U_2 = \tfrac{1}{2}\,(U_1 + U_4) \qquad \text{...(9.8)}$$

According to Equation 9.8, using this model, the wind velocity at the rotor plane is the average of the upstream and downstream wind speeds.

Here, we introduce the axial induction factor a, as the fractional decrease in wind velocity between the free stream and the rotor plane, then

$$a = (U_1 - U_2)/U_1 \qquad \qquad ...(9.9)$$

Now

$$U_2 = U_1(1 - a) \qquad \qquad ...(9.10)$$

and

$$U_4 = U_1(1 - 2a) \qquad \qquad ...(9.11)$$

In simple terms, the component of energy in the wind absorbed by the rotor disk over unit time is equal to the change in kinetic energy flowing through the rotor area over the same unit time. In other words, the power in the wind converted into the power in the rotor is given by

$$P = \tfrac{1}{2}\,\rho\,A_2\,U_2\,(U_1^2 - U_4^2) \qquad \qquad ...(9.12)$$

Substituting Equations 9.10 and 9.11 in Equation 9.12, we have

$$P = \tfrac{1}{2}\,\rho\,A_2\,U_1^3\,4a(1 - a)^2 \qquad \qquad ...(9.13)$$

Now, the control volume area A_2 at the rotor is same as the rotor area A, and if we replace the free-stream wind velocity u_1 by wind velocity u, we have

$$P = \tfrac{1}{2}\,\rho\,AU^3\,4a(1 - a)^2 \qquad \qquad ...(9.14)$$

For maximum power output, we set dP/da = 0, which yields

$$a = 1/3 \qquad \qquad ...(9.15)$$

Substituting the value of a in Equation 9.14, we have

$$P = 16/27\,(\tfrac{1}{2}\,\rho\,AU^3) \qquad \qquad ...(9.16)$$

The factor 16/27 or 0.592 is also called the Betz coefficient (Betz 1926). It represents the maximum fraction of power in air flow, which an ideal rotor, under the assumed conditions, can theoretically extract from the flow.

In the above analysis, we have assumed that no rotation was imparted to the flow. Now, this analysis can be extended to the case where the rotating rotor generates angular momentum. In case of a wind turbine rotor, the flow behind the rotor rotates in the opposite direction, which represents an extra loss of kinetic energy.

To incorporate this aspect in our analysis, we consider an annular stream tube with a ring radius r and a thickness dr as shown in Figure 9.9(b). The

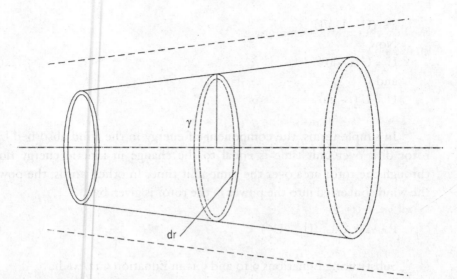

Figure 9.9b Annular tube of radius r

cross-sectional area of the annular tube becomes 2πrdr. Now, as the rotor turns, a pressure difference occurs over the blades. The relative angular velocity increases from Ω to Ω + ω, while the axial components remain unchanged.

Applying Bernoulli's equation, we have

$$p^+ - p^- = \tfrac{1}{2}\rho\,(\Omega + \omega)^2 r^2 - \tfrac{1}{2}\rho\,\Omega^2 r^2 \qquad \ldots(9.17)$$
or
$$p^+ - p^- = \rho\,(\Omega + \tfrac{1}{2}\omega)\,\omega\, r^2 \qquad \ldots(9.18)$$

Therefore, the resulting thrust on the annular element of the rotor is

$$dT = \rho\,(\Omega + \tfrac{1}{2}\omega)\,\omega\, r^2 2\pi\, r dr \qquad \ldots(9.19)$$

We introduce a tangential induction factor a′, where

$$a' = (\tfrac{1}{2}\omega)/\Omega \qquad \ldots(9.20)$$

The expression of thrust changes to

$$dT = 4a'(1 + a')\,\tfrac{1}{2}\,\rho\,\Omega^2 r^2 2\pi\, r dr \qquad \ldots(9.21)$$

Now, from Equation 9.7, we have

$$T = \tfrac{1}{2}\,\rho\, A_2\,(U_1^2 - U_4^2)$$

For the annular element as $A_2 \rightarrow 2\pi r dr$, and $U_1 \rightarrow U$, we have from Equation 9.11

$$U_4 = U(1 - 2a) \qquad \qquad ...(9.22)$$

Therefore,

$$dT = 4a(1 - a) \tfrac{1}{2} \rho U^2 2\pi \, r dr \qquad \qquad ...(9.23)$$

Similarly, one can derive the equation for torque as

$$dQ = 4a'(1 - a) \tfrac{1}{2} \rho U \, \Omega r^2 2\pi \, r dr \qquad \qquad ...(9.24)$$

Performance coefficients

The power P generated by the rotor is a function of the torque Q and the rotational speed of the rotor Ω. The three are related by the following equation

$$P = Q\Omega \qquad \qquad ...(9.25)$$

As the wind speed varies, the torque and/or rotational speed also changes, therefore, for each value of the wind speed U, density ρ, and radius R, separate curves would have to be drawn. This is a rather tedious approach to studying rotor characteristics.

It is more convenient to handle the dimensionless entities given below.

Power coefficient $C_p = P/(\tfrac{1}{2} \rho A U^3)$ = (Rotor power)/(Power in wind)

$$...(9.26)$$

Torque coefficient $C_Q = Q/(\tfrac{1}{2} \rho A U^2 R) \qquad \qquad ...(9.27)$

Tip speed ratio $\lambda = \Omega R/U \qquad \qquad ...(9.28)$

where A is the rotor area corresponding to radius R.

From our earlier discussion on the Betz coefficient, we know that 0.592 is theoretically the maximum value that C_p can have for an ideal rotor under assumed conditions.

Replacing the values of P and Q in Equation 9.25 from Equations 9.26 and 9.27

$$UC_p = \Omega R C_Q$$

From Equation 9.28, we have

$$C_p = C_Q \lambda \qquad \qquad \ldots(9.29)$$

With the help of Equations 9.26 and 9.27, the characteristics of rotors with different dimensions and at different wind speeds can be studied by means of $C_p - \lambda$ and $C_Q - \lambda$ curves. Typical plots of C_p and C_Q are shown in Figures 9.10 and 9.11.

As mentioned earlier, the Betz limit arrived at in this manner is based on an idealized rotor. Other factors that result in further reduction in practically achieving maximum C_p are given below.

- Rotation of the wake behind the rotor
- Finite number of blades and associated tip losses
- Non-zero aerodynamic drag

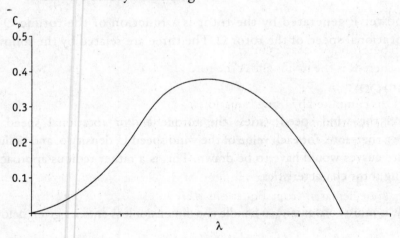

Figure 9.10 $C_p - \lambda$ curve

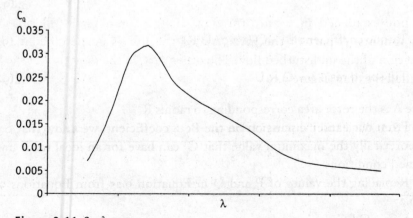

Figure 9.11 $C_Q - \lambda$ curve

However, new blade designs have been developed that have resulted in a C_p as high as 0.56 due to improvements carried out in surface, blade root section, and aerofoil design. Some technologists have also questioned the Betz limit.

Example 1

A fixed-speed, pitch-regulated, grid-connected wind turbine rated at 1000 kW has a rotor radius of 30 m. The cut-in wind speed[1] is 3 m/s and the rated wind speed is 13 m/s. The speed of the rotor is 45 rpm. The power output from the wind turbine for wind speeds between 3 and 13 m/s is given in Table 9.1 for air density of 1.121 kg/m³. Calculate, and plot C_p and C_Q for every interval in wind speeds. Also plot $C_p - \lambda$ and $C_Q - \lambda$ curves.

Solution

R = 30 m
Therefore, A = 2827.43 m²
where R is the radius and A the rotor area.
rpm = 45
C_p is computed by using Equation 9.26.
The rotational speed, Ω is computed as
Ω = 45 2π/60 rad/s = 4.71 rad/s

The tip speed ratio is computed by Equation 9.28, and the C_p and C_Q values are calculated from Equations 9.26 and 9.29, respectively. The results are presented in Table 9.2 and the curves are plotted in Figures 9.12–9.15.

Airfoils

All bodies placed in a uniform flow experience a force that has two components. A part of the force, called the lift, is perpendicular to the direction of the undisturbed flow. The other part, in the direction of the flow, is called the drag (Figure 9.16).

Table 9.1 Wind turbine power table

Wind speed (m/s)	3	4	5	6	7	8	9	10	11	12	13	14	15
Power (kW)		10	26.1	78	145	230.5	349.3	530.1	711	837	950	1000	1000	1000

---

[1] Wind speed at which the wind turbine starts generating electricity.

Table 9.2 Calculated values of rotor parameters for different wind speeds

U(m/s)	P(kW)	λ	C_p	C_Q
3	10	47.1	0.23	0.005
4	26.1	35.3	0.26	0.007
5	78	28.3	0.39	0.014
6	145	23.6	0.42	0.018
7	230	20.2	0.42	0.021
8	349	17.7	0.43	0.024
9	530	15.7	0.46	0.029
10	711	14.1	0.45	0.032
11	837	12.8	0.40	0.031
12	950	11.8	0.35	0.029
13	1000	10.9	0.29	0.026
14	1000	10.1	0.23	0.023
15	1000	9.4	0.19	0.02

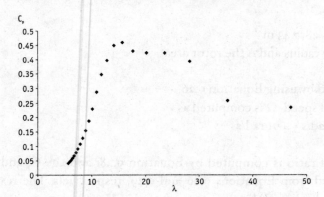

Figure 9.12 C_p–λ curve

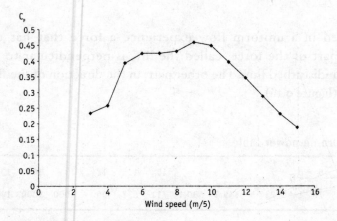

Figure 9.13 C_p–wind speed curve

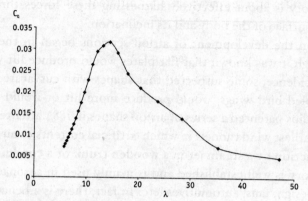

Figure 9.14 C_Q–λ curve

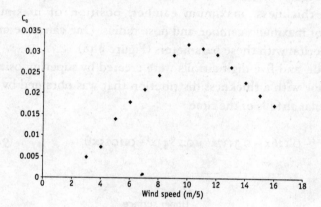

Figure 9.15 C_Q–wind speed curve

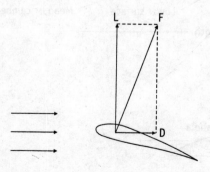

Figure 9.16 Lift and drag working on a body

The airfoil theory is about effectively harnessing these forces through the geometry of the surface of the body and its inclination.

Serious work on the development of airfoil sections began in the late 19th century. Although it was known that flat plates would produce lift when set at an angle of incidence, some suspected that shapes with curvature, that more closely resembled bird wings, would produce more lift or would do it more efficiently. Phillips patented a series of airfoil shapes in 1884 after testing them in one of the earliest wind tunnels in which 'artificial currents of air were produced from induction by a steam jet in a wooden trunk or a conduit'.

The airfoil theory is well established and is mainly used in aeronautical designs, helicopter design, fans, automotives, etc. In fact, there is a helicopter theory that deals with the design of rotor blades for helicopters.

Airfoil geometry can be characterized by the co-ordinates of the upper and lower surfaces (Figure 9.17). It is often summarized by a few parameters such as maximum thickness, maximum camber, position of maximum thickness, position of maximum camber, and nose radius. One can generate a reasonable airfoil section with these parameters (Figure 9.18).

The NACA four- and five-digit airfoils were created by superimposing a simple meanline shape with a thickness distribution that was obtained by fitting a couple of popular airfoils of the time

$$y = \pm (t/0.2)(0.2969x^{0.5} - 0.126x - 0.3537x^2 + 0.2843x^3 - 0.1015x^4) \qquad ...(9.30)$$

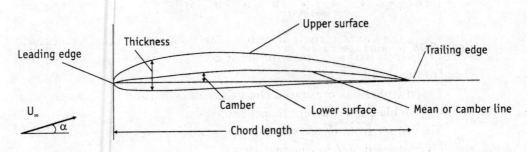

Figure 9.17 A typical airfoil

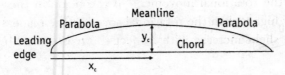

Figure 9.18 Airfoil description

The camberline of four-digit sections was defined as a parabola from the leading edge to the position of maximum camber, and then another parabola back to the trailing edge.

It is easier to describe the lift and drag properties of different airfoils using dimensionless lift and drag coefficients, which are given by

$$C_l = L/(1/2\rho\, AU^2) \qquad\qquad\qquad\qquad ...(9.31)$$

$$C_d = D/(1/2\rho\, AU^2) \qquad\qquad\qquad\qquad ...(9.32)$$

where L is the lift and D the drag in kg m/s², C_l is the lift coefficient, and C_d is the drag coefficient.

C_d/C_l is an important ratio and at its minimum value, lift, and consequently, the coefficient of performance, will be maximized.

The angle between the direction of the undisturbed wind speed and a reference line of airfoil is called the angle of attack α. For an airfoil, the reference line is a line connecting the trailing edge with the centre of the smallest radius of curvature at the leading edge. In case of a curved plate, it is simply a line connecting the leading edge and the trailing edge.

C_l–α curves enable a designer to find the angle of attack corresponding to maximum C_l under given conditions.

The values of C_l and C_d of a given airfoil vary with wind speeds and the Reynold's number Re that is given by

$$Re = Uc/\nu \qquad\qquad\qquad\qquad ...(9.33)$$

where c is the characteristic length of the body and ν is the kinematic viscosity, which is 15×10^{-6} m²/s for air at 20 °C. In case of airfoils, c is the chord of the airfoil.

Lysen (1982) has presented the values of α, C_l, and C_d/C_l as given in Table 9.3 for blades of different shapes and types.

We can see from Figure 9.19 that the relative wind speed w to the blade airfoil is composed of following two parts.

1 The original wind speed u, but slowed down to a value $(1 - a)$ u as a result of power extraction

2 A wind speed due to the rotational movement of the blade in the rotor plane, which is slightly higher than the rotational speed of the blade, Ωr at the cross-section. The slight increase with respect to Ωr is caused by the wake behind the rotor.

Table 9.3 Airfoil characteristics

	C_d/C_l	α	C_l
Flat plate	0.10	5	0.80
Curved plate (10% curvature)	0.02	3	1.25
Curved plate with tube on concave side	0.03	4	1.10
Curved plate with tube on convex side	0.20	14	1.25
Airfoil NACA 4412	0.01	4	0.80

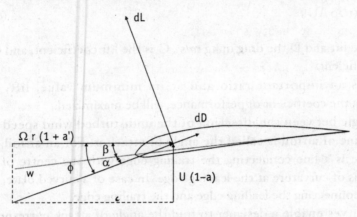

Figure 9.19 Airfoil geometry, orientation, and forces

The angle between the relative wind speed w and the rotor plane is ϕ. The lift force L, due to the action of the wind speed w on the blade cross-section is, by definition, perpendicular to the direction of w.

As a result, the angle between L and the rotor plane is $90° - \phi$. The forward component of the lift in the rotor plane is equal to $L\sin\phi$ and the component of the drag force in the rotor plane is $D\cos\phi$.

Example 2

What is the relationship between various forces when the rotational speed is constant?

Solution

At a cross-section at radius r on the blade, we have

Rotational speed = Ωr = constant.

Since the rotational speed is constant, the net resultant of forces in the rotational plane should be zero. Assuming negligible frictional force at the hub[2] of the rotor, we have

---

[2] Point at the root of the blade where it is fixed to the rotor shaft, also the centre of the rotor.

$$D\cos \phi + L\sin \phi = 0$$
or
$$L\sin \phi = -D\cos \phi$$

Therefore, in the situation of constant rotational speed, the components of the lift and drag forces in the rotor plane are equal and opposite.

Blade element theory

The axial momentum theory is not adequate to provide us with the necessary information to design a rotor blade. However, the blade element theory that allows computation of forces on each element of the blade together with the axial momentum theory enables us to work towards a rotor blade design.

The underlying assumptions in the blade element theory are as follows.

- There is no interference between adjacent blade elements along each blade.
- The forces acting on a blade element are solely due to the lift and drag characteristics of the sectional profile of a blade element.

The blade element theory involves computation of forces on an element of the blade, their integration over the entire blade length, and subsequently multiplication over the number of rotor blades. The cross-section of a blade element of thickness dr is given in Figure 9.20. The expressions for the lift and drag on the element are given below.

$$dL = C_l (\tfrac{1}{2})\rho w^2 c\,dr \qquad\qquad ...(9.34)$$

$$dD = C_d (\tfrac{1}{2})\rho w^2 c\,dr \qquad\qquad ...(9.35)$$
where c is the chord of the blade.

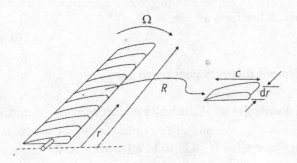

Figure 9.20 Airfoil elements

The thrust experienced by the blade element is the force normal to the plane of rotation, and is given by the sum of the components of the lift and the drag forces normal to the plane of rotation.

$$dT = dL \cos \phi + dD \sin \phi \qquad \qquad ...(9.36)$$

On the other hand, the torque is caused by the forces along the plane of rotation and is given by

$$dQ = r (dL \sin \phi - dD \cos \phi) \qquad \qquad ...(9.37)$$

For a rotor with B number of blades, using Equations 9.34 to 9.37, we have

$$dT = B (1/2)\rho w^2(C_l \cos \phi + C_d \sin \phi)cdr \qquad ...(9.38)$$

$$dQ = B (1/2)\rho w^2(C_l \sin \phi - C_d \cos \phi)crdr \qquad ...(9.39)$$

In axial momentum theory also (Equations 9.23 and 9.24), we had derived equations for thrust and torque on an element of the rotor as

$$dT = 4a(1 - a) \tfrac{1}{2} \rho U^2 2\pi rdr \qquad ...(9.40)$$

$$dQ = 4a'(1 - a) \tfrac{1}{2} \rho U \Omega r^2 2\pi rdr \qquad ...(9.41)$$

From Figure 9.21, we have

$$w = (1 - a)U/\sin \phi = (1 + a')\Omega r/\cos \phi \qquad ...(9.42)$$
or
$$\tan \phi = (1 - a)/ (1 + a')\lambda_r \qquad ...(9.43)$$

Introducing the local solidity ratio as
$$\sigma = Bc/2\pi r \qquad ...(9.44)$$

The results of the blade element theory transform to

$$dT = [(1 - a)^2 \sigma C_l \cos \phi/\sin^2 \phi][1+ (C_d/C_l)\tan \phi]\tfrac{1}{2} \rho U^2 2\pi\, rdr \qquad ...(9.45)$$

$$dQ = [(1 + a')^2 \sigma C_l \sin \phi/\cos^2 \phi][1 - (C_d/C_l)(1/\tan \phi)]\tfrac{1}{2} \rho\Omega^2 r^2 2\pi rdr \qquad ...(9.46)$$

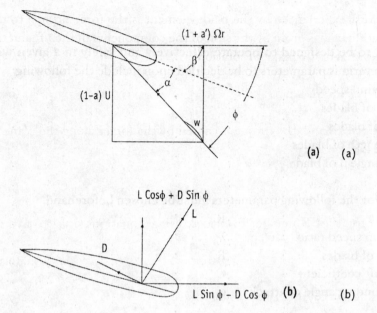

Figure 9.21 Airfoil velocity diagram

Combining Equations 9.40 and 9.45, we have

$$4a/(1 - a) = [\sigma\ C_l \cos \phi/\sin^2 \phi][1 + (C_d/C_l)\tan \phi] \qquad ...(9.47)$$

Similarly, using Equations 9.41, 9.43, and 9.46, we have

$$4a'/(1 + a') = [\sigma\ C_l/\cos \phi][1 - (C_d/C_l)(1/\tan \phi)] \qquad ...(9.48)$$

According to Lysen (1982), the drag terms can be omitted from Equations 9.47 and 9.48 because the profile drag does not induce velocities at the blade itself within the approximation of the small blade chords. However, he cites examples of researchers who continue to use drag terms. Here, if we drop the drag terms, we have

$$4a/(1 - a) = \sigma\ C_l \cos \phi/ \sin^2 \phi \qquad ...(9.49)$$

$$4a'/(1 + a') = \sigma\ C_l /\cos \phi \qquad ...(9.50)$$

Rotor design

A rotor has to be designed to operate effectively and safely in a given wind regime. The various parameters to be decided upon include the following.

- Design wind speed
- Number of blades
- Length of blades
- Chord length of blades
- Setting angle β of blade

The values for the following parameters must be chosen beforehand.

- Radius   R
- Design tip speed ratio λ_d
- Number of blades   B
- Design lift coefficient C_{ld}
- Corresponding angle of attack α_d

For wind turbine rotor design, radius can either be calculated from the rated power output at rated wind speed or from the expected annual energy output.

For example, assuming a coefficient of performance of 0.3 at rated wind speed U_r, we have

$$P = 0.3 \; \tfrac{1}{2}\rho \; \pi R^2 U_r^3 \qquad \qquad ...(9.51)$$

or

$$R = [P/(0.3 \; \tfrac{1}{2}\rho \; \pi \; U_r^3)]^{\frac{1}{2}} \qquad \qquad ...(9.52)$$

The values for λ_d and B can be chosen from Table 9.4 (Lysen 1982). Airfoil characteristics can be chosen from Table 9.3.

The following equations enable calculation of the other parameters.

Chord	c	$= 8\pi r \, (1 - \cos \phi)/B \, C_{ld}$	...(9.53)
Blade setting angle	β	$= \phi - \alpha$	...(9.54)
Flow angle	ϕ	$= \tan^{-1}[2/3\lambda_r]$	...(9.55)
Local design speed	λ_{rd}	$= \lambda_d r/R$	...(9.56)

Table 9.4 λ_d and B values for wind turbine blades

λ_d	1	2	3	4	5–8	8–15
B	6–20	4–12	3–6	2–4	2–3	1–2

Note Generally, for wind turbines, λ_d is in the range 5–8.

Example 3

Work out the dimensions of a 1000-W wind turbine that achieves its rated output at 12 m/s and is to operate in a wind regime with an average air density of 1.121 kg/m³.

Solution

> We have
> Rated power, P = 1000 W
> Air density, ρ = 1.121 kg/m³
> We assume C_p = 0.3
>
> Using Equation 9.52, we have
> Radius R = 1.04 m, say 1.05 m
> Assuming the number of rotor blades, B to be 3, and using Table 9.4, we

have

> Design tip speed ratio, λ_d = 7, corresponding to B = 3
> From Table 9.3, we select the blade profile of airfoil NACA 4412, and

correspondingly

> C_{ld} = 0.8
> α_d = 4°

We now have two design options, that is, either to keep C_{ld} constant, and vary chord c and setting angle β or to vary C_{ld} and keep the chord constant.

Assuming C_{ld} to be constant, and using Equations 9.53–9.56, we calculate the parameters given in Table 9.5.

It can be seen that the chord is larger closest to the root of the blade and minimal at the tip. The setting angle also changes from 29.6° to 1.44°.

The reader is advised to compute characteristics with constant chord.

Table 9.5 Blade parameters

r(m)	r/R	λ_{rd}	ϕ	σ_d	β	C (m)
0.15	0.142857143	1	33.69007	4	29.69007	0.301503
0.3	0.285714286	2	18.43495	4	14.43495	0.184247
0.45	0.428571429	3	12.52881	4	8.528808	0.128247
0.6	0.571428571	4	9.462322	4	5.462322	0.097702
0.75	0.714285714	5	7.594643	4	3.594643	0.078738
0.9	0.857142857	6	6.340192	4	2.340192	0.065879
1.05	1	7	5.440332	4	1.440332	0.056606

Evolution of the modern wind turbine and its sub-systems

The main sub-systems in a wind turbine are the prime mover (rotor), transmission or the drive train including shafts, couplings, and bearings, and the generation system including the generator, instrumentation and controls, yawing mechanism, tower and foundations or structure, and the grid connection system (Figure 9.22).

Rotor

It is the rotor of the wind turbine that makes it a unique machine. While other sub-systems such as generators, gearboxes, shafts, or couplings are used in different kinds of machines, the rotor and the entire engineering involved in its development and manufacture are dedicated to wind turbines.

The design approach of the modern wind turbine has consolidated around the three-bladed horizontal axis wind turbine. Till early 1990s, vertical axis wind turbines were also being experimented with. However, it now seems that there is a consensus among technologists and design engineers on a three-bladed horizontal axis rotor. All new wind turbines being developed in the

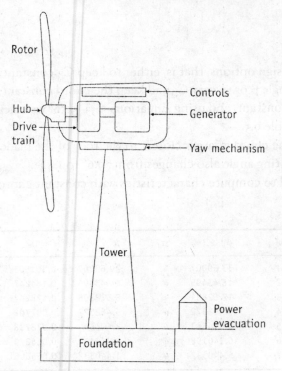

Figure 9.22 Sub-systems of a wind turbine system

world follow this basic approach. Further R&D is directed at achieving greater operational efficiency, higher reliability, and compatibility with the grid (Figures 9.23 and 9.24).

Drive train

Typically, a drive train includes a low-speed shaft on the same side as the rotor, a gear box, and a high-speed shaft on the generator side. In recent times, gearless wind turbines have also been introduced (Figure 9.25).

Figure 9.23 A typical rotor hub

Figure 9.24 A typical rotor hub assembly with blades bolted to it

(a)

(b)

Figure 9.25 Typical drive trains of wind turbines on display
(a) A 2-MW wind turbine (top); (b) another model (bottom)

Pitch- and stall-regulated wind turbines

Two different approaches of regulating rotor power in high wind speeds are pitch and stall regulation. These two approaches determine the blade structure and the choice of the aerofoil.

Functionally, stall implies breakdown of the lifting force when the angle of flow over an aerofoil becomes too steep. In case of aircrafts, stall is an undesirable state but in wind energy technology, it has been used effectively to limit power from the rotor.

A stall-regulated wind turbine runs at approximately constant speed in high wind, not producing excessive power while the rotor blades remain at a fixed angle with respect to the plane of rotation. As the wind speed increases, with constant rotor speed, the flow angles over the blade sections become steeper and this limits the power output from the rotor to acceptable levels without any provision for active control of the rotor blade angles. However, as mentioned, for the stall feature to function, the rotational speed of the rotor must be held constant, which is achieved by connecting the generator to electrical grid. In the grid-connected mode, wind speed fluctuations do not result in changes in the rotational speed of the rotor; rather these fluctuations are absorbed in the grid as fluctuations in electricity generated from the wind turbine generator. Stall regulation, therefore, is a subtle process from an aerodynamic and an electrical viewpoint (Figure 9.26).

The alternative to stall regulation is pitch regulation wherein the angle of individual or all the rotor blades can be changed with respect to the plane of the rotor. With variation in blade angle, and, hence, in the geometry of the rotor, the power output from the wind turbine gets limited. This approach enables more uniform and precise regulation of power (Figure 9.27).

There was a time when, most of the wind turbines used to be stall-regulated. Over a period of time, the pitch regulation approach has emerged as the winner. There are many advantages associated with pitch regulation, which include reduced loading, soft braking, and greater rotor efficiency. Most of the modern wind turbines are pitch-regulated, though there are exceptions (Figure 9.28).

Figure 9.26 Pitch motor and gear inside the rotor hub

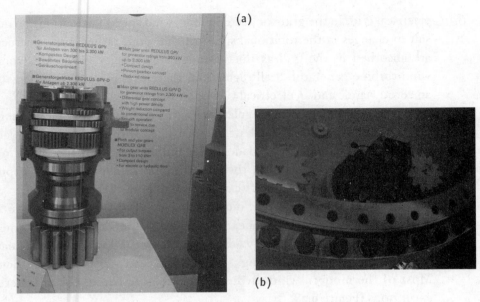

Figure 9.27 (a) Cut section of pitch gear assembly
(b) Pitch gear mechanism

Blade materials

The earlier wind turbines (till 1960s) used blades made up of steel. Later, even aluminium blades were used. However, since the 1970s, the trend has been increasingly towards the use of composites. Fibreglass in polyester resin has been the main composite used in the manufacture of blade. Wood epoxy laminates have also been widely used in modern wind turbines. Some of the modern wind turbines are now opting for carbon fibre and glass composites.

Figure 9.28 Pitch gear engagement

Generation system

Initially, most wind turbines operate at fixed rotor speed when producing power, and at high wind speeds, the stall or pitch regulation mechanism would regulate the power. These wind turbines worked with induction generators that had their rotational speed linked to the grid frequency, and this rotational speed fluctuated very little with respect to grid frequency. Fixed rotor speed, however, imposes some restrictions on the rotor in achieving a high coefficient of performance.

In order to achieve a high coefficient of performance, the rotor should be allowed to vary its rotational speed with changes in wind speed. In modern wind turbines, two different approaches are currently being followed. One of them is the variable speed approach, where the generator is delinked from grid frequency, allowing it to operate at variable speed. Such machines are called variable speed machines. In a variable speed wind turbine, output from the generators has variable frequency and voltage. Complex power electronics circuitry is used to convert the variable frequency and variable voltage electricity into constant voltage and frequency within the range acceptable to the electricity grid. In the second approach, the generation system is not delinked from the grid frequency but a limited amount of change in rotor speed is allowed (modified fixed speed wind turbine). This limited amount of variation is able to achieve the purpose of high performance without drastic changes in design philosophy as seen in case of variable speed wind turbines. Variation in rotor speed is achieved by pitch regulation. Therefore, whether it is variable speed wind turbine or modified fixed speed wind turbine, pitch regulation (also called active pitch control) plays an important role.

Yet another drastic modification of the variable speed concept is the gearless wind turbine, where a multi-pole generator, also called the ring generator, is directly connected to the rotor hub and runs at a low 50–60 rpm. Due to large number of poles, the ring generator is able to operate at a low rpm. So far the ring generator has, however, been incorporated into the wind turbine system design, by only one company in Germany.

Yaw system

Yaw is a term used to indicate the rotation of the rotor–nacelle assembly on a vertical axis to move in or out of the wind direction. The system is a necessity for all horizontal axis wind turbines. The yaw mechanism includes a large plate fixed to the top of the tower of the wind turbine having gear tooth on its periphery and a smaller gear fixed to the bottom of the nacelle. The nacelle sits on a yaw bearing that transfers the nacelle load to the structure. The yaw gear is driven by a yaw drive, which consists of an electric motor, speed reduction gears, and a pinion gear. The speed is reduced to slow down the rotation to an acceptable level, otherwise very large gyroscopic forces can overturn the turbine (Figure 9.29).

Sensory and control systems

A wind turbine has a large number of sensors in it, both for control purposes as well as for monitoring the health of the machine. Both wind speed and direction are measured at one or more points on the structure. Usually, wind speeds and direction sensors are mounted on top of the nacelle. Wind speed measurements are usually used in the control system for yawing the wind turbine into the wind, and for braking it and yawing it out of the wind when the wind speeds are very high. In modern wind turbines with active pitch

Figure 9.29 A yaw gear engagement and yaw motor

control, wind speeds measured are also used for active pitch control to regulate power output from the wind turbine.

Apart from the wind speed and direction measurements, thermometers are used to measure temperatures inside the gearbox, generator(s), brake pump, etc. The wind turbine is stopped when the temperatures recorded are above a set limit that indicates overheating of the system component.

Tower and foundation

Different kinds of towers being used for wind turbines are tubular steel towers, lattice steel towers, and tubular concrete towers. Tubular steel and concrete towers are common in India. Offshore wind turbines demand yet another approach (Figure 9.30).

Wind turbine size

The size of the commercial wind turbines has consistently increased with units of 50–100 kW appearing on the first windfarms in 1985/86 to the latest multi-megawatt wind turbines with rotor diameters of up to 90 m and a hub height of 100 m (as is the case with Suzlon's 2-MW wind turbine, tested in

Figure 9.30 A tripod design for offshore wind turbines on display

Tamil Nadu in 2004/05). In Europe, larger-sized wind turbines of up to 5 MW are evolving for offshore projects (Figure 9.31).

Worldwide, R&D efforts are continuing in these directions. At present, the modified fixed speed wind turbine is the most preferred one for land and offshore projects. It is interesting to note that across the world with the growth in the wind turbine size, there has also been a growth in the market size as well as the number of wind turbine manufacturers. We can also conclusively say that growth in the market, the size of wind turbines, modernization, and R&D are all related (Figure 9.32).

Windfarms

Normally, wind turbines are installed in large clusters, arranged in an array. The array of wind turbines is also called a windfarm. The position of each wind turbine in an array has to be selected carefully keeping in view the wind speed direction, terrain, and the influence of one wind turbine on another. The process of finalizing the positions of individual wind turbines in the windfarm is known as *micrositing*. Different kinds of softwares such as *windfarmer* (http://www.garradhassan.com/products/ghwindfarmer/) are also used to carry out micrositing in an optimal manner (Figure 9.33).

The influence of one wind turbine on the other results in reduced electricity generation. This is also known as array loss. The process of

Figure 9.31 Trends in wind turbine sizes
Source www.ewea.org

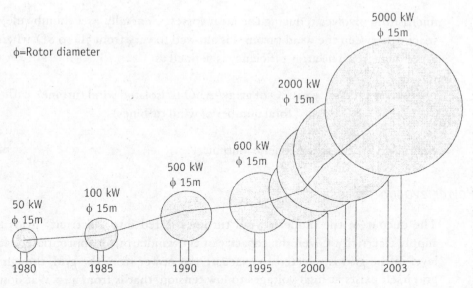

ϕ=Rotor diameter

5000 kW
ϕ 15m

2000 kW
ϕ 15m

600 kW
ϕ 15m

500 kW
ϕ 15m

100 kW
ϕ 15m

50 kW
ϕ 15m

1980   1985   1990   1995   2000   2003

Figure 9.32 Trends in wind turbine technology
Source www.ewea.org

Figure 9.33 Front view of a typical array of wind turbines at a windfarm in Denmark

micrositing involves reducing the array losses. Generally, as a thumb rule, the spacing between the wind turbines is allowed to vary from 5D to 8D, where D is the diameter. The array efficiency is defined as

$$\text{Array efficiency} = \text{AEO of array}/(\text{AEO of isolated wind turbine}) \times \\ (\text{Total number of wind turbines})$$

where AEO is the annual energy output.

Power evacuation aspects

The output of the modern wind turbines is fed into electricity grids that supply electricity. Given the capacity of the windfarm, an appropriate voltage level in the grid is selected to evacuate the power from the grid, though the grid itself exists at high-voltages to low tension, that is from 440 V at domestic supply level to 400 kV at the high-voltage transmission level. As each level of the grid has a certain current carrying capacity, an 11-kV grid (common at distribution level) may be able to carry current to meet a total load of 10 MVA, while a 110-kV grid may have a current-carrying capacity to meet a load of 100 MVA. Therefore, it may not be possible to connect a windfarm of 25 MW capacity rated output with an 11-kV grid.

Individual wind turbine generates 440 V of electricity. Depending on the design of the windfarms, output from each individual wind turbine or from a number of wind turbines is stepped up to the 11-kV or 33-kV level. The aggregate output from the windfarm is stepped up to an appropriate level and then transmitted to a utility sub-station where it is connected to the grid (Figure 9.34).

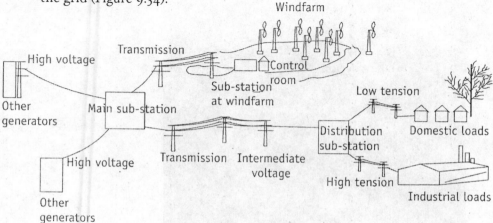

Figure 9.34 A typical windfarm–grid diagram

Problems

1 How old is the concept of windmills?
2 What was the type of the windmill used in Persia?
3 How is the traditional European windmill different from the Persian windmill?
4 What are the different design approaches in a wind turbine? Which type has emerged as commercially acceptable?
5 What are the main assumptions in axial momentum theory?
6 What is the significance of induction factors in axial momentum theory?
7 What is Betz limit? Derive it.
8 Discuss the performance coefficients and explain how they vary with tip speed ratio?
9 Carry out computations and plots of Example 1 assuming the rotational speed of the rotor to be 60 rpm. Compare the results with those computed at 45 rpm.
10 Discuss the working of a simple airfoil.
11 Work out the dimensions of a 100-W wind turbine that achieves its rated output at 10 m/s and is to operate in a wind regime with an average air density of 1.121 kg/m³. Assume constant chord length.
12 Work out the dimensions of a 1-MW wind turbine that achieves its rated output at 14 m/s and is to operate in a wind regime with an average air density of 1.121 kg/m³. Assume variable chord length.
13 What are the various sub-systems in a wind turbine? Briefly discuss each of them.
14 What is the significance of active pitch control in a modern wind turbine?

Nomenclature

a' Tangential induction factor
a Axial induction factor
A Area (m²)
C_{ld} Design lift coefficient
$\dot{m}$ Mass flow rate (kg/s)
p Pressure (N/m²)
P Power (watt)
Q Torque
r Radius of annular stream tube (m)
R Radius of the rotor (m)

T   Thrust (N)
U   Air velocity (m/s)
w   Incremental angular velocity
Ω   Angular velocity (rad/s)
ρ   Air density (m³/kg)

References

Betz A. 1926
Windenergie under Ihre Ausuntzung durch windmillen
Vandenhoeck and ruprecht
Gottingen, Germany
[cited by Manwell *et al.*]
London: John Wiley & Sons, Ltd

Cheremisinoff N P. 1978
Fundamentals of Wind Energy
Ann Arbor: Science Publishers Inc. 169 pp.

Lysen E H. 1982
Introduction to Wind Energy, 310 pp.
The Netherlands: SWD 82–1 Steering Committee Wind Energy Developing Countries.

Manwell J F, McGowan J G, and Rogers A L. 2002
Wind Energy Explained: theory, design, and application
London: John Wiley & Sons, Ltd. 577 pp.

Gipe P. 1995
Wind Energy Comes of Age, 536 pp.
London: John Wiley & Sons, Ltd.

Small hydro: resource and technology

10

Ashish V Kulkarni
Energy–Environment Technology Division, T E R I, New Delhi

Introduction

The word 'hydro' comes from the Greek word 'hydra' meaning water. The energy of water has been harnessed to produce electricity since long. It is the first renewable energy source to be tapped essentially to produce electricity. Hydropower currently accounts for one-fifth of the global electricity supply, along with improving the electrical system reliability and stability throughout the world. It also substantially avoids greenhouse gas emissions, thus complementing the measures against climate change issues.

Global scenario

Globally, hydropower accounts for about 16% (2809 TWh/year) of the world's total electricity. Its total technical potential is about 16 494 TWh/year, of which just over 9061 TWh/year is currently being considered as economically feasible for implementation. The total installed capacity according to a recent survey by the World Energy Council is about 730 GW, excluding the 100 GW under construction (WEC Survey 2004). The continent-wise break-up of the hydropower potential is given in Figure 10.1. Figure 10.2 gives regional technically exploitable potential and generation.

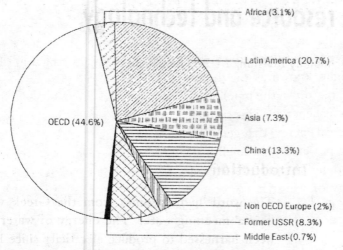

Figure 10.1 Regional distribution of installed hydropower capacity
Source World Energy Congress (2001)

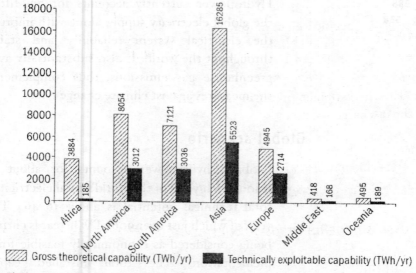

Figure 10.2 Regional technically exploitable potential and generation

Small hydro resource assessment

Resource

Flowing water falling from a sufficient gradient forms the resource for small hydropower. The difficulty arises in finding both the flow of water and the gradient.

Power

Power from water is a result of conversion of potential energy (the water body at a certain height, known as the 'head') to kinetic energy (a flow which is known as 'discharge' down the pipe), which is transferred to the buckets in the turbine (mechanical energy). This turbine, in turn, is the prime mover for the generator (electrical energy) that generates electricity.

Essentially, power from a small hydro potential site is derived from two parameters: head and discharge. 'Head' is the vertical height from which the potential energy of water is converted into electricity after a drop and 'discharge' is the flow rate of the water in the stream/river.

$$\text{Power (kW)} = H \times Q \times \gamma \qquad \qquad ...(10.1)$$

where,

H = Head in m

Q = Discharge in m³/s

γ = Specific weight of water, product of mass and acceleration due to gravity (9.81 kN/m³).

Example 1

Calculate the hydraulic power in a small stream if the head available is 60 m and the discharge is 0.02 m³/s.

Solution

Available power = 60 × 0.02 × 9.81

= 11.77 kW

Resource assessment methods for determining 'head' and 'flow'

Head measurements

The 'head' at a particular site can be measured by using two types of devices.

Abney level

The Abney level is a hand-held meter, also called the clinometer, which measures the slope. This requires a certain amount of skill and gives an accuracy of (±)5 m. This measurement requires two people and is based on simple trigonometry. Two posts, each about a metre high, are put up at two locations: the forebay and the turbine. One person stands at the post located

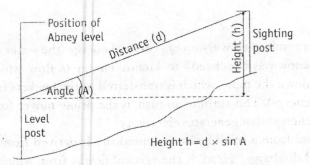

Figure 10.3 Head measurement by Abney level
Source Mayer and Smith (2001)

at the turbine location and measures the angle of the post at the forebay location using the Abney level. The distance between the two posts is also measured by a tape. Some of the Abney levels have it in-built, which makes the process easier. Now, the vertical height is measured using a trigonometry formula. Looking at the figure illustrated, we could determine the head as a product of sine of the angle measured and the distance between the posts. The limitation of the Abney level measurement is that there has to be a clear line of vision between the two strategic locations, otherwise the head has to be measured in steps to get the total head available (Figure 10.3).

Example 3
If the angle A measured by the Abney level is 34° and the distance d between the two posts is 30 m, then calculate the head.

Solution
$$\text{Head} = D \times \sin A \qquad\qquad ...(10.2)$$
$$= 30 \times \sin 34$$
$$= 16.77 \text{ m}$$

Flow measurements

There are three standard methods to measure the flow.

Bucket method

This method is used to measure small streams/flows usually during the summer season, when flow is generally very small. It is quite accurate and needs three to four measurements. The method is very simple. A bucket/container of known volume, say 5 litres or 10 litres, is used. A small weir is constructed using the boulders available at the site, which facilitates the water

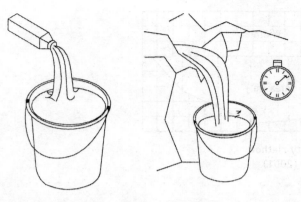

Figure 10.4 Flow measurement by bucket method
Source Mayer and Smith (2001)

into the container. Some natural spot where water could thus be collected has to be determined during the site study. The time taken to collect water, measured using a stopwatch, determines the flow of the stream. Three to four readings are taken (Figure 10.4).

Example 4
　　If it takes 10 s to fill a 20-litre container then
　　Flow = 20/10 = 2 litres/s = 0.002 m³/s.

Area velocity method (Float method)

Although this method is quite accurate, three to four readings are needed to ensure the same. Here, the area of the cross-section of the river/stream bed is calculated (m²). A person has to actually cross the whole river and measure the different sections of the river bed. Generally, local people are employed to do this. A known length along the river is determined. It is always better to have a long length for each reading as it ensures accurate measurements. A twig, leaf, or a small wooden block is made to float with the flow along the known length. This determines velocity (m/s). The product of the area of the cross-section and the velocity gives us the flow of the stream. The measurement of area is a crucial reading and a river with a smoother bed and sides provides an easier reading, than the one with a rocky bed. The velocity of the river/stream also gets affected in case of rocky beds. Therefore, a velocity correction factor is employed, which is to be multiplied with the measurement taken (Figures 10.5 and 10.6). The correction factor is usually 0.85 for smoother beds, 0.7 for smooth, deep, slow-moving streams, and 0.3 for rocky beds.

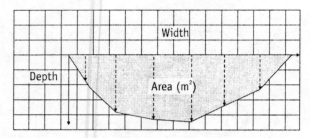

Figure 10.5 Area velocity method
Source Mayer and Smith (2001)

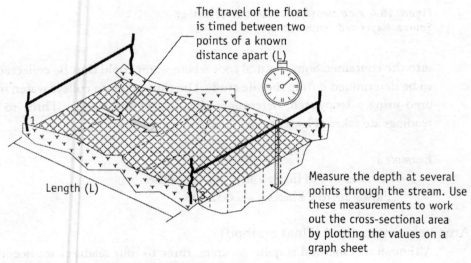

The travel of the float is timed between two points of a known distance apart (L)

Measure the depth at several points through the stream. Use these measurements to work out the cross-sectional area by plotting the values on a graph sheet

Length (L)

Figure 10.6 Flow measurement by area velocity method
Source Mayer and Smith (2001)

Example 5

Readings are taken with a float that is made to flow along a known length of 20 m along a river with smooth bed and sides; the width and depth of the river being 15 m and 2 m, respectively. Calculate the flow of the river if the time taken for the float to traverse 20 m is 25 s.

Solution

Calculation of cross-sectional area	= 15 × 2 = 30 m²
Velocity	= 20/25 = 0.8 m/s
Velocity correction factor	= 0.85 (since the bed and sides are smooth)
Flow	= 30 × 0.8 × 0.85
	= 20.4 m³/s

Salt gulp method

This method involves complex procedures and calculations as compared to other methods but gives an accurate result. It is mostly used for micro- and small-hydro projects. It can be applied to those rivers that have a very rocky bed. It involves the use of a conductivity meter that gives a digital reading. A known mass M of salt is first mixed with water to prepare a solution that is poured into the river. The amount of water used to make the salt solution need not be measured. The only precaution to be taken is that the salt solution after preparation should not be spilt so as to maintain accuracy in the mass of salt added, which is crucial for the reading to be taken. The mass of the salt depends on the flow of the stream/river. Usually, this comes through practice. A thumb rule is adding about 2 kg of salt for 1 m³/s of flow. The salt solution is then poured into the river. The probe of the conductivity meter is then dipped into the stream at a point that is 15–20 m from the point where the salt is added. It usually takes 10–15 minutes for the salt to reach the point where the conductivity is measured. The conductivity is measured every 5 s. A graph is plotted with time on the x axis and conductivity on the y axis (Figure 10.7). The area under the curve is then measured. The formula to calculate the flow is

$$Q = M \text{ (kg)} \times k^{-1} \text{ (ohm}^{-1}/\text{mg litre}^{-1})/A \text{ (s} \times 10^{-6} \times \text{ohm}^{-1}) \qquad \text{...(10.3)}$$

The conductivity is converted into salt concentration by multiplying with a conversion factor k^{-1}, which takes into account the temperature of the water in the stream. For 22 °C, k^{-1} is 2.04.

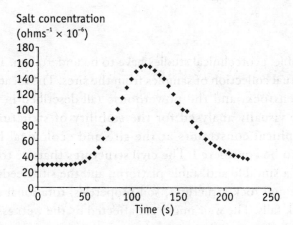

Figure 10.7 Flow measurement by salt gulp method
Source Mayer and Smith (2001)

A proper feasibility study would need at least five flow measurements in a day and at least 12 readings of 12 days, one in each month in a year. This is the minimum reading required for defining the hydrology of a site.

Site evaluation

Site selection is very important for designing a power plant. After assessing the head and the discharge, site evaluation has to be done to ensure that the available power could be generated. The decision about whether a site is suitable for a power plant depends on an iterative process in which the topography, environmental sensibility, and knowledge of the scheme to be set up are of utmost importance. Thus, site evaluation avoids the dangerous failures of operation of the plant. Surveying techniques are changing day by day. These need to be addressed and the knowledge of the same is required for site evaluation.

Cartography

Maps of locations to the required scale are generally available with the local or state government authorities, and they can be used for analysing the topography of the site. In countries where these maps are not available, site evaluation has to be done using engineering techniques like the GPS (global positioning system). Photographs of the sites help in evaluating the site and help in determining the geological structures, the type of rocks, etc. These also help engineers to determine stability of the civil structures to be built at the site.

Geotechnical studies

When no maps are available, geotechnical studies have to be undertaken. This involves site visits and actual collection of samples from the sites. The location of the weir, channel, penstock, and the powerhouse (all described in the next section) need to be visually analysed for the stability of structures. The design and geographical constraints at the site and ecological history of the place need to be considered. The civil structures that are to be erected at the site need a suitable and stable platform, and the site needs to be assessed for the slope and rock structures, soil properties, rotational and planar soil slides, and rock falls. The weir might get affected by the wetness of the soil, and the penstock during real time implementation may face many problems due to landslides or rocky surfaces. Sometimes due to the terrain, there may be a need to increase the length of the penstock. Therefore, site visits and geotechnical studies become very important. The methodologies used for such studies are photogeology, geomorphologic maps, laboratory analysis, geophysical studies, structural geological analysis, and physical studies.

Small hydro technology

Small hydro technology is mainly a 'run-of-the-river' scheme (see Figure 10.8) that operates by diverting part of the river flow or the entire river by means of constructing a weir and intake, through a penstock (or pipe) and a turbine that drives a generator to produce electricity. The water then flows back into the river through a civil construction known as the tail race. Recently, small hydro projects are also being built on irrigational canal drops in view of the fact that irrigational canal flows are already known and the power generated can be quantified. These projects do not envisage construction of reservoirs or dams. This is preferable from an environmental point of view as seasonal river flow patterns downstream are not affected and there is no flooding of valleys upstream of the system. A further implication is that the power output of the system may not be determined by controlling the flow of the river in case of micro- and pico-hydro projects; instead, the turbine operates when there is water flow and the output power is regulated or governed. The systems can be built locally at low cost, and their simplicity gives rise to better long-term reliability. Small hydro systems are thus suitable as remote area power supplies for rural and isolated communities, and as an economic alternative to extending the electricity grid. These are also commercially

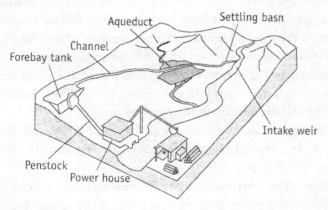

Figure 10.8 Small hydro project configuration

attractive if the power generated is fed into the grid. The main advantage of these systems is that they provide cheap, independent, and continuous power, without degrading the environment.

System components

The system with all its components could be divided into three major categories.
1 Civil works
2 Electro-mechanical equipment
3 Distribution network

A brief description of the components of a small hydro scheme is given below.

Weir/intake   A weir is a civil work done across the river to facilitate the diversion and intake of required discharge to the feeder channel. As the system is usually 'run-of-the-river' type, the weir may be built using the available boulders/minimum civil works.

Feeder channel   This is made of mild steel/PVC (polyvinyl chloride) along the length to the forebay with desilting tanks to remove the silt.

Forebay   This is a tank designed and built in accordance with the discharge available at the site. It also maintains a balance in the power output by maintaining the design head for the project. It facilitates the intake for the penstock, which feeds the discharge to the turbine. A feeder channel may be of the open type or made of PVC/steel.

Penstock   The penstock is a pipe designed to discharge potential at the site and is usually made up of mild steel. It feeds the discharge to the turbine.

Turbine   Turbines are selected according to the design head available. The selection is site-specific: for a typically high-head site, Pelton/Turgo impulse turbines are used; cross-flow impulse turbines are used at medium-head sites; and Kaplan/Francis turbines are used for low-head applications.

Generator   Synchronous generators are generally used for small hydro sites, whereas for very small hydro sites (in the range of 0–10 kW), induction motors which could be self-excited are used as generators.

Governor   Electronic load controllers can be used for small hydro projects up to 25 kW, but typical mechanical governors are mostly recommended for higher power projects.

Distribution system   The distribution system is mainly the overhead lines supported by locally available poles. In some cases, where the terrain is rough consisting of dense forest areas, underground cables are used.

We will now look into each of the components in detail.

Weir/intake

A weir, in terms of small hydro projects, is a civil construction built to divert some amount of the whole discharge to produce electricity. Weir structures all over the world are built using different methods and materials. Usually, they are designed such that the design discharge flows for power production and the rest overflows into the normal course of the river. The simplest of the weirs are constructed using boulders and stones available at the site (Figure 10.9). The new material being used for construction is rubber, which has the ability to inflate if there is need for more discharge. Weirs are usually designed with spillways that could actually be controlled using valves.

The different types of forces acting are the hydrostatic forces acting on the upstream and downstream faces; hydrostatic uplift acting on the base of the weir; forces due to the silt deposited at the base of the dam; and movements that are acting on the base of the dam due to the water movement. The contact stress of the weir with the base should be greater than the forces acting on the weir. It is also necessary that the weir so built should not slide away. Therefore, a static coefficient of 0.6–0.75 should be maintained between the vertical and horizontal forces acting on the weir. The weir should be built keeping all the above conditions in mind (Figure 10.10). To raise the water level, we could have a trapezoidal structure with spillways (Figure 10.10). New materials like rubber have also been used to raise the water level, as rubber can be inflated when the water level increases due to floods in the stream. These rubber dams are only in an experimental stage.

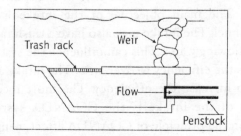
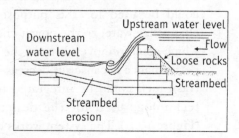

Figure 10.9 Different types of weirs using boulders
Source Penche and de Minas (1998)

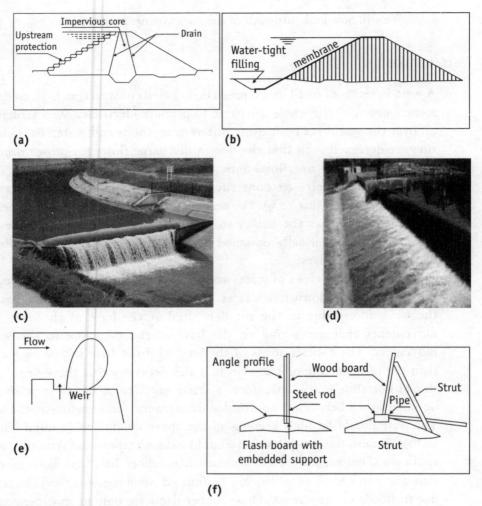

Figure 10.10 Different types of weirs
Source Penche and de Minas (1998)

The intake could possibly be adjacent to the weir. One of the spillways could be used for this purpose. The main objective of the intake is to channelize water to the power channel. The intake may also have a trash rack, which could remove silt from the flowing water. This can influence the system performance because silt decreases the efficiency to over 20%. The intake and trash rack could be integrated to increase the efficiency. The most recent technology used for the design of the intake is the COANDA system (Figure 10.11). The system works on the principle of COANDA effect, named after the person who invented it. In this system, water flows at the same angle as the surface on which it flows. Such an intake could be designed to enhance

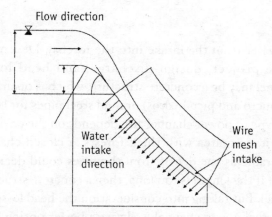

Figure 10.11 COANDA intake

the performance of the whole system. Two other types of intakes are the Siphon type and French or drop intake. Siphon type is very cost-effective (Figure 10.12) and is the simplest of all. It is not gated and at runaway conditions, is easier to shut off. The French or drop intake is a canal built along the length of the weir with a trash rack on top of the canal (Figure 10.13).

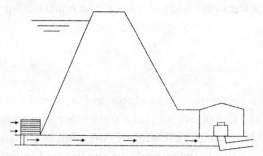

Figure 10.12 Siphon intake

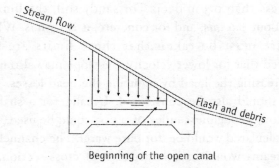

Figure 10.13 French or drop intake
Source Penche and de Minas (1998)

Power channel

The power channel feeds water from the intake into the forebay. Depending on the economics of the project, design discharge, and head losses calculated, the power channel may be a concrete structure, which is open, or a pipe made up of PVC (for micro and pico hydros) or mild steel pipes for larger hydros. In many cases, the type of power channel also depends on the environment of the entire site. If it is an area with thick forests, a closed channel made of mild steel is preferable as the debris from the trees could decrease the efficiency of the system. If the length is too long, then a concrete structure could be used for the channel. But taking into consideration the head losses to be accounted for, the mild steel power channel is always a better option. The length of the power channel is determined by geological surveys of the site and diameter or the width could be determined by the design discharge.

Recent designs of intakes also include a fish passage, which has emerged as one of the main issues of environmental concern worldwide. A lot of research is being done in this area so that the number of fishes being killed due to turbines could be reduced.

The channel is to be designed using two parameters, the velocity v of the discharge (m³/s) and the area of the cross-section A (m²). The equation for flow rate is

$$Q = v \times A \qquad \qquad ...(10.4)$$

The velocity of the flow depends on the slope and material used for construction of the channel. There are certain limits for the velocity depending on the material. There are possibilities of corrosion or damage to the channel if the velocity of the discharge goes above the limits. Limits have been set for velocity of different materials used for construction of channels that are shallow, that is, less than 0.3 m deep. For sandy soil, the limit is 0.4 m/s; for clay soil it is about 0.6 m/s; and for concrete, it is 1.5 m/s. While designing a channel, care must be taken that these limits are not exceeded. It should be noted that for lower velocity, the slope may also need be lower. This helps in increasing the head by reducing the head losses. The roughness of the channel should also be considered. Usually, for a shallow channel with some vegetation, a roughness index of 0.007 could be used. The area of cross-section thus calculated would be for pure water. The channel for river water and natural streams would have larger area of cross-section, by about 30%.

Example 6
To design a concrete channel having a discharge of 20 m³/s, the velocity is about 1.5 m/s. Therefore, the area of cross-section of water can be calculated as

$Q = v \times A$

$20 = 1.5 \times A$

$A = 20/1.5 = 13.3$ m², that is, height = width = 3.6 m, if a square cross-section of water is taken.

Therefore, if we take a square channel, we would have the width and height of the channel as

$3.6 \times 1.3 = 4.7$ m (approximately)

The sides of the channel should always be outwards to avoid any collapse.

Forebay

A forebay is a component of the system, which may or may not be present, depending on the site conditions, although it is better to have one to maintain stability of the output power in smaller hydro systems. The design of the forebay depends mainly on the time for which the power is to be regulated.

Example 7
Suppose, we need a forebay to maintain a water level for three minutes, with a discharge of 200 litres/s or 0.2 m³/s, then the forebay would have a volume of

$0.2 \times 60 \times 3 = 36$ m³

Thus, we could have the forebay constructed for that particular volume.

Penstock

A penstock could be installed underground or above the ground, depending on the geological studies undertaken at the site. The nature of the ground, temperature/climate of the area, and ecological studies undertaken determine the material of the penstock to be installed.

For pico/micro hydros, a PVC penstock may be installed. The rock and gravel in the ground provide good insulation if the penstock is designed to be buried under the ground. This also eliminates the use of anchor blocks and expansion joints that have to be used if the penstock is installed above ground. Penstock corrosion should be prevented by using paints and other material. Usually a penstock is designed to be straight or nearly straight with

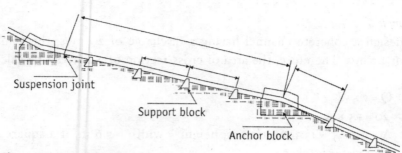

Figure 10.14 Penstock
Source Penche and de Minas 1998

the required amount of concrete blocks for support near the joints and bends to avoid the forces acting on the penstock (Figure 10.14). Rocks and other locally available materials should be used as support along the length of the penstock.

There is a wide range of materials to be chosen for the penstock. Usually for larger heads, fabricated welded steel is used; steel pipes could also be used for this purpose. For high heads, the usual choice is steel or ductile pipes. For lesser diameters, steel is the obvious choice and for low heads, PVC or plastic pipes could be used. A diameter of 0.4 m could be easily taken as a standard for heads up to 200 m. PVC pipes are usually installed underground and do not need any protection. These pipes are also less prone to corrosion. These should not be installed in rocky areas. Plastic pipes could be easily installed above ground. Concrete penstocks, both pre-stressed with high tensile strength or steel reinforced, constitute another solution but due to heaviness they are not competitive. Some of the properties of the materials used pressure pipes are given in Table 10.1.

Table 10.1 Some properties of penstock materials

Material	Young's modulus of elasticity E (N/m²)	Coefficient of linear expansion a (m/°C)	Ultimate tensile strength (N/m²)	η
Welded steel	206	12	400	0.012
Polyethylene	0.55	140	5	0.009
Polyvinyl chloride	2.75	54	13	0.009
Asbestos cement	NA	8.1	NA	0.011
Cast iron	78.5	10	140	0.014
Ductile iron	16.7	11	340	0.015
Wood-stave (new)	—	—	—	0.012
Concrete (smooth finish)	—	—	—	0.014

E – Young's modulus of elasticity in N/m²;
η – Manning's constant;   a – coefficient of linear expansion in m/°C
Source Penche and de Minas (1998)

Hydraulic design of penstock

The main parameters of design of the penstock are the diameter and the wall thickness. The diameter of a penstock is designed to reduce the frictional losses and thus minimize the head losses.

The diameter is selected keeping in view the fact that the total head loss should be less than four per cent of the total head available at any particular site. Equation 10.4 is applicable for flows in penstock also.

The diameter of the pipe can also be designed using the empirical formula.

$$D = (10.3 \times \eta^2 \times Q^2 \times L/h_f)^{0.1875} \qquad \qquad ...(10.5)$$

where η is the Manning's constant, Q is the discharge, L is the length of the penstock in m, and h_f is the friction loss. Table 10.1 gives the values of Manning's constant for different materials.

If we have to minimize the head losses due to friction in the pipe to four per cent

$$h_f = 4 \ H/100$$

where H is the gross head. Equation 10.5 can be then written as

$$D = 2.69 \ (\eta^2 \times Q^2 \times L/H)^{0.1875} \qquad \qquad ...(10.6)$$

Example 8

A system has a gross head of 80 m and a discharge of 3 m³/s and the length of the penstock is about 170 m. The diameter of the steel pipe, keeping in view the fact that the head losses through the steel pipe should be around four per cent, can be calculated as

$$D = 2.69 \ (\eta^2 \times Q^2 \times L/H)^{0.1875}$$
$$= 2.69 \ (0.012^2 \times 3^2 \times 170/80)^{0.1875}$$
$$= 0.89 \ m.$$

Therefore, we select a 1-m diameter steel pipe for the penstock.

The wall thickness of a penstock should be selected such that it takes into account the wear and tear due to corrosion.

The thickness of the pipe could be calculated as

$$t_{min} \ (in \ cm) = (\rho \times H \times D)/(2 \times \sigma \times \Phi) \qquad \qquad ...(10.7)$$

where, ρ = density of water = 0.001 kg/cm³

H = gross head (in cm)

D = diameter of the pipe (in cm)

σ = permissible stress in steel = 1500 kg/cm2

Φ = joint efficiency (0.85–0.9)

Example 9

A penstock of a system of head 80 m and diameter 1 m should have minimum thickness of

$$t_{min} \text{ (in cm)} = (\rho \times H \times D)/(2 \times \sigma \times \Phi)$$
$$= 0.001 \times 8000 \times 100/2 \times 1500 \times 0.9$$
$$= 0.296 \text{ cm}$$

Provision for corrosion

$$= 0.2 \text{ cm}$$
$$= 0.496 \text{ cm}$$
$$= 4.96 \text{ mm}$$

Therefore, the penstock thickness should be about 5 mm.

Turbine

A turbine is a mechanical component in the hydro system, which converts the potential energy of the falling water into rotating shaft power. Selection of a turbine is a very critical issue which mostly depends on the site characteristics. The head and flow at the site form the basic parameters to select a turbine. The other important parameters that can be considered are running speed of the generator, other loads, and part load conditions.

We may classify the turbines according to head as high-head, medium-head, or low-head machines (Table 10.2). They can also be classified by the type, that is, impulse or reaction. The normal range of efficiencies for impulse turbines is 90% and that for reaction turbines is 85%–90%.

The main difference between a reaction turbine and an impulse turbine is that the rotating element, called the 'runner', is fully immersed in water in a

Table 10.2 Impulse and reaction turbines

Type of turbine	High head	Medium head	Low head
Impulse turbines	Pelton	Cross-flow	Cross-flow
	Turgo	Multi-jet Pelton	—
Reaction turbines	–	Francis	Propeller
			Kaplan

reaction turbine, whereas in an impulse turbine, it operates by the impact of a jet of water in air.

In a reaction turbine, the runner blades are positioned in such a way that there occurs pressure difference across each of them, which creates lifting forces; the best example being an aircraft that causes the runner to rotate. The blades are enclosed in a pressure casing.

In an impulse turbine, a nozzle converts the low-velocity water into a high-speed jet, which hits the blades. The blades deflect in such a way as to maximize the momentum that, in turn, rotates the runner. Impulse turbines are more economical because of the absence of a specialized pressure casing as required by reaction turbines.

Impulse turbine
Impulse turbines have the following advantages over reaction turbines.
- More rugged
- Easier to operate and maintain
- No pressure points around the shaft
- Easier fabrication
- Better part-flow efficiency

Pelton turbine
A set of specially designed buckets mounted on the periphery of a circular disk forms the runner of the pelton turbine (Figure 10.15). The water jet gushing out from the nozzle/nozzles of the turbine strikes the buckets and the runner is turned. The buckets are designed in such a way that the water jet that strikes them does not deflect water away from the oncoming jet (Figure 10.16). The buckets are also split in the middle so as not to create a dead spot. The buckets are also designed at a particular angle taking into consideration the fluid mechanics. For example, the pelton bucket is designed at 165° from the incoming jet, which is the maximum permissible angle for the return jet so as not to interfere with the incoming jet.

In large hydro plants, pelton turbines are usually employed for heads above 150 m. At lower heads, they have slower rotational speeds (RPM). If the runner size is increased considerably and the generator could run at lower speeds, then pelton turbines could be effectively used for lower heads. The runner speeds should be higher so as to either increase the number of jets or have twin runners.

In case the number of jets is increased, a smaller runner could be employed because it would facilitate more flow of water and increase the

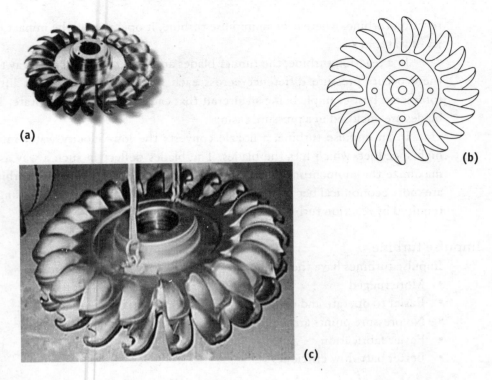

(a)

(b)

(c)

Figure 10.15 Runner of a pelton turbine
Sources Penche and de Minas (1998); http://www.jfccvilengineer.com/turbines.htm, last accessed on 1 August 2005

rotational speed to obtain the required power. The part-flow efficiency can be increased as the turbine can be operated with a lower number of nozzles and the jets in use would still have the optimal flow to run the turbine at a particular speed. In case of twin runners, which is rarely designed, the runners are placed on the shaft either on the opposite sides of the alternator or on the same side. This is designed only after maximizing the number of jets.

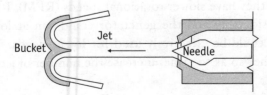

Figure 10.16 Bucket shape
Source Penche and de Minas (1998)

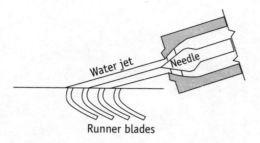

Figure 10.17 Turgo runner blades and water jet
Source Penche and de Minas (1998)

Turgo turbine

Turgo impulse turbines are designed for higher speeds and these could be used for middle-head applications. They are slightly different from pelton turbines. The jets are designed such that the water discharged from the buckets does not interfere with the incoming jet (Figure 10.17). This facilitates higher rotational speeds. Therefore, turgo impulse turbine could also be economical as compared to a pelton turbine for a required power because it can have a lesser runner diameter. Another significant feature is that a turgo turbine could possibly be coupled directly to the generator rather than using costlier speed-increasing transmission levers. A turgo impulse turbine could be mounted horizontally or vertically and is also very efficient at part loads.

Cross-flow turbine

The cross-flow turbine, also known as the Mitchell-Banki turbine, consists of a drum-shaped governor with two parallel disks that are connected near the rims by a series of curved blades (Figure 10.18). The cross-flow turbine is limited only to horizontal shafts, whereas other impulse turbines could have both horizontal and vertical shafts.

When in operation, a rectangular nozzle directs the jet of water onto the full length of the runner. The water strikes the blades and imparts most of its kinetic energy. It then passes through the runner and strikes the blades again on exit, impacting a smaller amount of energy before leaving the turbine. Although it is classified as an impulse turbine, hydrodynamic pressure forces are involved and a mixed-flow definition would be more accurate.

High part-flow efficiency can be maintained at less than a quarter of the full flow by the arrangement for flow partitioning. At low flows, the water can be channelled through either two-thirds or one-third of the runner, thereby sustaining relatively high turbine efficiency (Figure 10.19).

Figure 10.18 Cross-flow turbine
Source http://www.palangthai.org/images/microhydro/ewijo/6004.jpg

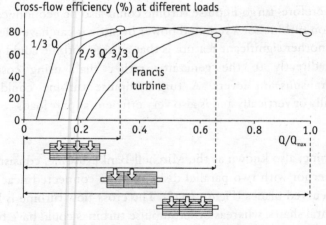

Figure 10.19 Part-flow efficiency of a partitioned cross-flow turbine
Source http://www.ossberger.de/cms/en/hydro/the-ossberger-turbine,
last accessed on 4 July 2005

Reaction turbines

Reaction turbines are of two types: Francis and propeller. The specific speed
of reaction turbines is always higher than that of impulse turbines and these
are, therefore, very popular amongst the medium- and low-head plants. These
turbines also reduce the cost to a large extent as mentioned earlier. They can
be coupled directly to the generator and many small-head systems of small
power have already been tested, commissioned, and are in operation in China

and Nepal. Such systems can also be used for small enterprise development like battery charging, etc. These turbines may also not need a complicated drive system that reduces the cost to a considerable extent.

These turbines require improved fabrication as they involve more profiled blades that are large in size and intricately arranged in specially designed casings. These involve extra expenses but ensure that these turbines would have high efficiency. Propeller machines are easier to handle than the Francis turbines. It is a known fact that reaction turbines have poor part-flow efficiencies.

Francis turbine

The presence of guide vanes that feed water to the blades of the runner at the most appropriate angle forms the basis of reaction turbines. A Francis turbine can either be volute-cased or open-flumed. It has a spiral casing that distributes water uniformly to the runner. The blades are complexly profiled so that the water exits through the centre of the turbine via draft tube after imparting all its pressure energy to the blades of the runner (Figure 10.20).

The adjustable guide vanes of these turbines regulate the water flow as it enters the runner and are usually linked to a governing system, which matches flow to turbine loading in the same way as a spear valve or a deflector plate does in a pelton turbine. When the flow is reduced, efficiency of the turbine drops (Figure 10.19).

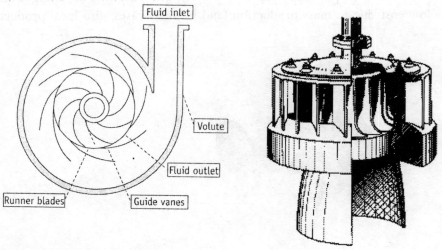

Figure 10.20 Francis turbine
Source Adapted from <http://lingolex.com/bilc/engine.html> and <http://www.tech.plym.ac.uk/sme/mech225/mom4pdf>, last accessed on 1 August 2005

Propeller turbine

The basic propeller turbine consists of a propeller fitted inside a continuation of the penstock tube. The turbine shaft passes out of the tube at the point where the tube changes position. The propeller usually has three to six blades, and the water flow is regulated by static blades. This kind of propeller turbine is known as a fixed-blade axial flow turbine because the pitch angle of the rotor blades cannot be changed. The part-flow efficiency of the fixed-blade propeller turbines tends to be poor.

Kaplan

Large-scale hydro sites make use of more sophisticated versions of the propeller turbines. Varying the pitch of the propeller blades together with the static plate adjustment enables reasonable efficiency to be maintained under part-flow conditions. Such turbines are known as variable pitch or Kaplan turbines (Figure 10.21).

Siphons like the guide vanes are nothing but a gate-closure mechanism. They are very effective but very noisy as well. Gate valves could also be used. The siphons do not usually allow the turbine to go through the runaway speed.

Reverse pump turbine

Centrifugal pumps can be used as turbines by passing water through them in reverse. Research is being carried out to enable the performance of pumps as turbines to be predicted more accurately. The potential advantages are the low cost due to mass production (and in many cases, also local production),

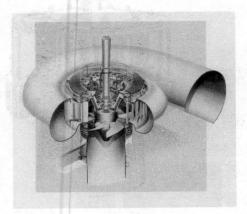

Figure 10.21 Kaplan turbine
Source http://www.tfd.chalmers.se/~hani/phdproject/proright.html,
last accessed on 4 July 2005

availability of spare parts, and the wider dealer/support networks. The disadvantages are poorly understood performance (efficiency 60%) characteristics and very poor part-flow efficiency. Many companies have used pumps as turbines at various times, but the technology remains unproven and relatively poor in performance.

Selection criteria of a turbine

The selection of a turbine for a particular location involves the following criteria.

- Net head available
- Range of discharge available
- Specific speed and rotor diameter
- Rotational speed and runaway speed

Net head available   Turbines are designed for specific heads (Table 10.3).

Discharge   The discharges are known when the system is designed for a particular capacity. The turbine can be selected by knowing the flow and the head available on the x and y axes, respectively (Figure 10.22).

The closure of these turbines and the speed increasers for the same are also very distinct (Table 10.4).

Specific speed and rotor diameter   The specific speed of the turbine constitutes the main criterion while selecting a turbine. Once this is fixed, the other dimensions and the turbine selected for a particular site could well be put in place (Penche and de Minas 1998).

The specific speed and the rotor diameter for a given turbine are given by

$$\text{Specific speed } (n_s) = n\sqrt{P}/H^{5/4} \qquad \qquad ...(10.8)$$
$$\text{Rotor diameter } (D) = 38.567 \times \sqrt{H}/n \qquad \qquad ...(10.9)$$

where n is the rotational speed, P is the output power of the turbine, and H is the head.

Table 10.3 Turbine types and applicable heads

Turbine type	Head range (in metres)
Kaplan and propeller	2 < H < 40
Francis	10 < H < 350
Pelton	50 < H < 1300
Michell-Banki	3 < H < 250
Turgo	50 < H < 250

H – head

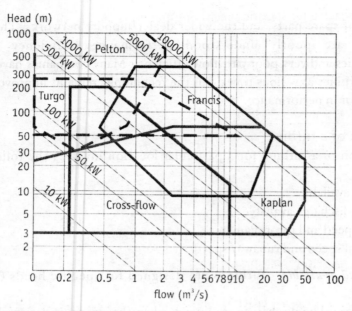

Figure 10.22 Selecting a turbine

Table 10.4 Configuration and selection of speed increasers

Configuration	Flow	Closing system	Speed increaser
Vertical Kaplan	Radial	Guide-vanes	Parallel
Vertical semi-Kaplan siphon	Radial	Siphon	Parallel
Inverse semi-Kaplan siphon	Radial	Siphon	Parallel
Inclined semi-Kaplan siphon	Axial	Siphon	Epicycloidal
Kaplan S	Axial	Gate valve	Parallel
Kaplan S right angle drive	Axial	Gate valve	Epicycloidal
Kaplan inclined right angle	Axial	Gate valve	Conical
Semi-Kaplan in pit	Axial	Gate valve	Epicycloidal

Source Penche and de Minas (1998)

Generally, all manufacturers specify the turbine-specific speed for their systems. A number of statistical studies have been done to determine the specific speed of the turbines. These studies have resulted in formulae determining the specific speed to be used for each type of turbine (Table 10.5). The equations given in Table 10.5 are illustrated in Figure 10.23. Turbines can be chosen using the above information.

Table 10.5 Formulae for specific speeds of different turbines

Pelton (1 jet)	$n_s = 85.49/H^{0.243}$
Francis	$n_s = 3763/H^{0.554}$
Kaplan	$n_s = 2283/H^{0.436}$
Cross-flow	$n_s = 513.25/H^{0.505}$
Propeller	$n_s = 2702/H^{0.5}$
Bulb	$n_s = 1520.26/H^{0.2837}$

n_s – specific speed
Source Penche and de Minas (1998)

Example 10

If the system has a head of 100 m and we wish to have a 850-kW turbine coupled to a generator of 1500 RPM, then the

$$\text{specific speed } (n_s) = n\sqrt{P}/H^{5/4}$$
$$= 1500 \times \sqrt{800}/100^{5/4}$$
$$= 134 \text{ RPM}$$

From Figure 10.23, we have Francis turbine as the obvious choice. We can also increase the speed by using more jets and a speed increaser ratio of 3:1 and also using a cross-flow or pelton having a specific speed of 45 RPM. The rotor diameter can be calculated as follows.

$$\text{Rotor diameter (D)} = 38.567 \times \sqrt{H}/n$$
$$= 38.567 \times \sqrt{100}/1500$$
$$= 0.257 \text{ m}$$

Rotational speed and runaway speed   The rotational speed of the turbine should always be the same as that of the generator. Therefore, standard generators with rotational speeds as given in Table 10.6 are preferred.

Table 10.6 Rotational speed of generators

Number of poles	Frequency		Number of poles	Frequency	
	50 Hz	60 Hz		50 Hz	60 Hz
2	3000	3600	16	375	450
4	1500	1800	18	333	400
6	1000	1200	20	300	360
8	750	900	22	272	327
10	600	720	24	250	300
12	500	600	26	231	277
14	428	540	28	214	257

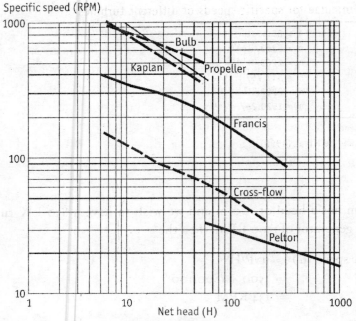

Figure 10.23 Graph to select the type of turbine based on the specific speed

The runaway speed of the turbine is the speed at which it would run if the whole electrical load was removed. This is usually two to three times the normal speed. Both the normal speed of the turbines and the runaway speed ratio of conventionally used turbines are given in Table 10.7.

Generators

Asynchronous generators

Taking into account their great similarity with the asynchronous motors currently used in industry, asynchronous generators are the most robust

Table 10.7 Normal speed to runaway speed ratio of different turbines

Turbine type	Normal speed n (RPM)	Runaway speed ratio (n_{max}/n)
Kaplan single-regulated	75–100	2.0–2.4
Kaplan double-regulated	75–150	2.8–3.2
Francis	500–1500	1.8–2.2
Pelton	500–1500	1.8–2.0
Cross-flow	60–1000	1.8–2.02
Turgo	600–1000	2

RPM – revolutions per minute

and cost-effective. While choosing an asynchronous generator in a small hydro project, the gap thickness between rotor and stator must be sufficiently large in order to avoid the mechanical failure that could lead to a certain extent of reduced overall efficiency of the machine and of the power factor.

The main limitation of the asynchronous generators is their need for reactive power. Reactive power necessary to magnetize the machine must be delivered by the public network and/or by static capacitors. They can never deliver a reactive power to an autonomous and/or public network and that is the reason for the net saleable energy reduction (power factor too weak). Their coupling to the network can easily be done at almost synchronous speed. However, when these machines are associated with capacitive compensation, the costs incurred because of components like capacitors must be considered. They also come with a plethora of functional constraints resulting from capacitive self-excitation, dielectric rupture of capacitors, and limited operating ranges.

Synchronous generators

The synchronous generator, having a rotor excitation winding, eliminates the limitations of the asynchronous generator with regard to reactive power. This results from the sufficient maintenance of thickness of the gap between the rotor and stator so as to avoid mechanical failures. Besides, its efficiency is higher than that of the asynchronous generator but its cost is also higher because the rotor configuration is more complex (windings on salient poles, excitation rings, or brushless excitation). The rotor excitation control (reactive power delivered to the network or absorbed by the machine) implies the use of a more complex regulating device, which would be more expensive. Permanent magnets can be replaced at the rotor excitation winding to keep the excitation control simple, which results in reduction in adjustment capacity. Table 10.8 gives different parameters of a synchronous generator.

Speed governors

The turbine and the generator in a given plant are designed for a certain speed and load. If the speed increases or decreases due to flow differences or if the output load increases or decreases, the turbine and the generator get affected and may fail under certain circumstances such as no-load conditions or heavy-flow conditions. This has to be controlled or governed using electro-mechanical devices. Control is achieved using two methods. First, the flow of water into the turbine can be controlled, and second, the flow could be left constant and the electrical output from the generator could be controlled

Table 10.8 Different parameters of a synchronous generator

Type (kW)	Power (V)	Voltage (A)	Current (RPM)	Speed (Hz)	Frequency (V)	Excitation voltage (Amp)	Excitation current (%)	Efficiency (kg)	Weight
sf 5-4/ 250SFW	5	400	9.02	1500	50	35.0	9.8	84.7	104
sf 8-4/ 250SFW	8	400	14.4	1500	50	34.5	15.0	86.0	113
sf 18-4/ 368SFW	18	400	32.5	1500	50	28.6	23.9	83.7	250
sf 26-4/ 368SFW	26	400	46.9	1500	50	35.7	23.9	85.5	280
sf 12-6/ 368SFW	12	400	21.7	1000	50	20.7	30.0	83.6	260
sf 18-6/ 368SFW	18	400	32.5	1000	50	26.5	30.0	84.9	290
sf 40-4/ 423SFW	40	400	72.2	1500	50	21.3	47.8	89.0	450
sf 55-4/ 423SFW	55	400	99.2	1500	50	25.7	48.5	87.8	520
sf 26-6/ 423SFW	26	400	46.9	1000	50	23.8	42.6	86.4	460
sf 30-6/ 423SFW	30	400	54.1	1000	50	23.9	48.5	86.9	460
sf 40-6/ 423SFW	40	400	72.2	1000	50	29.6	48.5	88.0	530
sf 75-4/ 493SFW	75	400	135.3	1500	50	22.0	42.0	88.9	710
sf 100-4/ 493SFW	100	400	180.4	1500	50	32.0	47.0	91.1	830
sf 55-6/ 493SFW	55	400	99.2	1000	50	32.0	36.0	89.3	750
sf 75-6/ 493SFW	75	400	135.3	1000	50	40.0	50.0	90.6	850
sf 40-8/ 493SFW	40	400	72.2	750	50	31.7	54.6	87.8	780
sf 55-8/ 493SFW	55	400	99.2	750	50	45.8	45.7	89.5	870
sf 100-6/ 590SFW	100	400	180.4	1000	50	24.0	120.0	90.1	1300
sf 75-8/ 590SFW	75	400	135.3	750	50	24.0	119.0	89.7	1320
sf 100-8/ 590SFW	100	400	180.4	750	50	29.0	122.0	90.9	1420

using electrical ballasts/dump loads at the output. The second method could be used for systems up to 100 kW capacities. But for higher capacities, mechanical governors are used.

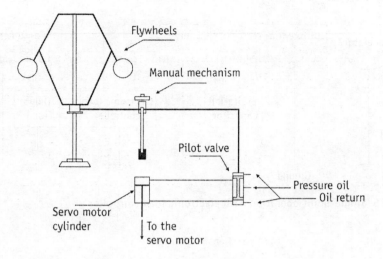

Figure 10.24 Mechanical governor operation

Mechanical speed governors

These governors have a gate that is connected to an electrical actuator, which in turn, is connected to the servo motor. As the flow or the electrical load increases or decreases, the actuator sends signals to the servo motor and this closes or opens the valve to control the flow into the turbine. Thus, the flow, and hence the load, is controlled (Figure 10.24).

Electronic load governors

Electronic load governors work on the single operating point principle. Here, the turbine is uncontrolled, and the electrical output is controlled using electronic switches that regulate the output power into the main load consumed by the end-users and the dump load (Figure 10.25). The switches used here are fast-acting power electronic devices such as thyristors or IGBTs (insulated gate bipolar transistors).

Switchgear

The switchgear to be used forms an important part of any system. It ensures safety and also maintains efficient performance of the system on the whole. Different countries adapt different standards for the switchgear and the distribution part of the plant. The main features that a plant should have are shown in Figure 10.26.

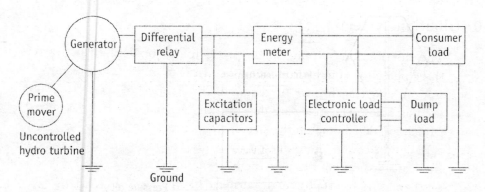

Figure 10.25 Single-point operation with electronic load controller

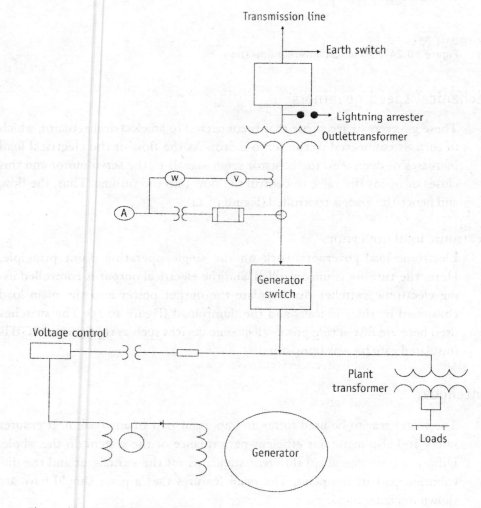

Figure 10.26 Switchgear system

Distribution system

The distribution system (single phase for pico/micro hydro systems and three phase for mini hydro schemes) consists of poles, transmission lines, sub-station, and transformer (if required). Pre-fabricated concrete columns are used as poles. The poles are dug into the ground to a limit as stated in the guidelines, depending on the size of the plant. The transmission lines may be aerial bunched, underground, or live wires. Guy wires are used as support for the transmission lines to give them strength against the various forces acting on them. The sub-station and transformers to step up and step down are used according to the size of the plant. Metering is also done wherever required. Necessary protection devices are installed and automated, if required.

Nomenclature

A	Area of cross-section, area under curve
D	Diameter of the pipe in cm
H	Head in m
h_f	Friction loss
k^{-1}	Conversion factor
L	Length of the penstock in m
M	Mass of salt
η	Manning's constant
n	Rotational speed
n_s	Specific speed
n_{max}/n	Runaway speed ratio
P	Output power of the turbine, kW
Q	Discharge in m³/s
s	Salt concentration
t_{min}	Minimum thickness of the pipe
v	Velocity in m/s
σ	Permissible stress in steel in kg/cm²
Φ	Joint efficiency
ρ	density of water

References

Mayer P and Smith N. 2001
Pico Hydro for Village Power, 106 pp.
London: UK Department for International Development

MNES (Ministry of Non-conventional Energy Sources). 2002
A Report on UNDP–GEF Hilly Hydro Project
New Delhi: MNES

Penche C and de Minas I. 1998
Layman's Guidebook on How to Develop a Small Hydro Site, 266 pp.
Bruselas: European Small Hydropower Association

Bibliography

Bevan G. 1999
Opportunities for Renewable Energy Technologies and Markets—costs and policies, 13 pp.
UK: UK Trade and Commerce

Croockewit J. 2004
Handbook for Developing Micro Hydro in British Columbia, 69 pp.
British Columbia: BC Hydro

Egrlea D and Milewski J C. 2002
The diversity of hydropower projects
Energy Policy **30:** 1225–1230

ESHA (European Small Hydro Association). 2004
BlueAGE (Blue Energy for A Green Europe) Report. A Strategic Study for the Development of Small Hydro Power in the European Union, 94 pp.
Belgium: ESHA

IEA (International Energy Agency). 2000
Economic Risk and Sensitivity Analysis for Small-scale Hydropower Projects
[IEA Technical Report], 40 pp.
USA: IEA

International Hydropower Association. 2000
Hydropower and the World's Energy Future
Sutton, Surrey: International Hydropower Association

Jiandong T. 2003
Small Hydro Power: China's practice, 4 pp.
Hangzhou: International Network for Small Hydro Power

Kueny J L. 1999
Objectives of Small Hydro Technology, 35 pp.
Grenoble Cedex 9, France: Institut National Polytechnique de Grenoble

Nouni M R, Mullick S C, and Kandpal T C. 2007
Techno-economics of micro-hydro projects for decentralized power supply in India
Energy Policy **35**: 2491–2506

Oud E. 2002
The evolving context for hydropower development
Energy Policy **30:** 1215–1223

Paish O. 2002
Small hydro power: technology and current status
Renewable and Sustainable Energy Reviews **6**: 537–556

Price T and Probert D. 1997
Harnessing hydropower: a practical guide
Applied Energy 57 (2/3): 175–251

Wilson E M. 2000
Assessment Methods for Small Hydro Projects
[IEA (International Energy Agency) Technical Report], 68 pp.
USA: IEA

WEC (Workd Energy Council). 2004
Survey of Energy Resources
Elsevier

Geothermal energy, tidal energy, wave energy, and ocean thermal energy*

V V N Kishore, Senior Fellow and Prashant Bhanware, Research Associate
Energy–Environment Technology Division, T E R I, New Delhi

Stuart L Ridgway#
537, 9th Street, Santa Monica, CA 90402, USA

Geothermal energy

When the earth was formed some 4600 million years ago, it was a mass of hot substances. After it cooled, the outer surface became a crust, but the inner mass remained hot and molten. The temperature at the centre of the earth is about 7000 °C and because of the large temperature difference between its interior and the surface, heat flows out at a steady rate. Also, the earth contains small quantities of long-lived radioactive isotopes, principally, thorium-232, uranium-238, and potassium-40, which liberate heat as they decay. The amount of heat flowing from the interior of the earth through the surface is about 10^{21} J per year, which, though small compared to about 5.4×10^{24} J per year of solar energy falling on it, is substantial for warranting exploration as a source of energy. Heat is transferred through the earth mainly by creep processes in hot solids.

The convective heat transfer process is quite efficient, resulting in small variations in temperature across the depth of the convecting layer. However, close to the surface (across the

---

* The sections on geothermal, wave and tidal power are primarily based on The Open University (1994) T521: Renewable Energy Resource Pack for Tertiary Education, subsequently republished as Boyle G (ed.) 2004 Renewable Energy: Power for a Sustainable Future, published by Oxford University Press, and reproduced with permission from The Open University.

outer 100 km of the earth), the material is too hard to convect, so heat is transported by conduction. Because of the low thermal conductivities, temperature increase with depth is much larger. The rigid outer boundary layer or shell is broken down into a number of fragments called the lithospheric plates that move around the surface at speeds of a few centimetres per year, their movement being controlled by the convective motions beneath. At the boundaries between the plates where they are in relative extension or compression, the heat flows are high (~300 mW/m^2) compared with a global mean of about 60 mW/m^2.

Along the plate margins, there are localized spots of very high heat flows because the rock material reaches the surface in molten form resulting in volcanic activity. The storage of molten or partially molten rock at about 1000 °C, just a few kilometres beneath the surface, strongly augments heat flow around even dominant volcanoes. These high heat flows result in high temperature gradients. Over the geological periods of time, these flows have resulted in large quantities of heat being stored in rocks at shallow depths, and these rocks form the sources of geothermal energy. The regions of highly concentrated heat flows are termed as 'high-enthalpy' regions. Heat is available in the form of steam and hot water at temperatures of 150–200 °C in the high-enthalpy regions. In areas of lower heat flow, where the convection of molten rock or water is reduced or absent, the temperature in the shallow rocks is lower (below 100 °C) and these are termed as the low-enthalpy regions. While high-enthalpy sites are suitable for power generation, low-enthalpy sites can be directly used for heating application, drying, etc.

The extraction of heat for useful purposes can aptly be termed as 'heat-mining'. Geothermal sources are not strictly renewable though at one time it was thought that many high-enthalpy resources were indeed renewable, in the sense that they could be exploited indefinitely. However, the experience of declining temperatures in steam fields and the simple calculations of heat supply and demand show that heat is being mined on a non-sustainable basis. The known and potential geothermal energy sources, however, can be exploited on a near-sustainable basis for several years to come.

Potential sites for geothermal energy utilization worldwide include the Himalayan geothermal belt, eastern China, Russia, Japan, the Philippines, Indonesia, New Zealand, Canada, United States, Mexico, Central American volcanic belt, the Caribbean, Ireland, eastern and southern Mediterranean, and the East Africa rift system (Figure 11.1).[1]

[1] Details available at <www.geothermal marin.org>, last accessed on 17 July 2005.

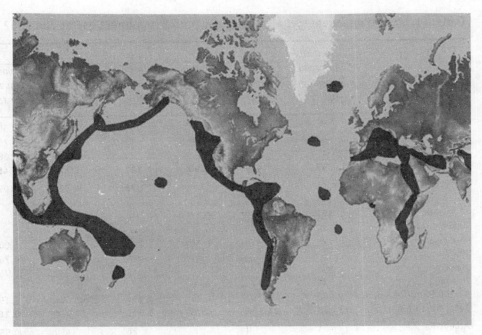

Figure 11.1 Potential sites geothermal global
Reproduced with permission from Geothermal Education Office, Tiburon, California
Source http://geothermal.marin.org

The exploited potential and the ultimate potential are given in Table 11.1 and Figure 11.2. (Gawell, Reed, and Wright 1999) (WEC Member Committees 2000). It can be seen that the potential is quite large compared to the current level of use.

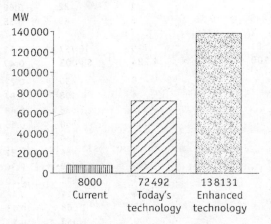

Figure 11.2 Geothermal energy potential and current utilization (in megawatts)
Source http://www.geo-energy.org/publications/reports/preliminary%20report.pdf

Table 11.1 Geothermal energy: electricity generation and direct use at end 2002

Country	Electricity generation			Direct use		
	Installed capacity (MW$_e$)	Annual output (GWh)	Annual capacity factor	Installed capacity (MW$_t$)	Annual output (TJ)	Annual capacity factor
Algeria				100	1 588	0.50
Kenya	57	447	0.90	1	11	0.25
Tunisia				20	173	0.28
Total Africa	**57**	**447**	**0.90**	**121**	**1 772**	**0.46**
Canada				378	1 022	0.09
Costa Rica	145	984	0.78			
El Salvador	161	940	0.67			
Guadeloupe	4	25	0.67			
Guatemala	28	175	0.71	3	108	1.00
Honduras				1	18	0.76
Mexico	853	5 398	0.72	164	3 920	0.76
Nicaragua	35	275	0.90			
United States of America	2 002	13 357	0.76	5 366	25 006	0.15
Total North America	**3 228**	**21 154**	**0.75**	**5 912**	**30 074**	**0.16**
Argentina	1	N		26	450	0.55
Chile				N	7	0.55
Colombia				13	266	0.63
Peru				2	50	0.65
Venezuela				1	14	0.63
Total South America	**1**	**N**		**42**	**787**	**0.60**
China	29	100	0.39	2 814	31 406	0.35
Georgia				250	6 307	0.80
India				80	2 516	1.00
Indonesia	807	6 238	0.88	7	43	0.19
Japan	547	3 431	0.72	258	5 836	0.72
Korea (Republic)				51	1 076	0.67
Nepal				1	22	0.66
The Philippines	1 931	10 248	0.61	3	40	0.38
Thailand	N	2	0.61			
Turkey	18	81	0.53	820	15 757	0.61
Total Asia	**3 332**	**20 100**	**0.69**	**4 284**	**63 003**	**0.47**
Austria	1	4	0.48	58	430	0.23
Belgium				4	108	0.87
Bulgaria				107	1 638	0.48
Croatia				114	550	0.15
Czech Republic				13	130	0.33
Denmark				3	84	0.81
Finland				300	3 200	0.34
FYR Macedonia				81	511	0.20
France				326	4 914	0.48
Germany				397	1 568	0.13
Greece				100	350	0.11
Hungary				250	3 600	0.46

Continued...

Table 11.1 (*Continued*)

Country	Electricity generation			Direct use		
	Installed capacity (MW$_e$)	Annual output (GWh)	Annual capacity factor	Installed capacity (MW$_t$)	Annual output (TJ)	Annual capacity factor
Iceland	202	1 433	0.81	1 800	24 700	0.44
Italy	862	4 660	0.62	680	8 916	0.42
Lithuania				41	681	0.53
The Netherlands				11	58	0.17
Norway				6	32	0.17
Poland				55	274	0.16
Portugal	16	105	0.75	2	3	0.25
Romania				191	2 128	0.35
Russian Federation	73	300	0.47	307	6 131	0.63
Serbia, Montenegro				86	2 415	0.89
Slovakia				75	340	0.14
Slovenia				103	1 080	0.33
Spain				70	1 051	0.47
Sweden				377	4 129	0.35
Switzerland				547	2 387	0.14
United Kingdom				3	36	0.38
Total Europe	**1 154**	**6 502**	**0.64**	**6 107**	**71 434**	**0.37**
Israel				63	1 714	0.86
Jordan				153	1 541	0.32
Total Middle East				**216**	**3 255**	**0.48**
Australia	N	1	0.68	10	295	0.90
New Zealand	448	2 715	0.69	308	7 081	0.73
Total Oceania	**448**	**2 716**	**0.69**	**318**	**7 376**	**0.74**
Total world	**8 220**	**50 919**	**0.71**	**17 000**	**177 701**	**0.33**

Source WEC Member Committees (2000)

Geothermal resources should, in general, have three important characteristics, as shown in Figure 11.3, viz an aquifer[2] containing water that can be accessed by drilling, a cap rock to retain the geothermal fluid,[3] and a heat source.

A geothermal aquifer must be able to sustain a flow of geothermal fluid so that even highly porous rocks are suitable only if the pores are interconnected. The velocity u of a fluid moving through a porous medium can be described by Darcy's law

$$u = K_w \left(\frac{H}{L} \right) \qquad \qquad \qquad ...(11.1)$$

[2] Porous rock that can store water and through which water can flow.
[3] Geyser or hot springs.

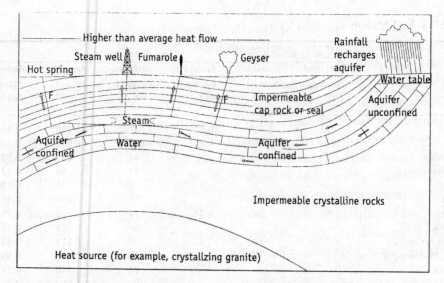

Figure 11.3 A schematic cross-section showing essential characteristics of a geothermal site
Permission granted by The Open University©
Source Open University (1994)

where K_w is the hydraulic conductivity and (H/L) is the hydraulic gradient, or change in the head H per metre of distance L along the direction of flow. The volumetric rate Q is given by

$$Q = AK_w \left(\frac{H}{L} \right)$$

...(11.2)

where A is the cross-sectional area. Some values of K_w for different rocks are given in Table 11.2.

Table 11.2 Porosities and hydraulic conductivities of selected materials

Material	Porosity (%)*	K_w (m/day)
Gravel	25–35	500–10 000
Sand, volcanic ash	30–40	1–500
Silt	40–50	10^{-2}–1
Clay	45–60	$<10^{-2}$
Mudrock	5–15	10^{-8}–10^{-6}
Sandstone	5–30	10^{-4}–10
Limestone	0.1–30	10^{-5}–10
Solidified lava	0.001–1	0.0003–3
Granite	0.0001–1	0.003–0.03
Slate	0.001–1	10^{-8}–10^{-5}

*Cavities present in rock
Source Open University (1994)

Cap rock consists of material that is relatively impermeable so that it acts like a seal for the geothermal fluid. The importance of cap rocks was discovered by the Italians in the early 1980s while exploring geothermal sources in an obvious place: the flanks of the Vesuvius volcano. Only small amounts of low-pressure fluid were discovered because the volcanic ashes that formed its flanks were apparently quite permeable. This proves that over time, gradual alteration in the uppermost deposits by hot water and steam creates clay or salt deposits that block the pores and thus seal the aquifer. This also explains why the youngest volcanic areas, like Vesuvius, are not necessarily the most productive geothermal sources.

The third pre-requisite for exploiting geothermal energy is the presence of a heat source. There are three different types of sources: volcano-related heat sources, sedimentary basins, and hot dry rocks.

Volcano-related heat sources

Several of the world's most advanced geothermal sites are located in the extinct volcanic areas. As rocks are good insulators, magmatic[4] intrusions may take millions of years to cool, and thus act as the focus for the 'hot fluid' in the permeable strata. The nature of the resource depends upon the local conditions of pressure and temperature in the aquifer, and especially on the P–T (pressure–temperature) depth profiles. Such profiles determine the extraction technology and economic viability of the site (Figure 11.4).

For a typical P–T depth curve shown in Figure 11.4, at shallow depths (~250 m), the temperature is too low for any water present to boil and pressure increases hydrostatically. However, in the depth interval 250–575 m, the temperature is high enough for water to vaporize so that the temperature curve lies slightly to the right of the boiling point curve. Pressure increase across this region is small because the pores are occupied by convecting water vapour rather than liquid water. Due to this reason, this is also an isothermal zone. Between 575 and 700 m, the rate of increase of temperature is proportional to the pressure corresponding to the boiling point so this is a liquid-dominated zone. Below 700 m, the pressure increases more rapidly and the P–T path deviates from the boiling point curve of water. High-enthalpy systems are subdivided into vapour- and liquid-dominated (that is, steam or liquid water) systems in the reservoir. The vapour-dominated systems are best and most suitable, mainly because steam is dry and has very high enthalpy.

---

[4] Magma is the partially molten rock.

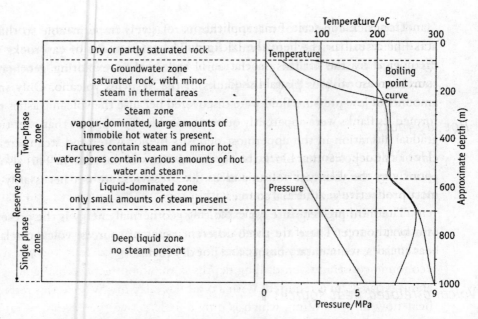

Figure 11.4 Variation of pressure and temperature with depth for a typical geothermal field
Source Open University (1994)

In the liquid-dominated systems, steam is often wet and of lower enthalpy, which adds to the technical problems of electricity production.

Sedimentary basins

These are heat sources where aquifers carry water to the depths where it becomes warm enough to exploit. Heat flow in rocks is usually governed by the conduction equation

$$q = k \left(\frac{\Delta T}{z} \right) \qquad \qquad ...(11.3)$$

where q is the heat flux in W/m², k is the thermal conductivity, and ΔT is the temperature difference over depth z.

The value of k is 2.5–3.5 W/m K for sandstones, limestones, and many crystalline rocks. However, mud rocks (clays and shales) have lower values of the order of 1–2 W/m K. These are also quite impermeable, so mud rocks act as impermeable cap rocks and enhance the geothermal gradient above aquifers in regions with otherwise normal heat flows. This has led to explorations aimed at locating natural warm waters in the areas of thick sedimentary rock sequences containing both mud rocks and permeable limestones or sandstones. Large-scale heat applications of geothermal energy worldwide take place in basins where the background heat flow is above average. The geological reasons for the association of high heat flow with sedimentary basins are reasonably well established.

Hot dry rocks

Hot dry rock resources refer to the heat stored within the impermeable or poorly permeable rock strata and to the process of extraction of heat. An artificial fracture zone is created within the suitably hot rocks and water is circulated through such a zone for extracting heat. Drilling is expensive, so only the top 6 km of the earth's crust is used for calculating the geothermal potential with the present technology. Given the current technical and economic constraints on drilling depths, a minimum geothermal gradient of about 0.025 °C/m is required. With a typical k value of 3 W/m k, this requires a heat flow of 75 MW/m^2, which is only a little above the earth's average. In practice, it is customary to look for rocks with much higher heat flows and the ideal targets are granite bodies.

Technologies for utilization of geothermal resources

The first stage in prospecting for geothermal resources in the volcanic areas involves a range of geographical studies aimed at locating rocks that have been altered chemically by hot geothermal brines. Surface thermal manifestations, such as hot springs[5] or mud pools, are also examined carefully. Chemical investigations and the release of gases through fractured rocks provide means of assessing the composition and resource potential of the trapped fluids. Geophysical prospecting, particularly by resistivity surveying and other electrical methods for detecting zones with electrically conducting fluids (brines), is probably the most effective technique in locating buried geothermal resources. Once a suitable geothermal aquifer has been located, exploration and production wells are drilled using special techniques to cope with the much higher temperatures and, in some cases, harder rock conditions compared to the oil and water wells. As fluid pressures in the aquifer can go up to 10 MPa, dense drilling muds are required to counter these pressures and to avoid a 'blow out' where an uncontrollable column of gas is discharged. The dimensional and constructional details of the wells to be dug are well established.

[5] Hot water only.

The technologies for utilizing geothermal heat for useful applications can be broadly divided into two categories: using the geothermal fluids directly and using a heat exchanger for the extraction of heat. The direct methods of using heat for power generation can further be classified as (1) dry steam plants, (2) single-flash steam power plants, and (3) double-flash power plants, shown in Figures 11.5 (a), (b), and (c), respectively.

The dry steam power plant is installed in sites where superheated steam at 180–185 °C and 0.8–0.9 MPa is available. The power plants of 1960s required about 15 kg of steam per saleable kWh, but modern plants with higher temperature steam and better turbine designs can achieve 6.5 kg steam/kWh. Plant efficiency is strongly affected by the presence of non-condensable gases, such as carbon dioxide and hydrogen sulphide.

In the single-flash system (Figure 11.5b), the fluid reaching the surface may be wet steam (water that has flashed within the well during ascent) or hot

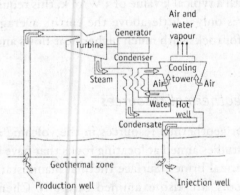

Figure 11.5 (a) Dry steam power plant

Figure 11.5 (b) Single-flash steam power plant

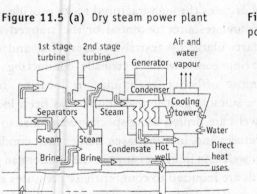
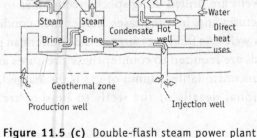

Figure 11.5 (c) Double-flash steam power plant
Permission granted by The Open University©
Source Open University (1994)

water at high pressure. It is often better to avoid flashing in the well because it can lead to scale deposits and plugging of the well. Conventional steam turbines are employed but these operate at lower efficiencies due to lower steam pressures and temperatures. The bulk of fluid mined (often up to 80%) may be in the form of unflashed brine, which is re-injected, unless there are local direct uses.

Double-flash systems (Figure 11.5c) are ideal where geothermal fluids contain low levels of impurities. The scaling and non-condensable gas problems are minimal in such cases. Unflashed liquid remaining after the initial high-pressure flashing flows to a low-pressure tank where another pressure drop provides additional steam. This steam is mixed with the exhaust from the high-pressure turbine to drive a second turbine, raising power output by 20%–25% for a 15% increase in plant cost.

The indirect method of using geothermal heat for power production employs a binary cycle power plant, also called the ORC (Organic Rankine Cycle) system (Figure 11.6). It uses a secondary working fluid with a boiling point lower than that of water, such as pentane or butane, which is vaporized and used to drive the turbine. The geothermal brine is pumped at reservoir pressure through a heat exchanger unit and is then re-injected. Though there are clear advantages, such as the utilization of low-temperature heat and protection from impurities, the capital costs are high. Also, the parasitic power consumption is high (~30%) due to the need for using large pumps for injection, secondary fluid circulation, etc. These kinds of power plants had been used in solar ponds and the OTEC (ocean thermal energy conversion) systems.

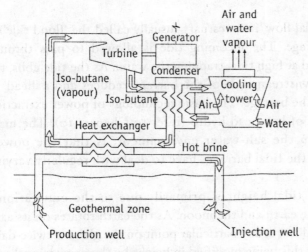

Figure 11.6 Binary cycle power plant
Permission granted by The Open University©
Source Open University (1994)

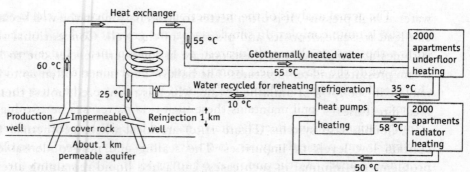

Figure 11.7 A scheme for utilization of geothermal energy for non-electrical application
Permission granted by The Open University©
Source Open University (1994)

An example of the indirect method of using geothermal energy for non-electrical application is shown in Figure 11.7. Here, heat pumps are used to enhance the system efficiency. Heat pumps, which work on the same principle as refrigeration, use electricity, and also increase the overall efficacy of a thermal application, such as residential heating. In the longer term, heat pumps may allow widespread economic development of even shallower, cooler geothermal aquifers.

Tidal power

The use of tides to provide energy has a long history. The idea of using tidal energy on a large scale to generate electricity was first considered in 1938 with turbines mounted in large barrages (essentially low dams) built across suitable estuaries.

The upstream 'tidal flow' in an estuary (usually called the 'flood tide') is trapped behind a barrage. The incoming tide is allowed to pass through sluices, which are closed at high tide, trapping the water. As the tide ebbs, the water level on the downstream side of the barrage reduces and a 'head' of water develops across the barrage. The basic technology of power extraction is then similar to that of low-head hydro (refer to Chapter 10). The main difference, apart from the salt-water environment, is that the power-generating turbines in the tidal barrages have to deal with regularly varying heads of water.

The variation in tidal height is primarily due to the gravitational interaction between the earth and the moon. As the earth rotates on its axis, gravitational forces produce, at any particular point on the globe, a twice-daily rise and fall in sea level; this being modified in height by the gravitational pull of the sun, and by the topography of landmasses and oceans.

The actual analysis of the interaction between the earth, the moon, and the sun is quite complex. In simple terms, the gravitational pull of the moon draws the seas on the earth 'nearest' to the moon into a bulge *towards* the moon while the seas furthest from the moon experience a less than average lunar pull and bulge *away* from the moon (Figure 11.8). As the earth rotates on its axis, the lunar pull maintains these high-tide patterns. The two high-tide configurations will in effect, be drawn around the globe as the earth rotates, giving, at any particular point, *two* tides per day occurring approximately 12.5 hours apart. As the moon also moves in an orbit around the earth, the timing of these high tides at any particular point will vary, occurring approximately 50 minutes later each day.

This basic pattern is modified by the pull of the sun, as shown in Figure 11.9. When the sun and the moon pull together (in line), the result is very high 'spring tides'. When they are 90° apart, the result is the lower 'neap tides'. The period between the neap and spring tides is approximately 14 days, that is, half the 29.5-day lunar cycle. The ratio between the height of the maximum spring and minimum neap tides can be more than two.

The above pattern is modified in reality by the fact that the moon's orbit is *elliptical*. There are other longer-term variations, for example, a semi-annual cycle caused by the inclination of the moon's orbit in relation to that of the earth. There are also forces due to spin of the earth, called the *Coriolis forces*,

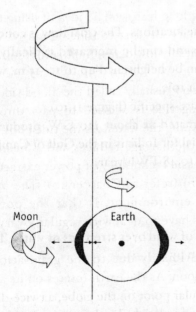

Figure 11.8 Relative rotation of the moon and the earth
Permission granted by The Open University©
Source Open University (1994)

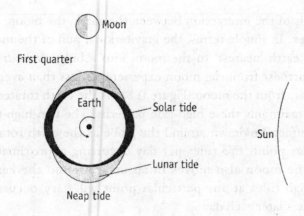

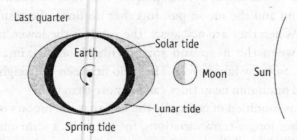

Figure 11.9 Influence of the sun on tides
Permission granted by The Open University©
Source Open University (1994)

(see Chapter 8) which modify tides in some locations. The tidal flow is concentrated as the tide approaches the shore and can be increased typically up to 3 m. In suitably shaped estuaries, it can be heightened up to 10–15 m, with complex 'resonance effects' playing a major role.

Tidal power availability is thus very site-specific (Figure 11.10).

The practical tidal potential is estimated at about 120 GW, producing approximately 190 TWh/year. The potential for India is in the Gulf of Cambay (16.4 TWh/year) and the Gulf of Kachchh (0.48 TWh/year).

Wave power

Ocean waves are generated by the passage of wind over stretches of water. The power P (in kW/m) of an ideal wave is given by

$$P = \frac{\rho\, g^2\, H^2\, T}{32\,\pi}$$

where H is the wave height in metres and T is the wave period in seconds.

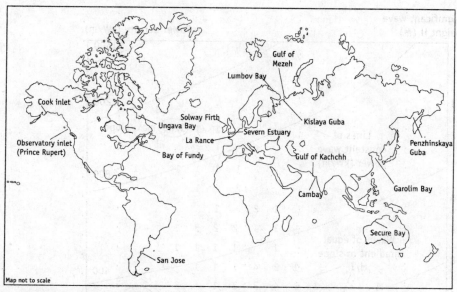

Figure 11.10 Some promising tidal power sites in the world
Permission granted by The Open University©
Source Open University (1994)

By deploying a wave-rider buoy, it is possible to record the variation in the surface level during a chosen period of time (Figure 11.11).

For a typical irregular sea, the average total power is given by

$$P_s\,(kW/m) = \alpha_s H_s^2 T_e$$

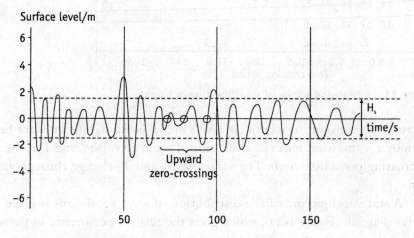

Figure 11.11 A typical wave record
Permission granted by The Open University©
Source Open University (1994)

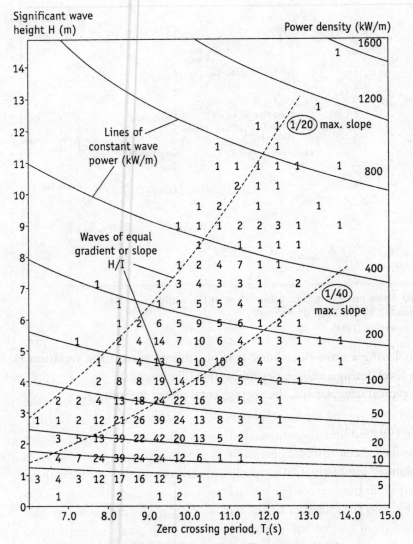

Figure 11.12 Scatter diagram of significant wave heights

where α_s is a constant equal to 0.49 kW/s m³, H_s is the significant wave height given by 4 × rms (root mean square) value of water elevation, and T_e is the zero up-crossing period (second). The values of H_s and T_e change throughout the year.

A statistical picture of the distribution of wave conditions is given by a scatter diagram (Figure 11.12), which gives the relative occurrence in parts per 1000 of the contributions of H_s and T_e.

The estimates of wave power density at various locations in the world are shown in Figure 11.13.

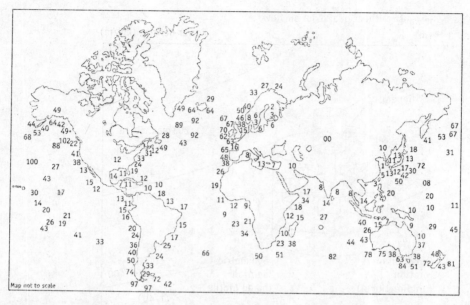

Figure 11.13 Annual average wave power in kW/m for various locations in the world

Wave energy technology

In order to capture energy from the sea waves, it is necessary to intercept the waves with a structure that can respond in an appropriate manner to the forces applied to it by the waves. The structures can be classified as fixed or floating. Fixed seabed and shore-mounted devices are the only wave energy converters to have been tested as prototypes at sea. The majority of devices tested and planned are of the OWC (oscillating water column) type. In these devices, an air chamber pierces the surface of water and the contained air is forced out of it into the chamber by the approaching wave crests and troughs. On its passage from and to the chamber, air passes through an air turbine generator and produces electricity. A novel axial flow air turbine, the Wells turbine, is proposed for many OWCs (Figures 11.14 and 11.15).

The Wells turbine rotates in the same direction, irrespective of whether the airflow is into or out of the chamber, and has aerodynamic characteristics, particularly suitable for wave applications.

Wave energy research has been going on in many places in the world but more so in Japan. The other countries doing research and development till the late 1980s were China, Denmark, India, Norway, Portugal, Spain, Sweden, and the UK. The status of various prototypes in the early 1990s is shown in Table 11.3.

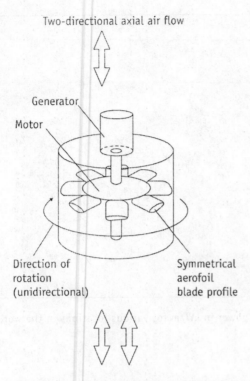

Figure 11.14 The Wells turbine for conversion of wave energy

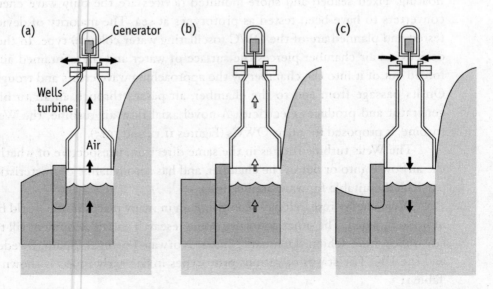

Figure 11.15 Principle of operation of an oscillating water column wave energy converter

Table 11.3 Summary of various prototypes of wave energy systems

Year	Type	Location	Owner	Installed capacity (kW)	Comments
1965	Navigation buoy OWC	Japan	Maritime agency	0.05	Several hundred deployed around the coastline of Japan
1978–86	Kaimei	Japan	IEA	375–1000	Vessel motion compromised the system performance. No further interest in energy testing but fundamental data on moorings and materials
1983	OWC	Sanze, Japan	Mitui and Fuji	40	Low output; decommissioned after one year
1983	Pendulor	Muroran, Japan	Muroran Institute of Technology	5	Still operational
1984 floating	Kaiyo Japan terminator	Okinowa, Japan	Institute of Ocean Environmental Technology	Not known completed	Research programme
1985	OWC Norway	Toftestallen, Brug	Kvaerner	600	Good performance. Destroyed by storms in December 1988
1985	Tapchan	Toftestallen, Norway	Norwave	350	Good performance, still operational
1985	Pendulor	Mashike, Japan	Mashike Port	5	Supplies hot water, still operational
1985	OWC	Neya, Japan	Taisei Corporation	40	Wells turbine driving a heat-generating eddy current-type device. Tests finished in 1988
1988	OWC array	Kujukuri, Japan	Takenaka Komuten Company	30	Array of 10 OWCs, with rectifying valves feeding a common high-pressure reservoir Planned to continue until 1995
1989	Hinged flap	Wakasa Bay, Japan	Kansai Electric Power Company	1	Under test
1989	Tethered float	Hanstholm, Denmark	Danish Wave Power APS	45	Problems with rubber seals. Further trials are planned.
1989	OWC	Sakata, Japan	Port and Harbour Research Institute	60	The OWC is an integral part of a new harbour wall. Now operational
1991	OWC	Islay, UK	Queen's University, Belfast	75	Still operational. Over 1000 hours of testing
1991	OWC	Trivandrum, India	IIT Madras	150	Latest available information in 1992 was that device was nearing completion
1992	FWPV	West Coast of Sweden	Sea Power AB	110 like Tapchan	Successful operation, not sensitive to tidal range
1992	OLAS 1000	West coast of Spain	Union Fenosa, Madrid		Array of seven hose pumps/buoys

OWC – oscillating water column;   FWPV – floating wave power plant;
IEA – International Energy Agency;   IIT – Indian Institute of Technology
Source Open University (1994)

Ocean thermal energy conversion[6]

The tropical oceans of the world are an enormous energy resource. Their surface water is a heat source typically at 25–27 °C, and the deep water below is available for a heat sink at a temperature of about 5 °C. This temperature difference used to make useful energy is called OTEC. Temperature difference between surface and depth of 1000 m for different zones is shown in Figure 11.16.

In 1994, William H Avery, Applied Physics Laboratory, Johns Hopkins University, and Chih Wu, United States Naval Academy, published 'Renewable Energy from the Ocean'. From their preface

'The upper layers of the tropical oceans are a vast reservoir of warm water that is held at a temperature near 27 °C by a balance between the absorption of heat from the sun and the loss of heat by evaporation, convection, and long-wavelength radiation. On an average day, the water near the surface absorbs more heat from the sun in one square mile (2.5 km²) of ocean area than could be produced by burning 7000 barrels of oil. For the whole tropical ocean area, the solar energy absorbed per day by the surface waters is more than 10 000 times the heat content of the daily oil consumption of the United States. A practical and economical technology for converting even 0.01% of the absorbed solar energy into electricity or fuel in a form suitable for delivery to consumers on land could have a profound impact on world energy availability and economics.'

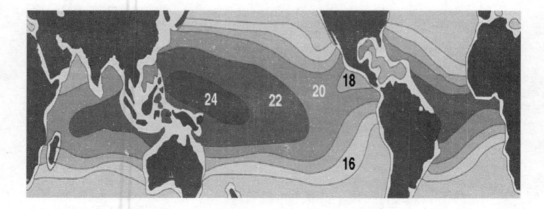

Figure 11.16 OTEC resource map
Source http://www.xenesys.com/english/otec/area.html

[6] Contributed by Stuart L Ridgway

OTEC needs supplies of warm and cold water with a temperature difference of at least 20 °C, an inexpensive and very efficient heat engine for extracting work from the heat supply heat sink combination, and reasonable methods for delivering the output to users.

History

The French physicist Jaques D'Arsonval suggested the use of OTEC resource over a hundred years ago. The Rankine (closed) cycle heat engines he suggested would have used a typical refrigerant as working fluid. For example, liquid ammonia would have been boiled using warm water heat, the vapour expanded through a turbine to do the cycle work, and then condensed back to liquid to be pumped back into the boiler. For modest yields of output work, large quantity of heat was required for which large boilers and condensers were needed.

Georges Claude, who had liquefied air and proposed separating neon by fractional distillation for the 'neon sign', attempted to build OTEC plants in the 1920s and 1930s, which used water vapour flashed from warm surface water in a vacuum as the working fluid. The warm and cold waters flowing through the apparatus were the essential heat exchange surfaces, saving heat exchanger costs. However, very low density of water vapour at OTEC temperatures makes it a poor working fluid for a power extraction turbine. The size and cost of a turbine adapted to this very low density vapour is a handicap to the Claude open cycle. He did achieve 22-kW output from his turbine, but the utilities supplying water and creating vacuum used more power.

Back then, the supply of low-cost fossil fuels was more and the environmental consequences of uninhibited combustion did not loom large. Forests dying of acid rain were rare. Global warming due to carbon dioxide emissions to the atmosphere was not an issue. So economic success eluded him.

The Arab oil embargo of 1973/74 emptied the Los Angeles freeways; low speed limits were made mandatory on the interstate highways; and there were long queues for fuel at the gas stations. There was a rush to buy locking lids for one's fuel tanks to frustrate the siphoners.

All this stimulated a substantial programme in the 1970s and early 1980s to develop the technology to exploit the OTEC resource.

Various attempts have been made to develop low-cost heat engines that can exploit this small temperature differential to provide useful energy. However, the heat engines seemed too costly, so it was back to inexpensive oil.

Now that oil is becoming more expensive again, and the climate experts are insisting on reducing carbon dioxide emissions, the possibilities of OTEC ought to be re-examined.

Closed-cycle development

Closed-cycle or Anderson cycle systems (Figure 11.17) use fluid with a low boiling point, such as ammonia, to rotate a turbine to generate electricity. Heat is transferred in the evaporator from the warm sea water to the working fluid. The working fluid exits from the evaporator as a gas near its dew point. The high-pressure, high-temperature gas is then expanded in the turbine. From the turbine exit point, the working fluid enters the condenser where it rejects heat to the cold sea water. Usually, the subcooled liquid leaves the condenser and finally, this liquid is pumped to the evaporator, completing a cycle. The T–S diagram of the closed system is shown in Figure 11.18.

The heat exchangers (evaporator and condenser) are a large and crucial component of a closed-cycle power plant. Much work has been done on materials for OTEC heat exchangers. Inexpensive aluminium alloys are used for this purpose.

This process can be modified to produce desalinated water as a by-product; the cold water (warmed only about 10 °C by the OTEC process) can

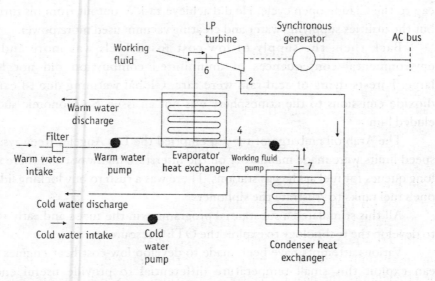

Figure 11.17 Schematic diagram of a closed cycle system for OTEC
Source http://www.answers.com/topic/ocean-thermal-energy-conversion

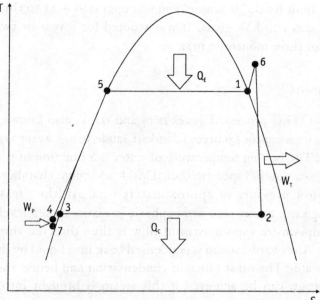

Figure 11.18 T–S diagram of the closed power plant
Source http://www.answers.com/topic/ocean-thermal-energy-conversion

condense large volumes of fresh water when it is passed through a heat exchanger in contact with a humid tropical atmosphere.

The world's first net power producing OTEC plant, 'Mini-OTEC', was a closed cycle plant and was deployed in 1979 on a barge off the Natural Energy Laboratory of Hawaii by the State of Hawaii, Lockheed Ocean Systems, and other private sector entities. It used ammonia as its working fluid.

This plant operated for three months, generating approximately 50 kW of gross power with net power of 10–17 kW (Ridgway 1984). Though only about 20% of the Mini-OTEC's gross power was available for export, the net-to-gross ratio was expected to approach 75% for closed-cycle plants larger than about 10 MW, making the process commercially more viable.

Other considerations associated with a closed-cycle OTEC power plant are the potential leakage of ammonia and the discharge of small amounts of chlorine that are added to the ocean water to prevent fouling of heat exchangers. Practices developed over the past 100 years in the refrigeration industry can minimize ammonia leakage. Experiments at the Natural Energy Laboratory of Hawaii (Ridgway, Hammond, and Lee 1981) have demonstrated that very small, environmentally benign levels of chlorine can successfully control the micro-fouling that could dramatically diminish the efficiency of the heat exchangers at the small ΔT available for the OTEC operation.

A plant was built by the Japanese, and was operated next to the island of Nauru. Power was 100 kW gross. It was studied for a year or two, and produced power for three months in 1981.

Open-cycle development

In the open-cycle OTEC process (Figures 11.19 and 11.20), also known as the Claude cycle after its inventor Georges Claude (Claude 1930), water vapour is the working fluid. The boiling temperature of water is a function of pressure. The warm surface sea water evaporates (boils) inside a vacuum chamber that is maintained at a low pressure of approximately 0.34 psi (the pressure at 80 000 feet [24.4 km], about 1/40 atmospheric pressure at sea level). The resulting low temperature vapour (steam) flow is then directed through a turbine generator. Afterwards, steam is condensed back into liquid by the flow of cold deep sea water. The most efficient condensation and hence, the highest electricity output, can be achieved if this steam is brought into direct contact with the cold sea water.

However, if steam flows through a surface condenser, in which it does not directly come in contact with the cold sea water, the resulting condensate is desalinated water. This pure fresh water 'by-product' is valuable for human consumption and agricultural purposes, especially in regions where natural freshwater supplies are limited. The reduced efficiency of the surface condenser, however, significantly reduces the production of electrical energy from the turbine. Designs that condense half the steam with a surface

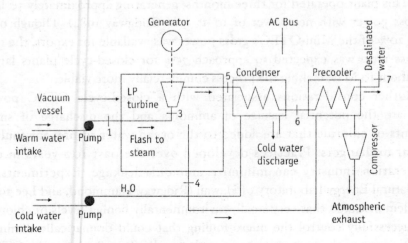

Figure 11.19 Schematic diagram of open-cycle OTEC
Source http://www.answers.com/topic/ocean-thermal-energy-conversion

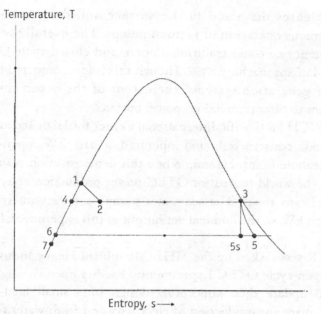

Figure 11.20 T–s diagram of open cycle power plant
Source http://www.answers.com/topic/ocean-thermal-energy-conversion

condenser, and then finish the steam condensation with the partially warmed sea water exiting the surface condenser can greatly improve electrical efficiency but reduce the freshwater output.

As the pressure drop across the turbine is the difference between the low pressure at which the water vaporizes and the lower pressure remaining after condensation, open-cycle systems require very large turbines to capture relatively small amounts of energy. Claude calculated that a 6-MW turbine would need to be about 10 m in diameter, and he could not design a realistic turbine larger than this. Recent re-evaluation of Claude's work (Parson, Bharathan, and Althof 1985) indicates that modern technology cannot improve significantly on his design, so it appears that the open-cycle turbines are limited to about 6 MW.

Less than half of the one per cent of the incoming warm ocean water becomes steam, so large amounts of water must be pumped through the plant to create enough steam to run the large, low-pressure turbine. However, this does not substantially reduce the surplus or net electrical power as pumping surface sea water requires little energy. In an ideal open-cycle plant, the vacuum pumps could be shut down after start-up, as all the water vaporized in the evaporator would be condensed in the condenser, leaving behind a vacuum. In the real world, however, both inevitable vacuum leaks and non-condensable gases dissolved in the surface and deep sea water necessitate continuous operation of vacuum pumps. The overall thermal-to-electrical efficiency of these traditional open- and closed-cycle OTEC plants is very similar, approaching 2.5%. Though this is low compared to the traditional power generation systems, the extent of the ocean thermal resource is sufficient to offer tremendous power outputs.

In 1993, the PICHTR (Pacific International Center for High Technology Research) designed, constructed, and operated a 210 kW open-cycle OTEC plant at Keahole Point, Hawaii. When this demonstration plant was operational, it set the world record for OTEC power production at 255 kW (gross) (Vega and Evans 1994). The sea water pumps and vacuum systems consumed about 170 kW, so the nominal net output of this experimental plant was about 40 kW.

The PICHTR was tasked by the MHI (Mitsubishi Heavy Industries) to maintain the open-cycle OTEC Experimental Facility open through December 1998, and update their conceptual design of a small land-based open-cycle OTEC plant for production of electricity and fresh water for Pacific Islands. In 1991, PICHTR, under the sponsorship of MoFA (Ministry of Foreign Affairs), documented the conceptual design for a 1.8-MW (gross) land-based open-cycle-OTEC plant for production of electricity and fresh water. The MHI task was to update the design using information gathered at the open-cycle OTEC Experimental Facility since 1992. The conceptual design was to include technical specifications for the major Open Cycle OTEC components that the MHI was able to manufacture. These are the low-density steam turbines and the pumps required to maintain the process vacuum by removing non-condensables (air and steam).

Following the successful completion of experiments, the 210-kW OTEC plant was shut down and demolished in January 1999.

Mist Lift

The mist lift process, a new open-cycle concept introduced in 1977, offers a way around the high cost difficulties of the previous OTEC engines. It avoids the giant heat exchangers of the 'closed cycle' originally proposed by D'Arsonval, and the enormous water vapour turbine required by Claude's 'open cycle'.

In the mist lift process, warm ocean water is sprayed upward from the bottom into an evacuated vertical duct. The ambient pressure is of the order of 2400 Pascals (0.348 psi). Vapour evaporates from warm water. Mist, a mixture of water droplets and water vapour, is formed. At a distance of 10–20 m above the bottom, cold water is sprayed upward into the duct. It condenses the vapour, and establishes a pressure of 1200 Pascals, which is lower than the bottom pressure. Driven by the pressure difference, the vapour flows upward from the bottom to the cold water spray-condensing region, dragging warm water droplets with it. The mist is thus accelerated to a substantial velocity. As the vapour condenses, the mist and cold water merge, forming a single-phase fluid, which goes to the top of the duct. The lifted water is then collected and passed through a hydraulic turbine to provide the output power of the plant.

The process uses the vapour flashed from a spray of very fine warm water droplets to lift these droplets to the height of Niagara Falls (140 feet [42.6 m]). Alternatively, water can be first dropped through a hydraulic turbine to provide the desired power output from the cycle, and then the mist is lifted and merged with the condensing cold water, and returned to the ocean. The mechanical coupling between the droplets and the lifting vapour depends upon the viscosity of the vapour, which does not diminish with lowering pressure, whereas the coupling between vapour and the turbine blades of the Claude-cycle depends on very low vapour density, which requires large turbines.

By placing the warm water and cold water injectors sufficiently below the sea level, one may dispense with cold and warm water supply pumps, which gives the concept an additional cost advantage over closed or Claude cycle OTEC.

The concept was initially tested on a 3.7-m high column erected in a laboratory at Dynamics Technology. The experiments demonstrated sufficient coupling to lift the mist to a height of 50 m with temperature differences typical of the tropical seas. These freshwater results were verified in experiments at the Natural Energy Laboratory of Hawaii in the early 1980s for droplet to vapour coupling with sea water.

A cost estimate of a conceptual design of a 4-MW mist lift OTEC power plant was prepared and published in 1984. This design was based on a modest extrapolation of the mist transport data acquired in the freshwater experiments in 1980/81 and ocean water experiments in 1983. It was optimized for minimum cold water use with a condenser effectiveness of 0.9, yielding an output of 450 kJ/m^3 of cold water. Allowances for cold water pumping power, mist generator loss, filter loss, hydraulic turbine efficiency, exit loss, and non-condensable removal power reduced this yield to a net value of 300 kJ/m^3 of cold water. The projected cost was 10 million dollars.

The two-stage mist lift

The cold and warm waters that emerge from the mist lift plant are mixed. The process uses a larger flow of cold water than warm water. The emerging water is cool and could accept more heat. Improved performance and lower costs could be achieved by adding a second stage that uses the cool water output from the first stage, and thus reduces the total cost (Tables 11.4 and 11.5).

A recent design analysis of this concept predicts that a two-stage mist lift plant can provide net power of 600 kW/m³/s of cold water.

Thus, the two-stage mist lift can yield twice the output per unit cold water supply of the present OTEC versions, and has many other potential economies. Further research and development in this direction promise large returns.

Appendix

The maximum thermodynamically possible workout from a flow W_h of warm water at temperature T_1 and flow W_c of cold water at temperature T_0 is given by

(1) $\text{Work}/c_p = W_h(T_1 - T_2) - W_c(T_2 - T_0)$ from conservation of energy, where c_p is the fluid-specific heat and T_2 is the common temperature of the exit water. If the heat engine is ideal and no entropy is created, T_2 can be found as follows

Table 11.4 Characteristics of 1.6-MW two-stage mist lift OTEC

Cold water flow	2.0 T/s (tonnes per second)
Head loss cold water pipe	5.0 m
Warm water flow stage 1	1.8 T/s
Warm water flow stage 2	1.8 T/s
Warm temperature	25 °C
Cold temperature	5 °C
Transfer temperature	14 °C
Output water temperature	17.8 °C
Stage 1 power	900 kW
Stage 2 power	720 kW
Input turbine	3.6 m³/s at 10 m head
Output turbine	5.6 m³/s at 18 m head

Table 11.5 Performance comparisons of recent plants and designs

Item	Warm/cold ratio	Power/cold flow (MW/T/s)
Closed cycles		
Nauru	1.03	0.257
GE 40 MW	1.15	0.217
John Hopkins	0.97	0.292
Open cycles		
NELH net power	1.5	0.246*
Mist Lift 1984	0.6	0.308
Two-stage Mist Lift	1.80	0.605

Source Avery and Wu (1994)

The theoretical maximum power, given $T_{warm} = 298°$, $T_{cold} = 278°$, and warm/cold flow ratio = 1.0, is 1.4 MW/T/s, and for warm/cold = 1.8 is 1.9 MW/T/s. Therefore, there is much possibility for substantial increase in the OTEC performance.

*The maximum output was 200 kW; the water was supplied by NELH pumps that are not optimized for the power plant; a 100-kW charge was rather arbitrarily taken for the pumping. Zero charge would make its power/cold = 0.5.

(2) $(W_h + W_c)*\ln(T_2) = W_h*\ln(T_1) + W_c*\ln(T_0)$ and substituted back into (1) to obtain the possible work.

Nomenclature

Geothermal energy

H Head
H/L Hydraulic gradient
k Thermal conductivity (W/mK)
K_w Hydraulic conductivity
L Distance
P Pressure
q Heat flux (W/m²)
T Temperature
u Velocity (m/s)
z Depth (m)

Wave power

H Wave height (m)
H_s Significant wave height (m)
P Power (kW/m)

T   Wave period (s)

T_c Zero-up-crossing period (s)

Appendix

c_p Specific heat

T   Temperature

W   Flow rate

References

Avery J H and Wu C. 1994
Renewable Energy from the Ocean: a guide to OTEC, 446 pp.
New York: Oxford University Press

Claude G. 1930
Power from the tropical seas
Mechanical Engineering **52**: 1039 pp.

Gawell K, Reed M, and Wright P M. 1999
Geothermal Energy: the potential for clean power from the earth
Geothermal Energy Association
Details available at http://www.geo-energy.org/aboutGE/potentialUse.asp,
last accessed on 17 July 2005

Harrison R, Mortimer N D, and Smarason O B. 1990
Geothermal heating: a handbook of engineering economics, 572 pp.
New York: Pergamon Press

Open University. 1994.
T521 Renewable Energy: a resource pack for tertiary education
Milton Keynes: The Open University

Parson B K, Bharathan D, and Althof J A. 1985
Thermodynamic Systems Analysis of Open-cycle Ocean Thermal Energy Conversion (OTEC)
[SERI TR-252-2234]
Golden, CO: Solar Energy Research Institute

Ridgway S L. 1984
Projected capital costs of a mist lift OTEC power plant 84-WA/Sol-33
[Winter meeting ASME, December 1984, New Orleans]

Ridgway S L, Hammond R P, and Lee C K B. 1981
Experimental demonstration of the feasibility of the mist flow ocean thermal energy process
[American Institute of Aeronautics and Astronautics, Terrestrial Energy Systems Conference, 2nd, Colorado Springs, CO, 1–3 December, 8 p, Organized by American Institute of Aeronautics and Astronautics]

Vega L and Evans D E. 1994
Operation of small open cycle OTEC experimental facility
[Proceedings of Oceanology International 1994, Vol. 5]
Beighton, UK

WEC Member Committees. 2000
Geothermal energy: electricity generation and direct use at end-1999

Bibliography

Baker C. 1991
Tidal power
Energy Policy **19**(18): 792–797

Charlier R H. 1982
Tidal Energy, 351 pp.
New York: Van Nostrand Reinhold

Deffeyes K. 2001
Hubbert's Peak: the impending oil shortage, 208 pp.
Princeton, NJ: Princeton University

Goodstein D. 2004
Out of Gas: the end of the age of oil, 128 pp.
New York: W W Norton & Company

Heinberg R. 2003
The Party is Over: oil, war and the fate of industrial societies, 288 pp.
East Sussex: Clairview Books

Salter S. 1992
Wave energy: some questions and answers
International Journal of Ambient Energy **14**(1): 17–23

12

Bio-energy resources

Kusum Lata, Fellow, and Sanjay P Mande, Fellow
Energy–Environment Technology Division, TERI, New Delhi

'Biomass' is a collective term used for all materials that are biogenic in origin, that is, derived from the products of photosynthesis (Orr and Govindjee 2005). Photosynthesis is a complex set of reactions in which carbon dioxide is converted into organic material by getting reduced to carbohydrates using light and water (Figure 12.1).

Three basic requirements of the photosynthetic reaction are sunlight, carbon dioxide, and water. Pigments of green plants (primarily chlorophylls and carotenoids) absorb sunlight and make energy available for chemical reactions. The process of photosynthesis uses this energy to make sugar from water and carbon dioxide. Water provides electrons for reduction and converts into

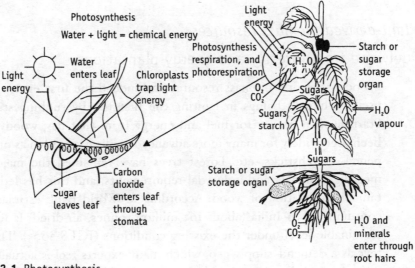

Figure 12.1 Photosynthesis

oxygen and protons. Sugar can be converted into starch for storage and is a potential source of energy. The basic reaction of photosynthesis can be given as

$$6CO_2 \text{ (carbon dioxide)} + 6H_2O \text{ (water)} \xrightarrow{\text{sunlight}} C_6H_{12}O_6 \text{ (glucose)} + 6O_2 \text{ (oxygen)}$$

Photosynthesis involves the massive conversion of sunlight into sugars and oxygen, which is the basis of energy for life on earth. It has been estimated that photosynthetic activity stores 17 times as much energy as is annually consumed by all nations in the world. Taking into account the energy requirements for collecting, processing, and converting, biomass still assures a bright future from the energy point of view. It holds the promise of addressing the energy needs of the world, if managed and used effectively and sustainably.

Biomass resources

Biomass being so versatile and scattered in nature, sufficient database and documentation, in terms of its availability and consumption/utilization patterns, are not always available. Although biomass meets a major part of the total energy requirements in many developing countries, it does not find an appropriate place in the overall energy balances, probably due to a lack of proper documentation and database. Biomass includes a wide range of plant and animal materials, available on a recurring basis. These can broadly be classified into wood, oil-bearing trees, residues, aquatic and marine biomass, and waste (Figure 12.2).

Plant-derived biomass resources

Fuelwood from forestry and energy plantations

Wood is the first biomass resource and indeed the first energy source to be used by human beings in igniting the first fire by rubbing sticks together. Starting with its use for fuel and energy from that day, woody biomass has been used widely for many more advanced applications such as pulp and paper, pencil, matchsticks, etc. Forest trees have remained the major source for meeting domestic and industrial requirements, and this has led to an unsustainable utilization of wood. According to World Bank estimates, fuelwood consumption in India (about 162 million tonnes, air-dried) is four times the sustainable level under the existing conditions (IGES 2005). This figure also reveals a demand–supply gap, which many experts feel is actually filled in by pilferage and illegal cutting down of trees in the country's forests (TEDDY

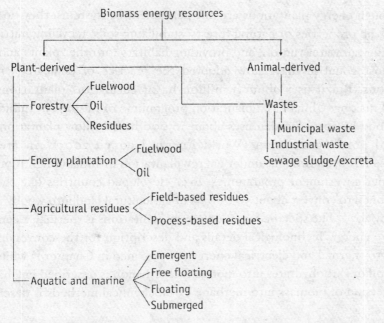

Figure 12.2 Classification of biomass resources on the basis of their origin

2002). However, the forest cover in recent years seems to be improving somewhat, which leads to the surmise that the demand gap is actually filled by 'trees outside forests', which include road-side plantations, orchards, trees growing in agro/farm forests and along canals, roads, railway lines, private lands, and wasteland, especially by *Prosopis juliflora* (*Vilayti babul*).

Growing energy plantations on wastelands, to meet the fuelwood needs, without affecting agricultural lands has been identified as one of the immediate economic solutions for harnessing solar energy. A plantation can be defined as a kind of forest created for commercial benefits, and the one dedicated to fuel/energy is called an energy plantation. The basic principle of energy plantation is that the amount of energy provided by the energy crop must exceed the energy used to produce it. These plantations are grown mainly in two ways.

- As energy crops in lands specifically dedicated to energy plantations.
- As an intercrop with non-energy crops.

The species for energy plantations are selected according to their wood properties, planted (usually as a single species), and grown to a useful size as quickly as possible. The selection of species depends on many factors such as adaptation to the local soil, irrigation needs, climate, cropping conditions, yields, pest tolerance, etc. (Appendix 1 [James 1983]).

Although energy plantations are not permanent ecosystems, they do have many of the properties of natural forests: stabilizing soils, recycling nutrients, controlling rainwater run-off, and providing habitats for other plants and animals. Most countries have now adopted the practice of dedicated energy plantations. Brazil uses about 3 million ha of eucalyptus plantations for charcoal making; China has a plantation programme of raising 13.5 million ha of fuelwood by 2010; Sweden uses about 16 000 ha of willow plantations for heat and power generation (World Energy Council 2005). An area of 17 000 ha has been brought under energy plantation in India (Singh 1996) till 1992 under government programmes. Even developed countries like the US have planned to convert about 50 000 ha of agricultural land into woody plantations by 2020. The specific target of energy plantations is thermal, electrical, or motive energy. Technological details and description for the conversion of biomass to thermal and electrical energy are provided in Chapter 13 while the conversion of carbohydrates into motive fuel (ethanol) is described in Chapter 15. Conversion of biomass into methane by biochemical methods is described in Chapter 14.

Oil from forestry and energy plantations

There are a number of plants that provide an alternative form of lighting/fuel. Some of these plants yield wax or oil that can be made directly into candles; some yield oil that can be burnt; and other plants can be used as wicks. Wax gets deposited on the plant's fruits, leaves, and catkins and is obtained by boiling the plant (usually only the fruit, which tends to have the greatest quantity of wax), allowing the liquid to cool, and then removing the wax as it solidifies. Wax can then be reheated and formed into candles and the liquid left behind is used as a blue dye. Some plants yield oil, which solidifies. Oil extracted from fruits and seeds of such plants has the consistency of tallow if allowed to stand so it can be formed into candles. Many plants produce seeds that are rich in oil, which can be extracted by pressing. Extraction of edible oil and fat from plant/tree parts (seeds, kernels, etc.) is an ancient practice employed to meet cooking oil requirements. This resource has an important role in the energy sector as it can also be used as a transportation and power generation fuel. Vegetable oil extracted from energy crops is an alternative to diesel fuel and can be blended with diesel. Biomass fuel oil industry is developing rapidly around the world. China has discovered 1554 species of oil plants, and among these, seeds of 154 plants contain more than 40% oil (Tao 2005). India has also carried out a detailed survey for identification of edible and non-edible oil-yielding varieties (Table 12.1). The chemistry of vegetable oils,

Table 12.1 List of some non-edible fuel-oil-yielding plants in India

Name of plant	Family name
Alangium salvifolium (Linn.f.) Wang	Alangiaceae
Aphanamixis piolystachya (Wall.) Parker	Meliaceae
Azadirachta indica A. Juss	Meliaceae
Barringtonia racemosa Roxb. (L.) Spreng	*Barringtoniaceae/Lecythidaceae*
Calophyllum inophyllum L. C. apetalum Willd	Clusiaceae
Calophyllum polyanthum Wall. ex Choisy	—
Canarium strictum Roxb.	Burseraceae
C. zeylanicum (Retz.) Bl. (Native of Sri Lanka)	Burseraceae
Cerbera manghas L.	Apocynaceae
Cirsum arvense (L.) Scop. (Native of Europe)	Asteraeae
Entada rheedei Spreng.	Mimosaceae
Argemone mexicana L.	Papaveraceae
Hernandia ovigera L.	Hernandiaceae
Hibiscus cannabinus L.	Malvaceae
Hydnocarpus alpina Wight	Flacourtiaceae
Jatropha curcas L.	Euphorbiaceae
Jatropha gossypifolia L.	Euphorbiaceae
Joannesia princeps Vell	Euphorbiaceae
Neolitsea umbrosa (Nees) Gamble	Lauraceae
Palaquimia ellipticum (Dalz) Baill	Sapotaceae
Pongamia glabra Vent.	Fabaceae
Derris indica (Lank.) Bennett.	
Putranjiva roxburghii Wall.	Euphorbiaceae
Drypetes roxburghii Wall. Hurusawa	
Rhus semialata Murr.	Anacardiaceae
Ricinus communis Linn.	Euphorbiaceae
Salvadora oleoides Decne	Salvadoraceae
Sterculia foetida Linn.	Sterculiaceae
Xylocarpus moluccensis (Lam) Roem.	Meliaceae
Carapa moluccensis Lam	
Garcinia rubro-echinata Kosterm.	Clusiaceae
Gymnacranthera canarica Warb.	Myristicaceae
Mesua ferrea L.	Clusiaceae
Tamarindus indica L.	Caesalpiniaceae
Cascabela thevetia (L.) Lippold	Apocynaceae
Thevetia peruviana (Pers.) Merr. (sometimes edible)	
Viburnum coriaceum Blume (Nepalese use it for edible purposes)	Caprifoliaceae
Celastrus paniculatus Willd.	Celastraceae
Neolitsea cassia (Linn.) Kostermans Syn.	
Sarcostigma kleinii Wight & Arn.	Icacinaceae
Vateria indica Linn. (non-edible)	Dipterocarpaceae

Source Botanical Survey of India (2004)

their processing, and an overview of different technologies of production are covered in Chapter 15.

Residues

Residue can be defined as a by-product of any plant-based activity that changes its shape, size, volume, state, and characteristics at each stage of processing. On the basis of their origin, residues can be classified as field- and process-based. Field-based residues are those plant materials that remain in the farm after removal of the main crop produce, for example, straw, stalks, sticks, leaves, fibrous materials, roots, branches, twigs, etc. Process-based residues (agro-industrial residues) are by-products of post-harvest processes of crops, namely, cleaning, threshing, sieving, crushing, etc., and can be in the form of husk, dust, or straws. Examples are groundnut shells, rice husk, bagasse, corn cobs, coconut shell, and coir pith (Dhingra, Mande, Kishore, *et al.* 1996). Effluents from the pulp and paper industry can also be considered under process-based residues but this chapter catagorizes industrial effluents under the waste category.

Field-based residues (wood pruning, stalks of mustard and cotton, straws of paddy, wheat, millet sorghum, pulses, etc.) have remained a major short- to medium-term source of bioenergy and can play a bigger role in the longer term. Residues from the agricultural sector are a large and under-exploited potential energy resource, and present many opportunities for better utilization. However, a number of important factors have to be addressed when considering the use of residues for energy. First, there are many other alternative uses, for example, as animal feed, as soil erosion control, as animal bedding, and as fertilizers (dung), etc. It has been estimated that the rural population in developing countries consumes about 1 tonne/person/year (15% moisture, 15 GJ/t) and in semi-urban and urban areas the consumption is about 0.5 tonne/person/year (including all types of biomass and end-uses) (World Energy Council 2005). Second, they are seasonally available and their availability is unpredictable. Along with other factors, availability depends on variability in productivity due to seasonal vagaries, the amount of residue deemed essential for maintaining soil organic matter, soil erosion control, efficiency in harvesting, losses, non-energy uses, etc. (Iyer, Rao, and Grover 2002). Although availability of residues at the time of harvest makes collection easy for small-scale utilization, it creates storage problems if residues have to be saved for use during other months of the year, especially due to their low bulk density. The amount of residues available depend upon harvesting time, their storage-related characteristics, storage facility, etc. Low bulk density residues can be densified for storage, easy handling, transportation, and further usage. The technologies utilized for residue densification and the combustion devices employed for thermal and electrical energy conversion are described in Chapter 13.

Aquatic and marine biomass

Aquatic macrophytes are a group of aquatic plants, which are macroscopic (can be seen by the naked eye) and include flowering vascular plants, mosses, ferns, and macroalgae. They have the characteristic feature of high growth rates (Table 12.2).[1] Generally, for fast-growing submerged plants, the doubling time is one to four days, which means that if the tank is one-fourth full of plants, it will be completely full in two to eight days. If we then remove three-fourth of those plants, the tank will again be completely full in another two to eight days, and so on. A few such varieties are Eurasian watermilfoil (*Myriophyllum spicatum*), elodea (*Elodea canadensis*), coontail (*Ceratophyllum demersum*), curly-leaf pondweed (*Potamogeton crispus*), water hyacinth (*Eichhornia crassipes*), hydrilla (*Hydrilla verticillata*), alligator-weed (*Alternanthera philoxeroides*), etc.

Aquatic macrophytes can be subdivided into four groups: emergent, floating-leaved, submerged, and free floating (Figure 12.3). Emergent plants have their tops elevated in the air, bottoms submerged in water, and are rooted to the base. These plants are mostly perennials[2] as the conditions are ideal for plant growth. The sediment provides nutrients and water and the overlying water and emergent portions of the plant gain carbon dioxide and sunlight.

Table 12.2 Aquatic plant production

Yield (tonne/ha/year)	Duckweed	Azolla	Water hyacinth	Salvinia
Fresh matter	521.0	569.0	2190.0	691.00
Dry matter	16.9	34.2	131.0	27.60
Crude protein	6.1	9.6	13.4	6.08

Source Chará, Pedraza, and Conde (1999)

[1] 'Growth' is a technical term, which refers to the rate at which plants increase their biomass. It is generally reported as grams produced per gram extant per day. A constant growth rate will, therefore, produce an exponential increase in biomass over time. An optimal or maximal growth rate is the maximum intrinsic growth rate of a given plant.

[2] Plants or plant parts living for more than a year.

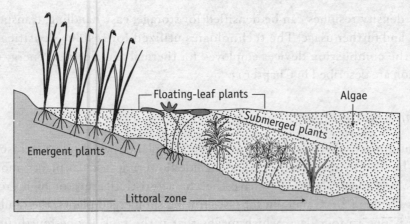

Figure 12.3 Different types of aquatic biomass

Being rooted to the bottom gives them strength to withstand the wind and waves in the shallow water zone. In northern climates, dry dead stems often supply oxygen for root respiration during winter when water bodies are locked under ice. Thus, cutting off dead stems below the water surface before the water body freezes is used as one of the management techniques as it limits oxygen supply and kills rhizomes. Common emergent macrophytes include plants such as reeds (*Phragmites* spp.), bulrushes (*Scirpus* spp.), cattails (*Typha* spp.), and spikerushes (*Eleocharis* spp.). Some emergents, wild rice (*Zizania* spp.) for example, have submerged or floating leaves before mature aerial leaves form.

Floating-leaved macrophytes are those plants that are rooted to the lake bottom with their leaves floating on the surface of the water. They generally occur in those areas of the water body that do not dry out frequently. Common representatives include water lilies (*Nymphaea* spp.), spatterdock (*Nuphar* spp.), and water shield (*Brasenia* spp.). Floating leaves are attached to the roots or rhizomes with a flexible, tough stem (actually in many cases a leaf stalk). Floating leaves survive in two extremely different habitats with water on the bottom and air on top. A thick, waxy coating protects the top of the leaf from the aerial environment. This makes herbicidal control difficult without the addition of special chemicals called adjuvants (wetting agents) to help the herbicide stick to and penetrate the waxy surface. Adjuvants are also used on many kinds of emergent and free-floating species when treating with herbicides for the same reason. Floating leaves can be ravaged by wind and waves so these plants are usually found in protected areas.

Submerged macrophytes are those plants that grow completely under water. These include diverse group of plants such as quillworts (*Isoetes* spp.), mosses (*Fontinalis* spp.), muskgrasses (*Chara* spp.), stoneworts (*Nitella* spp.), and numerous vascular plants. Submerged species face specific problems of obtaining light for photosynthesis and carbon dioxide from water, where it is available in much less quantities compared to air. Submerged species invest much less energy into structural support because they are supported by water, which accounts for about 95% of the weight of these species.

Free-floating macrophytes are those plants that typically float on, or just under, the water surface with their roots spread out in water and not in sediment. Small, free-floating plants include duckweeds (*Lemna* spp.), mosquito fern (*Azolla caroliniana*), and water fern (*Salvinia* spp.). Larger surface floating plants include water hyacinth (*E. crassipes*) and frog's bit (*Limnobium spongia*). Free-floating species are entirely dependent on water for their nutrient supply and are thus also used to remove excess nutrients from waste water, for example, water hyacinth. The wind, waves, and current change, so they are likely to be found in quiet locations and embayments.

Cultivation of several of these macroscopic algae and freshwater aquatic weeds is considered to have potential for large-scale growth and conversion into energy. Aquatic plants are considered for this purpose as some of them are known to be capable of exceptionally high levels of biological productivity and organic yields per unit of time and space. Second, they are grown naturally in the existing coastal waters and wetlands.

On the other hand, a very limited number of freshwater sources are appropriate for aquatic biomass production as these weeds often pose a hazard to them. Presently, their overgrowth in freshwater resources is a global problem. The most common methodology followed for the management of aquatic weeds is mechanical removal due to the various advantages associated with it, for example immediate availability of water for use, absence of dead and decaying organisms, non-toxicity, etc. This methodology is associated with high cost, low efficiency, and slow production of a large volume of plant residues. Economic utilization of the harvested plant residues for energy generation using biochemical/thermochemical conversion technologies (Mande and Lata 2005) can make mechanical removal of biomass a cost-effective option (Table 12.3).

Waste

Waste can be defined as solid, liquid, or gaseous material that is to be discarded by disposing it off, burning or incinerating, or recycling. Waste associated with the biomass processing industries and the domestic sector is considered under biomass resources. These include animal residues such as

Table 12.3 Characterization of a few aquatic weeds

Plant	Moisture (%)	TS (%)	VS (% of TS)	Calorific value (HHV, kcal/kg)
Paragrass				
Vegetative part	74.9	25.1	92.9	3731.1
Root part	89.5	10.5	82.8	3189.0
Root mat	89.3	10.7	62.4	2272.2
Salvania plant	92.1	7.9	62.3	2250.9
Phragmites				
Vegetative part	54.8	45.2	90.3	3734.7
Root part	89.1	10.9	85.2	3299.4
Cyprus	82.1	17.9	90.1	3561.9

TS – total solids; VS – volatile solids; HHV – higher heating value
Source TERI (2003)

cattle dung; poultry litter; MSW (municipal solid waste), including food and vegetable market waste; and industrial organic waste such as that from food-processing industries, sugar industry, plantation industries, such as coffee-processing industry, leather, distilleries, pulp and paper, and many more. In addition, agricultural waste and aquatic and wetland vegetation form an important category of bioresources (considered separately in the 'Residue/aquatic biomass' section), suitable for conversion into useful energy and manure through biological process.

Characterization of biomass

The properties of biomass feedstock are among the key factors that influence the selection of the processing technology and its design, operation, efficiency, maintenance schedule, etc. Therefore, characterization of biomass is the first step for efficient utilization of its energy potential. The processing technologies can broadly be categorized as combustion-based and biochemical-based (for biogas and biofuels). Characterization requirement of the feedstock and the impact of each parameter vary with the processing technology. Therefore, an understanding of the significance of each parameter with respect to the processing technology is also important.

Various properties taken into account for characterization of feedstock are presented in Figure 12.4. There are specific methods devised by the ASTM (American Society for Testing and Materials) (Appendix 2), APHA (American Public Health Association), and TAPPI (Technical Association of the Pulp and Paper Industry) for biomass characterization. Many quick analytical

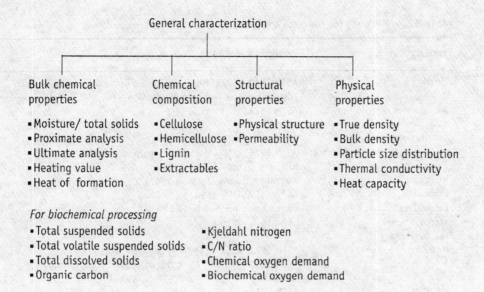

General characterization

Bulk chemical properties
- Moisture/ total solids
- Proximate analysis
- Ultimate analysis
- Heating value
- Heat of formation

Chemical composition
- Cellulose
- Hemicellulose
- Lignin
- Extractables

Structural properties
- Physical structure
- Permeability

Physical properties
- True density
- Bulk density
- Particle size distribution
- Thermal conductivity
- Heat capacity

For biochemical processing
- Total suspended solids
- Total volatile suspended solids
- Total dissolved solids
- Organic carbon
- Kjeldahl nitrogen
- C/N ratio
- Chemical oxygen demand
- Biochemical oxygen demand

Figure 12.4 Selective parameters of biomass characterization

procedures are also available now, which use modern techniques and instruments to carry out these analyses (Curvers and Gigler 1996) (Table 12.4). Broadly, the pathway that can be followed for biomass characterization is shown in Figure 12.5.

Bulk chemical properties

Moisture content

The moisture content of most biomass feedstocks depends on the category of biomass and its origin. Moisture content of the biomass is a combination of inherent moisture and surface moisture, and moisture analysis generally represents physically bound water only. It is an important property with regard to the energy conversion technology. Water released by chemical reaction during pyrolysis is incorporated into volatile matter during proximate analysis.

For thermochemical conversion technologies, moisture content of the biomass is an undesirable burden, which has to be removed to the extent feasible before usage. A high-moisture-containing fuel makes its ignition increasingly difficult and decreases the thermal efficiency of the system. This is especially important for gasification and pyrolysis, as heat is used to drive off the water and consequently this energy is not available for facilitating further chemical reactions to produce combustible gases/vapours.

Table 12.4 Methods for biomass analysis

Property	Analytical method
Heating value	ASTM E711, D2015, Modified Dulong formula, Tillman formula, Institute of Gas Technology equation
Particle size distribution	ASTM E828
Bulk density	ASTM E873
Constitutional analysis	
Lignin/ structural carbohydrate	ASTM E1758-01, Unnumbered method by LAP
Protein	Nitrogen factor
Nitrogen	Kjeldahl nitrogen (Jackson 1967)
Ethanol	LAP-011, ASTM E 1690
Extractives	Unnumbered method by LAP
Starch	LAP-016
Proximate analysis	
Moisture	ASTM E871, ASTM E1756-01, LAP-001, T-412, OM-02
Ash	ASTM E830 (575 °C), ASTM D1102 (600 °C), Unnumbered method by LAP, E1755-01 (550 °C–600 °C)
Volatiles	ASTM E872/E897
Fixed carbon	By difference
Ultimate analysis	
C(carbon), H(hydrogen)	ASTM E777
N(nitrogen)	ASTM E778
S (sulphur)	ASTM E775
Cl (chloride)	ASTM E776
Ash elemental	ASTM D3682, ASTM D2798, ASTM D4278
Ash fusibility	Microwave digest, ASTM E953/D1857, ash sinter test, fuel pellet test
Water-soluble alkali	ASTM C114
Chemical fractionation	ASTM Unnumbered method
Metals	ASTM E885
Digested slurry composition	
Proximate analysis	LAP-012
Insoluble solids	LAP-018

ASTM – American Society for Testing and Materials; LAP – Laboratory Analytical Prcedure

The quality of gas deteriorates, which, in turn, decreases the heat value of the gas (Figure 12.6). If gas is to be used for direct combustion purposes, low heating values can be tolerated and the use of feedstock with moisture content in the range 40%–50% (wet basis) is feasible. In downdraft gasification systems, high moisture content not only gives rise to low gas heating values but also leads to low temperatures in the reaction zone, which, in turn, leads to insufficient tar-converting capability, that is, low-grade gas for engine applications.

Efficient processes for low-moisture content biomass materials are: integrated gasification combined cycle, supercritical water gasification combined

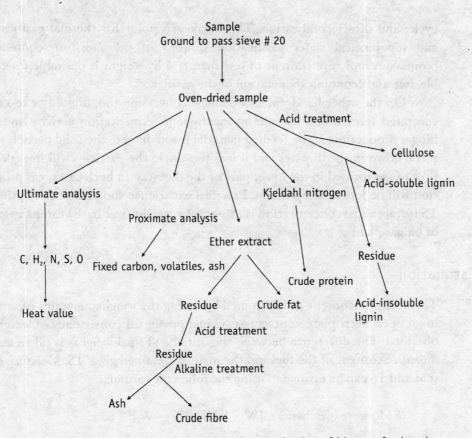

Figure 12.5 Broad pathway representing characterization of biomass feedstock

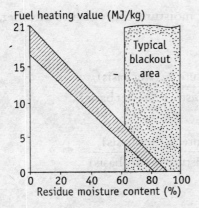

Figure 12.6 Effect of moisture content on heating value of fuel

cycle, and direct combustion. The breakeven point for thermal gasification and supercritical water gasification is 40% of the moisture content in biomass. A moisture content of less than 15% by weight is desirable for trouble-free and economical operation of the gasifier.

On the other hand, for biochemical conversion too, diluted or too concentrated fermentation results in insufficient fermentation activity and low biogas production. If the feeding material is too diluted, the solid particles will settle down in the digester and if it is too thick, the particles will impede the flow of gas formed in the lower part of the digester. In both cases, gas production will be less than optimum. Practical experience shows that the optimum TS (total solids) concentration is 8%–11% by weight (wet basis) for many types of biogas plants.

Estimation

The moisture content is determined by drying the weighed amount of sample in an open petri-plate kept at 105 °C in an oven till consistency of weight is obtained. The difference between the initial and final weights is taken as the moisture content of the fuel and the material retained gives TS. Moisture content and TS can be estimated using the following formula.

$$\% \text{ Moisture (wet basis)} = ([W_2 - W_f]) / ([W_2 - W_1]) \times 100$$

$$\% \text{ Moisture (dry basis)} = ([W_2 - W_f]) / ([W_f - W_1]) \times 100$$

where W_1 is the weight of empty crucible, W_2 is the weight of the crucible and sample, and W_f is the constant weight of crucible and sample after drying.

In order to convert one form of moisture content into another, the following formulae can be used.

$$\text{Moisture \% (dry basis)} = \frac{\text{Moisture \% (wet basis)}}{100 - \text{Moisture \% (wet basis)}}$$

$$\text{Moisture \% (wet basis)} = \frac{\text{Moisture \% (dry basis)}}{100 + \text{Moisture \% (dry basis)}}$$

Proximate analysis

Proximate analysis characterizes the biomass feedstock for volatile matter, fixed carbon, and ash content. The proximate analysis of several biomass materials is given in Appendix 3. Many a time, moisture analysis is also included as part of proximate analysis. The dry matter (residue after moisture removal) consists of combustible fraction and ash. Combustibles can further be subdivided into volatile matter that can be driven off during biomass pyrolysis (volatile carbon) and the remaining char (fixed carbon). The division of combustible matter into 'volatiles' and 'fixed carbon' depends upon the experimental conditions and hence, a method of determination has to be specified and standardized. A rapid heating rate normally yields more volatile matter.

$$\text{Biomass} = \text{moisture} + \underline{\text{volatile matter} + \text{fixed carbon} + \text{ash}}$$
$$\text{TS (total solids)}$$

TS content can be estimated as

% TS = 100 – % MC (wet basis)

Ash represents the transformed minerals after combustion of biomass at high temperatures. For exact ash content analysis, small corrections to the ash weight are necessary to correct it to a mineral matter basis.

Volatile matter

During combustion, volatile matter and inherently bound water in the fuel are released in the pyrolysis zone forming vapours consisting of water, tar, oils, and gases. These have to be burned again properly in order to reduce emissions and to realize high efficiencies of combustion equipment. Quantitative estimates of volatiles thus helps in designing equipment such as wood burning stoves (chullas), biomass fixed boilers, and gasifiers.

In the context of biochemical conversion processes, volatile matter is an indication of the amount of organic matter available as feed source for microbial consortia. The methane production in a biogas plant is directly proportional to the volatile solids. Thus, it is customary to report biogas yields in terms of $kgCH_4/kgVS$. For a bioreactor, loading rate of the reactor is expressed as the weight of total volatile solids added per day, per unit volume of the digester.

Estimation

For thermochemical processing, volatile matter is estimated by heating the oven-dried sample at 600 ± 25 °C for six minutes and then at 900 ± 25 °C for another six minutes in a pre-weighed open silica crucible in a muffle furnace. The amount of weight loss in the sample gives the volatile matter of the biomass sample. It can be estimated using the formula given below.

$$\% \text{ Volatile matter (dry basis)} = \frac{(W_2 - W_f)}{(W_2 - W_1)} \times 100$$

where W_1 is the weight of the silica crucible, W_2 is the weight of the crucible and the oven-dried sample, and W_f is the constant weight of the crucible and sample after heating.

Volatile matter estimation for biochemical processing is done in a somewhat different way by heating a known amount of oven-dried biomass sample in an open and pre-weighed silica crucible at 550 ± 25 °C till constant weight is achieved. It can be calculated using the same formulae as given above for the thermo-chemical process. The VS for the biochemical context is thus lower in comparison to the volatile matter in the context of combustion.

Ash content

Ash content and its properties have an impact on the combustion systems. The amount and nature of ash and its behaviour at high temperatures affect the design and type of ash-handling system employed in biomass-burning equipment. At high temperatures, ash becomes sticky and eventually forms molten slag, which results in slagging or clinker formation in the gasification reactor. Slagging can lead to excessive tar formation and/or complete blocking of the reactor. Excessive slag formation can occur if the ash melting point is low. Biomass feedstock with high ash content and low ash fusion temperatures are difficult to handle, especially in gasifiers.

This signifies that other than ash content, its melting behaviour is also an important characteristic to be studied for its suitability for gasification. The melting behaviour should be studied both in oxidizing and reducing atmospheres. The chemical composition of ash for some biomass materials is given in Appendix 4. The ash fusion temperatures of selected biomass materials are given in Table 13.10 of Chapter 13.

Estimation

The ash content in biomass samples is estimated by combusting a known quantity of oven-dried sample in a pre-weighed and closed silica crucible at

750 ± 25 °C for a minimum of four hours in a muffle furnace. The amount is estimated using the formula given below.

$$\% \text{Ash (dry basis)} = \frac{(W_f - W_1)}{(W_2 - W_1)} \times 100$$

where W_1 is the weight of the silica crucible, W_2 is the weight of the crucible and the oven-dried sample, and W_f is the constant weight of the crucible and the sample after combustion.

Fixed carbon

The fixed carbon is estimated on material balance basis using the following formula.

$$\% \text{ Fixed carbon (dry basis)} = 100 - \% \text{ volatile matter} - \% \text{ ash content}$$

The amount of FC present gives a rough indication of the charcoal yield. Also, a higher FC material is generally better suited for gasification than a lower FC material.

Example 1
The fixed carbon and ash content of cedar wood was found to be 13.1% and 0.2% on wet basis, respectively. Estimate the volatile matter of the wood on dry basis considering moisture content of the wood as five per cent.

Solution
Volatile matter + fixed carbon + ash + moisture = 100 (wet basis)
Volatile matter (wet basis) = 100 – 13.1 – 0.2 – 5 = 81.7
Volatile matter (dry basis) = volatile matter (wet basis)/(1 – moisture)
= 86%

Ultimate analysis

Ultimate analysis is basically an elemental analysis of biomass, involving measurement of carbon, hydrogen, oxygen, nitrogen, and sulphur content (by difference). Ultimate analysis of waste samples also includes chlorine estimation.

For a thermochemical process, carbon is the most important element in fuels as it has a direct influence on the heating value (Figure 12.7). Ultimate analysis is also necessary for calculating stoichiometric air requirements for combustion and for flue gas analysis (discussed in Chapter 13). A negligible

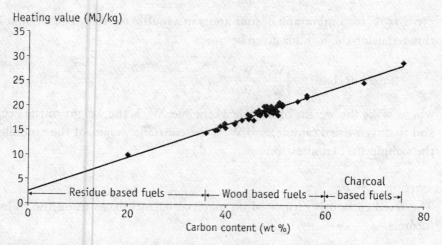

Figure 12.7 Effect of carbon content on heating value of biomass (generated from Appendices 5 and 7)

presence of nitrogen and sulphur in biomass signifies low NO_x and SO_x emissions. The bound oxygen content in biomass feedstock is high due to the presence of ether, acids, and alcohol groups in cellulose, hemicellulose, and lignin. The C/H/O ratios of most of the biomass materials are similar, despite different feedstocks, when expressed on a dry and ash-free basis (Figure 12.8). The H/C ratio in most of the biomass materials feedstock is 1.5, while it is unity for coal. The ultimate analysis of few biomass materials is given in Appendix 5.

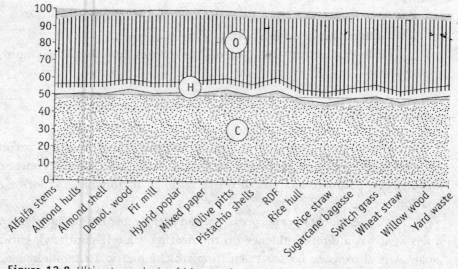

Figure 12.8 Ultimate analysis of biomass fuels on ash-and moisture-free basis

In biochemical systems, nitrogen, phosphate, and other elements are required by the microbial population for optimal growth. These microbes are responsible for waste conversion and stabilization during anaerobic digestion. Nitrogen is the second important major nutrient for the growth and proper functioning of microbial systems (Bryant 1979; Stafford, Hawkes, and Horton 1980). Nitrogen present in the feedstock has two benefits.

- It provides an essential element for the synthesis of amino acids, proteins, and nucleic acids.
- It is converted into ammonia, which, as a strong base, neutralizes the volatile acids produced by fermentative bacteria, and thus helps maintain neutral pH conditions essential for cell growth.

An overabundance of nitrogen in the substrate can lead to excessive ammonia formation, resulting in toxic effects. The nitrogen requirement of the anaerobic digestion process is considered to be about one-fifth of the aerobic process (Britz, Noeth, and Lategan 1983) and the nitrogen requirement for carbohydrate digestion has been found to be six times higher than that for protein and fatty acids (Speece and McCarty 1964). The synthesis of new microbial cells is, however, directly proportional to the amount of nutrients available. According to Speece, the anaerobic microbial cells have a nitrogen/cell ratio of 11% (w/w), and at a 5% cell synthesis ratio, the nitrogen requirement could be 6 kg nitrogen/1000 kg of COD (Speece 1987).

The composition of the organic matter (carbon-containing compounds) added to a digestion system has an important role to play in the growth rate of the anaerobic bacteria and the production of biogas.

The carbon, hydrogen, nitrogen, and sulphur content is determined using a modern CHNS analyser, and oxygen is calculated by difference basis. Most of these systems employ a catalytic combustion or pyrolysis step to decompose the sample into carbon dioxide, water, hydrogen sulphide, and nitrogen which are then determined quantitatively by gas chromatography using a flame ionization detector or thermal conductivity detector. It is important to mention that the hydrogen and oxygen reported in the ultimate analysis is in addition to the moisture determined during proximate analysis. Therefore, it is normally suggested to report all values on a dry basis for ultimate analysis.

Importance of C/N ratio for biochemical reactors

The ratio of organic matter to nitrogen in the feedstock affects their utilization rate by different bacteria present in the digester. Bacteria need a suitable

C/N ratio for their metabolic processes. A C/N ratio ranging from 20 to 30 is considered optimum for anaerobic digestion. If the C/N ratio is very high, nitrogen will be consumed rapidly by methanogens for meeting their protein requirements and will no longer react with the left-over carbon content of the material. The organic accumulation in the digester is inhibitory, as a result the gas production will be low. On the other hand, if the C/N ratio is very low, nitrogen will be liberated and accumulated in the form of ammonia. Its accumulation in the digester increases the pH and a pH higher than 8.5 has a toxic effect on the methanogen population. The C/N ratio of a few biomass samples is given in Appendix 6.

Use of ultimate analysis for fuel formula

Elemental analysis can be used to estimate the fuel formula. For deriving the fuel formula from the ultimate analysis, the weight per cent is converted into moles and then the number of moles of each element is divided by the number of carbon moles.

Example 2
Ultimate analysis (dry basis) of rice hulls is given as carbon: 38.5%, hydrogen: 5.7%, nitrogen: 0.5% , sulphur: 0.0%, oxygen: 39.8% and ash: 15.5%. Derive its fuel formula and fuel weight.

Solution
Basis: 100 g of dry rice hulls.
Dividing the weight of carbon, hydrogen, nitrogen, and oxygen by atomic weight, that is, 12, 1, 14, and 16, respectively, gives their number of moles as 3.21, 5.70, 0.04, and 2.49, respectively. Dividing the number of moles of each element by the number of moles of carbon gives the fuel formula as

$$CH_{1.78} N_{0.01} O_{0.78}$$

Combining the molecular weight of each element as per the fuel formula gives the molecular weight as 26.34.

Use of ultimate analysis for assessing quality of volatile matter

Ultimate analysis and simple stoichiometry can be used to assess the quality of volatile matter. If it is assumed that fixed carbon contains only carbon, then volatiles will constitute all hydrogen, oxygen, and some portions of carbon.

Example 3

The proximate analysis of rice hull gives 16.2% of fixed carbon (dry basis). Estimate the weight% of carbon in volatiles and the molar ratio of carbon, hydrogen, and oxygen in volatile matter, using results of Example 2.

Solution

Using the ultimate analysis results of Example 2 and assuming fixed carbon containing only carbon, the carbon content in the volatiles can be estimated using the same basis of 100 g, as

Volatile carbon = total carbon – fixed carbon
Volatile carbon = 38.5 – 16.2 = 22.3 g

For molar ratio estimation, now volatiles composition is 22.3 g of carbon, 5.7 g of hydrogen, and 39.8 g of oxygen (as per ultimate analysis).
Moles of carbon = 22.3/12 = 1.9
Moles of hydrogen = 5.7/1 = 5.7
Moles of oxygen = 39.8/16 = 2.5

Hence, molar ratio of carbon, hydrogen, and oxygen is 1, 3.0, and 1.3, respectively (obtained by dividing the moles of each element by the carbon moles).

Calorific value/heating value

Calorific value is the amount of heat generated by unit mass of the fuel on its complete combustion. It can be reported in two ways, namely HHV (higher heating value) and LHV (lower heating value). HHV considers moisture in liquid state after combustion. The LHV considers that water formed by combustion of the fuel remains in vapour form. The difference in the calorific values is the latent heat of water for the amount of water produced during combustion. HHV and LHV are usually reported for oven-dry biomass, that is, on a moisture-free basis.

The heating value of biomass feedstocks is determined by the bomb calorimeter which measures the enthalpy change between reactants and products at 25 °C. The heating value obtained is termed HHV because the water of combustion is present in the liquid state at the completion of the experimental determination. Occasionally the HHV is also reported on wet basis. The two are related by the formula.

$$HHV(dry\ basis) = \frac{HHV(wet\ basis)}{1 - moisture\ fraction}$$

Conversion between bases is simple in the case of HHV as the moisture present in the sample is in the same state before and after combustion.

In order to obtain the relation between LHV and HHV, the account of water formed during the reaction (w) should be known. One can then write

HHV (wet basis) = LHV (wet basis) + wλ

where λ is the latent heat of water.

A NHV (net heating value), which is often used in stove literature, is defined as the heat available in wet fuel after allowing for the moisture in fuel to evaporated. Thus,

NHV = HHV (dry basis) × (1–MC) – MC × λ

where MC is the moisture content in the fuel.

Calorific values are also reported as gross calorific value (GCV) and NCV. It is important to know how these are measured and reported, and on what basis the efficiencies are reported. The latter can show a large deviation depending on which number is taken for calorific value.

The bomb calorimeter is the most common device for measuring the heat of combustion or the calorific value of a material. With this apparatus, a test specimen of specified mass is burned under standardized conditions. The heat of combustion determined under these conditions is calculated on the basis of the observed temperature rise while taking into account the heat loss. The combustion process is initiated inside an atmosphere of oxygen in a constant volume container, the bomb, which is a vessel built to withstand high pressures (Figure 12.9). It is immersed in a stirred water bath, and the whole device is the calorimeter vessel.

Initially, benzoic acid is used for estimating the apparent heat capacity of the calorimeter. The temperature in water due to the combustion of a known amount of benzoic acid pellet (<1 g) in oxygen-pressurized (25 atmospheres) bomb gives the apparent heat capacity of the calorimeter. The heat liberated by the igniting materials in the bomb needs to be deducted during estimation.

$$A(\text{cal}/°C) = \frac{(W_b \times C_b) + \sum_i^n C_i W_i}{\Delta T_b}$$

where A is the apparent heat capacity of the calorimeter, W_b is the weight of benzoic acid, W_i is the weight of the igniting material, C_b is the calorific value of benzoic acid (6318 cal/g), C_i is the calorific value of the igniting material, and ΔT_b stands for the temperature rise due to combustion of benzonic acid.

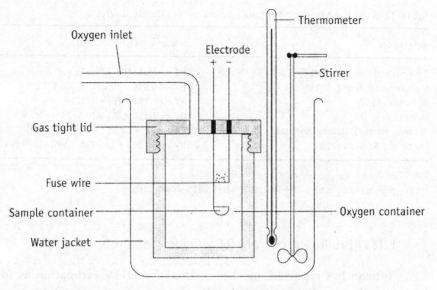

Figure 12.9 Schematic diagram of a bomb calorimeter

There are normally two materials which get used up in the calorimeter, namely thread and fuse wire. Thus n = 2 in the above formula for solids.

The value of apparent heat capacity is then used to estimate the calorific value of the biomass sample using the equation

$$C(cal/g) = \frac{(\Delta T_s \times A) - \sum_{i-1}^{n} W_i C_i}{W_s}$$

where W_s is the weight of the sample, W_i is the weight of the igniting material, C_i is the calorific value of the igniting material, and ΔT_s stands for temperature rise due to combustion of sample. Higher heating values of some biomass materials are given in Appendix 7.

Use of proximate analysis for heat value

Proximate analysis can be used for a quick estimation of HHV. Several correlations between HHV and proximate analysis values are given in Table 12.5.

Use of ultimate analysis for heat value estimation

Heating value is often estimated using the modified Dulong formula (wet basis). This equation has been found more accurate for feedstocks with a low moisture content and chars than for fresh biomass.

Table 12.5 HHV estimation formulae based on proximate analysis

Reference	Correlation (HHV, MJ/kg)
Jimenez and Gonzalez (1991)	HHV = −10.81408 + (0.3133 × (VM+FC))
Sheng and Azevedo (2005)	HHV = 19.914 − (0.2324 × Ash)
Demirbas (1997)	HHV = (0.196 × FC) + 14.119
Demirbas (1997)	HHV = (0.312 × FC) + (0.1534 × VM)
Cordero, Marquez, Rodriguez-Mirasol, *et al.* (2001)	HHV = (0.3543 × FC) + (0.1708 × VM)
Sheng and Azevedo (2005)	HHV = −3.0368 + (0.2218 × VM) + (0.2601 × FC)

VM, FC, Ash, are weight per cent on dry biomass basis.
HHV – high heating value; VM – volatile matter; FC – fixed carbon

$$HHV \text{ (Btu/lb)} = 145C + 610 \ (H - [O/8]) + 40S + 10N$$

Tillman has reported another method for HHV estimation as follows. This equation has been reported to be more accurate for fresh biomass than the Dulong-Berthelot equation.

$$HHV \text{ (Btu/lb)} = 188C - 718$$

The Institute of Gas Technology analysed 700 coal samples for their heating values and ultimate analysis. Comparing the heating values of these samples based on the ultimate analysis, a correction factor was incorporated in the equation. This equation has been reported to be more accurate than the Dulong and Tillman equations.

$$HHV \text{ (Btu / lb)} = 146.58C + 568.78H + 29.45 - 6.58 \ Ash - 51.53 \ (O + N)$$

Table 12.6 summarizes the various formulae derived by different authors for calculating HHV from the ultimate analysis.

HHV of the biomass can also be estimated from composition characterization. Table 12.7 gives some formulae, which have been derived by various researchers for HHV estimation using cellulose, lignin, and extractable matter content in the biomass. The composition characterization of biomass has been discussed in a separate section in this chapter.

Waste water analysis for solids

TS in waste water can be subcategorized as TSS (total suspended solids), VSS (volatile suspended solids), and TDS (total dissolved solids).

Table 12.6 Higher heating value estimation formulae based on ultimate analysis

Name of author	Correlation (HHV, MJ/kg)
Tillman (1978)	$HHV = 0.4373\,C - 1.6701$
Annamalai, Sweeten, and Ramalingam (1987)	$HHV = 0.3516\,C + 1.16225\,H - 0.11090 + 0.0628\,N + 0.10465\,S$
Institute of Gas Technology (1978)	$HHV = 0.341\,C + 1.322\,H - 0.120 - 0.12\,N + 0.0686\,S - 0.0153\,Ash$
Graboski and Bain (1981)	$HHV = 0.328\,C + 1.4306\,H - 0.0237\,N + 0.0929\,S - (1-Ash/100)$ $(40.11\,H/C) + 0.3466$
Channiwala and Parikh (2002)	$HHV = 0.3491\,C + 1.1783\,H + 0.1005\,S - 0.10340 - 0.0151\,N - 0.0211\,Ash$
Demirbas (1997)	$HHV = 0.335\,C + 1.423\,H - 0.154\,O - 0.145\,N$
Jenkins and Ebeling (1985)	$HHV = -0.763 + 0.301\,C + 0.525\,H + 0.064\,O$
Sheng and Azevedo (2005)	$HHV = -1.3675 + 0.3137\,C + 0.7009\,H + 0.03180*$

*O is the sum of the contents of oxygen and other elements (including S, N, Cl, etc.) in the organic matter, that is O = 100-C-H-ash.

Table 12.7 Higher heating value estimation formulae based on composition

Name of author	Correlation (HHV, MJ/kg)
Shafizadeh and Degroot (1976)	$HHV = 0.1739\,Ce + 0.2663\,L + 0.3219\,E$
Jimenez and Gonzalez (1991)[a]	$HHV = [1 - Ash/(100 - Ash)](0.1739\,Ce + 0.2663\,L + 0.3219\,E)$
Tillman (1978)[b]	$HHV = 0.1739\,Ce + 0.2663\,(1 - Ce)$
Demirbas (2001)[c]	$HHV = 0.0889\,L + 16.8218$

[a] Ce, L, and E are weight fraction of cellulose (including cellulose and hemicellulose), lignin, and extractives, respectively on dry biomass basis
[b] Ce is cellulose (cellulose and hemicellulose) on dry extractable-free basis
[c] HHV and L in this equation are on dry ash-free and extractable-free bases
HHH – high heating value; L – lignin; E – extractives

$$TS = TDS + VSS + ash$$

where VSS + ash = TSS

TSS is a measure of the mass of fine particles suspended in water. It is an important characterization of waste water for biochemical reactor operation. Suspended particulates form scum layers and foam due to the presence of insoluble components with floating properties, such as fats and lipids; retard or even completely obstruct the formation of sludge granules in UASB (upflow anaerobic sludge blanket) reactors; cause entrapment of granular sludge in a layer of adsorbed insoluble matter and sometimes also disintegration of granular sludge, and sudden washout of sludge; decline in overall methanogenic activity of the sludge due to accumulation of TSS, clog the reactor due to TSS settling, etc. A suitable TSS concentration for a UASB reactor is less than 500 mg/litre.

For its estimation, a well-mixed sample is filtered through a weighed standard paper filter or a glass-fibre filter. The residue retained on the filter is dried to a constant weight at 105 °C. The increase in the weight of the filter gives TSS.

VSS are the amount of suspended solids which are lost by ignition at 550 °C. This is a useful parameter for waste water assessment and biochemical reactor assessment because it roughly represents the amount of organic matter present in the solid fraction of waste water, activated sludge, and industrial waste.

TDS are those solids that pass through a filter with a pore size of 2.0 microns or smaller. After filtration, the filtrate (liquid) is dried and the remaining residue is weighed and calculated as milligram per litre. These methods have a detection limit of approximately 4 mg/litre for the TSS and 10 mg/litre for TDS.

Chemical composition of biomass*

The three main constituents are structural and non-structural carbohydrates (cellulose and hemicellulose), lignin, and extractable matter. The concentration of each class of compound varies depending upon the species, type of plant tissue, stage of growth, and growing conditions. The chemical composition of selected biomass samples is given in Table 12.8.

Due to the carbohydrate structure, biomass is highly oxygenated with respect to conventional fossil fuels, including HC (hydrocarbon) liquids and coals. Typically, 30–40 wt% of the dry matter in biomass is oxygen. The principal constituent of biomass is carbon, making up 30–60 wt% of dry matter. Of the organic components, hydrogen is the third major constituent, comprising typically five to six per cent dry matter. Nitrogen, sulphur, and chlorine usually comprise less than one per cent dry matter but occasionally well above this. These are important in the formation of pollutant emissions. Nitrogen is a macronutrient for plants and critical to their growth. Certain inorganic elements can be found in high concentration as well. In annual growth tissues, concentrations of the macronutrient potassium frequently exceed one per cent dry matter. In some of the gramineae (grasses and straws), silica is the third-largest component (in rice straw, silica is 10%–15% dry matter).

*See also chapter 15 for discussion on chemical structures of cellulose, etc.

Table 12.8 Cellulose, hemicellulose, and lignin content in common agricultural residues and wastes.

Agricultural residues	Cellulose	Hemicellulose	Lignin
Hardwood stem	40–50	24–40	18–25
Nut shells	45–50	35–35	25–35
Corn cobs	45	35	15
Grasses	25–40	35–50	10–30
Wheat straw	33–40	20–25	15–20
Rice straw	40	18	5.5
Leaves	15–20	80–85	0
Coastal Bermuda grass	25	35.7	6.4
Switch grass	30–50	10–40	5–20
Solid cattle manure	1.6–4.7	1.4–33	2.7–5.7
Swine waste	6.0	28	—
Paper	85–99	0	0–15
Newspaper	40–55	25–40	18–30

Source Prasad, Singh, and Joshi (2007)

Composition of solid waste (domestic, commercial, and industrial waste) is studied further for fractions of green waste, paper waste, plastics, building waste, textiles, glass, wood, metal, cardboard, kitchen waste, and recoverable goods.

Paper and organic waste (food and yard waste) constitute 50%–60% of the total residential waste generated (Table 12.9). The rest is made up of a wide variety of materials, including non-recycled materials such as ceramics, textiles, leather, rubber, batteries, ashes, rubble, fibre glass, and drywall. The industrial waste composition varies from one particular industry or commercial sector to another. Within the manufacturing sector, composition of the waste produced depends to a large extent on the products being made. For example, waste from an electronics manufacturer would have more plastics, metal, and paper than waste from a furniture manufacturer, whose waste has more wood. Similarly, waste from different retail establishments may vary. Restaurants and retail food establishments generate more organic materials than office establishments. However, some materials such as cardboard packaging are in such common use that they appear in the waste stream of virtually all industrial and commercial waste.

Cellulose

Cellulose is a linear polysaccharide of β-D glucopyranose units linked with (1-4) glycosidic bonds (Figure 12.10) while the hemicelluloses are polymers of

Table 12.9 Municipal solid waste composition (%) in different countries

Countries	Paper	Textiles	Plastic	Glass	Metals	Organic materials	Bulky wastes	Other wastes
USA (Geological and Mining Engineering and Sciences 2005)	37.00	4.0	11.0	5.00	8.00	29.00	—	6.0
Developing countries* (Geological and Mining Engineering and Sciences 2005)	11.00	4.0	11.0	4.00	2.00	56.00	—	12.0
Western EU** (Geological and Mining Engineering and Sciences 2005)	26.00	2.0	7.0	6.00	4.00	27.00	3.00	25.0
Candidate countries*** (Geological and Mining Engineering and Sciences 2005)	14.00	5.0	9.0	5.00	5.00	42.00	0.00	20.0
Phnom Penh (International Transfer Centre Environmental Technology 2005)	4.00	3.4	13.90	0.90	0.20	73.10	1.7-wood	2.8
Taiwan (Lee Shang-Hsiu 2005)	21.1	2.5	5.80	2.00	3.50	61.00	3.00	1.1

* Brazil, China, India, Iran, Laos, Lebanon, Malaysia, Mauritius, Mexico, Nepal, Pakistan, Singapore, Sri Lanka, Thailand, and Vietnam
**Norway, Switzerland, Iceland, Belgium, Denmark, Germany, Greece, France, Ireland, Italy, Luxembourg, the Netherlands, Austria, Portugal, Finland, Spain, Sweden, and United Kingdom, none of which are developing countries
***Bulgaria, Cyprus, Czech Republic, Estonia, Hungary, Latvia, Lithuania, Malta, Poland, Romania, Slovakia, Slovenia, and Turkey, all of which are developing countries except for Malta and Turkey

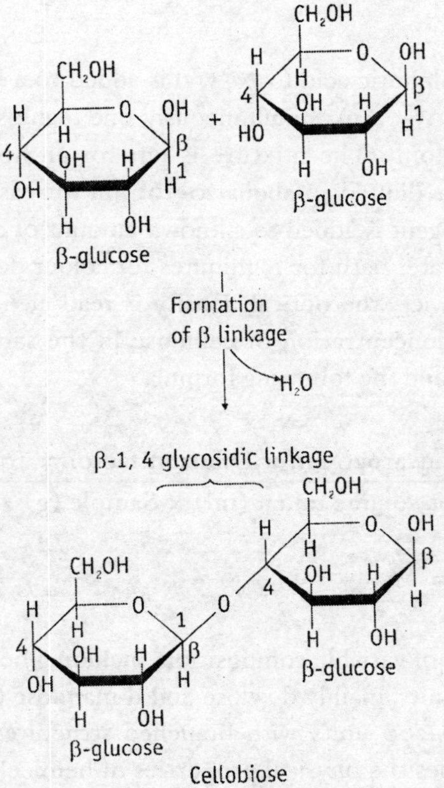

Figure 12.10 Linkage of glucose for cellulose formation

d-xylose and d-mannose. Plants producing large amounts of free sugars, such as sugar cane and sweet sorghum, and starch crops, such as maize and other grains are attractive as feedstocks for fermentation to produce biofuels. Therefore, the cellulose and hemicellulose fraction of biomass feedstock is extremely useful in biochemical conversion technologies (biogas and ethanol formation).

The dominant physical characteristic of cellulose is its extreme insolubility due to strong secondary bonding, which retards not only acid and enzymatic hydrolysis but also removes lignins and hemicelluloses interspersed throughout the cellulose structures. Cellulose can be dissolved by strong acids such as hydrochloric, sulphuric, and phosphoric acids. This property of cellulose is used for its estimation (Updegraff 1969) where it is degraded with concentrated sulphuric acid and is further estimated using calorimeter after colour development with anthrone reagent.

Estimation

For cellulose estimation, 10 ml sulphuric acid (67%, v/v) is added to a known quantity of oven-dried sample (0.05–0.1 g). Simultaneously, one blank sample is prepared with sulphuric acid only. The mixture is left for an hour for cellulose digestion. The aliquot is diluted to about 20–100 ml for easy handling. About 4 ml of anthrone reagent is added to a known amount of diluted aliquot (0.2–1 ml) and kept in a water bath for 15 minutes for colour development. After cooling the tubes in ice, the optical density is read at 620 nm against the reagent blank. The concentration of cellulose in the sample is calculated using glucose standard and the following formula.

$$\text{Cellulose }(\%, w/w) = \frac{\text{Conc. at } 620 \text{ nm} \times \text{Dilution factor} \times 100}{\text{Aliquot volume taken (ml)} \times \text{Sample (g)} \times 10^6}$$

Hemicellulose

Hemicellulose is a polysaccharide of variable composition, including both five and six carbon monosaccharide units, mainly d-xylose and d-mannose (Figure 12.11). Hemicellulose contains 50–200 units with branched structures. This structural characteristic determines the physical properties of hemicellulose. Hetero-polymerization decreases the ability to form regular, tight-fitting crystalline regions and thus makes hemicellulose more soluble than cellulose.

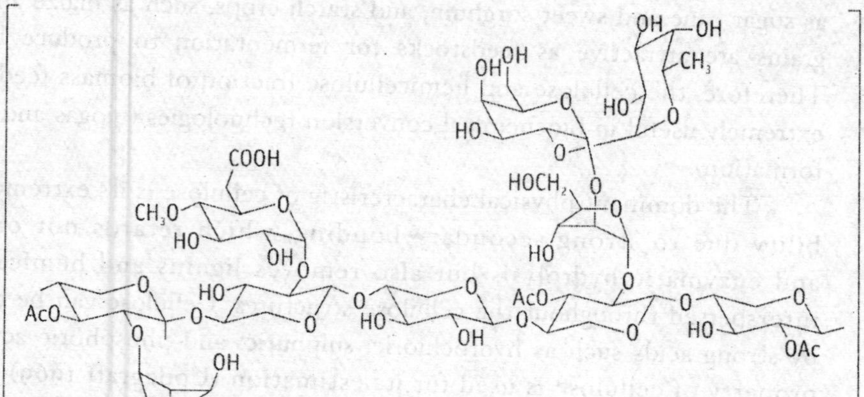

Figure 12.11 Structure of hemicellulose

Hence, it can be differentiated from cellulose by its solubility in weak acid and alkaline solutions and this property is used for its estimation (Deschatelets and Yu 1986) .

Estimation

For hemicellulose estimation, add 10 ml sulphuric acid (three per cent, w/v) to a known amount (0.1–0.5 g) of oven-dried sample. Simultaneously, one blank is prepared with sulphuric acid only. The sample tubes are autoclaved at 121 °C for 60 minutes. The digested sample is neutralized with 10 N KOH (potassium hydroxide) and HCl (hydrochloric acid) and diluted (50–100 ml) with distilled water. 1 ml p-bromoaniline reagent is added to a known quantity of aliquot (2–5 ml) and kept in a water bath at 70 °C for 10 minutes for colour development. Leaving the tubes in dark at room temperature for 70 minutes, the optical density is taken at 520 nm using a spectrophotometer against the reagent blank. The concentration of hemicelluloses in the sample is calculated using the xylose standard and the following formula

$$\text{Hemicelluloses (\%, w/w)} = \frac{\text{Concentration at 520 nm} \times \text{Dilution factor} \times 100}{\text{Aliquot volume taken (ml)} \times \text{Sample (g)} \times 10^6}$$

Lignin

Lignin is an irregular three-dimensional polymer of phenyl-propane units with heavy cross-linkages produced by random coupling of the hydroxyl group of monomers created by a non-reversible reaction by removal of water from sugars to create aromatic structures. A few of the monomers are shown in Figure 12.12. The hydroxyl groups (either the alcoholic hydroxyles on the chains or the phenolic hydroxyles on the aromatic rings) can react with each other or with the aldehyde or ketone groups. The reaction of one hydroxyl with the other forms ether linkage, while the reaction with aldehyde forms hemiacetal and the reaction with ketones forms ketals. An early stage in the condensation of various monomers to form lignin is shown in Figure 12.13. The hydroxyl groups can further react to extend the polymer or for cross-linking.

Thus, the structure of lignin is complex and not yet known fully. It deposits in an amorphous state surrounding the cellulose fibres and is bound to the cellulose directly by ether bonds.

The complexity of lignin resists attack by most microorganisms (aerobic and anaerobic). Thus, it is not considered fermentable or digestible.

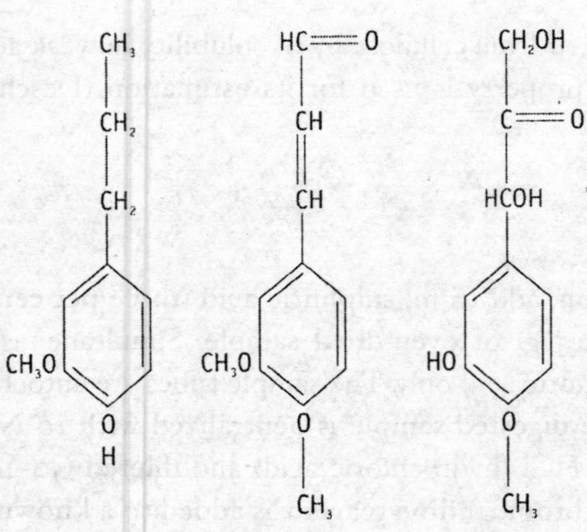

Figure 12.12 Monomers of lignin

As the complex structure of biomass does not have significant influence on combustion behaviour, thermochemical means are usually proposed for their conversion.

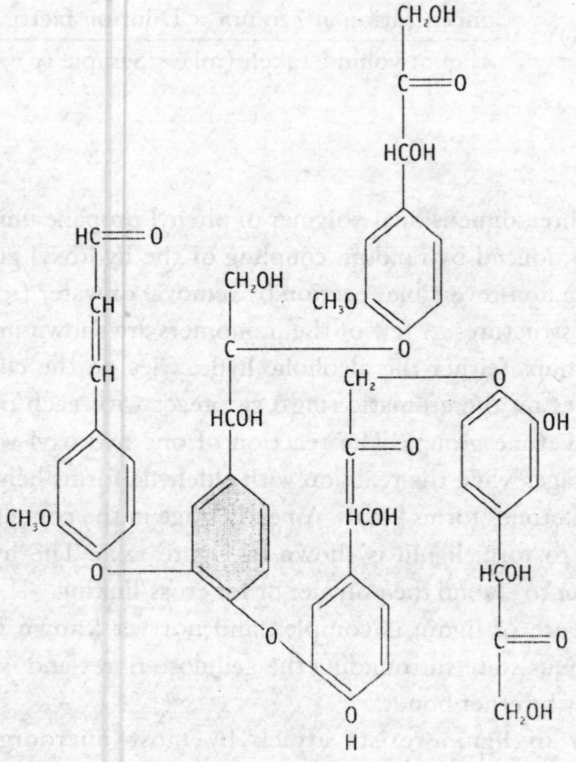

Figure 12.13 Condensation of lignin monomers

Lignin can be fractionated into acid-insoluble and acid-soluble materials. Acid-insoluble material may also include ash and protein, which must be accounted for during gravimetric analysis.

Estimation

For lignin estimation, 3 ml of 72% sulphuric acid is added to a known amount of oven-dried sample (300–500 mg). After mixing, the tubes are then placed in a water bath set at 30 ± 3 °C for 60 minutes for lignin digestion. The sample is stirred after 5–10 minutes, without removing it from the water bath, to ensure even acid to particle contact and uniform hydrolysis. Upon completion of the 60-minute hydrolysis, tubes are taken out from the water bath and 84 ml of water is added to dilute the acid to a concentration of four per cent. The tube is then screwed with Teflon caps and the sealed samples are then autoclaved for one hour at 121 °C. Hydrolysates are allowed to slowly cool to near room temperature before removing the caps. The sample is now vacuum filtered for all further analysis. The residue and filtrate are both used for AIL (acid-insoluble lignin) and ASL (acid-soluble lignin), respectively.

AIL (left after vacuum filtration) is oven-dried at 105 °C till a constant weight (as per the procedure for TS) is achieved and further combusted at 550 °C till a constant weight is achieved to get the AIL. Weighing the ash left after combustion, AIL can be estimated using the following formula.

$$AIL\,(\%) = \left(\frac{W_r}{W_s} \times 100 \right) - \left(\frac{W_a}{W_r} \times 100 \right) - \%P$$

where W_r is the weight of the oven-dried acid-insoluble residue, W_s is the weight of the sample, W_a is the weight of the ash left after combustion, and P is the protein content in biomass.

For ASL, absorbance of hydrolysis liquor aliquot is measured at 320 nm wavelength on a UV (ultraviolet) visible spectrophotometer against the blank. The sample is diluted to bring the absorbance into the range 0.7–1.0, recording the dilution. Deionized water or four per cent sulphuric acid may be used to dilute the sample, but the same solvent should be used as a blank. The concentration of ASL can be calculated using the following formula.

$$ASL = \frac{UV_{abs} \times volume_{filtrate} \times dilution}{\varepsilon \times W_s} \times 100$$

where UV_{abs} is the absorbance of sample at 320 nm, 'volume$_{filtrate}$' is the volume

Table 12.10 Absorptivity of few biomass feedstocks

Biomass type	Lambda max (nm)	Absorptivity at lambda max. (litre/g cm)	Recommended wavelength (nm)	Absorptivity at recommended wavelength (litre/g cm)
Pinus radiata	198	25	240	12
Bagasse	198	40	240	15
Corn stover	198	55	320	30
Populus deltiodes	197	60	240	25

of hydrolysed filtrate (87 ml), 'dilution' is the volume of sample of diluting solvent/volume of sample, ε is the absorptivity of biomass at specific wavelength (Table 12.10), and W_s is the oven-dried weight of sample. Total lignin content can be given as the sum of AIL and ASL concentrations.

Protein content in biomass

Biomass feed stocks containing protein and other nitrogen-containing materials are used as feedstock for conversion to fuels and chemicals. These constituents must be measured as part of a comprehensive biomass analysis. Protein concentration in biomass is not measured directly. In many cases, nitrogen content of the biomass sample is measured and protein content is estimated using an appropriate NF (nitrogen factor). Most standard methods recommend use of an NF of 6.25 for all types of biomass except wheat grains where an NF of 5.70 is recommended (Mossé 1990; Mossé and Huet 1985).

Protein content = Nitrogen content × NF

Extractables

Extractables can be defined as the fraction of biomass soluble in water, ether, and ethanol. They include resin, volatile oils, waxes and fatty acids (ether soluble), pigments (alcohol soluble), carbohydrates (water soluble), etc. As the quality and nature of extractables vary, the products after processing also vary.

Estimation

Estimation of water-soluble and ethanol-soluble extractables is done in a two-step soxhlet extraction process. Water-soluble materials may include inorganic material, non-structural sugars, and nitrogenous material, among others. Inorganic material in the water-soluble material may come from both the biomass and any soluble material that it is associated with, such as soil or fertilizer. Ethanol-soluble material includes chlorophyll, waxes, or other minor components. A few biomass may require both extraction steps, while other biomass may only require exhaustive ethanol extraction.

For a two-step process, take 2–10 g of oven-dried sample on an extraction thimble (less than the height of the soxhlet siphon tube height). Extraction is carried out in the soxhlet apparatus with 190 ml HPLC (high performance liquid chromatography) grade water for six to eight hours at a rate of about four to five siphon cycles per hour. Successive ethanol extraction is carried out in a new dried and weighed extraction flask with 190 ml proof ethanol. The water extraction flask and ethanol extraction flask are dried at 105 °C and 40 °C, respectively. Later, both the results can be added for cumulative extractables as per the given formula.

$$\text{Extractable (dry basis), \%} = \frac{W_{wr} + W_{er}}{W_s} \times 100$$

where W_{wr} is the weight of water-extracted residue after drying at 105 °C, W_{er} is the weight of ether extracted residue after drying at 40 °C, and W_s is the weight of oven-dried sample .

Chemical oxygen demand

The chemical composition of liquid effluents is most commonly assessed by the COD. The COD of waste water is an important parameter for deciding the operational parameters (feed rate) and designing parameters of a biochemical reactor. For example, the loading rate of a bioreactor is an important design parameter, and can be defined in terms of kg COD/m³ reactor/day. Its estimation is based on oxidation of organic matter in the sample with potassium dichromate under strong acid conditions. Silver sulphate and mercuric sulphate are used as catalyst and for removal of chloride interferences, respectively. After digestion, the remaining unreduced dichromate is titrated with ferrous ammonium sulphate using ortho-phenonthroline ferrous complex as an indicator. The consumed potassium dichromate is converted into the oxygen equivalent. For example, the chemical reaction involved in oxidation of potassium acid phthalate by dichromate can be given as

$2 KC_8H_5O_4$ (potassium acid phthalate) + $10 K_2Cr_2O_7$ (potassium dichromate) + $41 H_2SO_4$ (sulphuric acid) → $10 Cr_2(SO_4)_3$ (chromium (III) sulphate) + $11 K_2SO_4$ (potassium sulphate) + $16 CO_2$ + $46 H_2O$

As 10 mole of potassium dichromate has the same oxidation power as 15 mole of oxygen, the equivalent reaction is

$$2 KC_8H_5O_4 + 15 O_2 + H_2SO_4 \rightarrow 16 CO_2 + 6 H_2O + K_2SO_4$$

Thus, 2 moles of potassium acid phthalate require 15 moles of oxygen. The theoretical COD of potassium acid phthalate is 1.175 g of oxygen/g of potassium acid phthalate.

The COD concentration can be calculated using the following formula.

$$COD\,(mg/litre) = \frac{[(A-B) \times N] \times 8000}{S}$$

where A is the ferrous ammonium sulphate solution required for titration of the blank (ml), B is the ferrous ammonium sulphate solution required for titration of the sample (ml), N is the normality of the ferrous ammonium sulphate solution, and S is the sample used for the test (ml).

Structural properties

Physical structure

Woody biomass feedstocks are composed of cells of various sizes and shapes. Long pointed cells are known as fibres. The length of the fibres varies in hard and softwood. Hardwood fibres are about 1 mm in length while softwood fibres are 3–8 mm. Thickness of the cell wall determines the density and mechanical properties of woody biomass.

Particle size and size distribution

The size and size distribution of biomass are important in selection of equipment, pre-processing required, etc. Fluidized bed combustors require a small and uniform size. Biomass gasifiers accept a size range of 25–100 mm, whereas grated combustors can accept larger sizes. Biomass has to be ground to a fine size for purposes of densification. Large sizes are generally not desirable for biochemical reactors.

Permeability

Permeability is important for both the combustion and biochemical processes. Gases and liquids generated during combustion/microbial action have to pass through the porous structure to the surrounding. Low permeability of fuel in the combustion process increases the residence time of the primary pyrolysis products in the hot zone increasing their probability for the second reaction.

Permeability of wood varies from 10^{-4} to 10^4 cm³ (air)/cm atm. In biochemical reactors, low permeability of substrate obstructs the availability of the formed products for subsequent reactions.

Physical properties

Bulk density

Bulk density is defined as the weight per unit volume of loosely packed biomass. It includes the actual volume of biomass, pore volume, and void volume between particles. Biomass with high bulk density is advantageous for combustion systems because it represents a high energy value for smaller volumes and needs less storage space. Average bulk densities of wood and charcoal are in the range of 300–550 kg/m³ and 200–300 kg/m³, respectively. Straws and hulls have bulk densities as low as 70 kg/m³. Densification (for example, briquetting), can increase bulk densities up to 700–800 kg/m³.

Density is also reported as apparent density and skeletal density, also called true density. Apparent density includes particle solid volume and pore volume while skeletal density includes solid volume only. The three densities are related as

$$D_a = D_s(1-P_p)$$
$$D_b = D_a(1-P_b)$$

where D_a is the apparent density, D_s is the skeletal density, D_b is the bulk density, P_p is the particle porosity, that is, volume of pores/volume of pores and volume of solids, P_b is the bed porosity, that is, void volume/void volume and particle volume.

Thermal analysis

Thermal analysis generally covers three different experimental techniques: TGA (thermo gravimetric analysis), DTA (differential thermal analysis), and DSC (differential scanning calorimetry). TGA measures change in the mass of a sample as a function of temperature and time. Samples that do not undergo mass change with temperature are analysed using DTA and DSC. As biomass degrades with temperature, TGA is the most common methodology adopted for thermal analysis. Various reactions taking place during pyrolysis/gasification of biomass materials are given in Figure 12.14. Each reaction is assumed

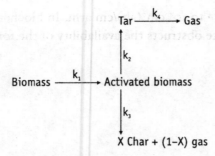

Figure 12.14 Reactions occurring during prolysis/gasification of biomass materials

to be irreversible and of the first order, with a rate k_i given by the Arrhenius Equation.

TGA gives weight loss of the biomass versus temperature/time data. Figure 12.15 gives a TGA graph obtained for pine wood. A weighed sample of the biomass is placed in the furnace chamber. Inert gas (nitrogen/argon) flow is started at least 15 minutes prior to furnace heat-up so as to maintain an oxygen-free atmosphere to restrict the combustion of biomass. Continuous data is generated for mass change of the sample with respect to temperature and time.

TGA gives important data of the activation energy E_a and the frequency factor A, which are further used to estimate the rate of the reaction of solid biomass using the Arrhenius Equation. The rate of reaction of solid biomass for the reaction

Biomass (solid) = Char (solid) + volatiles (gas)

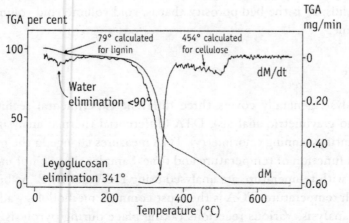

Figure 12.15 Thermo gravimetirc analysis graph of pine wood
Source <http://www.chamotlabs.com/ApplicationNotes/HTC/HTC34thMidwest.html>

is estimated using the following equation

$$\frac{db}{dt} = k(1-b)^n$$

where b is the fraction of biomass present at time t, k is the rate constant, and n is the order of reaction.

Using the Arrhenius Equation

$$k = Ae^{-E_a/RT}$$

where k is the reaction rate constant, A is the frequency factor (s^{-1}), E_a is the activation energy (kcal/mole), R is the universal gas constant, and T is temperature (Kelvin), the rate of reaction can be written as

$$\frac{db}{dt} = Ae^{-E_a/RT}(1-b)^n$$

if β is the linear heating rate dT/dt, the above equation can be written as

$$\frac{db}{dT} = \frac{A}{\beta}.e^{-E_a/RT}.(1-b)^n$$

Assuming a first order reaction (n=1), the kinetic parameters A and E_a can be obtained from weight loss data of TGA (Volker and Rieckmann, 2002; Raveendran and Ganesh, 1993).

TGA finds its application in combustion technologies to understand the fuel properties, reactivity, pyrolysis kinetics, etc. TGA data can be used in modelling of charcoal kilns, pyrolysis oil-producing equipment, cooking stoves, and gasifiers.

Properties of microbial biomass

Protein estimation in microbial biomass

Protein content of the microbial biomass is a qualitative and quantitative indicator of the microbial biomass in the reactor. For its estimation, approximately 1 ml of sample is harvested at 10 000 × g (g is the acceleration due to gravity) for 10 minutes and the pellet is resuspended in 0.5 ml phosphate buffer (100 mM, pH 7.0). To a 100 µl cell suspension, 1.9 ml of 100 mM phosphate buffer (pH 7.0) and 1 ml of Biuret reagent is added and left at room temperature for 15–20 minutes. Optical density is measured at 310 nm against a control (2 ml of phosphate buffer + 1 ml of Biuret reagent) and protein content is estimated (mg/litre) using the BSA (bovine serum albumin) as standard.

Estimation of dry cell weight

Dry cell weight is estimated in order to assess the total microbial biomass in the reactor. A sufficient amount of cellular mass is centrifuged at 8000 RPM for 15 minutes in order to get about 5 g of cellular pellet. The pellet is dried at 105 °C till constant weight is achieved, giving dry cell biomass.

Flocculating ability

The flocculating ability of the sludge can be determined as the reflocculation ability of sludge flocs after disruption (Jorand, Guicherd, Urbain, *et al.* 1994). About 80 ml of sludge sample with a suspended solid concentration of approximately 4 g/litre is transferred in a beaker placed in an ice bath and sonicated at 50 W for 15 seconds. This is sufficient to disrupt the flocs, but does not cause cell rupture. A 10-ml aliquot of the suspension is centrifuged at 1200 RPM for 2 minutes and the absorbance of the supernatant is measured at 650 nm (A). The rest of the sonicated suspension is stirred on a magnetic stirrer (set at a specific speed to keep the sludge flocs suspended) at ambient temperature for 15 minutes after which a 10-ml aliquot is analysed in the same way as before (B). The flocculating ability of the flocs can be calculated as

$$\text{Flocculating ability}(\%) = \frac{1-B}{A} \times 100$$

Relative hydrophobicity of sludge

The relative hydrophobicity of sludge flocs is measured as adherence to hydrocarbons (Chang and Lee 1998). About 30 ml of the sludge sample is washed and suspended in Tris buffer (0.05 mM at pH 7.1). The activated sludge suspension (thickened to approximately, 4 g/litre) is homogenized by sonication (50 W for 2 minutes) at 4 °C to disrupt the flocs into single cells and small microcolonies. The suspension is agitated uniformly for 5 minutes with 15 ml hexadecane in a separatory funnel. After 30 minutes, when the two phases are separated completely, the aqueous phase is transferred into glassware. The relative hydrophobicity is expressed as the ratio of suspended solid concentration in the aqueous phase after emulsification (TSS_e) to the concentration of suspended solids in the aqueous phase before emulsification (TSS_i).

$$\text{relative hydrophobicity}(\%) = \left[1 - \frac{TSS_e}{TSS_i} \right] \times 100$$

Sludge volume index

The SVI (sludge volume index) is given as the volume occupied by 1 g of suspension after 30 minutes of settling. A higher SVI implies lower settleability.

$$SVI = \frac{Sludge\ volume\ settled\ (ml/litre) \times 1000}{Total\ suspended\ solids\ (mg/litre)}$$

Biomass resource assessment

Estimation of the quantities of biomass available at specified geographical locations with the existing physical, technical, environmental, economic, policy, and social conditions is called biomass resource assessment. Here, 'physical' stands for type of soil and water availability and 'technical' for the quality of harvesting machinery available, etc; environmental consideration includes the minimum residues needed to prevent soil erosion, wildlife diversity, etc.; economic consideration includes land competition for food and fuel; policy consideration includes crop subsidies, energy policies, emission regulations, etc; and social factors include ethical and aesthetic values. So it can well be said that biomass resource assessment is a comprehensive study comprising many parameters other than estimation of the amount of biomass.

Biomass-based energy generation interventions are gaining importance with the increasing awareness about climate change, energy security, and other concerns (see Chapter 1). Assessment of biomass resources is an important tool for the planning and management of energy. For strategic planning and implementation, there is a need for reliable data about resources and their sustainable availability. The assessment gives information about nature, sustainability, and availability of biomass for domestic and industrial consumption, planning of the shortfall, and interventions. Focus is mainly on energy crops, wood from forest, agricultural residues, and animal manure due to their potential usage for energy generation. Periodic assessments enable adjustments to be made in management or silviculture practices so as to optimize sustainable yield on a sound economic and environmental basis for such interventions.

There are three basic approaches for biomass resource assessment, namely, the inventory approach, quantity and associated cost approach, and supply curve approach. While the inventory approach estimates the physical quantity of biomass, the second one estimates the average range of quantities with an associated range of production and collection costs, and the last approach estimates quantities as a function of the price that can be paid. The basic step for all approaches is to gather information about the occurrence of different kinds of biomass in the region, their growing patterns, for example their growth rate, rate of harvesting, biomass yield/ha, biomass residue/ha, rotation age, etc., and their current alternative uses.

Estimation of woody biomass

Biomass inventory preparation is preferable only for smaller areas as it becomes a costly option when very large areas are involved under the assessment study. A multistage approach is followed for large area assessment studies. With this methodology, initially low-cost assessment methods are used to identify areas with sustainable biomass supply problems and availability, which are then studied intensively. The multistage methodology for woody biomass assessment is as follows.

Stage I: Review of existing data

The first step of woody biomass assessment is to define the objectives so as to collect, review, and evaluate the existing data for the area concerned. The available data, objectives, and catchment area of study decide the intensity and methodology of the assessment study.

Stage II: Low spatial resolution imagery—AVHRR (resolution 1–8 km)

One of the low cost approaches is to use AVHRR (advanced very high resolution radiometry) with 1, 4, and 8 km resolution; limited ground verification; and a literature-sourced biomass database. These satellite imageries provide overview data with low level of accuracy for cover typing and type borders, that is, maps with (1:5 000 000) broad vegetation types and rough biomass estimates. These low accuracy data, along with limited ground truthing and existing demographically related wood fuel consumption data, enable the selection of areas. Being low cost and repeatable, the time sequential data provides valuable information for monitoring macro and seasonal changes in the woody biomass resource base. AVHRR data is used to identify and map vegetation/biomass classes by means of the GVI (global vegetation index).

Stage III: Imagery

This stage is more intensive and uses satellite imagery with relatively higher spatial resolution (30–80 m) for limited areas. The high resolution imagery is used for annual monitoring as more reliable interpretation of ground cover/ vegetation types is possible with the visible spectrum. The main precaution is to use imagery of the same season, as leafless periods or periods when both crops and trees have green leaves could confuse interpretation. The data needs to be supported by greater degree of ground truthing and aerial photos to identify vegetation/ground cover types discernable through the lower resolution imagery. The higher cost of aerial photography limits use for limited selected areas, hence, it is not feasible to use it for monitoring ground cover changes of large areas. The cost-efficient method of monitoring ground cover changes of large areas is to stratify the area on a broad climatic/ecological basis and randomly select a sample of imagery frame locations for which imagery is obtained over the same period.

Aerial photography is most commonly used for forest inventory usually at a scale of 1:20 000 to 1:50 000 and over a limited area (300 000 ha). Although it is expensive, it provides detailed information of vegetation types and sometimes even identification of individual tree species. The larger scale photography can even be used to measure tree height and/or crown diameter in more open stands. This measurement is facilitated by using low-level photography (100–200 m above-ground altitude) in a vertical position.

Stage IV: Map preparation

Ground cover/biomass-type paper maps are produced using the range of imagery and photography described above. The objective is to produce maps to determine the spatial extent of the various ground cover types and to enable the biomass types to be stratified so as to facilitate subsequent ground assessment of woody biomass.

The three stages of map preparation are (i) production of provisional type maps, (ii) field verification, and (iii) final type classification and maps preparation.

Stage V: Ground inventories

The imagery and maps provide spatial information on the types, but for wood fuel resource planning, development, and management, it is essential to know the quantity, condition, and growth rates of the resource within the types. These data are usually obtained using a ground inventory in forests and non-forest areas with woody vegetation, using the maps and imageries to stratify the sample so as to enable a cost-efficient assessment to a desired level of accuracy.

Principle of remote sensing

The principle behind recognizing vegetation by remote sensing systems is their operation in green, red, and near infrared regions of the electromagnetic spectrum. Sensors are sensitive enough to discriminate among the radiation absorption and reflectance by different types of objects and vegetation mainly due to the special characteristic features of the semi-transparent physiology of leaves. Chlorophylls absorb blue and red light and carotenoids absorb blue and green light. As green and yellow lights are not effectively absorbed by photosynthetic pigments in plants, these colours are either reflected by leaves or pass through them. Therefore, plants seem to be green. Due to the absorption of visible red and blue colour by the chlorophyll, the leaves reflect visible wavelengths that are concentrated in the green. Other than this, there is another strong reflectance between 0.7–1.0 μm (near infrared) by the spongy mesophyll cells located in the interior or back of a leaf. As the intensity of this reflectance is much higher than most inorganic materials, the presence of vegetation can be clearly observed on multispectral images by their specific tonal signatures: darker tones in the blue and especially red bands, and notably light in the near infrared bands.

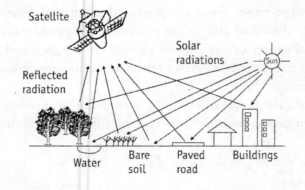

Before assessment, correlation between various parameters are established, such as D (diameter at breast height), stump diameter, total tree height or crown diameter, volume/weight of woody biomass to a utilization limit, etc. Development of appropriate tree-weight functions requires destructive sampling of trees. Measurements of tree height, crown diameter, and stem diameter at breast are taken prior to cutting them down and weighing.

Assessments are done using the Quadrant method. The plot size is deter-mined by terrain; stand characteristics; and type of sampling (random, stratified, phasing, parameters to be monitored, etc.). Generally, the plot size varies from 0.005 ha (7 m × 7 m) to 0.5 ha (50 m × 100 m). Plots may contain one or more sub plots. In order to obtain the time series data on woody biomass, permanent plots are used. These are particularly useful for growth data.

In a forestland quadrants of standard size, individual species are identi-fied and enumerated. In case of a pure forest, species-wise GS (growing stock) can be obtained, whereas in case of a heterogeneous forest, GS of miscellane-ous trees can be obtained. Standard productivity values can be used if they are available for the particular species. D and the height for the sample trees are measured and used for calculating the volume of the trees using the following equations (Chaturvedi 1984).

$$V = a + b \times D^2 H$$

where V is volume (m³), H is height (m), D is diameter at breast height (m), and a and b are regression constants.

GS is estimated using the following equation. The volume generated by the wood species under different land categories is aggregated to compute the total growing stock in sample area.

GS = Tree cover for each category of land (ha) ×
Annual productivity (m³/ha) × Density (tonne/m³)

Estimating the annual yield from GS is the key to determining an acces-sible sustainable supply, which can be estimated as follows.

$$\text{Sustainable yield (tonnes/year)} = 2 \times \frac{GS}{\text{Rotation}}$$

Here, the GS of a particular type of species represents the average of all the ages of trees. General rule to estimate the stock at maturity and the wood removed in thinning is to double the average stock. The rotation ages for vari-ous tree formations vary from two to three years for agroforestry trees to 60–100 years for high forest trees.

Another method of sustainable yield estimation is the NPP (net pri-mary production) of wood, which is defined as the total growth rate above ground plant (wood, leaves, grass, herbs, etc.), that is, the amount of vegeta-ble matter produced (net primary production) per day, week, or year. In closed formations, 50%–70% of annual NPP can be in the form of wood, whereas in woodlands, which are more open, 40%–60% of the annual NPP could be wood. The average values of NPP according to the ecosystem are summarized in Figure 12.16.

Stage VI: Estimation of surplus biomass availability

Although most studies estimate the maximum physical quantities of biomass, it is also necessary to estimate the quantities which would be available, including alternate uses, and technical, economic, and environmental factors. The factor which has been arrived at considering the limitations of soil type, temperature and water conditions, machinery efficiency, accessibility due to remoteness, and per cent needed to safeguard the GS and to ensure sustainable supply in the following years in case of adverse circumstances is multiplied with the annual sustainable yield.

ESY (extractable sustainable yield) =
60% (say) × ASY (annual sustainable yield)

Finally, the amount consumed for alternate uses of biomass is subtracted to estimate the surplus biomass availability.

Surplus biomass availability (tonne/year) =
ESY – biomass consumption by all sources.

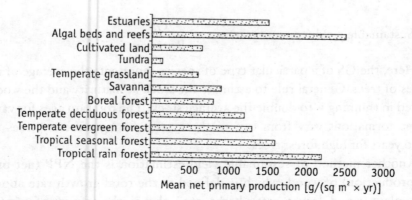

Figure 12.16 Mean net primary production by ecosystem
Source Holdgate (1996)

Example 4

A forest of 2705 ha has a productivity of 76.782 m³/ha and rotation age of 50 years and density of 0.8 tonnes/m³ (wet basis). Calculate the ESY of the forest.

Solution

The growing stock of the forest would be
2705 × 76.782 × 0.8 = 166 156 tonne

The ASY would be
2 × 166 156/50 = 6646.2 tonne/year

Extractable sustainable yield would be
0.6 × 6646.2 = 3987.5 tonne (wet basis)/year

Estimation of non-woody biomass

As discussed earlier, residues are generated from forestry, energy plantation, agricultural crops and fields, agro-industries, etc. One of the difficulties is to estimate the potential of residues that can be available for energy use on a national or regional basis due to their competing usage in assessment areas as fuels, in-situ burning, mulching, animal feed, house-building, etc. A demand survey is necessary to give directions about the residues usage for different purposes and in different periods. For example, rice and maize straw may be used as the principal fuel by the householders for a period immediately after the harvest. The rural industry may be dependent on crop residues to produce heat (for example, bagasse in jaggery making units), to fire pottery or bricks kilns with coffee husks, rice hulls, etc. If they are of minor importance then estimation may be done to the order of magnitude of standard conversion factors and the quantities of crop production (Table 12.11). However, the quantity of agricultural residues produced is also affected by seasons, soil type, irrigation conditions, and at times, with the variety of the seeds in one crop itself.

Residue production (tonnes) = Grain production (tonnes) × RPR (residue–product ratio)

It may be noted that with improved agricultural farming techniques, the production of crops has been increasing consistently in the past three decades. Correspondingly, availability of agricultural residues has also been changing with time. Improvements are also made to reduce residues. Alternately, residue production values may also be estimated by direct

Table 12.11 Air dry weight of residue produced per tonne of crop produced

Crop	Residue	RPR	Average residue production per ha
Barley	Straw	0.6–1.2	
	Residue		4.3 tonnes/ha
Cassava	Stem	0.2	
Coconut	Pith	246 g/nut	
	Fibre	164 g/nut	
	Frond	5.0	
	Husk	0.42–1.6	
	Shell	0.15–0.65 or 135 g/nut	
Coffee (wet process)	Cherry	0.75	
	Husk	0.25	
Coffee (dry process)	Cherry and husk	1	
Corn	Residue		10.1 tonnes/ha
Cotton	Residue		6.7 tonnes/ha
	Stalk		3.2–4.2 or 3 tonnes/ha
	Stem	3.5–4.0	
Cowpea	Stalk	2.9	
Groundnut	Shell	0.3–0.5	
	Straw	2.3–2.9	
	Residue		5.6 tonnes/ha
Jute	Stick		2.0 or 3 tonnes/ha
Maize	Straw	1.0	
	Cob	0.18–0.27	
	Husk	0.2	
	Stover and leaves	1.0–2.5	
Millet	Straw	1.4	
	Stalk	2.0–3.7	
Mungbean	Stalk and steam	3.2	
Oats	Straw	0.9–1.8	
	Residue		5.6 tonnes/ha
Palm kernel	Shell	0.35	
Palm Oil	Empty bunches	0.234	
	Fibre	0.18	
	Shell	0.073	
Peanuts			
Pigeon pea	Stalk	5.0	
Rice paddy	Husk	0.22–0.5	
	Straw	0.447–2.9	
	Bran	0.1	
	Residue		6.7 tonnes/ha
Rye	Straw	1.1–2.0	
Sesame	Stalk	2.3–5.3	
Sisal	Waste	1.2	

Continued

Table 12.11 *(Continued)*

Crop	Residue	RPR	Average residue production per ha
Sorghum	Straw	1.4	
	Leaves and stem	1.2	
	Stalk	0.9–4.6	
	Residue		8.4 tonne/ha
Soybean	Straw	1.1	
	Stalk	0.394	
Sugarbeet	Pulp	0.1–0.2	
	Residue		5.6 tonnes/ha
Sugar cane	Bagasse	0.1–0.3	
	Top and Trash	0.125	
Sunflower	Straw	2.5	
Tobacco	Residue		4.0 tonnes/ha
Wheat	Straw	0.7–1.8	
	Residue		5.0 tonnes/ha

Residue stands for plant parts left in the field after harvest
Sources Hall and Overend (1987); Watson and Ramstad (1987); FAO (1982); Cornelius (1983); PRESSEA (2005); Djevic (2005); Lal (2005); MoA (1999); Vimal, Tyagi, and Grover (1984); TERI (1987)

measurement in the fields during harvesting. An estimation of biomass residues in India is shown in Appendix 8. An estimation of world residue production is given in Lal (2005).

Estimation of waste

Like crop residues, animal waste, industrial waste, and household waste are also an important fuels. An accurate assessment of animal dung availability is essential, for example in rural energy planning, methane emissions, etc. Dung production depends upon the efficiency of animal digestive system. Some animals like pigs and cattle have an efficient digestive system, whereas horses have a poor system and their dung contains undigested leaves. Its production also varies as per the quality and quantity of feed. Dung has several other uses. Farmers and ecologists consider it to be too valuable to burn directly and prefer to use it as manure. It is also used as binding agent during house construction or as a wall and floor coating agent in the rural areas. Thus, all these facts have to be taken into consideration when assessing dung production and its availability for energy generation.

When measuring the potential supply of animal residues, it is important to have a realistic animal count by region or district, their average weight, and relevant conversion factors. While differences in animal numbers can account for some of the discrepancies, the largest discrepancy could be related to conversion factors. When animals are fully grown, dung production is in proportion to the food intake, which is more or less related to the animal size and differs from country to country and region to region (Table 12.12). As the moisture content of the dung also varies with respect to the feed intake, the production is compared on a dry basis for energy generation.

MSW is generally defined as the total waste from the domestic sector, that is, the total waste excluding industrial and agricultural waste, and sewage sludge. As defined by the USEPA (United States Environmental Protection Agency), it includes durable and non-durable goods; containers and packaging; food waste; yard waste; and miscellaneous inorganic waste from residential,

Table 12.12 Animal dung production in different countries

Country	Animal	Dung production (dry basis) kg/head/day	Volatile solid (kg/head/day)
North America	Dairy cattle	5.68	5.23
	Non-dairy cattle	2.55	2.35
Western Europe	Dairy cattle	5.52	5.08
	Non-dairy cattle	2.88	2.65
Eastern Europe	Dairy cattle	4.49	4.13
	Non-dairy cattle	2.91	2.68
Oceania	Dairy cattle	3.77	3.47
	Non-dairy cattle	3.29	3.03
Latin America	Dairy cattle	3.16	2.91
	Non-dairy cattle	2.70	2.48
Africa and Middle East	Dairy cattle	2.01	1.85
	Non-dairy cattle	1.68	1.54
Asia	Dairy cattle	3.07	2.82
	Non-dairy cattle	2.49	2.29
Indian subcontinent	Dairy cattle	2.87	2.64
	Non-dairy cattle	1.50	1.38
	Buffalo	2.65	2.43
Other livestock Developing countries	Swine		0.34
	Sheep		0.32
	Goat		0.35
	Camel		2.49
	Horse		1.72
	Mule/Ass		0.94
	Poultry		0.02

Source IPCC (1996)

commercial, and institutional sources. The MSW industry has four components: recycling, composting, landfilling, and waste-to-energy conversion via incineration. Depending upon the fraction of each component, its usage as fuel varies. The potential of MSW usage as energy thus varies from country to country and from region to region within a country. Various social and economic factors affecting the quality and quantity of MSW are urbanization level, economic status, food habits, culture, disposal habits, etc. One of the studies in Florida witnessed a variation in the MSW production between coastal and non-coastal areas too. Hence, whenever assessment of MSW is carried out for energy generation, all these factors should be considered before planning and management. The average per-capita waste generation from a few countries has been summarized in Table 12.13. This is an average figure for the country in consideration, as wide variations exists within each country.

Table 12.13 Per-capita municipal solid waste generation in different countries (2000)

Country	kg/capita/annum
Australia	400
Austria	380
Belgium	450
Canada	330
Denmark	560
Finland	190
France	360
Germany	340
India	146
Ireland	330
Japan	270
New Zealand	380
Norway	330
Switzerland	450
The Netherlands	530
United Kingdom	480
United States	460

Appendix 1 List of energy crops

Abies balsamea	*Bactris gasipaes*	*Croton tiglium*
Acacia albida Del.	*Bambusa arundinacea*	*Cynodon dactylon*
Acacia auriculiformis	*Beta vulgaris*	*Cyperus papyrus*
Acacia cyclops	*Betula lenta L.*	*Dactylis glomerata*
Acacia farnesiana	*Bothriochloa pertusa*	*Dalbergia sissoo*
Acacia mangium	*Bracheria mutica*	*Daucus carota*
Acacia mearnsii	*Brassica juncea*	*Dichanthium annulatum*
Acacia nilotica	*Brassica napus*	*Digitaria decumbens*
Acacia saligna	*Brassica nigra*	*Echinochloa crusgalli*
Acacia senegal	*Brassica rapa*	*Eichornia crassipes*
Acacia seyal	*Bromus inermis*	*Elaeis guineensis*
Acacia tortilis	*Bruguiera gymnorrhiza*	*Eleusine coracana*
Acroceras macrum	*Cajanus cajan*	*Eragrostis curvula*
Agropyron elongatum	*Calliandra calothyrsus*	*Erythrina berteroana*
Ailanthus altissima	*Camellia sinensis*	*Erythrina fusca*
Albizia falcataria	*Cannabis sativa*	*Erythrina poeppigiana*
Albizia lebbek	*Caragana arborescens*	*Eucalyptus*
Albizia procera	*Carica papaya*	*Eucalyptus camaldulensis*
Aleurites fordii	*Carthamus tinctorius*	*Eucalyptus citriodora*
Aleurites moluccana	*Cassia fistula*	*Eucalyptus globulus*
Aleurites montana	*Casuarina cunninghamiana*	*Eucalyptus gomphocephala*
Alnus glutinosa	*Casuarina equisetifolia*	*Eucalyptus grandis*
Alnus maritima	*Casuarina glauca*	*Eucalyptus microtheca*
Alnus nepalensis	*Casuarina junghuhniana*	*Eucalyptus occidentalis*
Alopecurus pratensis	*Cenchrus ciliaris*	*Eucalyptus robusta*
Anacardium occidentale	*Ceriops tagal*	*Eucalyptus saligna*
Ananas comosus	*Chloris gayana*	*Eucalyptus tereticornis*
Andropogon gayanus	*Chrysothamnus nauseosus*	*Eucalyptus viminalis*
Apios americana	*Cichorium intybus*	*Euphorbia lathyris*
Arachis hypogaea	*Citrullus colocynthis*	*Euphorbia tirucalli*
Arundo donax	*Cocos nucifera*	*Festuca pratensis*
Asclepias syriaca	*Coffea arabica*	*Gleditsia triacanthos*
Atriplex hortensis	*Coix lacryma-jobi*	*Gliricidia sepium*
Avena sativa	*Conocarpus erectus*	*Glycine max*
Avicennia germinans	*Copaifera langsdorfii*	*Gmelina arborea*
Avicennia marina	*Corchorus olitorius*	*Gossypium hirsutum*
Avicennia officinalis	*Crambe abyssinica*	*Guizotia abyssinica*
Axonopus affinis	*Crotalaria juncea*	*Helianthus annuus*

Helianthus tuberosus

Hevea brasiliensis

Hibiscus cannabinus

Hibiscus sabdariffa

Hordeum vulgare

Humulus lupulus

Hymenaea courbaril

Inga edulis

Inga vera

Ipomoea batatas

Jatropha curcas

Juglans nigra

Juglans regia

Juniperus virginiana

Laguncularia racemosa

Lesquerella fendleri

Leucaena leucocephala

Limnanthes alba

Linum usitatissimum

Lolium perenne

Macadamia

Malus sylvestris

Manihot esculenta

Medicago sativa

Melaleuca quinquenervia

Melinis minutiflora

Mimosa scabrella

Moringa oleifera

Morus alba

Oryza sativa

Panicum maximum

Papaver somniferum

Paspalum notatum

Pennisetum purpureum

Phalaris arundinacea

Phaseolus vulgaris

Phleum pratense

Phoenix dactylifera

Phragmites australis

Picea spp.

Pinus elliottii

Pithecellobium dulce

Pittosporum resiniferum

Pongamia pinnata

Populus deltoides

Prosopis alba

Prosopis chilensis

Prosopis cineraria

Prosopis glandulosa

Prosopis juliflora

Prosopis pallida

Prosopis tamarugo

Prunus dulcis

Pterocarpus indicus

Pueraria lobata

Rhizophora mangle

Rhizophora mucronata

Ricinus communis

Robinia pseudoacacia

Saccharum officinarum

Salsola kali

Samanea saman

Sapium sebiferum

Secale cereale

Sesbania bispinosa

Simmondsia chinensis

Sinapis alba

Sindora supa

Solanum tuberosum

Sorghum halepense

Sorghum sudanense

Sorghum X almum

Symphytum peregrinum

Theobroma cacao

Thuja occidentalis

Trifolium pratense

Triticum aestivum

Typha spp.

Vicia faba

Vigna unguiculata

Vitis vinifera

Zea mays

Zizania aquatica

Appendix 2 ASTM test methods for biomass

- E1126-94a Standard Terminology Relating to Biomass Fuels
- E1288-89(1994) Standard Test Method for the Durability of Biomass Pellets
- E1358-97 Standard Test Method for Determination of Moisture Content of Particulate Wood Fuels Using a Microwave Oven
- E1534-93(1998) Standard Test Method for Determination of Ash Content of Particulate Wood Fuels
- E1535-93(1998) Standard Test Method for Performance Evaluation of Anaerobic Digestion Systems

- E1757-01 Standard Practice for Preparation of Biomass for Compositional Analysis
- E1821-01 Standard Test Method for Determination of Carbohydrates in Biomass by Gas Chromatography
- E870-82(1998) Standard Test Methods for Analysis of Wood Fuels
- E871-82 (1998) Standard Method of Moisture Analysis of Particulate Wood Fuels
- E872-82 (1998) Standard Test Method for Volatile Matter in the Analysis of Particulate Wood Fuels
- E873-82(1998) Standard Test Method for Bulk Density of Densified Particulate Biomass Fuels
- PS121-99 Provisional Specification for Biodiesel Fuel (B100) Blend Stock for Distillate Fuels

ASTM standards test methods for ash content analysis

- C561-91(2000) Standard Test Method for Ash in a Graphite Sample
- D1119-05 Standard Test Method for Per cent Ash Content of Engine Coolants
- D1506-99 Standard Test Methods for Carbon Black-Ash Content
- D2415-98(2003) Standard Test Method for Ash in Coal Tar and Pitch
- D2866-94(2004) Standard Test Method for Total Ash Content of Activated Carbon
- D2875-00 Standard Test Method for Insoluble Ash of Vegetable-Tanned Leather
- D2974-00 Standard Test Methods for Moisture, Ash, and Organic Matter of Peat and Other Organic Soils
- D4422-03 Standard Test Method for Ash in Analysis of Petroleum Coke
- D4427-92(2002)e1 Standard Classification of Peat Samples by Laboratory Testing
- D4574-02 Standard Test Methods for Rubber Chemicals–Determination of Ash Content
- D4715-98(2003) Standard Test Method for Coking Value of Tar and Pitch (Alcan)
- D4868-00(2005) Standard Test Method for Estimation of Net and Gross Heat of Combustion of Burner and Diesel Fuels
- D4906-95(2001) Standard Test Method for Total Solids and Ash Content in Leather Finishing Materials
- D5347-95(2001) Standard Test Method for Determination of the Ash Content of Fats and Oils
- D5630-01 Standard Test Method for Ash Content in Thermoplastics
- D5667-95(2005) Standard Test Method for Rubber from Synthetic Sources–Total and Water Soluble Ash
- D586-97(2002) Standard Test Method for Ash in Pulp, Paper, and Paper Products

- D6511-00 Standard Test Methods for Solvent Bearing Bituminous Compounds
- E1131-03 Standard Test Method for Compositional Analysis by Thermogravimetry
- E1534-93(1998) Standard Test Method for Determination of Ash Content of Particulate Wood Fuels
- E830-87(2004) Standard Test Method for Ash in the Analysis Sample of Refuse-Derived Fuel

ASTM standards for moisture content analysis

- C566-97(2004) Standard Test Method for Total Evaporable Moisture Content of Aggregate by Drying
- D1037-99 Standard Test Methods for Evaluating Properties of Wood-Base Fibre and Particle Panel Materials
- D143-94(2000)e1 Standard Test Methods for Small Clear Specimens of Timber
- D2867-04 Standard Test Methods for Moisture in Activated Carbon
- D2974-00 Standard Test Methods for Moisture, Ash, and Organic Matter of Peat and Other Organic Soils
- D3466-76(1998) Standard Test Method for Ignition Temperature of Granular Activated Carbon
- D3737-04 Standard Practice for Establishing Allowable Properties for Structural Glued Laminated Timber (Glulam)
- D4442-92(2003) Standard Test Methods for Direct Moisture Content Measurement of Wood and Wood-Base Materials
- D4643-00 Standard Test Method for Determination of Water (Moisture) Content of Soil by the Microwave Oven Method
- D4931-92(2002) Standard Test Method for Gross Moisture in Green Petroleum Coke
- D4933-99(2004) Standard Guide for Moisture Conditioning of Wood and Wood-Base Materials
- D4959-00 Standard Test Method for Determination of Water (Moisture) Content of Soil by Direct Heating
- D5147-02ae1 Standard Test Methods for Sampling and Testing Modified Bituminous Sheet Material
- D6403-99(2004) Standard Test Method for Determining Moisture in Raw and Spent Materials
- D6420-99(2004) Standard Test Method for Determination of Gaseous Organic Compounds by Direct Interface Gas Chromatography-Mass Spectrometry
- D644-99(2002) Standard Test Method for Moisture Content of Paper and Paperboard by Oven Drying

- D6776-02 Standard Test Method for Determining Anaerobic Biodegradability of Radiolabelled Plastic Materials in a Laboratory-scale Simulated Landfill Environment
- D817-96(2004) Standard Test Methods of Testing Cellulose Acetate Propionate and Cellulose Acetate Butyrate
- D871-96(2004) Standard Test Methods of Testing Cellulose Acetate
- E1358-97 Standard Test Method for Determination of Moisture Content of Particulate Wood Fuels Using a Microwave Oven
- E410-98(2003) Standard Test Method for Moisture and Residue in Liquid Chlorine

ASTM standards for proximate analysis

- D3172-89(2002) Standard Practice for Proximate Analysis of Coal and Coke
- D3174-04 Standard Test Method for Ash in the Analysis Sample of Coal and Coke from Coal
- D3175-02 Standard Test Method for Volatile Matter in the Analysis Sample of Coal and Coke
- D5142-04 Standard Test Methods for Proximate Analysis of the Analysis Sample of Coal and Coke by Instrumental Procedures
- E830-87(2004) Standard Test Method for Ash in the Analysis Sample of Refuse-Derived Fuel

ASTM standards for ultimate analysis

- D3174-04 Standard Test Method for Ash in the Analysis Sample of Coal and Coke from Coal
- D3176-89(2002) Standard Practice for Ultimate Analysis of Coal and Coke
- D3177-02 Standard Test Methods for Total Sulphur in the Analysis Sample of Coal and Coke
- D3178-89(2002) Standard Test Methods for Carbon and Hydrogen in the Analysis Sample of Coal and Coke
- D3179-02e1 Standard Test Methods for Nitrogen in the Analysis Sample of Coal and Coke
- D5142-04 Standard Test Methods for Proximate Analysis of the Analysis Sample of Coal and Coke by Instrumental Procedures
- D5373-02 Standard Test Methods for Instrumental Determination of Carbon, Hydrogen, and Nitrogen in Laboratory Samples of Coal and Coke
- E775-87(2004) Standard Test Methods for Total Sulphur in the Analysis Sample of Refuse-Derived Fuel
- E778-87(2004) Standard Test Methods for Nitrogen in the Analysis Sample of Refuse-Derived Fuel

- E830-87(2004) Standard Test Method for Ash in the Analysis Sample of Refuse-Derived Fuel
- E856-83(2004) Standard Definitions of Terms and Abbreviations Relating to Physical and Chemical Characteristics of Refuse Derived Fuel
- E870-82(1998)e1 Standard Test Methods for Analysis of Wood Fuels

ASTM standards for carbon content

- D3178-89(2002) Standard Test Methods for Carbon and Hydrogen in the Analysis Sample of Coal and Coke
- E777-87(2004) Standard Test Method for Carbon and Hydrogen in the Analysis Sample of Refuse-Derived Fuel

ASTM standards for nitrogen content

- D2868-96(2001) Standard Test Method for Nitrogen Content (Kjeldahl) and Hide Substance Content of Leather
- D3179-02e1 Standard Test Methods for Nitrogen in the Analysis Sample of Coal and Coke
- D4795-94(2003) Standard Test Method for Nitrogen Content of Soluble Nitrocellulose—Alternative Method
- E258-67(2002) Standard Test Method for Total Nitrogen in Organic Materials by Modified Kjeldahl Method
- E778-87(2004) Standard Test Methods for Nitrogen in the Analysis Sample of Refuse-Derived Fuel
- UOP731-74 Total Sulphur in Heavy Oils, Tars and Solids by Microcoulometric Titration, Single Entry Boat Technique

ASTM standards for hydrogen content

- D2650-04 Standard Test Method for Chemical Composition of Gases by Mass Spectrometry
- D3178-89(2002) Standard Test Methods for Carbon and Hydrogen in the Analysis Sample of Coal and Coke
- E777-87(2004) Standard Test Method for Carbon and Hydrogen in the Analysis Sample of Refuse-Derived Fuel
- UOP735-73 Hydrogen Content of Gases by Gas Chromatography

ASTM standards for oxygen content

- D6733-01 Standard Test Method for Determination of Individual Components in Spark Ignition Engine Fuels by 50-Meter Capillary High Resolution Gas Chromatography
- D888-03 Standard Test Methods for Dissolved Oxygen in Water
- UOP649-74 Total Oxygen in Organic Materials by Pyrolysis-Gas Chromatographic Technique

ASTM standards for other parameters

- D4012-81(2002) Standard Test Method for ATP (Adenosine Triphosphate) Content of Microorganisms in Water
- D5271-02 Standard Test Method for Determining the Aerobic Biodegradation of Plastic Materials in an Activated-Sludge-Wastewater-Treatment System
- D6530-00 Standard Test Method for Total Active Biomass in Cooling Tower Waters (Kool Kount Assay; KKA)
- E1279-89(2001) Standard Test Method for Biodegradation By a Shake-Flask Die-Away Method
- E1534-93(1998) Standard Test Method for Determination of Ash Content of Particulate Wood Fuels
- E1535-93(1998) Standard Test Method for Performance Evaluation of Anaerobic Digestion Systems
- E1690-01 Standard Test Method for Determination of Ethanol Extractives in Biomass
- E1721-01 Standard Test Method for Determination of Acid-Insoluble Residue in Biomass
- E1755-01 Standard Test Method for Ash in Biomass
- E1756-01 Standard Test Method for Determination of Total Solids in Biomass
- E1757-01 Standard Practice for Preparation of Biomass for Compositional Analysis
- E1758-01 Standard Test Method for Determination of Carbohydrates in Biomass by High Performance Liquid Chromatography
- E1821-01 Standard Test Method for Determination of Carbohydrates in Biomass by Gas Chromatography
- E870-82(1998)e1 Standard Test Methods for Analysis of Wood Fuels
- E873-82(1998) Standard Test Method for Bulk Density of Densified Particulate Biomass Fuels

ASTM standards for combustion

- D1072-90(1999) Standard Test Method for Total Sulphur in Fuel Gases
- D240-02 Standard Test Method for Heat of Combustion of Liquid Hydrocarbon Fuels by Bomb Calorimeter
- D4485-05a Standard Specification for Performance of Engine Oils
- D4868-00(2005) Standard Test Method for Estimation of Net and Gross Heat of Combustion of Burner and Diesel Fuels
- D5485-99 Standard Test Method for Determining the Corrosive Effect of Combustion Products Using the Cone Corrosimeter
- E1131-03 Standard Test Method for Compositional Analysis by Thermogravimetry
- E1678-02 Standard Test Method for Measuring Smoke Toxicity for Use in Fire Hazard
- E870-82(1998)e1 Standard Test Methods for Analysis of Wood Fuels
- E955-88(2004) Standard Test Method for Thermal Characteristics of Refuse-Derived Fuel Macrosamples

Appendix 3 Proximate analysis of selected biomass

Biomass	Fixed carbon	Volatile matter	Ash	References
Alabama oakwood	21.90	74.70	3.30	Boley and Landers (1969)
Alfalfa stems	15.81	78.92	5.27	Jenkins, Baxter, Miles, et al. (1998)
Almond	21.54	76.83	1.63	Jenkins and Ebeling (1985)
Almond forest	13.8	76.40	1.00	González, González-Garcia, Ramiro, et al. (2005)
Almond hulls	20.07	73.80	6.13	Jenkins, Baxter, Miles, et al. (1998)
Almond peel	13.90	71.20	3.60	González, González-Garcia, Ramiro, et al. (2005)
Almond pruning	15.14	75.00	3.59	Gañán (2004)
Almond pruning	15.90	72.20	1.20	González, González-Garcia, Ramiro, et al. (2005)
Almond shell	15.80	80.30	0.60	González, González-Garcia, Ramiro, et al. (2005)
Almond shells	20.71	76.00	3.29	Jenkins, Baxter, Miles, et al. (1998)
Bagasse	7.00	70.90	22.10	Grover, Iyer, and Rao (2002)
Bagasse	11.90	86.30	1.80	Grover, Iyer, and Rao (2002)
Bagasse pith	10.60	86.60	2.80	Grover, Iyer, and Rao (2002)
Bamboo dust	9.30	74.20	16.50	Grover, Iyer, and Rao (2002)
Bamboo stick waste	47.70	12.70	39.60	Grover, Iyer, and Rao (2002)
Block wood	14.59	83.32	2.09	Grover, Iyer, and Rao (2002)
Cabernet Sauvignon	19.20	78.63	2.17	Jenkins and Ebeling (1985)
Canyon live oak	11.30	88.20	0.50	Rossi (1984)
Castor seed cake	25.20	67.90	6.90	Grover, Iyer, and Rao (2002)

Continued...

Appendix 3 Continued...

Biomass	Fixed carbon	Volatile matter	Ash	References
Ceder cones	28.10	70.40	1.50	Grover, Iyer, and Rao (2002)
Cherry stone	25.90	73.90	0.20	González, Encinar, Canito et al. (2003)
Cherry stone (gasified residual)	94.00	3.00	3.00	González, Encinar, Canito, et al. (2003)
Coconut coir	29.70	66.58	3.72	Jenkins and Ebeling (1985)
Coconut shell char: 750 °C	87.17	9.93	2.90	Parikh, Channiwala, and Ghosal (2005)
Coconut shell powder	20.58	79.07	0.35	Jimenez and Gonzalez (1991)
Coconut stem	23.10	74.40	2.50	Grover, Iyer, and Rao (2002)
Coffee chaff	19.60	75.80	4.60	Grover, Iyer, and Rao (2002)
Corncob	16.80	82.10	1.10	Grover, Iyer, and Rao (2002)
Corncob	18.54	80.10	1.36	Jenkins and Ebeling (1985)
Corncob	12.50	86.50	1.00	Demirbas (2001)
Cornstover	17.60	78.70	3.70	Demirbas (2001)
Cotton gin waste	14.97	83.41	1.61	Parikh, Channiwala, and Ghosal (2005)
Cotton shells	16.90	68.50	14.60	Jimenez and Gonzalez (1991)
Cotton shells briquettes	17.10	77.80	5.10	Grover, Iyer, and Rao (2002)
Cotton stalk	17.30	65.40	17.30	Jenkins and Ebeling (1985)
Cottongin trash	15.10	67.30	17.60	Jenkins and Ebeling (1985)
Dal lake weed	3.60	47.70	48.70	Grover, Iyer, and Rao (2002)
Demol. wood	12.32	74.56	13.12	Jenkins, Baxter, Miles, et al. (1998)
Douglas fir	25.80	73.00	1.20	Tillman (1978)
Eucalyptus	15.40	72.90	0.90	Ganan, Kassir Abdulla, Miranda, et al. 2005)
Eucalyptus	21.30	75.35	3.35	Parikh, Channiwala, and Ghosal (2005)
Eucalyptus (gasified residual)	90.80	5.60	2.70	Ganan, Kassir Abdulla, Miranda, et al. (2005)
Eucalyptus saw dust	16.20	83.60	0.20	Grover, Iyer, and Rao (2002)
Eucalyptus stalk	12.20	87.30	0.50	Grover, Iyer, and Rao (2002)
Eucalyptus-Grandis	16.93	82.55	0.52	Jenkins and Ebeling (1985)
Eucatlyptus bark	15.30	65.70	19.00	Grover, Iyer, and Rao (2002)
Fir mill	17.48	82.11	0.41	Jenkins, Baxter, Miles, et al. (1998)
Fly ash (Bagasse fuel)	19.80	8.10	72.10	Grover, Iyer, and Rao (2002)
Grewia optiva (Bhimal)	14.20	85.50	0.30	Grover, Iyer, and Rao (2002)
Hazelnut shell	28.30	69.30	1.40	Demirbas (1997)
Hybrid poplar	12.49	84.81	2.70	Jenkins, Baxter, Miles, et al. (1998)
Industrial waste (stalla)	20.10	75.10	4.80	Grover, Iyer, and Rao (2002)
Lantana briquettes	11.90	20.80	67.30	Grover, Iyer, and Rao (2002)
Loblolly pine	33.90	65.70	0.40	Risser (1981)
Macadamia shell	23.68	75.92	0.40	Jenkins and Ebeling (1985)
Mentha piperita	7.50	79.00	13.50	Grover, Iyer, and Rao (2002)
Miscalthus (elephanta grass)	12.40	87.20	0.40	Grover, Iyer, and Rao (2002)
Mixed paper	7.42	84.25	8.33	Jenkins, Baxter, Miles, et al. (1998)
Moringa oleifera (leaves)	10.70	67.80	21.50	Grover, Iyer, and Rao (2002)
Mulberry stick	22.80	75.10	2.10	Grover, Iyer, and Rao (2002)

Appendix 3 Continued...

Biomass	Fixed carbon	Volatile matter	Ash	References
Municipal solid waste –Bareilly	5.00	25.00	70.00	Grover, Iyer, and Rao (2002)
Municipal solid waste –Moradabad	4.00	35.70	60.30	Grover, Iyer, and Rao (2002)
Poultry pure waste	25.90	14.30	59.80	Grover, Iyer, and Rao (2002)
Press mud briquettes	8.60	54.70	36.70	Grover, Iyer, and Rao (2002)
RDF	0.47	73.40	26.13	Jenkins, Baxter, Miles, et al. (1998)
Redwood	19.92	79.72	0.36	Jenkins and Ebeling (1985)
Redwood char– 790–1020 °F	67.70	30.00	2.30	Boley and Landers (1969)
Rice hull	16.22	63.52	20.26	Jenkins, Baxter, Miles, et al. (1998)
Rice husk	13.20	65.30	19.20	Ganan, Kassir Abdulla, Miranda, et al. (2005)
Rice husk (gasified residual)	74.40	4.40	20.30	Ganan, Kassir Abdulla, Miranda, et al. (2005)
Rice husk char	41.20	5.90	52.90	Grover, Iyer, and Rao (2002)
Rice straw	15.86	65.47	18.67	Jenkins, Baxter, Miles, et al. (1998)
Rice straw (ground)	16.20	68.30	15.50	Grover, Iyer, and Rao (2002)
Rose apple char	12.20	22.20	65.60	Grover, Iyer, and Rao (2002)
Sal seed husk	28.06	62.54	9.40	Grover and Anuradha (1988)
Sawdust	25.00	72.40	2.60	Grover, Iyer, and Rao (2002)
Sawdust + mustard	11.90	55.40	32.70	Grover, Iyer, and Rao (2002)
Seed corn	12.90	85.60	1.50	Zhang, Cummer, Suby, et al. (2005)
Sena leaves	25.50	57.20	17.30	Grover, Iyer, and Rao (2002)
Sewage sludge (community)	8.60	53.00	38.40	Thipkhunthod, Meeyoo, Rangsunvigit, et al. (2005)
	6.70	51.20	42.00	Thipkhunthod, Meeyoo, Rangsunvigit, et al. (2005)
	7.00	50.00	43.00	Thipkhunthod, Meeyoo, Rangsunvigit, et al. (2005)
	4.00	47.60	48.40	Thipkhunthod, Meeyoo, Rangsunvigit, et al. (2005)
	6.00	42.20	51.80	Thipkhunthod, Meeyoo, Rangsunvigit, et al. (2005)
	3.70	34.50	61.80	Thipkhunthod, Meeyoo, Rangsunvigit, et al. (2005)
	5.00	39.00	56.00	Thipkhunthod, Meeyoo, Rangsunvigit, et al. (2005)
	3.20	33.30	63.50	Thipkhunthod, Meeyoo, Rangsunvigit, et al. (2005)
	3.10	32.90	64.00	Thipkhunthod, Meeyoo, Rangsunvigit, et al. (2005)
	1.80	30.60	67.60	Thipkhunthod, Meeyoo, Rangsunvigit, et al. (2005)
	2.20	24.80	72.90	Thipkhunthod, Meeyoo, Rangsunvigit, et al. (2005)
	2.40	23.40	74.20	Thipkhunthod, Meeyoo, Rangsunvigit, et al. (2005)
Sewage sludge	5.10	55.50	39.40	Thipkhunthod, Meeyoo, Rangsunvigit, et al. (2005)
	6.80	52.60	40.60	Thipkhunthod, Meeyoo, Rangsunvigit, et al. (2005)
	6.50	47.70	45.90	Thipkhunthod, Meeyoo, Rangsunvigit, et al. (2005)
	3.90	50.40	45.70	Thipkhunthod, Meeyoo, Rangsunvigit, et al. (2005)
	3.20	36.60	60.20	Thipkhunthod, Meeyoo, Rangsunvigit, et al. (2005)
Sewage sludge	3.20	54.50	42.30	Thipkhunthod, Meeyoo, Rangsunvigit, et al. (2005)
	2.80	45.60	51.60	Thipkhunthod, Meeyoo, Rangsunvigit, et al. (2005)
	3.00	38.20	58.80	Thipkhunthod, Meeyoo, Rangsunvigit, et al. (2005)

Continued...

Appendix 3 Continued...

Biomass	Fixed carbon	Volatile matter	Ash	References
Softwood	28.10	70.00	1.70	Demirbas (1997)
Spearmint	11.80	70.10	18.10	Jimenez and Gonzalez (1991)
Spruce wood	28.30	70.20	1.50	Demirbas (1997)
Subabul	13.80	85.20	1.00	Grover, Iyer, and Rao (2002)
Subabul wood	18.52	81.02	1.20	Parikh, Channiwala, and Ghosal (2005)
Sudan Grass	18.60	72.75	8.65	Jenkins and Ebeling (1985)
Sugar cane bagasse	11.95	85.61	2.44	Jenkins, Baxter, Miles, et al. (1998)
Sugarcane leaves	14.90	77.40	7.70	Grover, Iyer, and Rao (2002)
Sweet sorghum bagasse	5.00	75.00	20.00	Grover, Iyer, and Rao (2002)
Switch grass	14.34	76.69	8.97	Jenkins, Baxter, Miles, et al. (1998)
Tan oak	9.20	90.60	0.20	Rossi (1984)
Tannary waste	1.00	45.00	54.00	Grover, Iyer, and Rao (2002)
Tapero root skin scale scrapping	11.40	35.10	39.20	Grover, Iyer, and Rao (2002)
Tea bush	21.80	76.50	1.70	Grover, Iyer, and Rao (2002)
Tea waste	13.60	85.00	1.40	Demirbas (1997)
Vine pruning	14.60	71.40	3.70	Cuellar Borrego (2003)
Vine pruning (gasified residual)	91.2	3.20	2.80	Cuellar Borrego (2003)
Walnut	20.80	78.50	0.70	Grover, Iyer, and Rao (2002)
Water hyacinth	1.90	87.30	10.80	Grover, Iyer, and Rao (2002)
Wheat straw	17.71	75.27	7.02	Jenkins, Baxter, Miles et al. (1998)
Wheat straw	23.50	63.00	13.50	Demirbas (1997)
Wheat straw	24.00	69.60	6.40	Grover, Iyer, and Rao (2002)
Wheat straw	11.70	80.60	7.70	Grover, Iyer, and Rao (2002)
White fir	16.58	83.17	0.25	Jenkins and Ebeling (1985)
Willow wood	16.07	82.22	1.71	Jenkins, Baxter, Miles, et al. (1998)
Wood chips	23.50	76.40	0.10	Jenkins (1980)
Wood chips	15.40	83.40	1.20	Grover, Iyer, and Rao (2002)
Yard waste	13.59	66.04	20.37	Jenkins, Baxter, Miles, et al. (1998)

Appendix 4 Elemental composition of ash (%)

Biomass	SiO_2	Al_2O_3	TiO_2	Fe_2O_3	CaO	MgO	Na_2O	K_2O	SO_3	P_2O_5	Reference
Alfalfa stems	5.79	0.07	0.02	0.30	18.32	10.38	1.10	28.10	1.93	7.64	Jenkins, Baxter, Miles, et al. (1998)
Almond hull	9.28	2.09	0.05	0.76	8.07	3.31	0.87	52.90	0.34	5.10	Jenkins, Baxter, Miles, et al. (1998)
Almond shell	8.71	2.72	0.09	2.30	10.50	3.19	1.60	48.70	0.88	4.46	Jenkins, Baxter, Miles, et al. (1998)
Demol. Wood	45.91	15.55	2.09	12.02	13.51	2.55	1.13	2.14	2.45	0.94	Jenkins, Baxter, Miles, et al. (1998)
Fir mill	15.17	3.96	0.27	6.58	11.90	4.59	23.50	7.00	2.93	2.87	Jenkins, Baxter, Miles, et al. (1998)
Hazelnut shell	33.70	3.10	0.10	3.80	15.40	7.90	1.30	30.40	1.10		Kalac, Svoboda, and Havliekova (2004)
Hybrid poplar	5.90	0.84	0.30	1.40	49.92	18.40	0.13	9.64	2.04	1.34	Jenkins, Baxter, Miles, et al. (1998)
Mixed paper	28.10	52.56	4.29	0.81	7.49	2.36	0.53	0.16	1.70	0.20	Jenkins, Baxter, Miles, et al. (1998)
Olive pitts	30.82	8.84	0.34	6.58	14.66	4.24	27.80	4.40	0.56	2.46	Jenkins, Baxter, Miles, et al. (1998)
Pistachio shell	8.22	2.17	0.20	35.37	10.01	3.26	4.50	18.20	3.79	11.80	Jenkins, Baxter, Miles, et al. (1998)
RDF	33.81	12.71	1.66	5.47	23.44	5.64	1.19	0.20	2.63	0.67	Jenkins, Baxter, Miles, et al. (1998)
Rice hull	91.42	0.78	0.02	0.14	3.21	0.01	0.21	3.71	0.72	0.43	Jenkins, Baxter, Miles, et al. (1998)
Rice straw	74.67	1.04	0.09	0.85	3.01	1.75	0.96	12.30	1.24	1.41	Jenkins, Baxter, Miles, et al. (1998)
Spruce wood	49.30	9.40		8.30	17.20	1.10	0.50	9.60	2.60		Kalac, Svoboda, and Havliekova (2004)
Sugar cane bagasse	46.61	17.69	2.63	14.14	4.47	3.33	0.79	0.15	2.08	2.72	Jenkins, Baxter, Miles, et al. (1998)
Switch grass	65.18	4.51	0.24	2.03	5.60	3.00	0.58	11.60	0.44	4.50	Jenkins, Baxter, Miles, et al. (1998)
Wheat straw	55.32	1.88	0.08	0.73	6.14	1.06	1.71	25.60	4.40	1.26	Jenkins, Baxter, Miles, et al. (1998)
Wheat straw	48.00	3.50		0.50	3.70	1.80	14.50	20.00	1.90		Kalac, Svoboda, and Havliekova (2004)
Spruce wood	49.30	9.40		8.30	17.20	1.10	0.50	9.60	2.60		Kalac, Svoboda, and Havliekova (2004)
Sugar cane bagasse	46.61	17.69	2.63	14.14	4.47	3.33	0.79	0.15	2.08	2.72	Jenkins, Baxter, Miles, et al. (1998)
Switch grass	65.18	4.51	0.24	2.03	5.60	3.00	0.58	11.60	0.44	4.50	Jenkins, Baxter, Miles, et al. (1998)
Wheat straw	55.32	1.88	0.08	0.73	6.14	1.06	1.71	25.60	4.40	1.26	Jenkins, Baxter, Miles, et al. (1998)
Wheat straw	48.00	3.50		0.50	3.70	1.80	14.50	20.00	1.90		Kalac, Svoboda, and Havliekova (2004)
Willow wood	2.35	1.41	0.05	0.73	41.20	2.47	0.94	15.00	1.83	7.40	Jenkins, Baxter, Miles, et al. (1998)
Yard waste	59.65	3.06	0.32	1.97	23.75	2.15	1.00	2.96	2.44	1.97	Jenkins, Baxter, Miles, et al. (1998)

SiO_2 – silicon dioxide; Al_2O_3 – aluminium oxide; Fe_2O_3 – ferric oxide; CaO – calcium oxide; MgO – magnesium oxide; Na_2O – sodium oxide;
K_2O – potassium oxide; SO_3 – sulphite; P_2O_5 – phosphorus pentoxide

Appendix 5 Ultimate analysis of biomass materials

Biomass	C	H	O	N	S	Cl	Ash	References
Alabama oakwood waste	49.50	5.70	41.30	0.20	0		.	Boley and Landers (1969)
Alfalfa stems	47.17	5.99	38.19	2.68	0.200	0.50	5.27	Jenkins, Baxter, Miles, et al. (1998)
Almond	51.30	5.29	40.90	0.66	0.010			Jenkins and Ebeling (1985)
Almond forest	46.50	6.80		1.88	0	0.03		Gonzalez, Gonzalez-Garcia, Ramiro, et al. (2005)
Almond hulls	47.53	5.97	39.16	1.13	0.060	0.02	6.13	Jenkins, Baxter, Miles, et al. (1998)
Almond peel	43.00	5.70		3.28	0.008	0.01		Gonzalez, Gonzalez-Garcia, Ramiro, et al. (2005)
Almond pruning	43.10	5.60		3.30	0.008	0.01		Gañán (2004)
Almond Pruning	51.30	6.50		0.77	0.035	0.05		Gonzalez, Gonzalez-Garcia, Ramiro, et al. (2005)
Almond shell	50.50	6.60		0.21	0.006	0.05		Gonzalez, Gonzalez-Garcia, Ramiro, et al. (2005)
Almond shell	49.30	5.97	40.63	0.76	0.040	0.01	3.29	Jenkins, Baxter, Miles, et al. (1998)
Block wood	46.90	6.07	43.99	0.95	0			Parikh, Channiwala, and Ghosal (2005)
Block wood	46.90	6.07	43.99	0.95	0			Grover, Iyer, and Rao (2002)
Cabernet Sauvignon	46.59	5.85	43.90	0.83	0.040			Jenkins and Ebeling (1985)
Canyon live oak	47.84	5.80	45.76	0.07	0.010			Rossi (1984)
Cherrystone	51.10	6.50		0.38	0.020			González, Encinar, Canito, et al. (2003)
Coconut coir	50.29	5.05	39.63	0.45	0.160			Jenkins and Ebeling (1985)
Coconut shell char-750 °C	88.95	0.73	6.04	1.38	0			Parikh, Channiwala, and Ghosal (2005)
Corncob	46.58	5.87	45.46	0.47	0.010			Jenkins and Ebeling (1985)
Corncob	49.00	5.40	44.60	0.40				Demirbas (2001)
Cottongin waste	42.66	6.05	49.50	0.18	0			Parikh, Channiwala, and Ghosal (2005)
Cotton stalk	39.47	5.07	39.14	1.20	0.020			Jenkins and Ebeling (1985)
Cottongin trash	39.59	5.26	36.38	2.09	0			Jenkins and Ebeling (1985)
Dal lake weed	19.12	2.00	25.96	4.22	–			Grover, Iyer, and Rao (2002)
Demol. wood	46.30	5.39	34.45	0.57	0.120	0.05	13.12	Jenkins, Baxter, Miles, et al. (1998)
Douglas fir	56.20	5.90	36.70	0	0			Tillman (1978)
Eucalyptus	7.90	7.90		0.22	0.029	0.05		Ganan, Kassir Abdulla, Miranda, et al. (2005)
Eucalyptus	46.04	5.82	44.49	0.30	0			Parikh, Channiwala, and Ghosal (2005)
Eucalyptus sawdust	49.37	6.40	42.01	2.02	–			Grover, Iyer, and Rao (2002)
Eucalyptus-Grandis	48.33	5.89	45.13	0.15	0.010			Jenkins and Ebeling (1985)
Fir mill	51.23	5.98	42.10	0.06	0.030	0.19	0.41	Jenkins, Baxter, Miles, et al. (1998)
Hazelnut shell	52.90	5.60	42.70	1.40	–			Demirbas (1997)
Hybrid poplar	50.18	6.06	40.43	0.60	0.020	0.01	2.70	Jenkins, Baxter, Miles, et al. (1998)
Loblolly pine	56.30	5.60	37.70	0	0			Risser (1981)

Macadamia shell	54.41	4.99	39.69	0.36	0.010			Jenkins and Ebeling (1985)
Mixed paper	47.99	6.63	36.84	0.14	0.070		8.33	Jenkins, Baxter, Miles, et al. (1998)
Mulberry stick	44.23	6.61	46.25	0.51	-			Grover, Iyer, and Rao (2002)
Nutshell	6.90	6.90		0.25	0.025	0.06		Ganan, Kassir Abdulla, Miranda, et al. (2005)
Oak char: 820–1185 °F	67.70	2.40	14.40	0.40	0.200			Boley and Landers (1969)
Olive pitts	52.80	6.69	38.25	0.45	0.050	0.04	1.72	Jenkins, Baxter, Miles, et al. (1998)
Paddy straw	35.97	5.28	43.08	0.17	-			Grover, Iyer, and Rao (2002)
Peach pit	49.14	6.34	43.52	0.48	0.020			Rossi (1984)
Pine	6.80	6.80		0.90	0.031	0.05		Ganan, Kassir Abdulla, Miranda, et al. (2005)
Pistachio shells	50.20	6.32	41.15	0.69	0.220	0.01	1.41	Jenkins, Baxter, Miles, et al. (1998)
Pistachio shell	48.79	5.91	43.41	0.56	0.010			Jenkins and Ebeling (1985)
RDF	39.70	5.78	27.24	0.80	0.350		26.13	Jenkins, Baxter, Miles, et al. (1998)
Redwood	50.64	5.98	42.88	0.05	0.030			Jenkins and Ebeling (1985)
Redwood char: 790–1020 °F	75.60	3.30	18.40	0.20	0.200			Boley and Landers (1969)
Rice hull	38.83	4.75	35.47	0.52	0.050	0.12	20.26	Jenkins, Baxter, Miles, et al. (1998)
Rice husk	37.80	4.70		0.40	0.020	0.03		Ganan, Kassir Abdulla, Miranda, et al. (2005)
Rice straw	38.24	5.20	36.26	0.87	0.180	0.58	18.67	Jenkins, Baxter, Miles, et al. (1998)
Sal seed husk	48.12	6.55	35.93	0	0			Grover and Anuradha (1988)
Seed corn	45.82	5.96	1.21	0.14	45.370			Zhang, Cummer, Suby, et al. (2005)
Sena leaves	36.20	4.72	37.49	4.29	-			Grover, Iyer, and Rao (2002)
Sewage sludge (community)	31.10	4.20	24.30	3.30	1.100	0.80		Thipkhunthod, Meeyoo, Rangsunvigit, et al. (2005)
	27.50	4.10	23.30	4.00	1.100	1.20		Thipkhunthod, Meeyoo, Rangsunvigit, et al. (2005)
	26.40	4.10	23.70	4.30	0.900	0.50		Thipkhunthod, Meeyoo, Rangsunvigit, et al. (2005)
	23.90	3.90	21.80	3.80	1.300	0.60		Thipkhunthod, Meeyoo, Rangsunvigit, et al. (2005)
	20.90	3.40	21.70	3.30	0.900	0.40		Thipkhunthod, Meeyoo, Rangsunvigit, et al. (2005)
	18.00	2.90	16.70	2.30	0.800			Thipkhunthod, Meeyoo, Rangsunvigit, et al. (2005)
	19.50	3.20	19.40		3.100	0.80		Thipkhunthod, Meeyoo, Rangsunvigit, et al. (2005)
	14.50	2.60	18.10		2.600	1.20		Thipkhunthod, Meeyoo, Rangsunvigit, et al. (2005)
	15.30	2.50	17.70		2.300	0.50		Thipkhunthod, Meeyoo, Rangsunvigit, et al. (2005)
	12.70	2.00	17.50		1.800	0.60		Thipkhunthod, Meeyoo, Rangsunvigit, et al. (2005)
	10.60	2.00	15.70		1.600	0.40		Thipkhunthod, Meeyoo, Rangsunvigit, et al. (2005)
	9.00	2.20	18.20		1.500	1.60		Thipkhunthod, Meeyoo, Rangsunvigit, et al. (2005)
Sewage sludge (hospital)	26.70	4.00	27.50		4.300	0.70		Thipkhunthod, Meeyoo, Rangsunvigit, et al. (2005)
	29.60	4.60	21.50		5.000	1.00		Thipkhunthod, Meeyoo, Rangsunvigit, et al. (2005)
	25.50	3.90	21.70		4.200	1.00		Thipkhunthod, Meeyoo, Rangsunvigit, et al. (2005)
	25.00	3.80	24.30	3.70		0.80		Thipkhunthod, Meeyoo, Rangsunvigit, et al. (2005)
	19.00	3.00	16.80	2.70		1.20		Thipkhunthod, Meeyoo, Rangsunvigit, et al. (2005)

Continued...

Appendix 5 Continued...

Biomass	C	H	O	N	S	Cl	Ash	References
Sewage sludge (industrial)	25.10	4.00	26.10	3.80	0.90			Thipkhunthod, Meeyoo, Rangsunvigit, et al. (2005).
		22.60	3.20	20.30	2.90	2.00		Thipkhunthod, Meeyoo, Rangsunvigit, et al. (2005)
		18.30	3.40	18.70	1.80	1.80		Thipkhunthod, Meeyoo, Rangsunvigit, et al. (2005)
Softwood	52.10	6.10	41.00	0.20	–			Demirbas (1997)
Spearmint	37.23	5.34	33.38	5.95	–			Jimenez and Gonzalez (1991)
Spruce wood	51.90	6.10	40.90	0.30	–			Demirbas (1997)
Subabul wood	48.15	5.87	44.75	0.03	0			Parikh, Channiwala, and Ghosal (2005)
Sudan Grass	44.58	5.35	39.18	1.21	0.01			Jenkins and Ebeling (1985)
Sugarcane bagasse	48.64	5.87	42.82	0.16	0.04	0.03	2.44	Jenkins, Baxter, Miles, et al. (1998)
Sugarcane leaves	39.75	5.55	46.82	0.17	–			Grover, Iyer, and Rao (2002)
Switch grass	46.68	5.82	37.38	0.77	0.19	0.19	8.97	Jenkins, Baxter, Miles, et al. (1998)
Tan oak	48.67	6.03	44.99	0.06	0.04			Rossi (1984)
Tea bush	47.67	6.13	43.16	1.33	–			Grover, Iyer, and Rao (2002)
Tea waste	48.60	5.50	39.50	0.50	–			Demirbas (1997)
Vine pruning	41.80	5.90		0.72	0.01	0.13		Cuellar Borrego (2003)
Walnut	48.20	6.25	43.24	1.61	0.16	0.23	7.02	Grover, Iyer, and Rao (2002)
Wheat straw	44.92	5.46	41.77	0.44	–			Jenkins, Baxter, Miles, et al. (1998)
Wheat straw	45.50	5.10	34.10	1.80	–			Demirbas (1997)
White fir	49.00	5.98	44.75	0.05	0.01			Jenkins and Ebeling (1985)
Willow wood	49.90	5.90	41.80	0.61	0.07	0.01	1.71	Jenkins, Baxter, Miles, et al. (1998)
Wood chips	48.10	5.99	45.74	0.08	0			Jenkins (1980)
Yard waste	41.54	4.79	31.91	0.85	0.24	0.30	20.37	Jenkins, Baxter, Miles, et al. (1998)
Salad (g/kg)	383.10			40.60				Dignac, Houot, Francou, et al. (2005)
Zucchinis (g/kg)	381.60			31.00				Dignac, Houot, Francou, et al. (2005)
Carrot (g/kg)	397.70			10.60				Dignac, Houot, Francou, et al. (2005)
Orange (g/kg)	414.20			11.80				Dignac, Houot, Francou, et al. (2005)
Paper (g/kg)	405.50			<0.10				Dignac, Houot, Francou, et al. (2005)
Green wastes (g/kg)	251.50			16.10				Dignac, Houot, Francou, et al. (2005)

C – carbon; H – hydrogen; O – oxygen; N – nitrogen; S – sulphur; Cl – chlorine

Appendix 6 C/N (carbon/nitrogen) ratio of some biomass feedstock

Solid fuel	C/N ratio
Switch grass	58
Duck dung	8
Human excreta	8
Chicken dung	10
Goat dung	12
Pig dung	18
Sheep dung	19
Cow dung/ buffalo dung	24
Water hyacinth	25
Elephant dung	43
Straw (maize)	60
Straw (rice)	70
Straw (wheat)	90
Biogas plant slurry	80

Appendix 7 Higher heating values (constant volume)

Biomass	MJ/kg	Reference
Alabama Oakwood waste	19.228	Boley and Landers (1969)
Alfalfa stems	18.670	Jenkins, Baxter, Miles, et al. (1998)
Almond	20.010	Jenkins and Ebeling (1985)
Almond forest	18.400	Gonzalez, Gonzalez-Garcia, Ramiro, et al. (2005)
Almond hull	18.890	Jenkins, Baxter, Miles, et al. (1998)
Almond peel	16.200	Gonzalez, Gonzalez-Garcia, Ramiro, et al. (2005)
Almond pruning	16.200	Gañán (2004)
Almond pruning	18.200	Gonzalez, Gonzalez-Garcia, Ramiro, et al. (2005)
Almond shell	18.200	Gonzalez, Gonzalez-Garcia, Ramiro, et al. (2005)
Almond shell	19.490	Jenkins, Baxter, Miles, et al. (1998)
Bagasse	14.258	Grover, Iyer, and Rao (2002)
Bagasse	18.167	Grover, Iyer, and Rao (2002)
Bagasse pith	17.192	Grover, Iyer, and Rao (2002)
Bamboo dust	15.890	Grover, Iyer, and Rao (2002)
Bamboo stick waste	17.657	Grover, Iyer, and Rao (2002)
Block wood	18.261	Grover, Iyer, and Rao (2002)
Cabernet Sauvignon	19.030	Jenkins and Ebeling (1985)
Canyon live oak	18.981	Rossi (1984)
Castor seed cake	21.010	Grover, Iyer, and Rao (2002)
Ceder cones	21.097	Grover, Iyer, and Rao (2002)
Cherrystone	22.500	González, Encinar, Canito, et al. (2003)
Coconut coir	20.050	Jenkins and Ebeling (1985)

Continued...

Appendix 7 (*Continued...*)

Biomass	MJ/kg	Reference
Coconut shell char-750 °C	31.124	Parikh, Channiwala, and Ghosal (2005)
Coconut shell powder	19.675	Jimenez and Gonzalez (1991)
Coconut stem	19.436	Grover, Iyer, and Rao (2002)
Coffee chaff	17.686	Grover, Iyer, and Rao (2002)
Corncob	18.795	Grover, Iyer, and Rao (2002)
Corncob	18.770	Jenkins and Ebeling (1985)
Corncob	17.000	Demirbas (2001)
Cornstover	17.800	Demirbas (2001)
Cottongin waste	17.483	Parikh, Channiwala, and Ghosal (2005)
Cotton shell	16.376	Jimenez and Gonzalez (1991)
Cotton shell briquette	19.055	Grover, Iyer, and Rao (2002)
Cotton stalk	15.830	Jenkins and Ebeling (1985)
Cottongin trash	16.420	Jenkins and Ebeling (1985)
Dal lake weed	8.887	Grover, Iyer, and Rao (2002)
Demol. wood	18.41	Jenkins, Baxter, Miles, *et al.* (1998)
Douglas fir	22.098	Tillman (1978)
Eucalyptus	18.01	Ganan, Kassir Abdulla, Miranda, *et al.* (2005)
Eucalyptus	18.640	Parikh, Channiwala, and Ghosal (2005)
Eucalyptus saw dust	18.502	Grover, Iyer, and Rao (2002)
Eucalyptus stalk	19.097	Grover, Iyer, and Rao (2002)
Eucalyptus-Grandis	19.350	Jenkins and Ebeling (1985)
Eucatlyptus bark	15.195	Grover, Iyer, and Rao (2002)
Fir mill	20.42	Jenkins, Baxter, Miles, *et al.* (1998)
Fly ash (Bagasse fuel)	8.385	Grover, Iyer, and Rao (2002)
Grewia optiva (Bhimal)	18.000	Grover, Iyer, and Rao (2002)
Hazelnut shell	19.300	Demirbas (1997)
Hybrid poplar	19.02	Jenkins, Baxter, Miles, *et al.* (1998)
Industrial waste (stalla)	18.928	Grover, Iyer, and Rao (2002)
Lantana briquettes	7.687	Grover, Iyer, and Rao (2002)
Loblolly pine	21.772	Risser (1981)
Macadamia shell	21.010	Jenkins and Ebeling (1985)
Mentha piperita	15.153	Grover, Iyer, and Rao (2002)
Miscalthus (elephanta grass)	19.297	Grover, Iyer, and Rao (2002)
Moringa-oleifera (leaves)	14.232	Grover, Iyer, and Rao (2002)
Mulberry stick	18.356	Grover, Iyer, and Rao (2002)
Municipal solid waste-bareilly	5.630	Grover, Iyer, and Rao (2002)
Municipal Solid Waste–Moradabad	7.183	Grover, Iyer, and Rao (2002)
Mustard stalk	17.489	Grover, Iyer, and Rao (2002)
Nutshell	18.1	Ganan, Kassir Abdulla, Miranda, *et al.* (2005)
Oak char-820-1185 °F	24.796	Boley and Landers (1969)
Olive marc	21.055	Jimenez and Gonzalez (1991)
Olive pitts	21.59	Jenkins, Baxter, Miles, *et al.* (1998)
Olive twigs	18.699	Jimenez and Gonzalez (1991)
Paddy straw	14.522	Grover, Iyer, and Rao (2002)
Peach pit	19.423	Rossi (1984)

Appendix 7 Continued...

Biomass	MJ/kg	Reference
PhC 300	22.840	Cordero, Marquez, Rodriguez-Mirasol, *et al.* (2001)
Pine	17.900	Ganan, Kassir Abdulla, Miranda, *et al.* (2005)
Pine wood	16.644	Grover, Iyer, and Rao (2002)
Pistachio shells	18.220	Jenkins, Baxter, Miles, *et al.* (1998)
Pistachio shell	19.260	Jenkins and Ebeling (1985)
Plywood	19.720	Grover, Iyer, and Rao (2002)
Poultry pure waste	11.712	Grover, Iyer, and Rao (2002)
Press mud briquettes	11.972	Grover, Iyer, and Rao (2002)
Pyrolysed residual of almond shell	21.100	Gañán (2004)
Pyrolysed residual of cherrystone	32.000	Kalac, Svoboda, and Havliekova (2004)
Pyrolysed residual of eucalyptus	30.200	Ganan, Kassir Abdulla, Miranda, *et al.* (2005)
Pyrolysed residual of nutshell	25.400	Ganan, Kassir Abdulla, Miranda, *et al.* (2005)
Pyrolysed residual of pine	29.700	Ganan, Kassir Abdulla, Miranda, *et al.* (2005)
Pyrolysed residual of rice husk	21.200	Ganan, Kassir Abdulla, Miranda, *et al.* (2005)
Pyrolysed residual of vine pruning	26.900	Cuellar Borrego (2003)
RDF	15.540	Jenkins, Baxter, Miles, *et al.* (1998)
Redwood	20.720	Jenkins and Ebeling (1985)
Redwood char-790-1020 °F	28.844	Boley and Landers (1969)
Rice hull	15.840	Jenkins, Baxter, Miles, *et al.* (1998)
Rice husk	13.100	Ganan, Kassir Abdulla, Miranda, *et al.* (2005)
Rice husk char	14.944	Grover, Iyer, and Rao (2002)
Rice straw	15.090	Jenkins, Baxter, Miles, *et al.* (1998)
Rice straw (ground)	15.614	Grover, Iyer, and Rao (2002)
Rose apple char	7.577	Grover, Iyer, and Rao (2002)
Sal seed husk	20.600	Grover and Anuradha (1988)
Sawdust	20.930	Grover, Iyer, and Rao (2002)
Sawdust + mustard	13.727	Grover, Iyer, and Rao (2002)
Sena leaves	18.125	Grover, Iyer, and Rao (2002)
Sewage sludge (community)	13.900	Thipkhunthod, Meeyoo, Rangsunvigit, *et al.* (2005)
	13.200	Thipkhunthod, Meeyoo, Rangsunvigit, *et al.* (2005)
	12.600	Thipkhunthod, Meeyoo, Rangsunvigit, *et al.* (2005)
	11.000	Thipkhunthod, Meeyoo, Rangsunvigit, *et al.* (2005)
	10.100	Thipkhunthod, Meeyoo, Rangsunvigit, *et al.* (2005)
	9.400	Thipkhunthod, Meeyoo, Rangsunvigit, *et al.* (2005)
	8.700	Thipkhunthod, Meeyoo, Rangsunvigit, *et al.* (2005)
	6.900	Thipkhunthod, Meeyoo, Rangsunvigit, *et al.* (2005)
	6.500	Thipkhunthod, Meeyoo, Rangsunvigit, *et al.* (2005)
	5.700	Thipkhunthod, Meeyoo, Rangsunvigit, *et al.* (2005)
	4.300	Thipkhunthod, Meeyoo, Rangsunvigit, *et al.* (2005)
	3.500	Thipkhunthod, Meeyoo, Rangsunvigit, *et al.* (2005)
Sewage sludge (hospital)	13.300	Thipkhunthod, Meeyoo, Rangsunvigit, *et al.* (2005)
	12.800	Thipkhunthod, Meeyoo, Rangsunvigit, *et al.* (2005)
	12.400	Thipkhunthod, Meeyoo, Rangsunvigit, *et al.* (2005)
	11.100	Thipkhunthod, Meeyoo, Rangsunvigit, *et al.* (2005)
	8.200	Thipkhunthod, Meeyoo, Rangsunvigit, *et al.* (2005)

Continued...

Appendix 7 Continued...

Biomass	MJ/kg	Reference
Sewage sludge (industrial)	10.900	Thipkhunthod, Meeyoo, Rangsunvigit, *et al.* (2005)
	9.900	Thipkhunthod, Meeyoo, Rangsunvigit, *et al.* (2005)
	9.000	Thipkhunthod, Meeyoo, Rangsunvigit, *et al.* (2005)
Softwood	20.000	Demirbas (1997)
Spearmint	15.530	Jimenez and Gonzalez (1991)
Spruce wood	20.100	Demirbas (1997)
Subabul	16.660	Grover, Iyer, and Rao (2002)
Subabul wood	19.777	Parikh, Channiwala, and Ghosal (2005)
Sudan grass	17.390	Jenkins and Ebeling (1985)
Sugarcane bagasse	18.99	Jenkins, Baxter, Miles, *et al.* (1998)
Sugarcane leaves	17.410	Grover, Iyer, and Rao (2002)
Sweet sorghum bagasse	13.730	Grover, Iyer, and Rao (2002)
Switch grass	18.060	Jenkins, Baxter, Miles, *et al.* (1998)
Tan Oak	18.934	Rossi (1984)
Tannery waste	7.685	Grover, Iyer, and Rao (2002)
Tapero root skin scale scrapping	9.228	Grover, Iyer, and Rao (2002)
Tea bush	19.842	Grover, Iyer, and Rao (2002)
Tea waste	17.100	Demirbas (1997)
Vine pruning	16.620	Cuellar Borrego (2003)
Walnut	19.967	Grover, Iyer, and Rao (2002)
Water hyacinth	14.806	Grover, Iyer, and Rao (2002)
Wheat straw	17.940	Jenkins, Baxter, Miles. *et al.* (1998)
Wheat straw	17.000	Demirbas (1997)
Wheat straw	18.905	Grover, Iyer, and Rao (2002)
Wheat straw	17.355	Grover, Iyer, and Rao (2002)
White fir	19.950	Jenkins and Ebeling (1985)
Willow wood	19.590	Jenkins, Baxter, Miles, *et al.* (1998)
Wood chips	19.916	Jenkins (1980)
Wood chips	20.031	Grover, Iyer, and Rao (2002)
Yard waste	16.300	Jenkins, Baxter, Miles, *et al.* (1998)

Appendix 8 Estimated biomass residue production in India

Crop	Economic produce 1996–97			2010			Residue to final economic produce ratio	Type of residue	Moisture %	
	Gross cropped area (Mha)	Total economic production (Mt)	Total residue production Mt (air dry)	Gross cropped area (Mha)	Total economic production (Mt)	Total residue production Mt (air dry)			At harvest	At use
Rice	43.3	81.3	145.5	46.1	118.8	213.9	1.8	Straw+husk	30	10
Wheat	25.9	69.3	110.6	28.5	98.5	157.6	1.6	Straw	30	10
Jowar	11.6	11.0	22.3	5.3	6.1	12.2	2.0	Stalk	30	10
Bajra	10.0	7.9	15.8	8.6	6.8	13.6	2.0	Stalk+cobs	30	10
Maize	6.2	10.6	26.3	6.6	13.0	32.5	2.5	Straw	30	10
Other cereals	4.3	4.7	9.4	1.3	1.4	2.8	2.0	Stalk	30	10
Red gram	3.6	2.7	13.5	3.6	2.7	11.2	5.0	Waste	30	10
Gram	7.1	5.7	9.3	7.7	7.0	13.5	1.6	Waste	30	10
Other pulses	12.5	5.8	17.1	12.5	5.9	17.1	2.9	Shell+waste	30	10
Ground nut	7.9	9.0	20.7	9.3	12.2	28.1	2.3	Waste	30	10
Rape seed and mustard	6.9	6.9	13.8	10.7	12.0	24.1	2.0	Waste	30	10
Other oil seeds	12.1	9.1	18.2	18.0	13.5	27.1	2.0	Waste	30	10
Cotton	9.1	14.3	50.0	10.1	15.9	55.7	3.5	Seeds+waste	30	10
Jute	0.9	9.8	15.7	0.6	6.5	10.5	1.6	Waste	30	10
Sugar cane	4.2	277.2	110.8	5.5	463.5	185.4	0.4	Bagasse+leaves	30	10
Coconut + arecanut	2.0	—	20.0	2.8	—	28.2	10t ha^{-1}yr^{-1}	Fronds+shells	30	10
Mulbery	0.3	—	3.0	0.3	—	3.3		Sticks	30	10
Coffee + tea	0.7	0.8	3.42	0.8	1.0	3.9	4.0	Twigs+branch	30	10
Total	168.6		626.5	178.2		840.6				

Source Ravindranath et al 2005

References

Annamalai K, Sweeten J M, and Ramalingam S C. 1987
Estimation of gross heating values of biomass fuels
Transactions of ASAE **30**: 1205–1208

Boley C C and Landers W S. 1969
Entrainment Drying and Carbonization of Wood Waste
[Report of investigation 7282]
Washington, DC: Bureau of Mines. 15 pp.

Botanical Survey of India. 2004
BGRIS/BSI database

Britz T J, Noeth C, and Lategan P M. 1983
Nitrogen and phosphate requirements for the anaerobic digestion of a petrochemical effluent
Water Research **22**: 163–169

Bryant M P. 1979
Microbial methane production: theoretical aspects
Journal of Animal Sciences **48**: 193–201

Chamot E and Porankiewicz B. 1999
An imomo approach to calculate the thermal stability of polymers: application of quantum mechanics to a 'wear' problem
[Polymer Characterization Symposium, 27–29 October 1999]
Quincy: American Chemical Society

Chang I and Lee C. 1998
Membrane filtration characteristics in membrane-coupled activated sludge system: the effect of physiological states of activated sludge on membrane fouling
Desalination **120**: 221–233

Channiwala S A and Parikh P P. 2002
A unified correlation for estimating HHV of solid, liquid and gaseous fuels
Fuel **81**: 1051–1063

Chará J, Pedraza G, and Conde N. 1999
The productive water decontamination system: a tool for protecting water resources in the tropics
Livestock Research for Rural Development **11**
Details at <http://www.cipav.org.co/lrrd/lrrd11/1/cont111.htm>
last accessed on 12 July 2006

Chaturvedi A N. 1984
Assessment of biomass production
Indian Forester **110**: 726–728

Cordero T, Marquez F, Rodriguez-Mirasol J, Rodriguez J J. 2001
Predicting heating values of lignocellulosics and carbonaceous materials from proximate analysis
Fuel **80**: 1567–1571

Cornelius J A. 1983
Processing of Oil Palm Fruit and its Products
[Report G149]
London: Tropical Products Institute. 95 pp.

Cuellar Borrego S. 2003
Aplicaciones de las fracciones obtenidas en la pirólisis del sarmiento de la vid Proyecto Fin de carrera. EE. II. Universidad de Extremadura

Curvers A and Gigler J K. 1996
Characterization of biomass fuels. An inventory of standard procedures for the determination of biomass properties
ECN–C–96–032, pp. 66.

Czernik S, Koeberle P G, Jollez P, Bilodeau J F, Chornet E. 1993
Gasification of residual biomass via the biosyn fluidized bed technology
In *Advances in Thermochemical Biomass Conversion*, edited by A V Bridgwater, pp. 423–437
Glasgow: Blackie Academic and Professional

Demirbas A. 1997
Calculation of higher heating values of biomass fuels
Fuel **76**: 431–434

Demirbas A. 2001
Relationships between lignin contents and heating values of biomass
Energy Conversion and Management **42**: 183–188

Deschatelets L and Yu K C. 1986
A simple pentose assay for biomass conversion studies
Applied Microbiology and Biotechnology **24**: 379–385

Dhingra S, Mande S, Kishore V V N, Joshi V. 1996
Briquetting of biomass – status and potential
Proceedings of the International Workshop on Biomass Briquetting, New Delhi, Regional Wood Energy Development Programme in Asia, Report No 23
Bangkok: Food and Agriculture Organization, pp 24–30

Dignac M F, Houot S, Francou C, Derenne S. 2005
Pyrolytic study of compost and waste organic matter
Organic Geochemistry **36**: 1054–1071

Djevic M D Sc. 2005
Use of crop residues for fuel
Details available at <http://www.rcub.bg.ac.yu/~todorom/tutorials/rad37.html>, last accessed on 6 June 2006

Ergudenler A and Ghaly A E. 1992
Quality of gas produced from wheat straw in a dual-distributor type fluidized bed gasifier
Biomass and Bioenergy **3**(2): 419–430

FAO (Food and Agriculture Organization). 1982
Agricultural residues: bibliography 1975–81 and quantitative survey 1982
Agricultural Services Bulletin, 171 pp.

Gañán J. 2004
Aprovechamiento energético mediante combustión, pirólisis y gasificación de los residuos del almendro. Tesis doctoral. EE. II. Universidad de Extremadura

Ganan J, Kassir Abdulla A, Miranda A B, Turegano J, Correia S, Cuerda E M. 2005
Energy production by means of gasification process of residuals sourced in Extremadura (Spain)
Renewable Energy **30**: 1759–1769

Geological and Mining Engineering and Sciences. 2005

González J F, Encinar J M, Canito J L, Sabio E, Chacón M. 2003
Pyrolysis of cherry stones: energy uses of the different fractions and kinetic study
Journal of Analytical Applied Pyrolysis **67**: 165–190

Gonzalez J F, Gonzalez-Garcia C M, Ramiro A, Ganan J, Gonzalez J, Sabio E, Roman S, Turegano J. 2005
Use of almond residues for domestic heating. Study of the combustion parameters in a mural boiler
Fuel Processing Technology **86**: 1351–1368

Graboski M and Bain R. 1981
Properties of biomass relevant to gasification
In *A Survey of Biomass Gasification*, Volume II, *Biomass Gasification Principles and Technology*, edited by T B Reed
New Jersey: Noyes Data Corporation. 239 pp.

Grover P D and Anuradha G. 1988
Thermochemical Characterization of Biomass for Gasification
[Report on physico-chemical parameters of biomass residues]
Delhi: Indian Institute of Technology

Grover P D, Iyer P V R, and Rao T R. 2002
Biomass–thermochemical characterization, 3rd edn
Delhi: Indian Institute of Technology and Ministry of Non-conventional Energy Sources, 159 pp

Hall D O and Overend R P. 1987
Biomass: regenerable energy
London: John Wiley and Sons. 504 pp.

Holdgate. 1996
Mean net primary production by ecosystems
Details available at <http://rainforests.mongabay.com/03net_primary_production.htm>
last accessed on 6 June 2006

IGES (Institute of Global Environmental Strategies). 2005
Details available at <http://www. iges. or. jp/APEIS/RISPO/p_report_2nd/
15_3_4_2_promotion_of_biomass_energy. pdf>, last accessed on 6 June 2006

Institute of Gas Technology. 1978
Coal Conversion Systems Technical Data Book
Washington, DC: Department of Energy

International Transfer Centre Environmental Technology. 2005
Details available at <http://www. bionet. net>, last accessed on 6 June 2006

IPCC (Intergovernmental Panel on Climate Change). 1996
IPCC guidelines for national greenhouse gas inventories: reference manual
Geneva: IPCC, 6–30 pp.

Iyer P V R, Rao T R, and Grover P D. 2002
Biomass: thermo-chemical characterization, revised 3rd edn
[Published under Ministry of Non-conventional Energy Sources sponsored Gasifier
Action Research Project]
Delhi: Indian Institute of Technology

Jackson M L. 1967
Soil Chemical Analysis, 498 pp.
New Delhi: Prentice-Hall of India Pvt. Ltd

James A D. 1983
Handbook of Energy Crops
Details available at <http://www.hort.purdue.edu/hort/default.shtml>
last accessed on 24 October 2006

Jenkins B M. 1980
Downdraft gasification characteristics of major California residue derived fuels
Davis: University of California
[Doctoral thesis submitted to University of California, Davis]

Jenkins B M and Ebeling J M. 1985
**Correlations of physical and chemical properties of terrestrial biomass with
conversion**
[Symposium on energy from biomass and waste IX Institute of Gas Technology, Chicago,
IL] 271 pp.

Jenkins B M, Baxter L L, Miles J, Miles T R. 1998
Combustion properties of biomass
Fuel Processing Technology 54: 17–46

Jimenez L and Gonzalez F. 1991
Study of the physical and chemical properties of lignocellulosic residues with a view to the production of fuels
Fuel 70: 947–950

Jorand F, Guicherd P, Urbain V, Manem J, Block J. 1994
Hydrophobicity of activated sludge flocs and laboratory-grown bacteria
Water Science and Technology 30: 211–218

Kalac P, Svoboda L, and Havliekova B. 2004
Contents of detrimental metals mercury, cadmium and lead in wild growing edible mushrooms: a review
Energy Education Science and Technology 13, 33–38

Lal R. 2005
World crop residues production and implications as its use as a biofuel
Environment International 31: 575–584

Lee Shang-Hsiu. 2005
Waste-to-energy facilities in Taiwan
WTERT/Earth Engineering Center
Details available at <http://www.seas.columbia.edu/earth/wtert/sofos/Lee_TW.pdf>,
last accessed on 6 June 2006

Mande S. 2005
Biomass gasifier based power plants: potential, problems and research needs for decentralized rural electrification
In *Wealth from Waste*, pp. 1–28
New Delhi: The Energy and Resources Institute

Mande S and Lata K. 2005
Potential of converting phumdi waste of Loktak lake into briquettes for fodder and fuel use
In *Aquatic Weeds: problems, control and management*, pp. 100–108, edited by S M Mathur, A N Mathur, R K Trivedi, Y C Bhat, P Mohonot
New Delhi: Himanshu Publications

MNES (Ministry of Non-conventional Energy Sources). 1996
Background Information for Conference of Mayors and Municipal Commissioners, Urban and Industrial Energy Group
New Delhi: MNES. 30 pp.

MoA (Ministry of Agriculture). 1999
Agricultural Statistics at a Glance, March 1997
New Delhi: Department of Agriculture and Cooperation, Directorate of Economics and Statistics, MoA

Mossé J. 1990
Nitrogen to protein conversion factor for 10 cereals and 6 legumes or oilseeds: a reappraisal of its definition and determination – variation according to species and to seed protein content
Journal of Agriculture and Food Chemistry **38**: 18–24

Mosse J and Huet J C. 1985
The amino-acid composition of wheat-grain as a function of nitrogen-content
Journal of Cereal Science **3**: 115–130

Orr L and Govindjee. 2005
Photosynthesis and the web: 2005
Details available at <http://photoscience. la. asu. edu/photosyn/photoweb/default. html>, last accessed on 25 November 2005

Parikh J, Channiwala S A, and Ghosal G K. 2005
A correlation for calculating HHV from proximate analysis of solid fuels
Fuel **84**: 487–494

Prasad S, Singh A, and Joshi H C. 2007
Ethanol as an alternative fuel from agricultural, industrial and urban residues
Resources, Conservation and Recycling **50**(2007): 1–39

PRESSEA. 2005
Biomass: current and planned utilisation

Raveendran K, and Ganesh A. 1993
Estimation of kinetic parameters of biomass pyrolysis from non-isothermal TGA data—a combined numerical and optimization method
In *Fourth National Meet on Biomass Gasification and Combustion*, P J Paul and H S Mukunda (eds)
Bangalore: Interline Publishing

Ravindranath N H *et al.* 2005
Assessment of sustainable non-plantation biomass resources potential for energy in India
Biomass and Bioenergy **29**: pp. 178–190

Risser P G. 1981
Agricultural and forestry residues
In *Biomass Conversion Processes for Energy and Fuels*, pp. 25–56, edited by S S Soffer and O R Zaborsky
New York: Plenum Press

Rossi A. 1984
Fuel characteristics of wood and non-wood biomass fuels
In *Progress in Biomass Conversion*, vol. 5, edited by D A Tillman and E C Jahn
New York: Academic Press. 69 pp.

Shafizadeh F and Degroot W G. 1976
Combustion characteristics of celluosic fuels
In *Thermal Uses and Properties of Carbohydrates and Lignins*
New York: Academic Press

Sheng C and Azevedo J L T. 2005
Estimating the higher heating value of biomass fuels from basic analysis data
Biomass and Bioenergy 28: 499–507

Singh N P. 1996
Biomass programme in India: technology development, demonstration and commercialisation of biomass production and improvement practices
BUN Newsletter 1(1)

Speece R E. 1987
Nutrient requirements for anaerobic digestion
Symposium on Biotechnical Advances in Processing Municipal Waste for Fuels and Chemicals, Minneapolis, pp. 195–221, Noyes Data Co.

Speece R E and McCarty P L. 1964
Nutrient requirements and biological solids accumulation in anaerobic digestion
Advances in Water Pollution Research 2: 305–322

Stafford D A, Hawkes D L, and Horton R. 1980
Methane Production from Waste Organic Matter, 285 pp.
West Palm Beach, Boca Raton, FL: CRC Press

Tao W. 2005
40% of global energy consumption to come from organisms in 2015

TERI (Tata Energy Research Institute). 1987
Potential for Renewable Energy Utilisation in Andaman and Nicobar and Lakshadweep Islands
[Submitted to the Department of Non-Conventional Energy Sources, Government of India]
New Delhi: TERI. 183 pp.

TERI (The Energy and Resources Institute). 2003
TEDDY (TERI Energy Data Directory and Yearbook). 2002/03
New Delhi: TERI, 503 pp.

TERI (The Energy and Resources Institute). 2003
Feasibility of Developing Processing Technique to Convert Phumdi Waste of Loktak Lake into Wealth
New Delhi: TERI, 80 pp.

Thipkhunthod P, Meeyoo V, Rangsunvigit P, Kitiyanan B, Siemanond K, Rirksomboon T. 2005
Predicting the heating value of sewage sludges in Thailand from proximate and ultimate analyses
Fuel 84, 849–857

Tillman D A. 1978
Wood as an Energy Resource, 252 pp.
New York: Academic Press

Updegraff D M. 1969
Semi-micro determination of cellulose in biological materials
Analytical Chemistry 32: 420–425

Van den Aarsen, Beenackers A A C M, and Van Swaaij W O M Y. 1982
Performance of rice husk fuelled fluidized bed pilot plant gasifier
In *Producer Gas* 1982 edited by B Kjellstrom, H Stassen, and A A C M Beenackers,
pp. 381–391
Stockholm: Beijer Institute

Vimal O P, Tyagi, and Grover P D. 1984
Energy from Biomass
New Delhi: Agricol Publishing Company. 440 pp.

Volker S, and Rieckmann, Th. 2002
Thermokinetic investigation of cellulose pyrolysis – impact of initial and final mass on kinetic results
Journal of Analytical and Applied Pyrolysis, 62, pp. 165–177

Watson S A and Ramstad P E. 1987
Corn: Chemistry and Technology
St Paul: American Association of Cereal Chemists. 605 pp.

World Energy Council. 2005
Survey of energy resources

Zhang R, Cummer K, Suby A, Brown R C. 2005
Biomass-derived hydrogen from an air-blown gasifier
Fuel Processing Technology 86: 861–874

13

Thermochemical conversion of biomass

Sanjay P Mande, Fellow
Energy–Environment Technology Division, TERI, New Delhi

Biomass conversion routes

Biomass can be used for different purposes such as cooking, process heating, electricity generation, steam generation, and mechanical or shaft power applications. It also produces a variety of chemicals as by-products. Various biomass conversion processes used for achieving these objectives can broadly be classified as follows.

- Physical
  - Fuel processing: chopping, shredding, pulverizing, and densification into briquettes/pellets
- Thermochemical
  - Combustion
  - Pyrolysis
  - Gasification
  - Liquification
  - Ammonia production
- Chemical
  - Acid hydrolysis
- Biochemical
  - Anaerobic digestion to methane
  - Ethanol fermentation

The first two biomass conversion routes, namely physical and thermochemical (combustion, pyrolysis for charcoal production, and gasification), are covered in this chapter, while biochemical conversion routes will be covered in the next chapter.

Biomass densification technologies

Many non-woody biomass residues suffer from major disadvantage of having low bulk densities for their efficient utilization. For example, the bulk density of the majority of agro-residues lies in the low range of 50–200 kg/m³ (Table 13.1) as compared to 800 kg/m³ for coal of the same size. This results in huge storage space requirements, difficulty in handling, and higher transportation costs, which makes them uneconomical as a marketable commodity (Dhingra, Mande, and Kishore 1996). Also, low bulk densities and the loose nature of available biomass are associated with faster burning of fuels resulting in higher flue gas losses (lower operating thermal efficiencies) and emissions in the form of fly ash or particulates in the atmosphere. This makes them poor quality biomass fuels. In order to improve the marketability of the available loose biomass as fuel, pre-processing becomes necessary (TERI 2004).

Densification of biomass can be done by the briquetting or pelletizing technology that compresses loose biomass into densified forms. This reduces the transportation and storage costs, and improves the effectiveness of biomass for use as a combustible fuel (Mande and Lata 2005).

Densified briquettes/pellets produced from biomass are fairly good substitutes for coal, lignite, and firewood and have several advantages.

- They are renewable and sustainable sources of energy.
- They are cheaper than coal.
- They are of consistent quality and size.
- They have better thermal efficiency than loose biomass.
- High density (800–1200 kg/m³) compared to loose biomass (50–200 kg/m³).
- They provide value addition for rural biomass.

Table 13.1 Bulk density of selected agro-residues

Biomass material		*Bulk density* (kg/m³)
Saw dust	Loose	177
Saw dust	Briquetted	350–400
Straw	Loose	80
Straw	Bales	320
Coir pith	—	47
Jute dust	—	74
Groundnut shell	Pulverized	165
Bagasse pith	—	74
Sugar cane leaves	Pulverized	167

Source Iyer, Rao, and Grover (2002)

Appropriate biomass residue for briquetting

Almost all types of biomass can be densified. However, many factors need to be considered before biomass qualifies as an appropriate feedstock for briquetting. The main characteristics of an appropriate biomass residue for briquetting are discussed below.

Moisture content

Higher moisture content poses problems in grinding and requires higher energy for the drying operation. Therefore, biomass with lower moisture content, preferably below 10%–15%, is desirable.

Ash content and composition

The majority of biomass residues (except rice husk with 20% ash) have a low ash content but they contain higher percentages of alkaline (especially potash) minerals which contribute towards lowering the sintering temperature leading to ash deposition. Higher ash content thus increases the slagging tendency, which becomes more acute with biomass containing more than four per cent ash.

Flow characteristics

Fine granular material with uniform size flows easily in the fuel hoppers and storage bins/silos. Thus, some of the appropriate biomass materials for briquetting include sawdust, coffee husk, groundnut shell, pulverized mustard stalk, and cotton sticks.

Binding mechanism of densification

Briquetting is one of the several agglomeration techniques used for the densification of biomass residues. On the basis of compaction, briquetting technologies can be classified as follows.
- High-pressure compaction
- Medium-pressure compaction
- Low-pressure compaction

Normally, binders are not required in high- and medium-pressure compaction, but sometime pre-heating of biomass is used to enhance the compaction process. In all compaction processes, individual particles are pressed together in a confined volume. In case of biomass, the binding mechanism under pressure can be divided into cohesive and adhesive forces, Van der Waal's forces of attraction between solid particles, and mechanical interlocking bonds under pressure, which create strength in bonding during compaction. Binders, highly viscous bonding media such as tar and other molecular organic liquid or cow dung, are used in low-pressure compaction to enhance adhesion among biomass particles by creating solid–liquid bridges. The lignin present in the biomass helps in creating such bonds due to its softening at higher temperatures and its adsorption on solid particle layers. The strength of the resulting agglomeration depends on the type of interaction and the material characteristics. Some important parameters for agglomeration are discussed below.

Particle size

Granular biomass material of 6–8 mm size with about 10%–20% powdery (<4 mesh) material normally gives the best results. Though high-pressure (1000–1500 bar) compaction machines such as piston press or screw extruder can briquette larger particle-sized biomass, it can lead to choking of the entrance to ram or die portion of the briquetting machine. The condensation of vapour released from the larger particles onto finer particles can create lumps, which affect free flow. However, presence of only the finer material is not always good due to its low density and flowability. Presence of different-sized particles improves the packing dynamics contributing to higher strength.

Moisture

The level of moisture content is a very critical factor as presence of the right amount of moisture (7%–10%) leads to the development of self-bonding properties in lignocellulosic substances at elevated temperatures and pressures prevailing in piston press and screw extruder briquetting machines. A higher moisture content can result in poor and weak briquettes having cracks due to the escape of steam. It also results in erratic operation as the feed flow chokes due to steam formation. The briquettes produced should have a moisture content higher than the equilibrium value, otherwise they would regain moisture from the atmosphere, resulting in swelling during storage, transportation, and disintegration when exposed to humid conditions.

Biomass feed temperature

At higher temperatures, moisture of biomass gets converted into steam under higher prevailing pressures, which hydrolyses the hemicellulose and lignin portions of the biomass into lower molecular carbohydrates, lignin products, sugar polymers, and other derivatives. These act like in situ binding material. Better compaction gives higher briquette density and strength with higher biomass feed temperature and pressure (Figure 13.1). This also softens the fibres and their resistance to briquetting, which results in lower power requirements and reduction in the wear of the contact parts. However, the temperature should be kept lower than 300 °C, beyond which biomass starts pyrolysing.

Briquetting can be done with or without binder material. No binders are generally required in high-pressure briquetting. Prior to the briquetting process, the biomass has to be broken up into small pieces and then dried to a

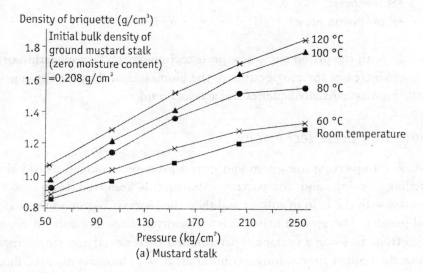

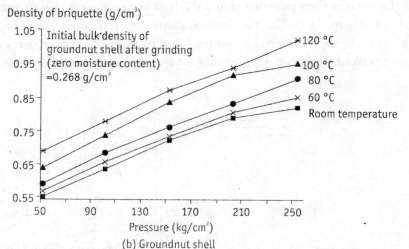

Figure 13.1 Variation in briquette density with pressure at different biomass feed temperatures. (a) Mustard stalk (b) groundnut shell
Reproduced with permission from FAO
Source FAO (1996a, b)

moisture content of about 12%–15%. The briquetting plants in India use saw-dust, bamboo dust, groundnut shell, mustard stalk, cotton stalk, coffee husk, baggasse, sugar mill waste mud (commonly called press mud), jute waste, coir pith, etc. as raw material. All these biomass briquettes, except for press mud, have good calorific values of the order of 3800–4000 kcal/kg.

Commercially, briquetting of biomass without binders is done by briquetting machines based on the following technololgies.

- Screw press
- Ram-piston press

In both the piston and screw-press technologies, the application of high pressure increases the temperature of the biomass, and thus the lignin present in the biomass partially liquefies and acts as a binder.

Ram and piston press technology

Biomass briquetting using ram and piston press technology involves drying, grinding, sieving, and compacting. Moisture is removed from the loose biomass with the help of a dryer and then the biomass is ground in a hammer mill grinder. The ground material is transported using pressurized air, sepa-rated from air using a cyclone separator, and then sieved and stored in a bin above the hopper for ensuring a continuous flow of biomass material into the press. Biomass is then punched into a die by a reciprocating ram to produce briquettes (Figure 13.2). The ram moves about 250–300 times per minute in the process. In the briquetting machine, due to wear of contact parts (here ram and piston), frequent maintenance and/or replacement is required. The

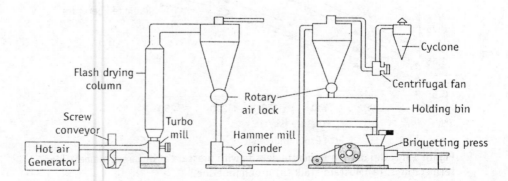

Figure 13.2 Ram and piston type briquetting press

average frequencies of replacement for some of the machine components are: 300 hours for ram, scrapper and wear rings, and 500 hours for taper, split die, and hammers.

In India, piston-press briquetting machines are commercially available. They are available in different capacity ranges, from 250 to 2250 kg/h. Table 13.2 gives the share of various input costs in the production of biomass briquettes. It can be seen that at higher production capacities, the raw material contributes more as input cost and the share of other costs diminishes due to the scale of production. Table 13.3 gives the power requirement for different capacity machines for briquetting various types of biomass materials. Power requirement increases with capacity of the machine. For given production capacity, it increases from fine to coarse to stalky biomass due to an increasing pre-processing requirement of biomass. For a given type of biomass, wet biomass requires higher power for an additional drying operation.

Table 13.2 Contribution of different heads (in %) for unit cost of briquettes

Input cost component	Briquetting machine production capacity (kg/h)					
	250	500	750	1000	1500	2250
Capital	19.2	14.6	12.1	9.7	8.8	7.7
Raw material	41.3	54.5	59.6	57.1	64.3	67.2
Operation	10.1	6.6	4.8	4.4	3.3	3.2
Electricity	25.5	20.3	19.9	24.7	19.9	18.0
Repair and maintenance	3.9	4.0	3.6	4.1	3.7	3.9

Source Tripathi, Iyer, and Kandpal (1998)

Table 13.3 Power requirement for briquetting units

Briquetting production capacity (kg/h)	Power requirement (kW)					
	Fine granular		Coarse granular		Stalky material	
	Dry	Wet	Dry	Wet	Dry	Wet
250	17.5	26.5	36.0	45.0	43.5	2.5
500	25.0	34.0	43.5	52.5	51.0	60.0
750	32.5	41.5	58.5	67.5	73.5	82.5
1000	50.5	65.5	101.0	116.0	116.0	131.0
1500	65.5	80.5	108.5	123.5	123.5	138.5
2250	98.0	113.0	141.0	156.0	156.0	171.0

Source Tripathi, Iyer, and Kandpal (1998)

Screw-press technology

In this process, biomass is dried to get an optimum moisture content value by passing hot air produced by burning part of the briquettes in the furnace. Using the heated oil obtained through a heat exchanger, biomass is further pre-heated to about 100–120 °C so as to minimize the wear of the dies and improve the true density of the briquettes formed. This pre-heated material is then fed to the screw extruder where a revolving screw (at about 600–700 RPM) continuously compacts the material through a tapered die, which is heated externally to reduce friction between the biomass and the die surface (Figure 13.3). The briquette obtained through this technology has a hole in the centre and its outer surface is partially carbonized.

The briquettes extruded using screw press are more homogeneous (as output is continuous and not in strokes) and have better crushing strength and combustion properties (due to larger combustion area per unit weight). Since the outer surface of the briquettes is carbonized, it facilitates easy and clean ignition, and also this impervious layer provides protection from moisture, thereby increasing its storage life. However, power consumption and wear of screw are higher than those in the ram and piston type of reciprocating machines (Table 13.4).

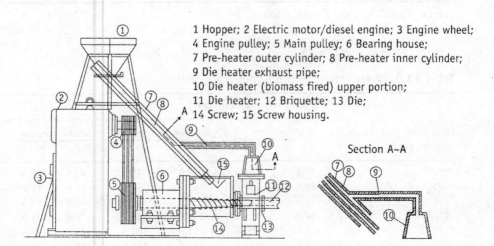

1 Hopper; 2 Electric motor/diesel engine; 3 Engine wheel;
4 Engine pulley; 5 Main pulley; 6 Bearing house;
7 Pre-heater outer cylinder; 8 Pre-heater inner cylinder;
9 Die heater exhaust pipe;
10 Die heater (biomass fired) upper portion;
11 Die heater; 12 Briquette; 13 Die;
14 Screw; 15 Screw housing.

Section A–A

Figure 13.3 Schematic of screw type briquetting machine with biomass fired die heater

Table 13.4 Comparison between piston press and screw extruder

Parameter	Piston press	Screw extruder
Optimum moisture content of raw material	10%–15%	8%–9%
Wear of contact parts	Lower	Higher
Output from machine	In strokes	Continuous
Power consumption	50 kWh/tonne	60 kWh/tonne
Density of briquette	1–1.2 g/cm^3	1–1.4 g/cm^3
Maintenance	Higher	Lower
Combustion performance of briquettes	Good	Better
Homogeneity of briquettes	Less homogenous	More homogenous

Source FAO (1996a)

Pelletizing machine

Pelletization produces somewhat lighter, and smaller pellets of biomass compared to briquetting. The pelletization machines are based on fodder making technology. Pelletizing generally requires conditioning of biomass material either by mixing with a binder or by raising its temperature through direct addition of steam or both. The material is dropped in the pressing chamber of the pellet mill where it forms a carpet on the die surface. The rollers roll over this layer and press it through tapered die holes. The pressing force keeps on increasing during rolling in the direction of the die holes. With each roll, a small disc is formed in the die hole that gets attached to the pressed piece already in the hole. The plugs are pushed forward uniformly and the hot pellets are ejected out at about 50–90 °C, which are cooled on conveyor belts before their storage (Figure 13.4). Compared to briquetting machines, pellet mills are simpler, and since they operate at lower pressures, the power consumption is lower. The processing requirements are also relatively less rigid for pelletizing.

Low-density briquettes using binder

In remote areas, for decentralized operation, a simple low-pressure briquettng technique using binder material like tar or cow dung can be adopted. T E R I has developed a simple screw extruder (Figure 13.5) coupled to a small motor, which produces medium density briquettes (400–600 kg/m³). The system was operated in village Dhanwas in Haryana for converting locally available agro-residue (mustard stalk) into briquettes (after pre-processing like chopping and pulverizing) using cow dung, biogas plant effluent slurry or clay as binder material. These briquettes were used as a gasifier fuel. The gasifier system provided electricity to the village community besides supplying parasitic power for the briquetting machine (Raman, Mande, and Kishore

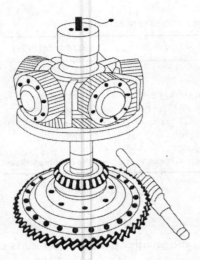

Figure 13.4 Schematic diagram of pelletizing machine

1993). Briquettes made from pyrolysed biomass were also sold as fuel substitutes for Hara (a local stove using dung cakes for simmering milk) in the village households and the road-side restaurants for cooking.

Recently, T E R I has successfully carried out a feasibility study of making briquettes from oil refinery waste sludge by mixing it with locally available loose biomass material. This will not only solve the waste disposal problem of hazardous waste from the oil refinery but can also yield easy-to-use briquettes for substituting coal in the surrounding region (T E R I 2001).

In IIT Delhi, a simple hand operated briquetting plant was developed for making beehive charcoal briquettes which can be used as a clean burning fuel. Locally available leafy biomass, lantana or pine needles are carbonized using a simple drum charring system. This is then mixed with a suitable binding material like clay or cow dung and pressed into a hand mould to form large

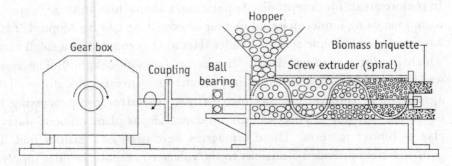

Figure 13.5 Low–density screw extruding briquetting machine

cylindrical briquettes of about 90 mm height and 125 mm diameter, with 19 parallel vertical holes in it (Figure 13.6). One person can make about 30 briquettes (weighing about 1/2 kg after drying) per hour if the charcoal–clay mixture is kept ready. These briquettes after drying can be used in an *angeethi* (a charcoal burning stove), which burns slowly for long durations without smoke. With about a 20% clay content in hardwood charcoal powder, the calorific value is around 18 MJ per kg or 9 MJ per briquette, and one briquette keeps burning for about an hour. A single briqutte is used in the Indian *chulha* but multiple, vertically stacked briquettes for long-duration operations are commonly used in the high altitude areas such as Tibet (Neinhuysm 2003).

Biomass combustion

Combustion is the most direct process of biomass conversion into energy that can be used for a variety of applications. The difficulty in combustion is in starting the process, as high temperatures – at least 550 °C (Quaak, Knowf, and Stassen 1999; TNO 1992) – are required for the ignition of biomass. However, once ignition starts, the combustion process will continue if sufficient air supply is available and if the moisture of biomass is not too high, till the biomass is completely converted into residual ash. Figure 13.7 shows the fire triangle delineating the components, namely, fuel, air, and heat, essential for combustion. Fire can be extinguished by breaking the triangle, that is, either by removing the fuel, by smothering (removing air), or by cooling (spraying with water).

Biomass combustion is employed for a variety of applications such as cooking, process heating, power generation and cogeneration. With the rising prices of fossil fuels, biomass is gaining importance and there is increasing sophistication of biomass combustion devices to increase efficiency and to reduce emissions. In order to harness biomass energy to the maximum extent, it is important to understand biomass combustion. This includes understanding properties of biomass fuels (see Chapter 12), and the fundamentals of numerous complex reactions associated with biomass combustion.

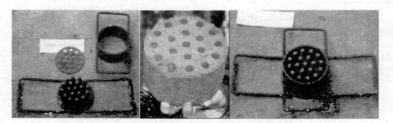

Figure 13.6 Low-density beehive briquetting mould

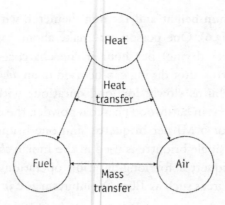

Figure 13.7 Fire triangle

Biomass combustion process

Combustion is a process whereby the carbon and hydrogen in the fuel react with oxygen ultimately to form carbon dioxide and water through a series of free radical reactions resulting in the liberation of heat. General combustion mechanisms have been postulated defining the various stages of chain reactions, namely, initiation, propagation, and termination. Generally, these stages include the following.

- Heating and drying
- Pyrolysis and reduction
- Gas phase pre-combustion and combustion reactions
- Char oxidation reactions
- Post-combustion reactions

Figure 13.8 depicts the processes associated with combustion.

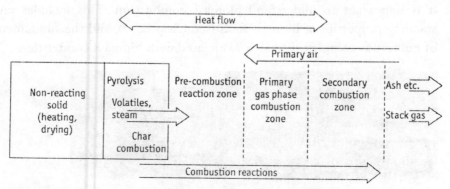

Figure 13.8 Representative model of biomass combustion mechanism

Biomass can be represented as $C_xH_yO_z$ and the overall combustion reaction can be written as

$$C_xH_yO_z + (x + 0.25y - 0.5z)O_2 \rightarrow xCO_2 + (0.5y).H_2O$$

Before the onset of combustion, physico-chemical processes such as drying, pyrolysis, etc., take place, as described below.

Heating and drying

In these processes, physical reactions dominate chemical reactions and hence there is a strong influence by the fuel particle size and moisture content. The presence of moisture enhances the ability to conduct heat to the centre of the fuel particle, but it also increases the energy requirement for heating and drying. About 2.8 MJ of energy is required to drive out one kilogram of moisture in the fuel.

Pyrolysis and reduction

Pyrolysis is of central importance in the flaming combustion of fuel though sufficient information is not yet available about this complex process for quantitative prediction of the pyrolysis kinetics. The carbon is left behind, as charcoal is the only solid produced in biomass pyrolysis. As the temperature of dried fuel is elevated to about 225–325 °C, pyrolysis of hemicelluloses begins. Cellulose gets pyrolized at a temperature range of 325–375 °C while lignin starts pyrolyzing at a temperature range of 350–500 °C (Shafizadeh and Chin 1997). Pyrolysis gases escape and char layer is formed on the fuel particle. This layer or reaction front hampers further pyrolysing of inner layers of fuel particles. Various pyrolysis pathways leading to gaseous products, tars, and char are heavily influenced by the fuel particle size, heating rate of the particle, and the ultimate pyrolysis temperature attained. Higher proportion of gaseous volatile products is obtained due to a faster heating rate of smaller fuel particles to higher temperatures, while the heating of larger particles, at a slower rate, to lower ultimate temperatures favours char formation (Wenzl 1970).

Gas phase reactions

The gaseous compounds produced during biomass pyrolysis are further fragmented before undergoing actual combustion sequences. These fragmentation reactions can be represented by decarboxylation and decarbonylation of acetic acid and acetaldehyde produced by pyrolysis of holocellulose as shown below.

$$CH_3COOH \rightarrow CH_4 + CO_2$$
$$CH_3CHO \rightarrow CH_4 + CO$$

Chain initiation commences after such fragmentation reactions with further breaking down of volatiles, and one such probable sequence begins with the ethane evolved through pyrolysis.

$$C_2H_6 + M \rightarrow 2CH_3 + M$$
$$2CH_3 + 2C_2H_6 \rightarrow 2CH_4 + 2C_2H_5$$
$$C_2H_5 + M \rightarrow C_2H_4 + H + M$$
$$H + C_2H_6 \rightarrow H_2 + C_2H_5$$

Chain propagation commences, once chain initiation occurs. Among almost infinite number of chain propagation reactions, the most commonly cited ones are as follows.

$$CH_3 + O_2 + M \rightarrow CH_3O_2 + M$$
$$CH_3O_2 \rightarrow CH_2O + OH$$

CH_2O is a key combustion intermediate whose concentration reaches the maximum in flames at 1050 °C, that later forms HCO and in turn reacts with the hydroxyl radical OH to form carbon monoxide (Palmer 1974). Among the post-combustion or chain termination reactions which follow the dominant sequences are

$$HCO + OH \rightarrow CO + H_2O$$
$$CO + OH \rightarrow CO_2 + H$$

These fast reactions are complemented by slower oxidation reactions of carbon monoxide.

$$CO + O_2 \rightarrow CO_2 + O$$

Final concluding reactions forming carbon dioxide and water are as follows.

$$H + OH \rightarrow H_2O$$
$$CO + O \rightarrow CO_2$$

Char oxidation reactions

Typical biomass chars, having empirical formulae $C_{6.7}H_{3.3}O$, are highly reactive (Bradbury and Shafizadeh 1980a,b; Mulcahy and Young 1975) and proposed char oxidation mechanisms are as follows.

$$2OH + C \rightarrow CO + H_2O$$
$$OH + CO \rightarrow CO_2 + H$$

The hydroxyl radical (OH) required can come from the dissociation of water (H_2O) or from the hemolytic cleavage of hydroxyl functional groups during pyrolysis. Gasification reactions such as the Bouduard and the steam-carbon reactions, as shown below, are the other two final char oxidation mechanisms used in grate-fired, pile-burning systems, inclined grate furnaces, and gasifiers.

$$C + CO_2 \rightarrow 2CO$$
$$C + H_2O \rightarrow CO + H_2$$

Combustion stoichiometry

The stoichiometric equation for combustion of any fuel is the one that represents a balanced equation of complete combustion of all reactants in the fuel with no excess air. The principal reactions for the combustion of any fuel are summarized below.

$$C + O_2 \rightarrow CO_2$$
$$CO + \frac{1}{2}O_2 \rightarrow CO_2$$
$$H_2 + \frac{1}{2}O_2 \rightarrow H_2O$$
$$S + O_2 \rightarrow SO_2$$
$$C_mH_n + \left(m + \frac{n}{4}\right)O_2 \rightarrow mCO_2 + \frac{n}{2}H_2O$$

where C_mH_n represents a volatile hydrocarbon present in the fuel.

During the normal combustion process, air is a common oxidizer, which is a mixture of 21% oxygen and 79% nitrogen on volume basis (that is, 1 mole of oxygen is accompanied by 3.76 (79/21) moles of inert nitrogen). Thus, the combustion of carbon with air can be rewritten as

$$C + O_2 + 3.76N_2 \rightarrow CO_2 + 3.76N_2$$

Stoichiometric air requirement for fuel combustion is the minimum theoretical air required for complete combustion based on the chemical composition of fuel. In reality, some excess air is always required for complete combustion to occur, value of which depends on the design of the combustion chamber and the type of fuel. Natural gas-fired boilers operate with about five per cent excess air, coal-fired boilers operate with about 20% excess air, while gas turbines operate with very lean mixtures with excess air levels of the order of 400%. Rich mixtures (with less excess air) result in incomplete combustion, leading to emissions in the form of PICs (products of incomplete combustion) such as carbon monoxide, methane, and NMOC (non-methane organic carbon), which is also associated with loss of energy.

If fuel composition is known, the theoretical or stoichiometric air can be calculated using the above-mentioned equations by the mass balance method or the mole method.

Example 1

Calculate the stoichiometric air requirement to burn 1 kg fuel, if the biomass fuel composition on mass basis is carbon 44%, hydrogen 15%, nitrogen 1%, oxygen 28%, moisture 10%, sulphur 0.5%, and ash 1.5%.

Solution

On mass basis

For 1 kg fuel, the weight of oxygen required to burn its various combustible fuel constituents is as follows.

For carbon:	$0.44 \times (32/12)$	$= 1.173$ kg
For hydrogen:	$0.15 \times (16/2)$	$= 1.200$ kg
For sulphur:	$0.005 \times (32/32)$	$= 0.005$ kg

Thus, the total oxygen required is 2.378 kg per kg of fuel, of which 0.28 kg is already present in the fuel. Therefore, the net oxygen required from the air supply is 2.098 kg (0.378 – 0.280). As air contains 23% oxygen by weight, the stoichiometric air required to be supplied for complete combustion is 9.123 kg (2.098 × 100/23) per kg of fuel.

Considering the molecular weight of air as 29, the air density at NTP (normal temperature and pressure) works out to be 1.295 kg/m³ (29/22.4, where 22.4 is molecular volume of air at NTP). Thus, the minimum amount of air required to be supplied for complete combustion is 7.045 m³ (9.123/1.295).

Example 2

Calculate the composition of dry flue gas if 20% excess air is supplied for the combustion of fuel with the composition given in Example 1.

Solution

Let us first convert the fuel composition from the mass basis to the mole basis for 100 kg fuel. Then from the basic combustion reactions given earlier, calculate the amount of oxygen required in kmol basis (1 kmol oxygen is required for 1 kmol of carbon and sulphur, while 0.5 kmol oxygen is required for 1 kmol of hydrogen).

Fuel constituent	Mass (kg)	kmol	kmol O_2 required
Carbon (C)	44	44/12 = 3.667	3.667
Hydrogen (H_2)	15	15/2 = 7.5	3.750
Nitrogen (N_2)	1	1/28 = 0.036	
Oxygen (O_2)	28	28/32 = 0.875	
Moisture (H_2O)	10	10/18 = 0.556	
Sulphur (S)	0.5	0.5/32 = 0.016	0.016

Thus, the stoichiometric oxygen requirement is 7.432 kmol, the net minimum oxygen required to be supplied is 6.557 kmol (7.432 – 0.875), and the stoichiometric air requirement is 31.225 kmol (6.557 × 100/21) per 100 kg of fuel or 699.4 m³ of air (as 22.4 m³ is the volume per kmol at NTP).

As 20% excess air is supplied, the actual amount of air supplied is 37.470 kmol (31.225 × 120/100) per 100 kg fuel. Of this, oxygen is 21%, that is 7.869 kmol and the balance is nitrogen. Since the stoichiometric oxygen requirement is 6.557 kmol, the excess 1.331 kmol (7.869 – 6.557) oxygen will appear as it is in flue gases. Thus, the dry flue gases after the combustion of 100 kg fuel with 20% excess air would be

Carbon dioxide   = 3.667 kmol from combustion of carbon
Sulphur dioxide   = 0.016 kmol from combustion of sulphur
Nitrogen   = 29.637 kmol (29.601 and 0.036 kmol from air and fuel, respectively)
Oxygen   = 1.331 kmol as excess oxygen

Thus, the total volume of dry flue gases is 34.651 kmol. Therefore, the volumetric composition of dry flue gases works out as follows.

Carbon dioxide   = 10.58% (3.667/34.651)
Sulphur dioxide   = 0.05% (0.016/34.651)
Nitrogen   = 85.53% (29.637/34.651)
Oxygen   = 3.84% (1.331/34.651)

Biomass combustion for useful heat production

Direct combustion systems are used to produce heat, which can be either used directly (for example, brick making) or transferred to a working fluid, such as steam, for further use in process heat or in steam engines or turbines for power production. The combustion efficiency can be defined as follows

$$\eta_{comb} = \frac{\text{Thermal energy available in the product gases}}{\text{Chemical energy in the supplied fuel}}$$

The combustion efficiency is mainly determined by the completeness of the combustion process. The flame temperature plays an important role in deciding the overall efficiency of the combustion device. For boiler, the following simplified regression equation can be used to calculate the approximate adiabatic flame temperature (T_{ad} in °C) (Tillman and Anderson 1983).

$$T_{ad} = 420 - 10.1(MC)_w + 1734\lambda + 0.6(T_{in} - 25)$$

where $(MC)_w$ is the moisture content on wet basis, λ is the excess air factor (ratio of actual fuel:air ratio to stoichiometric fuel:air ratio) and T_{in} is the temperature of the combustion air (°C) entering the combustor. It can be seen that excess air has more influence than the moisture content on the adiabatic flame temperature (Figure 13.9). Theoretically, the highest temperature can be achieved with stoichiometric air supply ($\lambda = 1$) but in practice, excess air is

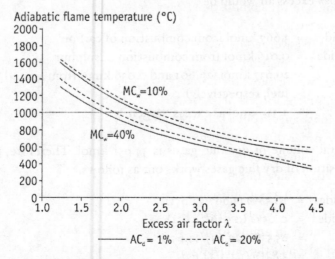

Figure 13.9 Effect of moisture (MC_w), ash content (AC_d), and excess air (λ) on adiabatic flame temperature
Reproduced with permission from the World Bank
Source Quaak, Knowf, and Stassen (1999)

always supplied to ensure complete combustion. The optimal values of λ depend on the furnace design, fuel type, and fuel feeding/firing system used. For well-designed furnaces/combustion devices, the λ values are in the range 1.6–2.5, while in poorly designed furnaces, value of λ reaches as high as 4–5.

The first law efficiency can typically be calculated using the heat loss method and for typical biomass-fired boilers, efficiencies are in the range 60%–80%, depending on the ultimate analysis of fuel, moisture content, and excess air factor (which normally ranges from 25% to 100%, depending on combustor design and fuel quality). The boiler efficiency can also be approximated (to an accuracy of ± 2%) using the following regression equation.

$$\eta = 96.84 - 0.28MC_g - 0.064T_s - 0.065EA$$

where MC_g is the moisture content of flue gas on wet basis, T_s is the flue gas temperature in °C, and EA is the excess air (percentage) that has a strong influence on the efficiency (Tillman 1981).

Emissions during biomass combustions

Biomass combustion for producing useful heat is associated with airborne emissions, which are undesirable but at the same time unavoidable too. Particulate emission has long been considered as a major problem associated with biomass combustion, while NO_x has emerged recently with the development of higher temperature combustion devices using fuels such as rice hulls, cotton processing wastes, etc.

Fly ash is largely governed by fuel type (fines in fuel) used, fuel feed rate, excess air used, and its distribution. Particulate emissions are minimized in staged combustion where the stoichiometric air is supplied under the grate while excess air (of the order of 50%–60%) is supplied as over-fire air. Particulate emissions range from a sub-micron size to 2 mm with typical concentrations of the order of 30–100 g/m³ (Jung 1979). The fly ash problem can be controlled by using cyclone separators, dry scrubbers, electrostatic precipitators, and bag-house filters.

Two sources of NO_x exist during biomass combustion, namely fuel NO_x and thermal NO_x. Fuel NO_x is governed by the concentration of nitrogen in fuel, which is mainly in amine form (Cowling and Kirk 1976). Fuel NO_x is formed by the oxidation of the reduced form of nitrogen contained in the fuel and is generally not sensitive to temperature (Edwards 1974). Since nitrogen contained in biomass fuels volatilizes preferentially, fuel NO_x can be controlled by staged combustion.

Thermal NO_x is formed by the oxidation of nitrogen in the combustion air, which is a high temperature reaction that is not favoured at flame temperatures below 1500–1600 °C. With the development of advanced combustion devices with refractory lining and low excess air combustors, thermal NO_x has become a significant problem. Thermal NO_x can be regulated by proper control of excess air and by not over-emphasizing the pre-heating of combustion air. Normally, NO_x emissions are observed in the range 0.4–1.2 kg/MJ, which are significantly lower than the emissions from coal combustion (Kester 1980; Munro 1983).

Types of combustors

Biomass combustors are designed in such a way that the combustion mechanisms/processes described earlier are controlled to release heat through the oxidation process of various chemical constituents of the fuel in the most practical optimum manner. Various types of combustors are:

- Fixed-bed or grate-fired combustors
- Suspension burners
- Fluidized bed systems

Fixed-grate systems

Fixed-grate systems were, for many years, the most common biomass combustion devices. Fixed-bed systems are mainly distinguished by the type of grate used and the mechanism used to supply or transport fuel through the furnace. Fixed-bed systems include manually fed systems, spreader-stoker systems, under-screw systems, through-screw systems, static grate, or inclined grates, and travelling grate systems. In the simplest form of fixed-grate systems, primary air for combustion of char is supplied under the grate while secondary air for combustion of volatiles is supplied above the grate. Primary air continues the combustion of char on the grate and the heat released during this process enables the pyrolysis of the fresh fuel added, releasing volatiles in it. The secondary air completes the combustion process to exploit the heat content in the released volatile gases. Typical combustion bed temperatures in fixed-grate systems are of the order of 900–1400 °C, and ash is removed below the grate. As compared to coal, biomass has a higher volatile fraction, and hence larger combustion space above the grate is provided. Therefore, biomass requires a higher proportion of secondary to primary air supply as compared to coal.

In inclined grate systems (developed for coal during the 1920s–30s), fuel is supplied at the top and gradually moves downwards during the combustion process (Figure 13.10). The first moving sloping grate system was introduced in 1940s. In this system, the residence time of fuel is fixed by the speed of the moving grate. The combustion chamber can be made further compact (higher heat release rate per unit grate area) for uniform-size biomass and with complete combustion occurring in different stages.

For small- to medium-size particle fuels, screw feeder systems are developed to push the fuel to the centre of the combustion and to take out the ash from other side. For large-size fuels and fuels with high ash content, a through-screw system is used. Here, the fuel is burned while being screw-fed through the combustion zone and the remaining ash is deposited into the ash pit at the end of the screw. For relatively small operations, special feed systems have been developed consisting of screw and spreader stokers. Fuel particles are spread above the reaction zone with spreader stokers, which resembles like suspension burning. Part combustion occurs when the fuel particle is in suspension while moving through the gas above the grate (Figure 13.11). A comparison of various fixed bed combustion systems is given in Table 13.5.

Grate-fired systems can handle fuels of larger particle sizes and with higher moisture contents (up to 50% wet basis). They are also capable of fuel utilization up to 600 GJ/h with complete combustion in various stages.

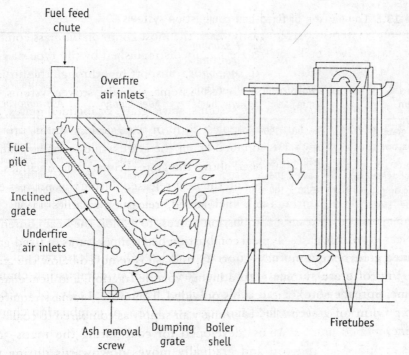

Figure 13.10 Sloping grate combustion chamber

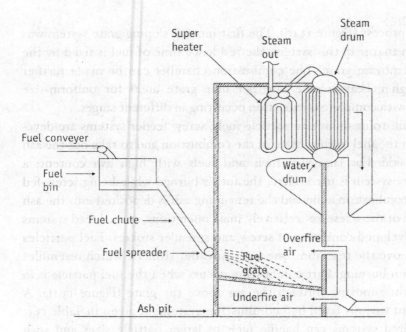

Figure 13.11 Spreader–stoker grate combustor

Table 13.5 Comparison of fixed-bed combustion systems

System	Fuel size (mm)	Maximum moisture content (per cent wet basis)	Fuel supply	Ash removal
Static grate	<φ100 × 300	50	Manual/automatic	Manual/automatic
Under screw	<40 × 30 × 15	40	Automatic	Manual/automatic
	>20 × 20 × 10	40	Automatic	Manual/automatic
Through screw	<φ50 × 100	40	Automatic	Automatic
Moving or inclined grate	<300 × 100 × 50	50	Automatic	Automatic
Spreader–stoker	<40 × 40 × 40	50	Automatic	Manual/automatic

Inclined grate systems normally operate with lower heat release rates of about 3.5 GJ/m² of grate surface and about 500 MJ/m³ of combustion chamber volume. Spreader–stoker can achieve higher heat release rates of the order of 10.2 GJ/m² of grate surface and 500–750 MJ/m³ of combustion chamber volume.

Suspension burner

These are special purpose burners, similar to pulverized coal-fired burners, developed for biomass. They have increased specific capacity (per volume of reactor) and produce an oil-type combustion flame, but require more extensive fuel preparation and storage than grate-fired systems (Figure 13.12). The heat release rate of suspension burners is in the range 500–600 MJ/m³ of combustion volume. They require fine biomass particulate size (< 2 mm) with less than 15% moisture (wet basis). The main drawback of the suspension burner is a low operating efficiency as high level of excess air (more than 100%) is required to prevent the build-up of slag in the burner/combustor. In the slagging mode, with low excess air levels, higher temperatures of the order of 1600–1700 °C can be achieved. Another drawback is that in the absence of staged combustion, suspension burning results in high fly ash and also higher fuel nitrogen conversion to oxides of nitrogen as compared to spreader–stoker firing.

Fluidized bed systems

In a fluidized bed combustor, fuel is burned in a hot (800–1000 °C) turbulent bed of non-combustible material (sand, limestone, etc.), which acts as a medium of heat transfer. The bed is fluidized by using fans to blow air through

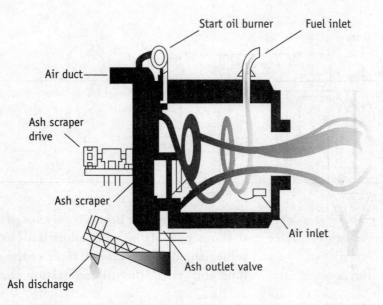

Figure 13.12 Suspension burner

a perforated bottom plate. Fluidization facilitates high heat transfer rates by creating a large heat transfer surface. This helps in complete combustion with low excess air levels (25%–35%), resulting in a high overall efficiency. A high thermal mass of inert material also enables good combustion of very wet fuels. Fluidized bed combustors are gaining increased acceptance, especially for loose biomass combustion, due to the several advantages associated with it. These are:

- Flexibility to accommodate changes in fuel properties, size, and shape
- Capability to handle high moisture (up to 65% wet basis) content fuels
- Capability to handle high ash content (up to 50%) fuels like rice husk

Depending on the air velocity, either a BFB (bubbling fluidized bed) or a CFB (circulating fluidized bed) is created. In a BFB, the combustor is divided into two zones, namely a zone containing free-moving sand-bed particles supported by air flowing upward giving the resemblance of bubbling fluid, and a free board zone above the fluidized bed (Figure 13.13). In a CFB, the velocity is so high that the lighter bed and fuel particles get carried away with the flow in circulating motion and get separated in cyclone and later return to the reactor (Figure 13.14). Thus, light fuel particles burn during circulation while larger/heavier particles burn until they become light enough to join the circulating stream. BFB has a heat release rate of about 5.6 GJ/m² grate equivalent and 470 MJ/m³ of reactor volume (Envirosphere 1980). Higher rates can be achieved with CFB systems.

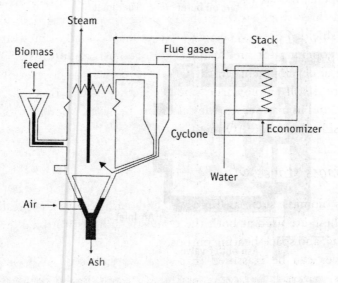

Figure 13.13 Bubbling fluidized bed combustion system

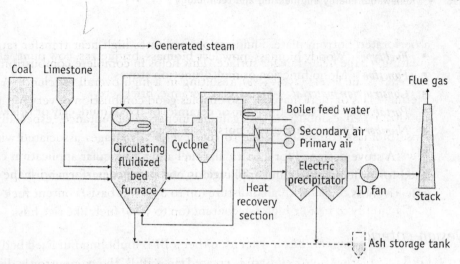

Figure 13.14 Circulating fluidized bed combustion system
Source <www.nedo.go.jp/sekitan/cct/eng_pdf/2_1a2.pdf> last accessed on 20 April 2007

Biomass stoves

Archaeological findings indicate that the use of fire was known to mankind 400 000 years ago. The use of fire for cooking dates back to 100 000 years ago, and cooking was presumably done over an open fire (mainly for cooking/roasting meat) with the fuel arranged in a pyramid shape. Despite lack of control on the fire and the smoky conditions, the open fire had the benefits of preserving food, protecting against animals, and providing warmth. A major development in open fire was the evolution of different-shaped vessels and later of a shielded three-stone stove for holding the pot over the open fire. Subsequently, the shielded fire was changed to a U-shaped mud or mud/stone stove with a front opening for fuel feeding and combustion air entry. Since then, despite several developments in the wood stoves, a large population in the developing world still employs the traditional three-stone or U-shaped shielded fire stove, and in many cases, they alternate between wood, cow dung cakes, and agro-residue for fuel (FAO 1993).

Classification of biomass stoves

Stoves that burn biomass such as firewood and agro-residues are called biomass stoves. These are used at both the domestic and institutional levels for cooking, heating, and space-heating purposes.

Biomass stoves can be classified in several ways, based on their attributes, functions, material, fuel types, etc. (FAO 1993). Based on various characteristics, stoves are classified as follows.

- *Fuel type*   Woody biomass, powdery biomass, briquettes, cow dung, etc.
- *Function*   Mono function or multi-function stoves
- *Construction material*   Metal, clay, ceramic, brick, etc.
- *Portability*   Portable (metallic or ceramic) or fixed (mud, clay or brick)
- *Number of pots*   Single, two pots, three pots, etc.

A stove designed for a particular fuel and a particular application can be used for different fuels and applications but may not perform with the same effectiveness.

Design criteria

A stove is a consumer-specific device. The effort on developing improved stoves is mainly aimed at improving energy efficiency (saving fuel) or reducing emissions (improving working conditions and reducing an adverse impact on health). While designing a stove, both engineering and non-engineering or social parameters are required to be considered.

Social factors

The interlinking of various criteria for stove design is shown in Figure 13.15. User need and availability of local biomass are the two important social factors that need to be taken into account while designing a stove. User needs would include factoring into the design various cooking operations (boiling, frying, baking, grilling, steaming, pressure-cooking, etc.) that have to be performed on the stove. Apart from this, it is also important to know the cooking time and the process heat requirements, which would determine the power range for the stove. Availability of local construction material, desired portability, seasonal availability of local biomass, etc. also need to be considered while designing a stove for the target group (Verhaart 1983).

Technical factors

A high-performance stove should be efficient from both the efficiency and the emission points of view. A fuel-efficient stove could reduce the drudgery of collecting fuel while reduction in emissions could save the users from the harmful impact due to exposure to smoke. However, as mentioned earlier, the general strategy adopted in designing an improved stove is improving the energy efficiency (by enhancing heat transfer) and providing a chimney for the removal of smoke. Though this strategy improves the fuel efficiency and also

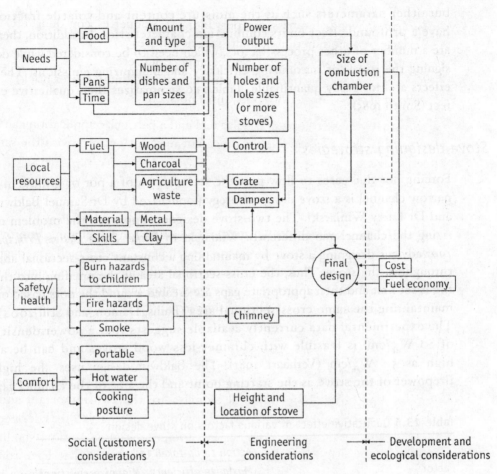

Figure 13.15 Design considerations for a stove

the quality of the indoor air, the overall reduction in emissions is questionable. Therefore, in addition to improving the heat transfer efficiency, it is necessary to ensure improvement in the combustion efficiency whereby unburned harmful pollutants can be minimized. Complete combustion can be ensured as follows.

- Maintaining higher temperatures in the combustion chamber
- Providing sufficient air and ensuring proper mixing for complete combustion
- Ensuring sufficient residence time for the completion of combustion reactions

Process factors

Composition of several biomass materials in terms of C–H–O content is quite similar on ash- and moisture-free basis (as mentioned in Chapter 12), but other parameters such as the moisture content and volatile fraction have a profound effect on its combustion characteristics. In addition, there are a number of other process factors, which need to be considered while designing the stove for maximizing efficiency and minimizing emissions. Their effects are not easily quantifiable. Table 13.6 summarizes their qualitative effect (Smith 1985).

Stove designing strategies

Forcing hot flue gases to flow past the surface area of a pot or griddle in a narrow channel is a stove design strategy popularized by Dr Samuel Baldwin and Dr Larry Winiarski. The two stove designers approach the problem of sizing the channel gap differently. Winiarski in *Rocket Stove Design Principle 1997* advices designing a stove by maintaining a constant cross-sectional area throughout the stove. Thus, the cross-sectional area at the opening into the fire is set first and then appropriate gaps are created around the pots based on maintaining the same cross-sectional area (Bailis, Damon, and Still 2004). The experimental data currently available suggests that a power density of 20 W_{th}/cm^2 is feasible with chimney-less wood stoves and can be as high as 50 W_{th}/cm^2 (Verhaart 1981). The Baldwin design uses the high firepower of the stove as the starting point and the size of the channel gap

Table 13.6 Qualitative effect of various factors on stove design

Factor	Action to be taken to	
	Maximize efficiency	Minimize emissions
Stove factors		
Combustion confinement	Maximize	Minimize
Temperature of combustion chamber	Minimize	Maximize
Excess air	Optimize	Optimize
Thermal mass		
- for short cooking time	Minimize	Minimize
- for long cooking time	Maximize	Maximize
Operational factors		
Fuel burning rate	Minimize	Maximize
Fuel charge size	Minimize	Minimize
Ratio of charge size to burn rate	Minimize	Minimize
Volume to surface ratio	Maximize	Maximize
Fuel actors		
Moisture content	Optimize @10%	Optimize @25%
Volatile matter content	Minimize	Minimize
Ash content	Minimize	Minimize

is then determined accordingly. Table 13.7 shows the values of various stove gaps from Baldwin's findings, which is an approximation meant to serve as a guide to the relationship between fire power, burning rate, length and width of gap size, and stove efficiency.

Performance testing of biomass stoves

The overall thermal efficiency η_o of the stove is defined as the ratio of the amount of useful heat absorbed for cooking food to the amount of energy content in the fuel used (Khummomgkol 1986; Mande and Lata 2001). It is a combination of partial efficiencies such as the combustion efficiency η_c (fraction of energy content of fuel converted into heat through its combustion), heat transfer efficiency η_h (fraction of heat generated that is transferred to the pot), and pot efficiency η_p (fraction of useful energy actually used for cooking food). Thus,

$$\eta_o = \frac{\text{Useful heat absorbed by food}}{\text{Heat content of biomass fuel used}} = \eta_c \times \eta_h \times \eta_p$$

Tabel 13.7 Baldwin's suggested channel gap sizes for stove design

Burning rate (kg/h)	Skirt gap[#] (mm)	Length of gap[#] (cm)	Stove thermal efficiency (%)	Firepower[+] (kW)
0.50	8	20	40	2.8
0.75	10	20	35	4.1
1.00	11	20	30	5.5
1.25	12	20	28	6.9
1.50	13	20	26	8.3
1.75	14	20	25	9.6

Source Bailis, Damon, and Still (2004)
[#] See the accompanying sketch; [+] Burning rate x calorific value

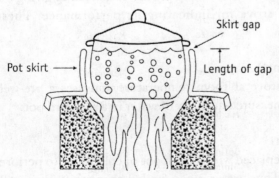

Sketch of stove for parameters in Table 13.7

The following three types of standard testing methods were evolved after critical review in the workshop organized by VITA (Volunteers in Technical Assistance).

- WBT (water boiling test) under laboratory conditions
- CCT (controlled cooking test) involving actual cooking carried out in the laboratory
- KPT (kitchen performance test) involving actual cooking in an actual kitchen

Water boiling test

WBT recommended by VITA is the most commonly used methodology to measure the power output of the stove and its efficiency (VITA 1985). It tries to simulate the boiling and simmering operations commonly used during cooking. A known quantity of water W is filled up to two-third level in a vessel of known weight V and is heated from ambient temperature T_a to boiling point T_b. The lid is kept closed, but is partially open during simmering. It is boiled for 15 minutes and then kept simmering (maintaining temperature within 2 °C of boiling point) for 60 minutes. The stove efficiency η is computed using the following equation:

$$\eta = \frac{W \times C_{pw} \times (T_b - T_a) + V \times C_{pv} \times (T_b - T_a) + dW \times L_w}{F \times CV}$$

where dW is the amount of water evaporated, L_w is the latent heat of water, F is the amount of fuel burnt, CV is the calorific value of fuel, C_{pw} and C_{pv} are specific heat values of water and the vessel, respectively.

Recently, a modified WBT version 1.5 is being developed under Shell HEH (Household Energy and Health) Programme (Bailis, Damon, and Still 2004). This consists of separate measurements under three different phases, simulating various commonly used cooking processes instead of giving a single efficiency number for a stove for indicating its performance. These are explained as follows.

Phase 1: High power (cold start)
Testing begins with the stove at room temperature and uses a pre-weighed bundle of wood to boil a measured quantity of water in a standard pot.

Phase 2: High power (hot start)
The boiled water is then replaced with a fresh pot of cold water to perform the second phase of the test. Thus, water is now boiled on a hot stove in order to

identify the differences in the performance of the stove in its hot and cold stages.

Phase 3: Low power (simmering)

A measured amount of water is boiled and then, using a pre-weighed bundle of wood, the water simmers at just below the boiling point for a measured period of time (45 minutes). This phase tries to simulate the common cooking process that entails cooking of legumes or pulses. For a multi-pot stove, only the primary pot will be assessed for its performance during simmering of the water.

The combination of tests is intended to measure the stove's performance at both high and low power outputs P_H and P_S, respectively, which are important indicators of the stove's ability to conserve fuel. Thus, rather than reporting a single number indicating the thermal efficiency of the stove, which alone cannot predict stove performance, this test is designed to yield several numerical indicators including the following.

- Time to boil
- Fuel burning rate (kilogram per hour)
- Specific fuel consumption (kilogram per task)
- Turndown ratio (ratio of high to low power output)

A well-designed stove should ideally perform with the same efficiency over the entire power level range. However, most stoves do not have such ideal performance, making it necessary to evaluate stove performance under maximum and minimum power levels.

Controlled cooking test

This test essentially gives the fuel consumption of a given stove for carrying out a typical cooking operation. It is done under laboratory conditions on a stove with a typical vessel size and shape, normally used for cooking a typical food of the region, using a commonly used cooking operation. The amount of fuel used up for cooking a known quantity of food and the time required for cooking are measured. The test is repeated at least three to five times to get average values. It can also be used to compare two different stoves for the same cooking operation or compare stove performance for the various cooking operations.

Kitchen performance test

This performance test is carried out in the field under real-life conditions. The various cooking operations performed by the family during cooking are monitored and recorded to arrive at the total fuel consumption per meal per person. The test is performed for several days to get realistic average values. This test overcomes the drawback of stove performance for making a typical meal and takes into account the stove performance for a variety of cooking operations such as boiling, frying, roasting, etc. encountered during meal preparation. Though, for theoretical analysis, this is too gross a test, but it gives a realistic measure of stove performance comparison, namely, whether the improved stove actually saves fuel under field conditions (Prasad, Sangen, and Visser 1985).

All these tests can only be used for a relative comparison of stoves for a given task under given operating conditions. These tests cannot be accepted universally for defining stove efficiencies as a small variation in the cooking practice or operating conditions would significantly affect the performance. The procedure for determining the efficiency of the gas stoves includes operating the stove at different constant power levels, using the water boiling test for a turndown ratio (ratio of maximum to minimum power output level). Though it is easy to operate gas stoves at different constant power output levels, it is difficult and impractical to achieve such conditions for wood burning stoves (Bussmann 1988).

Emissions from biomass stoves

Among the numerous pollutants emitted from biomass stoves, the most important ones are carbon monoxide, TSP (total suspended particulates), PAH (polycyclic aromatic hydrocarbons), and formaldehyde. Several design and operating parameters resulting in incomplete combustion contribute to these emissions. These include: insufficient supply of combustion air, lower temperatures in the combustion zone, improper mixing of air and fuel, etc. Many times, some design modification to improve the thermal performance or efficiency, such as reducing the gap between the pan and the stove mouth as well as between the pan and the grate, may actually result in increased emissions. Therefore, there is need to take into consideration the effect of any design modification on the stove as a whole. Generally, emissions are seen to increase with increasing power levels. Similarly, higher emissions are observed with smaller stoves. Often, this is due to the lower residence time of fuel in the combustion chamber of the stove resulting in incomplete combustion and hence higher emissions.

Emission testing of biomass stoves

Emissions from the stoves with chimney are measured by placing sensors in the flue gas path. For the majority of stoves without chimney, the sensors are kept attached to the operator as emissions are measured to assess their impact on the health of the operator. However, these are non-standard techniques and depend upon the local conditions of ventilation, position of doors and windows, direction of the wind, and so on. The following two methods available for monitoring emissions from a stove without chimney are commonly used

- Hood method
- Chamber method

Hood method

This is also called a direct method in which the stove is kept in the enclosed hood. The monitoring of air supply is done through pre-defined vents and the flue is extracted from the hood exit under isokinetic sampling conditions (Figure 13.16). The measurement of flow rates through vents is quite expensive and not very accurate with natural draft flows. If air flow is induced using an exhaust fan then it can affect the normal operation of the stove. Since it is not always possible to construct hood in situ under field conditions, this method is normally used for laboratory experiments.

Chamber method

The chamber method, first proposed by Ahuja, Joshi, Smith, et al. (1987) is also called a simulated kitchen method or indirect method (Figure 13.17). This

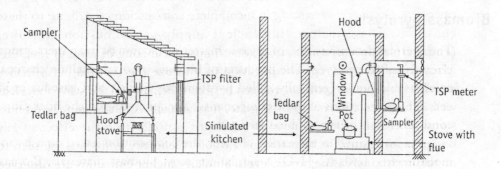

Figure 13.16 Hood method
Source Smith et al. (2000)

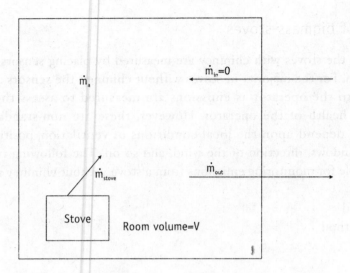

Figure 13.17 Chamber method
Source Ballard–Tremeer and Jawurek (1999)

method can also be used in the field. However, for accurate measurements of emission performance using this method, it is necessary to have relatively constant ventilation conditions throughout the experimentation period. Pollutants near the stove and in the background room conditions are measured for the particulate levels (using filter paper) and carbon monoxide (using a sampler) at regular intervals throughout the experiment. The air exchange rate in the room is monitored using the standard exponential carbon monoxide decay method (where the carbon monoxide level is raised up to a certain level and its decay after its source is removed is monitored with time). Normally, emissions from various stoves are compared for a given standard task rather than the emissions per unit fuel or heat output.

Biomass pyrolysis

The thermal decomposition of organic matter in vacuum or in an inert atmosphere is called pyrolysis. The products of biomass pyrolysis include charcoal, condensable liquid (generally called pyroligneous liquid), and gaseous products. The proportions of these components vary depending on the operational conditions and the type of biomass used.

As mentioned in the earlier chapter, the composition based on ash- and moisture-free basis (C–H–O) is similar for most biomass materials. Biomass consists of three major constituents, namely cellulose, hemicellulose, and lignin. In most biomass materials, cellulose is in the form of glucan polymer with an average molecular weight of 100 000. Hemicellulose has a molecular weight (<30 000) lower than that of cellulose and is a mixture of polysaccharides glucose, mannose, galactose, xylose, and galacturonic acid residues. Lignin is more abundant and has a higher degree of polymerization in soft woods than in hard woods. It is a randomly linked amorphous phenolic compound with high molecular weight.

Pyrolysis process and product chemical composition

During wood pyrolysis at a low heating rate for charcoal production, generally the following sequence of events occurs. Initially, drying (removal of moisture) occurs below 200 °C. Till about 300 °C, reduction in the molecular weight (or degree of polymerization) of mainly amorphous cellulose occurs with the formation of carbonyl and carboxyl group radicals, giving rise to the outgassing of carbon monoxide and carbon dioxide. The resultant crystalline cellulose gets decomposed with the temperature rising beyond 300 °C, resulting in the formation of char, tar, and gaseous products. Simultaneously, hemicellulose first gets decomposed into soluble polymer fragments followed by further decomposition into volatile gases and gives less tar and char. Lignin decomposes between 300–500 °C, giving more char (about 55%)' yield as compared to cellulose or hemicellulose. It also gives about 20% aqueous distillate mainly consisting of methanol, acetic acid, acetone, and water. The first two steps till 300 °C are endothermic while the decomposition between 300 and 500 °C is exothermic. Therefore, during charcoal making, once the temperature builds up to 300 °C, further external heating is not required and the temperature rises to about 400–450 °C. For achieving higher temperatures during pyrolysis for maximizing gas or liquid yield, additional external heat is required.

Process conditions

The process conditions affect the products of pyrolysis, namely, char, tar, and gas, apart from volatile fraction and moisture content of fuel. The influence of three major parameters, the heating rate (thermal flux), highest temperature used, and residence time on the products of biomass pyrolysis is shown in Figure 13.18 (Roy, De Caumia, and Menard 1983). It is observed that with an increase in the heating rate with lower residence time, the yield of volatiles increases, but at higher temperature, these volatiles get cracked resulting in more gas yield. The char yield is maximized at lower heating rates and temperatures but with longer residence time (Wen and Dutta 1979).

In practice, it is easy to control the highest temperature achieved and the residence time in the reactor, but it is not easy to control the heating rate to maximize either the char or the tar yields. The heating rate depends on two main factors, namely, particle size and reactor type. For a given reactor, the larger the particle size, the lower is the heating rate (due to low conductivity of wood) which results in more char yield (Figure 13.19). For pyrolysis reactors, where the temperatures are normally below 1000 °C radiation heat transfer does not play a major role and the dominant heat transfer mode is conduction and convection. Some reactors like fluidized bed can handle only smaller/finer particles (Villermaux 1982). The following observations can be made from Figures 13.18 and 13.19.

- Charcoal can be produced by using stacking kilns, multiple hearth kilns, and rotary kilns with low heating rates, low temperature, and high residence time.
- Pyrolytic liquid can be obtained using transported bed reactors with high heating rates, low temperature, and shorter residence time.
- Gas can be obtained by means of fluidized bed, circulating bed, or even cyclone reactors with high heating rates and higher temperatures.

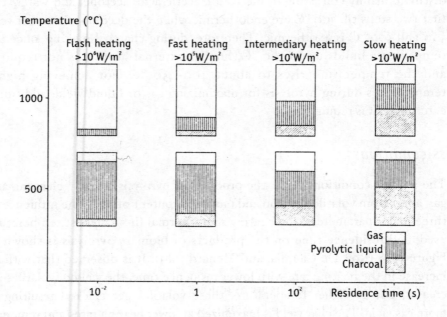

Figure 13.18 Effect of various operating parameters on products of pyrolysis

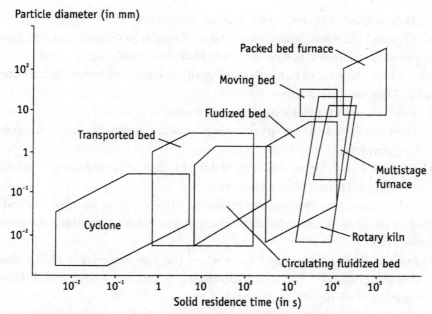

Figure 13.19 Various operating zones of different pyrolysis reactors

Charcoal production

Pyrolysis processes are generally called carbonization processes when the objective is to produce charcoal from the pyrolysis of biomass. Nowadays, though charcoal is mainly used for industrial and chemical applications in developed countries, it is still an important fuel for cooking and heating in developing countries. It is estimated that through the control of pyrolysis reactions it is possible to obtain 30–40 kg each of char and pyroligneous liquid, and 15–20 kg gases per 100 kg. The major parameters for charcoal making are as follows.

- *Yield* The ratio of the weight of charcoal obtained to the dry weight of feed material. Normally, it ranges from 1:3 to 1:5, depending on the quality of feed material and the operating conditions maintained in the reactor used.
- *Volatile content* The weight loss of charcoal when it is heated up to 900 °C per unit weight of charcoal.
- *Fixed carbon content* The weight of charcoal minus volatiles and uncombustible ash fraction per unit weight of dry charcoal.

Charcoal is used for a variety of applications, depending on its varying properties. Some applications are as follows.

- *Domestic* for cooking and heating
- *Agricultural* for drying and processing

- *Metallurgical* copper, bronze, nickel, aluminium, etc.
- *Chemical* carbon monoxide, carbon dioxide, activated carbon, calcium carbide, absorbent, soil conditioner, pharmaceutical, etc.

More than 100 charcoal-making processes are known today (Emrich 1985). They can be classified as follows.

- *Kilns* Used for producing charcoal alone.
- *Converter* Used for pyrolyzing smaller particles and recovering by-products.
- *Retort* Used for pyrolyzing billets or logs reduced in size to about 30 cm length and 18 cm diameter.

In modern technology, the following three types of heating methods are used in order to initiate the carbonization process and maintain the required temperature.

- *Internal heating* Normally used when the raw material is cheap. Part of the material used for charcoal making is burned within the kiln using controlled air supply.
- *External heating* Normally used in retorts to indirectly heat the material to be carbonized. Fuel for this can be obtained from the gases or liquid released during the pyrolysis process itself.
- *Heating with gas recirculation* Normally expensive and used for large systems. Part of the gas produced is burned and directed inside the reactor for its carbonization under controlled conditions.

Traditional charcoal-making kilns

Most of the charcoal-making processes adopted in developing countries are governed by low investment and low operating cost, with limited range of biomass material available locally. As a result, these processes are generally labour intensive, have low yield, and take a longer operating time. These can be classified basically as stacked bed kilns and are generally operated in batches.

A huge quantity of locally available biomass is cut and stacked on the ground or inside a shallow pit. It is then covered with layers of leafy material and soil layers. It is slowly ignited through the holes provided, and once the ignition spreads throughout the fuel bed the air supply is cut off by closing these holes and gradually the entire fuel bed gets carbonized in the absence of air for oxidation. The charcoal is then removed after the cooling down of the kiln (Figure 13.20).

Figure 13.20 Traditional charcoal-making kiln

Modern charcoal Missouri kiln

The Missouri kiln is widely used in developed countries (Massengale 1985). Here pieces of wood of approximately 20 cm are neatly stacked and about 15 cm of air space is ensured at regular intervals. The wood-holding capacity of these kilns is in the range 100–180 tonnes and it takes about 8–20 days for completing a batch of charcoal-making with 20% yield. Unlike the earth mounted traditional charcoal kiln, it is a permanent structure made up of brick or concrete construction and the same kiln can be used for several batches with minor maintenance (Figure 13.21).

Lambiotte retort

It is a continuous carbonization process suitable for industrial-scale application requiring huge investments, power, water, etc. The retort is normally characterized by automation, energy, as well as labour-saving techniques (Carre, Herbert, Lacrosse, *et al.* 1985) and is a multi-stage retort (Figure 13.22). Uniform-size wood of 10 cm diameter and 30 cm length is fed from the top into the cylindrical retort. It gets dried in the first stage followed by carbonization in the second stage and cooling in the third stage. Pyroligneous gases are taken out from the carbonization zone and burned with controlled air supply in the drying zone. For cooling, gases are removed from the carbonization zone and are re-injected through the bottom of the retort after the tars get cracked and gases get cooled by a water-cooling system.

Fluidized bed carbonizer converter

This is a rapid pyrolysis (short residence time) process for forming fine charcoal powder from particles. Biomass particles (less than 6 mm in size) are directed to the glowing charcoal bed, which is fluidized with an oxygenated air–gas mixture. A cyclone removes dust and pyrolytic liquid from the gases

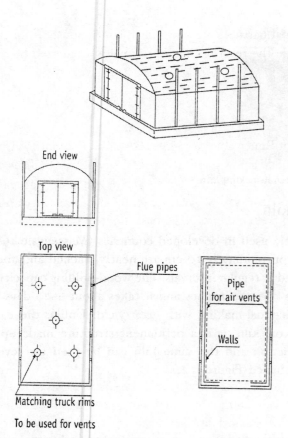

End view

Top view

Flue pipes

Pipe for air vents

Walls

Matching truck rims

To be used for vents

Figure 13.21 Missouri kiln

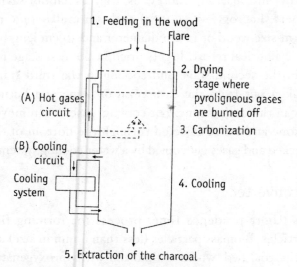

1. Feeding in the wood

Flare

2. Drying stage where pyroligneous gases are burned off

(A) Hot gases circuit

3. Carbonization

(B) Cooling circuit

Cooling system

4. Cooling

5. Extraction of the charcoal

Figure 13.22 Lambiotte: multi-stage pyrolyser retort

for further recovery. Charcoal particles are withdrawn continuously from the overflow pipe situated inside the reactor and are normally briquetted before further use (Figure 13.23).

Flash carbonization

Professor Michael Antal Jr at Renewable Resources Research Laboratory, University of Hawaii, has developed a new pyrolysis process of flash carbonization to convert green biomass waste into liquid fuel and solid charcoal (Várhegyi, Szabó, and Antal 1994). This process involves the ignition of flash fire at an elevated pressure in the packed bed of green biomass. Because of the elevated pressure, the fire spreads quickly throughout the biomass bed, triggering the transformation of biomass into biocarbon within a short time span of 30–45 minutes. In order to minimize carbon monoxide and other pollutant emissions, an after burner is provided where temperature levels of 1200 °C are maintained (Figure 13.24).

Charring drum pyrolyzer

For use in developing countries, a simple charring drum had been used to carbonize loose biomass, which can later be briquetted for densification to use it as fuel for charcoal substitution. A used oil barrel/drum of 200-litre capacity is

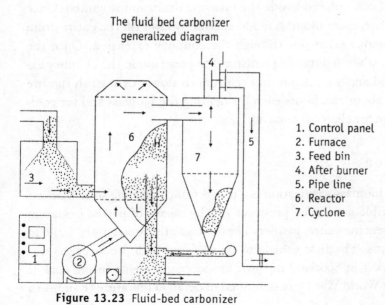

The fluid bed carbonizer
generalized diagram

1. Control panel
2. Furnace
3. Feed bin
4. After burner
5. Pipe line
6. Reactor
7. Cyclone

Figure 13.23 Fluid-bed carbonizer

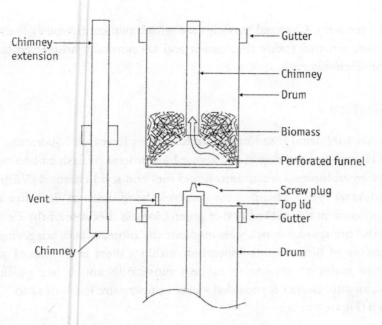

Figure 13.24 Schematic of drum pyrolyser

used for this purpose (Figure 13.24). Loose biomass is placed inside it and the air is distributed either by providing controlled air entries at various levels (TERI 1990) or by placing an inverted perforated cone inside the drum with the chimney (Neinhuysm 2003). Loose dry biomass is placed around the inverted perforated cone placed inside the charring drum and is ignited. Once the layer catches fire, more biomass is added layer by layer till the entire drum is filled. White smoke vents out through the chimney extension. Once the smoke turns from white (containing moisture) to grey colour, the chimney extension is removed and a lid is put on the drum to slowly extinguish the fire inside it. It takes about two hours each for igniting the biomass and for cooling down the drum for charcoal removal.

Biomass gasification

Thermochemical biomass gasification is a process of converting solid biomass fuel into combustible gas (called producer gas) by means of partial oxidation carried out in a reactor called gasifier. The first gasifier units were built in France during 1850s. The first vehicle to be powered with producer gas was made by J W Parker in Scotland in 1901. Producer gas plants (using coal or peat), used during World War I, are described in detail in literature (Rambrush

1923). In 1921, George Imbert attracted considerable attention in France by driving a car equipped with a gasifier from Strassbourg to Paris. Despite a number of public rallies, a public test run by the French government, automobile club, and research institutes, little interest was generated in gasifiers. The application of gasifiers assumed significance and attention during World War II due to the scarcity of petroleum products. By the end of the war in 1945 more than a million vehicles were in operation using gasifiers throughout the world (Kaupp and Goss 1987). However, after the end of the war, they were largely decommissioned as petroleum products once again became widely available at cheap rates.

The energy crisis in 1970s brought renewed interest in gasification. The technology was perceived as a relatively cheaper option in developing countries having sufficient sustainable biomass for small-scale industrial as well as power-generation applications. In the beginning of the 1980s, more than 10 European manufacturers offered gasifier-based power plants with capacities up to 250 kW$_e$. Developing countries like Brazil, India, Indonesia, and the Philippines started a gasifier implementation programme based on locally developed technologies. Gasification of biomass looks simple in principle and many types of gasifiers have been developed. The production of gaseous fuel from solid fuel offers easy handling, better control on combustion, and the possibility of using it in internal combustion engines for shaft power or electricity production which makes gasification very appealing, especially for small decentralized options. However, biomass fuels used in gasifiers vary widely in physical and chemical properties, making gasifier design much more complicated than what was envisaged by early developers.

Principle of gasification

Biomass gasification occurs through a sequence of complex thermochemical reactions. In the first stage, partial combustion of biomass to produce gases and char occurs along with the generation of heat. This heat is utilized in the drying of biomass to evaporate its moisture as well as for pyrolysis reactions to bring out volatile matter and to provide the heat energy necessary for further endothermic reduction reactions to generate producer gas. This gas consists of a mixture of combustible gases such as carbon monoxide, hydrogen, and traces of methane and other hydrocarbons. Normally, air is used as a gasifying agent; however, the use of oxygen can produce higher calorific value gas but is not usually preferred due to the cost implications.

Gasification reactions

Gasification is a complex thermochemical process, which is difficult to understand. Splitting the gasifier into strictly distinct zones is not realistic, but nevertheless conceptually essential. Various gasification reactions occur simultaneously in different parts of the gasifier. The broad stages involved in gasification are described below.

Drying

Biomass fuels contain moisture ranging from 5% to 35%. At a temperature of about 100 °C, water from the fuel gets converted into steam. Fuels do not experience any kind of decomposition during the drying process.

Pyrolysis

Pyrolysis is the thermal decomposition of biomass fuels in the absence of oxygen. It involves release of three kinds of products, namely, solid charcoal, liquid tars, and gases whose proportion depends on the fuel type and the prevailing operating conditions. It is to be noted that in any gasifier system, there will be always a low temperature zone in which pyrolysis takes place generating condensable hydrocarbons (normally called tars).

Oxidation

Air introduced in the oxidation zone contains inert gases such as nitrogen besides oxygen and water vapour, which are considered to be non-reactive with fuel constituents during the gasification process. The oxidation takes place at about 700–1400 °C . A heterogeneous reaction takes place in the oxidation zone between solid carbonized fuel and oxygen in the air producing carbon dioxide and releasing a substantial amount of heat.

$$C + O_2 = CO_2 + 406 \text{ kJ/g.mole}$$

This formula means that 12 kg of carbon burns completely with 32 kg of oxygen to produce 406 MJ of heat. The plus sign indicates the release of heat energy during the reaction. The hydrogen in the fuel reacts with the oxygen in the air blast producing steam.

$$H_2 + \tfrac{1}{2}O_2 = H_2O + 242 \text{ kJ/g.mol}$$

Reduction

In the reduction zone of all types of gasifiers, a number of high-temperature chemical reactions take place in the absence of oxygen or under a reducing atmosphere. The principal reduction reactions that take place are:

Bouduard reaction

$$CO_2 + C = 2CO - 172.6 \text{ kJ/g.mole}$$

Water–gas reaction

$$C + H_2O = CO + H_2 - 131.4 \text{ kJ/g.mol}$$

Shift reaction

$$CO_2 + H_2 = CO + H_2O - 41.2 \text{ kJ/g.mol}$$

Methane production reaction

$$C + 2H_2 = CH_4 + 75 \text{ kJ/g.mol}$$

The sequence of these reactions in a typical downdraft gasifier is shown in Figure 13.25. The equilibrium composition of a given solid fuel depends upon the air supply per unit weight of the biomass. A dimensionless parameter, known as ER (equivalence ratio), is applied to characterize the air supply conditions, and is usually defined as follows.

$$ER = \frac{(\text{Weight of oxygen/weight of dry fuel})_{actual}}{(\text{Weight of oxygen/weight of dry fuel})_{stoichiometric}}$$

The denominator in the above equation is the oxygen required for complete combustion of the fuel, and it varies from fuel to fuel. The mole fraction (or volume fraction) of the various components of the producer gas as a function of ER is shown in Figure 13.26. It is generally observed that for effective gasification, the ER should be in the range 0.2–0.4. If the ER value is less than 0.2, pyrolysis predominates the process and if it is above 0.4, combustion predominates (Kaupp and Goss 1984).

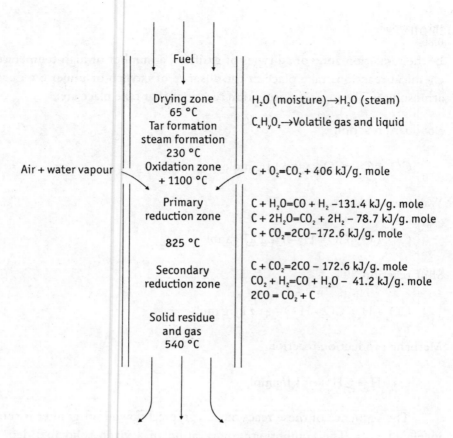

Figure 13.25 Sequence of reactions in a downdraft gasifier

Prediction of gas composition

It can be observed from the earlier discussions that the Bouduard and the water–gas reactions are the main reduction reactions requiring heat and thereby result in the lowering of the gas temperature. The water shift reaction describes the so-called water–gas equilibrium. For a given reaction temperature, the reaction between the products of concentration of carbon monoxide and water and the products of concentration of carbon dioxide and hydrogen are fixed by the value of the water gas equilibrium constant K_{WE}.

$$K_{WE} = \frac{[CO] \times [H_2O]}{[CO_2] \times [H_2]}$$

The equilibrium composition of gas can only be achieved in practice in cases having a sufficient reaction rate and reaction time. Normally, the reaction rate decreases with a fall in temperature. For example, in the case of the water–gas reaction, the reaction is said to be frozen below 700 °C as the

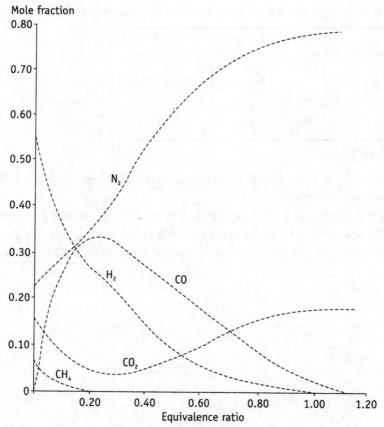

Figure 13.26 Effect of air–fuel ratio on gas composition
Reprinted from GTZ publications
Source Kaupp and Goss (1984)

reaction rate drops significantly to a low level. The typical values of K_{WE} for various reaction temperatures are given in Table 13.8.

Equilibrium gas composition from a gasifier can theoretically be calculated through the mass balance of the four major in-going elements (carbon, hydrogen, oxygen, and nitrogen) and by assuming that the water–gas equilibrium is reached at a given reactor temperature. An additional set of relations can be obtained by further assuming that a constant amount of methane is obtained in the producer gas per kilogram of dry fuel which happens under prevailing operating conditions. This can help in calculating the gas composition for a wide range of input parameter such as the fuel moisture content and the fraction of heat loss from the reactor (through conduction, convection, and radiation). Figures 13.27 and 13.28 show typical equilibrium gas composition highlighting the effect of various operating

Table 13.8 Water–gas equilibrium constant at various reaction temperatures

Reaction temperature (°C)	Equilibrium constant (K_{WE})
600	0.38
700	0.62
800	0.92
900	1.27
1000	1.60

Source FAO (1986)

conditions like the fuel moisture content and the fraction of heat loss from the reactor, respectively.

In a real gasifier, however, the composition of the gas differs greatly from the equilibrium value and is generally dependent on the gasifier design. The factors affecting the gas composition are temperature distribution in the fuel

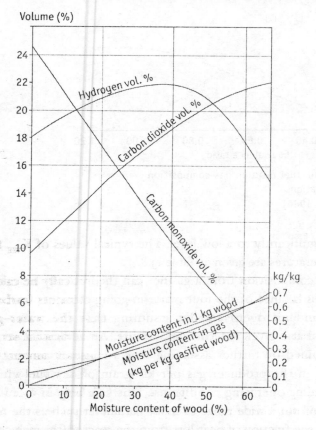

Figure 13.27 Effect of wood moisture content on producer gas composition
Reproduced with permission from FAO
Source FAO (1986)

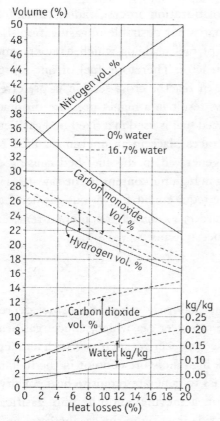

Figure 13.28 Effect of heat losses on producer gas composition
Reproduced with permission from FAO
Source FAO (1986)

bed, average gas residence time, and the residence time distribution. These are, in turn, dependent upon the mode of air entry, dimensions of the gasifier, and the quantum of heat loss to the surroundings. The modelling of the various processes occurring in the gasifier requires not only knowledge of kinetics but also an understanding of the heat and mass transfer processes occurring in various zones of the gasifier. A brief description of some important factors that affect the quality of producer gas generation in gasifiers is given below.

Fuel size

Fuel size affects the fuel movement within the reactor as well as the rate of reaction and the energy intensity per unit volume. Large wood pieces provide a smaller surface area per unit volume of the reactor, which, in turn, affects the quality of gas as volatilization or pyrolysis becomes less intense. Too small a size of biomass leads to an intense volatilization process leading to the formation of significant pyrolytic liquid that is not desirable in gasification. Larger fuel sizes also increase the chances of fuel bridging which hampers smooth fuel movement within the gasifier reactor. Therefore, fuel of one-fourth or one-fifth of the smallest dimension of reactor cross-section is preferred to avoid fuel bridging. Smaller fuel sizes result in a higher pressure drop across the fuel-bed in the reactor. If pulverized fuel is used in a fixed-bed gasifier, the fuel movement is not uniform and takes place in an uneven manner. Many times, tunnelled gas pathways are generated in the fuel bed causing a tar carryover without cracking, resulting in high tar content in the raw gas. Figure 13.29 shows the range of particle size requirement for various gasifier reactor types. The top three designs in Figure 13.29, namely counter-current, cross-flow and co-current pertain to fixed bed systems.

Fuel moisture content

With high moisture content, the net calorific value of the fuel decreases and so does the calorific value of the producer gas, which reduces gasification efficiency. Also, the tar fraction in the producer gas increases with an increase in the moisture in the biomass. With high moisture content in the fuel, more air is required to combust the biomass to generate requisite heat for drying–heating of biomass as well as for achieving endothermic gasification reactions. As a result, the carbon dioxide fraction in the gas increases and the

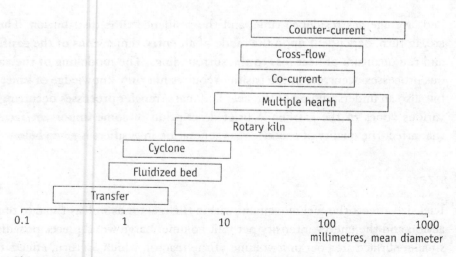

Figure 13.29 Fuel size requirements for various gasifier reactor types

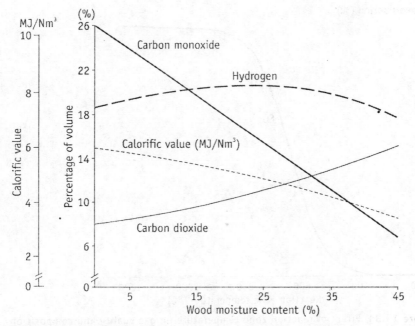

Figure 13.30 Effect of fuel moisture on producer gas composition and quality
Reprinted from GTZ publications
Source GTZ (1986)

carbon monoxide reduces. Also, it becomes difficult to maintain high fuel-bed temperatures with biomass having very high moisture content. The hydrogen percentage in the gas increases slightly initially till about 20% moisture content, but then starts reducing again (Figure 13.30).

Fuel-bed temperature

A high fuel-bed temperature (above 800 °C) is preferred to achieve a high carbon conversion of biomass and low tar content in the resultant product gas. Figure 13.31 shows the importance of the presence of an active char bed and higher temperature in the reactor zone for a theoretical situation in which carbon dioxide reacts with char to produce carbon monoxide. As temperatures increase, CO_2 fraction will decrease and CO fraction will increase. In reality, however, the shift reaction does not allow for zero levels of CO_2. Temperature affects not only the amount of tar formed but also the composition of tar by influencing the chemical reactions involved in the gasification. High temperatures help in achieving tar cracking under a reducing environment. It is reported that higher temperatures favour the formation of fewer aromatic tar species without substituting groups like benzene, naphthalene, etc.

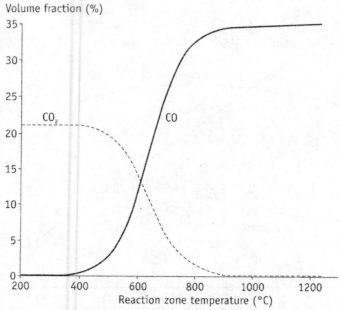

Figure 13.31 Effect of reaction zone temperature on gas quality and composition
Reprinted from GTZ publications
Source Kaupp and Goss (1984)

(Kinoshita, Wang, and Zhou 1994). Besides, the temperature also influences the formation of ammonia and nitrogen, the levels of which depend on various temperature-dependent thermochemical reactions occurring inside the gasifier reactor. With sawdust fuel, more than 50% reduction in ammonia was reported when the gasifier reactor temperature was increased from 750 to 900 °C (Wang, Padban, Ye, *et al.* 1999; Zhou, Masutani, Ishimura, *et al.* 2000). However, several other factors like moisture content and properties of the feed material limit the gasifier operating temperature that largely influences the entire gasification process. Figure 13.32 shows a typical temperature range for different feed materials that influence various important factors such as the gas heating value, char conversion, and the risk of sintering apart from tar content in the producer gas.

Residence time and gas superficial velocity

Often the effect of residence time is overlooked while designing a gasifier. In order to generate good quality gas at various fuel burning rates (or gas-producing rates), there is a need to ensure sufficient time for the completion of gasifier reactions within the reactor zone. Gas residence times of 0.1–0.5 s have been used for designing the combustion–reduction zone lengths.

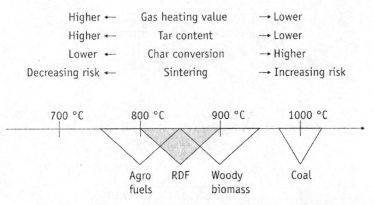

Figure 13.32 Typical gasification temperatures for various feedstocks and their influence on some critical factors

The SV (superficial velocity), is one of the most important design parameters controlling the gas energy content, production, or fuel consumption rate as well as the tar and char production rate. The gasifier superficial velocity V_s is defined as

$$\text{Superficial velocity} = \frac{\text{Gas production rate}}{\text{Cross sectional area}} \quad (\text{Nm}^3/\text{h m}^2)$$

Other terms which are used to describe the superficial velocity are hearth load and specific gasification rate, expressed as kg/(h)(m²) or Nm³/(cm²)(h). It controls the rate at which air and then gas pass through the gasifier reactor which, in turn, has an effect on the heat transfer around each particle during flaming pyrolysis of volatiles, combustion, tar cracking, and charcoal gasification. At low superficial velocities, particles are heated slowly to a low pyrolysis temperature (about 600 °C). These particles essentially remain isothermal producing more charcoal and large quantities of volatile tars. At high superficial velocities, the outer surface of particles may become incandescent (>800 °C) and still pyrolysis can continue at a cooler particle centre. Thus, the escaping gases can react with charcoal (simultaneous pyrolysis and gasification), reducing the charcoal yield and increasing the gas yield with lower tar content (Figures 13.33 and 13.34).

For heat applications, the tars in the producer gas are useful fuels having a high calorific value (provided they are not allowed to condense before reaching the burner). Thus, a gasifier with low SV as in the case of inverted downdraft gasifier stove can provide a simple design for cooking applications and can also produce more charcoal. However, for power generation applications, the gasifier needs to be operated with a high SV for minimizing tar and charcoal production. Since SV is independent of gasifier size, it can also be a

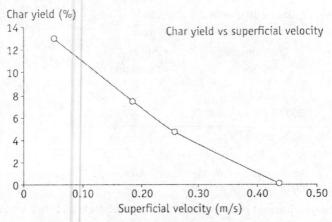

Figure 13.33 Charcoal yield from gasifier as function of superficial velocity
[Reproduced with permission from Biomass Energy Foundation]
Source Reed, Walt, Ellis, *et al.* (1999)

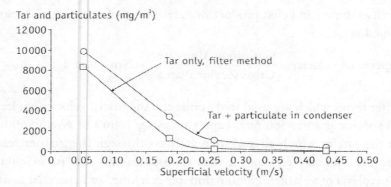

Figure 13.34 Tar yield from gasifier as function of superficial velocity
[Reproduced with permission from Biomass Energy Foundation]
Source Reed, Walt, Ellis, *et al.* (1999)

good parameter for comparing the performance of gasifiers with different dimensions.

Ash content and ash fusion temperatures

Mineral content in the fuel that remains in oxidized form even after the completion of combustion is generally called ash. However, in the case of gasifiers, the ash also contains some amount of unburnt fuel in char form. Thus, biomass having a high ash content lowers the amount of energy available from the gasifier and also requires a larger volume for accommodating ash or frequent removal of ash during gasifier operations. Materials with high ash content are problematic to handle as they can lead to the problem of clinker forming in the fuel-bed, especially if the ash has a low fusion point. Also, several inorganic matters such as mud, sand, and grit are picked up during collection and harvesting. It is not unusual considering the practice of handling agro-based biomass. This may not pose any problems in combustion devices/furnaces but can cause serious blockage of the entire reactor due to ash fusion inside at higher temperatures, thereby hampering the fuel movement and the gasification process. The common belief that wood is low in ash is not correct and there are several types of wood and agro-residues that have a higher ash content than even some varieties of coal. Based on the ash content, biomass can be classified as low ash (<5%), medium ash (5%–10%), and high ash (>10%) material (Table 13.9).

The melting point of ash is more responsible for the slagging behaviour of the fuel and the degree of slagging is more severe in biomass fuels having a high ash content. High-ash content biomass residues such as rice husk and similar fuels having a high ash fusion temperature generally do not pose the problems of clinker formation. However, high ash content retards heat and mass transfer rates, resulting in incomplete conversion of carbon in the case of insufficient residence time within the reactor and/or lack of agitation of the

Table 13.9 Biomass classification based on ash content (%)

Low ash type	Ash content	Medium ash type	Ash content	High ash type	Ash content
Powdery material					
Eucalyptus sawdust	0.4	—	—	Industrial bamboo dust	9.8
Saw-mill dust	1.3	—	—	Jute dust	19.9
				Sugarcane leaves	9.5
				Coir pith	15.4
Granular material					
Bagasse	1.8	Arecanut shell	5.1	Sweet sorghum bagasse	20.0
Coconut shell	1.9	Cotton shell	4.6	Rice husk	22.4
Groundnut shell	3.6	Coffee husk	5.8	Tea waste	19.8
Acacia	0.6	Sugarcane leaves	7.7	(decaffeinated)	
Pine needle	1.5	—	—	MSW	27.5–70.0
Stalk-like material					
Arhar stalks	3.4	Ragi stick	7.1	Jowar stalk	9.5
Corn cob	1.2	Sweet sorghum stalk	7.4	Rice straw	20.0
Jute stick	1.2	Castor stick	5.4	Ficus	10.9
Lantana camara	3.5	Congress grass	4.2		
Mulberry sticks	2.5	(*Parthenium hystophorus*)			

Source Iyer, Rao, and Grover (2002)

fuel bed. On the other hand, fuels such as corncobs, groundnut shell, and coconut coir have a low to medium ash content but also have low ash deformation temperatures in the range of 800–1200 °C that can lead to clinker formation. Materials such as cotton stalk, arhar stalk, acacia, bamboo, and pine needle are more suitable as gasifier fuels due to their low to medium ash content and very high ash deformation temperatures (Table 13.10). Methods of determining ash deformation and ash fusion temperatures are described in Iyer, Rao, and Grover (2002).

Gasifier reactor types

While the various reactors used for biomass gasification can be classified in many different ways, the density factor (ratio of dense biomass phase to total reactor volume) is a simple and effective method of classification. Thus, gasifiers can be classified as
- dense phase reactors and
- lean phase reactors

  In dense phase reactors, such as fixed bed reactors (updraft, downdraft, cross-draft, etc.), the biomass or feedstock occupies maximum reactor volume with typical density factors of around 0.3–0.8. On the contrary, in lean phase reactors, such as fluidized bed reactors, the biomass occupies very little reactor volume—of the order of 0.05–0.2 (Table 13.11).

Table 13.10 Ash fusion and deformation temperatures of biomass

Biomass fuel	Ash deformation temperature (°C)	Ash fusion temperature (°C)
Arhar stalk	1250–1300	1460–1500
Bagasse	1300–1350	1420–1450
Coffee husk	950	1020
Coir pith	850	930
Cotton stalk	1320–1380	1400–1450
Corn cob	800–900	950–1050
Groundnut shell	1180–1200	1220–1250
Jute stick	1300–1350	1400–1450
Kikar (acacia)	1300–1350	1380–1400
Mustard shell	1350–1400	1400–1450
Pine needle	1250–1300	1350–1400
Rice husk	1430–1500	1650–1680
Rice straw	1350	1450

Source Iyer, Rao, and Grover (2002)

Table 13.11 Comparison of dense and lean phase gasifier reactors

General advantages

Dense phase gasifiers
- High carbon conversion
- Low ash carryover
- Large residence time of solids
- Relatively simple construction

Lean phase gasifiers
- Very good gas–solid contact and mixing
- High throughputs
- High specific capacity

Specific advantages

Updraft
- Low gas outlet temperature
- High thermal efficiency
- Suitable for direct firing

Downdraft
- Low tar yield

Fluidized bed
- Good temperature control
- Low-medium tar yield
- Tolerates variations in fuel quality
- Can operate at partial load
- Easily started and stopped

Entrained flow
- Produces low-tar gas and little methane
- High feedstock utilization due to high reaction rates

General disadvantages

Dense phase gasifiers
- Low specific capacity
- Poor turn-down capability
- Uniformily sized feedstock required with minimum of fines
- Ash fusion and clinker formation on grate

Lean phase gasifiers
- Low feedstock inventory
- Extensive particulates clean-up required

Specific disadvantages

Updraft
- High tar yield

Downdraft
- Unsuitable for high moisture fuels

Fluidized bed
- Carbon loss with ash
- Low operating temperature

Entrained flow
- High outlet gas temperature
- Slagging

Dense phase reactors

These are the most common gasifier types having a long history and were the first to be produced commercially due to the advantage of simplicity in construction as well as operation. One of the important characteristics of these reactors is that they have fairly distinct reaction zones within the reactor such as drying, pyrolysis, combustion, and reduction. However, the relative position of these zones varies with the fuel characteristics and operating conditions.

Updraft or counter-current gasifier

In the counter-current moving bed reactor, also called the updraft gasifier, the air flows counter to the downward fuel flow and enters into the gasifier from below the grate and flows in the upward direction within the gasifier. The different zones for partial combustion, reduction, and distillation in an updraft gasifier are shown in Figure 13.35. The gas produced in the reduction zone leaves the gasifier reactor together with the pyrolysis products and the steam from the drying zone. The resulting combustible producer gas is rich in hydrocarbons (tars) and, therefore, has a relatively higher calorific value. Therefore, it is more suitable for thermal applications, such as direct heating in industrial furnaces, as it gives higher operating thermal efficiencies. If it is to be used for electricity generation by IC (internal combustion) engines, it has to be cleaned thoroughly.

In some updraft gasifiers, steam is injected or evaporated into the hot partial combustion zone, which has a beneficial effect on gas quality and also helps in preventing the lower portion of the gasifier from overheating. Generally, sensible heat of the producer gas or radiative heat from the gasifier shell is used for generating the necessary amount of steam.

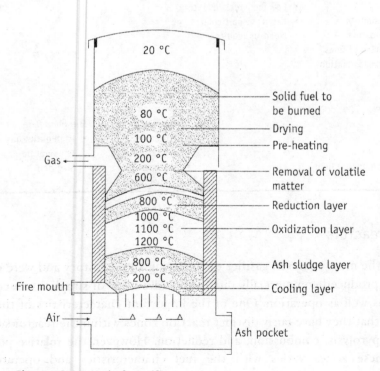

Figure 13.35 Updraft gasifier

The characteristics of the updraft gasifier are as follows.

- It has reasonably well defined zones for various reactions.
- Its efficiency is high because hot gases pass through the entire fuel bed and leave at lower temperatures. The sensible heat of hot gas is used for the reduction, pyrolysis, and drying processes.
- Products from pyrolysis and drying, containing water vapour, tar, and volatiles, leave the gasifier without passing through high temperature zones and, therefore, do not get cracked. Hence, elaborate gas cleaning is required before use, making it less suitable for engine operation.
- Unsuitable for high volatile fuels such as cashew nut shells.

Downdraft or co-current gasifier

In the co-current moving bed reactor or downdraft gasifier, the air enters at the middle level of the gasifier above the grate and the resulting gases flow down cocurrently through the gasifier reactor. All the decomposition products from the pyrolysis and drying zones are forced to pass through the oxidation zone (Figure 13.36). This leads to thermal cracking of the volatiles, resulting in reduced tar content in the producer gas. Therefore, it is very attractive to use this gas for engine applications. There is a constriction in the older designs at

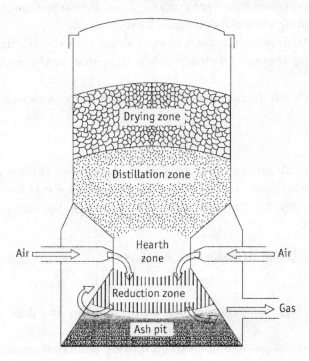

Figure 13.36 Downdraft gasifier

the level of the oxidation zone to force the pyrolysis products through a concentrated high temperature zone to achieve a high degree of cracking. This concentrated oxidation zone can cause sintering or slagging of ash, resulting in clinker formation and consequent blocking of the constricted area and/or channel formation. Continuously rotating grates or other mechanical shaking may be required to avoid this problem. Compared to the updraft gasifier, the disadvantages of a downdraft gasifier are higher gas outlet temperatures and somewhat lower thermal efficiency. Additional steam injection is not common in downdraft gasifiers. They can be operated at a very high specific gasification rate (or hearth load) in the range 2900–3900 kg/h m² which corresponds to about 1 Nm³/h m². The highest reported was 5020 kg/h m². The starting ignition time required for downdraft gasifiers is lower as compared to updraft gasifiers but is still perceived to be long as it is in the range 15–20 minutes. It also has relatively better load following capability (ability to quickly extend the partial combustion zone to produce the requisite higher gas quantity).

The characteristics of the downdraft gasifier are as follows.
- Tars and volatiles pass through the high temperature bed in the oxidation zone before leaving and hence get cracked. Thus, the output gases are relatively clean requiring a less elaborate gas-cleaning system.
- The gas leaves at relatively high temperatures of about 400–500 °C, hence the gasifier operating thermal efficiency is less than that in the updraft gasifier.
- Not suitable for high-ash, high-moisture, and low-ash fusion temperature fuels.

Cross-draft gasifier

In a cross-draft gasifier, air enters from one side of the gasifier reactor and leaves from the other side (Figure 13.37). Cross-draft gasifiers have very few applications and can hardly be credited with any advantage beyond good permeability of the bed.

The characteristics of this type of gasifier are as follows.
- Grate not required
- Single air tier
- Ash formed due to high temperature; falls to the bottom and does not hinder operation
- High exit temperature of gas and low carbon dioxide reduction results in poor quality of the gas and low efficiency

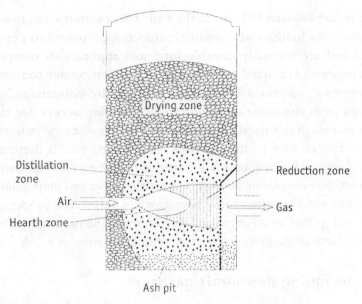

Figure 13.37 Cross-draft gasifier

- Not much research occured. However, it probably produces lesser tar than updraft gasifier and has a narrow turn-down ratio.

Fluidized bed and other lean phase reactors

Lean phase reactors have recently been adopted from pulverized coal gasifiers. In lean phase gasifiers, distinct reaction zones do not exist like in dense phase reactors. All the reactions, namely drying, oxidation, pyrolysis, and reduction take place effectively in the same region. Among the lean phase reactors, the most commonly used reactor types are the fluidized bed and entrained flow reactors. Recently, other reactor designs like cyclonic reactors have been experimented with.

Fluidized bed reactors are more attractive than the dense phase reactors due to their high heat and mass transfer rates and good mixing of the solid phase giving a uniformly high (800–1000 °C) bed temperature. The fluidized bed reactor/gasifier is essentially a hot bed of sand particles agitated constantly by the gasifying agent, steam, and inert gas. The fluidizing gas is distributed through nozzles located at the bottom of the bed. Although it has a somewhat higher throughput per unit of reactor volume than the moving bed, its main disadvantages are a high outlet gas temperature and an entrainment of charcoal fines, and it requires a complex control system because of the low biomass hold-up in the bed. These systems are, however, most appropriate for biomass whose particle sizes range from 0.1 to 1 cm (see Figure 13.29) and are normally suitable for small applications comprising high carbon loss with entrained ash. The loss of fluidization due to sintering of ash is a common problem, which can be controlled by maintaining higher bed temperatures of the order of 800–900 °C. Sintering occurs due to the agglomeration of alkali metals from biomass ash with the silica in the sand.

In the entrained flow gasifier reactors, no inert material is present but finely reduced fuelstock is required. They are normally used for large-capacity (30 tonnes/hour) fast-circulating bed gasifiers for the paper and pulp industry.

Tables 13.12 and 13.13 give typical operational data and the producer gas composition and quality in different gasifier reactors. The performance characteristics of various air gasifier reactors are compiled in Annexe 1.

Guidelines for designing downdraft gasifiers

This section gives a general review of the design characteristics of an Imbert type downdraft gasifier on the basis of the Swedish experience.

The design of an Imbert type downdraft gasifier is based on specific gasification rate, also called the hearth load G_H. It is defined as the amount of

Table 13.12 Typical operational data for different types of gasifiers

| Parameter | Downdraft | Updraft | Fluid bed | |
			Conventional	Circulating
Grate energy release (GJ/h.m²)	1.5–4	2.5–5	6–9	40
Offgas temperature, (°C)	600–800	75–150	650–850	800–900
Oils and tar (kg/kg dry feed)	0.001–0.01	0.05–0.15	0.01–0.05	—
Char loss (kg/kg dry feed)	0.02	0.01–0.02	0.02–0.05	—

Table 13.13 Producer gas characteristics from different gasifiers

| Gasifier reactor type | Gas composition, dry, vol % | | | | | HHV MJ/m³ | Gas quality | |
	H_2	CO	CO_2	CH_4	N_2		Tars	Dust
Fluid bed air-blown	9	14	20	7	50	5.4	Fair	Poor
Updraft air-blown	11	24	9	3	53	5.5	Poor	Good
Downdraft air-blown	17	21	13	1	48	5.7	Good	Fair
Downdraft oxygen-blown	32	48	15	2	3	10.4	Good	Good
Multi-solid fluid bed	15	47	15	23	0	16.1	Fair	Poor
Twin fluidized bed gasification	31	48	0	21	0	17.4	Fair	Poor

HHV – higher heating value

producer gas to be obtained per unit cross-sectional area of the throat, which is the smallest area of cross-section in the reactor. It is normally expressed in terms of $Nm^3/h\ cm^2$, where N indicates that the gas volume is calculated at normal pressure and temperature conditions. It is reported that the gasifier can be operated with G_H in the range 0.1–0.9 $Nm^3/h\ cm^2$. Normal Imbert gasifiers show a minimum value of G_H in the range 0.30–0.35, resulting in a power turndown ratio of about 2.5–3. With better insulating materials, modern gasifiers can now be operated at lower tar levels with G_H below 0.15. Based on the maximum value of the hearth load G_H, the throat diameter d_t can be calculated. Other dimensions such as the height h of the nozzle plane above the throat, nozzle area, and the diameter of nozzle top ring (see Figure 13.38 for details) can then be calculated from the graph given by the Swedish Academy of Engineering Sciences reproduced in Figures 13.39 to 13.41 (FAO 1986).

The comparison of various types of gasifiers available during the World War II indicates that the maximum specific gasification rate (or hearth load) is of the order of 0.09, 0.3, and 0.9 $Nm^3/h\ cm^2$ for 'no throat', 'single throat', and 'double throat' gasifiers, respectively (FAO 1986).

The comparison also shows the following.
- Nozzle air-blast velocities should be of the order of 22–33 m/s.
- Throat inclination should be about 45°–60°.
- Hearth diameter at air inlet should be 10 cm and 20 cm larger than the smallest cross-section (throat) in the case of single- and double-throat design, respectively.

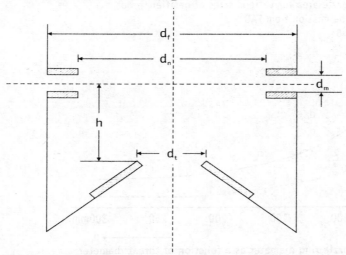

Figure 13.38 Design parameters for Imbert-type gasifiers

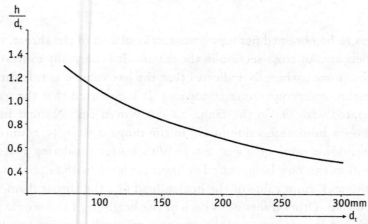

Figure 13.39 Height of nozzle plane above throat for various throat diameters
Reproduced with permission from FAO
Source FAO (1986)

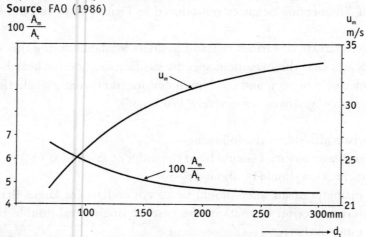

Figure 13.40 Nozzle area for various sizes of gasifier throat
Reproduced with permission from FAO
Source FAO (1986)

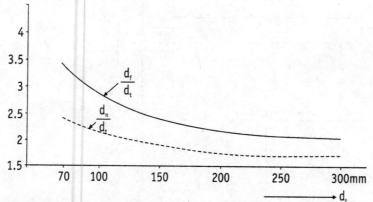

Figure 13.41 Nozzle ring diameter as a function of throat diameter
Reproduced with permission from FAO
Source FAO (1986)

- Reduction zone height should be more than 20 cm.
- Air inlet nozzle plane should be located more than 10 cm above the throat section.

It should be emphasized that all these empirical design guidelines are based on the experience with different wood gasifiers.

Example 3

Design an Imbert-type downdraft gasifier reactor zone for operating a three-cylinder four-stroke engine on 100% producer gas with the following specifications.

Piston diameter (D)	: 110 mm
Piston stroke (s)	: 100 mm
Number of cylinders (n)	: 3
Engine RPM	: 1500
Volumetric efficiency (f)	: 80%

Solution

For designing a gasifier, the first step is to find out the required gas production rate. Engine swept volume can be calculated as

$$V_s = \tfrac{1}{2} \times rpm \times N \times \tfrac{\Pi}{4} \times D^2 \times S$$

$$= \tfrac{1}{2} \times 1500 \times 3 \times \tfrac{\Pi}{4} \times 0.11^2 \times 0.1 = 2.137 \, m^3 / min = 128.23 \, m^3 / h$$

For stoichiometric air-fuel (gas) ratio 1.1:1.0, the air requirement for m^3 of gas is 1.1. Thus, if V_g is the gas intake rate, the air + fuel intake will be 2.1 V_g. Hence

$$V_g = f \times \frac{V_s}{2.1} = 0.8 \times \frac{128.23}{2.1} = 48.85 \, m^3 / h$$

For maximum hearth load G_H of 0.9 N m^3/h cm^2, the throat area A_t is

$$A_t = \frac{V_g}{G_{H\,max}} = \frac{48.85}{0.9} = 54.28 \, cm^2$$

Thus, the throat diameter d_t for circular cross-section works out to be

$$d_t = \sqrt{\frac{4 \times A_t}{\pi}} = \sqrt{\frac{4 \times 54.28}{\pi}} = 8.32 \, cm$$

Height h of the nozzle plane above the throat cross-section can be determined using the graph in Figure 13.39. $h/d_t = 1.2$ for throat diameter

$d_t = 83$ mm. Thus, h works out to be

$$h = d_t \times (h/d_t) = 83 \times 1.2 = 99.6 \, mm$$

The diameter of firebox d_f and the diameter of nozzle top ring d_n can be determined using the graph in Figure 13.41 by noting the ratio $d_f/d_t = 3.2$ and $d_n/d_t = 2.3$, respectively, for throat diameter $d_t = 83$ mm. Thus, d_f and d_n work out to be

$$d_f = 83 \times 3.2 = 265.6 \, mm$$

$$d_n = 83 \times 2.3 = 190.9 \, mm$$

Assuming that five nozzles are used for supplying the required amount of air for gasification and noting the ratio of $100(A_m/A_t)$ as 6.3 for calculated throat diameter from the graph given in Figure 13.40, the nozzle diameter can be calculated as follows.

$$A_m = 6.3 \times \frac{A_t}{100} = 6.3 \times \frac{54.28}{100} = 3.42 \, cm^2$$

$$\text{Area of each nozzle} = \frac{A_m}{5} = 0.684 \, cm^2$$

thus nozzle diameter d_m works out to be

$$\sqrt{\frac{4 \times A_n}{\pi}} = \sqrt{\frac{4 \times 0.684}{\pi}} = 0.93 \, cm = 9.3 \, mm$$

The air blast velocity u_m can also be found from Figure 13.40 as 25 m/s.

Tar formation and reduction

One of the major problem areas in biomass gasification is dealing with the tar formed during the process, namely minimizing its formation and the methods of its reduction or removal once it is formed. Tar is a complex mixture of condensable hydrocarbons and consists of both aromatic and PAHs (poly-aromatic hydrocarbons). The major tar components are toluene, naphthalene, and with process temperatures lower than 800 °C , phenol. Besides these, a number of other compounds occur as trace elements. Figure 13.42 shows the structures of some typical tar components. Macromolecular components with up to seven benzene rings may occur, particularly at temperatures over 800 °C (Spliethoff 2001). In the EU/IEA/US-DoE meeting on tar measurement protocol held in Brussels in 1999, various research groups agreed to define tar as all organic contaminants with a molecular weight

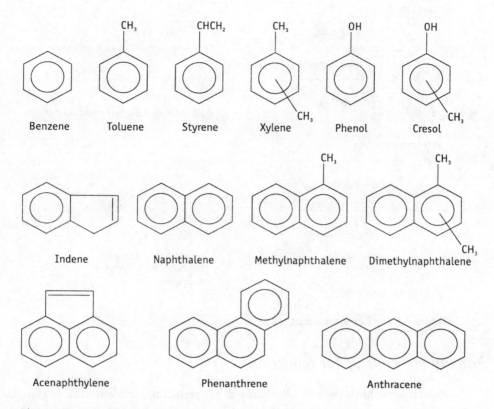

Figure 13.42 Structure of various components of tar

greater than that of benzene (Neeft, Knoef, and Onaji 1999). Tar is an undesirable product of the gasification process because of the various problems associated with its condensation, formation of tar aerosols, and polymerization to form more complex structures that cause problems in the process equipment as well as engines used in the application of producer gas.

Tar removal methods

Several methods of tar reduction/removal have been reported and can broadly be categorized into two types depending on the location where the tar is reduced.

- In gasifier itself; known as the primary methods
- Outside the gasifier; known as the secondary methods

Figure 13.43 shows the schematic of primary and secondary methods of tar removal. Thus, an ideal primary method concept will eliminate the use of secondary methods.

The following sections describe both methods with emphasis on the primary method.

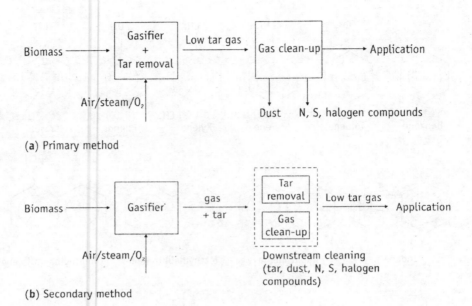

(a) Primary method

(b) Secondary method

Figure 13.43 Tar removal methods

Primary methods of tar reduction

Primary methods include all measures for reducing tar formation in the gasification process occurring in the gasifier by preventing or converting the tar formed.

Gasifier operating conditions

The operating conditions play an important role during biomass gasification to obtain good quality gas with regard to its composition and tar content. As discussed earlier, ER of about 0.25–0.3, high uniform fuel bed temperatures (above 800 °C), high superficial velocity, and sufficient residence time can help in achieving a better carbon conversion of biomass and low impurity levels in the producer gas obtained.

Recently, several researchers have investigated pressurized biomass gasification in an effort to achieve higher biomass carbon conversion and lower tar content. Almost complete elimination of phenols was observed at high pressures of the order of 21.4 bars; however, the fraction of PAH was found to increase with higher pressures (Knight 2000). A decrease in the amount of light hydrocarbons (lower than naphthalene) as well as that of tar was observed with an increase in ER for pressurized gasification with almost 100% carbon conversion (Wang, Padban, Ye, *et al.* 1999).

Bed additives

Several catalysts have been tried for tar reduction which include nickel-based catalysts, calcined dolomites, magnesites, zeolites, olivine, and iron catalysts out of which only a few have been tried as active bed additives inside the gasifier. There is a great potential of in-bed additives in terms of tar reduction as they act as in situ catalysts during several gasification reactions. Limestone was one of the first additives used in fluidized bed gasification and steam gasification, and it was found that its use could prevent agglomeration of the bed (Walawender, Ganesan, and Fan 1981). Dolomite is one of the most popular active in-bed additives studied for tar cracking in bed as well as in the secondary reactor (Karlsson, Ekstrom, and Liinaki 1994). Addition of three per cent of calcined dolomite is reported to result in 40% reduction in the tar levels (Narvaez, Orio, and Aznar 1996). Nickel-based catalysts are reported to increase the hydrogen content in the gas with considerable reduction in the methane content. They are also very effective in reducing/decomposing ammonia along with the conversion of light hydrocarbons. However, they have a major problem of fast deactivation due to carbon deposition and poisoning due to the presence of hydrogen sulphide (Wang, Ye, Padban, *et al.* 2000).

Gasifier design modification

The reactor design is very important for a gasifier system with respect to its efficiency and gas quality (composition and tar content). A two-stage gasifier design has been reported to be very effective in producing clean gas. The basic concept of this design is to separate the pyrolysis zone from the reduction zone. A two-stage gasifier is equivalent to two single-stage gasifiers. Tars formed during the first (pyrolysis) stage are decomposed in the second stage (reduction zone). Figure 13.44 represents the two-stage gasification concepts applied by AIT (Asian Institute of Technology), Bangkok, and TERI, New Delhi, resulting in significant (more than 75%–80%) reduction in tar content in the raw gas (Bhattacharya, Siddique, and Pham 1999; TERI 2005). The successful operation of this gasifier type depends on the stabilization of the pyrolysis zone which, in turn, depends on the balance between downward solid fuel movement and upward flame propagation.

An open-top gasifier design, originally proposed by Tom Reed, has been developed by the IISc (Indian Institute of Science), Bangalore, in which primary air is supplied from the open top of the gasifier and secondary air is supplied through the nozzles in the reduction zone as shown in Figure 13.45. It is claimed that the air supply from the top helps in the upward propagation of

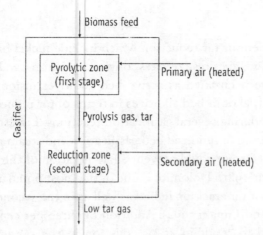

Figure 13.44 Two-stage gasification principle in TERI gasifier design

reaction front so as to get a higher residence time. It also helps in releasing the volatiles into gaseous form before entering the high-temperature zone where they get cracked to reduce the tar level (Dasappa, Mukunda, Paul, *et al.* 2003).

At DTU (Danish Technical University), another two-stage biomass gasifier design has been developed in which the gasification process occurs in two different systems connected to each other. Pyrolysis of the biomass feed occurs in the horizontal screw-feed pyrolyser, which is externally heated by the engine exhaust and heat recovered from the hot gas followed by partial oxidation of volatile products in the presence of a hot charcoal bed in

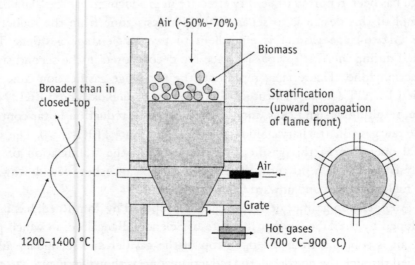

Figure 13.45 Open top gasifier design of IISc, Bangalore
Source Dasappa, Mukunda, Paul, *et al.* (2003)

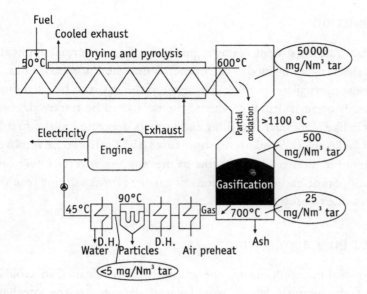

Figure 13.46 Two-stage gasification principle of DTU, Denmark
Reproduced with permission from Elsevier
Source Henriksen and Christensen (1994)

vertical reactor (Figure 13.46) (Henriksen and Christensen 1994). Very low tar (<25 mg/Nm³) is reported even in raw gas in this type of gasifier design.

Secondary methods of tar removal

As mentioned earlier, the quality of gas with regard to impurities such as tar, dust, and water vapour varies with different reactors, and even for a particular reactor type, it varies with the fuel used as well as the operating conditions. Depending on the application, the gas needs to be cleaned and conditioned. There is always a possibility of tar and vapour condensing along with the dust particles to form a semi-solid mixture, which gets deposited and blocks various filters and gasifier system components. In order to prevent this from happening, the gas is cleaned and conditioned. Normally, dust particulates are removed in hot conditions (tar and water vapour remaining in gaseous form) and then the gas is cooled to condense the tar and water vapour to form a liquid condensate, which is then trapped and removed. Further conditioning of gas includes the removal of mist and fine particles (<0.1 µm) from the gas. If the gas is to be used for chemical synthesis, it is necessary to remove all unnecessary chemicals such as heavy metals, sulphur, etc., which can poison the catalyst. The extent of treatment required varies, depending on the gas applications.

For direct combustion

Generally elaborate treatment is not required for direct thermal applications of gas as all the tar gets burned if the burner is designed properly. Thus, only removal of dust particulates is done. However, if the gas is to be used directly for drying seeds or agricultural products, the tar has to be removed, else the seed layer acts like a granular bed filter and tar gets deposited in the first layer (Barrett and Jacko 1984). Similarly, in some cases, the tar has to be removed in pottery or ceramic industry applications to prevent any adverse effect on the product quality. Some earlier applications of gasifiers for tea drying resulted in tainting of tea due to presence of tar.

For engine or turbine application

For engine or turbine applications, the gas needs to be cleaned to avoid condensation of the deposit which would cause damage to the mechanical components moving at high speed. There is no consensus about what should be the gas quality limit for tar and particulates. Various manufacturers indicated that the particulate content should be less than 3 mg/Nm³ for turbines and should lie in the range 50–100 mg/Nm³ for engines. However, according to the SERI manual (translation of Swedish GENGAS manual), the particulate content ranges between 10 and 20 mg/Nm³ (Buekens, Bridgewater, Ferrero, *et al.* 1990). Normally, the gas is not allowed to cool down below 80 °C in turbines as below this temperature tar condensation can occur (dew point temperature of producer gas is reported to be about 70 °C). In order to improve the breathing or volumetric efficiency of engine, it is necessary to cool down the gas closer to ambient temperature. Generally, it is observed that for every 10 °C hike in gas temperature, the engine efficiency drops by one per cent.

Various gas treatment techniques used for high temperature particulate removal are as follows (Figure 13.47).

- Cyclones and multi-cyclones (not effective with very small-size particulate distribution, large turndown ratios and if tar droplets are present in the gas)
- Hydraulic dust removal (cleaning fluid must be a liquid)
- Electrostatic precipitator (costly and high power consumption)
- Granular bed (effective if regeneration is not required or easy to organize)
- Porous layer filters (costly but may soon become viable with technological advances in ceramics and polymers)

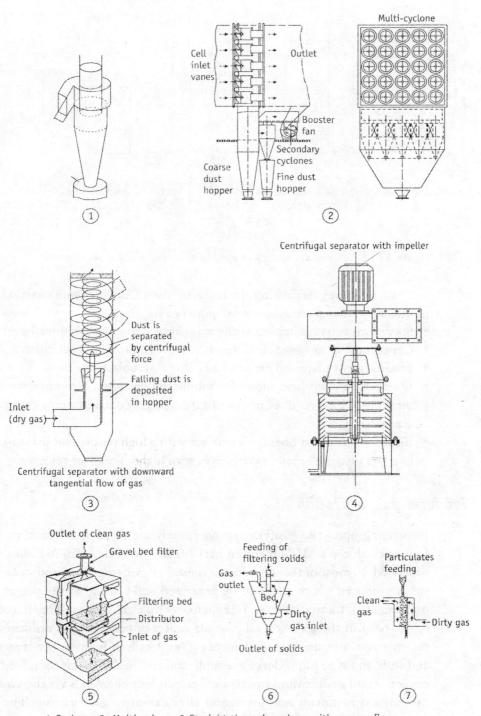

1-Cyclone; 2- Multicyclone; 3-Straight through cyclone with reserve flow;
4-Centrigugal separator with impeller; 5-Fixed bed filters;
6-Fluidized bed filters; 7-Mobile bed filters

Figure 13.47 Various gas treatment techniques for removal of particulates

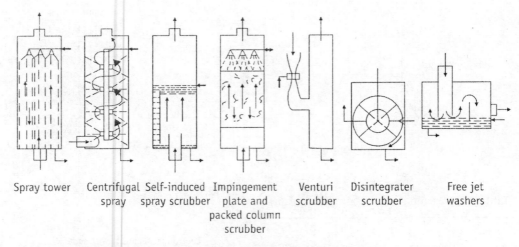

Spray tower   Centrifugal   Self-induced   Impingement   Venturi   Disintegrater   Free jet
               spray   spray scrubber   plate and   scrubber   scrubber   washers
                                      packed column
                                       scrubber

Figure 13.48 Various wet scrubbers used in gas treatment for tar removal

Various wet scrubbing techniques used for tar condensation and particulate removal are as follows (Figure 13.48).

- Spray towers (very simple, excellent for large particulate removal)
- Centrifugal spray towers (efficiently removes up to 1 μm particulate)
- Impingement plate and packed-bed column scrubbers
- Disintegrator scrubber (good for submicron particles, high energy requirement: 5–7 kJ/m³, requires pre-purification to get concentrations <2–3 g/m³ of size <10 μm)
- Ejector venturi scrubber (efficient but with a high permanent pressure drop of 0.4–1 kPa, the efficiency improves with higher liquid jet velocity.

Producer gas utilization

Depending upon the gasifying agent, namely, air or oxygen/steam, an LHV (calorific value 3–7 MJ/Nm³) or an MHV (calorific value 12–16 MJ/Nm³) gas is produced. In most of the gasifiers, air is used as gasifying agent and so an LHV gas is produced which needs to be generated and utilized at the site as compressing and transporting it is expensive due to high nitrogen content (50%–55%) in the gas. Several possible routes of producer gas utilization are summarized in Figure 13.49. Table 13.14 gives details of biomass pre-treatment, and scale and type of producer gas applications. The MHV gas (obtained from oxygen/steam gasification) can be used to produce chemicals via the synthesis gas (carbon monoxide and hydrogen) after extensive gas cleaning. The units for producing ammonia from biomass gasification tend to be too large to make it economically viable and are not expected to gain attention in near future.

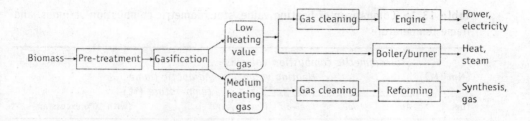

Figure 13.49 Possible routes for producer gas utilization

Table 13.14 Producer gas (air gasification) applications

Biomass pretreatment			Gasifier reactor	Type of application			Scale of application		
							<100 kWe	100–500 kWe	>1 MW
Simple	Medium	Extensive		Boiler	Burner	Engine			
¤			Downdraft	¤	¤	¤	¤	¤	
	¤		Open core	¤	¤	¤	¤	¤	
¤			Updraft	¤	¤		¤	¤	
	¤		Fluidized bed	¤	¤	¤		¤	¤
		¤	Entrained bed	¤	¤				¤

Direct combustion

Various gas characteristics (physical and chemical parameters) and velocity affect the performance of the gas in the burner. These include the following.

- Gas density as compared to air density
- Heating value of gas
- Stoichiometric air requirement (for complete combustion)
- Flame-speed coefficients of gas constituents

A variation in these parameters can result in a flame lift off (excessive gas velocity resulting in the loss of contact with the burner), internal burning of flame back (lower gas velocity than flame propagation resulting in internal burning), or unstable flame. Flame back can be dangerous, as the flame starts moving upstream in a pre-mixed gas–air mixture, leading to explosion. This can be prevented by a flame arrester but can result in extinguishing the flame. While changing the type of gaseous fuel used in a gas burner, there is a need to give attention to the calorific value (heating value), stoichiometric air requirements, and its effect on the adiabatic flame temperature as well as the flue gas volumes. Table 13.15 gives a comparison for the usage of LHV, MHV, and natural gas in the burner.

Table 13.15 Comparison of heating value, stoichiometric combustion volumes, and flame temperature

(Nm³/MJ)	Stoichiometric combustion volumes		Heating value (MJ/Nm³)			Adiabatic flame temperature (°C)
Fuel	Air	Fuel	Fuel+air	Flue gases		(with 10% excess air)
LHV	0.20	0.231	0.43	0.39	4.3	1480
MHV	0.20	0.104	0.30	0.27	10.0	1870
Natural gas	0.25	0.027	0.28	0.28	38.0	1860

It can be observed from Table 13.15 that for the generation of 1 MJ, the producer (LHV) gas requirement is 8.5 times higher than the natural gas. As a result, larger amounts of flue gases are generated when producer gas is used instead of natural gas and, therefore, modifications in flue gas paths might become necessary for compensating higher head losses. A lower flame temperature obtained with producer gas can have limitations on its use and at higher temperatures like metal melting or cement kilns, dual firing may be necessary. Recently, some successful attempts have been made to increase the flame temperature by supplying pre-heated air into the gas for high temperature applications like crematoria (Mande, Lata, and Kishore 2001) and steel re-rolling.

Combustion in an engine

While using producer gas in an engine, combustion limits, knock resistance and compression ratios are important parameters apart from the heating value and the stoichiometric ratio, as discussed earlier. The combustion limits of any fuel–air mixture are the two mixing ratios between which it is possible to ignite the mixture by means of an ignition source (spark in the case of an Otto engine), and the power output depends on the energy content of the fuel-air mixture per unit volume and the average cylinder pressure. The mean effective pressure also depends on the compression ratio and ignition timing. Higher the compression ratio, higher is the thermodynamic cycle efficiency. The compression ratio is normally limited by knowing the tendency of the fuel in the engine under operating conditions. Knocking is an untimely, jerky undesirable combustion process leading to a sharp rise in pressure (up to 50 000 bars/s) and oscillations giving rise to a hammering sound causing mechanical damage to the engine. Continued knocking results in reduced output, overheating, and finally mechanical damage or piston jam/seize. The knocking tendency is highly influenced by the change in the compression ratio and ignition timing. As a thumb rule, a reduction in the compression ratio by 1 lowers the octane requirement of Otto (spark ignition or SI) engine by 10–15 octanes and increases the start cetane requirement for diesel CI (compression ignition) engine by 3–10 cetane units.

When producer gas is used in an SI engine, the ignition time has to be advanced to gain maximum heating value of slow-burning carbon monoxide (flame speed of 0.52 m/s compared to 2.83 m/s for hydrogen) which, in turn, reduces the mean effective pressure and hence there is a reduction in power output resulting in derating of the engine. Due to the lower compression in the SI engine (about 6–8 as compared to 16–17 in the case of a CI engine), the derating in a SI engine with LHV producer gas is much higher, of the order of 30%–40%, as engine cycle efficiency decreases with lower compression ratio (refer to the section on thermodynamics in Chapter 3). Excessive advancement of ignition timing can at times cause backfiring due to the presence of hot spots. However, producer gas has a lower tendency for knocking due to its LHV and presence of a large fraction of inert nitrogen and carbon dioxide.

Therefore, normally diesel engines are preferred for the use of gas in engine operation as they operate with a higher compression ratio and excess air levels, resulting in lowering of the derating (of the order of 10%). Existing diesel engines can easily be operated on dual fuel (diesel + gas) mode with 15%–30% pilot diesel injection for ignition. Also, it gives the option of an easy switch back to 100% diesel operations and the governor system of diesel injection takes care of fluctuations in operating load conditions (Mande 2005).

Recently, development of modified diesel engines capable of operating on 100% producer gas has been initiated at several research institutes. The engine modifications basically involve

- modifying piston and/or cylinder head to reduce the compression ratio;
- replacing diesel injectors with spark plugs;
- using a diesel pump governing mechanism for spark distribution; and
- adjusting the spark ignition timing.

Inverted downdraft gasifier stove

Dr T B Reed developed a very small gasifier meant to act on an efficient cook stove, called the IDD (inverted downdraft) natural convection gasifier stove (Figure 13.50). The stove has a high efficiency – of the order of 30%–35%. He later worked on sizing the stove and trying to use it for charcoal making. After much testing and many publications, but with no real success in applications, the work was stopped in 1995. However, later in 1998, Dr Reed began work on

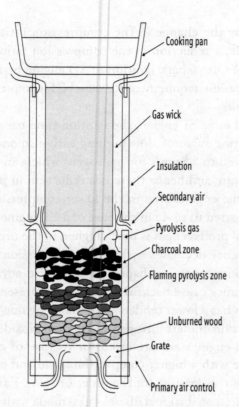

Figure 13.50 Natural draft gasifier stove operating on inverse down draft principle [Reproduced with permission from the Biomass Energy Foundation]
Source Reed and Walt (1999)

a smaller, forced convection model with a fan, and in 2001, a prototype forced-air gasifier stove was operated on a kitchen table. Figure 13.51 shows the schematic diagram of gasifier turbo-stove.

A series of modifications and improvizations resulted in the gasifier turbo-stove concept. These modifications included different stackable units in a heat column over a gasifier unit with an air pipe, with smaller holes for the entry of secondary air, pre-heated secondary air, a tapered chimney, and with independent structural components for the stove body. The gasifier chamber was removable and, therefore, could be emptied to save the resultant charcoal, re-loaded with biomass, re-lighted, and re-inserted into the heat column. Later in 2002, the 'Wood Gas Camp Stove' with battery-powered fan and the ability to produce an impressive flame for sustained periods was developed with forced-air designs and with the intention of making a stove for the affluent North American camper market. Variations of these stoves, have been developed and are being disseminated in Sri Lanka and India.

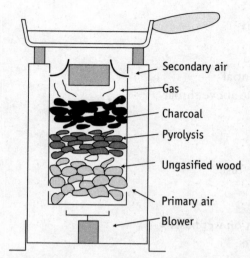

Figure 13.51 Gasifer turbo-stove
Source Reed and Walt (1999)

Nomenclature

A	Temperature of combustion air
A_m	Total nozzle area
A_n	Nozzle area
A_t	Throat area
BFB	Bubbling fluidized bed
C_{pv}	Specific heat of vessel
C_{pw}	Specific heat of water
CI	Compression ignition
CV	Calorific value of fuel
CCT	Controlled cooking test
CFB	Circulating fluidized bed
D	Diameter
d_f	Firebox diameter
d_m	Nozzle diameter
d_n	Diameter of nozzle top ring
d_t	Throat diameter
dW	Amount of water evaporated
EA	Excess air (percentage)
ER	Equivalence ratio
f	Volumetric efficiency

F	Amount of fuel burnt
G_H	Hearth load
G_{Hmax}	Maximum hearth load
GWP	Global warming potential
h	Height of nozzle plane above throat
HHV	Higher heating value
IDD	Inverted downdraft
K_{WE}	Equilibrium constant
KPT	Kitchen performance test
L_w	Latent heat of water
LHV	Lower heating value
MC_g	Gas moisture content on wet basis
MHV	Medium heating value
N	Number of cylinders
NTP	Normal temperature and pressure
PAH	Poly-aromatic hydrocarbons
RPM	Revolutions per minute
S	Stroke
SI	Spark ignition
SV	Superficial velocity
SGR	Specific gasification rate
T_a	Ambient temperature
T_{ad}	Adiabatic temperature
T_b	Boiling point
T_s	Flue gas temperature
u_m	Air blast velocity
V	Weight of water
V_g	Maximum gas intake
V_s	Swept volume
VS	Gasifier superficial velocity
W	Water quantity
WBT	Water boiling test
λ	Excess air factor
ϕ	Diameter
η	Efficiency
η_o	Overall efficiency
η_c	Combustion efficiency
η_h	Heat transfer efficiency
η_p	Pot efficiency

References

Ahuja D R, Joshi V, Smith K R, Venkatraman C. 1987
Thermal performance and emission characteristic of unvented biomass burning cookstoves: a proposed standard method for evaluation
Biomass 12: 247–270

Bailis R, Damon O, and Still D. 2004
The Water Boiling Test Version 1.5
[With input from Kirk Smith and Rufus Edwards]
Household Energy Health Programme, Shell Foundation

Ballard–Tremeer G and Jawurek H H. 1999
Evaluation of the dilution chamber method for measuring emissions of cooking devices
Biomass and Bioenergy 17, pp. 481–494

Barrett J R and Jacko R B. 1984
Environmental aspects of biomass furnaces used in agriculture—air pollution and grain contamination
In *Bioenergy 84 Conference Proceedings*, vol. 4, edited by H E Grens and A Ellogard
London: Applied Science Publishers

Bhattacharya S C, Siddique A H M R, and Pham H L. 1999
A study on wood gasification for low tar gas production
Energy 24: 285–296

Bradbury A G W and Shafizadeh F. 1980a
Chemisorption of oxygen on cellulose char
Carbon 18: 109–116

Bradbury A G W and Shafizadeh F. 1980b
Role of oxygen chemisorption in low-temperature ignition of cellulose
Combustion and Flame 35: 85–89

Bryden M, Still D, Scott P, Hoffa G, Ogle D, Balilis R, Goyer K (eds). 2004
Design Principles for Wood Burning Cook Stoves, edited by
Oreland: Aprovecho Research Centre, Shell Foundation Partnership for Clean Indoor Air. 40 pp.

Buekens A G, Bridgewater A V, Ferrero G L, Maniatis K. 1990
Commercial Marketing of Aspects of Gasifiers
Luxumberg: Commission of European Communities

Bussmann P J T. 1988
Wood Stoves: theory and applications in developing countries
Eindhoven: Eindhoven University of Technology

Carre J, Hebert J, Lacrosse L, Lecicq A. 1985
The Lambiotte continuous carbonization process
In *Proceedings of Symposium on Forest Products Research International,
Achievements and the Future,* vol. 5
Pretoria

Cowling E B and Kirk T K. 1976
**Properties of cellulosic and lignocellulosic materials as substrate for
enzymatic conversion process**
In *Enzymatic Conversion of Cellulosic Materials: technology and applications,* pp. 95–123
New York: Wiley Inter Science

Dasappa S, Mukunda H S, Paul P J, Rajan N K S. 2003
*Biomass to Energy: the science and technology of the IISc bio-energy
systems*
Bangalore: Advanced Bio-residue Energy Technology Society, Indian Institute of
Science. 154 pp.

Dhingra S, Mande S, and Kishore V V N. 1996
Briquetting of Biomass in India—status and potential
[FAO's Regional Wood Energy Development Programme, Report No. 23]
Edited by P D Grover and S K Misra, pp. 24–30
Bangkok: FAO

Edwards J. 1974
Combustion: formation and emission of trace species
Michigan: Ann Arbor Science

Emrich W. 1985
Handbook of Charcoal Making
Dordretcht/Boston/Lancaster: Rendel Publishing Company

Envirosphere. 1980
*Program Negative Declaration for the Biomass Energy Demonstration
Program of the California Energy Commission*
Santa Ana, California: Envirposphere Company

FAO. 1986
Wood Gas as Engine Fuel
[FAO Forestry Paper 72]
Bangkok: FAO. 139 pp.

FAO. 1993
Improved Biomass Burning Cookstoves: a development manual
[Regional Wood Energy Development Programme Field Document No. 44]
Bangkok: FAO

FAO. 1996a
Biomass Briquetting: technology and practices, edited by P D Grover and
S K Misra
[Field Document No. 46]
Bangkok: FAO. 48 pp.

FAO. 1996b
Proceedings of the International Workshop on Biomass Briquetting
[New Delhi, 3–6 April 1995], edited by P D Grover and S K Misra
[Regional Wood Energy Development Programme Report No 23]
Bangkok: FAO. 193 pp.

GTZ. 1986
Status Report: energy from biomass-direct and reduced combustion
[A publication of GATE (Deutsches Zentrum für Entwicklungstechlogien) in GTZ
(Deutsche Gesellschaft für Technische Zusammenarbeit GmbH)]
Braunschweig: Friedr Vieweg & Sohn Verlagsesellschaft.

Henriksen U and Christensen O. 1994
Gasification of straw on two-stage 50 kW gasifier
In *Proceedings of the Eighth European Conference on Biomass for Energy, Environment,
Agriculture and Industry,* edited by P H Charter, pp. 1568–1578
Vienna

Iyer P V R, Rao T R, and Grover P D (eds). 2002
Biomass: thermo-chemical characterization
New Delhi: Indian Institute of Technology Delhi. 131 pp.

Jung D C. 1979
*Design Guideline Handbook for Industrial Spreader Stoker Boilers Fired
with Wood and Bark Residue Fuels*
Corvallis, Oregon: Oregon State University

Karlsson G, Ekstrom C, and Liinaki L. 1994
The development of a biomass IGCC process for power and heat production
In *Proceedings of the Eighth European Conference on Biomass for Energy, Environment,
Agriculture and Industry,* edited by P H Chartie, pp.1538–1549
Vienna

Kaupp A and Goss J R. 1984
Small-scale Gas Producer Engine System
Germany: GATE/GTZ, Gmbh Publication

Kaupp A and Goss J R. 1987
History of Small Gas Producer Engine Systems
Wiesbaden: GATE Vieweg

Kester R. 1980
Nitrogen Oxide Emissions from a Pilot Plant Spreader Stoker Bark Fired Boiler
Seattle, Washington: University of Washington

Khummomgkol P. 1986
Review of standard methods of testing stove energy
In *Proceedings of ASEAN Conference on Energy from Biomass*, pp.118–130

Kinoshita C M, Wang Y, and Zhou J. 1994
Tar formation under different biomass gasification conditions
Journal of Analytical and Applied Pyrolysis 29: 169–181

Knight R A. 2000
Experience with raw gas analysis from pressurized gasification of biomass
Biomass and Bioenergy 18: 67–77

Mande S. 2005
Biomass gasifier based power plants: potential, problems and research needs for decentralized rural electrification
In *Wealth from Waste*, 2nd edn
New Delhi: The Energy and Resources Institute. 495 pp.

Mande S and Lata K. 2001
Experimental analysis of various cookstoves for efficiency and emission characteristics
SESI Journal 11 (11): 29–36

Mande S and Lata K. 2005
Potential of converting phumdi waste of Loktak lake into briquettes for fodder and fuel use
In *Aquatic Weeds: problems, control and management*, pp 100–108, edited by S M Mathur, A N Mathur, R K Trivedi, Y C Bhat, and P Mohonot
New Delhi: Himanshu Publications

Mande S, Lata K, and Kishore V V N. 2001
Biomass gasifier based crematorium: an efficient and eco-friendly way of cremation
In *Renewable Energy Technologies for New Millennium*, pp. 316–320
[Proceedings of NREC-2000, 20–22 Dec 2000, Mumbai]
Mumbai: Indian Institute of Technology

Massengale R. 1985
The Missouri charcoal kiln
In *Proceedings of Symposium on Forest Products Research International, Achievements and the Future*, vol. 5
Pretoria

Mulcahy M F R and Young C C. 1975
The reaction of hydroxyl radicals with carbon at 298 K
Carbon **13** (2): 115–124

Munro J M. 1983
Formation and Control of Pollutant Emissions in Spreader-Stoker-Fired Furnaces
Salt Lake City, Utah: University of Utah

Narvaez I, Orio A, and Aznar M P. 1996
Biomass gasification with air in an atmospheric bubbling fluidized bed. Effect of six operational variables on the quality of produced raw gas
Industrial and Engineering Chemistry Research **196** (35): 2110–2120

Neeft J P A, Knoef H A M, and Onaji P. 1999
Behaviour of tar in biomass gasification systems: tar-related problems and their solutions
[Novem Report No 9919 Energy from Waste and Biomass]
The Netherlands

Neinhuysm S. 2003
The Beehive Charcoal Briquette Stove in the Khumbu Region of Nepal
Kathmandu: SNV (Netherlands Development Organization). 22 pp.

Palmer H B. 1974
Equilibria and chemical kinetics in flames
In *Combustion Technology: some modern developments*, pp. 2–33, edited by H B Palmer and J M Beer
New York: Academic Press

Prasad K K, Sangen E, and Visser. 1985
Wood burning cookstoves
Advances in Heat Transfer **17**: 159–310

Quaak P, Knowf H, and Stassen H. 1999
Energy from Biomass: a review of combustion and gasification technologies
[World Bank Technical Paper Series No. 422]
Washington, DC: The World Bank. 178 pp.

Raman P, Mande S, and Kishore V V N. 1993
Multifuel, multipurpose biomass gasifier system to meet rural energy needs and to promote rural industries
Urjabharti **3** (3)

Rambursh N E. 1923
Modern Gas Producers
London: Benn Brothers

Reed T B and Walt R. 1999
The turbo wood-gas stove
In *Proceedings of Fourth Biomass Conference of Americas on 29 September 1999*
Oakland, CA. 7 pp.

Reed T B, Walt R, Ellis S, Das A, Deutch S. 1999
Superficial velocity—the key to downdraft gasification,
In *Proceedings of Fourth Biomass Conference of Americas on 29 September 1999*
Oakland, CA. 8 pp.

Roy C, De Caumia B, and Menard H. 1983
Production of liquids from biomass by vacuum pyrolysis—development of database for continuous process
In *Symposium on Energy from Biomass and Wastes*, vol. VII, pp. 1147–1170
Lake Buena Vista, Florida

Shafizadeh F and Chin P S. 1997
Thermal deterioration of wood
In *Wood Technology: chemistry aspects*, pp. 57–81, edited by I S Goldstein
Washington, DC: ACS Press

Smith K R. 1985
Biomass Fuels, Air Pollution and Health: a global review
New York: Plenum Publishing Co.

Smith K R *et al.* 2000
Greenhouse gases from small-scale combustion devices in developing countries: household stoves in India
EPA–600/R–00-052, United States Environmental Protection Agency Report

Spliethoff H. 2001
Status of biomass gasification for power production
[Article No 200109 November 2001]
IFRF – Combustion Journal: 25 pp.

TERI (Tata Energy Research Institute). 1990
Design, fabrication, testing and field demonstration of biomass gasifier system for irrigation pumping
[TERI Project report No. 1986/SG/61]
New Delhi: TERI

TERI (Tata Energy and Research Institute). 2001
Feasibility Study to Manufacture Briquette Fuel from Refinery Sludge
[TERI project report No. 2000/BE/62]
New Delhi: TERI

TERI (The Energy and Resources Institute). 2004
Feasibility of Utilizing Coir Pith as a Replacement of Fuelwood to Minimize its Adverse Impact on Local Environment
[TERI Project report No 2003/BE/61]
New Delhi: TERI

TERI (The Energy and Research Institute). 2005
Livelihood Improvement through Biomass Energy in Rural Areas (LIBERA)
[TERI project report No. 2003/BE/67]
New Delhi: TERI

Tillman D A. 1981
Review of mechanisms associated with wood combustion
Wood Science **13**(4): 177–184

Tillman D A and Anderson. 1983
Computer modelling of wood combustion with emphasis on adiabatic flame temperature
Journal of Applied Polymer Science: Applied Polymer Symposium **37**: 761–774

TNO (the Dutch Institute on Applied Scientific Research). 1992
Kleinschalige verbranding van schoon afvalhout in Netherland
[Prepared for NOVEM, Utrecht, The Netherlands]

Tripathi A K, Iyer P V R, and Kandpal T C. 1998
A techno-economic evaluation of biomass briquetting in India
Biomass and Bioenergy **14**: 479–488

Uma R, Kandpal T C, and Kishore V V N. 2004
Emission characteristics of an electricity generation system in diesel alone and dual fuel modes
Biomass and Bioenergy **27**: 195–203

Várhegyi G, Szabó P, and Antal M J. 1994
Kinetics of the thermal decomposition of cellulose under the experimental conditions of thermal analysis. Theoretical extrapolations to high heating rates
Biomass and Bioenergy **7**: 69–74

Verhaart P. 1981
On designing wood stoves
In *A Woodstove Compendium,* edited by G Delepelire, K Krishna Prasad, P Verhaart, and P Visser
Eindhoven: Woodburning Stove Group

Verhaart P. 1983
On designing wood stoves
In *Wood Heat for Cooking,* edited by K Krishna Prasad and P Verhaart
Bangalore: Indian Academy of Sciences

Villermaux J. 1982
Conception et Functionnement des reacteurs
Paris: Editions Lavoisier

VITA (Volunteers in Technical Assistance). 1985
Testing the Efficiency of Wood Burning Cook Stoves—International Standards
Arlington: VITA

Walawender W P, Ganesan S, and Fan L T. 1981
Steam gasification of manure in fluid bed, influence of limestone as bed additive
In *Symposium Papers on Energy from Biomass and Wastes,* pp. 517–527
Chicago: IGT

Wang W Y, Padban N, Ye Z C, Andersson A, Bjerle I. 1999
Kinetics of ammonia decomposition in hot gas cleaning
Industrial and Engineering Chemistry Research **38**: 4175–4182

Wang W Y, Ye Z C, Padban N, Olofsson G, Bjerle I. 2000
Gasification of biomass/waste blends in a pressurized fluidized bed gasifier
In *Proceedings of the First World Conference on Biomass in Energy and Industry*, edited by S Kyritsis, A A C M Neenackers, P Helm, A Gassi, D Chiaramonti, pp. 1698–1701
Seville

Wen C Y and Dutta S. 1979
Rates of coal pyrolysis and gasification reactions
In *Coal Conversion Technology*, edited by C Y Wen and E S Lee, pp. 57–170
Reading: Addison-Wesley

Wenzl H. 1970
The Chemical Technology of Wood
New York: Academic Press

Zhou J, Masutani S M, Ishimura D M, Turn S Q, Kinoshita C M. 2000
Release of fuel-bound nitrogen during biomass gasification
Industrial and Engineering Chemistry Research **39**: 626–634

Annexe 1 Performance characteristics of various air gasification reactors

Characteristics	Updraft gasifiers	Downdraft gasifiers	Single fluid bed gasifiers	Fast fluid bed gasifiers
Capacity				
Fuel feed rate (d.a.f.t/h)	0.2–40	0.01 0.75	0.3–15	2–27
Combustible gas output				
– hot (GJ/h)	5–200	0.11–12	5.5–275	36–500
– cold (GJ/h)	4–170	0.1–10	4.5–225	30–420
– power (kWe equivalent)	150–10000	10–750	300–15000	2000–27000
Specific capacity (kg/hm^2)	150–250	100–800	300–450	400–600
Grate energy release (GJ/hm^2)	2.5–5	1.5–4	6–9	8–12
Input				
Fuel size	10–200		3–50	0–15
Open core (mm)		5–15		
Throated (mm)		20–80		
Moisture content (% db)	0–100	0–30	0–100	0–70
Ash content (% db)	0–50	0–15	0–25	Low
Air: feed ratio (kg/kg)	1.4–1.9	1.25–1.75	1.2–2.2	1.4–2.2
Steam: air (kg/kg)	0–0.25	0	0	0
Operation				
Pressure (bar)	1–5	1 – 10	1 – 10	1–30
Temperature (°C max)	900–1200	1000–1200	700–900	700–1100
Turndown ratio	2–4:1		5–8:1	3–5:1
– open core		1.5 : 1		
– throated		3 : 1		
Operating power requirement (total system unless specified) (kWh/GJ)	2–4	2–4	3–6	3–6
Output				
Gas exit temperature (°C)	75–500		500–900	700–1100
– open core		300–700		
– throated		500–1000		
Gas composition (dry and cold)				
H_2 (mol%)	13.5	16	8	18
CO (mol%)	22	20	16	20
CO_2 (mol%)	9	12	15	14
CH_4 (mol%)	3	2	6	3
C_{2+} (mol%)	0.5	–	2	1
N_2 (mol%)	52	50	53	44
Higher heating value (MJ/m^3)	4.5–6.0	4.5–6	4–6	5.0–6.5
Particulars before clean up (ppm)	Low	Low	Moderate	Moderate
Tars and oil before clean up (ppm)	High	Low	Low–moderate	Moderate
Gas yield				
(Nm^3/kg daf)	1.9–2.2	2.0–2.2	2.1–2.4	2.2–2.7
(Nm^3/kg daf)	2.4–3.4	2.4–2.8	2.6–3.0	2.7–3.4
Condensable yield (excluding H_2O) (kg/kg daf)	0.01–0.1	0–0.02	0.01–0.05	0.01–0.05
Ash and char yield (kg/kg daf)	0.01–1.0	0.01–0.18	0.02–0.33	0.02–0.1

(Continued)

Annexe 1 (*Continued*)

Characteristics	Updraft gasifiers	Downdraft gasifiers	Single fluid bed gasifiers	Fast fluid bed gasifiers
Energy balance				
Fuel input (%)	100.0	100	100	100
Cold clean gas (%)	60–86	75–89	69–87	78
Sensible heat or hot gas				
(% recoverable)	0–5	5–10	7–10	7–10
(% non–recoverable)	10–20	3–7	3–12	3–12
Tars and oils (%)	2–8	1	1–4	1–4
Heat losses (%)	2–7	2–7	2–7	2–7

Biochemical methods of conversion

K V Rajeshwari, Fellow and Malini Balakrishnan, Fellow
Energy–Environment Technology Division, TERI, New Delhi

Introduction

Organic wastes are important resources that can be converted into useful products by adopting suitable bio-processes. Waste materials considered to be bio-resources include animal residues such as cattle dung, poultry litter, municipal solid wastes (including food and vegetable market wastes), industrial organic wastes such as those from food-processing industries, sugar, and plantation industries (for example, tea and coffee processing), leather, distilleries, pulp and paper, and many more (Dhussa and Tiwari 2000). In addition, agricultural wastes and wetland vegetation such as water hyacinth also form an important category of bio-resources.

Basics of biochemical processes

A biochemical process refers to a number of complex chemical reactions occurring in the living organisms or involving living organisms.

The basis of any biochemical reaction is (i) hydrolysis of complex materials in organic matter, such as carbohydrates, proteins, and fats, into simple nutrients or (ii) the synthesis of complex molecules from simple substances. The process is accompanied by the release of energy used for cell growth and maintenance and useful byproducts.

Food + Micro-organisms → Energy + Products

For example, photosynthesis is a biochemical process in which plants utilize solar energy for converting carbon dioxide and water into sugar and oxygen (Figure 14.1).

In the process of respiration, organic compounds are converted into complex molecules such as starch, proteins, and fats, which have energy stored in bonds. This energy is released in the form of ATP (adenosine triphosphate) during cellular respiration

Glucose + O_2 → CO_2 + H_2O + 36ATP

H_2O + ATP → ADP (adenosine diphosphate) + P + Energy

Fermentation is a biochemical process commonly used in industries for converting organic substrates (for example, carbohydrates) into useful products such as alcohols, acids, and so on. A common example of biochemical conversion is fermentation of milk with lactic acid bacteria. A list of starter cultures in fermentation of milk is given in Table 14.1.

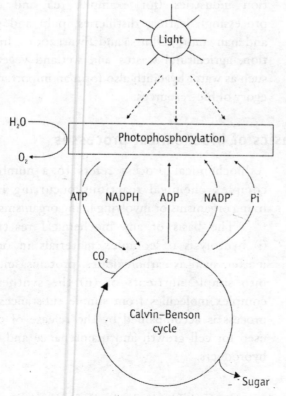

Figure 14.1 Photosynthesis process

Table 14.1 List of starter cultures

Streptococcus cremoris	*Lactobacillus bulgaricus*
Streptococcus lactis	*Leuconostoc cremoris*
Streptococcus thermophilus	*Leuconostoc dextranicum*
Streptococcus diacetylactis	*Lactobacillus acidophilus*

Various food products are obtained from fermentation processes. Most of the industrial food processes are based on fermentation by yeast, fungi, and mould. Alcohol in beer and wine is attributed to fermentation by yeast. Most dairy products are produced by fermentation of milk by microbes, namely lactococcus, lactobacillus, and streptococcus. Beer, a product of alcoholic fermentation of cereal by yeasts, can trace its origin to the Babylonian Empire (Mesopotamia) before 6000 BC. By 2000 BC, many types of beer existed in Babylon. The Egyptians were the first to discover that dough ferment resulted in light delicious bread. Some examples of food fermentation are given in Table 14.2.

Classification of micro-organisms

Micro-organisms can be broadly classified into plants, animals, and higher and lower protista. Micro-organisms belonging to the plant and animal kingdoms are multicellular, whereas those under protista can either be unicellular or multicellular. Microbes, which are of importance to waste treatment, are protists such as algae, protozoa, fungi, blue-green algae, and bacteria. These can be further classified as eukaryotic cells containing well-defined nucleus with a membrane or prokaryotic cells without a nuclear membrane. Bacteria and blue green algae are prokaryotic, whereas the algae, protozoa, and fungi are eukaryotic.

Table 14.2 Examples of food fermentation

Product	Substrate	Micro-organism
Chocolate	Cacao bean	*Saccharomyces cerevisiae*
		Candida rugosa
		Kluyveromyces marxianus
Bread	Flour	*S. cerevisiae*
Coffee	Coffee bean	*Erwinia dissolvens*
Sauerkraut	Cabbage	*Leuconostoc plantarum*
Soy sauce	Soyabean	*Aspergillus oryzae*

Bacteria can be classified on the basis of their shape and can be cylindrical, rod-shaped, spherical, and helical. In waste utilization, various groups of micro-organisms act on the substrate to be treated by one of the four processes, namely, aerobic, anoxic, anaerobic, and a combination of the aerobic/anoxic or anaerobic processes. Each process can be further subdivided into attached or suspended growth or a combination of both. The biochemical action of the microbes by any of the above routes results in the degradation of carbonaceous and nitrogenous matter.

Classification of micro-organisms can also be based on the use of carbon for growth. *Autotrophic* organisms are those that utilize carbon dioxide as a source of carbon, while organisms utilizing organic carbon are known as *heterotrophic*. Autotrophic organisms are further classified into photosynthetic or chemosynthetic, depending on their source of energy, that is, the sun or inorganic oxidation–reduction reactions.

In addition to classification based on source of energy and carbon, the oxygen requirement for growth leads to another category of classification of micro-organisms, namely aerobic (those that require oxygen), anaerobic (those that can survive in the absence of oxygen), and facultative (organisms that can exist under both aerobic and anaerobic conditions). Organisms that require the presence of oxygen for their metabolism are also known as obligate aerobes. Almost all animals, fungi, and several bacteria are obligate aerobes. Organisms for which the presence of molecular oxygen is toxic are known as obligate anaerobes. These are involved in acid fermentations, acetogenesis, and methanogenesis. Facultative anaerobes can use available oxygen and also other electron acceptors such as nitrate in absence of oxygen. Yeast and human cells are examples of facultative micro-organisms. Table 14.3 lists the organisms that are obligate aerobes, obligate anaerobes, and facultative anaerobes.

Table 14.3 Examples of different kinds of aerobic and anaerobic micro-organisms

Obligate aerobes	Facultative anaerobes	Obligate anaerobes
Bacillus *subtilis*	Bacillus *anthracis*	Clostridium botulinum
Bdellovibrio spp.	Corynebacterium diphtheriae	Clostridium tetani
Bordetella pertussis	Escherichia coli	
Legionella spp.	Lactobacillus spp.	
Mycobacterium leprae	Klebsiella spp.	
Mycobacterium tuberculosis	Salmonella sp.	
Pseudomonas spp.	Staphylococcus aureus	
	Streptococcus spp.	

Source Abedon (1998)

Enzymes in biochemical processes

Cell growth and multiplication of micro-organisms are facilitated by enzymes that are extracellular or intracellular. These enzymes act as catalysts during the absorption of substrate or nutrient by the cell and also during the conversion of substrate into product. The reaction is represented by the following equation.

$$[E] + [S] \rightarrow [E][S] \rightarrow [P] + [E]$$

where [E] is the enzyme, [S] the substrate, [E][S] the enzyme–substrate complex, and [P] is the product.

The activity of enzymes depends on the substrate, its concentration, pH, and temperature. The optimum pH for enzymatic activity is 7 and the optimum temperature is 35–40 °C. At higher temperatures, enzymes get denatured. The action of enzymes can be explained in a simplified manner by the lock-and-key mechanism (Figure 14.2). Each enzyme is specific to a specific substrate and has an active site where the substrate binds and reaction occurs. The conversion of a substrate into a product can be a single-step or a multi-step reaction, catalysed by a series of enzymes. For example, rennet, an enzyme preparation containing chymosin, is used to catalyse changes in milk proteins for gel formation during preparation of cheese. Metabolic pathways of various substrate conversions in a cell are very complex due to multiple reactions. For example, formation of ethanol from pyruvate in yeast and other micro-organisms is catalysed by two types of enzymes. In the first step, pyruvate is converted into acetaldehyde, which is catalysed by pyruvate decarboxylase. The second step is the reduction of acetaldehyde to ethanol, catalysed by alcohol dehydrogenase.

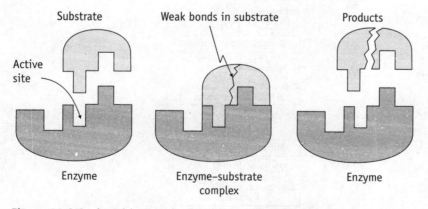

Figure 14.2 Lock-and-key mechanism of enzyme and substrate

$$\text{Pyruvate} + H^+ \rightarrow \text{Acetaldehyde} + CO_2$$

$$\text{Acetaldehyde} + NADH + H^+ \rightleftharpoons \text{Ethanol} + NAD^+$$

The glycolytic pathway for conversion of glucose into pyruvate along with energy production in the form of ATP is catalysed by different enzymes for each step (Stryer 1986; Figure 14.3). For details on glycolysis, Krebs' cycle and other biochemical path ways, any standard book on biochemistry (e.g. Stryer 1986) can be referred. Enzymatic activity is maximum at optimum temperature and pH. An acidic or alkaline pH affects the shape of enzyme molecules and inactivates the site at which reaction occurs (Figure 14.4).

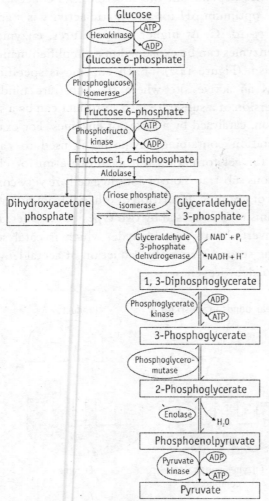

Figure 14.3 Glycolytic pathway

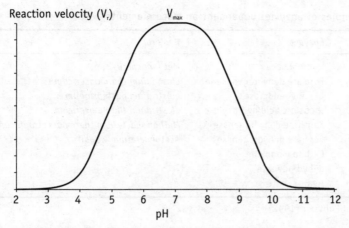

Figure 14.4 Effect of pH on enzymatic reaction

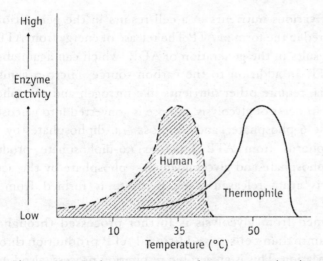

Figure 14.5 Optimum temperature in enzyme-catalysed reaction

At low temperatures, molecules of the substrate do not have enough energy for the reaction to take place. At high temperatures, molecules are highly energetic and cause degeneration of enzyme shape and its activity. Most enzymes involved in the metabolic reactions of the human body work at body temperature. Enzymes involved in the thermophilic reactions work at a higher temperature (Figure 14.5). In addition to the optimum temperature and pH, enzyme activity is also affected by the presence of various elements. The enzyme systems of acetogenic and methanogenic bacteria, for example, formate dehydrogenase and hydrogenase, involved in waste treatment processes require the presence of selenium, tungsten, and nickel (Table 14.4).

Table 14.4 Examples of enzymes dependent on trace elements

Trace element	Enzyme	Bacteria
Selenium	Hydrogenase	*Methanococcus* spp.
	Formate dehydrogenase	*Clostridium thermoaceticum*
	Glycine reductase	*Clostridium pasteurianum*
	Nicotinic acid hydroxylase	*Clostridium thermoaceticum*
	Xanthine dehydrogenase	*Methanobacterium thermoautotrophicum*
Tungsten	Formate dehydrogenase	*Acetobacterium woodii*
Nickel	Carbon monoxide dehydrogenase	
	Hydrogenase	

Source Malina and Pohland (1992)

Energy requirement for growth and maintenance

The conversion of various nutrients in a cell results in the generation of energy, which is stored in the form of ATP. The release of energy from ATP for various purposes results in the generation of ADP, which can again absorb energy to form ATP. In addition to the carbon source, micro-organisms, particularly bacteria, require other nutrients like nitrogen and phosphorus (Figure 14.3). In the process of glycolysis, glucose is converted into glucose-6-phosphate, fructose-6-phosphate, and fructose 1,6-diphosphate by the addition of phosphorus from ATP. Fructose 1,6-diphosphate produces dihydroxyacetone phosphate and glyceraldehyde-3-phosphate by the action of aldolase. ATP is again released when pyruvate is formed from 1,3-diphosphoglycerate.

Pyruvate formed from glycolysis is further processed through the Kreb's cycle in mammalian cells for additional ATP production through oxidative phosphorylation. This is an aerobic respiration process, also known as the citric acid cycle, in which pyruvate is converted into simpler compounds and citrate is formed as the first intermediate.

Figure 14.6 shows the Kreb's cycle depicting the conversion of pyruvate through TCA (tricarboxylic acid) pathways. The oxidative decarboxylation of pyruvate to form acetyl CoA (coenzyme A) is the link between glycolysis and the citric acid cycle.

$$\text{Pyruvate} + \text{CoA} + \text{NAD}^{\cdot} \rightarrow \text{Acetyl CoA} + \text{CO}_2 + \text{NADH}$$

In Kreb's cycle, decarboxylation of pyruvate occurs in the presence of CoA, resulting in the formation of acetyl CoA. This is catalysed by pyruvate dehydrogenase complex. Acetyl CoA condenses with oxaloacetate to form citrate that isomerizes to isocitrate, which is further oxidized and

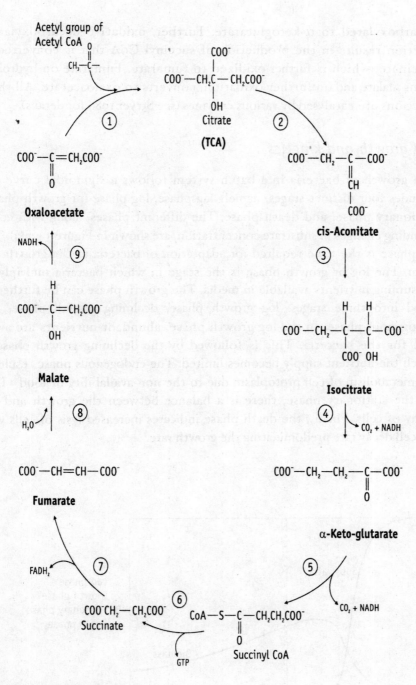

Figure 14.6 Kreb's cycle

decarboxylated to α-ketoglutarate. Further, oxidative decarboxylation reaction results in the production of succinyl CoA that is converted to succinate, which is further oxidized to fumarate. Fumarate on hydrolysis forms malate and on further oxidation, converts to oxaloacetate. All these reactions are catalysed by various enzymes (see Stryer 1986 for details).

Bacterial growth and kinetics

The growth of bacteria in a batch system follows a sigmoidal curve and includes four distinct stages, namely lag phase, log phase (or growth phase), stationary phase, and death phase. The different phases, along with corresponding changes in substrate concentration, are shown in Figure 14.7 (a). The lag phase is the time required for adaptation of bacteria to the growth medium. The log or growth phase is the stage in which bacteria multiply by consuming nutrients available in media. The growth phase can be further divided into three stages: log growth phase, declining growth phase, and endogenous phase. In the log growth phase, abundant nutrients are available for the bacteria. This is followed by the declining growth phase in which the nutrient supply becomes limited. The endogenous phase results in the metabolism of cell protoplasm due to the non-availability of food. During the stationary phase, there is a balance between the growth and the decay of cells. Finally, the death phase indicates increased lysis of cells with the cell decay rate predominating the growth rate.

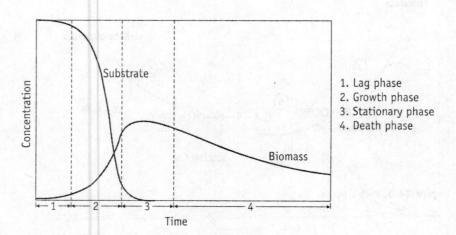

Figure 14.7 (a) Schematic diagram of bacterial growth curve

As mentioned in earlier sections, in addition to the formation of various products through biochemical conversion of resources, a major application of biological processes is in the utilization of waste water and solid wastes. The following section focuses on the kinetics and types of biochemical reactions for waste degradation.

Mixed cultures

Mixed cultures are microbial cultures, involving two or more micro-organisms. The bacterial consortium used in the biological waste treatment is a typical example of mixed cultures. This is a complex system as each micro-organism has a distinct growth pattern depending on the external parameters like pH, temperature, substrate, availability of oxygen, and presence of other micro-organisms. Growth rates of different micro-organisms in a mixed culture are shown in Figure 14.7(b). Growth of the mixed cultures in batch systems simultaneously follows the three steps: oxidation of organic matter, auto-oxidation of cells for energy, and growth of cells.

These three processes are depicted using the terms COHNS for organic matter and $C_5H_7NO_2$ to represent cell tissue as follows (Metcalf and Eddy 1993).

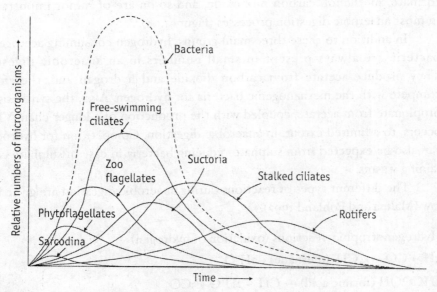

Figure 14.7(b) Growth curves of different microorganisms in mixed cultures

Oxidation

COHNS + O_2 + bacteria → CO_2 + NH_3 + other end products + energy

Synthesis

COHNS + O_2 + bacteria + energy → $C_5H_7NO_2$ (New cell tissue)

Endogenous respiration (auto-oxidation)

$C_5H_7NO_2$ + $5O_2$ → $5CO_2$ + NH_3 + $2H_2O$ + energy

As an example, anaerobic digestion of waste involves a consortium of bacteria, which can be broadly divided into three main groups.

Fermenting bacteria (also termed acidifying or acidogenic bacteria) These cause hydrolysis and acidogenesis of the substrate. The exo-enzymes released from these micro-organisms hydrolyse polymeric matter like proteins, fats, and carbohydrates into smaller units. These, in turn, undergo an oxidation–reduction process resulting in the formation of the VFA (volatile fatty acids) as well as some carbon dioxide and hydrogen.

Acetogenic bacteria These break down the products of the acidification step to form acetate. In addition, hydrogen and carbon dioxide (in the case of odd-numbered carbon compounds) are also produced.

Methanogenic bacteria These convert acetate or carbon dioxide and hydrogen into methane. Other possible methanogenic substrates like formate, methanol, carbon monoxide, and so on are of minor importance in most anaerobic digestion processes (Figure 14.8).

In addition to these three main groups, hydrogen-consuming acetogenic bacteria are always present in small numbers in an anaerobic digester. They produce acetate from carbon dioxide and hydrogen and, therefore, compete with the methanogenic bacteria for hydrogen. Also, the synthesis of propionate from acetate, coupled with the production of a longer chain VFA, occurs, to a limited extent, in anaerobic digestion. Competition for hydrogen can also be expected from sulphate-reducing bacteria in case of sulphate-containing wastes.

The different types of reactions during anaerobic digestion are given below (Malina and Pohland 1992).

Hydrogenotrophic reactions (oxidation of hydrogen)

$4H_2$ + CO_2 → CH_4 (methane) + $2H_2O$

$4HCOOH$ (formic acid) → CH_4 + $2H_2O$ + $3CO_2$

$4(CH_3CHOHCH_3)$ (2-propanol) + CO_2 → CH_4 + $4CH_3COCH_3$ + $2H_2O$

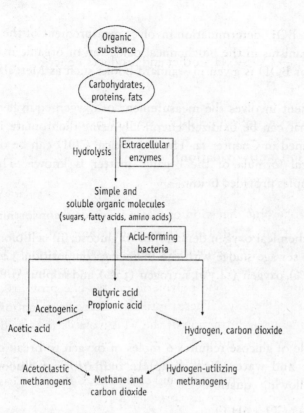

Figure 14.8 Steps in anaerobic digestion of waste
Source Malina and Pohland (1992)

Aceticlastic reaction (breakdown of acetate)

CH_3COOH (acetic acid) $\rightarrow$ $CH_4 + CO_2$

Disproportionation reactions (oxidation of substrate)

$4CH_3OH$ (methanol) $\rightarrow$ $3CH_4 + CO_2 + 2H_2O$

$4CH_3NH_2$ (methylamine) $+ 3H_2O \rightarrow 3CH_4 + CO_2 + 4NH_4^+$ (ammonium ion)

$2(CH_3)_2S$ (dimethyl sulphide) $+ 2H_2O \rightarrow 3CH_4 + CO_2 + H_2S$ (hydrogen sulphide)

$CH_4 + 2O_2 \rightarrow CO_2 + 2H_2O$

Organic wastes typically consist of proteins (40%–60%), carbohydrates (25%–50%), and oils and fats (8%–12%). At the elemental level, the organic matter consists of carbon, hydrogen, oxygen, and nitrogen. The aggregate organic matter in wastes and waste water can be quantified by measuring BOD (biochemical oxygen demand), COD (chemical oxygen demand), and TOC (total organic carbon). BOD determination involves measurement of the oxygen used by microorganisms in the biochemical oxidation of organic matter. The test procedure for BOD is given in standard books such as Metcalf and Eddy 1993.

COD measurement involves the measurement of oxygen-equivalent of the organic matter that can be oxidized chemically using dichromate in an acid solution, as outlined in Chapter 12. The theoretical COD can be determined if the chemical formula of the organic matter is known. This is illustrated in the examples provided below.

Example 1
Estimate the COD (chemical oxygen demand) of (i) glucose, (ii) cell biomass, (iii) methane, and (iv) sewage sludge with the following composition. Carbon (31.1%), hydrogen (4.2%), oxygen (24.3%), nitrogen (3.3%), and sulphur (1.1%).

Solution
(i) Glucose
A single molecule of glucose requires 6 moles of oxygen to break down into carbon dioxide and water. The complete oxidation of glucose is represented by the following equation.

$$C_6H_{12}O_6 + 6O_2 \rightarrow 6CO_2 + 6H_2O$$

Hence, COD of glucose is calculated as
$6 \times 32/180$ g of O_2/g of glucose = 1.07 g O_2/g glucose

(ii) Cell biomass
For oxidation of the cell biomass, 5 moles of oxygen are required as represented by the following equation

$$C_5H_7NO_2 \text{ (cell biomass)} + 5O_2 \rightarrow 5CO_2 + NH_3 + 2H_2O$$

Thus, COD of the cell biomass = $5 \times 32/113$ g of O_2/g of cell biomass = 1.42 g O_2/g cells.

(iii) Methane
The complete oxidation of methane is represented as

$$CH_4 + 2O_2 \rightarrow CO_2 + 2H_2O$$

The COD of methane, which is the amount of oxygen required for oxidation of each mole of methane = $(2 \times 32)/16$ = 4 g O_2/g of methane

(iv) Sewage sludge

The molecular formula is estimated by dividing the weight per cent of each element by the atomic weight and dividing the resulting number by the moles of carbon.

From the elemental composition given for sewage sludge, the number of moles can be calculated as

Carbon $31.1/12$ = 2.59
Hydrogen $4.2/1$ = 4.2
Oxygen $24.3/16$ = 1.51
Nitrogen $3.3/14$ = 0.235
Sulphur $1.1/32$ = 0.15

For molecular formulae
Carbon $2.59/2.59$ = 1
Hydrogen $4.2/2.59$ = 1.62
Oxygen $1.51/2.59$ = 0.58
Nitrogen $0.235/2.59$ = 0.09
Sulphur $0.15/2.59$ = 0.06

Hence, the molecular formula is

$$CH_{1.62} N_{0.09} O_{0.58} S_{0.06}$$

Molecular weight = $12 \times 1 + 1 \times 1.62 + 14 \times 0.09 + 16 \times 0.581 + 32 \times 0.06 = 26.096$
For estimating the COD of sewage sludge, the balancing equation for the oxidation of the sewage sludge is

$$CH_{1.62} N_{0.1} O_{0.6} S_{0.06} + 0.995 O_2 \rightarrow CO_2 + 0.1 NH_3 + 0.59 H_2O + 0.06 H_2S$$

Hence, COD = $(0.995 \times 32)/26.096 = 1.22$ g of O_2/g of sewage sludge.

Example 2
Estimate the theoretical methane yield for 1 g COD of organic matter.

Solution
Density of methane is 0.7167 g/litre.
Hence, 1 g of methane will occupy a volume of 1.4 litres.
As the COD of methane is 4 g [Example 1 (iii)], 1.4 litres of methane will be equivalent to 4 g of COD.

Thus, 1 g of COD of methane is equal to 1.4/4 litre or 0.35 litre of methane. Theoretical methane yield is thus 0.35 litre/g COD.

Growth kinetics of micro-organisms

It is important to understand the kinetics of bacterial growth for arriving at an optimal design of the reactor size and hence the cost. It is possible to have an optimum growth of micro-organisms through proper control of the environmental conditions such as pH, temperature, aerobic or anaerobic conditions, substrate level, additional micronutrients, and so on. The digestion period and hence the retention time is dependent on the growth rate of micro-organisms and the rate at which the substrate can be metabolized to give stable products.

Simple kinetics in batch culture (growth phase)

The microbial growth rate can be described by the equation

$$r_m = \frac{dX}{dt} = \mu X \qquad \qquad \text{...(14.1)}$$

where r_m is the rate of growth of micro-organisms (g/m³/day), μ is the specific growth rate (day⁻¹), and X is the microbial concentration. Microbial concentration, also called biomass concentration, is usually measured by VSS (volatile suspended solids). See Chapter 12 for measurement of VSS.

In case of substrate limitation, where the growth rate is affected by the limited concentration of nutrients or substrate, the cell growth is represented by the Monod equation as

$$\mu = \mu_{max} \times \frac{S}{k_s + S} \qquad \qquad \text{...(14.2)}$$

where μ_{max} is the maximum specific growth rate (g new cells/g cells/day), S is the limiting substrate concentration (g/m³), and k_s is the half-velocity constant (g/m³).

When $\mu = \mu_{max}/2$

Equation 14.2 becomes

$$\frac{\mu_{max}}{2} = \mu_{max} \times \frac{S_{1/2}}{k_s + S_{1/2}} \qquad \qquad \text{...(14.3)}$$

Rearranging the equation, we get

$$\frac{S_{1/2}}{k_s + S_{1/2}} = \frac{1}{2}$$

Or

$$k_s = S_{1/2}$$

The graphical representation of the variation of specific growth rate with substrate concentration is shown in Figure 14.9.

Other kinetic models proposed include

$$r_m = \mu_{max} \qquad \text{(constant)}$$
$$r_m = \mu_{max} \, XS \qquad \text{(first order)}$$
$$r_m = \mu_{max} \left[\frac{S/S_o}{k + (1-k)(S/S_o)} \right] X \quad \text{(mixed)}$$

where S_o is the influent substrate concentration and k is a dimensionless kinetic parameter.

$$r_m = \frac{\mu_{max}}{1 + (k/s) + \sum_{j=1}^{n}(S/k_j)} X \qquad \text{(mixed)}$$

Substituting for μ from Equation 14.2 in Equation 14.1, the overall rate expression can be represented as

$$r_m = \frac{dX}{dt} = \mu_{max} \times X \times \frac{S}{k_s + S}$$

...(14.4)

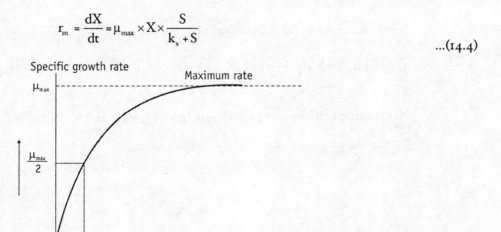

Figure 14.9 Variation of growth rate with substrate concentration
Reproduced with permission of The McGraw-Hill Companies
Source Metcalf and Eddy (1993)

The growth of microbes is accompanied by depletion of the substrate. The microbial growth rate and substrate depletion rate are thus related by the equation

$$r_m = -Y\, r_{sub} \qquad \qquad \ldots(14.5)$$

where Y is the yield coefficient.

The substrate concentration is usually measured in COD or BOD. Thus the yield coefficient is expressed in gVSS/gCOD removed.

As the concentration of substrate decreases with time, the rate is indicated as negative.

Substituting dX/dt from Equation 14.5 in Equation 14.4 gives

$$r_{sub} = -\frac{1}{Y}\mu_{max} \times X\, \frac{S}{k_s + S} \qquad \qquad \ldots(14.6)$$

The term (μ_{max}/Y) is also denoted k, the maximum specific substrate utilization rate. The decay of cells during the endogenous or death phase is given as

$$r_d = -k_d \times X \qquad \qquad \ldots(14.7)$$

where r_d is the rate of biomass decay (g VSS/m³/day) and k_d is the decay constant (g VSS/g VSS/day). The net rate of increase in biomass concentration will thus be

$$r_{m,net} = -r_m - (k_d \times X) \qquad \qquad \ldots(14.8)$$

Substituting for r_m from Equation 14.1, we get

$$r_{m,net} = \left(\mu_{max} \times X \times \frac{S}{k_s + S}\right) - k_d X \qquad \qquad \ldots(14.9)$$

Net specific cell biomass growth rate (g VSS/g VSS/day) is

$$\mu_{net} = \frac{1}{X}\, r_{m,net} \qquad \qquad \ldots(14.10)$$

or

$$\mu_{net} = \left(\mu_{max} \times \frac{S}{k_s + S}\right) - k_d \qquad \qquad \ldots(14.11)$$

Some indicative values of kinetic constants are given below.

	Acid producing bacteria	Methane producing bacteria
μ_{max}	≥ 1.33 day^{-1}	≥ 1.33 day^{-1}
Y	0.54 gVSS/gCOD utilized	0.14 gVSS/gCOD utilized
k_d	0.87 day^{-1}	0.02 day^{-1}

The VSS in any bioreactor includes the cell biomass, lysed cells, or cell debris due to the decay of the cells, and non-biodegradable VSS present in the feed or substrate.

Effect of temperature

Temperature affects the biological reactions through its influence on microbial metabolism, gas transfer rates, and settleability of solids. The microbial metabolic reactions are catalysed by enzymes that are influenced by temperature as explained earlier. The gas transfer rates are influenced by temperature due to its effect on diffusivity and viscosity of the media. Temperature also affects the settleability of solids by causing deflocculation of sludge.

The effect of temperature on reaction rate can be derived from the Vant Hoff–Arrhenius relation

$$\frac{d(\ln k)}{dT} = \frac{E}{RT^2} \qquad \ldots(14.12)$$

where k is the reaction rate constant, T is the temperature (°C), E is the activation energy constant in J/mole, and R is the ideal gas constant (8.314 J/mol °C).

Integration of Equation 14.12 gives

$$\ln \frac{k_2}{k_1} = \frac{E(T_2 - T_1)}{RT_1 T_2} \qquad \ldots(14.13)$$

where k_2 is the rate constant at temperature T_2 and k_1 is the rate constant at T_1.

As the waste treatment reaction occurs at ambient temperature, the term E/RT_1T_2 can be considered constant C. This transforms Equation 14.13 into

$$\ln \frac{k_2}{k_1} = C(T_2 - T_1) \qquad \ldots(14.14)$$

$$\text{or} \quad \frac{k_2}{k_1} = e^{C(T_2 - T_1)} \qquad \ldots(14.15)$$

$$\text{or } \frac{k_2}{k_1} = \eta^{(T_2 - T_1)}$$

where $\eta = e^C$ ...(14.16)

Choosing a reference temperature of 20 °C we get

$$k_T = k_{20} \times \eta^{(T-20)}$$ ...(14.17)

where k_T is the reaction rate constant at a temperature T °C, k_{20} is the reaction rate constant at a temperature 20 °C, and η is the temperature activity co-efficient.

Effect of nutrients

Certain inorganic elements are required for growth, including nitrogen, sulphur, phosphorus, magnesium, calcium, iron, sodium, and chlorine. Micronutrients such as zinc, manganese, molybdenum, selenium, cobalt, and nickel, and growth promoters, namely amino acids and vitamins, may also be required in certain cases.

A note on terminology

As mentioned earlier, substrate concentration is quantified as BOD and COD in waste waters. But as waste waters contain varieties of substrates some more terms are used to quantify, such as biodegradable COD (bCOD) or biodegradable soluble COD (bs COD). The waste waters contain colloidal and particulate matter, part of which may be biodegradable. Bacteria cannot consume the particulate substrates directly. These are first hydrolysed enzymatically to convert into soluble substrates. The bCOD can be a measure of particulate substrates also.

The bacterial concentration, also termed often as active biomass, is commonly measured as TSS (total suspended solids) and VSS (volatile suspended solids). In waste water treatment plants in which part of the sludge produced is recycled, the term MLVSS (mixed liquor volatile suspended solids) is used to define biomass concentration in the reactor input. The solids consist of bacteria, nbVSS (nonbiodegradable volatile suspended solids) and iTSS (inert total suspended solids).

The biomass yield Y can be estimated from stoichiometry considerations. For example, if we assume that the soluble substrate can be represented as glucose ($C_6H_{12}O_6$), and bacterial cells can be represented as $C_5H_7NO_2$, a stoichiometric equation can be written as

$$3C_6H_{12}O_6 + 8O_2 + 2NH_3 \rightarrow 2C_5H_7NO_2 + 8CO_2 + 14H_2O$$

The yield then can be obtained as

$$Y = \frac{2 \text{ moles of bacterial cells} \times 113}{3 \text{ moles of glucose} \times 180}$$

$$= 0.42 \text{ g.cells/g.glucose consumed}$$

In the above, 113 and 180 are the molecular weights of bacterial cells and glucose respectively.

The COD of glucose can be determined by the oxidation reaction

$$C_6H_{12}O_6 + 6O_2 \rightarrow 6CO_2 + 6H_2O$$

$$COD = \frac{6 \text{ moles } O_2 \text{ required} \times 32}{1 \text{ mole of glucose} \times 180}$$

$$= 1.07 \text{ gO}_2/\text{g. of glucose}$$

The yield in terms of COD will then be

$$Y = \frac{0.42}{1.07} = 0.39 \text{ g.cells/g COD used.}$$

The actual yield in any given biological treatment process will be less than the figures given above.

Biomass yield can also be estimated from considerations of bioenergetics (see Metcalf and Eddy 2003).

Aerobic and anaerobic systems for treatment of liquid wastes

Different types of bioreactors for waste water treatment

Batch reactor

It is a simple vessel, with continuous mixing of contents of the reactor with no inflow or outflow of materials. The H/D (height/diameter) ratio is generally kept at 1:3 (Figure 14.10).

Figure 14.10 Batch reactor

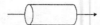

Figure 14.11 Plug flow reactor

Plug flow reactor

This type of reactor has a high length–width ratio with minimal or negligible longitudinal dispersion. The outflow of materials follows the same sequence as the influx, and each particle is retained in the reactor for the duration equivalent to the retention time (Figure 14.11).

CSTR (Continuous flow stirred tank reactor)

There is a complete mixing in the reactor that results in immediate dispersion of particles entering the reactor. For completely mixed conditions, the residence time is given as the volume of reactor divided by the volumetric flow rate of feed. Complete mixing is often not achieved in practical reactors and hence there is a residence time distribution, depending on the shape of reactor, degree of mixing, etc. (Figure 14.12).

Packed bed reactor

The reactor is packed with external media such as rock, plastic, and so on. This can be operated either as an anaerobic filter in absence of oxygen by complete loading of reactor or as a trickling filter by intermittent loading (Figure 14.13).

Fluidized-bed bioreactor

The reactor is similar to the packed bed reactor but differs in the characteristics of the packing medium, which is usually sand or powdered charcoal that has fluidizing properties. The bed expands due to the flow of liquid or air (Figure 14.14).

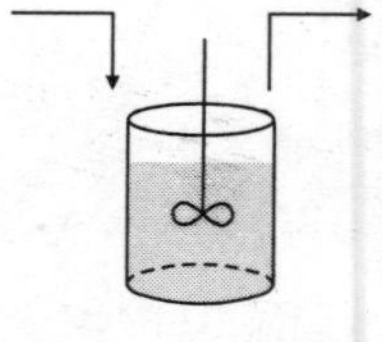

Figure 14.12 Continuous flow stirred tank

Packing medium

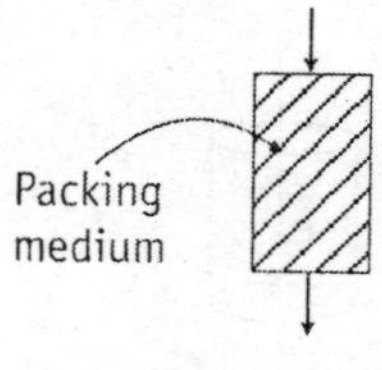

Figure 14.13 Packed bed

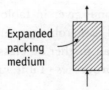

Expanded packing medium

Figure 14.14 Fluidized-bed reactor

The conversion of nutrients to final products and energy by complete oxidation of the decomposition products results in a higher energy production in the case of aerobic processes compared to anaerobic systems. However, most of the energy production goes towards the growth of cells resulting in a higher level of sludge production as compared to the anaerobic systems. Another advantage with the anaerobic systems is the low energy and maintenance requirement for the process resulting in higher net energy production. The cycle of conversion to the intermediate and final products of decomposition is shown in Figure 14.15.

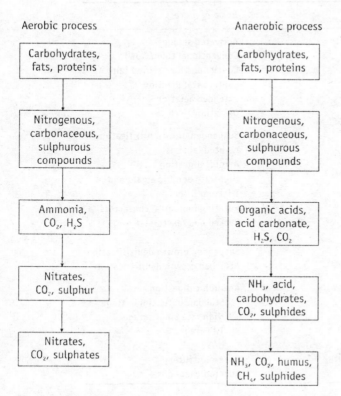

Aerobic process

Carbohydrates, fats, proteins

Nitrogenous, carbonaceous, sulphurous compounds

Ammonia, CO_2, H_2S

Nitrates, CO_2, sulphur

Nitrates, CO_2, sulphates

Anaerobic process

Carbohydrates, fats, proteins

Nitrogenous, carbonaceous, sulphurous compounds

Organic acids, acid carbonate, H_2S, CO_2

NH_3, acid, carbohydrates, CO_2, sulphides

NH_3, CO_2, humus, CH_4, sulphides

Figure 14.15 Decomposition of organic compounds in aerobic and anaerobic cycles

The comparison of aerobic and anaerobic digestion is given in Table 14.5. Biological treatment processes can be classified as aerobic, anaerobic, and anoxic as stated earlier. Each of these processes can further be classified as suspended growth or attached growth treatment processes (Table 14.6).

Table 14.5 Comparison of aerobic and anaerobic processes

Aerobic	*Anaerobic*
Electrical power required of power	No requirement of aeration hence less use
High maintenance and operational cost	Low operational and maintenance cost
High sludge production necessitating further treatment adding to the disposal cost	Less organic components converted to biomass
No useful energy production	Useful energy production in the form of methane

Table 14.6 Biological processes for waste water treatment

Type of process	*Name of process*
Aerobic suspended growth	Activated sludge
	Conventional (plug flow)
	Continuous flow stirred tank
	Contact stabilization
	Extended aeration
	Oxidation ditch
	Suspended growth nitrification
	Aerated lagoons
	Aerobic digestion
	High-rate aerobic algal ponds
Aerobic attached growth	Trickling filters
	Rotating biological contactors
	Packed bed reactors
Anoxic	
Suspended growth	Suspended growth denitrification
Attached growth	Attached growth denitrification
Anaerobic suspended growth	Anaerobic digestion
	Standard rate, single stage
	High rate single stage
	Two stage
	UASB
Anaerobic attached growth	Anaerobic filter
	Anaerobic lagoons

Source Metcalf and Eddy (2003)

Aerobic processes

Some of the aerobic suspended growth and attached growth processes are described below.

Activated sludge process

It is the most commonly used process for waste water treatment. The feed is pumped into the reactor (known as mixed liquor tank) containing acclimatized aerobic cultures. A mechanical stirrer along with aeration provides an aerobic environment for the cultures. After the desired retention time, the cells are separated from the treated waste water and partially recycled. The quantity of cells recycled and the retention time depend on the type of waste water and treatment efficiency. The mean cell retention time in the reactor depends on the flocculation and the settling property of micro-organisms.

A higher retention time results in better floc formation. This is because of the formation of a slimy material due to the ageing of cells, which facilitates the floc formation that ultimately settles down due to gravity.

Kinetics of activated sludge system in a continuous flow stirred tank

CSTR without recycle
Referring to Figure 14.16(a), the material balance equation for a CSTR can be written as

$$V\frac{dS}{dt} = QS_o - QS + V.\, r_{sub}$$

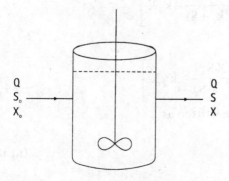

Figure 14.16(a) CSTR without recyle

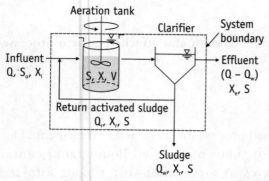

Figure 14.16(b) Schematic of the activated sludge process
Reproduced with permission from The McGraw-Hill Companies
Source Metcalf and Eddy (2003)

where V is the volume of reactor, Q is the volumetric feed rate and S and X are the substrate and biomass concentrations. Substituting for r_{sub} from Equation 14.6, we get

$$V\frac{dS}{dt} = Q(S_o - S) - V. \frac{\mu_{max}}{Y}. \frac{XS}{k_s + S}$$

At equilibrium dS/dt = 0 and one can write

$$0 = Q(S_o - S) - V. \frac{\mu_{max}}{Y}. \frac{XS}{k_s + S} \qquad \ldots(14.18)$$

A mass balance on bacterial biomass can be similarly written as

$$V\frac{dX}{dt} = QX_o - QX + V. r_{m,net}$$

Substituting for $r_{m,net}$ from Equation 14.9, we get

$$V\frac{dX}{dt} = Q(X_o - X) + V\left[\mu_{max}. \frac{XS}{(k_s + S)} - k_d X\right]$$

And at equilibrium,

$$0 = Q(X_o - X) + V\left[\mu_{max}. \frac{XS}{(k_s + S)} - k_d X\right] \qquad \ldots(14.19)$$

Equations 14.18 and 14.19 can be written as

$$0 = (S_o - S) - \theta \frac{\mu_{max}}{Y}. \frac{XS}{(k_s + S)} \qquad \ldots(14.18a)$$

$$0 = (X_o - X) - \theta \mu_{max}. \frac{XS}{(k_s + S)} - \theta k_d X \qquad \ldots(14.19a)$$

where $\theta = V/Q$

θ is called the HRT (hydraulic retention time) or simply retention time.

Assuming $X_o = 0$ and manipulating the above equations, one can get

$$X = \frac{Y(S_o - S)}{1 + \theta k_d} \qquad \qquad ...(14.20)$$

$$\text{and } S = \frac{k_s(1 + \theta.k_d)}{\left[\theta(\mu_{max} - k_d) - 1\right]} \qquad \qquad ...(14.21)$$

These equations can be used to predict the performance of CSTR without recycle.

Activated sludge system using a CSTR with recycle

The system is shown in Figure 14.16(b). Mass balance on the system boundary can be written as

$$V.\frac{dX}{dt} = QX_o - [(Q - Q_w)X_e + Q_wX_r] + V.r_{m,net} \qquad \qquad ...(14.22)$$

(Accumu = (input) – (output)   + (net growth)
lation)

where Q_w = waste sludge flow rate

X_r = biomass concentration in waste sludge

and X_e = biomass concentration in effluent stream.

Assuming $X_o = 0$ and substituting for $r_{m,net}$ from Equation 14.9, we get, at equilibrium (dX/dt = 0),

$$\frac{(Q - Q_w)X_e + Q_wX_r}{VX} = \mu_{max}\frac{S}{(k_s + S)} - k_d \qquad \qquad ...(14.23)$$

The left hand term has units of (day)$^{-1}$, and hence the inverse will have the units of time. This is called the SRT (solids rentention time), which can be written as

$$SRT = \frac{VX}{(Q - Q_w)X_e + Q_wX_r} \qquad \qquad ...(14.24)$$

For no recycle, $Q_w = 0$ and $X_e = X$ (assuming no clarification) and SRT then equals V/Q, which is nothing but HRT described earlier. Thus, for a complete mix CSTR with no recycle, the hydraulic and solid (or cell) retention times are the same. For increasing reaction rates (and to reduce reactor volumes), it is thus imperative that SRT be increased, either by solids recycling or by other processes such as suspended growth. Equation 14.23 can be written as

$$\frac{1}{SRT} = \mu_{max} \frac{S}{(k_s + S)} - k_d \qquad \qquad ...(14.25)$$

The above equation can be rearranged as

$$S = \frac{k_s[1 + (SRT).k_d]}{[(SRT)(\mu_{max} - k_d)] - 1} \qquad \qquad ...(14.26)$$

which is similar to Equation 14.21 with θ replaced by (SRT).

A substrate mass balance on the reactor gives

$$V\frac{dX}{dt} = (QS_o + Q_r S) - (Q + Q_r)S + V.r_{sub} \qquad \qquad ...(14.27)$$

Accumu = Input   – output   + generation
lation

Substituting dX/dt = 0 for steady state and for r_{sub} from Equation 14.6, we get

$$S_o - S = \theta \frac{\mu_{max}}{Y} \cdot \frac{XS}{(k_s + S)} \qquad \qquad ...(14.28)$$

where θ is the hydraulic retention time defined earlier. Substituting for $S/(k_s + S)$ from Equation 14.25 and rearranging, we get

$$X = \frac{(SRT)}{\theta} \left[\frac{Y(S_o - S)}{1 + k_d.(SRT)} \right] \qquad \qquad ...(14.29)$$

The above equation can be used to estimate the biomass concentration in the reactor, which is seen to be a function of both SRT and HRT besides the yield coefficient and decay coefficient. Further analysis of the activated sludge system is given in Metcalf and Eddy (2003).

Analysis of a plug flow reactor with recycle

Plug flow or tubular reactors are more complicated for analysis compared to CSTRs. These are however, industrially important. The Kompo gas system (described later) for biomethanation of solid organic wastes relies on a plug flow reactor. The leafy waste system of ASTRA, Bangalore and a few other anaerobic digestion designs in India are also based on plug-flow reactors, though it is difficult to distinguish between plug-flow and CSTR reactors when the L/D ratio is not very high, as is the case for some Indian reactor designs. For a true plug-flow reactor, all the particles entering the reactor spend the same time in the reactor before exiting. In comparison, CSTRs have a residence time distribution which may not be desirable, especially if mixing is not adequate, as in family size biogas plants (Raman, Sujatha, Dasgupta, *et al.*, 1989).

Analysis of plug-flow designs under certain simplifying assumptions is reported by Metcalf and Eddy (2003). The resulting expression for SRT is

$$\frac{1}{SRT} = \left[\frac{\mu_{max}(S_o - S)}{(S_o - S) + (1 + \alpha)k_s.\ln(S_{in}/S)} \right] - k_d \qquad ...(14.30)$$

where S_o = inlet concentration
S = outlet concentration
$S_{in} = (S_o + \alpha S)/(1 + \alpha)$
and α = recyle ratio

A number of CSTRs in series approximate the plug-flow reactor model.

Example 3

Waste water with an initial COD of 3 g/l is treated in a continuous flow reactor. A COD reduction of 75% is achieved. The inlet and outlet bacterial cell masses are 0 and 450 mg/l respectively. Calculate the yield coefficient Y.

Solution

For a COD reduction of 75%, the outlet substrate concentration would be 3 x (1–0.75) or 0.75 g/l. The yield coefficient is obtained as

$$Y = \left[\frac{X - X_o}{(S - S_o)} \right] = -\frac{450/1000}{(3 - 0.75)}$$
$$= 0.2g \text{ VSS/gCOD consumed}$$

Example 4

A CSTR without recycle treats waste water with an inlet COD of 450 mg/l and an inlet biomass concentration of 0. The outlet substrate concentration and biomass concentration for different values of HRT are as follows.

HRT (days)	S(mg COD/l)	X(mg VSS/l)
1.0	50	120
1.5	47	125
1.9	35	128
3.5	28	132

Calculate the maximum specific substrate utilization rate $k(=\mu_{max}/Y)$ and k_s.

Solution

From Equation 14.18(a), we can write

$$S_o - S = \theta.k.\frac{XS}{(k_s+S)}$$

$$\text{or } \frac{S_o - S}{\theta X} = \frac{k.S}{(k_s+S)}$$

$$\text{or } \frac{\theta X}{S_o - S} = \frac{k_s+S}{kS} = \left(\frac{k_s}{k}\right).\frac{1}{S} + \frac{1}{k}$$

Thus, plotting $\theta X/(S_o-S)$ vs $(1/S)$ should result in a straight line with an intercept equal to $(1/k)$ and a slope equal to (k_s/k).

From the given values of S and X for different θ, we can get the different co-ordinates as follows.

S	(1/S)	(S_o-S)	θX/(S_o-S)
50	0.02	400	0.3
47	0.021	403	0.465
35	0.028	415	0.58
28	0.035	422	1.09

A plot of $\theta X/(S_o-S)$ vs $(1/S)$ gives a straight line with an intercept of 0.1683 and a slope of 0.14. Hence

$$\frac{1}{k} = 0.1683, \text{ or } k = 5.94 \text{ day}^{-1}$$

$$\frac{k_s}{k} = 0.14 \text{ or } k_s = 0.14 \times 5.94 = 0.83 \text{ (mg VSS/l)}$$

Example 5

A CSTR has to be designed to treat waste water of COD 10 000 mg/l. The desired COD reduction is 85%. Calculate the hydraulic retention times for reactor temperatures of 25 °C and 35 °C. The various rate constants are given in the following table.

Temperature °C	μ_{max}, g/g.day	k_s, g/l	k_d, g/g.day
25	0.2	0.9	0.03
35	0.35	0.16	0.03

Solution

The outlet concentration S is obtained as $10(1 - 0.85) = 1.5$ g/l

From Equation 14.21,

$$S = \frac{k_s(1 + \theta k_d)}{\left[\theta(\mu_{max} - k_d) - 1\right]}$$

θ can be calculated from the given rate parameters as

θ (25 °C) = 10.5 days

θ (35 °C) = 3.5 days

Aerated lagoon

This is similar to the conventional activated sludge process and has a retention time of 10 days. The reactor content is aerated through the use of surface or diffused aerators with microbial cells in suspension. The different types of aerated lagoons include facultative partially mixed lagoons, aerobic flow-through partially mixed lagoons, and aerobic lagoons with solid recycling.

Attached growth aerobic systems

Attached growth systems involve the immobilization (binding of microbes, substrate, enzymes, and extracellular polymers) on support material such as plastic, sponge, activated carbon, and so on. This layer of micro-organisms, substrate, and extracellular material is known as biofilm. The substrate consumption occurs in the biofilm and there is diffusion of substrates, products, nutrients, and oxygen across a stagnant liquid layer, which separates the biofilm from the bulk liquid layer. The thickness of a biofilm varies between 100 µm and 10 mm. The substrate concentration within the biofilm is lower than that in the bulk liquid zone. The concentration within the biofilm varies with the depth of the film and rate of consumption. The layer of biofilm is not uniform, which makes the attached growth reactor kinetics more complex.

Anoxic processes

Although degradation of organic carbon is important in waste water, some of the inorganic compounds such as ammonia reduce the dissolved oxygen in the effluent by oxidizing to nitrate. The nitrification process is carried out either in the same bioreactor as the organic carbon degradation or in a suspended growth reactor following the activated sludge process. The design of the settling tank and the reactor for nitrification process is similar to that of the activated sludge process. Oxygen or air is used for nitrification.

Anaerobic reactors for waste water treatment

All modern high-rate anaerobic processes are based on the concept of retaining a high viable biomass by some mode of bacterial sludge immobilization. This is achieved by one of the following methods.

- Entrapment of sludge aggregates between the packing material inside the reactor, for example, downflow/upflow AFFR (anaerobic fixed film reactor).
- Formation of highly settleable sludge aggregates combined with gas separation and sludge settling, for example, UASBR (upflow anaerobic sludge blanket reactor) and ABR (anaerobic baffled reactor).
- Bacterial attachment to high density particulate carrier materials, for example AFBR (anaerobic fluidized bed reactors) and AEBR (anaerobic expanded bed reactors).

Some of the reactor types in use are described in subsequent sections (Kansal, Rajeshwari, Balakrishnan, *et al.* 1998).

Anaerobic fixed film reactor

This reactor employs a biofilm support structure (media) such as activated carbon, PVC (polyvinyl chloride) supports, hard rock particles, or ceramic rings for biomass immobilization, and can be operated in either the upflow or downflow mode. The advantages and disadvantages of an anaerobic fixed film reactor are given below.

Advantages	Disadvantages
<ul><li>Simple configuration and easy to construct</li><li>No mechanical mixing</li><li>Better stability at higher loading rates</li><li>Can withstand toxic and organic shock loads</li><li>Quick recovery after starvation period</li></ul>	<ul><li>High reactor volume</li><li>Susceptibility to clogging due to increased film thickness and high suspended solid concentration in waste water</li></ul>

In stationary fixed film reactors (Figure 14.17), cells are deliberately attached to a large-sized solid support. The reactor has (i) a biofilm support structure (media) for biomass immobilization, (ii) distribution system for uniform distribution of waste water above/below the media, and (iii) effluent draw off and recycle facilities (if required). The reactors can process different waste streams with little compromise in capacity and can adapt readily to changes in temperature. This is important for installations where waste water characteristics change rapidly. The reactor start-up can be very quick after a period of starvation (one or two days to reach maximum capacity after three weeks of starvation).

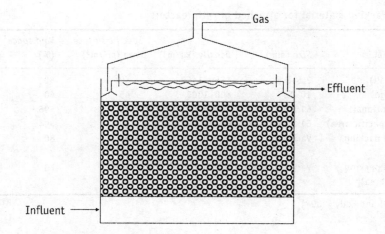

Figure 14.17 Schematic of stationary fixed film reactor

The common problem associated with stationary fixed film reactors is clogging due to non-uniform growth of biofilm thickness and/or high suspended solid concentration in waste water. Non-uniform growth and consequent clogging occurs especially at the influent entry. Some measures to combat this problem include (i) recirculation of effluent and gas for developing a relatively thin film and sloughing of biomass, (ii) provision for a relatively thin layer of media near the inlet to accumulate the excess biofilm, and (iii) improvement in the flow distribution system to avoid very low liquid velocity. Activated carbon, PVC supports, hard rock particles, and ceramic rings are the various types of film support that have been tried. Reactor configuration and operation (upflow or downflow mode of operation) have a marked effect on the performance of the reactor. With wastes containing large amounts of hard-to-digest suspended solids, recirculation aids in degradation as it keeps the solids in suspension (Figure 14.17).

The specific surface area of the packing averages about 100 m²/m³ and higher packing densities do not have any improvement in the process efficiency.

Some of the common packing materials and their properties are shown in Table 14.7. The performances of fixed bed reactors for different types of waste water are given in Table 14.8.

Estimation of design parameters for a packed-bed reactor

The contact time of liquid with the biofilm is related to the filter depth as follows.

Table 14.7 Packing material for attached growth reactors

Packing material	Size (mm)	Density (kg/m³)	Specific surface area (m²/m³)	Void space (%)
River rock (small)	2.5–7.5	1250–1450	60	50
River rock (large)	10–13	800–1000	45	60
Plastic (conventional)	61 x 61 x 122	30–80	90	>95
Plastic (high specific area)	61 x 61 x 122	65–95	140	>94
Plastic random packing (conventional)	Variable	30–60	98	80
Plastic random packing (high specific area)	Variable	50–80	150	70

Source Metcalf and Eddy (2003)

Table 14.8 Performance of fixed-bed reactor for different types of waste water

Type of waste water	Packing type	Loading rate (kg COD/m³ d)	Retention time (days)	COD reduction (%)
Guar gum	Pall rings	7.7	1.2	61
Chemical processing	Pall rings	12–15	0.9–1.3	80–90
Domestic	Tubular	0.2–0.7	0.5–0.75	90–96
Landfill leachate	Cross flow	1.5–2.5	2.0–3.0	89
Food canning	Cross flow	4–6	1.8–2.5	90

Source Metcalf and Eddy (2003)

$$t = \left(\frac{c\,h}{q^n}\right)$$

...(14.31)

where t is the contact time of the liquid (min), c is the constant associated with the packing material, h is the depth of packing (m), q is the hydraulic loading rate (litre/m² min), and n is the hydraulic constant for the packing material and is assumed to be 0.5.

$$q = \left(\frac{Q}{a}\right)$$

where Q is the influent feed rate (litre/min) and a is the cross-section area of the filter (m²).

Hence, the liquid contact time decreases with an increase in the flow rate. This can be attributed to the increase in the film thickness with increase in feed rate.

Substituting for q in Equation 14.31 gives

$$t = \frac{ch}{(Q/a)^n} \qquad \ldots(14.32)$$

The rate of substrate utilization in film can be represented as

$$\left(\frac{dS}{dt}\right) = -kS$$

Integrating the above expression

$$[\ln S]_{S_i}^{S_e} = -kt$$

Substituting for t from Equation 14.31,

$$\ln \frac{S_e}{S_i} = -\frac{kch}{q^n} = -kchq^{-n}$$

Upflow anaerobic sludge blanket reactor

The main feature of a UASB (upflow anaerobic sludge blanket) reactor is the formation of granular sludge, which is an agglomeration of microbial consortia. The system consists of a gas–solid separator (to retain the anaerobic sludge within the reactor), a feed distribution system, and effluent draw-off facilities (Figure 14.18). With a sophisticated feed distribution system installed in a UASB reactor, effluent recycle (to fluidize the sludge bed) is not necessary as sufficient contact between waste water and sludge is guaranteed even at low organic loads. The following table lists out the advantages and disadvantages of a UASB reactor.

Advantages	Disadvantages
■ Retention of the anaerobic sludge within the reactor	■ Long start-up period
■ High mean cell retention time due to formation of granular sludge with bigger particle size and high sludge settling and thickening property	■ Requires seed sludge for faster start-up
■ Highly cost-effective design	■ Requirement of controlled and skilled operation to maintain efficiency of the reactor
■ Low initial investment as compared to an anaerobic filter or a fluidized bed system	

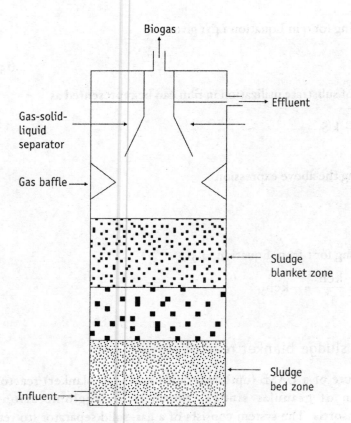

Figure 14.18 Schematic of a UASB reactor

UASB processes are applied for treating high strength and low/medium strength waste water and a variety of other substrates. The process has been applied to the waste water generated from a wide cross-section of industries, such as distilleries, food-processing units, tanneries, and so on, in addition to municipal waste water.

Expanded granular sludge bed reactor

The EGSB (expanded granular sludge bed) reactor is a modified form of UASB where a slightly higher superficial liquid velocity can be applied (5–10 m/h compared to 3 m/h for soluble waste water and 1–1.25 m/h for partially soluble waste water in the UASB). This results in the accumulation of granular sludge, and a part of the granular sludge bed is in an expanded or fluidized state in the higher regions of the bed. Advantages and disadvantages of an EGSB reactor are listed below.

Advantages	Disadvantages
▪ Sufficient contact between waste water and sludge	▪ Possibility of sludge loss
▪ A good flow of substrate into the sludge aggregates	▪ Requirement of effluent recycling
▪ High velocity	

In the expanded bed design, micro-organisms are attached to an inert support medium such as sand, gravel, or plastics as in a fluidized bed reactor. However, the diameter of the particles is slightly bigger as compared to that used in fluidized beds. The principle used for expansion is also similar to that for the fluidized bed, that is, high upflow velocity and recycling.

Anaerobic fluidized bed reactor

In AFBR (anaerobic fluidized bed reactor), the media for bacterial attachment and growth – generally small-sized particles of sand, activated carbon, and so on – is kept in a fluidized state by the drag forces exerted by the upflowing waste water. The fluidization of the media facilitates the movement of microbial cells from the bulk to the surface and thus enhances the contact between micro-organisms and the substrate. Increasing the recycling makes the process similar to a completely mixed system. Design of a fluidized bed thus consists of a feed distribution system, a media support structure, media, head space, effluent draw off, and recycle facilities. Thickness of the biofilm depends on the size and density of the inert media, bed regeneration, and upflow velocity. There can be a periodical removal of excess sludge from the zones of the fluidized bed where the thickness of the biofilm is maximum. It is possible to operate the reactor at lower retention times and/or higher loading rates. The stationary packed bed technology is adequate for the treatment of easily biodegradable waste water or water in which high COD removal is not required, while the fluidized bed technology is suitable for treatment of high strength complex waste water with components that are difficult to degrade (Figure 14.19). Advantages and disadvantages of an AFBR are given below.

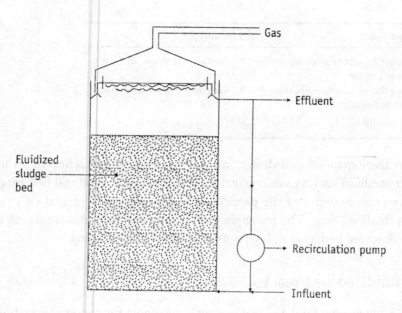

Figure 14.19 Fluidized bed reactor

Advantages	Disadvantages
▪ Elimination of bed clogging	▪ Need for recycling of effluent to achieve bed expansion
▪ Low hydraulic head loss combined with better hydraulic circulation	
▪ Greater surface area per unit of reactor volume	
▪ Lower capital cost due to reduced reactor volumes	

Hybrid reactor

The hybrid reactor (Figure 14.20) is a combination of the UASB system and fixed film reactors. By choosing a suitable highly porous packing material with a large specific surface, the adhesion of microbes can be greatly improved and concentration of activated sludge in the reactors can be considerably enhanced.

Advantages of a hybrid reactor are listed below.
▪ The reactor can withstand disturbances such as large fluctuations in the loading rate

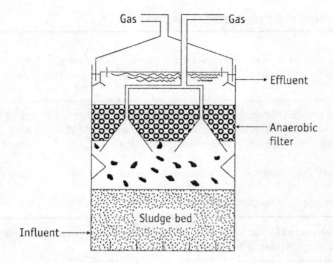

Figure 14.20 Schematic of hybrid reactor

- As microbes are held by the support media, there is no wash out even at a very high upflow velocity of effluent
- Successful phase separation (in terms of acidogenic and methanogenic phases) can be achieved
- Provides the best environment for growth of both acidogens and methanogens
- Can be used for a wide variety of industrial effluents, and it is possible to maintain the desired pH conditions for both the acidogens and methanogens
- There is high rate of mass transfer because of the increased contact time between the feed and the microbes
- Minimization of sludge loss due to immobilization
- Simplicity in operation and design
- More economical than fixed-bed system at the industrial scale

The characteristics of different types of reactors discussed have been summarized in Table 14.9.

Growth kinetics in anaerobic processes

The kinetic expression describing the anaerobic process is based on Monod's model, which is described earlier for aerobic processes in Equation 14.2.

$$\mu = \mu_{max} \frac{S}{k_s + S}$$

where μ and μ_{max} represent the specific growth rate and maximum

Table 14.9 Characteristics of different types of reactors

Anaerobic reactor type	Start-up period (days)	Channelling effect	Effluent recycle	Gas–solid separation device	Carrier packing	Typical loading rates (kg COD/m³ day)	HRT (days)
CSTR	—	Not present	Not required	Not required	Not essential	0.25–3	10–60
UASB	4–16	Low	Not required	Essential	Not essential	10–30	0.5–7
Anaerobic filter	3–4	High	Not required	Beneficial	Essential	1–40	0.5–12
Expanded bed	3–4	Less	Required	Not required	Essential	1–50	0.2–5
AFB	3–4	Very less	Required	Beneficial	Essential	1–100	0.2–5

COD – chemical oxygen demand; HRT – hydraulic retention time; CSTR – continuous stirred tank; UASB – upflow anaerobic sludge blanket; AFB – anaerobic fluidized bed;
Source Rajeshwari, Balakrishnan, Kansal, *et al.* (2000)

specific growth rate, respectively (g VSS/gVSS/d), S is the limiting substrate concentration (g/m³), and k_s is the half-velocity constant (g/m³).

SRT can be derived from Equation 14.25

$$\frac{1}{SRT} = \left[\frac{\mu_{max} \times S_e}{k_s + S_e} \right] - k_d$$

where S_e is the substrate concentration in the effluent (g/m³). Typical values of kinetic parameters for the anaerobic suspended growth process are: Y = 0.08 g VSS/g COD; k_d = 0.03 g/g.day; μ_{max} = 0.20 g/g.day (at 25 °C); k_s = 900 mg/L (at 25 °C), and methane production = 0.4 m³/kg COD utilized (at 35 °C).

Estimation of methane yields in an anaerobic reactor

If the composition of waste is known, estimation of CH_4, CO_2, NH_3, and H_2S can be made by the method proposed by Buswell and Boruff (1932). The over-all equation of biomethanation can be written as

$$C_cH_hO_oN_nS_s + \left(c - \frac{h}{4} + \frac{o}{2} + \frac{3n}{4} + \frac{s}{2} \right) H_2O \rightarrow$$

$$\left(\frac{c}{2} + \frac{h}{8} + \frac{o}{4} + \frac{3n}{8} + \frac{s}{4} \right).CH_4 + \left(\frac{c}{2} - \frac{h}{8} + \frac{o}{4} + \frac{3n}{8} + \frac{s}{4} \right)CO_2 \qquad ...(14.33)$$

$$+ nNH_3 + s.H_2S$$

The mole fractions of the evolved gases can be obtained as

$$x_{CO_2} = \frac{4c - h + 2o - 5n + 2s}{8(c - n + s)} \qquad \qquad ...(14.34)$$

$$x_{CH_4} = \frac{4c + h - 2o - 5n - 2s}{8(c - n + s)} \qquad \qquad ...(14.35)$$

$$x_{H_2S} = \frac{s}{8(c - n + s)} \qquad \qquad ...(14.36)$$

Critical design parameters for UASB process

Waste characteristics, volumetric organic loading rates, and upflow velocity are the important parameters for designing UASB reactors. Upflow velocity is the ratio of the feed flow rate and cross-sectional area of reactor. The reactor volume and height are determined by organic loading, superficial velocity, and effective liquid volume of reactor.

$$V = \frac{Q \times S_o}{OLR} \qquad \qquad ...(14.37)$$

where V is the effective liquid volume of reactor including the volume occupied by sludge blanket and active biomass (m³), Q is the feed rate (m³/day), S_o is the influent concentration (kg/m³), and OLR is the organic loading rate (kg COD/m³ d).

The total liquid volume V_e after excluding the space for gas collector is given as V/0.8. This is assuming 0.8 as the effectiveness factor indicating the space occupied by the sludge blanket.

Upflow superficial velocity u_s (m/h) is expressed as the ratio of the flow rate to the reactor area.

$$u_s = Q/A$$

or the area of the reactor $A = Q/u_s$
Height of the reactor $H_e = V_e/A$
Total height of the reactor $H = H_e + H_g$
where H_g is the additional height due to the gas collector.

The desired volumetric COD loading and upflow velocities for different waste water strengths are given in Tables 14.10 and 14.11, respectively.

Feed distribution is an important parameter while designing a UASB reactor. A gas–solid separator has to be designed to prevent sludge washout and

Table 14.10 Desired volumetric COD loading for 85%–95% removal at a temperature of 30 °C for a UASB reactor with granular sludge and little loss of TSS

Waste water COD (mg/litre)	Fraction as particulate COD	Volumetric loading (kg COD /m³d)
1000–2000	0.1–0.3	8–12
	0.3–0.6	8–14
	0.6–1.0	
2000–6000	0.1–0.3	12–18
	0.3–0.6	12–24
	0.6–1.0	
6000–9000	0.1–0.3	15–20
	0.3–0.6	15–24
	0.6–1.0	
9000–18 000	0.1–0.3	15–24
	0.3–0.6	
	0.6–1.0	

Source Metcalf and Eddy (2003)

Table 14.11 Desired upflow velocity and reactor height for different strengths of waste water

Waste water	Upflow velocity (m/h)	Reactor height (m)
COD 100% soluble	1.0–3.0	6–10
COD partially soluble	1.0–1.25	3–7
Domestic waste water	0.8–1.0	3–5

Note COD – chemical oxygen demand
Source Metcalf and Eddy (2003)

facilitate maximum separation of liquid and solids from the gas. The gas–solid–liquid separator is an important feature in a UASB reactor. It has an inverted V-shape and the slope of the settler in the gas–solid–liquid separation device should be between 45 and 60°. The preferred height of the gas collector is 1.5–2 m at 5–7 m reactor height. Table 14.12 gives the recommended parameters for feed distribution. Further design details are provided in Metcalf and Eddy (2003).

Table 14.12 Recommended parameters for feed distribution

Sludge category	Organic loading (kg COD/m³d)	Area of reactor distributed inlet (m²)
Dense flocculent	< 1	0.5–1
sludge (>40 kg TSS/m³)	1–2	1–2
	>2	2–3
Medium flocculent		
sludge, (20–40 kg TSS/m³)	<1–2	1–2
	>3	2–5
Granular sludge	1–2	0.5–1
	2–4	0.5–2
	>4	>2

Note TSS – total suspended solids; COD – chemical oxygen demand
Source Metcalf and Eddy (2003)

Importance of granular sludge and sludge volume index in designing UASB reactor

Development of granular sludge is necessary for efficient operation of the UASB reactors as well-formed granules with good settling property help in retaining the biomass even at higher upflow velocities. A good settling property of granules is judged based on the settling velocity. The settling velocity or terminal velocity of a sediment particle is the rate at which the sediment settles in a still fluid. It is dependent on granule size, shape, and density as well as on the viscosity and density of liquid. The settling velocity, u_t (m/s), can be calculated as follows (Harris 2003; Jimenez and Madsen 2003).

$$u_t = \sqrt{\left(\frac{\rho_g - \rho}{\rho}\right)\left(\frac{g}{C_d/2}\right)\left(\frac{V_g}{A_g}\right)} \qquad \ldots(14.38)$$

where ρ_g is the density of the granules (kg/m³), ρ is the density of the liquid (kg/m³), g is the acceleration due to gravity (m/s²), C_d is the drag coefficient dependent on size and shape of granule and viscosity of liquid, V_g is volume of granule particle (m³), and A_g is the cross-sectional area of granule (m²).

Granules with a settling velocity of up to 20 m/h are categorized as poor settling fraction, those between 20 and 50 m/h as moderate, and those over 50 m/h as good settling fraction (Schmidt and Ahring 1996). The shape and composition of the granular sludge is variable. It generally has a spherical form with diameter ranging from 0.14 mm to 5 mm. Depending on the type of waste water being treated, the size and inorganic composition of granules (calcium, potassium, and iron) differ. The structure and stability of granules is governed by extracellular polymers that vary between 0.6% and 20% of the VSS and consist of proteins and polysaccharides. An increase in the C/N ratio improves production of extracellular polysaccharide, thus promoting attachment of bacteria to the surface. Certain substrates, such as high concentration of fatty acids, particularly propionate, are found to inhibit the activity of granular sludge. *Methanosaeta* spp. and *Methanosarcina* spp. are important for initial granulation. In addition, granules are also found to contain *Methanobacterium formicicum*, *Methanobacterium thermoautotrophicum*, and *Methanobrevibacter* spp. (Figure 14.21).

Example 6

(i) Determine the height and diameter of a UASB reactor treating 100 m³/day of waste water with a soluble COD of 8000 mg/litre. (ii) Also estimate the HRT and methane gas production assuming COD degradation to be 75%. The effectiveness factor (fraction of the total liquid volume occupied by the sludge blanket) is 0.8. The other parameters are as follows.

Upflow velocity = 1.5 m/h
Volumetric loading rate = 18 kg s (soluble) COD/m³/day
Gas storage height = 2.5 m
CH_4 production = 0.35 litre/g COD (removed)

Solution

(i) From Equation 14.37

$$V = Q \times \left(\frac{S_o}{OLR} \right)$$

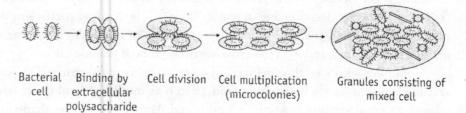

Bacterial cell   Binding by extracellular polysaccharide   Cell division   Cell multiplication (microcolonies)   Granules consisting of mixed cell

Figure 14.21 Granule formation
Source Schmidt and Ahring (1996)

V = Effective liquid volume of the reactor = 100 m³/day ÷ 8 kg COD/m³ × 18 kg COD/m³/day = 44.4 m³

Total liquid volume of reactor, V_e = V/effectiveness factor = 44.4/0.8 = 55.5 m³
Area of the reactor = feed rate/upflow velocity
    = 100 m³/day/(1.5 m/h × 24 h/day) = 100 m³/day/36 m/day = 2.77 m²
    A = 3.14 × D²/4 = 2.77 m²

    or D² = 3.53 m²

    D = 1.88 m

Liquid height of the reactor = V_e/A = 55.5/2.77 = 20 m
Total height = 20 + 2.5 = 22.5 m

(ii) HRT = liquid volume of reactor/Feed rate
    = (55.5 m³/100 m³/day) × 24 h/day = 13.32 h

(iii) Amount of soluble COD degraded = 8000 × 0.75 = 6000 mg/litre
    = 6 kg/m³
Effluent COD concentration = 8000 – 6000 = 2000 mg/litre
COD utilized = 6 kg/m³ × 100 m³/d = 600 kg COD/day
Methane produced = 600 kg COD × 0.35 m³/kg COD (removed) = 210 m³/day

Aerobic and anaerobic systems for solid waste treatment

The treatment of the organic solid wastes can be carried out either in the presence or absence of oxygen. Three common methods of processing include composting, vermicomposting, and biomethanation/anaerobic digestion.

Composting

Composting is a biological process in which the organic matter present in waste is converted into enriched inorganic nutrients. The manure obtained has high nitrogen, phosphorus, and potassium content. Heterotrophic microorganisms act upon the organic matter and by the action of enzymes, convert organic compounds first into simpler intermediates like alcohol or organic acids and later into simple compound like sugars. This produces humic acid and available plant nutrients in the form of soluble inorganic minerals like nitrates, sulphates, and phosphates. The quality of compost depends upon the waste being composted. The presence of high nitrogen, phosphorus, and potassium contents in the organic waste facilitates production of high-quality manure after composting. The average composition of different constituents in the compost is given in Table 14.13. Table 14.14 gives concentration limits of heavy metals in compost.

There are different methods of composting various kinds of organic wastes such as agro residues, animal waste, household waste, and so on. Some of these are heap, pit, lagoon, chamber, and Berkeley and Nadep methods (Agarwal and Saxena 2001).

Vermicomposting

This is a process whereby food materials, kitchen wastes – including vegetable and fruit peelings – papers, and so on, can be converted into compost by the

Table 14.13 Characteristics of compost

Constituent	Typical value (%)
Total nitrogen	1.3
Total phosphorus	0.2–0.5
Total potassium	0.5
Organic phosphorus	0.054
pH	8.6
Moisture	40–50
Organic matter	30–70

Sources US Composting Council (2002); Bhardwaj (1995)

Table 14.14 Limits of heavy metals in compost

Constiuents	Concentration not to exceed (mg/kg dry basis)
Arsenic	10.00
Cadmium	5.00
Chromium	50.00
Copper	300.00
Lead	100.00
Mercury	0.15
Nickel	50.00
Zinc	1000.00

Note *C/N ratio not to exceed 20–40 and pH not to exceed 5.5–8.5
Source The Gazette of India notification (2000)

action of earthworms. An aerobic condition is created due to the exposure of organic waste to air. Many Asian countries are adopting the process of vermi-composting for waste disposal. Although there are thousands of species of earthworms, *Eisenia foetida* and *Eisenia andrei* are effective in decomposition of organic wastes. Certain biochemical changes in the earthworm's intestine result in excretion of cocoons and undigested food known as vermicastings that are an excellent manure due to the presence of rich nutrients—vitamins and enzymes, nitrates, phosphates, and potash. The enzymes produced also facilitate the degradation of different biomolecules present in solid waste into simple compounds for utilization by micro-organisms. The different stages of vermicomposting are shown in Figure 14.22.

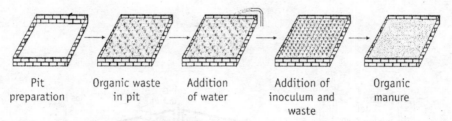

| Pit preparation | Organic waste in pit | Addition of water | Addition of inoculum and waste | Organic manure |

Figure 14.22 Different stages of vermicomposting

Anaerobic digestion systems for solid waste treatment

Anaerobic digestion involves breakdown of organic matter in biomass such as animal dung, human excreta, leafy plant materials, and so on, by micro-organisms in the absence of oxygen to produce biogas, which is a mixture of methane and carbon dioxide with traces of hydrogen sulphide. The optimum temperature for the anaerobic digestion process is 37 °C and the optimum pH is 7. In addition to waste treatment, the process of anaerobic degradation is advantageous due to the generation of biogas that can be used for various thermal applications or for power generation. Besides, digested sludge can be used as manure in place of chemical fertilizers. Biogas is a clean fuel as it is smokeless and thus does not cause health hazards of eye, throat, and lung. In some states in India like West Bengal, biogas slurry is also used for fish feed in pisciculture.

In India, biogas technology is more than five decades old. Different models of biogas plants in use are discussed below.

Biogas plants for animal wastes

KVIC (Khadi and Village Industries Commission) biogas plant

The KVIC has been a pioneering agency in the field of biogas in India. The first plant model, Gramalakshmi, was developed in 1950. This served as a prototype for the KVIC floating dome model that is being disseminated in the country since 1962. The KVIC biogas plant is a composite unit of a masonry digester and a metallic dome that acts as a gas holder wherein gas is collected and delivered at a constant pressure through a pipeline. A constant pressure is maintained due to the upward and downward movement of the lid of the digester along the central guide pipe fitted in a frame in the masonry (Figure 14.23).

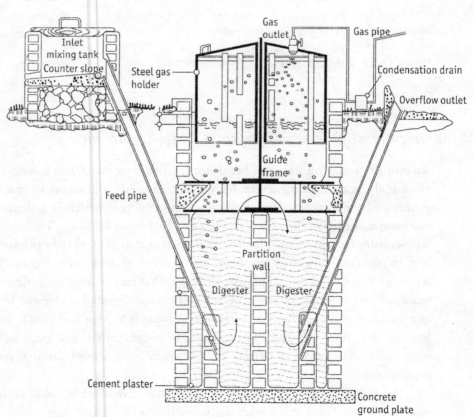

Figure 14.23 Schematic of a KVIC biogas plant
Source Eggeling, Guldager, Hilliges, *et al.* (1979)

Janata biogas plant

The first fixed dome plant (originated in China) was introduced in India in the form of Janata biogas plant by Gobar Gas Research Station, Ajitmal (a wing of Planning, Research, and Action Division, Lucknow) in 1978. In addition to the low installation cost as compared to the KVIC plants, the Janata model plants have the advantages of using locally available building materials and skills in construction. The design consists of a masonry structure, which integrates the reactor and gas-holding space. The inlet and outlet tanks are connected to the digester through displacement chambers. The gas outlet is located on top of the dome and the digested slurry is withdrawn from the opening in the outlet displacement chamber (Figure 14.24).

Deenbandhu biogas plant

The Deenbandhu plant was developed by AFPRO (Action for Food Production), New Delhi, in the early 1980s. In 1981, AFPRO constructed eight biogas plants of 2 m³ capacity of different designs and shapes as part of a training programme in collaboration with FAO/UNDP. The design was standardized after minor modifications in 1984. The design consists of segments of two spheres of different diameters joined at their bases, which acts as the digester as well as the gas storage chamber. To prevent short-circuiting, sufficient distance between the inlet and outlet tanks is maintained by incorporating segments of spheres with base diameters not less than the diameter of conventional Janata plants of the same capacity (Figure 14.25).

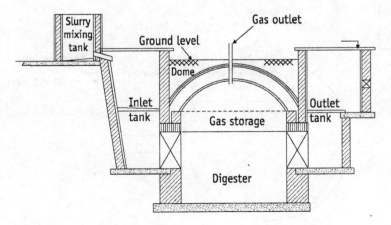

Figure 14.24 Schematic digram of Janata biogas plant

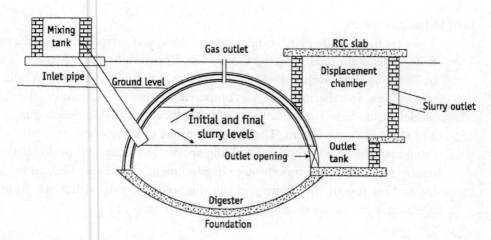

Figure 14.25 Schematic diagram of Deenbandhu biogas plant
Reproduced with permission from The McGraw-Hill Companies
Source Singh, Myles, and Dhussa (1987)

TERI biogas plant

The TERI biogas plant, first developed at its field research unit in Pondicherry, was introduced in 1985 in the village Dhanawas, Gurgaon district (Haryana). The model was introduced as a field prototype, and after improvements and modifications, the final design has been disseminated in different parts of India since 1987. About 173 TERI model biogas plants are installed in 46 villages spread over seven states of India (Figure 14.26).

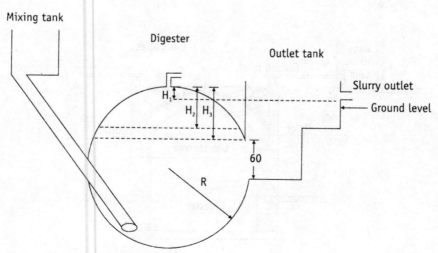

Figure 14.26 TERI fixed dome biogas digester

The TERI spherical type of biogas plant had evolved in an effort to reduce the construction cost and to improve biogas yield in the fixed dome biogas models. Some salient features of the TERI plant are listed below.

- Completely spherical shape, requiring lowest material consumption and amenable for fast construction
- Increased gas storage, prevention of slurry from entering gas pipes due to sufficient distance between slurry level and gas outlet
- Reinforcement of dome with tiles for better durability
- Costs and performance comparable to the Deenbandhu model
- Tangential input or a diffuser box results in an effective residence time of 32 days compared to an HRT of 40 days for conventional plants (Kishore, Raman, and Ranga Rao 1987; Kishore, Raman, Pal, *et al.* 1998).

In addition to the above there are several other models such as the VINCAP, Pragati and Krishna models which are popular in some regions.

Design calculations for fixed dome biogas digester

The biogas digester design is based on quantity of dung available. The important parameters in designing the digester are gas production rate G, active slurry volume V_s, and dome radius. The biogas plant consists of the digester, inlet, and outlet. The digester consists of active slurry volume, gas storage space, and buffer space.

Referring to Figure 14.26, the following parameters are considered for designing the biogas plant.

- Radius of the plant R (m)
- Distance between the gas outlet and slurry exit level of the outlet tank H_1 (m)
- Distance between the gas outlet and initial slurry level (level at nil pressure) in the biogas plant H_2 (m)
- Distance between gas outlet and final slurry level (level at full pressure) in the biogas plant H_3 (m)
- Volume of buffer space in dome V_1 (m³)
- Volume of spherical segment corresponding to height H_2 V_2 (m³)
- Volume of spherical segment corresponding to height H_3 V_3 (m³)

The volumes can be expressed as

$$V_1 = (\pi/3)H_1^2(3R - H_1) \qquad\qquad ...(14.39)$$
$$V_2 = (\pi/3)H_2^2(3R - H_2) \qquad\qquad ...(14.40)$$
$$V_3 = (\pi/3)H_3^2(3R - H_3) \qquad\qquad ...(14.41)$$

The volume of slurry displaced within the digester is equal to the volume of slurry displaced in the outlet tank, which is in turn equal to the required gas storage space. Cooking is generally done twice a day. Hence, the gas storage space would be roughly equal to half the volume of gas produced per day.

$$V_3 - V_2 = A_o \times \text{ height of outlet tank} \qquad \text{...(14.42)}$$

where A_o is the cross-sectional area of the outlet tank (m^2).

Assuming a gas storage space of 60% of the daily gas production, G (m^3),

$$(V_3 - V_2) = 0.6 \, G \qquad \text{...(14.43)}$$

Based on the field experience, to prevent gas line choking, the following values are chosen

$$H_1 = 0.3$$
$$H_3 - H_1 \geq 0.5$$

For an HRT of 40 days, the biogas production from 1 kg of cattle dung is about 0.04 m^3 of gas. With a dilution factor of 1:1 for slurry preparation from dung, the active slurry volume is (G/0.04) × (2/1000) × 40 or 2G m^3.
Hence

$$(4/3)\pi R^3 - V_3 = 2G \qquad \text{...(14.44)}$$

The ratio L/B for the outlet tank is taken as 2
where L is the length and B is the breadth.
Substituting for L = 2B

$$B = \sqrt{(A_o/2)}$$

The equations are solved using the EUREKA software resulting in various dimensions for different gas production volumes (Table 14.15).

Table 14.15 Design values for different gas production rates

Gas volume G, m^3	Radius R	h_1	h_2	h_3	A_o
1	0.98	0.3	0.78	0.98	1.24
2	1.18	0.3	0.78	1.07	2.52
3	1.34	0.3	0.83	1.17	3.38
4	1.43	0.3	0.7	1.14	5.93

Source Kishore, Raman, Pal, *et al.* (1998)

Bioreactors for other organic wastes

Organic wastes such as fruit and vegetable wastes, leafy wastes, food wastes, food-processing industry wastes, and organic fraction of municipal solid wastes after pre-sorting and segregation have a high organic and moisture content. Thus anaerobic digestion is a suitable process of disposal. However, the digester design is not as simple as that for liquid effluents and animal dung. This is due to the heterogeneous nature of the substrate, which affects the efficiency of the reactor during digestion. The use of conventional biogas plant design described above would result in operational problems such as use of large amounts of water for slurry preparation, extensive pre-treatment/shredding and pre-processing, handling of large amounts of slurry, and clogging of the gas outlet due to floating of partially digested material. The different types of biodigester designs used for handling solid wastes of different composition at different concentrations are described below.

Single-stage system

In this system, the entire process of hydrolysis, acidification, and methanogenesis occurs simultaneously in a single reactor. The single-stage system can be subdivided into low-solid (<15%) and high-solid (>20%) systems on the basis of the total solid content in reactor. The low solid process requires pre-treatment to prepare a homogeneous slurry through screening, pulping, and so on. To maintain homogeneity and to prevent the settling of heavier particles and floating of the lighter layer, which can affect the mechanical parts, mixing and periodical removal of scum may be required. Although simple in operation, due to low-solid content, slurry preparation requires the addition of water, which results in increased volume and cost, in addition to the increased drying and maintenance. Short-circuiting of feed material is one of the problems faced in this system.

Another classification is dry or wet system depending on the total solid concentration of slurry. Different configurations in each category are briefly described below.

Dry systems

Dranco (dry anaerobic composting) process

This comprises a thermophilic, single-phase anaerobic fermentation step, which is followed by the aerobic maturation phase. A wide range of solid wastes can be handled at different total solid concentration. Complete digestion of the residue is ensured due to the post-aerobic process.

Kompogas

The process involves thermophilic fermentation for microbial conversion of organic substance present in the material into compost and biogas. The process occurs at a temperature of 55–60 °C and the retention time is 15–20 days (Figure 14.27).

Valorga designs

The Valorga process was developed and patented by the French company Steinmuller Valorga to treat mixed solid waste. It is a single-stage plug flow type process without any mechanical mixing for the treatment of mixed municipal waste resulting in energy production. (Singh 2002) (Figure 14.28).

Wet systems

Single-stage wet systems

WAASA process

The process operates at both thermophilic and mesophilic temperatures, with the thermophilic process having an HRT of 10 days compared to 20 days in the mesophilic design. It has been used for various types of wastes, including municipal solid waste and bio-solids, and the concentration range is 10%–15% of total solids. The digester consists of a pre-digestion chamber and the contents are mixed by biogas circulation.

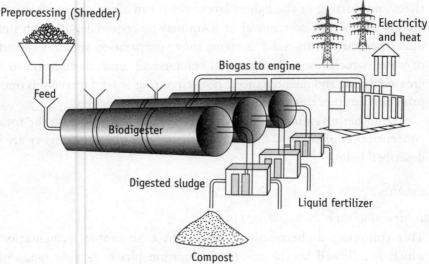

Figure 14.27 Kompogas process
Source <http://www.kompogas.ch/>, last accessed on 24 July 2006

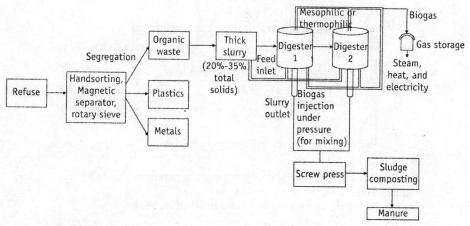

Figure 14.28 Valorga process
Source <www.valorgainternational.fr/en>, last accessed on 10 October 2006

Linde process

The process involves automatic separation of contaminants in the wet preparation stage. The digesters are designed such that there is gas recirculation resulting in high biogas yield. The digested residue has a high compost value due to complete decomposition. The process can also be used for combined digestion of bio-waste and sewage sludge and/or agricultural waste (manure) (<http://62.27.58.13/en/p0052/p0054/p0054.jsp#1>) (Figure 14.29).

Plug flow digester for leafy biomass wastes

The ASTRA centre at the Indian Institute of Science, Bangalore, has been working on various designs to convert leafy biomass to biogas. Based on experience with various digester designs, the plug flow digester design, where the

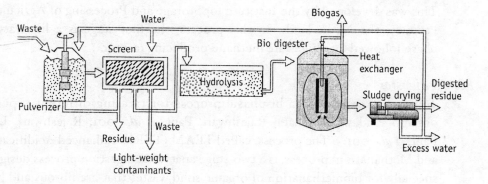

Figure 14.29 Linde process
Reproduced with permission from Elsevier

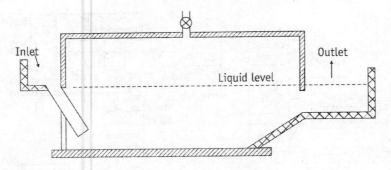

Figure 14.30 Plug flow digester for biogas generation
Reproduced with permission from Elsevier
Source Jagadish, Chanakya, Rajabapaiah, *et al.* (1998)

solid waste movement occurs horizontally, was favoured. A 5 m³/day biogas plant based on the plug flow design was set up for digestion of leafy wastes (Figure 14.30). Feeding of 50 kg/day resulted in gas production varying between 45 and 100 litres/kg of biomass with an average production of 50 litres/kg (Jagadish, Chanakya, Rajabapaiah, *et al.* 1998)

Two-stage systems
HITACHI design

The process includes a thermochemical pre-treatment of waste under alkaline conditions at 60 °C for three hours and a two-phase digestion process consisting of a liquefaction phase at a temperature of 60 °C. This results in reduction in the processing time to eight days.

IBVL design

This was developed by the Institute for Storage and Processing of Agriculture Produce, the Netherlands. In this process, the first stage is a liquefaction phase followed by a high-rate methane-producing reactor.

TEAM digester

TERI has developed a bi-phasic process for treating different types of organic solid wastes (Lata, Rajeshwari, Pant, *et al.* 2001; Rajeshwari, Lata, Pant, *et al.* 2001;). The process, called TEAM (TERI's Enhanced Acidification and Methanation) process, is a two-stage anaerobic digestion process designed specially for biomethanation of organic solid wastes that are fibrous and have light floating materials. Operating between 35 °C and 40 °C, the first stage of the process extracts the organic content from the solid wastes while the second stage generates biogas. The system has six acidification reactors operating in series and a single UASB reactor for methane production from volatile fatty acids. The retention time is six days for the acidification process and one day for the UASB reactor. The process does not involve any agitation mechanism resulting in low maintenance requirement. The digested sludge has a high NPK content and the treated effluent from the UASB is reused for extraction of the organic contents in the acidification phase (Figure 14.31).

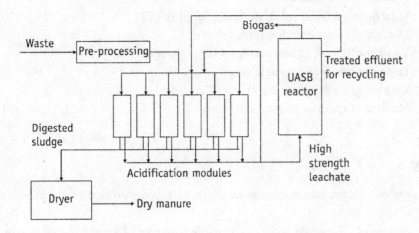

Figure 14.31 TEAM process for organic solid wastes

Nomenclature

A	Cross-section area of the filter (m²)
COD	Chemical oxygen demand (g/m³)
d_e	Diffusivity coefficient in biofilm (m²/day)
d_w	Diffusivity coefficient of substrate in water (m²/day)
h	Depth of packing (m)
H_e	Height of the reactor (m)
k_d	Decay constant (g VSS/g VSS/day)
k_s	Half-velocity constant (g/m³)
k_T	Reaction rate constant at a temperature of T °C (g/g/day)
k_{20}	Reaction rate constant at a temperature of 20 °C (g/g/day)
q	Hydraulic loading rate (litre/m²/min)
Q_r	Sludge recycle flow rate (m³/day)
Q_w	Sludge removal flow rate (m³/day)
Q	Feed rate of the substrate (m³/day)

r_m	Growth rate of micro-organisms (g/m³/day)
r_{sub}	Substrate comsumption rate
S	Substrate concentration (g/m³)
t	Contact time of liquid (min)
i-TSS	Concentration of total suspended solids in the influent (g/m³)
u_s	Superficial upflow velocity (m/hour)
u_t	Settling velocity (m/s)
V	Volume of the reactor (m³)
VSS	Volatile suspended solids in the influent (g/m³)
X	Microbial concentration (g/m³)
Y	Yield coefficient (g biomass/g COD removed)
η	Temperature activity coefficient
μ	Specific growth rate (time⁻¹)
μ_{max}	Maximum specific growth rate (time⁻¹)

Glossary

Aerobic treatment Degradation of organic matter by micro-organisms in presence of oxygen

Anaerobic treatment Degradation of organic matter by micro-organisms in absence of oxygen

Anoxic Condition of depleted oxygen level

ADP (adenosine diphosphate) A chemical converted to ATP

ATP (adenosine triphosphate) High-energy phosphate molecule that stores and releases energy for various metabolic reactions in the cell

Biodegradable substances Organic substances that can be degraded by micro-organisms into simple stable substances

Biogas A combustible gas composed of methane, carbon dioxide, and hydrogen sulphide, produced by anaerobic decomposition of organic material and can be used instead of natural gas as a source of energy

BOD (biochemical oxygen demand) Amount of oxygen consumed by bacteria to decompose organic matter in water or in effluents under aerobic conditions.

Cell The basic structural and functional unit of all organisms

COD (chemical oxygen demand) Amount of oxygen required for chemical oxidation of organic matter in water. A higher COD value indicates a higher level of water pollution. It accounts for oxidation of both biodegradable and non-biodegradable organic matter and hence, COD value is always higher than the BOD value

Enzyme A protein produced by a living organism; acts as catalyst enhancing the rate of a specific chemical or biochemical reaction, without affecting the nature of the reaction

Eukaryotes Unicellular or multicellular organism with a well-defined membrane-bound nucleus

HRT (hydraulic retention time) Average span of time for which waste water or organic slurry remains in a reactor

Media Any liquid or solid preparation containing essential nutrients for growth and maintenance of micro-organisms

Micro-organism Living organism too small to be seen with the naked eye; can be examined only microscopically

OLR (organic loading rate) Amount of organic load in the form of BOD, COD, or total solids applied to the reactor processing waste water/waste per unit time, per unit volume

Prokaryotes Organisms, mostly unicellular, without a well-defined nucleus

Respiration Process by which organisms consume organic matter and oxygen and release energy and carbon dioxide

Sludge Concentrated cell biomass in water or digested waste obtained in semi-solid form during biological treatment processes

SS (suspended solids) Refers to solids greater than 1 mm in size, which can be separated from the waste water by filtration. SS is an important parameter to determine the strength of waste water and efficiency of digestion

Substrate Substance on which an organism lives/grows and obtains nutrients

SVI (sludge volume index) Volume of 1 g of sludge after 30 min of settling; expressed as the ratio of the settled volume of sludge to the suspended solid content

TDS (total dissolved solids) A measure of all soluble solids in water

TOC (total organic carbon) Represents the total carbon atoms in organic molecules present in the organic resource

Total solids Total amount of solids in water including dissolved, suspended, and settleable solids

Total suspended solids A measure of total organic and inorganic suspended matter in liquid

VSS (volatile suspended solids) Amount of suspended solids that burn off at $550 \pm 50\,°C$

References

Abedon S T. 1998
Supplemental Lecture (98/04/06 update)
Ohio State University
Details available at <http://www.mansfield.ohio-state.edu/~sabedon/
biol2020.htm#obligate_aerobe>, last accessed on 29 June 2005

Agarwal S K and Saxena L M. 2001
Composting and Vermicomposting
New Delhi: Society for Environment and Development. 40 pp.

Bhardwaj K K R. 1995
**Improvements in microbial compost technology: a special reference to microbiology
of composting**
In *Wealth from Waste*, pp. 115–135, edited by S Khanna and K Mohan
New Delhi: Tata Energy Research Institute. 280 pp.

Buswell A W and Boruff C B. (1932)
**The relationship between chemical composition of organic matter and the
quality and quantity of gas production during digestion**
Sewage Works Journal 4(3): p. 454

Dhussa A K and Tiwari R C. 2000
Waste-to-energy in India
Bioenergy News 4(1): 12–14

Eggeling G, Guldager H U R, Hilliges G, Sasse L, Tietjen C, Werner U. 1979
Biogas Manual for the Realization of Biogas Programmes
Bremen: Ubersee Museum

Harris C K. 2003
**MS698–3, sediment transport processes in coastal environments Lecture 2:
settling velocities January 2003**
Details available at <http://www.vims.edu/~ckharris/MS698_03/lecture_02.pdf>,
last accessed on 29 June 2005

Jagadish K S, Chanakya H N, Rajabapaiah P, Anand V. 1998
Plug flow digesters for biogas generation from leaf biomass
Biomass and Bioenergy 14(5/6): 415–423

Jimenez J A and Madsen O S. 2003
A simple formula to estimate settling velocity of natural sediments
Journal of Waterway, Port, Coastal and Ocean engineering [ASCE] **129(2)**: 70–78
Details available at <http://www.ocean.washington.edu/people/faculty/parsons/
OCEAN542/jimenez-madsen-jwpcoe-04.pdf>, last accessed on 15 September 2005

Kansal A, Rajeshwari K V, Balakrishnan M, Lata K, Kishore V V N. 1998
**Anaerobic digestion technologies for energy recovery from industrial
wastewater: a study in Indian context**
TERI Information Monitor on Environmental Science 3(2): 67–75

Kishore V V N, Raman P, and Ranga Rao V V. 1987
Fixed Dome Biogas Plants: a design, construction and operation manual
New Delhi: Tata Energy Research Institute. 44 pp.

Kishore V V N, Raman P, Pal R C, Sharma S P. 1998
The TERI fixed dome biogas plant model: a case study of development through user interaction and field research
International Journal of Ambient Energy **19**(4): 199–210

Lata K, Rajeshwari K V, Pant D C, Kishore V V N. 2001
TEAM process: conceptualisation of efforts to meet the challenge of vegetable market waste management problem
Bioenergy News **5**(1): 21–23

Malina J F and Pohland F G. 1992
Design of Anaerobic Processes for the Treatment of Industrial and Municipal Wastes
USA: Technomic Publication Co. Inc. 213 pp

Metcalf and Eddy. 1993
Wastewater Engineering: treatment, disposal, reuse (2nd edn) (12th reprint)
New Delhi: Tata McGraw Hill Publishing Company. 895 pp.

Metcalf and Eddy. 2003
Wastewater Engineering: treatment and reuse, 4th edn
New Delhi: Tata McGraw-Hill, pp. 1819

Rajeshwari K V, Lata K, Pant D C, Kishore V V N. 2001
A novel process using enhanced acidification and a UASB reactor for biomethanation of vegetable market waste
Waste Management and Research **19**: 292–300

Rajeshwari K V, Balakrishnan M, Kansal A, Lata K, Kishore V V N. 2000
State of the art of anaerobic digestion technology for industrial wastewater treatment
Renewable and Sustainable Energy Reviews **4**: 135–156.

Raman P, Sujatha K, Dasgupta S, Kishore V V N. 1989
Residence time distribution studies in noncontinuous flow unstirred tank reactors with reference to biogas digesters
SESI Journal, **3**(2): pp. 1–12

Schmidt J E and Ahring B A. 1996
Granular sludge formation in UASB reactors
Biotechnology and Bioengineering **49**: 229–246

Singh J B, Myles R, and Dhussa A. 1987
Manual on Deenbandhu Biogas Plant
New Delhi: Tata McGraw Hill Publishing Company. 895 pp.

Stryer L. 1986
Biochemistry (First Indian edn)
New Delhi: CBS Publishers and Distributors. 927 pp

The Gazette of India notification. 2000
Ministry of Environment and Forests, Notification, New Delhi, 25 September 2000

US Composting Council. 2002
Evaluating composting quality
Details available at <http://www.compostingcouncil.org/section.cfm?id=39>,
last accessed on 17 September 2005

Yim G and Glover C.
Food microbiology: background, basics and the details of cheese production
BioTeach
Details available at <http://www.bioteach.ubc.ca/Bioengineering/FoodMicrobiology/#Fig 1>,
last accessed on 25 October 2005

Bibliography

Technology Innovation Management and Entrepreneurship Information service, a joint
project of NSTEDB, Department of Science and Technology and Federation of Indian
Chambers of Commerce and Industry
Site accessed on November 2004

Linde-KCA-Dresden

Singh R. 2002
The 1.5 million ton success: biomethanation of MSW by the Valorga process
Bioenergy News **6** (4)

The Kompogas process
Details available at <http://www.kompogas.ch/>, last accessed on 14 September 2005

Liquid fuels from biomass: fundamentals, process chemistry, and technologies

15

Linoj Kumar N V, Research Associate
Energy–Environment Technology Division, T E R I, New Delhi

Introduction

The development of biofuels can be traced back to early 19th century. In fact, development of diesel engines and biofuels has a concurrent history of technological advancements and economic struggle. The first authentic internal combustion engine in the US, developed in 1826, ran on alcohol and turpentine (Kovarik 1998). In 1896, Henry Ford built his first automobile to run on ethanol. In 1898, Rudolf Diesel operated an engine fuelled by peanut oil (Peterson 1986; Peterson and Auld 1991; Peterson, Wagner, and Auld 1983). During World War II, biofuels emerged as a major source to power the war machinery. Synthetic methanol was also used extensively to fuel vehicles. However, after the war, with a steady supply of petroleum fuels coupled with low prices, there was a decline in production and use of biofuels.

Liquid biofuels re-appeared in the 1970s, when there was substantial increase in the production and use of biofuels in many countries triggered by the oil shock. However, biofuels became less competitive vis-à-vis fossil fuels after the collapse of oil prices in the mid-1980s, and interest in biofuels in the transport sector diminished noticeably in the 1990s and the beginning of the 21st century. However, the upsurge in oil prices in the international market once again focussed attention on biofuels, particularly in countries with heavy dependence on oil import.

Today, concerns of energy security and climate change are driving the development of biofuels globally. Interest has grown significantly along with favourable market responses. World production of fuel ethanol reached 49 billion litres in 2007.[1] Bio-diesel production is also growing rapidly. Recent advancements in technologies are also catalysing the continued growth of biofuels.

Biofuels offer several advantages in terms of resource potential and applications. Primarily, these are renewable as long as they are properly managed. In contrast to fossil fuels, where reserves are restricted to a few pockets, biomass is evenly distributed all over. More importantly, biofuels provide enhanced employment opportunities and livelihood generation, leading to regional as well as national self-sufficiency. From the environmental point of view, biofuels are carbon-neutral fuels. Provided that an equivalent quantity of vegetation is replanted, use of biomass as fuel will not result in a net change in atmospheric carbon dioxide levels. Emission levels are also considerably less as compared to fossil fuels because of comparatively less nitrogen and sulphur in biofuels. As biofuels are oxygenated, these are clean burning fuels.

Liquid biofuels, as their name suggests, are fuels derived from biomass that is processed to produce a combustible liquid fuel. The liquid biofuels significant in the current context are alcohol fuels — methanol, ethanol, butanol, and vegetable oils — derived from plant seeds. The first section of the chapter deals with basic chemistry of the biomass components that are relevant to the production of liquid fuels. The remaining sections deal with the fundamentals of conversion of these constituents into liquid fuels, covering the principles of equilibrium, kinetics, etc. The chapter also gives an overview of different process stages and existing technologies for production of these fuels.

Biomass constituents relevant to production of liquid fuels

All organic materials produced by photosynthesis in the plant matter are of relevance to biomass energy utilization in one way or other. The major classes of organic derivatives, significant to liquid biofuel production, are triacylglycerols, carbohydrates, and lignin. While triglycerides are the key components for production of vegetable oil and bio-diesel, carbohydrates are a source of sugars that undergo fermentation to produce ethanol. Lignin is a complex biopolymer, which plays a critical role in both thermochemical and biochemical conversion of lignocellulosic biomass. Before going into the details of the individual liquid fuel, it is useful to have a basic understanding of the chemistry of these biomass components.

[1] US Renewable Fuels Association, details available at <http//www.ethanolrfa.org/resource/facts/trade> last accessed on 3 September 2008.

$$
\begin{array}{lll}
CH_2\!-\!\!-OH & HCOO\!-\!\!-R^{i} & CH_2\!-\!(HO\ \ H)\!-\!COO\!-\!\!-R^{i} \\[4pt]
CH\!-\!\!-OH \ + & HCOO\!-\!\!-R^{ii} \longrightarrow & CH\!-\!(HO\ \ H)\!-\!COO\!-\!\!-R^{ii} \\[4pt]
CH_2\!-\!\!-OH & HCOO\!-\!\!-R^{iii} & CH_2\!-\!(HO\ \ H)\!-\!COO\!-\!\!-R^{iii}
\end{array}
$$

(Glycerol)   (Fatty acid)   ...(15.1)

$$
\xrightarrow{-3H_2O}
\begin{array}{l}
CH_2\!-\!\!-O\!-\!\!-CO\ R^{i} \\[4pt]
CH\!-\!\!-O\!-\!\!-CO\ R^{ii} \\[4pt]
CH_2\!-\!\!-O\!-\!\!-CO\ R^{iii}
\end{array}
$$

Triglyceride (fatty acid esters of glycerol)

Triacylglycerols

Triacylglycerols commonly, known as triglycerides, are the main energy storage molecules in vegetable oils and animal fats. They are the predominant components of lipids present in the seeds or in the fleshy part of fruits. These tissues are, therefore, important commercial sources of fats and oils (Gurr 1980).

Chemically, triglycerides are the fatty acid esters of glycerol in which each of the three hydroxyl groups of glycerol are esterified with a fatty acid (Equation 15.1). Triglycerides have lower densities compared to water. At room temperature, they may exist as solid or liquid. When they exist as solid, they are called fat and when as liquid, they are called oil.

Partial glycerides have one or two hydroxyl groups esterified and are mono or diacylglycerols, respectively (Figure 15.1). Sometimes, additional fatty acids may be esterified to these mid-chain hydroxyl groups so that tetra, penta, and hexa acid glycerides occur in some plant oils (Gurr 1980).

Physical and chemical characteristics of fats/oils have a degree of correlation with the type and percentage of the fatty acids and their positioning with respect to the glycerol molecules. The component fatty acids can be saturated or unsaturated. While saturated fatty acids have only carbon–carbon single bonds, unsaturated fatty acids have carbon–carbon double/triple bonds. The melting point of saturated fatty acids increases with the chain length. For example, decanoic acid and longer chain molecules are solid at room temperature. Also, unsaturated fatty acids have a lower melting point than saturated fats, and most of them are liquid at room temperature.

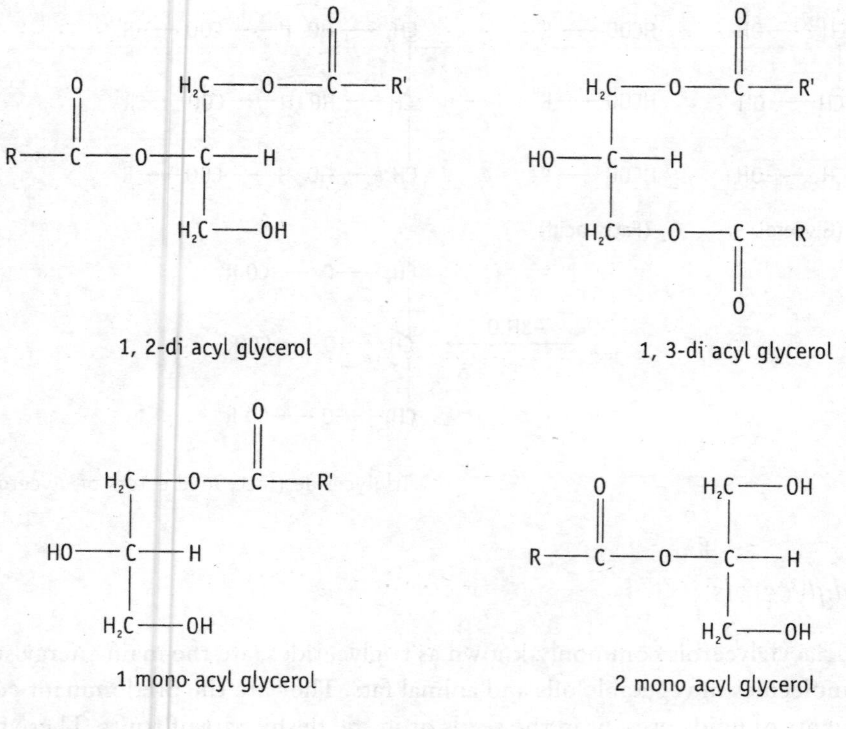

Figure 15.1 Structure of mono, di, and triglycerides

Mainly, five fatty acids dominate in plant triacylglycerols: palmitic, stearic, oleic, linoleic, and α-linolenic. Butyric, caproic, and vaccenic acids are commonly found in butterfats (Table 15.1). There are hundreds of others, which may also contribute a major proportion in individual seed oils. Examples are erucic acid in rapeseed and ricinoleic acid in castor oil (see Gurr (1980) for more information).

Carbohydrates

Chemically, carbohydrates are poly-hydroxy aldehydes and poly-hydroxy ketones or compounds that can be hydrolysed to them (Berg, Tymoczko, and Stryer 2003). All carbohydrates are saccharides, that is, either sugars or polymers of sugars. Biochemical conversion of carbohydrates present in biomass is one of the main routes for production of different value-added chemicals such as alcohols. Therefore, it is necessary to have an idea about the different types of carbohydrates present in biomass and their properties.

Table 15.1 Chemical names and descriptions of some common fatty acids

Common name	Carbon atoms	Double bonds	Scientific name (IUPAC)	Melting point, °C	Sources
Butyric acid	4	0	Butanoic acid	−7.9	Butterfat
Caproic acid	6	0	Hexanoic acid	−3.4	Butterfat
Palmitic acid	16	0	Hexadecanoic acid	62.9	Palm oil
Palmitoleic acid	16	1	9-Hexadecenoic acid	0	Animal fats
Stearic acid	18	0	Octadecanoic acid	69.6	Animal fats
Oleic acid	18	1	9-Octadecenoic acid	16.3	Olive oil
Vaccenic acid	18	1	11-Octadecenoic acid	44	Butterfat
Linoleic acid	18	2	9,12-Octadecadienoic acid	−6.5	Grapeseed oil
α-Linolenic acid	18	3	9,12,15-Octadecatrienoic acid	−12	Flaxseed (linseed) oil
γ-Linolenic acid	18	3	6,9,12-Octadecatrienoic acid		Borage oil

Based on the major functional group present, carbohydrates can be classified as aldoses and ketoses, which contain aldehyde and ketone groups, respectively. Depending on the size of molecules and difference in linkages, carbohydrates can be generally categorized as monosaccharides, disaccharides, and polysaccharides.

Monosaccharides

Monosaccharides are the simplest carbohydrates, which cannot be further hydrolysed. All monosaccharides, whether they are aldose or ketose, are reducing sugars because of the presence of hemiacetal linkages.[1] Depending on the number of carbon atoms present, monosaccharides can be classified as triose, tetrose, pentose, hexose, and so on. Because these molecules have multiple asymmetric carbon atoms, they exist as diasterioisomers as well as enantiomers (Berg, Tymoczko, Stryer 2003). For example, $C_6H_{12}O_6$ represents several sugars, including glucose, mannose, and fructose (Figure 15.2)

The most vital property of sugar relevant to biomass conversion is its fermentability. A few examples of fermentable sugars are D-glucose, D-mannose, D-fructose, and D-galactose.

Disaccharides

As sugars contain many hydroxyl groups, glycosidic bonds can join one monosaccharide to another. In other words, the alcohol component of a

---

[1] Many sugars in solution are not open chains. Open chains of these sugars cyclize into a ring in which joining of an aldehyde and alcohol results in hemiacetal linkages.

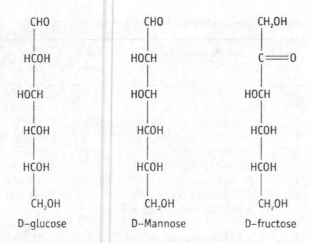

Figure 15.2 Structure of D-glucose, D-mannose, and D-fructose

glycoside is provided with the hydroxyl function of another monosaccharide, and the resulting compound is called a disaccharide. For example, in maltose, two D-glucose residues are joined by a glycosidic linkage between the α-anomeric form of C_1 on one sugar and the hydroxyl oxygen atom of C_4 on the adjacent sugar (Figure 15.3). Such a linkage is called α-1,4 glycosidic bond.

In short, disaccharides are composed of two monosaccharides. Conversely, disaccharides can be converted into simple fermentable sugars by hydrolysis using either chemicals or enzymes. For example, sucrose, which can be thought of simply as a combination of two sugars, glucose and fructose, is hydrolysed by the enzyme invertase present in yeast.

$$C_{12}H_{22}O_{11}(\text{sucrose}) + H_2O(\text{water}) \xrightarrow{\text{Invertase}} C_6H_{12}O_6 \,(\text{glucose}) + C_6H_{12}O_6 \,(\text{fructose})$$

$$...(15.2)$$

The resultant hydrolysed sugars can then be fermented to ethanol.

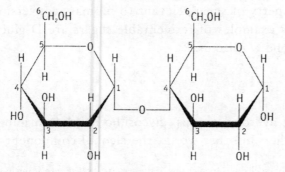

Figure 15.3 Glycosidic linkages in maltose

Example 1
While most of the disaccharides are reducing sugars, why is sucrose not?
(A sugar with hemiacetal/hemiketal group designated by COCOH is a reducing sugar)

Solution
Sucrose is formed by the combination of α-D glucose and β D fructose units.

The *glycosidic bond* for sucrose is sometimes referred to as an α-β-1-2 bond. The glucose and fructose units are joined by an acetal oxygen linkage in the alpha orientation. It contains a six-member ring of glucose and a five-member ring of fructose. The alpha acetal linkage is part of a double acetal, where hemiacetal of glucose and the hemiketal of fructose are joined together. Therefore, there is no hemiacetal remaining in the sucrose and hence no reducing groups are present.

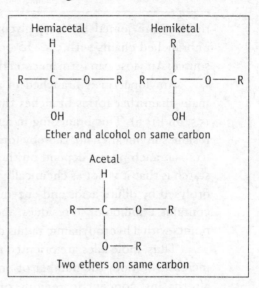

Polysaccharides

Polysaccharides are naturally occurring carbohydrate polymers formed from a combination of multiple monosaccharides by elimination of water molecules between adjacent ones. The most important compounds in this class are cellulose, hemicellulose, and starch. As these are abundantly available in nature, they are considered as the potential commercial sources of ethanol. In this context, base sugars are of extreme significance during biomass processing (Table 15.2).

Starch

Starch is considered to be the nutritional reservoir in plants. It is a granular polysaccharide, which accumulates in the storage organs of plants such as seeds, tubers, and roots. Starch consists of 10%–20% of ∝-amylose and 80%–90% amylopectin. Amylose and amylopectin are inherently incompatible molecules. α-Amylose is a linear molecule in which D-glucose

Table 15.2 Important polysaccharides and their composition

Polysaccharides	Approximate representation	Monomer building block	Molar mass
Cellulose	$(-C_6H_{10}O_5-)_n$	D- glucose	>100 000
Starch	$(-C_6H_{10}O_5-)_n$	D-glucose	35 000–90 000
Hemicellulose		Various sugars	10 000–35 000
	(mainly pentoses)		

residues are joined by α-1,4 glycosidic bonds. It consists of single, mostly unbranched chains with 500–20 000 α-(1→4)-D-glucose units dependent on source. Amylose can form an extended shape (hydrodynamic radius 7–22 nm.

Amylopectin is branched in a way that α-1,4 glycosidic bonds form a main chain that forms branches through α-1,6 glycosidic bonds (Figures 15.4, 15.5, and 15.6). This branching in amylopectin occurs for every 24–30 glucose residues in linear chain, thereby forming huge, but compact molecules and the -(1,6)- branch points depend on the source of starch. The major advantage of starch is that it is not as chemically resistant as cellulose and can be readily hydrolysed by dilute acids and enzymes to glucose. Each amylopectin molecule contains a million or so residues, about five per cent of which form the branch points with a hydrodynamic radius of 21–75 nm.

Thus, molecules are oriented radially in the starch granule. As the radius increases so does the number of branches, with the consequent formation of alternating concentric regions of amorphous and crystalline structures.

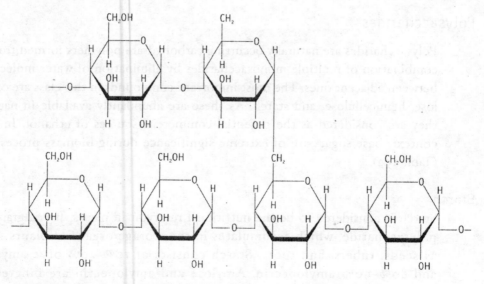

Figure 15.4 Glycosidic linkages in amylopectin

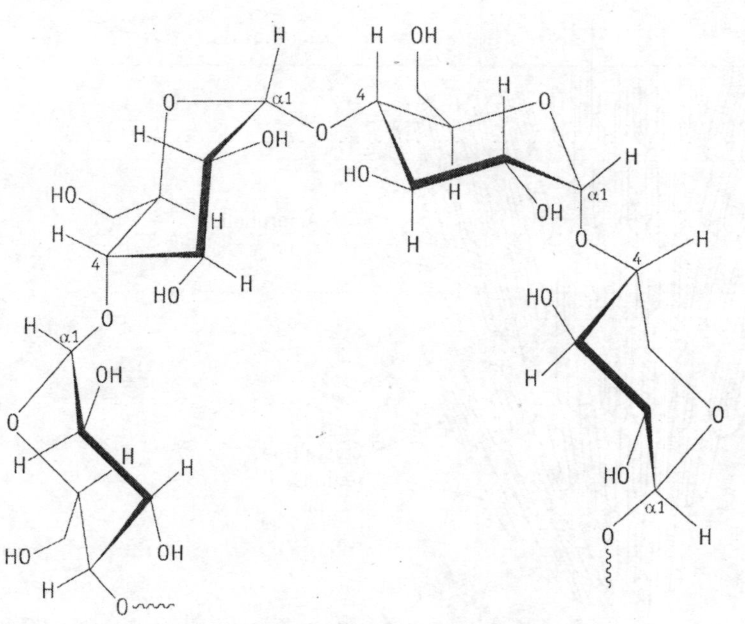

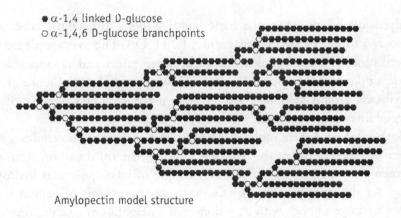

Figure 15.5 Representative partial structure of amylose

● α-1,4 linked D-glucose
○ α-1,4,6 D-glucose branchpoints

Amylopectin model structure

Figure 15.6 Representation of amylopectin branching pattern

Figure 15.7 shows the organization of amorphous and crystalline regions of the structure generating concentric layers that contribute to the growth rings that are visible by light microscopy.

Cellulose

Cellulose is the most abundant natural organic polymer in the biosphere, having a linear structure in which the glucose molecules are joined together by

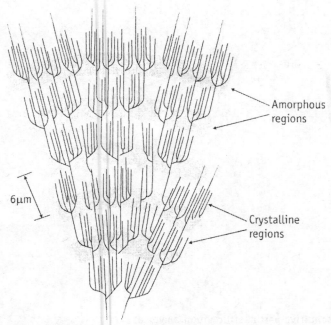

Figure 15.7 Molecular orientation of a starch granule

β-1,4 glycosidic bonds. Thus, the basic building block of cellulose is actually cellobiose, a dimer of two glucose units ($C_6H_{10}O_5$). The average degree of polymerization of plant cellulose varies between 7000 and 15 000 glucose units, depending on the source (Fengel and Wegener 1983). Cellulose chains have hydroxyl groups at both ends and have a reducing end because of the hemiacetal linkage (Figure 15.8).

Both inter- and intra-molecular hydrogen bonds exist in cellulose. The hydrogen bonding between the hydroxyl group of the third carbon atom and the oxygen atom in pyranose ring is an example of intra-molecular hydrogen bonding. As shown in the Figure 15.9, each monomer unit of cellulose can form two hydrogen bonds with a monomer in a neighbouring chain, resulting in an inter-molecular interaction. While the hydrogen bonds are formed between the chains in each layer (Figure 15.10a), the layers are connected with weak Van der Waals forces (Figure 15.10b). Thus, small units of cellulose formed through side-by-side hydrogen bonding are layered by Van der Waals forces. This causes the chains to be packed together into larger units called micro fibrils. This results in a stable configuration, which is essentially free of interstitial spaces, making it anhydrous. The important chemical property of cellulose which affects the biochemical conversion, is its extreme insolubility in water and recalcitrance towards hydrolysis reactions. However, in

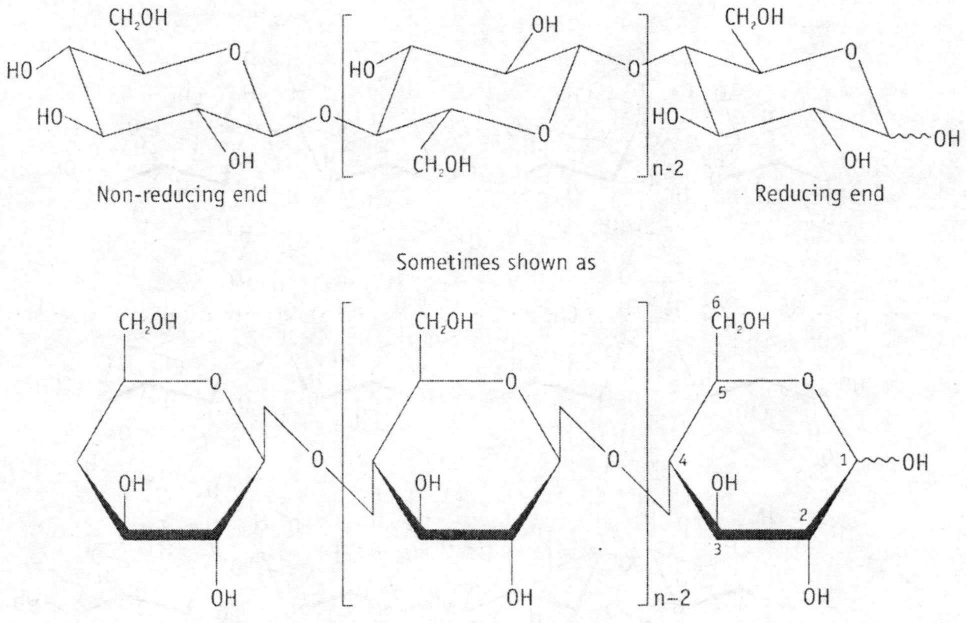

Figure 15.8 Structural representation of cellulose $(-C_6H_{10}O_5-)_n$

addition to the highly crystalline regions, native cellulose contains less ordered amorphous regions that can be hydrolysed rather easily by chemicals or enzymes providing a beginning to the cellulose saccharification.

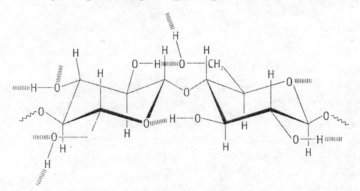

Figure 15.9 Hydrogen bonding in cellulose (intra- and intermolecular hydrogen bonding)

Hemicellulose

Hemicelluloses belong to a group of heterogeneous polysaccharides that are formed through biosynthetic routes different from those of celluloses. Hemicelluloses occur in association with cellulose, but are chemically different from it in several aspects. Hemicellulose is an amorphous polymer. It has

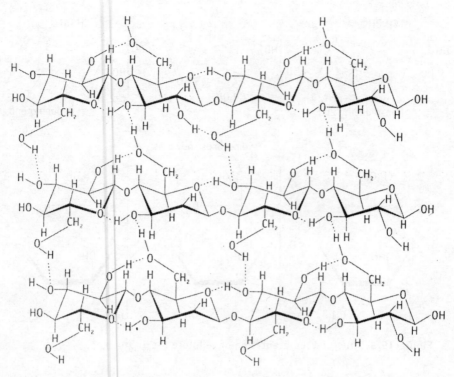

Figure 15.10(a) Linkage between the linear cellulose molecules by hydrogen bonding

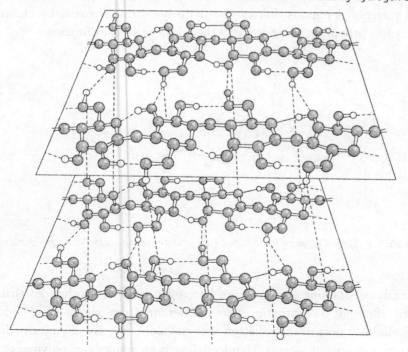

Figure 15.10(b) Staggered layers of cellulose sheets linked by Van der Waals forces

much lower molar masses than those of cellulose. It can be dissolved in dilute alkaline conditions. It can be relatively easily hydrolysed by acids to monomer components consisting of glucose, mannose, galactose, xylose, arabinose, and small amounts of rhamnose, glucuronic acid, methyl glucuronic acid, and galacturonic acid. They usually contain two to four different sugars as building blocks. Hemicelluloses of softwoods and hardwoods have different structures and composition.

D-glucose is a component of some hemicelluloses, although xylose is the dominant sugar in hardwood hemicellulose, and mannose is important in softwood hemicellulose. Most hemicelluloses have a DP (degree of polymerisation) of about 200.

Lignin

Lignin is a complex biopolymer having a heterogeneous three-dimensional structure made up of phenyl propanoid units. It forms a matrix surrounding the cellulose and hemicellulose, and acts as a natural glue, providing the plant with an additional mechanical strength. Biosynthetically, lignin arises from glucose via the formation of three precursor alcohols, namely, coniferyl, sinapyl, and p-coumaryl alcohols (Figure 15.11). In addition, lignins contain minor amounts of various aromatic acids (for example, vanillic, ferulic, p-coumaric, etc.) in ester-like combinations. The typical functional groups are methoxyl, phenolic hydroxyl, benzyl alcohol, and carbonyl groups (Figure 15.12). Chemical bonds have been reported between lignin and hemicellulose and even cellulose (Watanabe 2003). The possible linkages are ester, ether, acetal, and glycosidic bonds. Ether linkages are more common and more

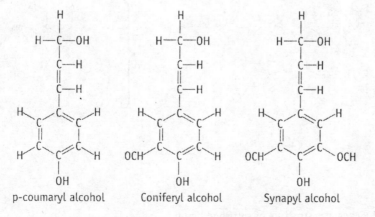

Figure 15.11 Three basic units of lignin: p-coumaryl alcohol, coniferyl alcohol, and synapyl alcohol

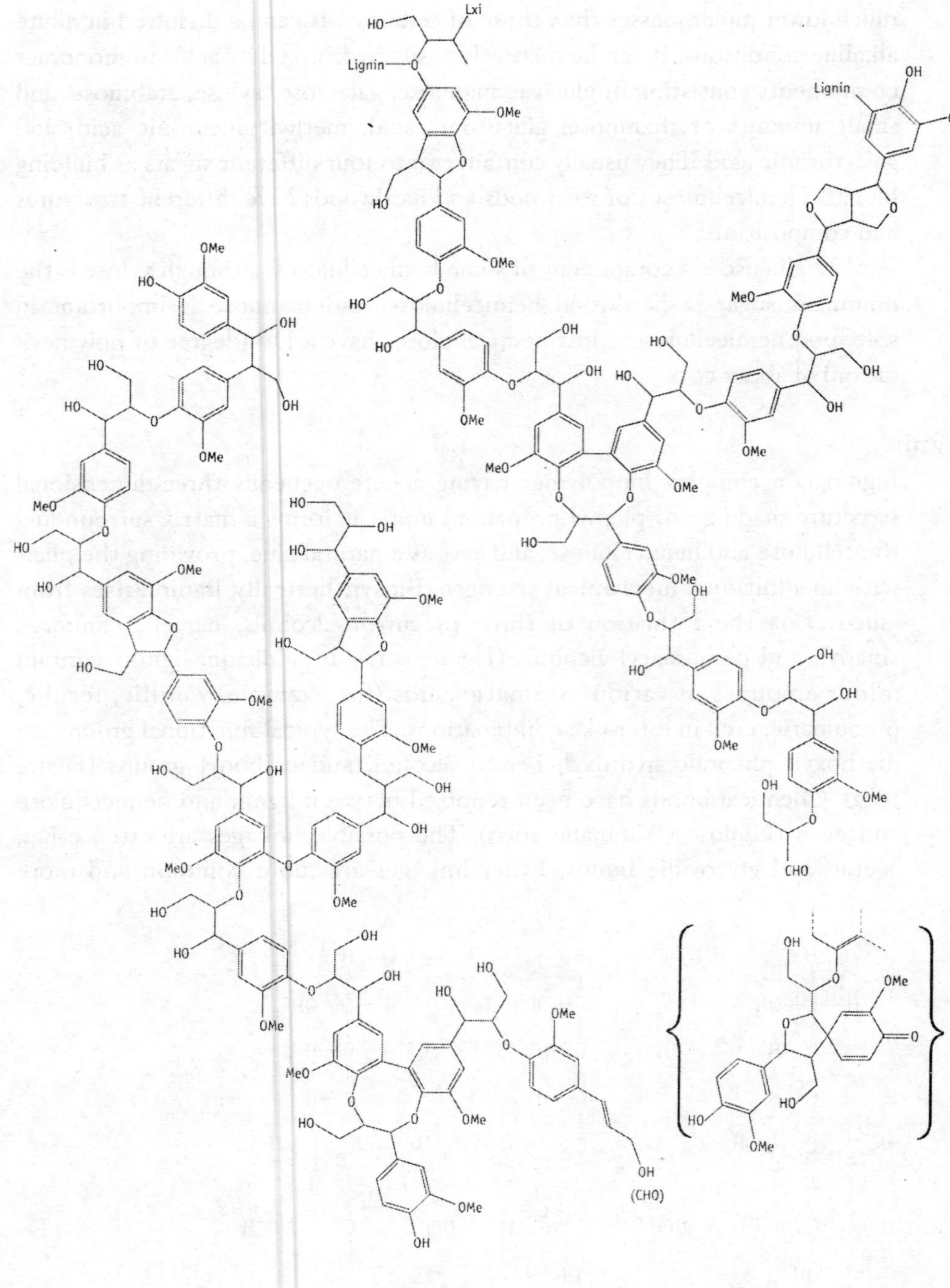

Figure 15.12 Schematic structure of softwood lignin
Source <http://www4.ncsu.edu/~dsargyro/>, last accessed on 31 March 2007

stable than ester bonds between lignin and carbohydrates. These units and linkages make lignins extremely resistant to chemical and enzymatic degradation. This recalcitrant nature of lignin is one of the major problems involved in the biochemical conversion of biomass. Biodegradation can mainly be accomplished by fungi and also by certain actinomycetes.

On the other hand, lignin plays a major role in the thermal conversion of biomass. It is the main contributor to the energy content of a fuel and its reduction may lead to calorific value reduction as well. It is well proven that HHV (higher heating values) of lignocellulosic fuels increase with an increase in their lignin content. A linear correlation exists between the HHV of the biomass fuel and its lignin content. HHV of the biomass fuel can be calculated as a function of lignin content using the following equations.

$$HHV\ (kJ/g) = 0.0893L + 16.9742 \quad \text{for woody biomass} \qquad ...(15.3)$$
$$HHV\ (kJ/g) = 0.0877L + 16.4951 \text{ for non-woody biomass} \qquad ....(15.4)$$

where L is the lignin content expressed in percentage.

The correlation coefficient r for first and second equations is 0.9658 and 0.9302, respectively (Demirbas 2001).

Liquid biofuels: alcohols

Methanol, ethanol, and bio-diesel are the three most common biomass-derived liquid fuels. The potential production processes considered here are conversions through chemical, biochemical, and thermochemical routes. While biochemical and thermochemical conversion processes are more appropriate for production of ethanol and methanol, respectively, chemical synthesis reactions, such as transesterification, hold promise for effective use of vegetable oils in the form of bio-diesel.

Ethanol is a type of alcohol produced by fermentation of sugars and starches or cellulosic biomass. Commercial production of ethanol is largely from sugar cane or sugar beet, as starches and cellulosic biomass usually require expensive pre-treatment before fermentation. The substitution of gasoline with ethanol in passenger cars and light vehicles is an extensive biofuels programme in Brazil and the US. Engines that run strictly on gasoline are no longer available in Brazil, having been replaced by neat-ethanol engines and gasohol engines that burn a mixture of 78% gasoline and 22% ethanol by volume (Newton 2003). Recent technological advances, including more efficient production and processing of multiple feedstock, are responsible for the availability of ethanol at comparatively low prices. Apart from being used as a fuel, ethanol finds numerous applications in the manufacture of cosmetics, pharmaceuticals, and also in the production of alcoholic beverages.

Methanol is produced by the process of thermochemical conversion. It can be produced from any biomass through the process of gasification followed by the catalytic hydrogenation of carbon monoxide/dioxide. As with ethanol, it can either be blended with gasoline to improve the octane rating of the fuel or used in its neat form as a fuel. Methanol is used extensively as an organic solvent and also for production of several chemicals, including formaldehyde, chloromethanes, methylamines, DMT (N,N-Dimethyltryptamine), pesticides, various bulk drugs, methyl acrylate, and dye intermediates.

Another way of extracting energy from biomass is through the use of vegetable oils as fuel. Many edible and non-edible oil crops are being grown in some countries for use as possible petroleum fuel substitutes. Because of problems associated with the high viscosity of the straight vegetable oil, they are also processed into a less viscous esterified oil known as bio-diesel. Bio-diesel has fuel characteristics comparable to diesel.

The details of all these fuels, their fundamentals, process chemistry, and an overview of the production technologies, are discussed in the following sections.

Methanol

Definition and properties

Methanol is a colourless, odourless, volatile, and inflammable alcohol, obtained by the destructive distillation of wood (Table 15.3). It is the simplest

Table 15.3 Properties of methanol

Physical properties	
Boiling point	64.7 °C
Freezing point	−97.68 °C
Density at 25 °C	780 kg/m^3
Vapour pressure at 25 °C (kPa)	127.2 mm Hg
Specific heat capacity (C_p)	79.5 J/mol·K
Fuel properties	
Flash point	11 °C
Auto ignition temperature	455 °C
Viscosity of liquid at 25 °C (cP)	0.541
Gross calorific value	23 MJ/kg

Source DSIR (1991)

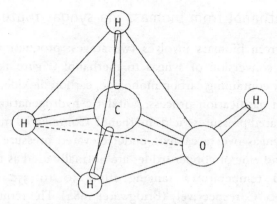

Figure 15.13 Tetrahedral structure of methanol

alcohol and has chemical formula CH_3OH (Figure 15.13). It is also known as wood alcohol or carbinol. It has a pleasant characteristic odour and pungent taste. It is highly poisonous and causes blindness, madness, and death, depending on the volume consumed. It is highly miscible in all proportions with water.

Methanol is the first in the series of aliphatic monohydric alcohols. As it contains only one carbon atom, it cannot undergo elimination of the hydroxyl group and beta hydrogen to form analogous olefin[2] unlike many other higher alcohols (DSIR 1991). Otherwise, it undergoes reactions that are typical of alcohols as a chemical class.

Production of methanol from biomass

Destructive distillation of wood

Destructive distillation is an ancient method for production of alcohol from biomass and it was the only source for commercial production of methanol till 1927 (Paul 1978). Hardwoods such as birch, oak, etc., thermally decompose during pyrolysis at temperatures ranging from 160 °C to 430 °C forming charcoal, tar, non-condensable gases, and a watery distillate known as pyroligneous acid. Pyroligneous acid, when distilled, gives a crude distillate of methanol having an offensive odour. The yields in this process are comparatively very small, ranging from 1% to 2% (six gallons per tonne) methanol (Paul 1978).

---

[2] Any class of unsaturated open-chain hydrocarbons, such as ethylene, having general formula C_nH_{2n}; an alkene with only one carbon–carbon double bond.

Synthetic production of methanol from biomass via syngas route

Production of methanol from biomass involves two stages: production of syngas from biomass and conversion of syngas to methanol (Figure 15.14). Syngas is a gaseous product containing carbon monoxide, carbon dioxide, and hydrogen and results from gasification process. Catalytic hydrogenation of these carbon oxides at elevated pressures yields methanol. Commercial manufacture involves passing syngas over a catalyst under elevated pressure and temperature. Chromium and zinc or copper oxides are normally used as catalysts with pressures and temperatures ranging from 50 to 350 atm (atmosphere) and up to 400 °C, respectively (Bridgwater 1984). The requisite temperature and pressure vary with composition of syngas. Average yield of methanol by this synthetic route is 100 gallons of methanol per tonne of feed material (Bridgwater 1984; Paul 1978).

Production of syngas from biomass

As mentioned above, syngas can be derived from gasification of biomass. Gasification is the partial oxidation of biomass with a sub-stoichiometric amount of oxygen. Product gas can either be producer gas or syngas, depending on the source of oxygen used. If air is used as a gasification agent, the products are diluted with nitrogen, generating fuel gas or producer gas. If pure oxygen or O_2/steam is used instead of air as a gasification agent, syngas is produced, which is rich in hydrogen and carbon monoxide. If only steam is used, no pure oxygen is required, but the reaction is strongly endothermic and heat has to be supplied resulting in a complex reactor system.

Many gasification reactors are available for syngas production. Based on the throughput, cost, complexity, and efficiency issues, only circulated fluidized bed gasifiers are preferred for large-scale fuel gas production. Direct gasification with air strongly increases the downstream equipment size as it results in dilution with nitrogen. Therefore, directly air-blown gasifiers are not

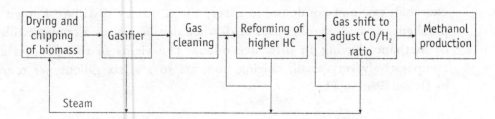

Figure 15.14 Process diagram for the production of methanol from biomass

usually preferred. Gasifiers such as GTI (Gas Technology Institute) pressurized direct-oxygen-fired gasifier, and the BCL (Battelle Columbus) atmospheric indirectly fired gasifiers are suitable for this process as both the gasifiers can produce medium calorific value gas undiluted by atmospheric nitrogen (Hamelinck and Faaj 2002).

GTI uses oxygen to reduce the downstream equipment size. It produces carbon-dioxide-rich gas, which lowers overall yield of methanol. However, the carbon monoxide/hydrogen ratio is still better for methanol production though the gasifier efficiency is lower and much more steam is needed. The indirectly fired BCL gasifier is fired by air but without any risk of nitrogen dilution, and there is no need of oxygen production and use. It produces a gas with low carbon dioxide content, but contains heavier hydrocarbons. Therefore, reforming process is necessary to enhance the carbon monoxide and hydrogen content.

Syngas cleaning

The produced gaseous mixture contains tars, dust, alkali compounds, and halogens, which can deactivate the catalysts downstream. The gas can be cleaned using the available conventional technology consisting of gas cooling, low-temperature filtration, and water scrubbing at 100–200 °C. Alternatively, hot gas cleaning can also be considered using ceramic filters and chemical reagents at 350–800 °C (Hamelinck and Faaj 2002).

As can be seen in Figure 15.15, in low-temperature cleaning, particulate matters are completely removed by the cyclone, bag filter, and the scrubbers. Essentially, all alkali and the bulk of sulphuric and nitrogenous compounds are removed by consecutive scrubbers. The zinc oxide bed or solvent absorption unit further brings down the sulphur concentration. In the hot gas cleaning system, granular bed and candle filters remove 99.8% of the particles (Hamelinck and Faaj 2002). Nitrogen and sulphur oxides are removed by the adsorbents. Alkali removal via physical adsorption or

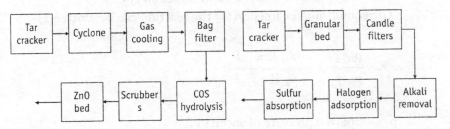

Figure 15.15 Conventional low-temperature wet cleaning (left) and advanced high temperature dry-cleaning systems (right)

chemisorption can be implemented at 750–900 °C. Sulphur is further removed by chemisorption. Thereafter, in absence of hydrogen sulphide, 99.5% of ammonia can be decomposed over a nickel catalyst. However, hydrogen cyanide cannot be sufficiently removed by hot gas cleaning, leading to shorter catalyst life in downstream reactors.

Syngas processing

Production of synthesis gas that meets the purity levels for methanol synthesis is a pre-requisite before it goes to the methanol reactor. The water gas shift is a common process operation to shift the energy value of carbon monoxide to hydrogen by reaction with steam. It brings the ratio of gases closer to the stoichiometric ratio. The required stoichiometric ratio of hydrogen and carbon monoxide is 2:1 and that of hydrogen to carbon dioxide is 3:1. The reaction is as given in Equation 15.5.

$$CO + H_2O(g) \rightleftharpoons CO_2 + H_2 \qquad \Delta H°_{298} = -41.2 \text{ kJ/mol} \qquad ...(15.5)$$

The equilibrium constant for the water gas shift reaction increases as the temperature decreases.

Also, sometimes, biomass-derived syngas contains methane and other light hydrocarbons, which represent a significant part of the heating value of gas. Steam reforming over nickel catalyst converts these compounds to carbon monoxide and hydrogen. Auto thermal reforming is another alternative to steam reforming, which combines partial oxidation in the first part of the reactor with steam reforming in the second part, thereby optimally integrating the heat flows.

Example 2

Find the equilibrium composition of a gas initially comprising 0.4 mol CO and 1 mol H_2O, at T = 1200 °C and P = 2 atm. (K = 0.371 at 1200 °C).

The water gas shift is as shown in Equation 15.5.
The equilibrium expression for the reaction can be written as

$$K = \frac{p_{H_2} \cdot p_{CO_2}}{p_{CO} \cdot p_{H_2O}} = \frac{x_{H_2} P \cdot x_{CO_2} P}{x_{CO} P \cdot x_{H_2O} P}$$

where

p_i = partial pressure of species i
x_i = mole fraction of gaseous species i
and P = total pressure (in atm)

Now we need an expression for the equilibrium mole fractions of all the species. First, we define the extent y as the number of moles of CO that react. Clearly, at equilibrium, we have

$$n_{CO} = 0.4 - y$$
$$n_{H_2O} = 1 - y$$
$$n_{H_2} = y$$
$$n_{CO_2} = y$$
$$n_{total} = 1.4$$

From the above, the equilibrium mole fractions can be obtained as

$$x_{CO} = (0.4 - y)/1.4$$
$$x_{H_2O} = (1 - y)/1.4$$
$$x_{H_2} = y/1.4$$
$$x_{CO_2} = y/1.4$$

The equilibrium mole fractions are now substituted into equation, which can be simplified to give

$$0.371 = \frac{y^2}{(0.4 - y)(1 - y)}$$

This equation can be re-arranged into quadratic form.

$$0.629\,y^2 + 0.5194\,y - 0.1484 = 0$$

The solution to this equation is $y = 0.225$. (The other root to the equation is -1.050, which is not physically possible.) This value for the extent of the reaction at equilibrium can now be used to calculate the number of moles of each of the species present at equilibrium.

The equilibrium composition is therefore

0.175 moles of CO		12.5% CO
0.775 moles of H_2O		55.3% H_2O
0.225 moles of H_2	or	16.1% H_2
0.225 moles of CO_2		16.1% CO_2

Chemical equilibrium and kinetics of syngas production

Gasifiers are commonly operated in the temperature range 800–1500 K. In the reaction zone, carbon monoxide, carbon dioxide, hydrogen, methane, and water are most likely to be present. As supplied oxygen is almost completely utilized for gasification, it is reasonable to assume that ideally, no oxygen is present in the equilibrium mixture. This requires that the amount of steam and oxygen fed to the gasifier are sufficient for the complete conversion of carbon. In this special case, by eliminating carbon and oxygen, the number of equations can be reduced to two independent equations (Probstein and Hicks 1982).

$$CH_4 + H_2O \leftrightarrow CO + 3H_2 \qquad \qquad ...(15.6)$$
$$CH_4 + CO_2 \leftrightarrow 2\,CO + 2H_2 \qquad \qquad ...(15.7)$$

Equilibrium conditions can be written as

$$K_I = \frac{(y_{co})\,(y_{H_2})^3\,P^2}{(y_{CH_4})(y_{H_2O})} \qquad \qquad ...(15.8)$$

$$K_{II} = \frac{(y_{co})^2\,(y_{H_2})^2\,P^2}{(y_{CH_4})(y_{CO_2})} \qquad \qquad ...(15.9)$$

where y's are the mole fractions and P is the total pressure of the system. As equilibrium chemistry can be expressed in terms of formation reactions, the equilibrium constant K can also be expressed in terms of these reactions. In this case, the relevant equilibrium constants in terms of formation reactions can be expressed as

$$K_I = \frac{K_{co}}{K_{H_2O} \cdot K_{CH_4}} \qquad \qquad ...(15.10)$$

$$K_{II} = \frac{K_{co}^2}{K_{CO_2} \cdot K_{CH_4}} \qquad \qquad ...(15.11)$$

where K_{CO} etc., are the equilibrium constants for formation reactions. Now K_I and K_{II} can be evaluated by using equilibrium constants for these formation reactions. Formation reactions and the corresponding equilibrium constants are given in Table 15.4.

If n is the total number of moles of all species present in the system and $n^{in}_{H_2}$, $n^{in}_{O_2}$, etc., are the number of moles of H_2, O_2, etc., fed into the reactor, one can write

$$(2y_{CH_4} + y_{H_2} + y_{H_2O})\,n = n^{in}_{H_2} \qquad \qquad ...(15.12)$$

Table 15.4 Equilibrium constants for formation reactions at different temperatures

| Reaction | $(\log_{10} K)$ | | | |
	+298K	700K	1000K	1500K
$C + \frac{1}{2} O_2 \rightarrow CO$	+24.065	+12.968	+10.483	+8.507
$C + O_2 \rightarrow CO_2$	+69.134	+29.502	+20.677	+13.801
$C + 2H_2 \rightarrow CH_4$	+8.906	+0.958	-0.999	-2.590
$2H_2 + O_2 \rightarrow 2H_2O$	+40.073	+15.590	+10.070	+5.733

Source Probstein and Hicks (1982)

$$(\tfrac{1}{2} y_{CO} + y_{CO_2} + \tfrac{1}{2} y_{H_2O}) \, n = n_{O_2}^{in} \qquad \qquad ...(15.13)$$

$$(y_{CO} + y_{CO_2} + y_{CH_4}) \, n = n_c^{in} \qquad \qquad ...(15.14)$$

$$\text{Also, } y_{CO} + y_{CO_2} + y_{CH_4} + y_{H_2O} = 1 \qquad \qquad ...(15.15)$$

These elemental mass balance equations can be used to obtain expressions for mole fractions of individual compounds in terms of y_{CO_2} and n.

Equilibrium constants K_I and K_{II} will vary with the temperature as shown in Table 15.4. If we know the gasifier pressure and the input rates of carbon, hydrogen, and oxygen, then it is possible to calculate the mole fractions of carbon monoxide, carbon dioxide, methane, and water.

$$(y_{co}) = \frac{-n_{H_2}^{in} + n_c^{in} + n}{(2n)} - y_{CO_2} \qquad \qquad ...(15.16)$$

$$(y_{CH_4}) = \frac{n_c^{in} + n_{H_2}^{in} - n}{(2n)} \qquad \qquad ...(15.17)$$

$$y_{H_2} = \frac{3n - 4n_{O_2}^{in} - n_c^{in} - n_{H_2}^{in}}{(2n)} + y_{CO_2} \qquad \qquad ...(15.18)$$

$$y_{H_2O} = \frac{4n_{O_2}^{in} + n_{H_2}^{in} - n_c^{in} - n}{(2n)} - y_{CO_2} \qquad \qquad ...(15.19)$$

Substituting these equations in Equations 15.8 and 15.9 results in a quadratic equation in y_{CO_2} and n. Substituting mole fractions in the Equation 15.8 results in an implicit equation for n (Probstein and Hicks 1982).

$$K_I(n_{H_2}^{in} + n_C^{in} - n)[4n_{O_2}^{in} + n_{H_2}^{in} - n_C^{in} - n(1+2y_{CO_2})] -$$

$$\left(\frac{P^2}{2n^2}\right)[n_c^{in} - n_{H_2}^{in} + n(1-2y_{CO_2})][n(3 + 2y_{CO_2}) -$$

$$n_c^{in} - 4n_{O_2}^{in} - n_{H_2}^{in}]^3 = 0 \qquad \qquad ...(15.20)$$

where

$$y_{CO_2} = \frac{b - \{b^2 - 16n^2(K_I - K_{II})K_I[4n_{H_2}^{in} + n_{H_2}^{in} - n_c^{in} - n][n_C^{in} - n_{H_2}^{in} + n]\}^{\frac{1}{2}}}{8n^2(K_I - K_{II})} \qquad ...(15.21)$$

and $b = 8nK_I n_{O_2}^{in} + 2nK_{II}(3n - n_c^{in} - 4n_{O_2}^{in} - n_{H_2}^{in})$ ...(15.22)

Example 3

Consider that an input into the gasifier at 1000 K and 1 atm pressure is 0.5 moles of steam and 0.25 moles of oxygen per mole of carbon. Calculate the equilibrium compositions of the gaseous mixture. K_I and K_{II} are given as 25.82 and 19.41 respectively.

Solution

$n^{in}_C = 1$

$n^{in}_{O_2} = 0.50$

$n^{in}_{H_2} = 0.50$

Substituting these values in Equation 15.20 and solving for n and other mole fractions, one can get

$n = 1.378$

$y_{H_2} = 0.264$

$y_{CH_4} = 0.044$

$y_{CO_2} = 0.034$

$y_{CO} = 0.648$

$y_{H_2O} = 0.010$

Effect of temperature on syngas production

The effect of temperature is related to changes in the heat of the reaction according to the van't Hoff equation.

$$\frac{d\ln K}{dT} = \frac{\Delta H_T^0}{RT^2} \qquad \qquad ...(15.23)$$

The above equation shows that equilibrium constants for endothermic reactions can increase with an increase in temperature. Equilibrium constants of a particular reaction can be evaluated by using this equation if data on temperature dependence of $\Delta H°_T$ is available. For a particular reaction, it is possible to plot $\log_{10} K$ versus $1/T$. This plot is also called van't Hoff plot.

Production of methanol from syngas

Fundamental chemistry

Formation of methanol from syngas is accompanied by the following reactions and energy changes (Gullu and Demirbas 2001; Hamelinck and Faaj 2002).

$$CO + 2H_2 \rightarrow CH_3OH \text{ (g)} \qquad \Delta H°_{298} = -90.8 \text{ kJ/gmol} \qquad ...(15.24)$$

$$CO_2 + 3H_2 \rightarrow CH_3OH \text{ (g)} + H_2O \qquad \Delta H°_{298} = -53.6 \text{ kJ/gmol} \qquad ...(15.25)$$

Effect of temperature and pressure on methanol production

Composition of the equilibrium mixture can vary with pressure. In the reaction of carbon monoxide and hydrogen to form methanol, as the total volume of gas is always proportional to the number of molecules present, the reaction results in a decrease in the volume by 67%. Therefore, a shift in equilibrium composition with a change in pressure can be thought of as a compensatory effect as stated in Le Chatlier's principle: 'If a change occurs in one of the factors, such as temperature or pressure, under which the system is in equilibrium, the system tends to adjust itself to annul as far as possible the effect of that change'. In general, whenever a gas reaction takes place, with a decrease in the number of molecules, a rise in pressure increases the fraction of the reactants converted.

As the reaction is exothermic, an increase in temperature can decrease the yield of methanol. At about ambient temperature, the equilibrium shifts towards the right and hence, a further drop in temperature would be of no use there. In spite of the adverse effect on equilibrium, exothermic reactions may mostly be carried out economically at elevated temperatures to permit efficient recovery of the heat released. In practice, methanol production is carried out at about 550–650 K. However, even at pressures up to 100 atm, methanol yields are very low and unreacted gases have to be recycled and separated. But commercial production of methanol is not feasible unless equilibrium is reached at acceptable rates. It can be accomplished through a change in the reaction path so that the activation energy is effectively lowered. This is done by catalysis.

Effect of catalyst on methanol synthesis

Adsorption of a reagent to a solid surface followed by the reaction of the adsorbed species with either another adsorbed species or with unadsorbed reactants is a well-known catalysis processes. If, as a consequence, the activation energy is effectively lowered, the reaction rate will be increased according to the Arrhenius law ($k = A \exp [-E/RT]$). The catalyst itself undergoes no net change, although it takes part in the reaction. The rate expression then becomes more complex, as suggested by simple stoichiometric dependence.

The most widely used catalyst is a mixture of copper, zinc oxide, and alumina, first used by ICI (Imperial Chemical Industries) in 1966. At 50–100 atm and 250 °C, it can catalyse production of methanol from carbon monoxide and hydrogen with a high degree of selectivity. A variety of catalysts are capable of causing conversion, including reduced nickel-oxide-based preparations, reduced copper/oxide shift preparations, copper/silicon dioxide and palladium/silicondioxide, and palladium/zinc oxide.

Methanol production technologies

Syngas is mixed with recycled unreacted gas, pressurized in the compressor until the operational pressure is reached, and then pre-heated in a heat exchanger until a temperature close to the reaction temperature is obtained. It is subsequently fed into the synthesis reactor. A large part of the heat of condensation of methanol is transferred to the reactor feed in a feed effluent heat exchanger before the gas is cooled to room temperature in a water cooler. Approximately, 95% of the methanol condenses here. Separation of the gas phase takes place in a separator. After passing through the recycle compressor, the gas is mixed with fresh syngas and subsequently reintroduced into the reactor. A fixed percentage of the gas leaving the separator is continuously purged. This is to prevent the build-up of unreacting gases such as nitrogen and methane. Crude methanol is purified by distillation. In the topping column, gases and impurities having low boiling points are removed. A second refining column removes the heavier products and water (Figure 15.16).

Temperature control is a critical factor in any reactor development for methanol production, as excessive temperature can lead to a reduced catalyst life. The catalyst deactivates primarily because of loss of active copper due to physical blockage of the active sites by large by-product molecules, poisoning by halogens or sulphur in the synthesis gas, and sintering of the copper crystallites into larger crystals. Mainly two types of technologies – fixed bed and slurry – are used for methanol production from syngas.

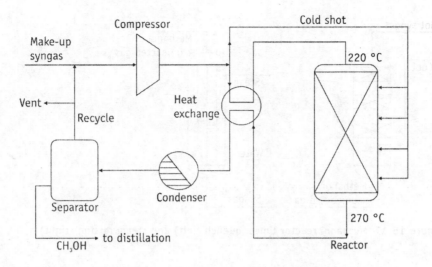

Figure 15.16 Schematic representation of a typical integrated methanol synthesis loop

Fixed-bed reactor

Conventional methanol reactors use fixed beds of catalyst pellets and operate in the gas phase. Two types of reactors predominate; ICI low-pressure systems and Lurgi systems. In ICI low-pressure process, an adiabatic reactor with cold unreacted gas injected between the catalyst beds is used. Subsequent heating and cooling leads to inherent inefficiency but the reliability is high.

The Lurgi system, with the catalyst loaded into the tubes and a cooling medium circulating on the outside of the medium, allows for near isothermal operation (Gullu and Demirbas 2001). The catalyst still deactivates, decreasing the overall yield of methanol. Conversion to methanol is limited by equilibrium considerations and the high-temperature sensitivity of the catalyst. However, it is important to note that catalyst productivity is a function of process pressure. Therefore, an increase in pressure can compensate for the loss in activity. Nowadays, large methanol plants exclusively use centrifugal compressors to bring the syngas to the desired operating pressure, which varies from 70 to 100 bars, depending on the activity of the catalyst. Higher the activity of the catalyst, lower will be the pressure used. Also, lower the activity of the catalyst, higher will be the pressure applied to compensate for the loss in activity.

Temperature moderation is also achieved by recycling large amounts of hydrogen gas, utilizing its higher heat capacity and the higher velocities to enhance heat transfer. Typically, a gas phase reactor is limited to the presence

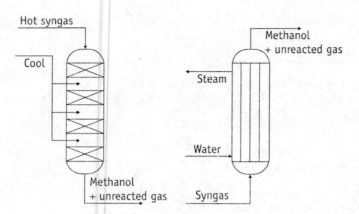

Figure 15.17 Methanol reactor types: quench (left) and steam raising (right)

of about 16% carbon monoxide in the inlet to the reactor in order to limit the conversion per pass to avoid excess heating (Figure 15.17).

Slurry bubble column reactor

A shift in equilibrium to the product side to achieve higher conversion per pass depends on the reaction conditions, catalysts, solvent, and space velocity. Experimental results show 15%–40% conversion for carbon-monoxide-rich gases and 40%–70% for hydrogen-rich gases.

The Eastman's chemicals LPMEOH™ (liquid phase methanol) demonstration unit uses an innovative mechanism for removal of heat and maintenance of uniform temperature (USDoE 1999). The SBCR (slurry bubble column reactor) of the LPMEOH process was invented in the late 1970s, and further developed and demonstrated in the 1980s. The process uses fine catalyst particles suspended in an inert mineral oil. The mineral oil acts as a temperature moderator, transferring the heat of reaction from the catalyst surface via liquid slurry to the boiling water in an internal tubular heat exchanger. As a result of maintaining a constant uniform temperature throughout the reactor, SBCR can achieve higher conversion rates (Figure 15.18).

The LPMEOH™ unit has three feed-gas streams: a balanced gas stream, which is diverted from the feed to a pre-existing gas-phase methanol unit, a high-pressure carbon monoxide stream, and a hydrogen stream from the exit of the gas-phase unit. The fresh feed is mixed with the recycled gas and spurged into the bottom of the reactor. Upon contact with the catalyst, methanol synthesis occurs. Separation of the methanol vapour and unconverted syngas from the slurry occurs in the free board volume in the reactor above the slurry catalyst bed. The exit gas is cooled and condensed

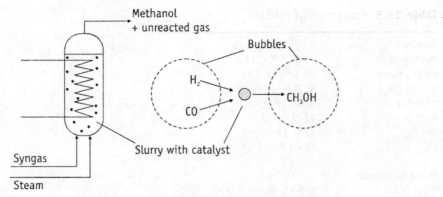

Figure 15.18 Schematic representation of LPMEOH reactor

into liquid methanol, which is collected in a product separator. Part of the overhead stream from the separator is recycled to the reactor and the rest is sent to the flue gas header.

The raw methanol stream is sent to a two-distillation-column recovery section. In the first column, dissolved gases are removed and sent to the flue header. The underflow from this column is sent to the second distillation column where purified methanol is recovered overhead. The bottom stream consisting of some methanol, higher alcohols, water, and mineral oils is sent to the distillation system of the pre-existing gas phase unit for methanol recovery (USDoE 1999).

Ethanol

Definition and properties

Ethanol is a colourless, flammable, oxygenated hydrocarbon liquid produced by fermentation of sugars. It is also known as ethyl alcohol, alcohol, grain-spirit, or neutral spirit. Its chemical formula is C_2H_5OH (Figure 15.19) and has a boiling point of 78.5 °C in the anhydrous state. However, it forms a binary azeotrope with water, with a boiling point of 78.15 °C at a composition of 95.57% by weight ethanol. Properties of ethanol are summarized in Table 15.5.

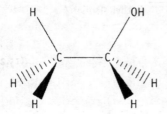

Figure 15.19 Ethanol molecule

Table 15.5 Properties of ethanol

Density	0.789 g/cm^3
Melting point	–114.3 °C (158.8 K)
Boiling point	78.4 °C (159 K)
Acidity (pK$_a$)	15.9 (H$^+$ from OH group)
Viscosity	1.200 cP at 20 °C
Flash point	17 °C
$\Delta H^0_{f, liquid}$	–277.38 kJ/mol
S^0_{liquid}	159.9 J/mol·K
C_p	112.4 J/mol·K
Phase behaviour	
Triple point	159 K (–114°C, –173.47 °F)
Critical point	514 K (241°C, 465.53 °F), 63 bar
Acid-base properties	
pH	7.0 (Neutral)

Source Morrison and Boyd (1987)

Production of ethanol

Basics of fermentation

Fermentation is defined as the energy-yielding anaerobic metabolic break-down of a nutrient molecule, such as glucose, without net oxidation (Stryer 1995). Fermentation results in production of simple organic compounds, such as acetic acid, lactic acid, ethanol, etc.

In aerobic respiration, a molecule of glucose is broken down through the process of glycolysis into pyruvate. In the first half of glycolysis, two ATP (adenosine triphosphate) molecules provide two phosphate groups to one molecule of glucose to form two molecules of G3P (glyceraldehyde-3-phosphate). In the next step, an NAD$^+$ (nicotinamide adenine dinucleotide) molecule replaces a hydrogen atom from G3P, converting G3P to 3-biphosphoglycerate with simultaneous conversion of NAD$^+$ to NADH (nicotinamide adenine dinucleotide).

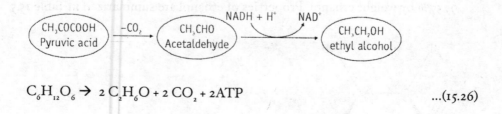

$$C_6H_{12}O_6 \rightarrow 2\,C_2H_6O + 2\,CO_2 + 2ATP \qquad \qquad ...(15.26)$$

When oxygen is present, it reacts with the hydrogen carried by NADH to produce water. When oxygen is not present in sufficient quantity to take up hydrogen, the cell's supply of NAD⁺ is converted to NADH, and G3P can no longer be converted to 3-biphosphoglycerate. Consequently, generation of ATP by the cell ceases unless another substance can be used to remove the hydrogen from NADH (Figures 15.20 and 15.21).

Fermentation is a biochemical pathway that provides such a substance. In ethanol fermentation used by yeasts, the ionized COO⁻ (carboxyl group) is removed from the pyruvate to generate a molecule of carbon dioxide, which is released by yeast into its surroundings. The resulting molecule, CH_3CHO (acetaldehyde), takes the place of oxygen as the chemical that accepts hydrogen from NADH (Lehninger 1981). This hydrogen, together with an H⁺ ion released during an earlier stage of glycolysis, is added to acetaldehyde, producing ethanol as shown in Equation 15.26.

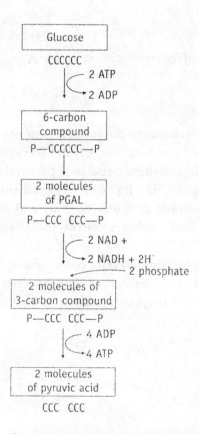

Figure 15.20 Glycolysis with the net gain of 2ATP molecules

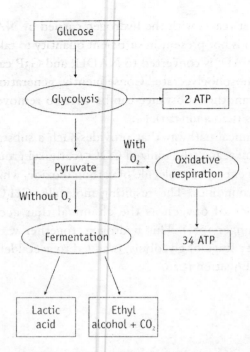

Figure 15.21 Conversion routes of pyruvate in the presence/absence of oxygen

Ethanol from molasses

Initially, the term 'molasses' referred specifically to the final effluent obtained in preparation of sucrose by repeated evaporation, crystallization, and centrifugation of juices from sugar cane and sugar beet. Today, several types of molasses are recognized and, in general, any liquid feed ingredient that contains in excess of 43% sugars is termed as molasses (Earl and Brown 1979). Examples are cane molasses, beet molasses, citrus molasses, and starch molasses.

During the process of ethanol production, molasses containing approximately 43% fermentable sugars are diluted and fermented. The essential steps involved in conversion of sugars to alcohol by yeast are given in Figure 15.22 and are outlined as follows.

1 Preparation of mash
2 Preparation of culture
3 Fermentation
4 Separation of products
5 Distillation

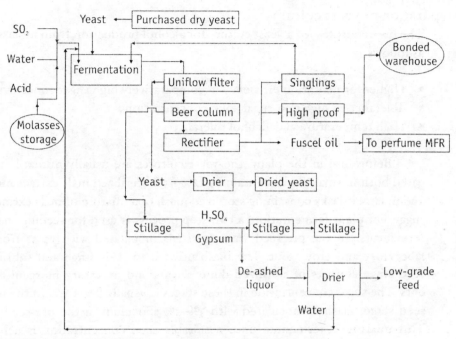

Figure 15.22 Different stages in conversion of molasses to ethanol

Preparation of mash

Mash is the diluted aqueous solution of sugar whose characteristics and composition are favourable to the fermenting organism in the efficient production of ethanol. This solution usually has 25% total solids with a sugar content of approximately 10%–12%. The pH of the mash is about 4.8 for maximum anti-bacterial action. However, optimum pH for maximum efficiency varies with the type of sugar and yeast used. pH adjustment is usually done by adding sulphuric acid, or hydrochloric acid or lactic acid. During fermentation, pH may drop up to 4.2–4.5 with no undesirable effects. If an increase in pH during fermentation is necessary, addition of aqueous ammonia is an option to effect this adjustment.

Nutrient requirement for the fermenting organism is another indispensable factor for maintaining an optimum growth rate. It is always desirable to add small quantities of ammonium salts as nitrogen source to the mash to increase the rate and efficiency of fermentation. This is exceptional for cane sugar molasses, which usually contain enough yeast nutrients to give fast and efficient fermentation. The optimum amount of nutrients may be determined by laboratory test fermentations. Other yeast nutrients, which have been used commercially are urea and occasionally, phosphates such as ammonium and potassium phosphates.

Preparation of yeast culture

The pre-requisites for a yeast culture for alcohol production from molasses are as follows.

- Higher fermentation efficiency in comparatively high sugar concentrations
- Tolerance to high non-sugar solid concentrations
- High temperature and alcohol tolerance

Before use in the plant, the yeast cultures are usually primarily propagated both in small flasks containing 12%–15% sterilized malt extract medium and in larger flasks containing sterilized mash of a mixture of malt extract and sugar. For plant fermentation, a two-step process is used for seeding the large fermentors. In the pre-seed stage, mash is inoculated with yeast from the laboratory and grown out. The mash utilized for this stage and subsequent seed stage consists of sterilized dilute sugars and necessary inorganic nutrients. The sugar concentration in these stages is usually 8%–10%. In the second seed stage, mash is inoculated with 3%–5% inoculum in the pre-seed stage. This mixture is fermented until a suitable seed concentration is achieved. When the concentration of yeast cells is $5–30 \times 10^7$ yeast cells/ml, the contents are ready for use as inoculum for one or more large fermentation tanks.

Fermentation

The actively growing yeast is added to the final fermentor. The usual volume of inoculum is two to four per cent of the final volume. Active yeast is added to the fermentation tank and at the same time, filling of the mash is also initiated to allow the development of yeast during the entire filling period, which may be as long as a couple of hours (Earl and Brown 1979). This provides additional fermentation time and also rules out the possibility of contamination during this period.

By the time the fermenting vessel is filled, actual fermentation proceeds at a rapid pace. At this stage, cooling of the fermentor is required for maximum fermentation efficiency and final product quality. If, for instance, initial temperature of the final fermentation mash is 26–28 °C then within hours of the initiation of the fermentation, the temperature will rise up to 35–38 °C. At this temperature, a significant loss of yeast activity results, giving a low yield of alcohol. In ordinary fermentation, temperatures are usually set between 21 °C and 27 °C in the beginning, and are held at 32–33 °C by the circulation of mash through external coolers.

The time necessary to complete the fermentation varies with the type of sugar used, the variety of yeast, and temperature. However, fermentation usually takes 36–48 hours. The course of fermentation can be checked by taking a specific gravity reading every hour while fermentation is in progress. During fermentation, the specific gravity of the mash decreases and when the specific gravity ceases to drop over a period of three hours, fermentation is considered complete. After this interval, the fermented mash is known as beer, which contains approximately, five to six per cent alcohol.

Separation of products

Once fermentation stops, the fermentation vessels are allowed to rest for six to nine hours to enable any suspended solids in the fermentor, such as yeast cells and insoluble solids, to settle at the bottom of the fermentor. The supernatant of the fermented mash is drawn off and pumped to the distillation house. Alternatively, prior to settling, the total content of the fermentation tank is passed through a centrifuge with the clarified beer being pumped to the overhead mash feed tank and the yeast and mud being diverted to a separate tank for conversion to yeast for cattle feed. This solid product collected in the centrifuge can be washed, collected, and dried further in a drum drier. This dried product is used both as fertilizer and animal feed.

Distillation

Ethanol is recovered from the fermentation broth by distillation and dehydration for the production of anhydrous alcohol. This is accomplished in two columns – the distillation and rectification columns – coupled with molecular sieves in which a mixture of nearly azeotropic water and ethanol is purified into pure ethanol. The distillation bottom stream is concentrated by evaporation using waste heat. The evaporated condensate is returned to the process while the concentrated syrup is combusted in a fluidized bed combustor to make steam for process heat while excess steam is converted into electricity for use in the plant. Part of the evaporator condensate along with waste water is treated by anaerobic and aerobic digestion, and the biogas resulting from anaerobic digestion is sent to the burner for heat recovery, while the treated water is recycled and returned to the process.

Ethanol from sweet sorghum

Sweet sorghum is a C_4 crop in the grass family belonging to the genus *Sorghum bicolor* L (Gnansounou, Dauriat, and Wyman 2005). Sweet sorghum is often considered to be one of the most drought-resistant agricultural crops as it has the capability of remaining dormant during the driest periods. Like other sorghum types, sweet sorghum probably originated from East Africa and spread to other African regions, southern Asia, Europe, Australia, and the United States. Of the many crops being investigated for energy industry, sweet sorghum is found to be one of the most promising candidates, particularly for ethanol production and especially, in the developing countries like India and China.

The technology of juice extraction from sweet sorghum is usually mechanical extraction, which involves a series of tandem roller mills with counter-current juice flow to leach the solubles. Efficiency of the extraction depends on the feedstock as well. For example, when one compares sugar cane and sweet sorghum, the extraction efficiency will be relatively low for sweet sorghum (85%–87%) because of its high fibre content. Production of ethanol from sweet sorghum juice is an established process. Once the juice is extracted, it is diluted to the desirable extent of fermentation. Fermentation and recovery of ethanol are same as described in the case of molasses.

Ethanol from starch-based resources

Starch-based resources include all cereals, such as corn, wheat, and rice. Unlike sugar-cane and sweet-sorghum-based ethanol production, starch-based feedstock needs pre-treatment and hydrolysis of starch constituents for the preparation of starch molasses, which can then be fermented to ethanol. Two types of processes are usually envisaged for hydrolysis and fermentation of starch—the dry milling and wet milling processes (Figure 15.23).

In dry milling, the process is divided into three stages: liquefaction, saccharification, and fermentation for ethanol production. The feedstock is first ground into flour and water is added to create a mash. For liquefaction, the mash is then mixed with hot water and alpha-amylase enzymes. Usually, a temperature of 88 °C is maintained during the process. The optimum pH for hydrolysis is 6, which is maintained by the addition of caustic soda. During liquefaction, the alpha-amylase enzymes attack the starch polymer randomly, producing maltose (dimer of glucose) and higher oligomers (Wallace, Ibsen, McAloon, *et al.* 2005).

After liquefaction, the mash is heated to 110 °C for 20 minutes and cooled to 60 °C. Continuous saccharification takes place in a stirred tank reactor in the presence of glucoamylase enzymes, and sulphuric acid is used for pH control. Residence time for saccharification is normally six hours and the optimum pH for saccharification is 4.4 (Wallace, Ibsen, McAloon, *et al.* 2005). During saccharification, the glucoamylase attacks the non-reducing end of maltose-producing glucose molecules. Commercial enzyme preparations usually contain a small fraction of pollulanases, which specifically attacks α 1-6

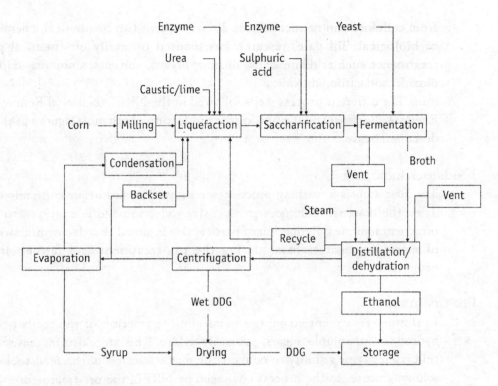

Figure 15.23 Different stages in conversion of starch to ethanol

linkages. The saccharified mash is then cooled to 32 °C and then fed to fermentors for production of ethanol. The by-products of the dry milling process are DDG (distillers gried grain) and carbon dioxide.

The wet milling process is used to obtain as many usable by-products as possible from different components of the feedstock. For example, in the case of corn, it is first steeped in water to separate the kernel. Following this, corn is ground and washed to further separate the kernel. When corn is processed, germ is removed to form corn oil and the starch is used for ethanol production. The remaining solids, gluten meal, and gluten feeds are sold as high-protein livestock feed.

Cellulose-based ethanol production

Cellulose-based ethanol technology converts agricultural waste, wood waste, and even some types of municipal solid wastes to ethanol. Unlike traditional ethanol conversion, sugars must be formed from the cellulosic material as a first step. Once formed, sugars can be fermented to ethanol. Like starch-based feedstock, cellulosic materials have to be pre-treated for production of sugars from cellulose and hemicelluloses. Pre-treatment can be physical, chemical, or biological. Till date, research has focused primarily on chemical pre-treatments, such as dilute acids, alkaline, organic solvents, ammonia, sulphur dioxide, and carbon dioxide.

The different process steps followed in the NREL (National Renewable Energy Laboratory) pilot plant for the production of ethanol (Figure 15.24) are described below.

Feedstock handling

The first step is a washing process to remove dirt and heavy contaminants. Then the material undergoes particle size reduction and is conveyed to the pre-treatment area. Reduction of particle size is aimed at reducing limitations of mass and heat transfer during the pre-treatment and fermentation processes.

Pre-treatment

In the pre-treatment stage, the hemicellulose fraction of the feedstock is hydrolysed to soluble sugars, primarily xylose. This step also increases the cellulase enzyme's ability to convert the major fraction of the feedstocks to soluble glucose. In the process envisaged by NREL, the pre-treatment mixes the feedstock with sulphuric acid and water (at approximately, one per cent acid in the resulting solution), and then raises the temperature of the slurry (20%–25% solids) to reaction temperatures (160–200 °C) with steam (Nguyen, Dickow, Duff, *et al.* 1996). The mixture is held at the reaction temperature for a pre-determined time (2–20 minutes), then flashed into a tank maintained at near atmospheric pressure. Because of the sudden drop in pressure, a fraction of the steam condensate and volatile compounds formed during the heating are evaporated and removed as flash tank overhead, which is condensed and sent to waste treatment. Lime is added to the remaining slurry to adjust the pH to 4.

Fermentation

The neutralized pre-treated material from the flash tank is taken to fermentors, where simultaneously saccharification and fermentation occur. While cellulase enzyme acts on the cellulose substrate to convert it to hexoses, xylose and generated hexose sugars are fermented to ethanol. Alternatively, some of the pre-treated material can be sent to seed fermentors and hydrolysed with cellulase to produce glucose for inoculum growth.

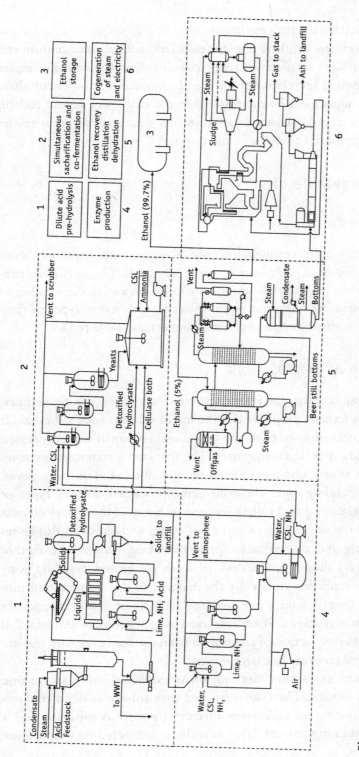

Figure 15.24 Schematic representation of NREL bio-ethanol process from ligno-cellulose

Distillation and downstream processing

After fermentation, the broth is sent for distillation. The product from the bottom of the distillation column is sent to a solid separator step. The liquid effluent can be used as recycle water for the process and sent to a neutralization step for pH adjustment before disposal in the city sewer system (Schell, Riley, Dowe, *et al.* 2004). The solids are disposed separately and sent to the neutralization step for pH adjustment and disposal.

Liquid biofuels: vegetable oils

Definition and properties

Vegetable oil is liquid fat, consisting mainly of triglycerides (generally more than 95%), extracted from plant sources, especially seeds. The fatty acids constituting triglycerides vary in their carbon chain length and in the number of double bonds. Therefore, different types of oils have different types of fatty acids, and consequently, different triglyceride molecules (Table 15.6).

Process of extraction of vegetable oils

Generally, mechanical extraction is used to extract oil from seeds. Sometimes, solvent extraction route is adopted, with the use of solvents such as hexane. It can aslo be a combination of both the processes. For small-scale processes, usually mechanical extraction is employed. The first step is to remove extraneous matter (chaff, stones, and metal) from the incoming seed utilizing sieves, magnets, etc. Depending upon the seed being processed, various types of grinding, cracking, flaking, and rolling equipment are used to rupture oil cells to allow egress of the oil. In case of soybeans, husks and hulls can also be removed by cracking and air separation prior to flaking. In order to further improve oil recovery during the process, all seeds are heated, typically using indirect steam, to condition and/or dry the material to the required moisture level prior to expelling (or flaking in case of soybean). Some products complete their processing at this stage and are marketed as expelled cake and expelled oil. However, further extraction of residual oil from the seed cake with the solvent can increase oil recovery to a large extent.

In the solvent extraction method, oil is extracted using an organic solvent, generally hexane. The cake is washed with solvent under a counter-current flow, producing an oil/solvent mixture (termed as miscella) and a residual meal-containing solvent. The miscella is subjected to evaporation

Table 15.6 Fatty acid profile of different vegetable oils

Type of seed	Oil content	Composition
Coconut	50%–70 %	Lauric acid content is up to 52%
Soybean	16.6%–21.6%	Ratio between mono- and poly-unsaturated fatty acids varied widely (1.7–3.9)
Sunflower	—	Palmitic – 6.0%, stearic – 4.6%, oleic – 17.8%, linoleic – 69.2%
Linseed	34%	Linolenic acid – more than 50%
Neem	33%–45%	Palmitic acid – 19.4%; stearic acid – 21.2%; oleic acid – 42.1%; linoleic acid – 14.9%; arachidic acid – 1.4%
Oil palm	50% (palm kernel)	96% neutral lipids, 2.4% phospholipids, 1.4% glycolipids
Olives	75% (pulp), 12%–28% (kernel)	Oleic acid (53%–86%)
Pongamia	27.5%	Oleic acid – 71.3%, linoleic acid –10.8%
Jatropha	35%–40%	21% saturated fatty acids and 79% unsaturated fatty acids
Rapeseed	33.2%–47.6%	5% saturated fat, 57% mono-unsaturated acid (oleic acid), 23% polyunsaturated omega-6 and 10%–15% polyunsaturated omega-3 and Erucic acid constituting 40%–50% fatty acid
Safflower	26%–32%	Linoleic acid: 74%–78 %
Camelina	30%–40%	Linoleic acid: 75%–80%, stearic 5%–10%
Castor beans	50%	Ricinoleic acid: 90%, oleic acid: 3%–4%, linoleic acid: 3%–4%

Sources Goering, Schwab, Daugherty, *et al.* (1982); Pryde (1983); Sonntag (1979)

and steam stripping to recover hexane from oil. The hexane is then separated from the water mixture by physical settlement and is recovered for reuse in the extraction operation. The solvent-laden meal is processed by steam distillation in a de-solventizer to recover most hexane content. The meal is then dried and cooled before storage in silos or loading, and may be sold as animal feed in this state or be processed further by grinding prior to sale. Table 15.7 compares decentralized cold pressing with the industrial solvent extraction process.

After oil extraction, the next step is the refining process, which generally includes degumming, neutralization, bleaching, deodorization, and further refining. The degumming process aims to remove seed particles, impurities, and (partly) phosphatides, carbohydrates, proteins, and traces of metals. In the process, the crude oil is treated with food grade processing acids and water, which leads to hydration of the main part of the phosphatides, proteins, etc. The hydrated material precipitates from the oil and can be removed by centrifugation.

Degumming is followed by alkali neutralization, which reduces the free fatty acids, oxidation products of free fatty acids, residual proteins, carbohydrates, traces of metals, and a part of the pigments. The treatment consists of a reaction with an alkali solution. This treatment results in the formation of a second phase of soap stock and the phase is separated and removed mostly by centrifuging the mixture. The recovered soap stock may be sold or acid-split on site to produce by-products for use in the soap industry and animal feed respectively. The neutral oil is usually washed after initial separation to remove any residual soap and is then dried.

The purpose of bleaching is to reduce the levels of pigments such as carotenoids and chlorophyll, heavy polycyclic aromatic hydrocarbons, and also residues of phospholipids, soaps, traces of metals, and oxidation products. These substances are removed by adsorption with activated clay, silica, and/or activated carbon. The bleaching clay containing all these substances is separated by filtration. The deodorization/stripping process is used to reduce/strip out free fatty acids and to remove odours, off-flavours, and other volatile components such as pesticides and light polycyclic aromatic hydrocarbons.

Table 15.7 Comparison between mechanical pressing and solvent extraction methods

Decentralized cold pressing	Industrial extraction with solvent
Production of cold pressed plant oil and press cake	Production of refined and half-refined plant oil and press cake
Low energy consumption (80 kWh/tonne seed)	High energy consumption (470 kWh/tonne seed)
No use of chemical solvents	Solvent extraction and thermal pre-conditioning
No waste water	Waste water from the refining (approximately 50 litres/litre of oil)
Small-scale capacities are possible (25 tonnes/day)	Higher capacities (Usually > 500 tonnes/day)
Low investment cost	High investment cost
Directly connected to agricultural production and additional value creation to the rural areas	Separated from agricultural production with presence of middle man

Source http://www.folkecenter.dk

Properties of vegetable oils

It has been found that neat vegetable oils can be used as diesel fuels in conventional diesel engines, but this leads to a number of problems as the injection, atomization, and combustion characteristics of vegetable oils in diesel engines are significantly different from those of diesel (Table 15.8).

It can be seen that kinematic viscosity of these oils varies between 30 and 40 cSt at 38 °C, which is due to their large molecular mass (600–900). The high viscosity of vegetable oils interferes with the injection process and leads to poor fuel atomization. The inefficient mixing of oil with air contributes to incomplete combustion, leading to heavy smoke emission, and the high flash point attributes to lower volatility characteristics. These disadvantages, coupled with reactivity of unsaturated vegetable oils, do not allow efficient engine operation for long periods of time. These problems can be solved if the vegetable oils are chemically modified to bio-diesel, which is similar in characteristics to diesel.

Bio-diesel

Definition and properties

Bio-diesel is defined as a mono alkyl ester of long chain fatty acids derived from renewable lipid feedstock, such as vegetable oils and animal fats, for use in compression ignition (diesel) engines and produced by transesterification of triglycerides or free fatty acid feedstocks by an alcohol.

Table 15.8 Comparative evaluation of the fuel properties of vegetable oil and diesel

Vegetable oil	Kinematic viscosity (cS)	Cetane number	Heating value (MJ/kg)	Cloud point (°C)	Pour point (°C)	Flash point (°C)	Density (kg/litre)
Jatropha	50.7	51.0	39.65	—	8	240	0.9186
Pongamia	27.8	—	34.60	—	—	205	0.912
Linseed	27.2	34.60	39.30	1.7	−15.0	241	0.9236
Peanut	39.6	41.80	39.80	12.8	−6.7	271	0.9026
Rapeseed	37.0	37.60	39.70	−3.9	−31.7	246	0.9115
Saffflower	31.3	41.30	39.50	18.3	−6.7	260	0.9144
Soybean	32.6	37.90	39.60	−3.9	−12.2	254	0.9138
Sunflower	33.9	37.10	39.60	7.2	−15.0	274	0.9161
Palm	39.6	42.00	—	31.0	—	267	0.9180
Diesel	3.06	50.00	43.80	—	−16	76	0.8550

Source Compiled from various sources

Bio-diesel is a clear amber-yellow liquid with a viscosity similar to petro-diesel. With a flash point of about 150 °C, bio-diesel does not ignite as readily as petroleum diesel (76 °C) and far less so than the explosively combustible gasoline (45 °C). Indeed, it is classified as a non-flammable liquid by OSHA (Occupational Safety and Health Administration), US, though it will of course burn if heated to a high enough temperature. Therefore, it is far safer to use it as a transport fuel. Bio-diesel gels at higher temperatures (about 0 °C) than petroleum diesel, which limits its pure form use in cold climates.

Unlike petro-diesel, bio-diesel is biodegradable and non-toxic, and it significantly reduces toxic and other emissions when burned as fuel. Properties of bio-diesel are summarized in Table 15.9.

Production of bio-diesel

Fundamentals

Bio-diesel is produced by transesterification of triglycerides and fatty acids present in oil. Transesterification is the process of exchanging the alkoxy group of an ester compound by another alcohol. Both acid and base can catalyse the reaction—while acid donates a proton to the alkoxy group, base removes a proton from alcohol, thereby making it more reactive. Transesterification of triglycerides takes place as given below.

Table 15.9 Properties of bio-diesel from different feedstock

Vegetable oil methyl esters (bio-diesel)	Kinematic viscosity (mm²/s)	Cetane number	Lower heating value (MJ/kg)	Cloud point (°C)	Pour point (°C)	Flash point (°C)	Density (kg/litre)
Jatropha	4.84	52	41	6	1	163	0.88
Pongamia	4.80	55.8	36.1	4	4	141	0.876
Neem	8.80	51	40.1	—	—	—	0.82
Peanut	4.90	54	33.6	5	—	176	0.883
Soybean	4.50	45	33.5	1	-7	178	0.885
Babassu	3.60	63	31.8	4	—	127	0.875
Palm	5.70	62	33.5	13	—	164	0.880
Sunflower	4.60	49	33.5	1	—	183	0.860
Tallow	—	—	—	12	9	96	—
Diesel	3.06	50	43.8	—	-16	76	0.855
20% bio-diesel	3.20	51	43.2	—	-16	128	0.859

Source Compiled from various sources

1 mole of triglyceride + 3 moles of methanol → 3 moles of ester + 1 mole of glycerin

$$CH_2 - O\text{-}CO\ R^I$$
$$|$$
$$CH - O\text{-}CO\ R^{II} \quad + 3\ CH_3OH \rightarrow$$
$$|$$
$$CH_2 - O\text{-}CO\ R^{III}$$

$$CH_2 - OH \qquad\qquad R^{III}\ COO\ CH_3$$
$$|$$
$$CH - OH \qquad + \qquad R^{II} - COO\ CH_3$$
$$|$$
$$CH_2 - OH \qquad\qquad R^{I} - COO\ CH_3$$

where R^I, R^{II}, R^{III} are the long chains of fatty acids. ...(15.27)

The most important process parameters that influence transesterification of vegetable oil are as follows.

1 Purity/composition of oil
2 Choice of alcohol
3 Quantity of alcohol
4 Reaction temperature
5 Catalyst

Purity/composition of oil

Efficiency of any oil to form esters through alkali-based transesterification mainly depends upon the triglyceride content of oil as it determines the stoichiometry of the reaction. Impurities in oil affect the conversion levels considerably. It is reported that the 65%–84% conversion can be increased up to 94%–97%, if oil is refined (Barnwal and Sharma 2005; Van Gerpan 2005; Van Gerpan, Shanks, Pruszko, *et al.* 2004). Bio-diesel yield and recovery will diminish in the presence of a significant amount of free fatty acid. The free fatty acid can react with the catalyst and form soap, which makes the separation of bio-diesel difficult. This problem can be solved by pre-treatments such as catalytic refining or acid-based transesterification. High temperature (240 °C) and pressure (9000 kPa) reactions can also avoid the formation of soap. Presence of water can also create problems by hydrolysing triglycerides and thereby increasing the amount of free fatty acids in the reaction mixture.

Choice of alcohol

Both methanol and ethanol can be used for the transesterification reaction. However, according to the stoichemistry, the quantity of ethanol required will be higher as compared to methanol. Also, in the present context, use of methanol is advantageous as it follows the simultaneous separation of glycerol. But reaction, using ethanol, is more complicated as it forms an azeotropic mixture which makes the separation rather difficult. Normally, methanol is used as it is the cheapest available alcohol, but in countries like Brazil, where there is abundant availability of ethanol, it is advantageous to use anhydrous ethanol. However, in this case, the optimum conditions and the choice of catalyst favoured will be different from methanol-based transesterification process.

From the environmental and energy security point of view, ethanol as a renewable fuel deserves priority compared to methanol. Once bio-diesel is available in the market at a reasonable price, the demand for methanol can be expected to jump up proportionately followed by the rising import bills. Against this background, ethanol promises more energy security as a domestic fuel and there is an urgent need in India to provide research and development support to develop and promote ethanol production technologies. However, currently, cheaper prices make methanol the better choice.

Quantity of alcohol

The stoichiometry of the reaction demands three molecules of alcohol for each mole of triglyceride reacted. On a weight basis, it corresponds to adding methanol at the rate of approximately 10% by weight per mass of oil processed and adding ethanol at the rate of about 15% of the oil processed. But at the stoichiometric molar ratio of 3:1, the reaction rate is slow and rarely proceeds to completion. For shifting the reaction forward, either the reaction should be conducted in steps or excess alcohol should be added. In the former case, a part of the alcohol and the catalysts are added at the beginning of each step and glycerol is removed at the end of each stage.

In the latter case, excess alcohol added depends upon the type of esterification. An acid-based process requires comparatively large amounts of excess alcohol to bring the reaction to completion. Significant amount of partially reacted mono/diglycerides of saturated fatty acids is present when the alcohol/oil ratio is low (Figure 15.25). The highest yield of 78%, using three per cent sulphuric acid as the reaction catalyst, has been reported at a methanol and triglyceride ratio of 23:1 (Schuchardta, Serchelia, and Vargas 1998). By extrapolation, the highest molar ratio needed to achieve complete methylation was found to be between 35:1 and 45:1 mol/mol. However, in base-catalysed transestrification, methanol quantity or twice the stoichiometric requirement, seems to be sufficient to get 90% yield. In both cases, excess alcohol was recycled at the end of the process.

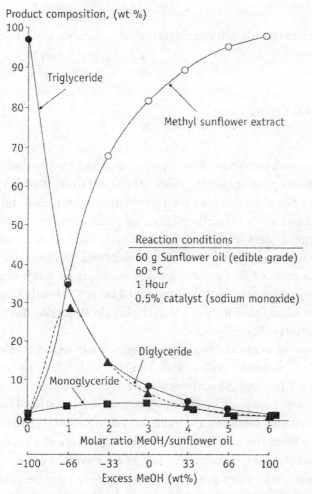

Product composition, (wt %)

Figure 15.25 Effect of alcohol-to-oil ratio on product composition for transesterification
Reproduced with permission from Elsevier
Source Van Gerpan (2005)

Reaction temperature

The rate of transesterification depends strongly on temperature. However, reaction is conducted close to a temperature below the boiling point of methanol at atmospheric pressure. Generally, transesterification reaction proceeds to completion after one hour at 60 °C, but if the temperature is reduced to about 32 °C, reaction takes four hours for completion (Freedman, Pryde, and Mounts 1984).

Types of catalyst

Several catalysts are used in transesterification of vegetable oils. These are listed below.

- Acid catalyst
- Alkali catalyst
- Non-ionic-base catalyst
- Lipase catalyst

Acid catalysis   The transesterification process is catalysed by Bronsted acids, preferably by sulphonic and sulphuric acids. These catalysts give very high yields in alkyl esters but the reactions are slow, requiring, typically, temperatures above 100 °C and more than three hours to reach complete conversion (Schuchardta, Serchelia, and Vargas 1998). It has been reported that the methanolysis of soybean oil, in the presence of 1 mol% of sulphuric acid, with an alcohol/oil molar ratio of 30:1 at 65 °C, takes 50 hours to reach complete conversion of the vegetable oil (> 99%). Butanolysis (at 117 °C) and ethanolysis (at 78 °C), using the same quantities of catalyst and alcohol, take three hours and 18 hours, respectively (Pryde 1984).

The mechanism of acid-catalysed transesterification of vegetable oil is well established for monoglycerides and can be extended to di- and triglycerides (Stoffel, Chu, and Ahrens 1959). The reaction proceeds through protonation of carbonyl group of the ester and leads to carbocation II, which, after a necleophilic attack, produces a tetrahedral intermediate III that eliminates the glycerol to form the new ester IV and to regenerate the catalyst H·. According to this mechanism, carboxylic acid can be formed by the reaction of carbocation II with the water present in the reaction mixture. This suggests that an acid-catalysed transesterification must be carried out in the absence of water in order to avoid competitive formation of carboxylic acid, which reduces the yield of alkyl esters (Schuchardta, Serchelia, and Vargas 1998) (Equations 15.28 and 15.29).

Ionic alkali base catalysis   Mainly three types of ionic alkali catalysts are used: alkaline metal alkoxides, hydroxides, and carbonates. The mechanism of base-catalysed transesterification of vegetable oils starts with reaction of a base with an alcohol-producing alkoxide and a protonated catalyst. The necleophilic attack of the alkoxides at the carbonyl group of the triglyceride generates a tetrahedral intermediate from which the alkyl group and the corresponding anion of the diglycerides are formed. The latter deprotonates the catalyst, regenerating the active species, which is now able to react with a second molecule of the alcohol, starting another catalytic cycle. Diglycerides

$$...(15.28)$$

$$...(15.29)$$

R" = [O / OH / OH] ; glyceride

R' = carbon chain of fatty acid
R = alkyl group of alcohol

Mechanism of the acid-catalysed transesterification of vegetable oils
Source Schuchardta, Serchelia, and Vargas (1998)

and monoglycerides are converted by the same mechanism to a mixture of alkyl esters and glycerol.

Alkaline metal hydroxides (potassium hydroxide and sodium hydroxide) are cheaper than metal alkoxides but less active. Though they serve as good alternatives in increased concentrations, their reaction with methanol in the esterification reactor leads to formation of some water, which hydrolyses the produced ester with consequent soap formation. This undesirable saponification reaction reduces the ester yields and considerably reduces the ease of recovery of glycerol due to formation of emulsions (Equations 15.30–15.33).

Though the use of potassium carbonate reduces soap formation, it also greatly reduces the yield. Even after adding 3 mol%, the yield is only 92.4%. Lack of soap formation can be explained on the basis of formation of bicarbonate instead of water. Enzyme-catalysed transesterification has not reached the commercial stage as a full-fledged technology. However, new results have been reported in recent articles and patents. Common to all these studies is the optimization of the reaction conditions, such as the solvent, pH, temperature, and type of microorganisms, which generate the enzyme in order to establish suitable characteristics for an industrial application. However, the reaction yields as well as reaction times are still unfavourable compared to base-catalysed reaction systems. It has also been found that amines, such as triethyl amines, piperidines, pyridines, and guanidines can also be used as catalysts in transesterification of oils. It was found that TBD (1,5,7 triazabicyclo [4,4,0] dec-5-ene) even if applied in 1 mol% produces more than 90% of

$$ROH + B \rightleftharpoons RO^- + BH^+ \quad \quad ...(15.30)$$

$$...(15.31)$$

$$...(15.32)$$

$$...(15.33)$$

Mechanism of the base-catalysed transesterification of vegetable oils
Source Schuchardta, Serchelia, and Vargas (1998)

methyl esters after one hour. Using other organic bases, under the same experimental conditions, the yields were not higher than 66%.

Alkaline metal alkoxides are the most active catalyst as they give very high yields in short reaction times even if they are applied in very low molar concentrations. In a study by Vicente, Martinez, and Aracil (2004) on transesterification with respect to sunflower oil, it was found that sodium methoxide with a concentration of 1 wt% of oil was the most efficient catalyst, yielding 99.33 ± 0.036 of bio-diesel with a purity of 99.72 ± 0.03. The findings of Vicente, Martinez, and Aracil (2004) on the effect of the catalyst on yield and the purity of bio-diesel from sunflower oil are given in Table 15.10.

The speed of the base-catalysed reaction, together with the comparatively less corrosive effect of alkaline catalyst in comparison with acid catalysis, makes the industries favour alkaline-catalysed reactions.

Non-ionic base catalysis Most of the organic bases for catalysing organic synthesis belong to the category of amidines, guanidines, and triamino (imino)

Table 15.10 Effect of different inorganic catalysts on bio-diesel yield and purity

Parameter	Sodium hydroxide	Potassium hydroxide	Sodium methoxide	Potassium methoxide
Bio-diesel purity	99.71 ± 0.04	99.76 ± 0.05	99.72 ± 0.03	99.52 ± 0.10
Bio-diesel yield	86.71 ± 0.28	91.67 ± 0.27	99.33 ± 0.36	98.46 ± 0.16

Source Vicente, Martinez, and Aracil (2004)

phosphoranes. The activity and efficiency of such non-ionic bases as catalysts for transesterification of vegetable oils were studied. Table 15.11 compares the catalytic performance of certain prominent organic catalysts (Schuchardta, Serchelia, and Vargas 1998).

The order of the catalytic activity is not directly related to the relative basicity of these compounds, as BEMP (tert-butylimino-2-diethylamino-1,3-dimethyl-perhydro-1,3,2 diazaphosphorane) and Me7P (tris [dimethylamino] methyliminophosphorane) should have been the more efficient catalysts, followed by TBD. However, guanidines are more active catalysts and the activity follows their relative basicity. The better performance of TBD as compared to BEMP and Me7P is also related to its kinetic activity. The catalytic site (unshared electron pair of the sp^2 N) of TBD is practically unhindered, allowing an easy access of methanol for proton transfer, while the steric hindrance shown by the triamino(imino)phosphoranes is so significant that they are practically inert to alkylating agents, such as isopropyl bromide, as well as extremely resistant to react with thionyl chloride and thiophosgene. Other

Table 15.11 Effect of non-ionic base catalysts on yield of bio-diesel *

Catalyst Chemical family	Chemical name	Yield (%) after 1 h	Relative basicity
Guanidines	TBD (1,5,7- triazabicyclo [4.4.0] dec-5-ene)	91	150.00
	MTBD (7-Methyl-1,5,7-triazabicyclo [4.4.0] dec-5-ene)	47	43.65
	TMG (1,1,3,3-tetramethylguanidine)	18	0.95
Triamino (imino) phosphoranes	BEMP (tert-butylimino-2-diethylamino-1, 3-dimethyl-perhydro-1,3,2 diazaphosphorane)	66	6873.00
	Me7P (Tris [dimethylamino] methylimino-phosphorane)	63	4762.00
Amidines	DBU (1,8- diazabicyclo [5.4.0] undec-7-ene)	32	3.40
	DBN (1,5-diazabicyclo [4.3.0] non-5-ene)	4.5	1.00

Source Schuchardta, Serchelia, and Vargas (1998)

*Reactions are conducted between 27.2 mmol of rapeseed oil with 62.5 mmol of methanol for one hour in presence of 1 mol% catalyst at 70 °C.

bases such as DMAP (4–methyl aminopyridine), pyridine, and triethylamine were also tested. However, even at 5 mol%, these amines did not give satisfactory yields. DMAP was the most active within this series, producing only 20% of methyl esters after one hour.

When compared to the typical industrial catalysts (sodium hydroxide and potassium carbonate), performance of TBD is good. Yields with TBD were close to those observed with sodium hydroxide and no undesirable by-products such as soaps (easily formed when alkaline metal hydroxides are used) were formed. When compared to potassium carbonate, TBD more active, even at low molar concentrations. A comparison data on efficiency of guanidine catalysts with inorganic catalysts is given in Table 15.12 (Schuchardta, Serchelia, and Vargas 1998).

Catalytic performance of alkyl guanidines follows the order TBD > TCG > DCOG > MTBD > PMG, (see Table 15.12 for chemical names) which corresponds to their relative base strength. This is increased by structural factors such as number and type of constituents. The excellent performance of DCOG and TCG is also attributed to the high symmetry of their guanidinium cations, as observed earlier for the symmetric 1,2,3-trimethylguanidine. Results obtained by transesterification of soybean oil with methanol show that PCBG (1,2,3,4,5-penta-cyclohexylbiguanidine) is even more active than TCG, as an 82% yield of methyl esters is obtained with PCBG after one hour, compared to 69% with TCG under similar conditions.

Lipase catalysis Due to their ready availability and ease with which they can be handled, hydrolytic enzymes have been widely applied in organic synthesis. They do not require any co-enzymes, are reasonably stable, and often tolerate

Table 15.12 Catalytic effect of different guanidines during bio-diesel production

Chemical family	Catalyst Chemical name	Yield (%) after one hour
Guanidines	TBD (1,5,7- triazabicyclo [4.4.0] dec-5-ene)	90.0
	DCOG (1,3-dicyclohexyl- 2-n-octylguanidine)	74.0
	TCG (1,2,3-tricyclohexylguanidine)	64.0
	MTBD (7-methyl-1,5,7-triazabicyclo [4.4.0] dec-5-ene)	47.0
	PMG (1,1,2,3,3- pentamethylguanidine)	49.0
Inorganic catalysts		
	NaOH (sodium hydroxide)	98.7
	K_2CO_3 (potassium carbonate)	84.0

Source Schuchardta, Serchelia, and Vargas (1998)

organic solvents. Although the enzyme-catalysed transesterification processes are not yet commercially developed, new results have been reported in recent articles and patents. The common aspects of these studies consist in optimizing the reaction conditions (solvent, temperature, pH, type of micro-organism which generates the enzyme, etc.) in order to establish suitable characteristics for an industrial application. However, the reaction yields as well as the reaction times are still unfavourable compared to the base-catalysed reaction systems.

Production processes

In simple terms, for transesterification to take place, alcohol, catalyst, and oil are made to react in a reactor and agitated for approximately one hour at 60 °C. The reaction is sometimes done in a two-stage system. In the first stage, 80% alcohol and catalyst are added and the reacted stream then undergoes a glycerol-removal step before entering into a second reactor in which the remaining alcohol and catalyst are added. The system has the potential of using less amounts of alcohol as compared to the single-stage systems. Following the reaction, glycerol is removed from the esters. Due to the very low solubility, separation between esters and glycerol occurs rather easily and can be accomplished with either a settling tank or a centrifuge. However, excess methanol can act as a solubilizer for both the layers and slow down the separation process. Excess methanol cannot be removed in advance because of the possibility of reversing transesterification. Water may be added to the reaction mixture after completion of the transesterification to improve the transesterification of glycerol (Figure 15.26).

Both batch and continuous transesterification processes are being used commercially. While batch processes are used for small-scale systems, continuous processes are widely used for large-scale bio-diesel manufacturing.

Batch process

In the batch process, oil is first released to the system followed by the catalyst and methanol. The system is agitated to bring the oil, catalyst, and alcohol into immediate contact. Towards the end of the process, less vigorous mixing can help in increasing the extent of separation of glycerol from the ester. After the reaction, the mixture is allowed to settle in the reactor to give the initial separation of glycerol and esters and it can further be separated by centrifuging. Alcohol is removed from both the glycerol and the ester stream using an evaporator or a flash unit. The esters are neutralized and washed gently using

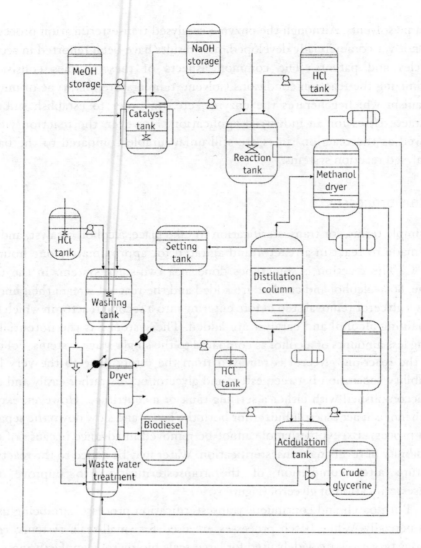

Figure 15.26 Process flow diagram for the production of bio-diesel

warm, slightly acidic water to remove the residual methanol and salts and then dried. The finished bio-diesel is then transferred to storage and the glycerol stream is neutralized and washed with soft water. Thereafter, the glycerol is sent to the refining section.

Continuous process

In continuous processes, generally CSTRs (continuous flow stirred tank reactors) in series are used. After process completion in the first reactor, glycerol is decanted, thus increasing the reaction rate in the second reactor.

An essential element in the design of the CSTR is to ensure sufficient mixing so that the composition throughout the reactor is constant. A plug-flow reactor system behaves as a series of small CSTRs chained together. In this reactor, the reaction mixture moves in a continuous plug with little mixing in the axial direction. The result is a continuous system that requires a very short residence time: as low as 6–10 minutes for completion of the reaction.

The maximum amount of free fatty acids acceptable in a base-catalysed system is generally less than 2%. Feedstocks having free fatty acids higher than this need special treatments. The processes generally adopted to process the free fatty acids are

- caustic stripping and direct esterification of free fatty acids, and
- glycerolysis of the oil

Caustic stripping

Caustic stripping is also referred to as catalytic refining. In this process, before transesterification of the triglycerides, a pre-treatment is given in which the free fatty acid content of the feedstock is made to react with caustic soda. The resulting soaps are stripped out using a centrifuge. Some triglycerides are lost during caustic stripping. The refined oils are dried and sent to a transesterification unit for further processing. If the soap mixture is disposed of, it will result in negative economy of production. Therefore, the soap mixture is acidulated and esterified using acid catalysis in a separate reactor. However, direct esterification requires a high alcohol-to-FFA ratio (between 20:1 to 40:1) and rather large amounts of acid catalyst. In addition, the process produces water that must be removed.

Glycerolysis

In this process, the triglyceride components of feedstock are completely hydrolysed to free fatty acids and glycerol. This is typically done in a counter-current reactor using sulphuric and/ sulphonic acids and steam (Van Gerpan, Shanks, Pruszko, et al. 2004). After separation of glycerol, pure free fatty acids are acid-esterified in another current reactor to transform them into methyl esters. The methyl esters are then neutralized and dried, and the yield can exceed 99%. However, the equipment needs to be acid-resistant. As the process can uptake very high free fatty acid feedstock, the cost of feedstock is extremely low.

While most bio-diesel producers use alkali-catalysed processes, technology advancements are also made using non-catalysed systems. Two well-known processes that do not use catalyst, are super-critical process and processing using a co-solvent.

Example 4
Consider a vegetable oil feedstock having free fatty acid content of 36% (15% palmitic, 18% stearic, and 3% oleic). How much amount of sodium hydroxide is required to neutralize the free fatty acid present in the oil?

Solution
1000 g oil contains 150 g palmitic, 180 g stearic, and 30 g oleic acid.

Molecular masses of palmitic, stearic and oleic acids are 256, 284 and 282 g, respectively, that is, 0.586, 0.634, and 0.106 moles, respectively.

Total number of moles of free fatty acid = 1.326

According to the stoichiometry,

1 mole of free fatty acid + 1 mole of sodium hydroxide =
  1 mole of soap + 1 mole of water.

Therefore 1.326 moles of sodium hydroxide, that is, 53 g is additionally required for every kg of oil reacted.

Co-solvent process

One of the major problems associated with transesterification of triglycerides is the extremely low solubility of methanol in triglycerides. Therefore, using a co-solvent, which facilitates the solubility of methanol, not only reduces the reaction time but also has the advantage of operating at a relatively low temperature. One such process nearing commercialization is the Biox co-solvent process. The process uses tetra hydro furan as the solvent, which has a boiling point close to that of methanol. The system operates at 30 °C and the reaction takes 5–10 minutes for completion (Boocock, Konar, Mao, *et al.* 1996). In this process, ester–glycerol phase separation is clean and the final by-products are the catalyst and water-free glycerol. At the end of the reaction, methanol and co-solvent are recovered in a single step.

Supercritical process

Supercritical fluids are highly compressed gases that have the intriguing aspect of having the properties of both gases and liquids at a temperature and pressure in excess of their critical point (Figure 15.27).

In the supercritical state, there is no longer a distinct liquid or vapour phase; only a single-fluid phase exists (Sahu 2003). The problems related to the phase separation of the vegetable oil and methanol mixture are not encountered and a single phase exists due to increase in the dielectric constant of methanol in the supercritical state. As a result, the reaction is completed in a very short time—within 2–4 minutes. The oil-to-alcohol ratio is 1:42 at

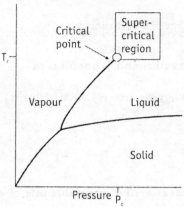

Figure 15.27 Phase diagram indicating supercritical state of matter

pressures greater than 80 atm and temperature of from 350–400 °C (Madras, Kolluru, and Kumar 2004).

References

Barnwal B K and Sharma M P. 2005
Prospects of bio-diesel production from vegetable oils in India
Renewable and Sustainable Energy Series **9**: 363–378

Berg C. 2004
World fuel ethanol analysis and outlook
Details available at <http://www.distill.com/World-Fuel-Ethanol-A&O-2004.html>,
last accessed on 4 January 2009

Berg J M, Tymoczko J L, and Stryer L. 2003
Biochemistry, 5th edn
New York: W H Freeman and Company

Boocock D G B, Konar S K, Mao V, Sidi H. 1996
Fast one-phase oil-rich processes for the preparation of vegetable oil methyl esters
Biomass and Bioenergy **11**(1): 43–50

Bridgwater A V. 1984
Thermo-chemical Processing of Biomass, 1st edn
Butterworths

Demirbas A. 2001
Relationship between lignin contents and heating values of biomass
Energy Conversion and Management **42**: 183–188

DSIR (Department of Scientific and Industrial Research). 1991
Technology in Indian Methanol Industry
[A status report prepared under the National Register of Foreign Collaborations]

Earl W B and Brown W A. 1979
Alcohol fuels from biomass in New Zealand: the energetics and economics of production and processing, vol. 1, pp. 1–11
[Third International Symposium on Alcohol Fuels Technology, Asilomar, CA, May 28–31]

Fengel D and Wegener G. 1983
Wood: chemistry, ultrastructure, reactions
Berlin: Walter de Gruyter & Co. 613 pp.

Freedman B, Pryde E H, and Mounts T L. 1984
Variables affecting the yields of fatty esters from transesterified vegetable oils
JAOCS **61**: 1638–1643

Gnansounou E, Dauriat A, and Wyman C E. 2005
Refining sweet sorghum to ethanol and sugar: economic trade-offs in the context of North China
Bioresource Technology **96**(9): 985–1002

Goering C E, Schwab A W, Daugherty M J, Pryde E H, Heakin A J. 1982
Fuel properties of eleven oils
Transactions of the ASAE **25**: 1472–1483

Gullu D and Demirbas A. 2001
Biomass to methanol via pyrolysis process
Energy Conversion and Management **42**: 1349–1356

Gurr M I. 1980
The Biosynthesis of Triacylglycerols, Lipids: structure and function
New York: P K Academic Press

Hamelinck C N and Faaj A P C. 2002
Future prospects for production of methanol and hydrogen from biomass
Journal of Power Sources **111**: 1–22

Kovarik B. 1998
Henry Ford, Charles F Kettering and the fuel of the future
Automotive History Review **32**: 7–27

Lehninger A I. 1981
Biochemistry, 2nd edn
New York: Worth Publishers Inc.

Madras G, Kolluru C, and Kumar R. 2004
Synthesis of biodiesel on super critical fluids
Fuel **83**: 2029–2033

Morrison R R and Boyd R N. 1987
Organic Chemistry, 5th edn
Boston, MA: Allyn & Bacon

Newton R. 2003
Biofuels are the future
Chemistry and Industry **11**: pp. 14–15

Nguyen Q A, Dickow J H, Duff B W, Farmer J D, Glassner D A, Ibsen K N, Ruth M F, Schell D J, Thompson I B, Tucker M P. 1996
NREL/DOE ethanol pilot-plant: current status and capabilities
Bioresource Technology **58**(2): 189–196

Paul J K. 1978
Methanol Technology and Application in Motor Fuels
New Jersey: Noyes Data Corporation

Peterson C L. 1986
Vegetable oil as a diesel fuel status and research priorities
Transactions of ASAE **29**(5): 1413–1422

Peterson C L and Auid D L. 1991
FACT - vol. 12, solid fuel conversion for the transport sector
ASME

Peterson C L, Wagner G L, and Auld D L. 1983
Vegetable oil substitutes for diesel fuel
Transactions of the ASAE **26**(2): 322

Probstein R F and Hicks R E. 1982
Synthetic Fuels
McGraw-Hill International Book Company

Pryde E H. 1983
Vegetable oil as diesel fuel: overview
JAOCS **60**: 1557–1558

Pryde E H J. 1984
American Oil Chemist's Society **61**: 1609

Sahu S. 2003
Super critical fluid extraction: a cleaner technology option for the industry
In *Green Chemistry: Environment Friendly Alternatives*
edited by R Sanghi and M M Srivastava
New Delhi: Narosa Publishing House

Schell D J, Riley C J, Dowe N, Farmer N, Ibsen K N, Ruth M F, Toon S T, Lumpkin R E. 2004
A bioethanol process development unit: initial operating experiences and results with a corn fibre feedstock
Bioresource Technology **91**(2): 179–188

Schuchardta U, Serchelia R, and Vargas R M. 1998
Transesterification of vegetable oils: a review
Journal of the Brazilian Chemical Society **9**(1): 199–210

Shay E G. 1993
Diesel fuel from vegetable oils: status and opportunities
Biomass and Bioenergy 4: 227–242

Sonntag N O V. 1979
Composition and characteristics of individual fats and oils
In *Bailey's Industrial Oil and Fat Products*, edited by D Swern, vol. 1, 4th edn
New York: John Wiley and Sons

Stoffel W, Chu F, and Ahrens E H. 1959
Analysis of long chain fatty acids by gas–liquid chromatography
Journal of Analytical Chemistry **31**: 307–308

Stryer L. 1995
Biochemistry, 4th edn
New York: W H Freeman and Co.

USDoE (US Department of Energy). 1999
Commercial-scale Demonstration of the Liquid Phase Methanol (LPMEOHTM) Process: clean coal technology
[Topical Report #11]
USDoE

Van Gerpan J. 2005
Biodiesel processing and production
Fuel Processing and Technology **86**: 1097–1107

Van Gerpan J, Shanks B, Pruszko R, Clements D, Knothe G. 2004
Biodiesel Production Technology
[Report No NREL/SR-510-36244]
Golden, Colorado: National Renewable Energy Laboratory

Vicente G, Martínez M, and Aracil J. 2004
Integrated biodiesel production: a comparison of different homogeneous catalysts systems
Bioresource Technology **92**(3): 297–305

Wallace R, Ibsen K, McAloon A, Yee W. 2005
Feasibility Study for Co-locating and Integrating Ethanol Production Plants from Corn Starch and Lignocellulosic Feedstocks
[Report No: NREL/TP-10-37092 prepared for US Department of Energy and US Department of Agriculture]

Watanabe T. 2003
Analysis of native bonds between lignin and carbohydrate by specific chemical reactions
In *Association Between Lignin and Carbohydrates in Wood and Other Plant Tissues*, pp. 91–130, edited by T Koshijima and T Watanabe
Heidelberg: Springer-Verlag

Index